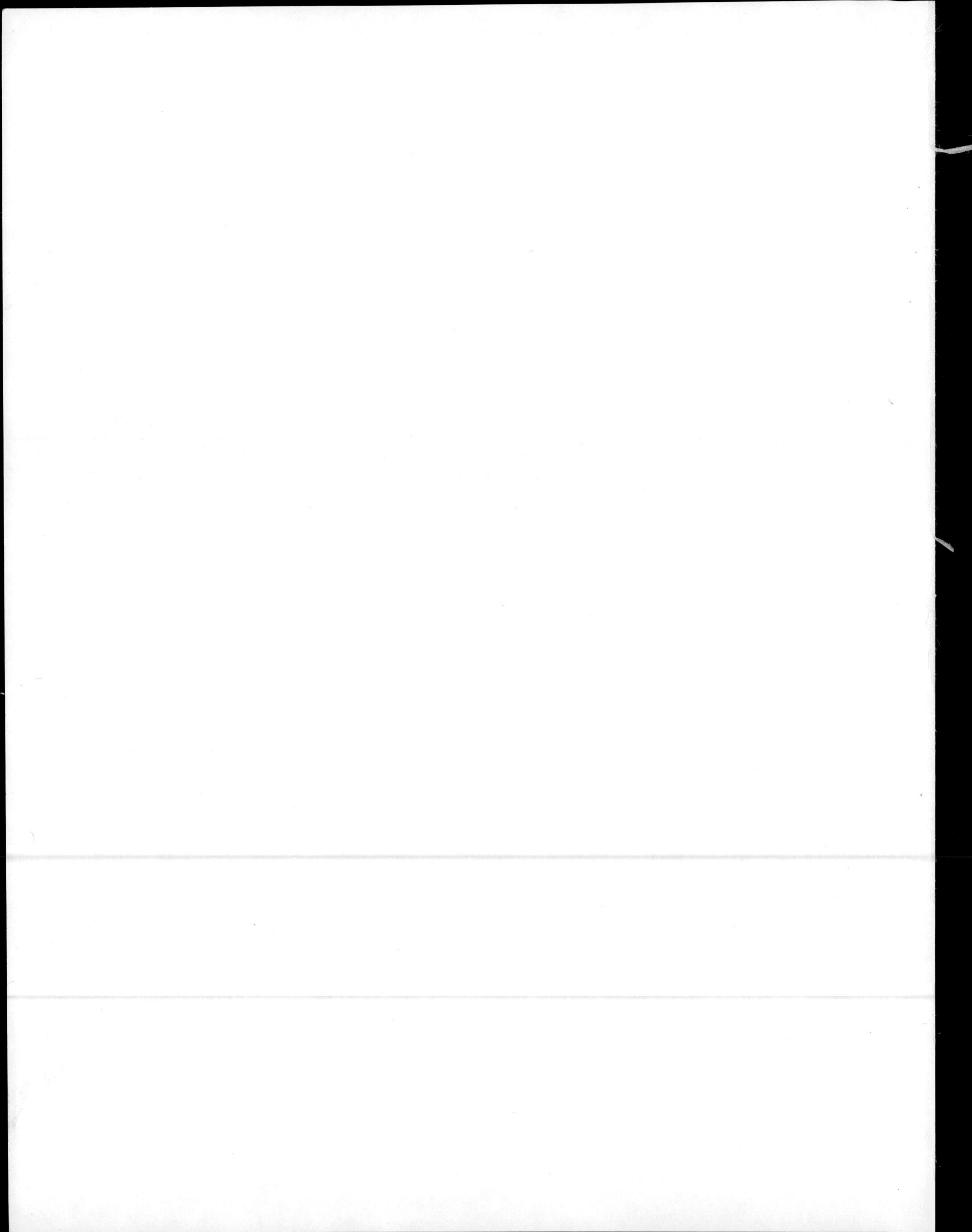

Encyclopedia of

ELECTRONIC CIRCUITS

Volume 5

Patent notice

Encyclopedia of

ELECTRONIC CIRCUITS

Volume 5

Rudolf F. Graf
&
William Sheets

TAB Books
Division of McGraw-Hill, Inc.
New York San Francisco Washington, D.C. Auckland Bogotá
Caracas Lisbon London Madrid Mexico City Milan
Montreal New Delhi San Juan Singapore
Sydney Tokyo Toronto

Published by TAB Books, a division of McGraw-Hill, Inc.

pbk 2 3 4 5 6 7 8 9 10 11 FGR/FGR 9 9 8 7 6 5
hc 2 3 4 5 6 7 8 9 10 11 FGR/FGR 9 9 8 7 6 5

Library of Congress Cataloging-in-Publication Data
(Revised for vol. 5)

Graf, Rudolf F.
The encyclopedia of electronics circuits.

Authors for v. 5– : Rudolf F. Graf & William Sheets.
Includes bibliographical references and indexes.
1. Electronic circuits—Encyclopedias. I. Sheets, William. II. Title.
TK7867G66 1985 621.3815 84-26772
ISBN 0-8306-0938-5 (v. 1)
ISBN 0-8306-1938-0 (pbk. : v. 1)
ISBN 0-8306-3138-0 (pbk. : v 2)
ISBN 0-8306-3138-0 (v. 2)
ISBN 0-8306-3348-0 (pbk. : v. 3)
ISBN 0-8306-7348-2 (v. 3)
ISBN 0-8306-3895-4 (pbk. : v. 4)
ISBN 0-8306-3896-2 (v. 4)
ISBN 0-07-011077-8 (pbk. : v. 5)
ISBN 0-07-011076-X (v. 5)

Acquisitions Editor: Roland S, Phelps
Editorial team: Andrew Yoder, Book Editor
Joanne Slike, Executive Editor
Production Team: Katherine G. Brown, Director
Jan Fisher, Coding
Lisa M. Mollott, Coding
Rose McFarland, Layout
Linda L. King, Proofreading
Nancy K. Mickley Proofreading
Joann Woy, Indexer
Design team: Jaclyn J. Boone, Designer
Brian Allison, Associate Designer
Cover design: Stickles Associates, Bath, Pa.

EL1
0110778

Contents

Introduction

The *Encyclopedia of Electronic Circuits, Volume V* adds approximately 1000 new circuits to the treasury of carefully chosen circuits that cover nearly every phase of today's electronic technology. These five volumes contain a wealth of new ideas and up-to-date circuits garnered from prestigious industry sources. Also included are some of the authors' original designs.

Each circuit is accompanied by a brief explanation of how it works, unless the circuit's operation is either obvious or too complex to describe in a few words. In the latter case, the reader should consult the original source listed in the back of the book. The index includes all entries from Volumes I to V. This provides instant access to about 5000 circuits, which make up the most extensive collection of carefully categorized modern circuits available anywhere.

Once again, the authors wish to extend their thanks to Ms. Loretta Gonsalves, whose virtuoso performance at the word processor contributed so much to the successful completion of the manuscript for this work. We look forward to the pleasure of working with her on Volume VI, which is now under development.

Rudolf F. Graf and William Sheets

Introduction

1

Alarm and Security Circuits

The sources of the following circuits are contained in the Sources section, which begins on page 675. The figure number in the box of each circuit correlates to the entry in the Sources section.

HIGH-POWER ALARM DRIVER

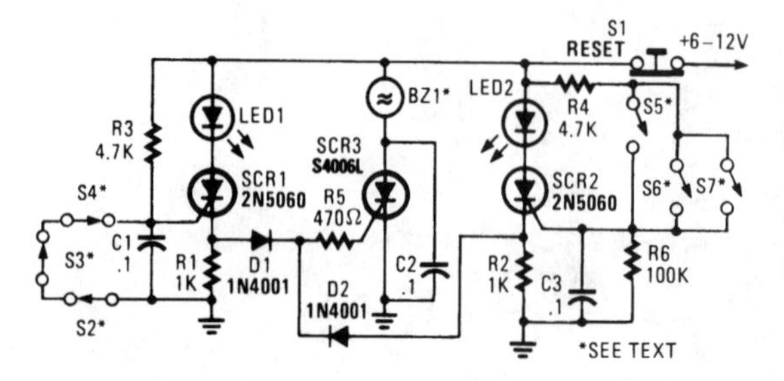

POPULAR ELECTRONICS

FIG. 1-1

In this circuit, a low-powered SCR is used to trigger a higher powered SCR. When a switch is opening (S2, S3, S4) or closing (S5, S6, S7), either SCR1 or SCR2 triggers. This triggers SCR3 via D1, D2, and R5. BZ1 is a high-powered alarm of the noninterrupting type.

MULTI-LOOP PARALLEL ALARM

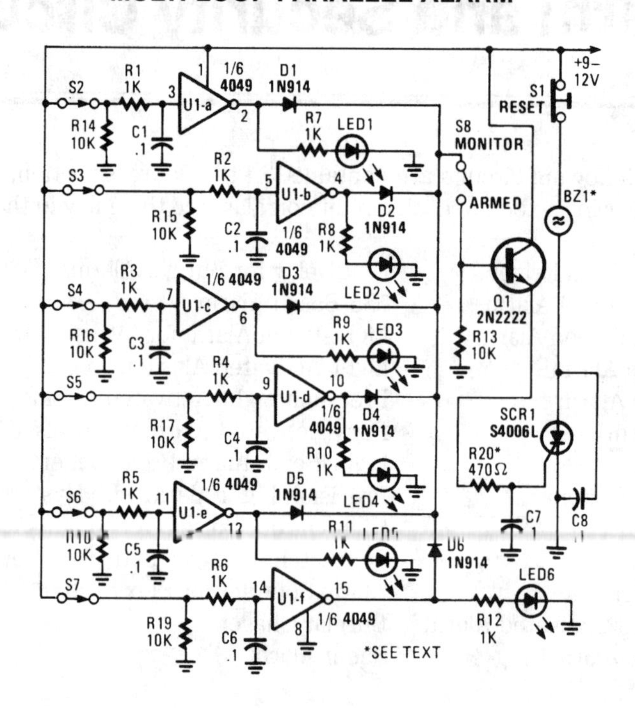

POPULAR ELECTRONICS

FIG. 1-2

This alarm has status LEDs connected across each inverter output to indicate the status of its associated sensor. S8 is used to monitor the switches via the LEDs, or to trigger an alarm via Q1 and SCR1. BZ1 should be a suitable alarm of the noninterrupting type.

SERIES/PARALLEL LOOP ALARM

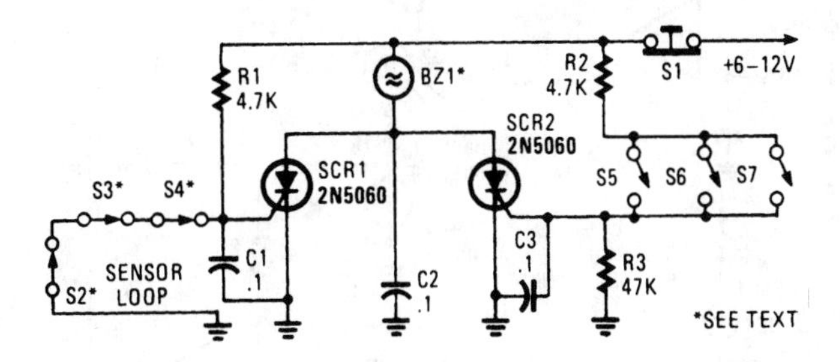

POPULAR ELECTRONICS *FIG. 1-3*

Two SCRs are used with two sensor loops. One loop uses series switches, the other loop parallel switches. When a switch actuation occurs, the SCR triggers. The alarm should be a noninterrupting type.

PARALLEL LOOP ALARM

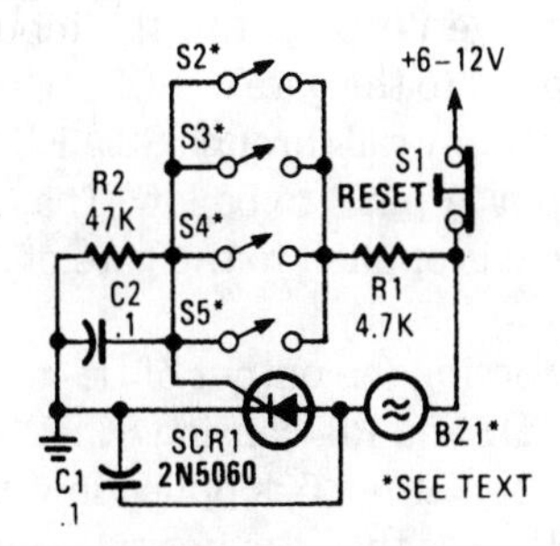

POPULAR ELECTRONICS *FIG. 1-4*

Four parallel switches are used to monitor four positions. When a closure occurs on any switch, SCR1 triggers, which sounds the alarm. The alarm should be of the noninterrupting type.

CLOSED-LOOP ALARM

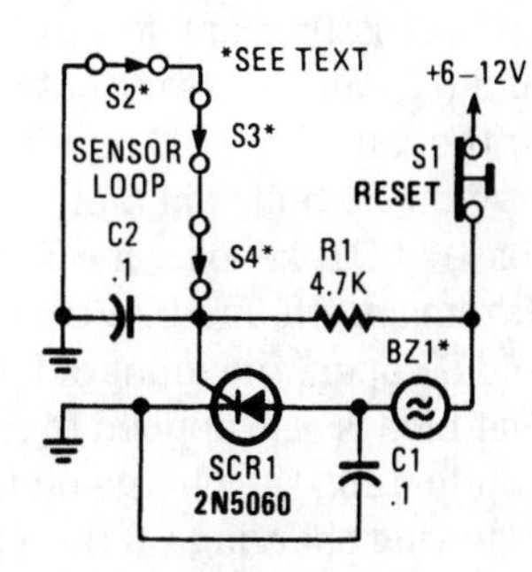

POPULAR ELECTRONICS *FIG. 1-5*

A string of three series-connected, normally closed switches are connected across the gate of an SCR. When one opens, the SCR triggers via R1, sounding an alarm. The alarm should be of the noninterrupting type.

DELAYED ALARM

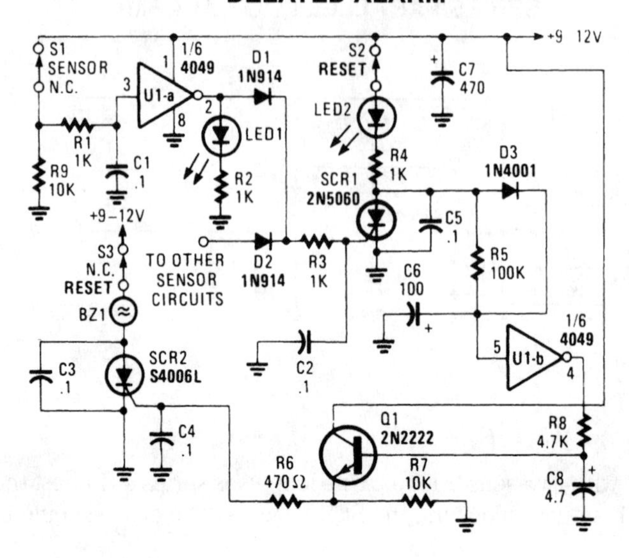

POPULAR ELECTRONICS *FIG. 1-6*

The alarm/sensor circuit shown is built around two SCRs, a transistor, a 4049 hex inverter, and a few support components, all of which combine to form a closed-loop detection circuit with a delay feature. The delay feature allows you to enter a protected area and deactivate the circuit before the sounder goes off.

Assuming that the protected area has not been breached (i.e., S1 is in its normally-closed position), when power is first applied to the circuit, a positive voltage is applied to the input of U1-a through S1 and R1, causing its output to go low. That low is applied to the gate of SCR1, causing it to remain off. At the same time, C6 rapidly charges toward the +*V* supply rail through S2, LED2, R4, and D3. The charge on C6 pulls pin 5 of U1-b high, causing its output at pin 4 to be low. That low is applied to the base of Q1, keeping it off. Because no trigger voltage is applied to the gate of SCR2 (via Q1), the SCR remains off and BZ1 does not sound.

But should S1 open, the input of U1-a is pulled low via R9, forcing the output of U1-a high, lighting LED1. That high is also applied to the gate of SCR1 through D1 and R3, causing SCR1 to turn on. With SCR1 conducting, the charge on C6 decays, the input of U1-b at pin 5 is pulled low, forcing its output high, slowing charging C8 through R8 to a voltage slightly less than the positive supply rail.

Transistor Q1 remains off until C8 has charged to a level sufficient to bias Q1 on, allowing sufficient time to enter the protected area and disable the alarm before it sounds. Once C8 has developed a sufficient charge, Q1 turns on and supplies gate current to SCR2 through R6, causing the SCR to turn on and activate BZ1. If the circuit is reset before the delay has timed out, no alarm will sound.

The delay time can be lengthened by increasing the value of either or both C6 and R5; decreasing the value of either or both of those components will shorten the delay time.

All of the switches used in the circuit are of the normally-closed (NC) variety. Switch S1 can be any type of NC security switch. Switch S2 can be either a pushbutton or toggle switch. Because S3 is used to disable the sounder (BZ1) only, anything from a key-operated security switch to a hidden toggle switch can be used.

DOOR MINDER

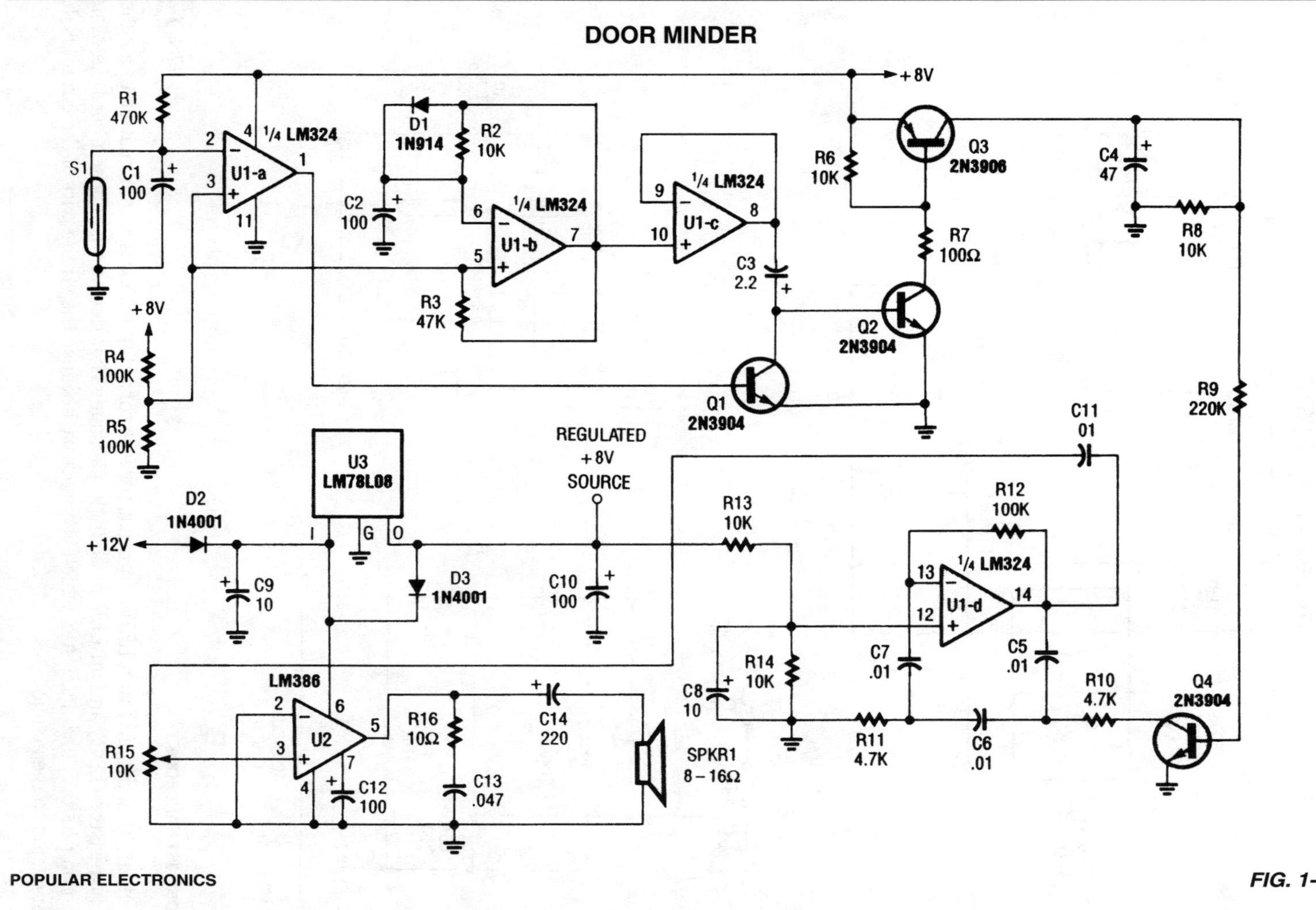

POPULAR ELECTRONICS

FIG. 1-7

This circuit monitors a door to determine if it has been left open. After 24 seconds, the alarm sounds. S1 is a magnetic sensor. The alarm is an electronic chime sound that is struck once per second.

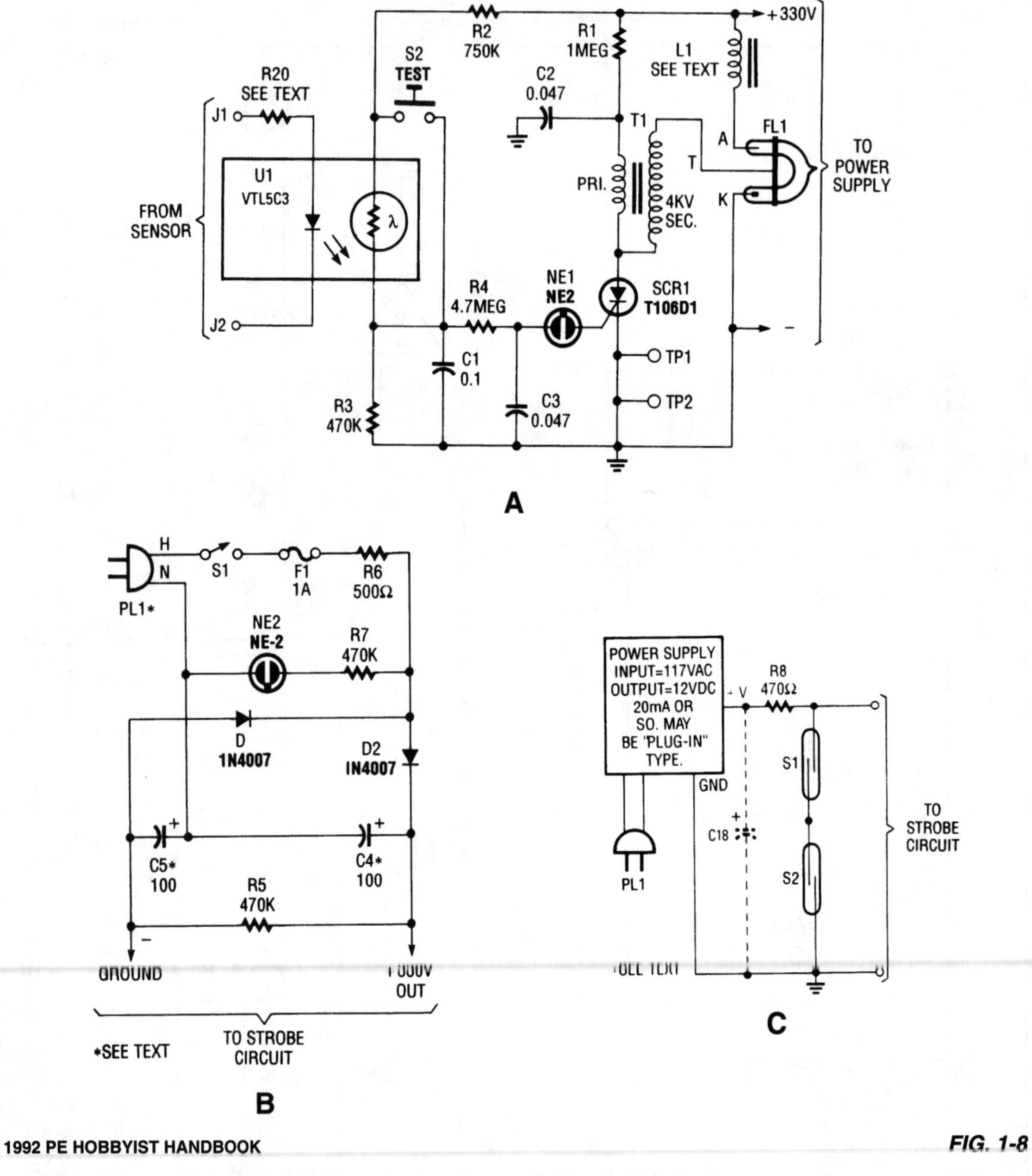

1992 PE HOBBYIST HANDBOOK *FIG. 1-8*

The circuit is activated by an LED/photoresistor isolator (U1), which is a combination of a light-dependent resistor (LDR) and an LED in a single package. That device was chosen because of its high isolation (2000 V) characteristic, which is necessary because the strobe part of the circuit is directly connected to the ac line.

STROBE ALERT SYSTEM (*Cont.*)

The voltage divider is formed by R2, U1's internal resistance, and R3. When U1's internal LED is off, U1's internal LDR has a very high resistance—on the order of 10 MΩ. The voltage applied to NE1 is considerably below its ignition voltage of approximately 90 Vdc.

The optoisolator's internal LED is activated by a dc signal supplying 20 mA. The external sensor(s) that supply the signal are connected to the strobe part of the circuit at J1 and J2.

When the internal LED lights, the LDR's resistance decreases to around 5 kΩ. Under that condition, about 125 Vdc is applied across C1, R4, and C2. The neon lamp periodically fires and extinguishes as capacitor C3 charges through R4, and discharges via NE1 and the SCR gate.

Resistor R4 restricts the current input to C3, and thereby controls the firing rate of NE1—about three times per second. The discharge through NE1 is applied to the gate of SCR1.

SCR1, a sensitive-gate unit, snaps on immediately when NE1 conducts, which completes the ground circuit for transformer T1 (a 4-kV trigger transformer). As SCR1 toggles on and off in time with the firing of NE1, capacitor C2 (connected in parallel with T1's primary) charges via R1, and then discharges very rapidly through T1's primary winding. A voltage pulse is applied to the trigger input of FL1, a Xenon flash lamp.

It is important to remember that the circuit is connected directly to the ac line. Resistor R6 is included to limit the amount of line current available to the circuit. The value of R6 can be decreased if you intend to modify the circuit for more flash power.

Warning: Even though the circuit is fuse-protected, it can still be dangerous if handled carelessly.

WARBLE ALARM

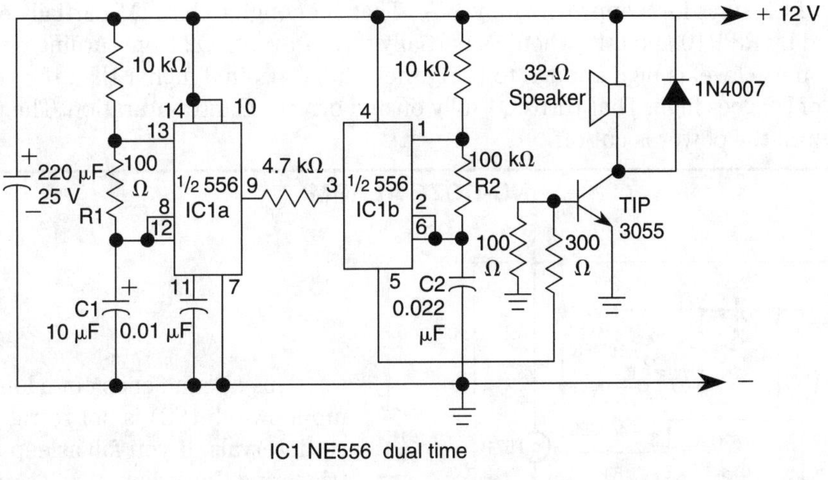

WILLIAM SHEETS

FIG. 1-9

This circuit uses a 556 to first generate a low frequency square wave, that is modulated to produce two alternate tones of about 400 and 500 Hz. Circuit generates warble alarm of European emergency vehicles. The frequencies of the oscillators are determined by the values of R1, C1 and R2, C2.

AUDIO ALARM

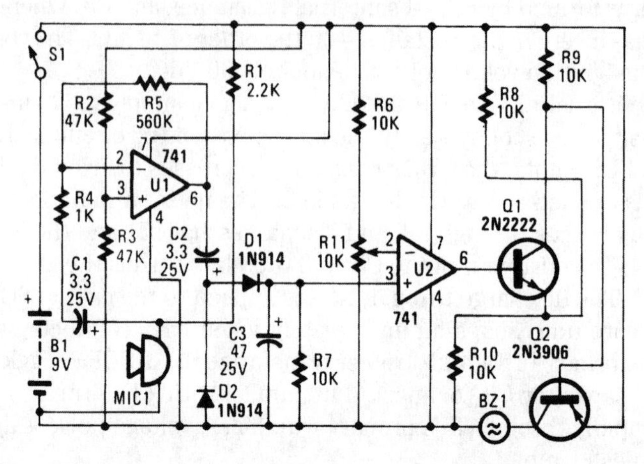

POPULAR ELECTRONICS *FIG. 1-10*

In the circuit, U1 amplifies the audio picked up by the condenser microphone. Resistor R1 limits its current, while R2 and R3 center the output of the amplifier to ½*B*+ to allow a single-ended supply to be used. Diodes D1 and D2 rectify the output of U1, and C3 filters the resulting pulsing dc. Thus, a dc voltage that is proportional to the ambient sound level is produced.

That voltage is presented to the noninverting input of U2. The inverting input is provided with a reference voltage of between 0 and ½*B*+, which is set by R11.

As long as the noise level is low enough to keep the voltage at pin 3 lower than the voltage at pin 2, the output of U2 stays low (approximately 1 V). That is enough to bias Q1 partially on. A voltage divider, formed by R8/R10 and Q1 (when it's partially on), prevents Q2 from turning on.

When the noise level is high enough to bring the voltage at pin 3 higher than the voltage at pin 2, the output of U2 goes high. That turns Q1 fully on and drives Q2 into saturation. The piezo buzzer then sounds until the power is cut off.

NO-DOZE ALARM

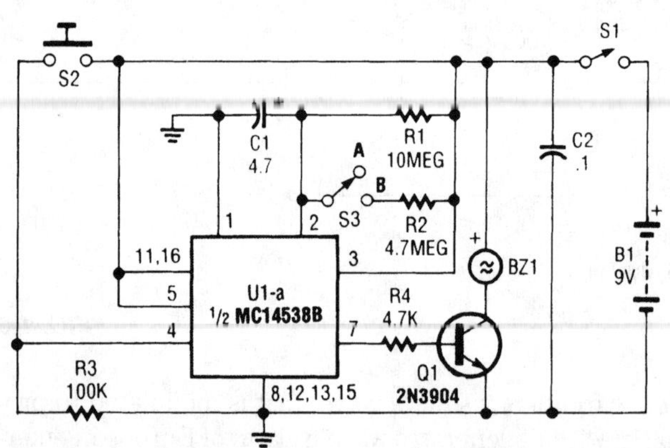

This circuit sends out a loud tone if the input switch (S2) is not retriggered at preset intervals. If you fall asleep and miss retriggering the circuit, it will sound until you press S2.

POPULAR ELECTRONICS *FIG. 1-11*

HEAT- OR LIGHT-ACTIVATED ALARM

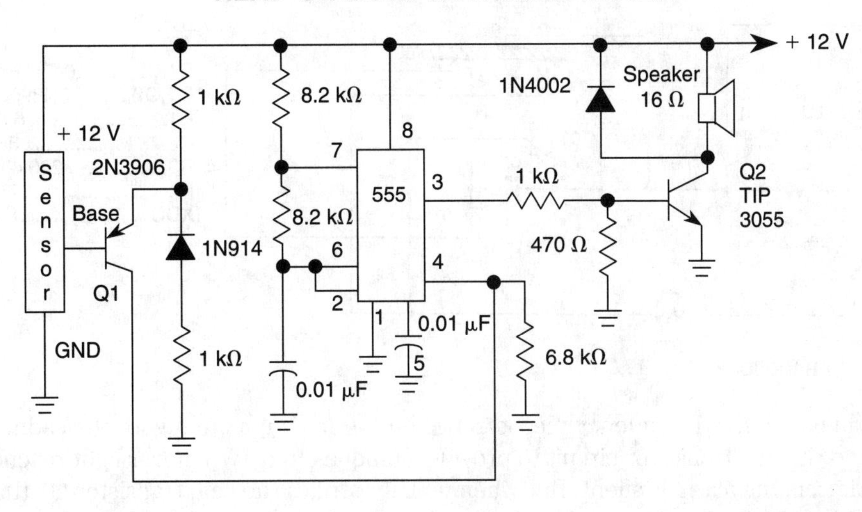

SENSOR CIRCUITS

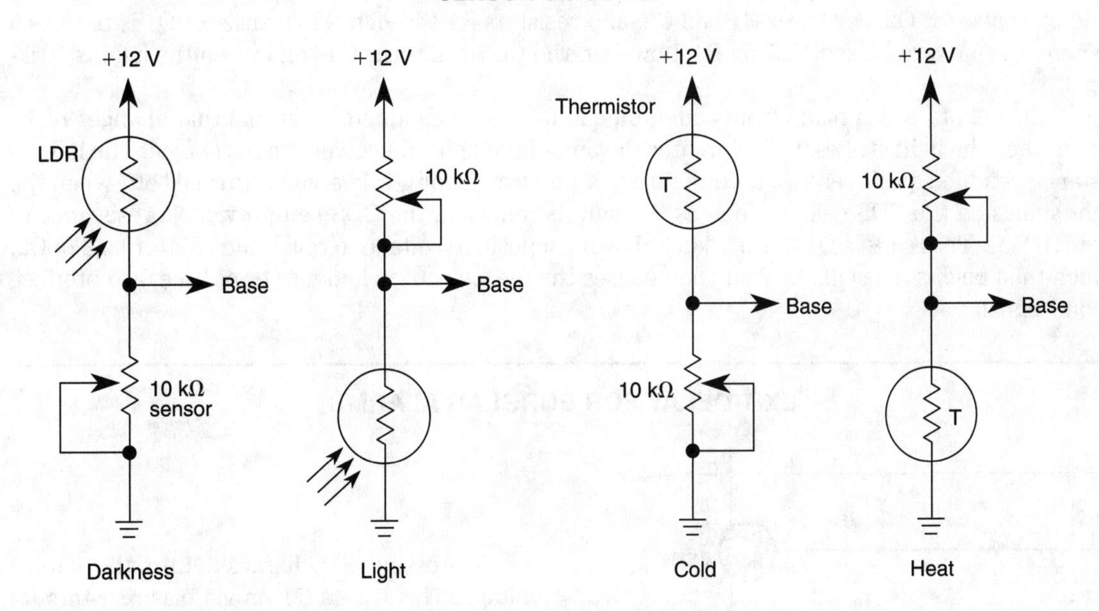

WILLIAM SHEETS ***FIG. 1-12***

The tone generated by a 555 oscillator can be turned on (activated) by heat or light. That causes Q1 to conduct transistor W2 (TIP 3055).Q2 (TIP 3055) acts as an audio amplifier and speaker driver.

PIEZOELECTRIC ALARM

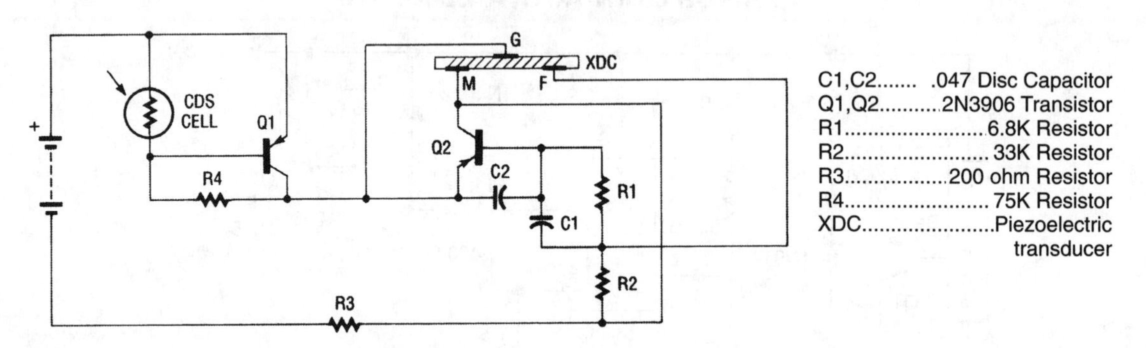

1991 PE HOBBYIST HANDBOOK ***FIG. 1-13***

The alarm uses a fixed-frequency piezoelectric buzzer in conjunction with the cadmium-sulfide (CDS) cell and the two-transistor circuit to provide a unique effect. Whenever light reaches the CDS photo-electric cell, the alarm is silent. But when no light strikes the cell, transistor Q1 turns on, and the circuit emits a high-pitched tone.

The alarm consists of a piezoelectric disk that oscillates at the fixed frequency of 3.137 kHz, created by transistor Q2, capacitor C1 and C2, and resistors R1 through R3. Transistor Q1 is used as a switch. It is forward-biased "on" by R4; however, the CDS cell turns Q1 "off" when the light is striking it.

A CDS photo cell is made from cadmium sulfide, a semiconductor material that changes resistance when the light strikes it. The greater the amount of light, the lower the resistance. The low resistance conducts positive voltage to the base of pnp transistor Q1, keeping it turned "off" when the light shines on the CDS cell. As soon as the light is removed, the CDS cell provides a resistance of over 100 kΩ. That causes Q1 to turn "on," allowing a positive voltage to reach the emitter lead of Q2, which then begins to oscillate. That then causes the piezoelectric element (transducer) to produce a loud signal.

EXIT DELAY FOR BURGLAR ALARMS

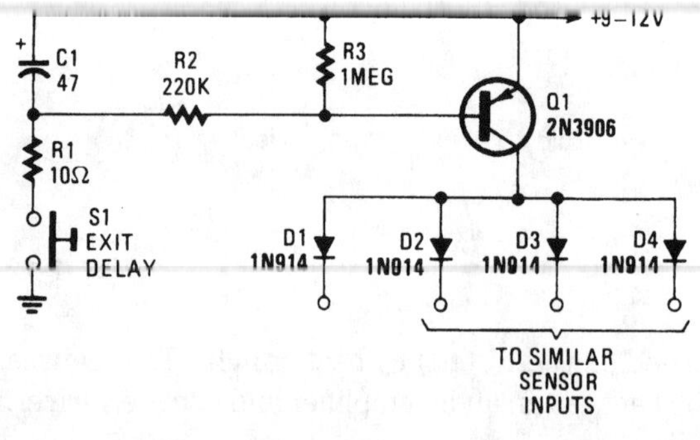

POPULAR ELECTRONICS ***FIG. 1-14***

Depressing S1 charges C1 to the supply voltage. This biases Q1 on via bias resistors R2 and R3. A voltage is available for the duration of the delay period, to hold off the alarm circuit. C1 can be increased or decreased in value to alter the delay times.

555-BASED ALARM

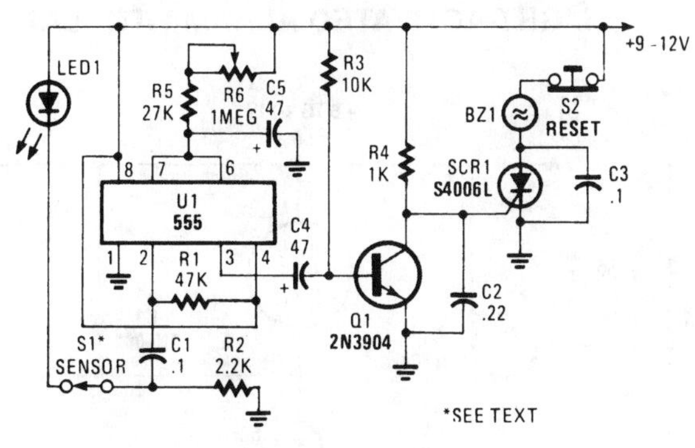

POPULAR ELECTRONICS *FIG. 1-15*

The alarm circuit has a single 555 oscillator/timer (U1) performing double duty; serving both in the alarm-trigger circuit and the entry-delay circuit. In this application, the trigger input of U1 at pin 2 is held high via R1. A normally-closed sensor switch, S1, supplies a positive voltage to the junction of R2 and C1, and lights LED1. With both ends of C1 tied high, there is no charge on C1. But when S1 opens, C1 (initially acting as a short) momentarily pulls pin 2 of U1 low, triggering the timed delay circle.

At the beginning of the timing cycle, U1 produces a positive voltage at pin 3, which charges C4 to near the positive voltage at pin 3, which charges C4 to near the positive supply voltage. Transistor Q1 is heavily biased on by R3, keeping its collector at near ground level. With Q1 on, SCR1's gate is clamped to ground, holding it off. When the delay circuit times out, pin 3 of U1 goes low and ties the positive end of C4 to ground. That turns Q1 off.

When Q1 turns off, the voltage at the gate of SCR goes positive, turning on the SCR and sounding the alarm. The delay time is adjustable from just a few seconds (R6 set to its minimum resistance) to about one minute (R6 adjusted to its maximum resistance).

LIGHT-BEAM ALARM FOR INTRUSION DETECTION

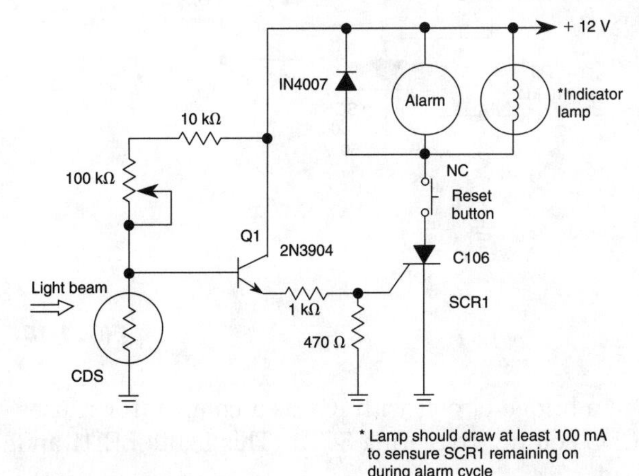

When the light beam that falls in the CDS photocell is interrupted, transistor (EN3904) conducts thereby triggering SCR1 (C106) and activating alarm bell. S1 resets the SCR. The alarm bell should be a self-interrupting electromechanical type.

WILLIAM SHEETS *FIG. 1-16*

LIGHT-ACTIVATED ALARM WITH LATCH

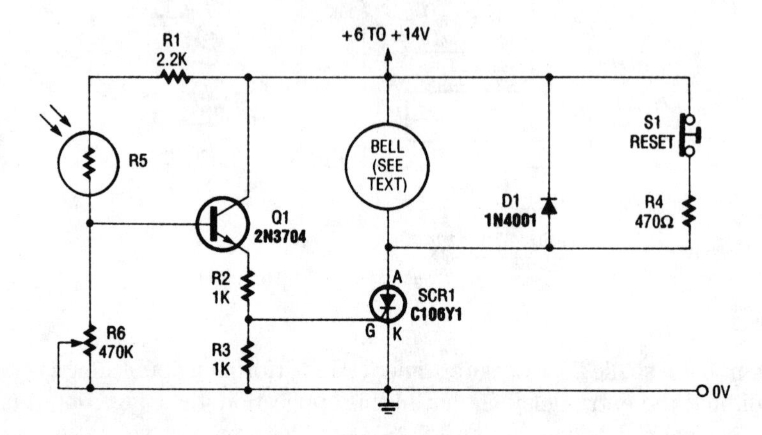

ELECTRONICS NOW

FIG. 1-17

In this circuit, light causes R5 to conduct forward-biasing Q1. R6 sets sensitivity. SCR1 is triggered from the emitter voltage on LQ1, sounding the alarm bell. When S1 is depressed, SCR1 unlatches. Be sure that a self-interrupting alarm (electromechanical buzzer or bell) is used.

PRECISION LIGHT-ACTIVATED ALARM

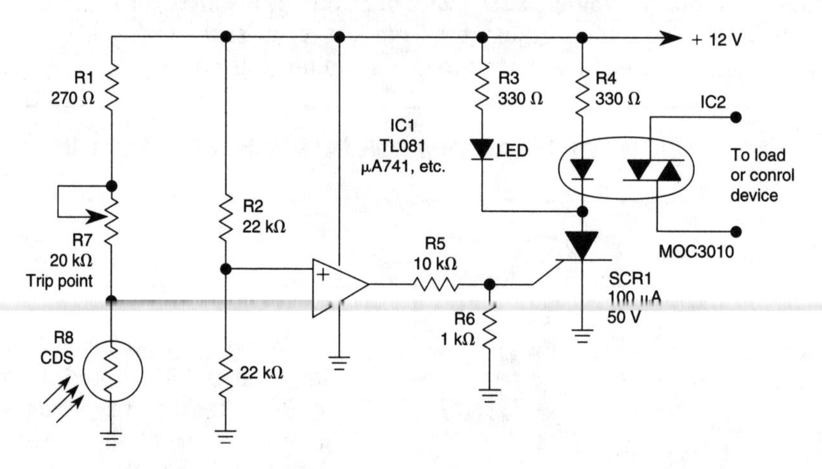

WILLIAM SHEETS

FIG. 1-18

The light-sensitive CDS cell R8 configured in a bridge circuit with IC1 as a comparator causes IC1's output to go high when light strikes the CDS cell R8, triggering SCR1. This lights LED1 and turns on opto isolator IC2, which switches the load.

DARK-ACTIVATED ALARM WITH PULSED TONE OUTPUT

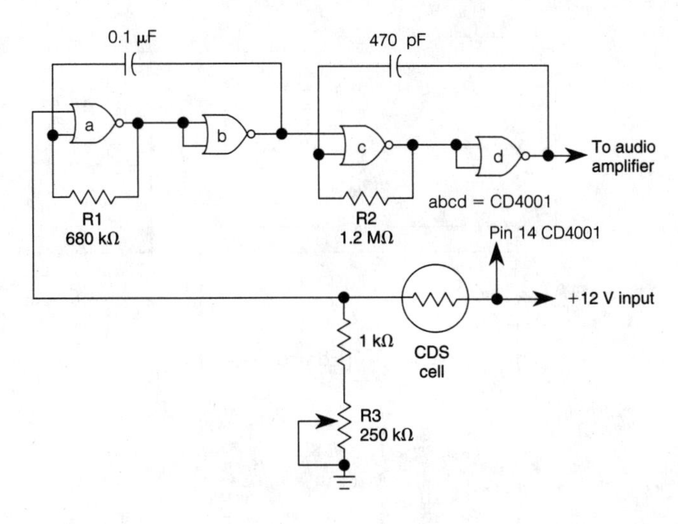

WILLIAM SHEETS

FIG. 1-19

NOR gates a and b form a low-frequency oscillator that is activated when the CDS cell, under dark conditions, causes NOR gate a to see a logic zero at one input. This low-frequency (10 Hz) gates a high-frequency oscillator (c and d) to oscillate at around 1000 Hz. R1 can be varied to change the pulse rate and R2 to change the tone. R3 sets the trigger point.

LIGHT-BEAM ALARM PREAMPLIFIER

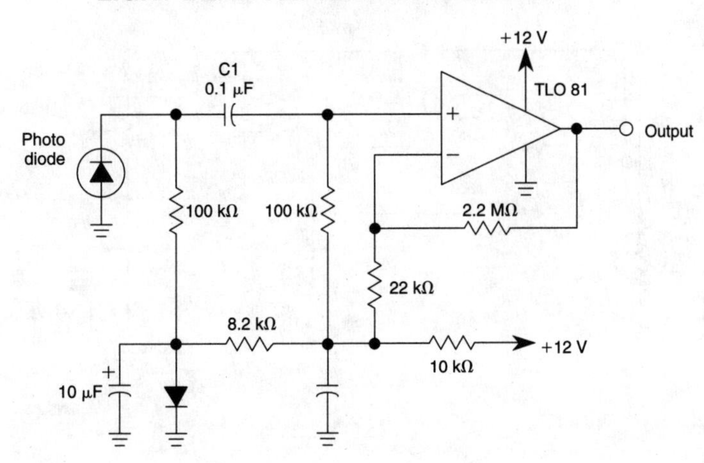

WILLIAM SHEETS

FIG. 1-20

This circuit can be used for light beams to 20 kHz. The gain of the operational amplifier is set for a 40-dB gain.

PRECISION LIGHT ALARM WITH HYSTERESIS

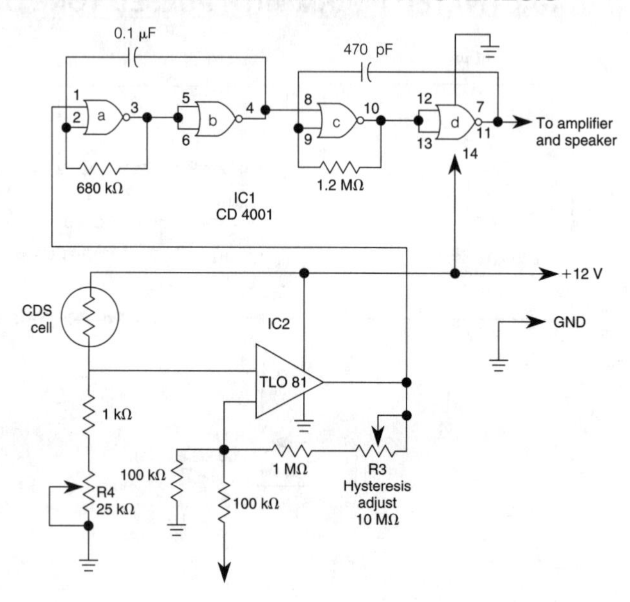

WILLIAM SHEETS **FIG. 1-21**

The TL081 is used as a comparator in a Wheatstone bridge circuit. When the CDS cell resistance decreases due to exposure to light, the output from IC2 cause the low-frequency oscillator (a) and (b) to generate a 10-Hz square wave, gating the 1000 Hz oscillator (c) and (d) on and off. This signal drives an amplifier. R3 controls hysteresis, which reduces on-off triggering near the threshold set by R4.

HIGH-OUTPUT PULSED-TONE/LIGHT-ACTIVATED ALARM

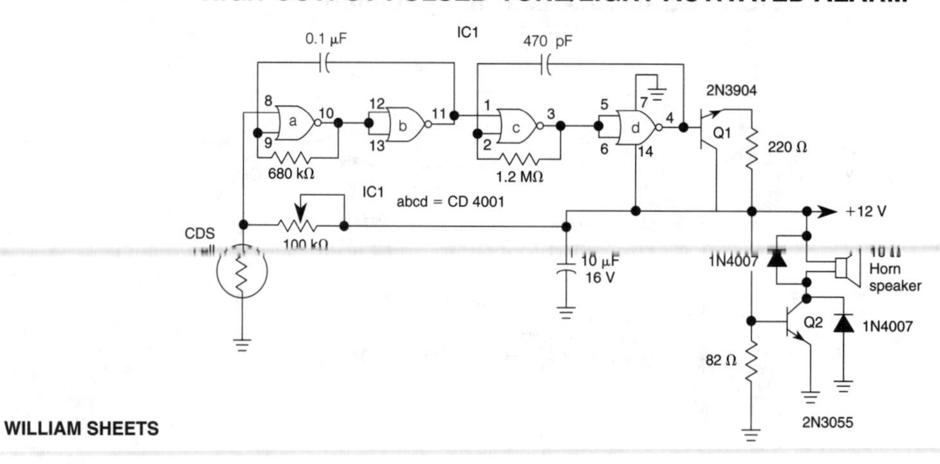

WILLIAM SHEETS **FIG. 1-22**

This circuit can produce up to 1 W of audio power to drive a speaker or horn. When the CDS cell is struck by light, its resistance decreases thus activating NOR gate (a) thereby causing (a) and (b) to produce a low-frequency (10-Hz) square wave. This pulses the 1-kHz oscillator (c) and (d), causing it to generate a pulsed 1-kHz tone at a 10-Hz rate. Q1 and Q2 amplify this signal. Q2 (2N3055) drives the speaker.

SELF-LATCHING LIGHT ALARM WITH TONE OUTPUT

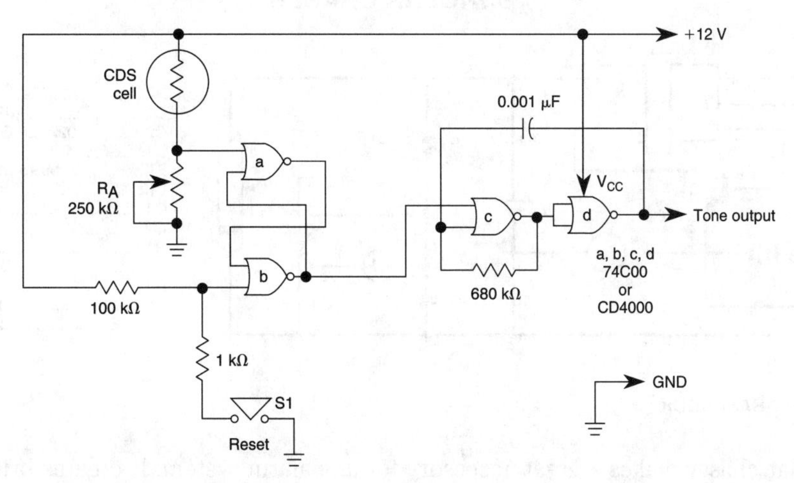

WILLIAM SHEETS *FIG. 1-23*

A decrease in the resistance of the CDS cell when light strikes it activates latch a and b, enabling tone oscillator c and d which produces an output of about 1000 Hz. R_A sets the trip level. S1 resets the circuit.

ALARM SOUNDER FOR FLEX SWITCH

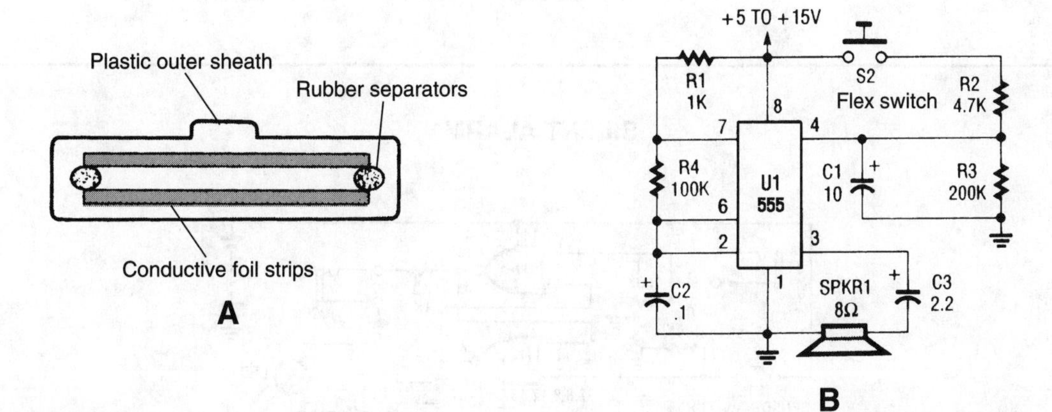

POPULAR ELECTRONICS *FIG. 1-24*

This is a cross-sectional diagram of a flex switch. They can be used as pushbutton or even position sensors. This schematic diagram shows an oscillator, which is used as an alarm sounder, triggered by a flex switch.

BURGLAR CHASER

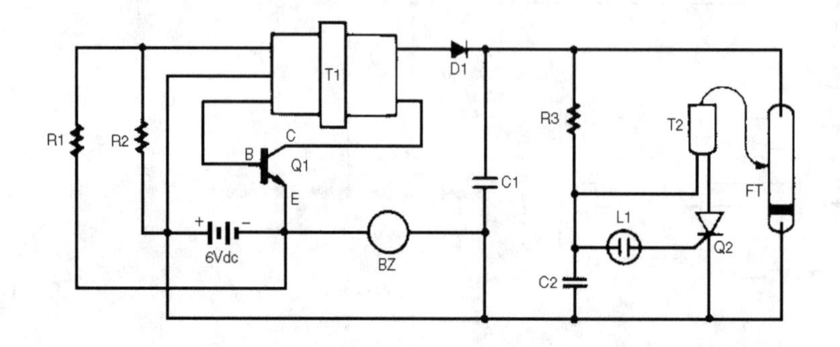

BZ	Metal Horn Buzzer
C1	.5 µF 250 volts Capacitor
C2	.022 µF Green Cap (223 K5K)
D1	1N4007 Diode
FT	Micro Strobe Tube/Reflector
L1	Neon Lamp
Q1	C1740 SW Transistor
Q2	106 SCR
R1	200 ohm Resistor
R2	820 ohm Resistor
R3	10 meg Resistor
T1	Inverter Transformer
T2	4 kV Trigger Coil

1991 PE HOBBYIST HANDBOOK ***FIG. 1-25***

The burglar chaser makes a great accessory for any alarm system. It creates brilliant flashes of white light and a loud, irritating sound from a metal horn buzzer. Transformer T1 is connected to Q1, R1, and R2 to form a blocking oscillator. This creates a 6-Vac signal on the primary of T1. Because of T1's large ratio of turns from primary to secondary, the 6-Vac signal is stepped up to a level of over 200 Vac, which is then rectified by D1. The resultant dc voltage is applied to storage capacitor C1 and the neon relaxation oscillator made up of R3, C2, and L1. Each time C2 charges up to a sufficient level, it ionizes L1, which causes SCR Q2 to fire. The firing SCR causes the charge on C2 to be applied to the trigger coil. The trigger coil converts the 200 V into the 4000-V pulse that is needed to fire micro xenon strobe tube/reflector FT. The cycle repeats itself after the strobe tube flashes.

SILENT ALARM

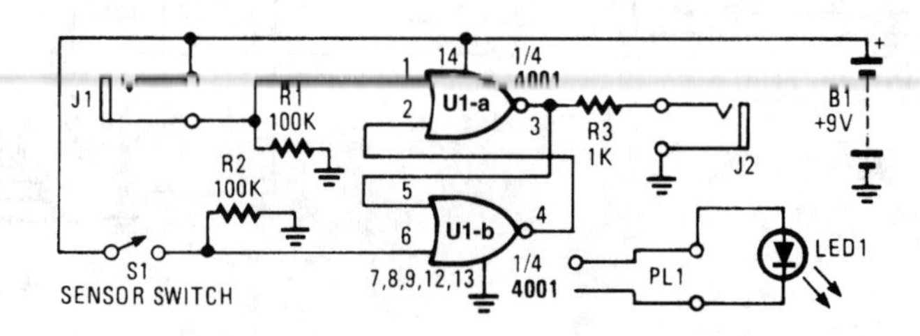

POPULAR ELECTRONICS ***FIG. 1-26***

A sensor switch triggers a set-reset flip flop and lights an LED.

2

Amplifier Circuits

The sources of the following circuits are contained in the Sources section, which begins on page 675. The figure number in the box of each circuit correlates to the entry in the Sources section.

DIFFERENCE AMPLIFIER

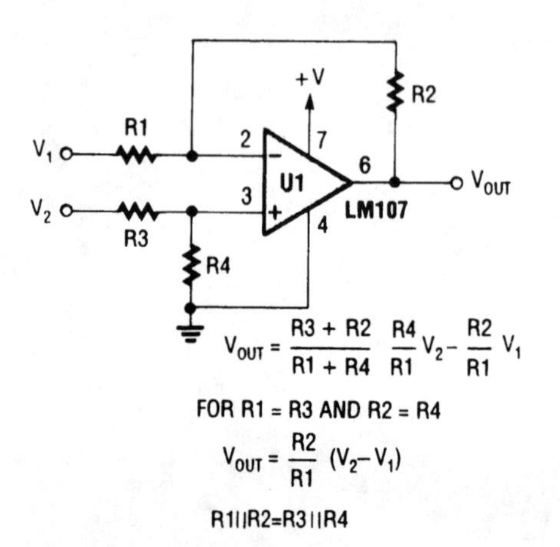

POPULAR ELECTRONICS FIG. 2-1

By using two inputs as shown, a difference amplifier yielding the differential between U1 and U2, times a gain factor results.

FAST-INVERTING AMPLIFIER WITH HIGH INPUT IMPEDANCE

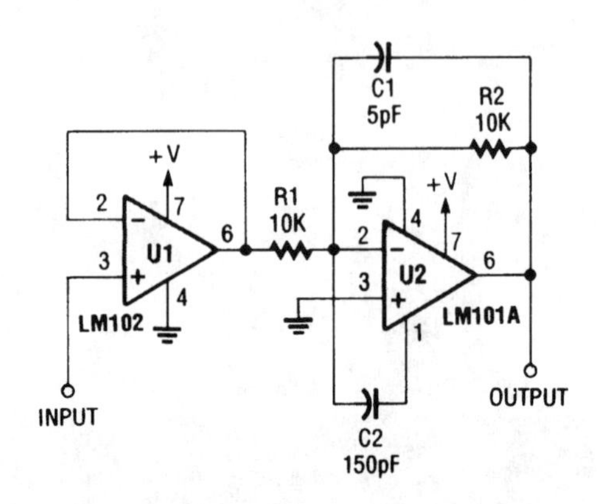

POPULAR ELECTRONICS FIG. 2-2

U1 is used as a voltage follower to feed inverter U2. Because U1 is in the voltage-follower configuration, it exhibits a high input impedance.

NONINVERTING ac AMPLIFIER

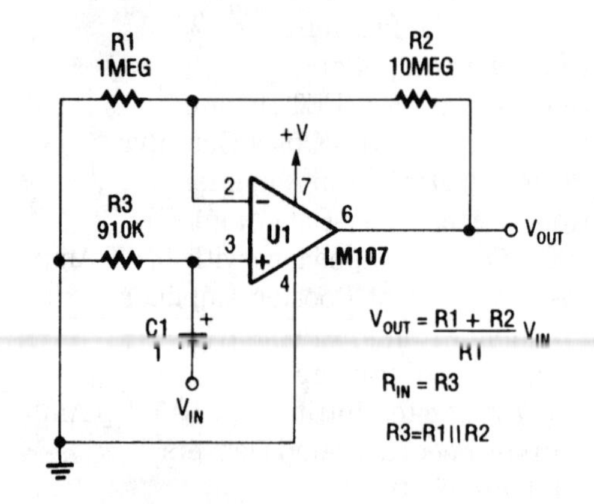

POPULAR ELECTRONICS FIG. 2-3

A general-purpose noninverting ac amplifier for audio of other low-frequency applications is shown. Design equations are in the figure. Almost any general-purpose op amp can be used for U1.

INVERTING SUMMING AMPLIFIER

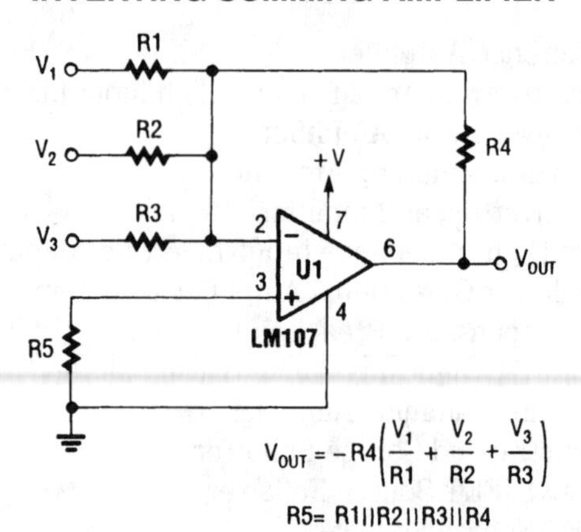

POPULAR ELECTRONICS FIG. 2-4

The output of U1 is the sum of V_1, V_2, and V_3, multiplied by R_1/R_4, R_2/R_4, and respectively. R1, R2, R3 are selected as required for individual gains. R4 affects gain of all these inputs.

NONINVERTING ac AMPLIFIER

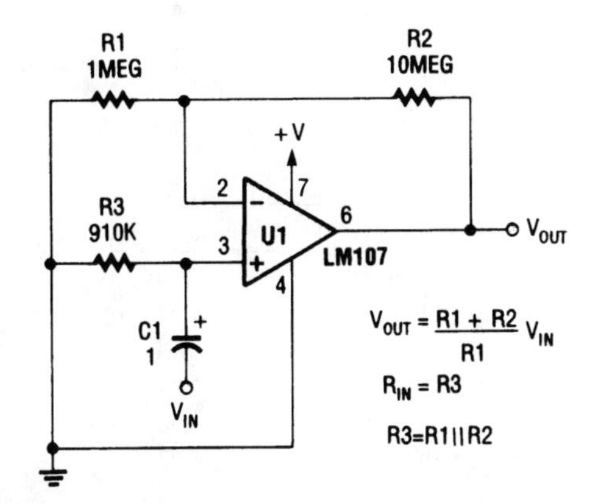

POPULAR ELECTRONICS *FIG. 2-5*

FAST HIGH-IMPEDANCE INPUT-INVERTING AMPLIFIER

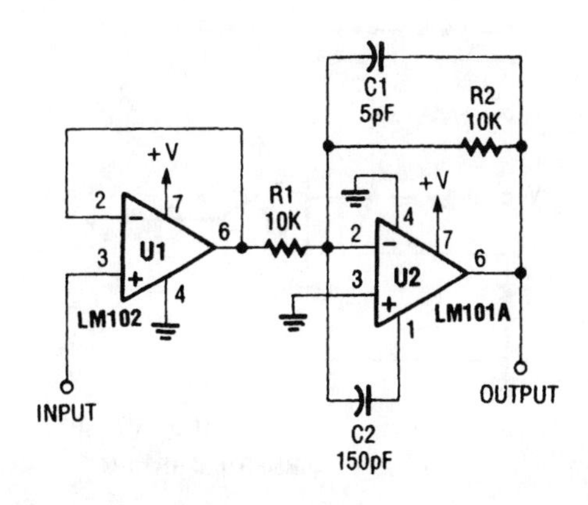

POPULAR ELECTRONICS *FIG. 2-6*

NONLINEAR OPERATIONAL AMPLIFIER WITH TEMPERATURE COMPENSATED-BREAKPOINT

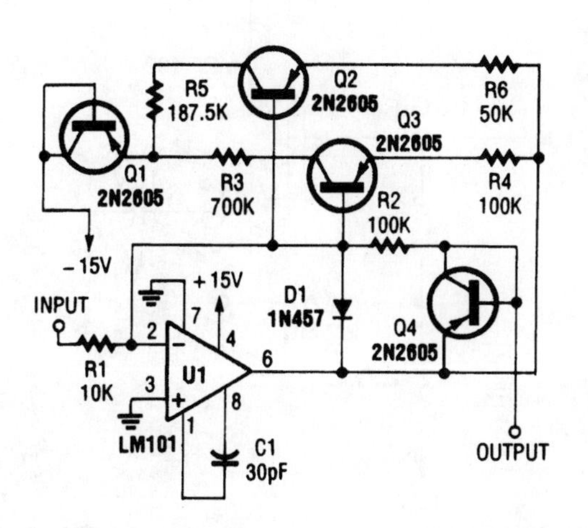

POPULAR ELECTRONICS *FIG. 2-7*

MOSFET HIGH-IMPEDANCE BIASING METHOD

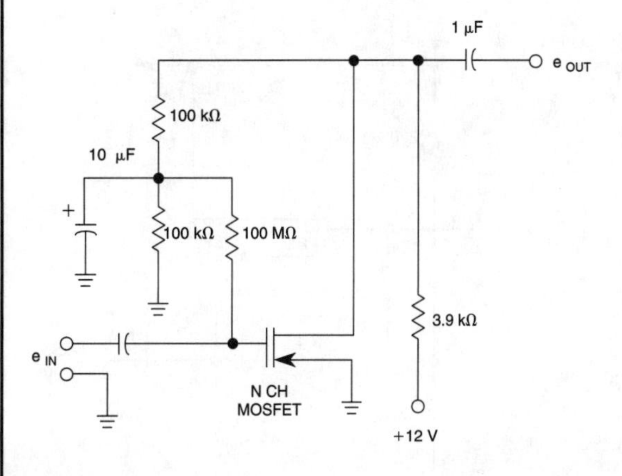

WILLIAM SHEETS *FIG. 2-8*

High-impedance biasing method for an N-channel MOSFET to form a linear-inverting amplifier.

INVERTING SUMMING AMPLIFIER

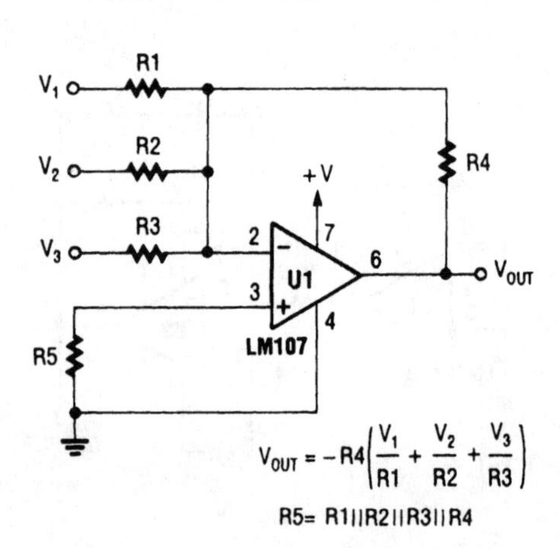

POPULAR ELECTRONICS *FIG. 2-9*

BOOTSTRAPPED SOURCE FOLLOWER

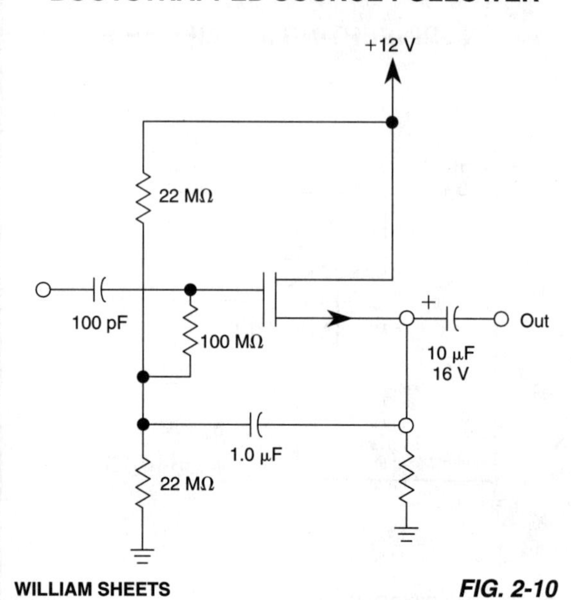

WILLIAM SHEETS *FIG. 2-10*

This bootstrapped source follower uses an N-channel MOSFET. It has a high input impedance.

30-MΩ JFET SOURCE FOLLOWER

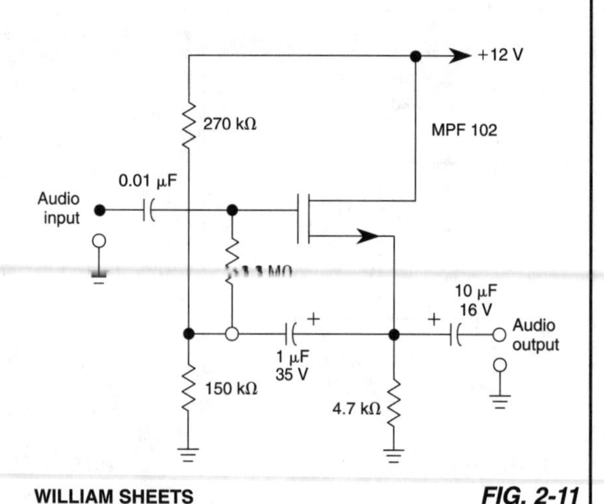

WILLIAM SHEETS *FIG. 2-11*

This JFET source-follower uses an MPF102 with offset biasing. It has an input impedance of >30 MΩ.

JFET SOURCE FOLLOWER

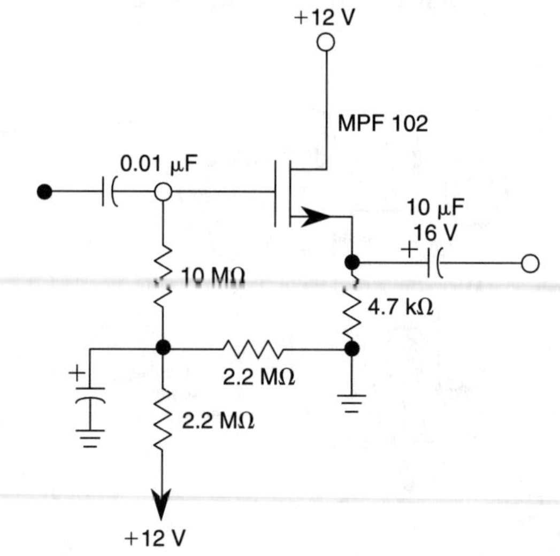

WILLIAM SHEETS *FIG. 2-12*

The circuit uses positive gate bias to improve the operating point for better dynamic range.

UNITY-GAIN NONINVERTING AMPLIFIER

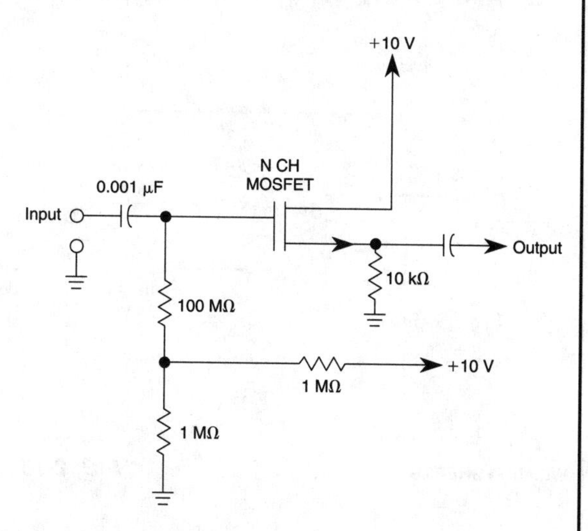

WILLIAM SHEETS *FIG. 2-13*

Biasing methods for an N-channel MOSFET to form a unity-gain noninverting amplifier or source-follower.

JFET AMP WITH CURRENT SOURCE BIASING

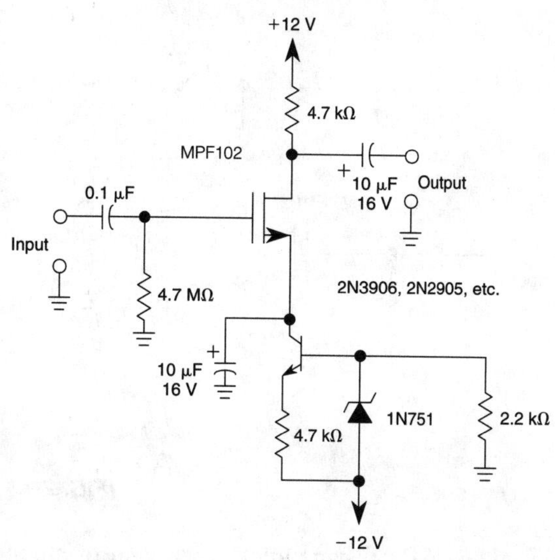

WILLIAM SHEETS *FIG. 2-14*

A current source (MPF102) in the source lead of bipolar transistor 2N3906 permits accurate control of drain current.

ELECTRET MIKE PREAMP

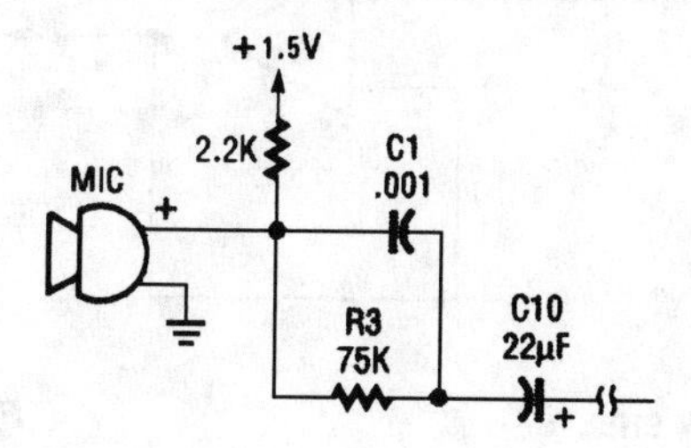

ELECTRONICS NOW *FIG. 2-15*

This circuit is suitable for using an electret microphone for many applications. A 1.5-V battery is used. C1 and R3 provide treble boost/bass cut; they can be eliminated, if desired.

DIFFERENCE AMPLIFIER

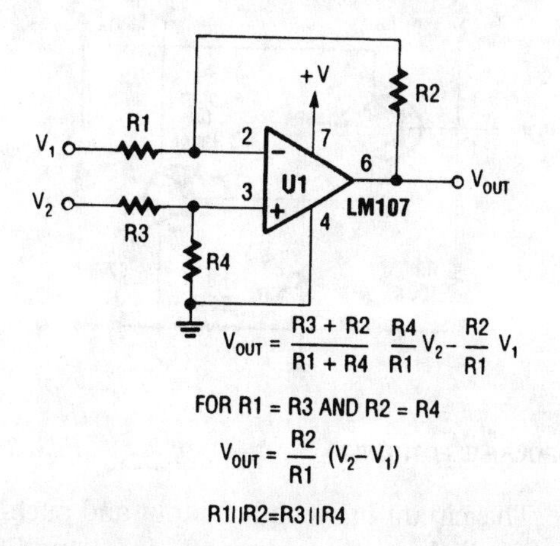

$$V_{OUT} = \frac{R3 + R2}{R1 + R4}\,\frac{R4}{R1}\,V_2 - \frac{R2}{R1}\,V_1$$

FOR R1 = R3 AND R2 = R4

$$V_{OUT} = \frac{R2}{R1}\,(V_2 - V_1)$$

R1||R2=R3||R4

POPULAR ELECTRONICS *FIG. 2-16*

GENERAL-PURPOSE JFET PREAMP

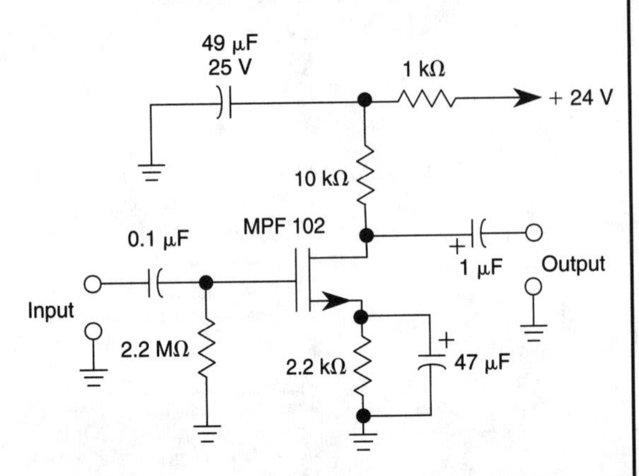

WILLIAM SHEETS

FIG. 2-17

This JFET preamplifier has a gain of about 20 dB and a bandwidth of over 100 kHz. It is useful as a low-level audio amplifier for high-impedance sources.

FET AMPLIFIER WITH OFFSET GATE BIAS

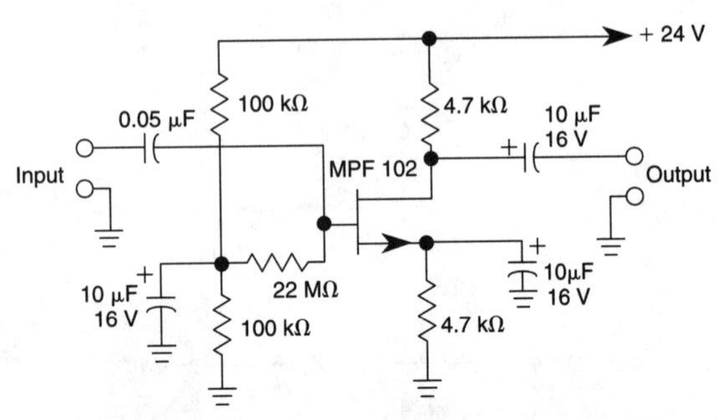

WILLIAM SHEETS

FIG. 2-18

In this amplifier circuit, the gate of the MPF102 is biased with an external voltage. This circuit achieves tighter control of the operating point and biasing conditions.

PUSH-PULL DARLINGTON AMPLIFIER

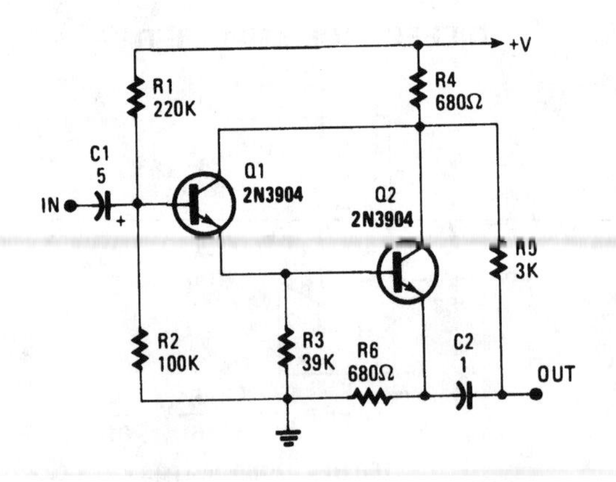

POPULAR ELECTRONICS

FIG. 2-19

This circuit has a high-*Z* input and push-pull output via the output taken across R4 and R6.

NONINVERTED UNITY-GAIN AMPLIFIER

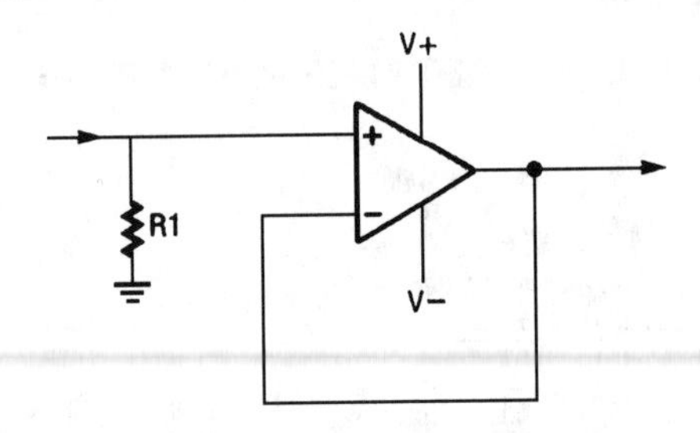

WILLIAM SHEETS

FIG. 2-20

An op amp can be used as a unity gain amplifier by connecting its output to its inverting input as shown. R1 should be low enough so the bias current of the op amp does not cause an appreciable offset.

500-MΩ INPUT IMPEDANCE WITH JFET AMP

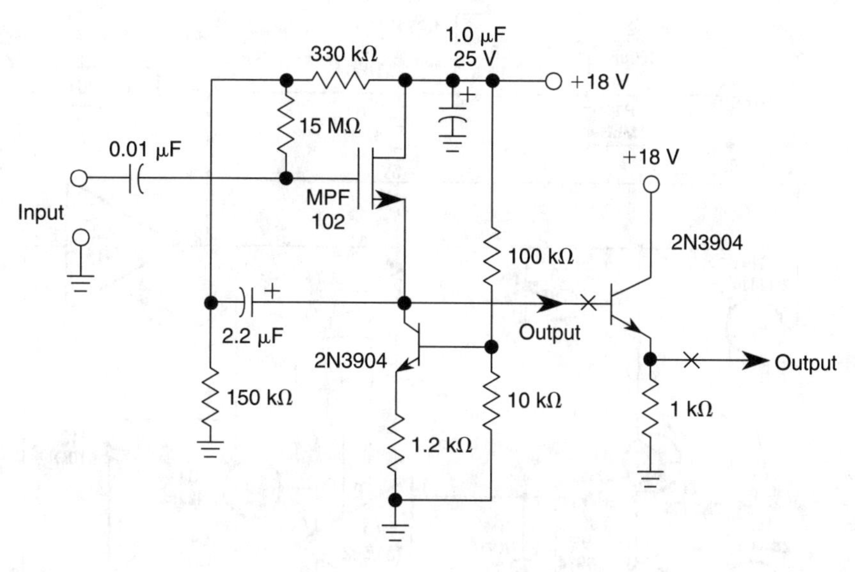

WILLIAM SHEETS

FIG. 2-21

A current source using a 2N3904 transistor plus bootstrapping, achieves an input impedance of 500 MΩ. A second 2N3904 transistor can be added at X to lower the output impedance.

DISCRETE CURRENT-BOOSTER AMPLIFIER

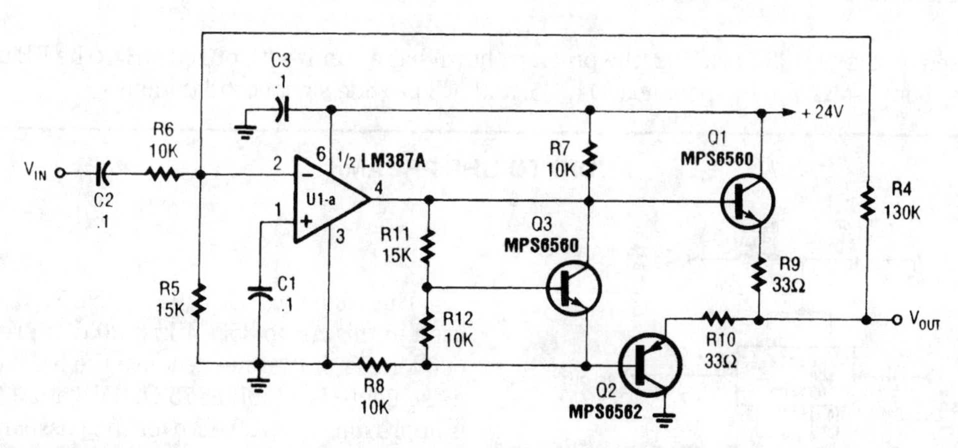

POPULAR ELECTRONICS

FIG. 2-22

Suitable as a line driver, this circuit is useable in many similar audio applications.

FREQUENCY COUNTER PREAMP

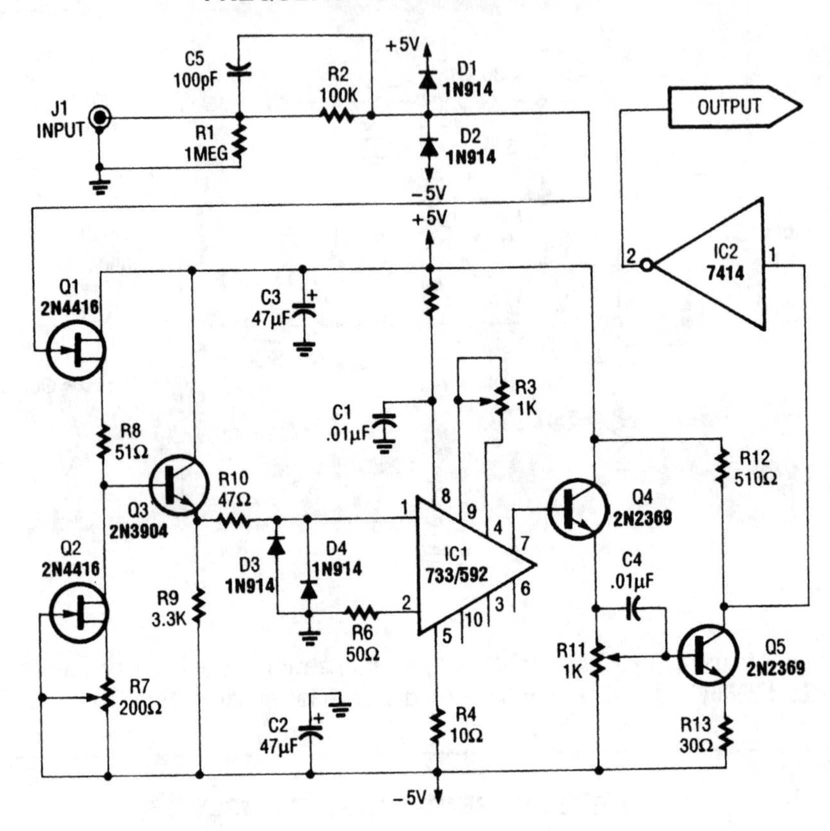

RADIO-ELECTRONICS

FIG. 2-23

Based on the LM733 or NE592, the preamp shown has a bandwidth of 100 MHz. The FET inputs provide about 1-MΩ input impedance. Q4, Q5, and IC2 provide signal conditioning.

AUDIO TO UHF PREAMP

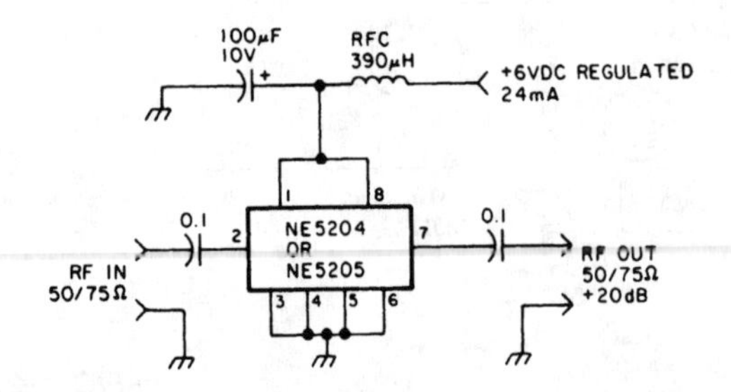

The Signetics NE5204 or NE5205 can be used in this AF to 350-MHz (–30 dB) preamp. If 600 MHz @ 3 dB is needed, use the NE5205. The noise figure is 4.8 dB at 75 Ω, 6 dB at 50 Ω. Gain is approximately +20 dB over the passband.

73 AMATEUR RADIO TODAY

FIG. 2-24

V- & I-PROTECTED INTRINSICALLY SAFE OP AMP

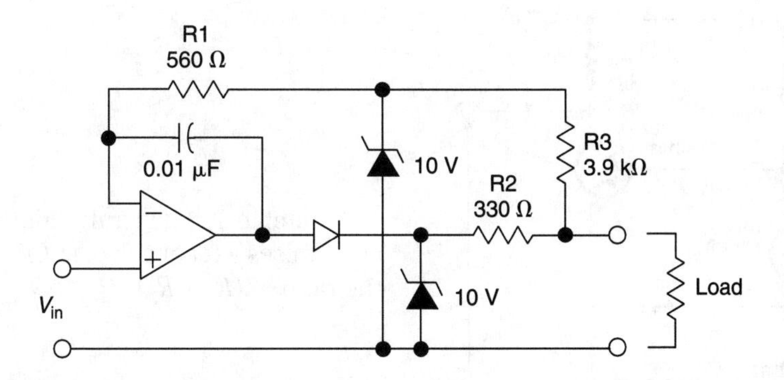

WILLIAM SHEETS

FIG. 2-25

The circuit is designed to drive an external load. A fault condition in the external load circuit could feed excessive current or voltage back into the line drive circuit. If excessive voltage appears from the load, the two zener diodes will clamp that voltage to a safe level, which in this case is 10 V.

The current in the zener diodes, op amp, and the remainder of the circuitry is limited to a safe level by resistors R1, R2, and R3. D1 protects the op-amp output stage from 10 V appearing across the clamp diodes under a fault condition.

The advantage of this circuit is that, although it's designed as unity gain buffer, the same techniques can be applied to inverting, noninverting, or differential gain stages.

CURRENT FEEDBACK AMP DELIVERS 100 mA @ 100 MHz

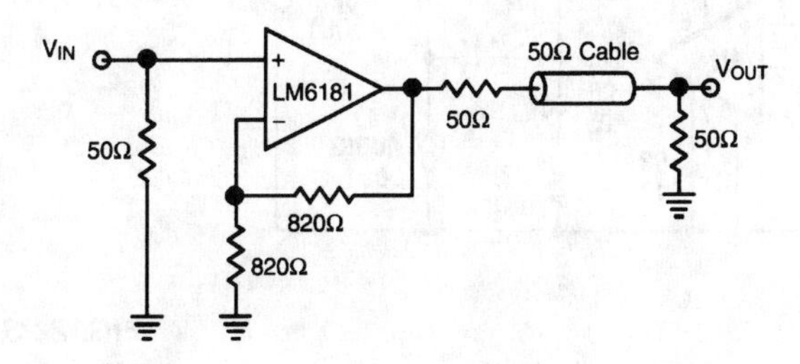

Using a NS LM6181, this IC is useful in cable drivers. The supply voltage is ±5 V to ±15 V.

NATIONAL SEMICONDUCTOR

FIG. 2-26

GENERAL-PURPOSE PREAMPLIFIER

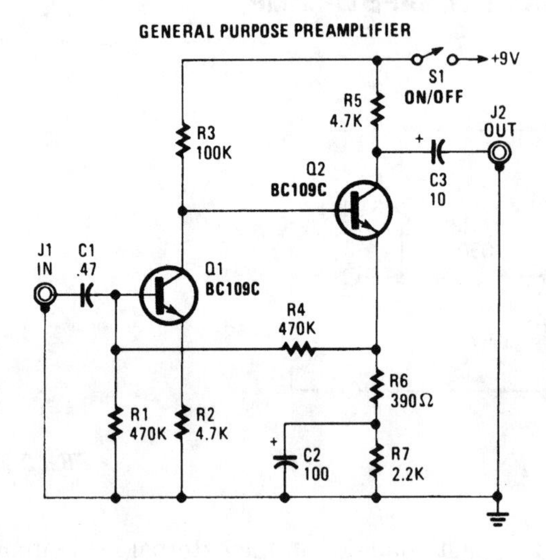

POPULAR ELECTRONICS *FIG. 2-27*

Suitable for general audio use, the preamp circuit uses a feedback pair. Current gain is set by the ratio of $(R_4 + R_6)/R_4$.

TEST BENCH AMPLIFIER

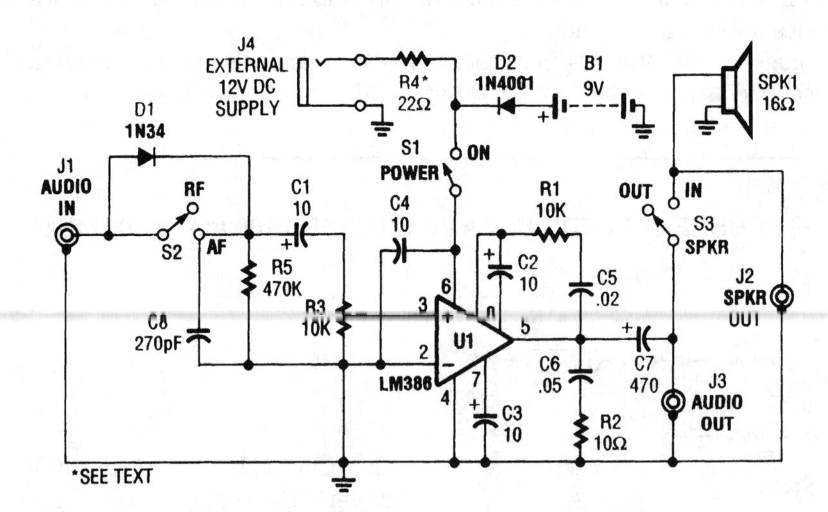

POPULAR ELECTRONICS *FIG. 2-28*

This amplifier might be useful in servicing or bench testing as a signal tracer or as a building block in various systems.

3

Analog-to-Digital Converter Circuits

The sources of the following circuits are contained in the Sources section, which begins on page 675. The figure number in the box of each circuit correlates to the entry in the Sources section.

ADC POLLER

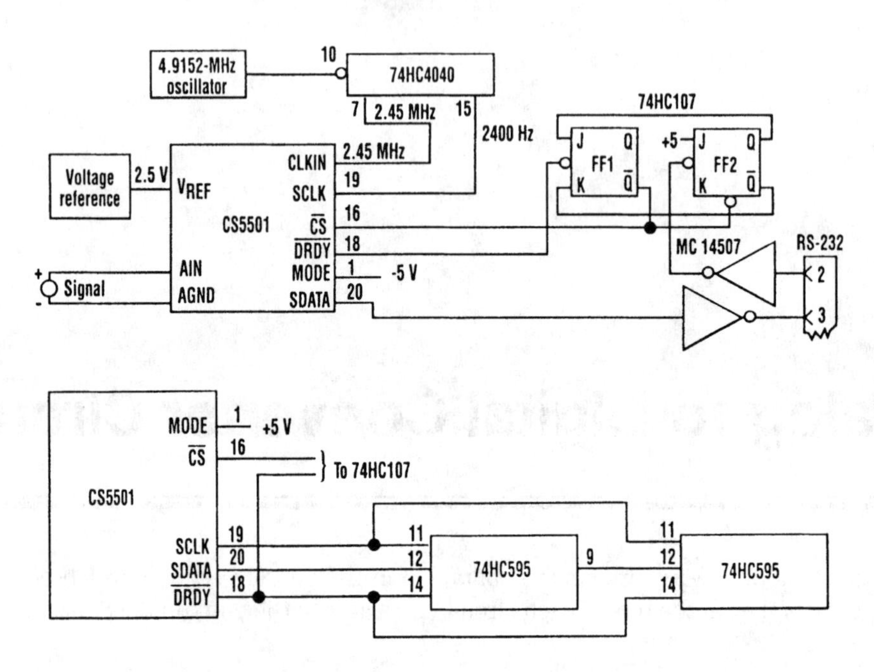

ELECTRONIC DESIGN *FIG. 3-1*

Because the CS5501 16-bit-delta-sigma analog-to-digital converter lacks a "start convert" command, it converts continuously, outputting conversion words to its output register every 1024 cycles of its master clock. However, by incorporating a standard dual J-K flip-flop into the circuit, the ADC can be configured to output a single-conversion word only when it is polled.

The CS5501 converter can be operated in its asynchronous communication mode (UART) to transmit one 16-bit conversion word when it is polled over an RS-232 serial line (see figure). A null character (all zeros) is transmitted to the circuit and sets the flip-flop FF2. The CS5501 can then output a single-conversion word, which is transmitted over the RS-232 line as two bytes with start and stop bits.

The baud rate can be chosen by selecting the appropriate clock divider rate on the 74HC4040 counter/divider as the serial port clock (SLCK) for the ADC. This type of polled-mode operation is also useful when the ADC's output register is configured to operate in the synchronous-serial clock (SSC) mode. In this case, the converter will load one output word into a 16-bit serial-to-parallel register (two 74HC595 8-bit registers) when polled to do so (see figure).

8-CHANNEL A/D CONVERTER FOR PC CLONES

The following program causes the A-D converter to perform eight sequential conversions and display the result. It's written in Turbo BASIC/Power BASIC source code, but it will run under the GW-BASIC interpreter if you replace the delay statements with FOR/NEXT loops, and add line numbers as shown in the second listing. These programs are available on the 73 BBS under the filenames ADC Turbo.BAS and ADCGW.BAS.

```
INITIALIZE:  'remarks follow the apostrophe
   screen 0                          'text mode 80 columns
   color 14,0                        'yellow on blue
   cls                               'clear the screen
   clear                             'clear all variables
   toggle%=2                         'initialize variables
   oddsign%=0

MINORLOOP:
   while not instat                  'keep going until a key is pressed
     out 888,1                       'regulator line high
   delay 1                           wait 1 second before next sample
     out 888,0                       'light up the regulator
     delay .054                      'wait 54 milliseconds to stabilize
   for ch%=0 to 7                    'scan 8 channels
     out 888,8                       'CS high pin 5
     out 888,0                       'CS low
     out 888,2                       'start bit is always high DI line
     out 890,0                       'clock high pin 1 of DB 25 printer
       for slow%=0 to 1:next slow%   'stretches clock pulse
     out 890,1                       'clock low
     out 888,2                       '8 single ended measurements selected
     out 890,0                       'clock high
       for slow%=0 to 1:next slow%   'stretches clock pulse
     out 890,1                       'clock low
     out 888,oddsign%                'part of the channel selection string
     swap oddsign%,toggle%           'toggles between high and low
     out 890,0                       'clock high
       for slow%=0 to 1:next slow%   'stretches clock pulse
     out 890,1                       'clock low
     out 888,select1%                'part of the channel selection string
     out 890,0                       'clock high
       for slow%=0 to 1:next slow%   'stretches clock pulse
     out 890,1                       'clock low
     out 888,select0%                'part of the channel selection string
     out 890,0                       'clock high
       for slow%=0 to 1:next slow%   'stretches clock pulse
   out 890,1                         'clock low
READBITS:
     for bit%=7 to 0 step -1         'MSB is first out
     out 890,0                       'clock high
       for slow%=0 to 1:next slow%   stretches clock pulse
     out 890,1                       'clock low
     ad%=inp(889)                    'port 889 pin 10  7-low135-high
     if ad%<120 then byte%=byte%+(2^bit%)
     next bit%
     if  ch%=0  then  select1%=0 : select0%=0 : ch0volts=byte%/51
     if  ch%=1  then  select1%=0 : select0%=2 : ch1volts=byte%/51
     if  ch%=2  then  select1%=0 : select0%=2 : ch2volts=byte%/51
     if  ch%=3  then  select1%=2 : select0%=0 : ch3volts=byte%/51
     if  ch%=4  then  select1%=2 : select0%=0 : ch4volts=byte%/51
     if  ch%=5  then  select1%=2 : select0%=2 : ch5volts=byte%/51
     if  ch%=6  then  select1%=2 : select0%=2 : ch6volts=byte%/51
     if  ch%=7  then  select1%=0 : select0%=0 : ch7volts=byte%/51
     byte%=0
   next ch%
   print using
   "##.#";ch0volts, ch1volts, ch2volts,ch3colts,ch4volts,ch5volts,ch6volts,ch7volts
     wend
```

GWBASIC Version

```
10 '  The following program causes the A-D converter to perform eight
20 '  sequential conversions and display the result.
30 SCREEN 0                          'text mode 80 columns
40 COLOR 14,0                        'yellow on blue
50 CLS                               'clear the screen
60 CLEAR                             'clear all variables
70 TOGGLE%=2                         'initialize variables
80 ODDSIGN%=0
90 IF INKEY$< >"" THEN END           'keep going until a key is pressed
100 OUT 888,1                        'regulator line high
110 OUT 888,0                        'light up the regulator
120 FOR W%=0 TO 500:NEXT W%          'wait 54 mseconds to stabilize
130 FOR CH%=0 TO 7                   'scan 8 channels
140 OUT 888,8                        'CS high  pin 5
150 OUT 888,0                        'CS low
160 OUT 888,2                        'start bit is always high DI line
170 OUT 890,0                        'clock high pin 1 of DB 25 printer
180 FOR SLOW%=0 TO 1:NEXT SLOW%      'stretches clock pulse
190 OUT 890,1                        'clock low
200 OUT 888,2                        '8 single ended measurements selected
210 OUT 890,0                        'clock high
220 FOR SLOW%=0 TO 1:NEXT SLOW%      'stretches clock pulse
230 OUT 890,1                        'clock low
240 OUT 888,ODDSIGN%                  'part of the channel selection string
250 SWAP ODDSIGN%,TOGGLE%            'toggles between high and low
260 OUT 890,0                        'clock high
270 FOR SLOW%=0 TO 1:NEXT SLOW%      'stretches clock pulse
280 OUT 890,1                        'clock low
290 OUT 888,SELECT1%                 'part of the channel selection string
300 OUT 890,0                        'clock high
310 FOR SLOW%=0 TO 1:NEXT SLOW%      'stretches clock pulse
320 OUT 890,1                        'clock low
330 OUT 888,SELECT0%                 'part of the channel selection string
340 OUT 890,0                        'clock high
350 FOR SLOW%=0 TO 1:NEXT SLOW%      'stretches clock pulse
360 OUT 890,1                        'clock low
370 REM                              'read output bits
380 FOR BIT%=7 TO 0 STEP -1           'MSB is first out
390 OUT 890,0                        'clock high
400 FOR SLOW%=0 TO 1:NEXT SLOW%      if ad%'stretches clock pulse
410 OUT 890,1                        'clock low
420 AD%=INP(889)                     'port 889 pin 10  7-low  135-high
430 IF AD%<120 THEN BYTE%=BYTE%+(2^BIT%)
440 NEXT BIT%
450 IF  CH%=0  THEN  SELECT1%=0 : SELECT0%=0 : CH0VOLTS=BYTE%/51
460 IF  CH%=1  THEN  SELECT1%=0 : SELECT0%=2 : CH1VOLTS=BYTE%/51
470 IF  CH%=2  THEN  SELECT1%=0 : SELECT0%=2 : CH2VOLTS=BYTE%/51
480 IF  CH%=3  THEN  SELECT1%=2 : SELECT0%=0 : CH3VOLTS=BYTE%/51
490 IF  CH%=4  THEN  SELECT1%=2 : SELECT0%=0 : CH4VOLTS=BYTE%/51
500 IF  CH%=5  THEN  SELECT1%=2 : SELECT0%=2 : CH5VOLTS=BYTE%/51
510IF  CH%=6  THEN  SELECT1%=2 : SELECT0%=2 : CH6VOLTS=BYTE%/51
520IF  CH%=7  THEN  SELECT1%=0 : SELECT0%=0 : CH7VOLTS=BYTE%/51
530 BYTE%=0
540 next ch%
550 PRINT USING"##.#";CH0VOLTS, CH1VOLTS, CH2VOLTS,CH3COLTS,
   CH4VOLTS,CH5VOLTS,CH6VOLTS,CH7VOLTS
```


FIG. 3-2

8-CHANNEL A/D CONVERTER FOR PC CLONES (*Cont.*)

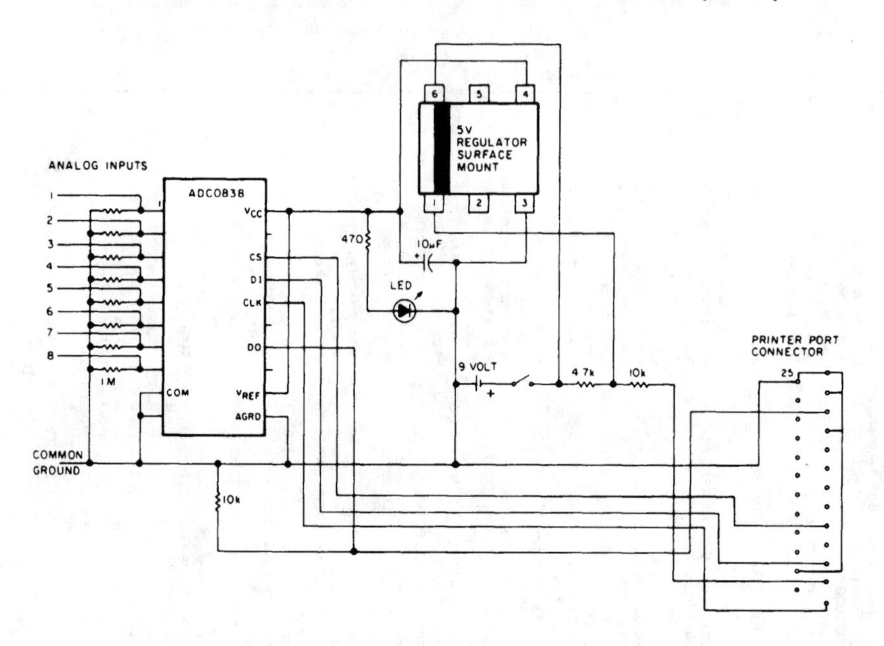

An A/D converter by National Semiconductor (ADC0838), converts 0- to 5-V analog inputs to a digital data format. A 9-V battery is used. The converter connects to the pointer port connector via a 25-pin connector.

4

Antenna Circuits

The sources of the following circuits are contained in the Sources section, which begins on page 675. The figure number in the box of each circuit correlates to the entry in the Sources section.

DUAL-BAND LOOP ANTENNA FOR 80 & 160 m

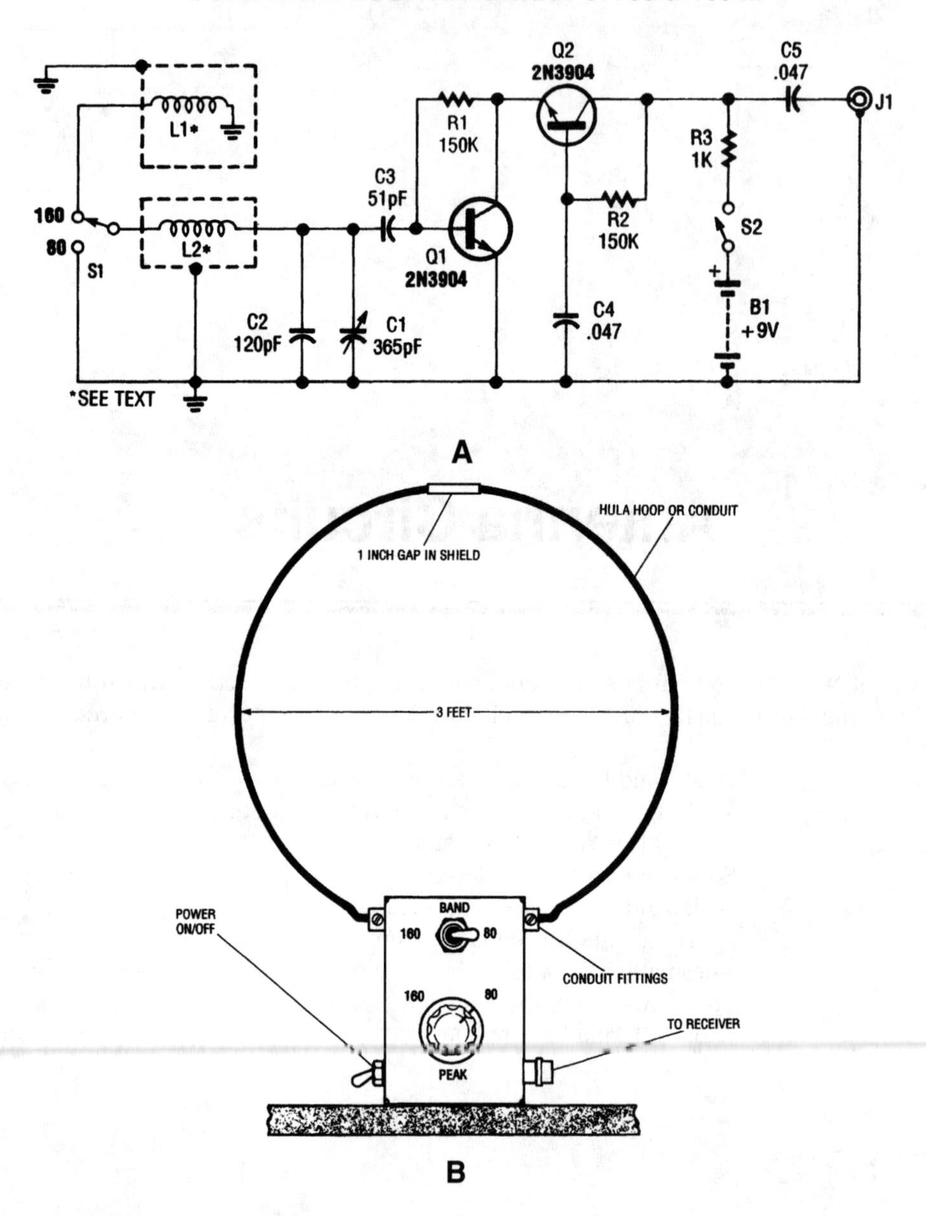

POPULAR ELECTRONICS *FIG. 4-1*

This antenna might help to reduce power-line noise. A plastic "hula hoop" or conduit 3 feet in diameter, covered with aluminum foil as a shield is used for L1 and L2. L1 is two turns and L2 is one turn, threaded through the loop. S1 selects 160- or 80-m operation. Q1 and Q2 form a preamplifier for the loop antenna. Do not transmit with this antenna—it is for receiving only.

VLF/VHF WIDEBAND LOW-NOISE ACTIVE ANTENNA

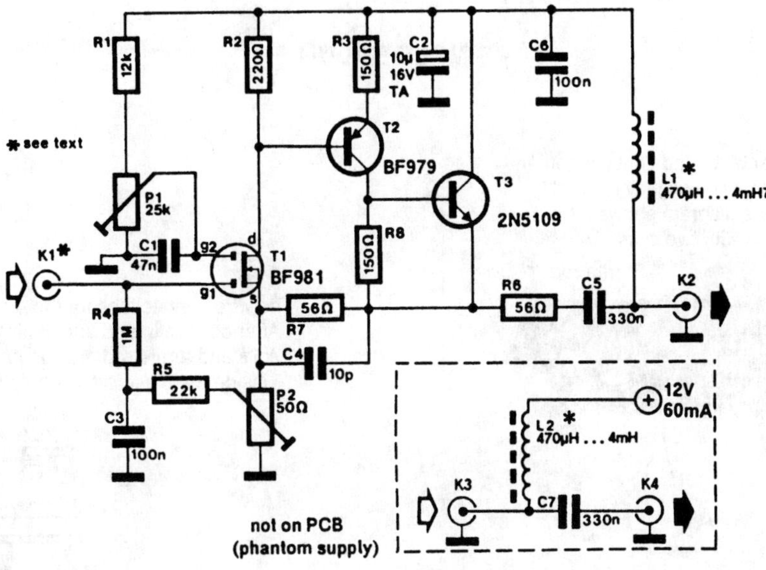

ELECTOR ELECTRONICS USA *FIG. 4-2*

A 30- to 50-cm whip antenna provides reception from 10 kHz to over 220 MHz. T1, a dual-gate MOSFET, provides low noise, high-input impedance, and high gain. The circuit is powered via the coaxial cable used to connect the antenna to a receiver.

VLF 60-kHz ANTENNA/PREAMP

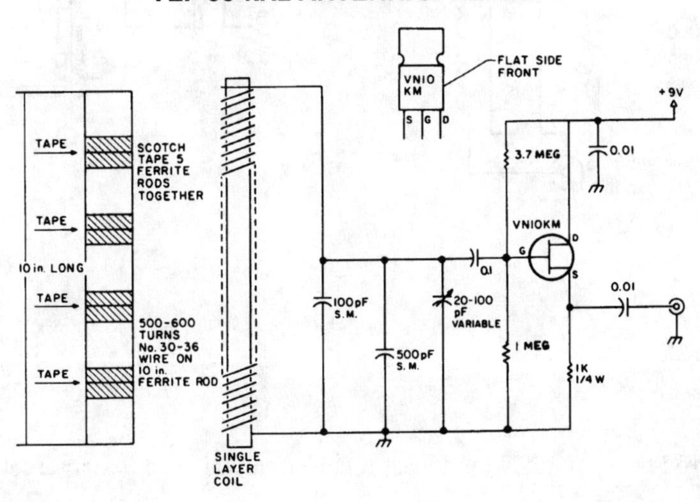

73 AMATEUR RADIO TODAY *FIG. 4-3*

Suitable for 60-kHz standard frequency reception, here is a schematic for a FET preamp and antenna.

SIMPLE BALUN

The wires must be bound tightly together, but windings may be slightly spaced if necessary. The diagram shows a bifilar balun with two coils.

Ferrite rod

A

An example of a 4:1 bifilar (a), and (b) a 1:1 trifilar balun.

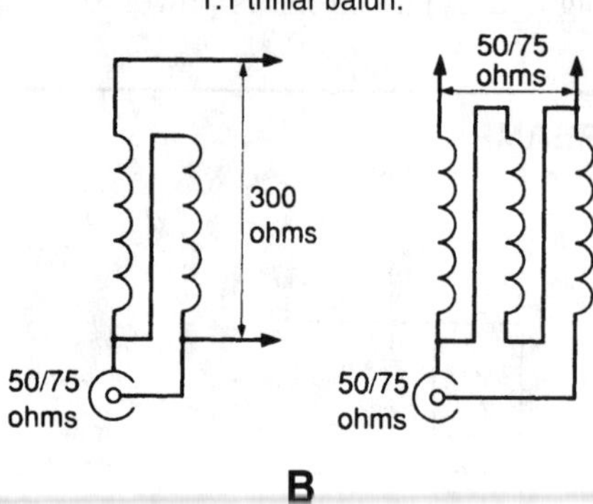

The wire connections for the 4:1 balun. After connecting up and testing, the coils and ferrite rod may be located inside the plastics film container.

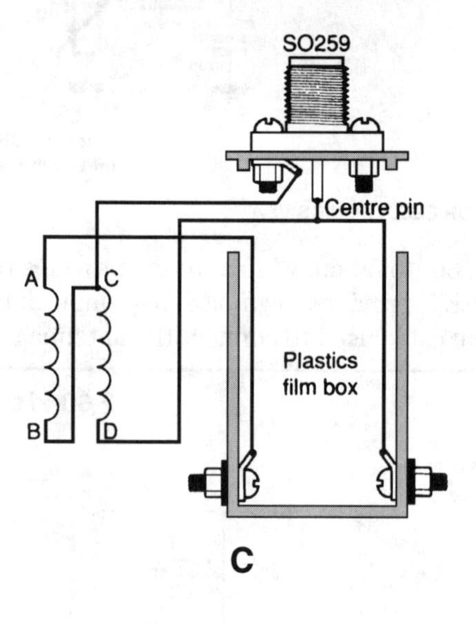

PRACTICAL WIRELESS ***FIG. 4-4***

An old ferrite rod from a junked broadcast receiver can be used to construct an antenna balun, as shown.

WIDEBAND ANTENNA PREAMPLIFIER

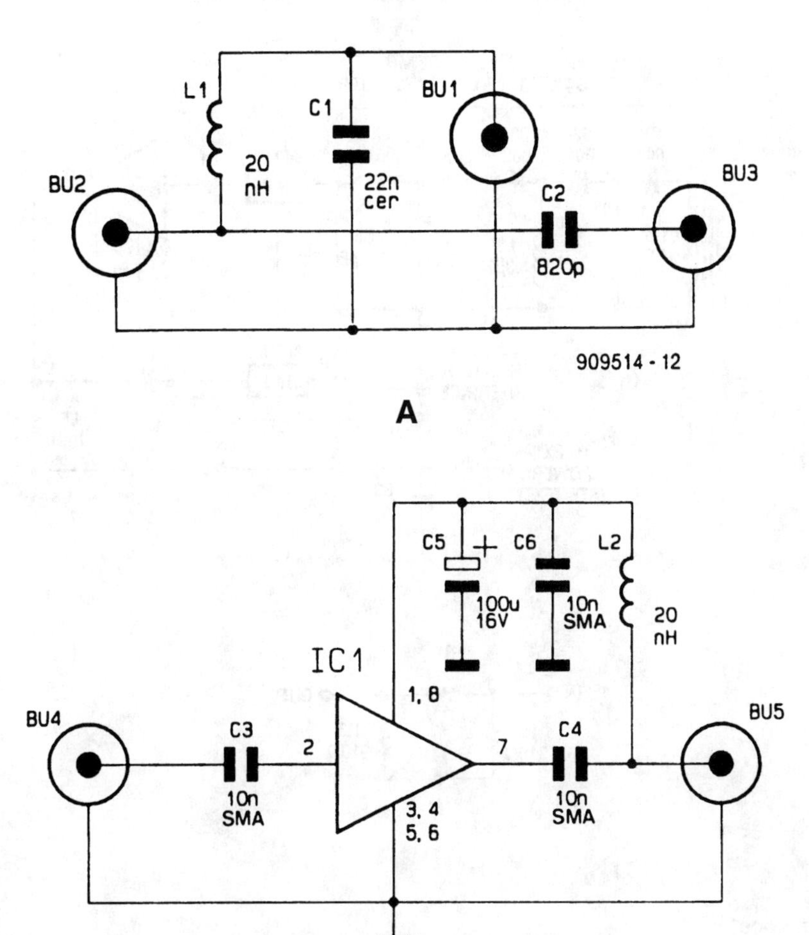

ELEKTOR ELECTRONICS ***FIG. 4-5***

This wideband antenna preamplifier has a gain of around 20 dB from 40 to 860 MHz, covering the entire VHF, FM, commercial, and UHF bands. A phantom power supply provides dc to the preamp via the coaxial cable feeding the unit.

HF BROADBAND ANTENNA PREAMP

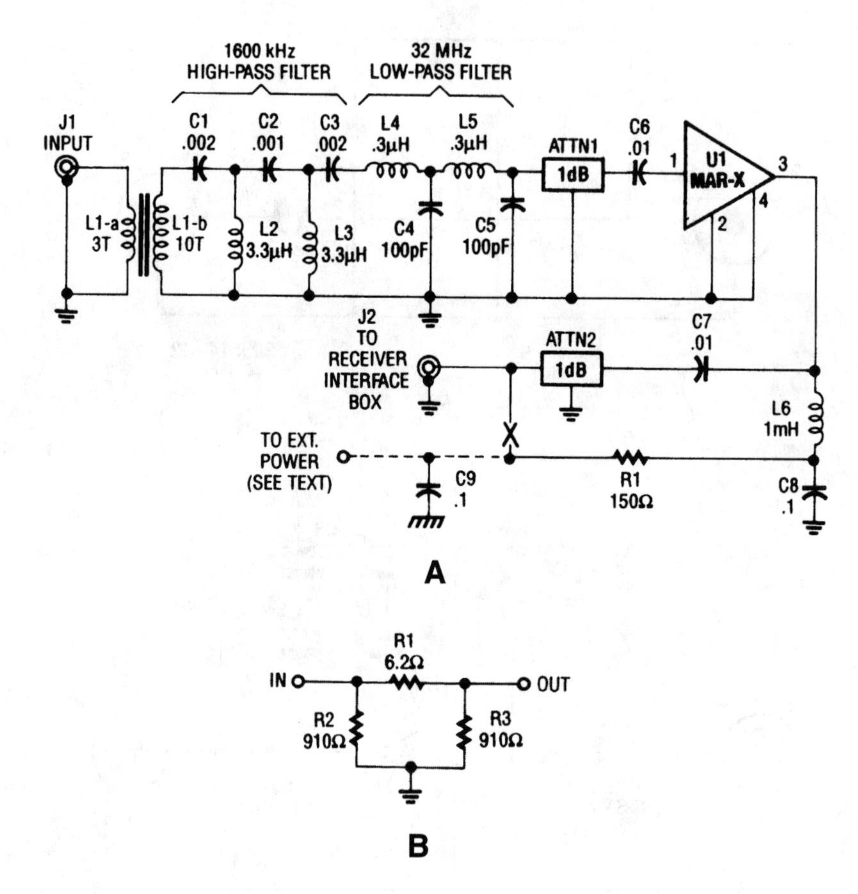

POPULAR ELECTRONICS **FIG. 1-6**

The HF/SW receiver preamplifier is comprised of a broadband toroidal transformer (L1-a and L1-b), LC network (comprised of a 1600-kHz, high-pass filter and a 32-MHz, low-pass filter), L2 and L3 (26 turns of #26 enameled wire wound on an Amidon Associates T-50-2, red, toroidal core), a pair of resistive attenuators (ATTN1 and ATTN2), and a MAR-x device.

Shown here is the composition of a basic 1-dB pi-network resistor antenuator. This is the method of supplying dc power to a preamplifier using only the RF coax cable.

AUTOMATIC TR SWITCH

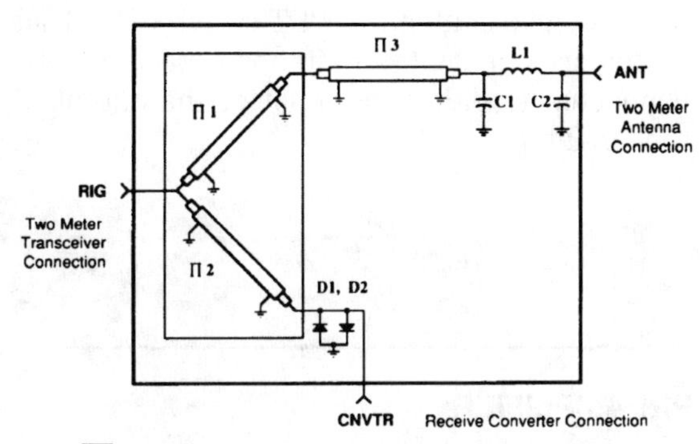

Indicates Wilkinson Hybrid section - See text for discussion

C1, C2 - 39pF mica caps

D1, D2 - 1N914, 1N4148 Si Diodes

L1 - 2 turns # 18 tinned wire, 1/4 inch ID, 0.2 inch long

Π1, Π2 and **Π3** consist of 75 Ω coax sections, 1/4 wave at the center of the transceiver transmitter band typically 147 Mhz.
Π1 and **Π3** are combined in one continuous length of cable - 1/2 wavelength total.
See text for additional discussion.

A

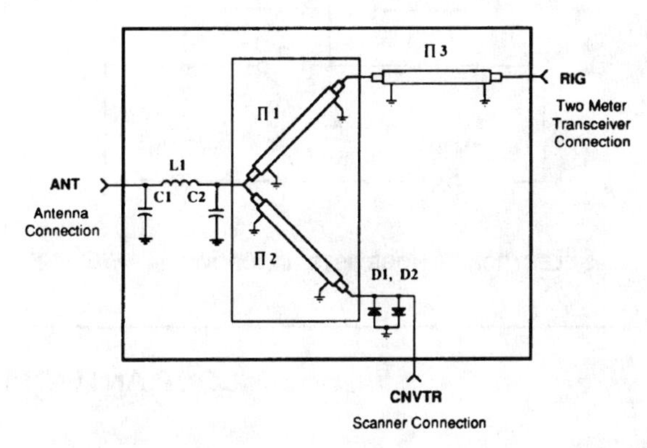

C1, C2 - 39pF mica caps

D1, D2 - 1N914, 1N4148 Si Diodes

L1 - 2 turns # 18 tinned wire, 1/4 inch ID, 0.2 inch long

Π1, Π2 and **Π3** consist of 75 Ω coax sections, 1/4 wave at the center of the transceiver transmitter band typically 147 Mhz.
Π1 and **Π3** are combined in one continuous length of cable - 1/2 wavelength total.
See text for additional discussion.

B

73 AMATEUR RADIO TODAY ***FIG. 4-7***

A pair of diodes and a quarter-wave transmission line are used as an automatic TR switch. D1 and D2 conduct during transmit periods, short-circuiting the scanner input. In this mode, the ¼-wave line appears as an open circuit. In receive, the circuit acts as a Wilkinson power divider.

LOW-POWER ANTENNA TUNER

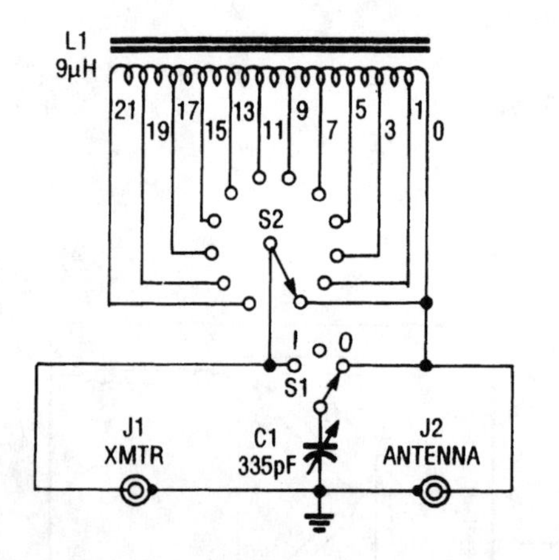

This antenna tuner is suitable for use with low-power (less than 5 W) transmitters or SW receivers. S2 selects inductance and S2 connects the 365-pF capacitor to either the transmitter or the side of the inductor. The tiny tuner is comprised of a tapped inductor (L1) and a variable capacitor (C1), which is connected to the inductor through a center-off SPDT switch (S1). That switch arrangement permits the capacitor to be connected to either the input or the output of the circuit.

1993 ELECTRONICS HOBBYISTS HANDBOOK *FIG. 4-8*

LOOP ANTENNA PREAMPLIFIER

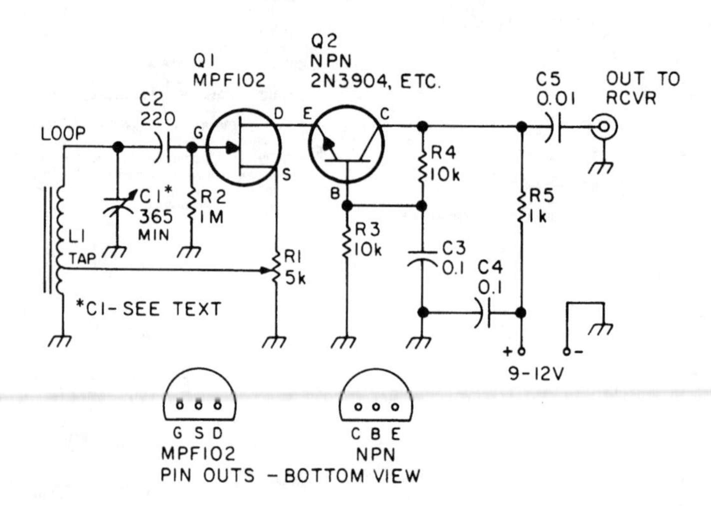

73 AMATEUR RADIO TODAY *FIG. 4-9*

This preamplifier has a built-in regeneration control boost gain selectivity. C1 is a single or multi-gang AM broadcast-band tuning capacitor. L1 is a ferrite loop antenna, tapped at about 15 to 25% of total turns. This circuit should prove useful for low-frequency (up to 3 MHz) reception, where a loop would be advantageous to reduce man-made noise pickup.

5

Audio Power Amplifier Circuits

The sources of the following circuits are contained in the Sources section, which begins on page 675. The figure number in the box of each circuit correlates to the entry in the Sources section.

20-W + 20-W STEREO AMPLIFIER

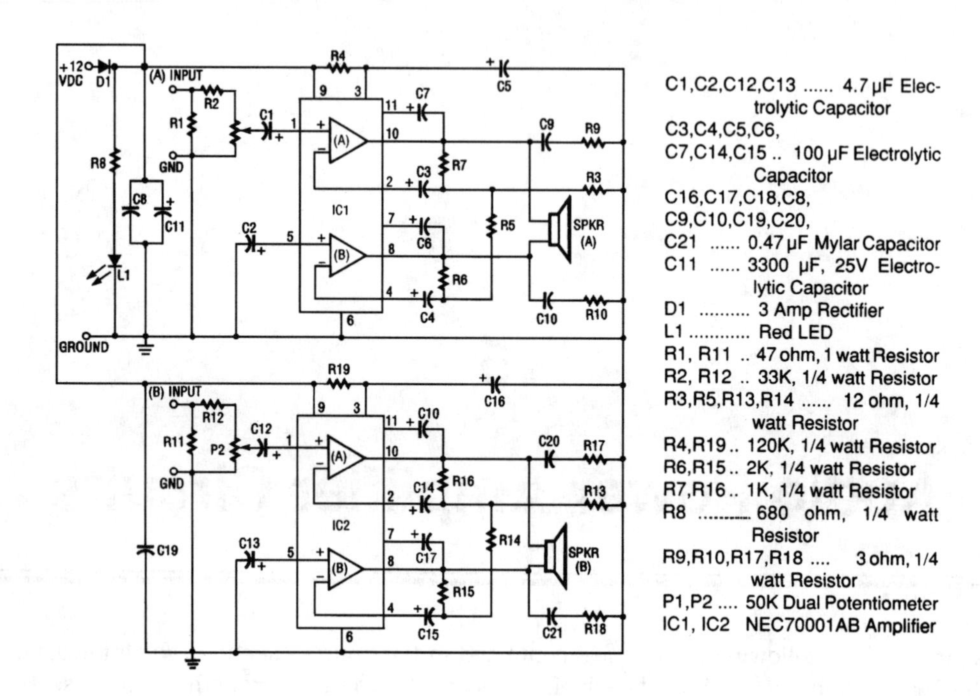

C1,C2,C12,C13 ...... 4.7 μF Electrolytic Capacitor
C3,C4,C5,C6,
C7,C14,C15 .. 100 μF Electrolytic Capacitor
C16,C17,C18,C8,
C9,C10,C19,C20,
C21 ...... 0.47 μF Mylar Capacitor
C11 ...... 3300 μF, 25V Electrolytic Capacitor
D1 .... 3 Amp Rectifier
L1 .... Red LED
R1, R11 .. 47 ohm, 1 watt Resistor
R2, R12 .. 33K, 1/4 watt Resistor
R3,R5,R13,R14 ...... 12 ohm, 1/4 watt Resistor
R4,R19 .. 120K, 1/4 watt Resistor
R6,R15 .. 2K, 1/4 watt Resistor
R7,R16 .. 1K, 1/4 watt Resistor
R8 .... 680 ohm, 1/4 watt Resistor
R9,R10,R17,R18 .... 3 ohm, 1/4 watt Resistor
P1,P2 .... 50K Dual Potentiometer
IC1, IC2 NEC70001AB Amplifier

1991 PE HOBBYIST HANDBOOK ***FIG. 5-1***

The 20-W + 20-W stereo amp consists of two complete, separate 20-W RMS bridge-type amplifiers. The input signal source is brought into the amplifier through the voltage divider network, which is made up of R1, R2, and P1. Resistor R1 provides a load impedance between the signal source and ground. Resistor R2 couples that signal to potentiometer P1.

The signal is coupled by capacitor C1 to the noninverting (+) input (pin 1) of internal amplifier (A) of IC1, where the signal is greatly amplified. Capacitor C2 couples the (+) input of the other (B) internal amplifier of IC1 to ground. That causes the input signal, which is referenced to ground, to be coupled to both amplifiers because both the inputs and outputs of IC1 (A) and IC1 (B) are connected in a bridge configuration. Notice that the output of IC1 (A) from pin 10 is connected to one side of the speaker and the output of IC1 (B) from pin 8 is connected to the other side of the speaker. That is why the speakers used cannot have one side connected to ground. Resistors R6 and R7 set the gain of the amplifier. Resistors R9 and R10 and capacitors C9 and C10 provide frequency stability and prevent oscillation. Capacitors C6 and C7 provide "bootstrapping," which prevents distortion at low frequencies. LED L1 lights up by way of a series resistor connected from the anode to +12 Vdc when power is applied.

Power for both IC1 and IC2 is brought in through D1 (to protect amplifiers from reverse polarity). Capacitor C11 provides additional power supply line filtering. This booster is capable of producing 20 W RMS output out of each channel.

40-W AMPLIFIER

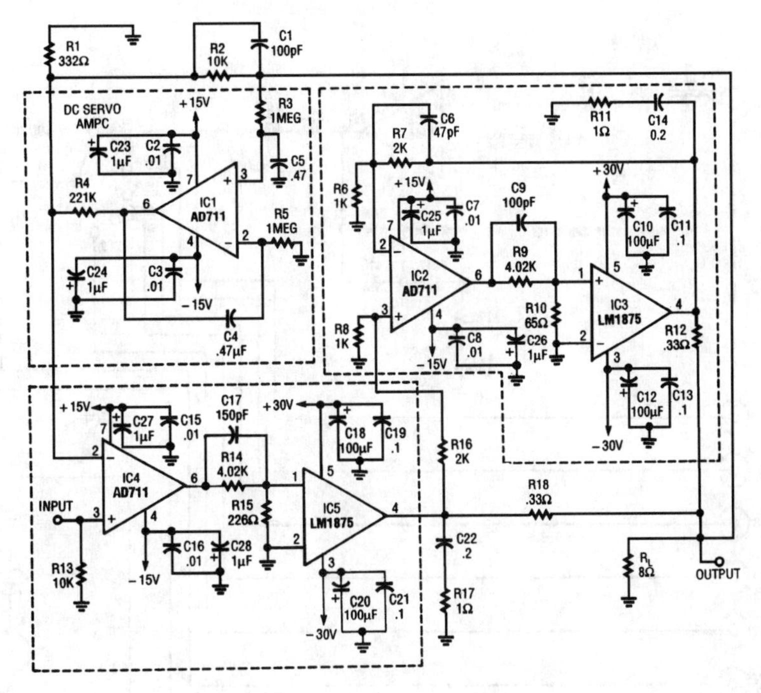

ELECTRONICS NOW *FIG. 5-2*

This circuit uses two LM1875 devices and a dc servo loop. This circuit provides 40-W output. IC3 and IC5 must be heatsinked.

HALF-WATT SINGLE-CHANNEL AUDIO AMPLIFIER

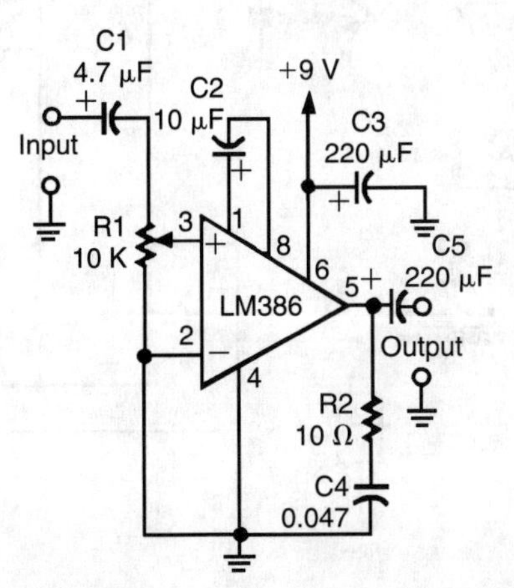

This circuit uses an LM386 IC and will work from 6- to 12-V battery sources. Output is about 0.5 W into 8 Ω.

ELECTRONICS NOW *FIG. 5-3*

DUAL AUDIO AMPLIFIER

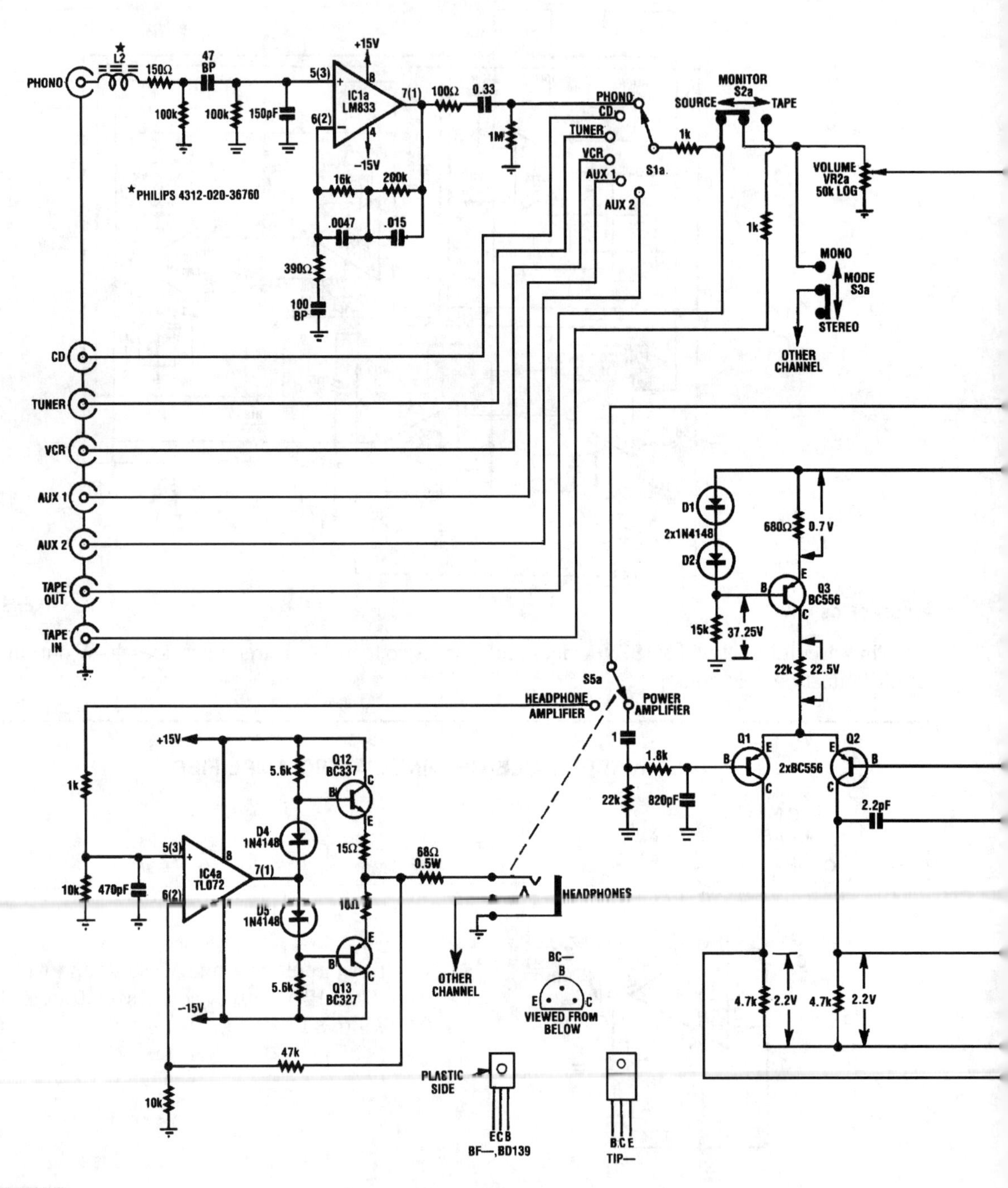

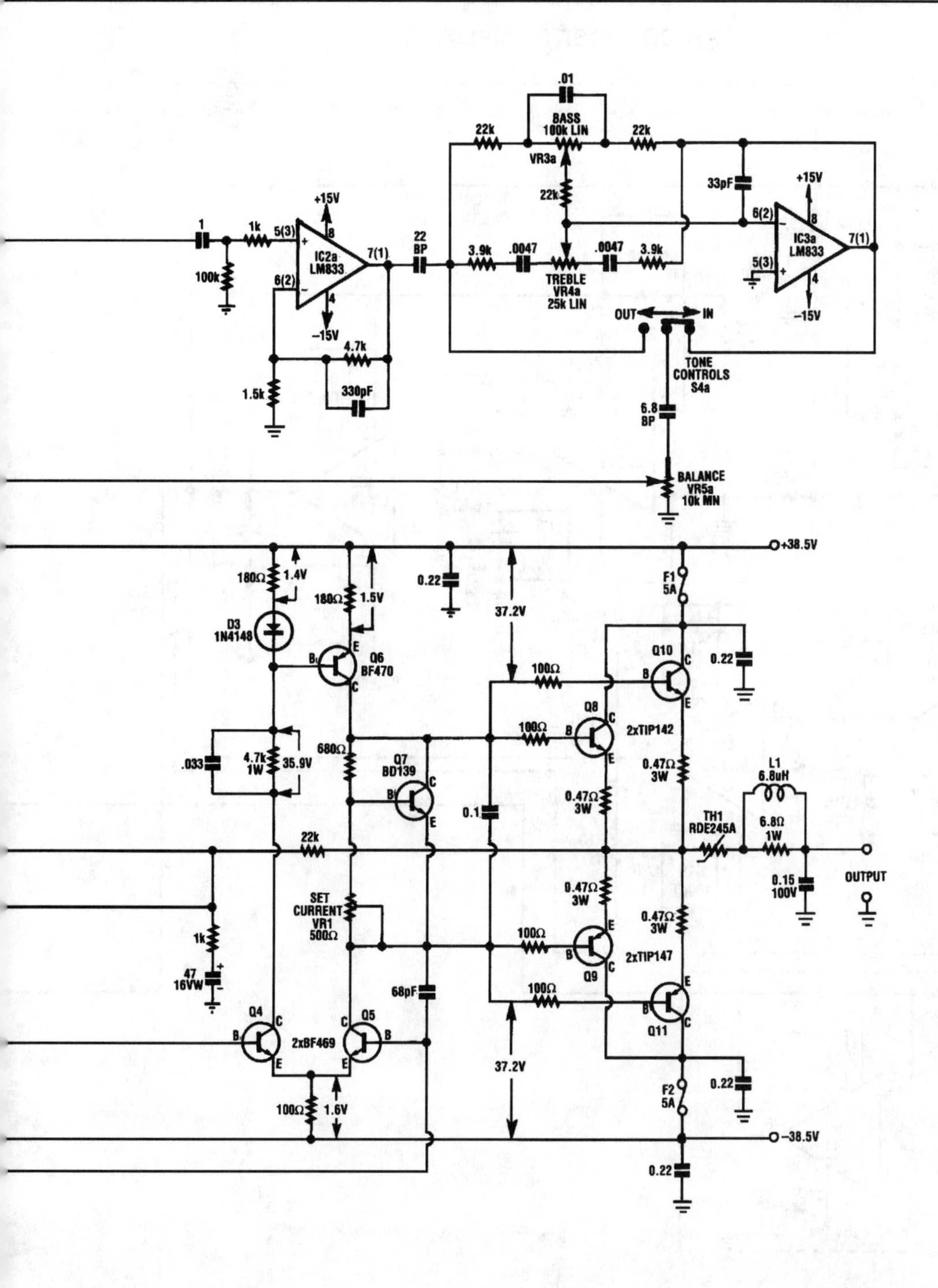
.01
BASS
100k LIN
22k
22k
VR3a
33pF
+15V
22k
6(2)
8
IC3a
LM833
7(1)
+15V
1
1k
5(3)
8
IC2a
LM833
7(1)
22
BP
3.9k
.0047
.0047
3.9k
5(3)
100k
6(2)
4
TREBLE
VR4a
25k LIN
4
OUT
IN
-15V
-15V
4.7k
TONE
CONTROLS
S4a
1.5k
330pF
6.8
BP
BALANCE
VR5a
10k MN
+38.5V
180Ω
1.4V
180Ω
1.5V
0.22
37.2V
F1
5A
D3
1N4148
Q6
BF470
Q10
0.22
100Ω
Q8
100Ω
2xTIP142
680Ω
.033
4.7k
1W
35.9V
Q7
BD139
0.47Ω
3W
L1
6.8uH
0.47Ω
3W
0.1
TH1
RDE245A
6.8Ω
1W
22k
0.15
100V
OUTPUT
SET
CURRENT
VR1
500Ω
0.47Ω
3W
0.47Ω
3W
100Ω
2xTIP147
Q9
1k
47
16VW
68pF
100Ω
Q11
Q4
Q5
2xBF469
37.2V
0.22
F2
5A
100Ω
1.6V
-38.5V
0.22

FIG. 5-4

A 70-W COMPOSITE AMPLIFIER

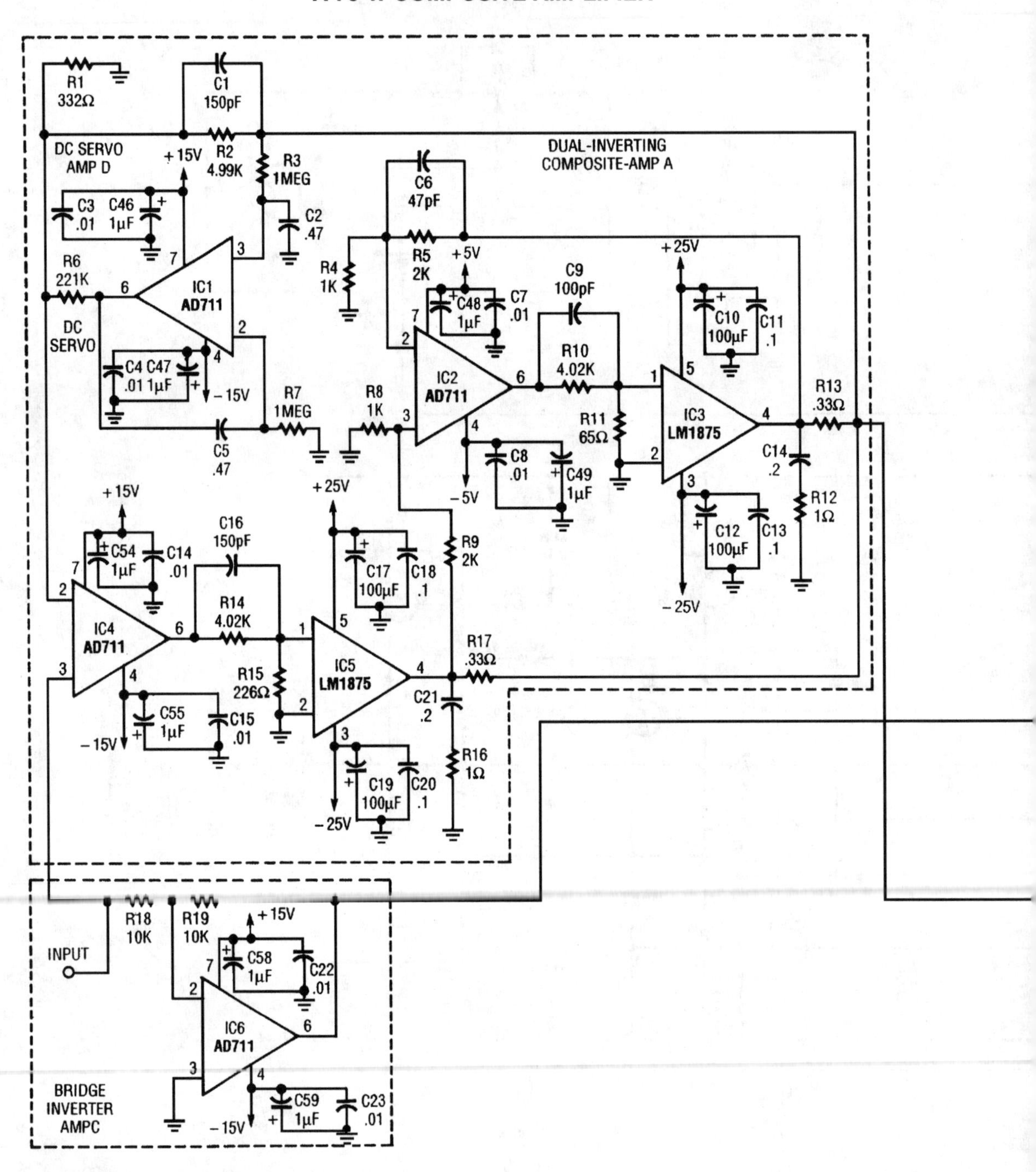

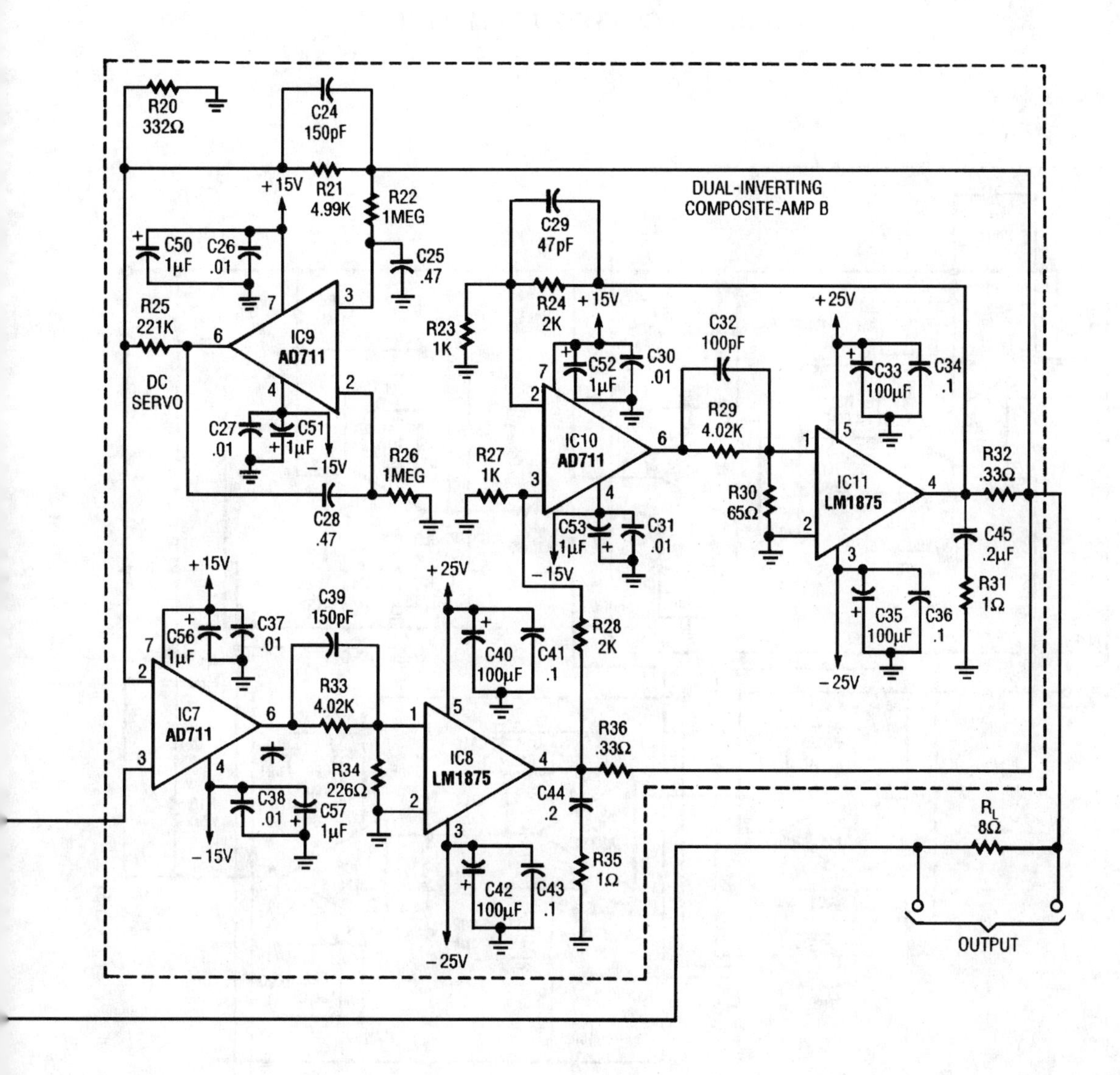

Four LM1875 devices, suitably heatsinked, and a ±25-V supply, 70 W of output are available from this circuit. IC6 is a phase inverter.

FIG. 5-5

A 33-W BRIDGE COMPOSITE AMPLIFIER

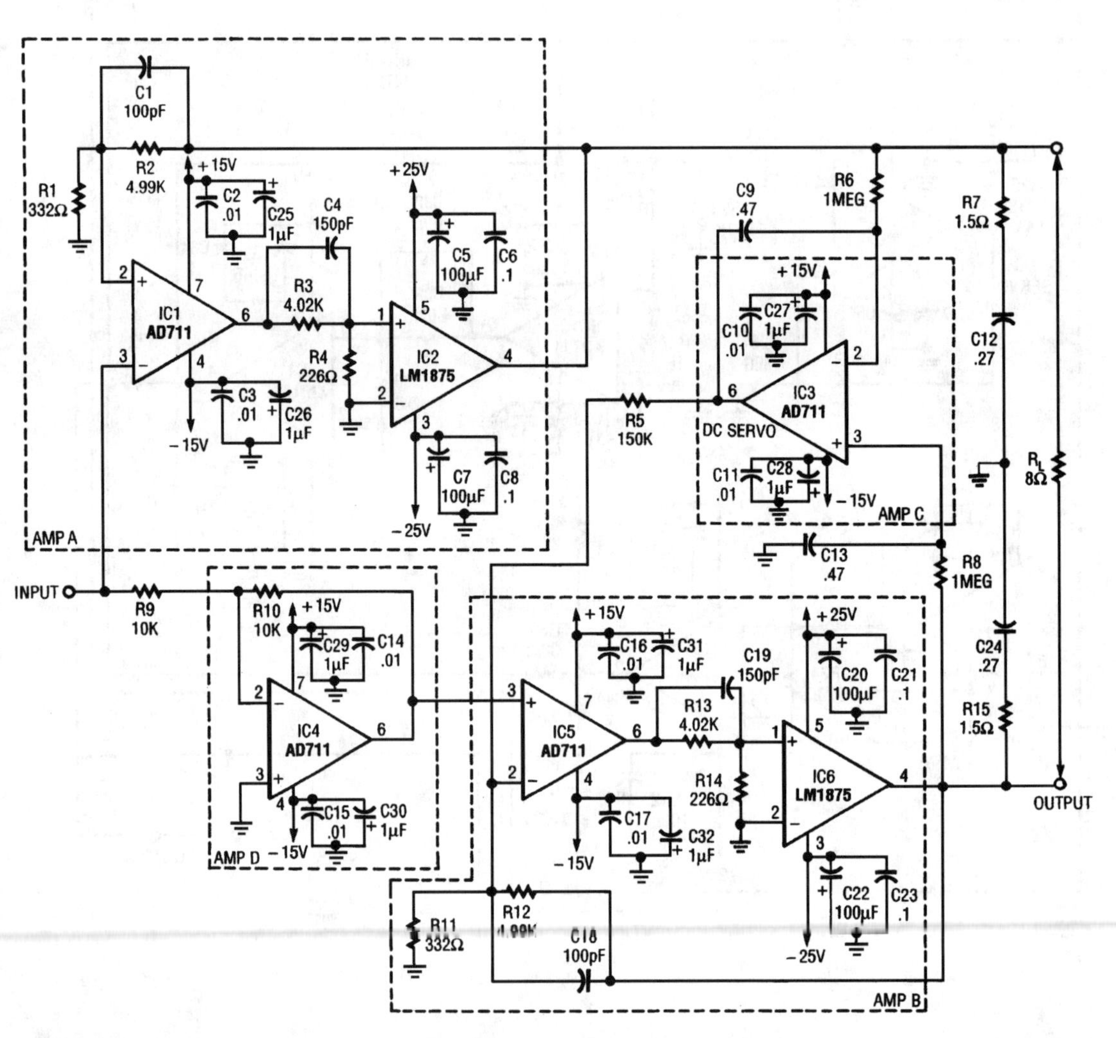

ELECTRONICS NOW

FIG. 5-6

Two LM1875 ICs provide 33 W of audio. IC4 is used as a phase inverter. IC6 and IC2 must be heatsinked.

MOSFET POWER AMPLIFIER

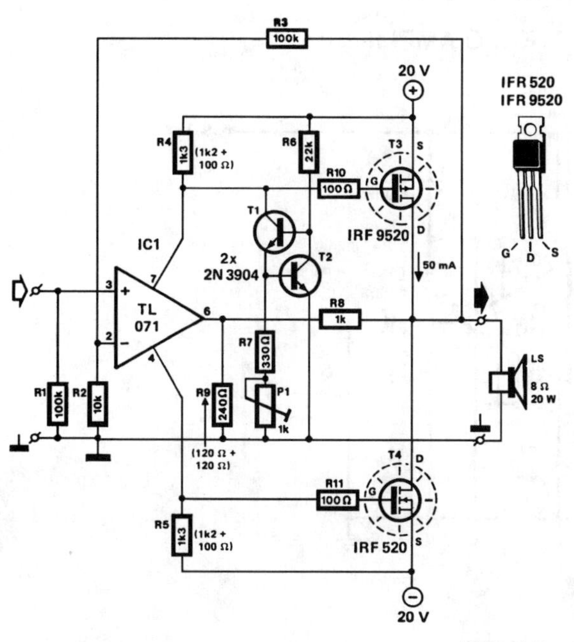

Two complementary MOSFETs are used to deliver 20 W into 8 Ω. A TL071 op amp is used as an input amplifier. The MOSFETs should be heatsinked with a heatsink of better than 5°C/W capability. THD is less than 0.15% from 100 Hz to 10 kHz.

303 CIRCUITS

FIG. 5-7

10-W NONINVERTING COMPOSITE AMPLIFIER

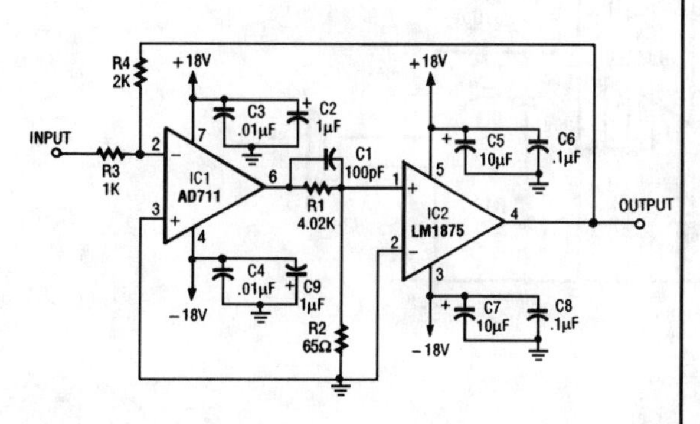

ELECTRONICS NOW

FIG. 5-8

By using an LM1875, suitably heatsinked, a 10-W amplifier that uses two IC devices can be built. IC2 must be heatsinked.

10-W INVERTING COMPOSITE AMPLIFIER

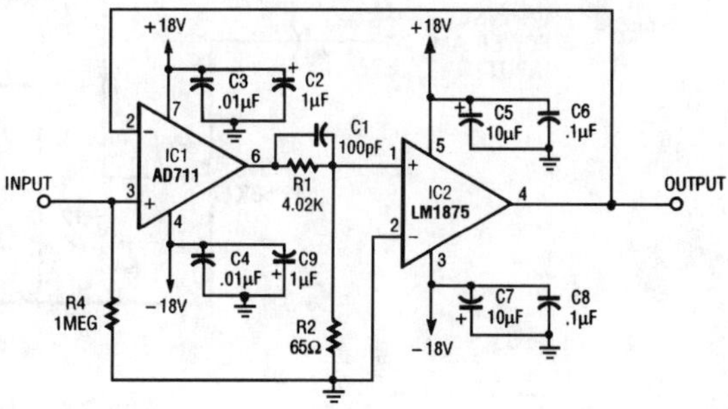

ELECTRONICS NOW

FIG. 5-9

Using an LM1875, a 10-W amplifier can be build using just two IC devices. The gain = R_4/R_3. Note that IC12 must be heatsinked.

LM380 PERSONAL STEREO AMPLIFIER

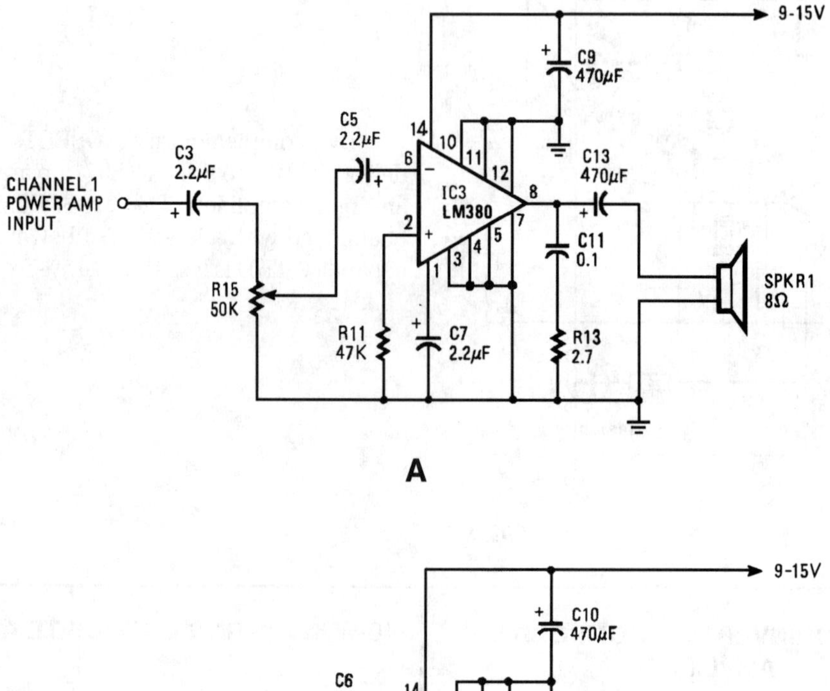

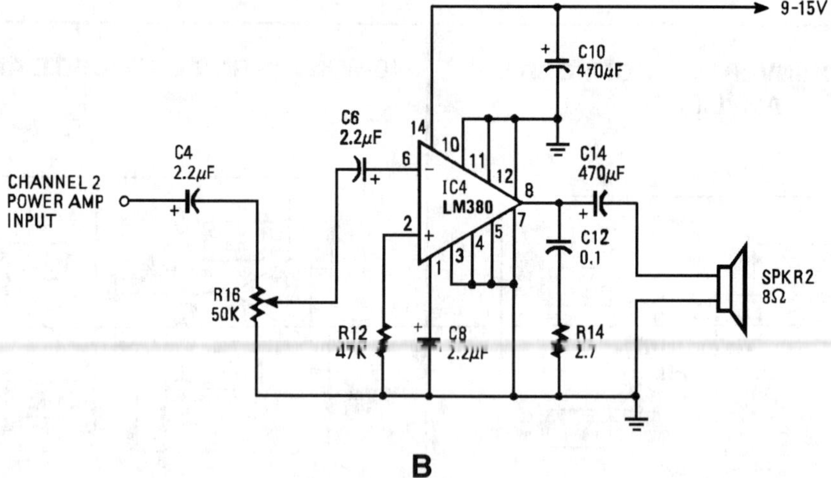

1987 R-E EXPERIMENTERS HANDBOOK ***FIG. 5-10***

With the simple circuit, you can use your personal stereo to drive standard 8-Ω speakers.

SUBWOOFER AMPLIFIER

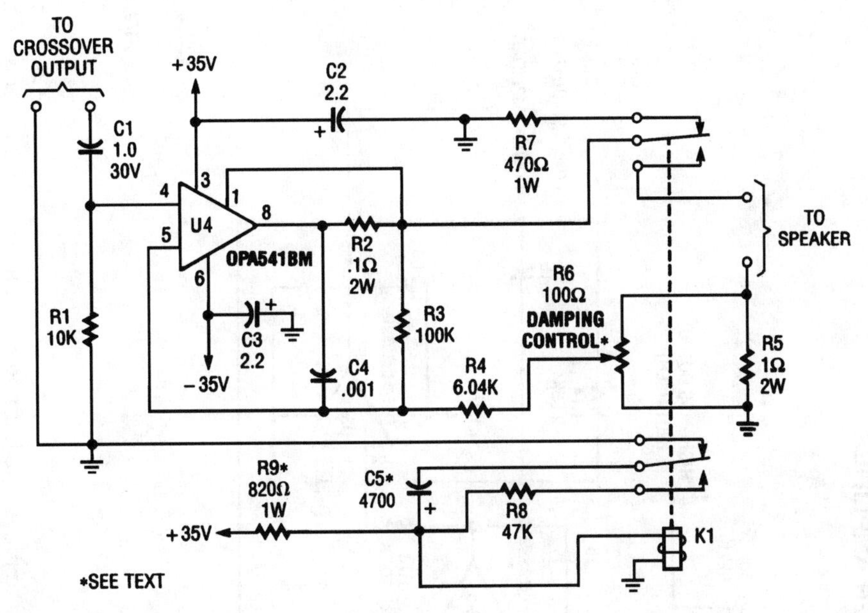

POPULAR ELECTRONICS *FIG. 5-11*

Designed to feed a low-frequency subwoofer speaker system, the amplifier is capable of up to 100 W into an 8-Ω load. The OPA541BM op amp requires heatsinking and is manufactured by Burr-Brown Corporation. A damping control and a relay to eliminate turn-on and turn-off thump in the speaker is included.

18-W BRIDGE AUDIO AMPLIFIER

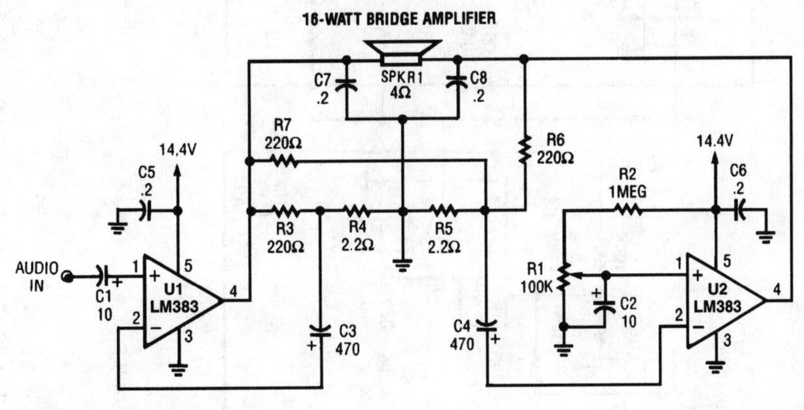

POPULAR ELECTRONICS *FIG. 5-12*

Two LM383 IC devices are used in a bridge circuit that is useful for auto sound applications.

SUBWOOFER CROSSOVER AMPLIFIER

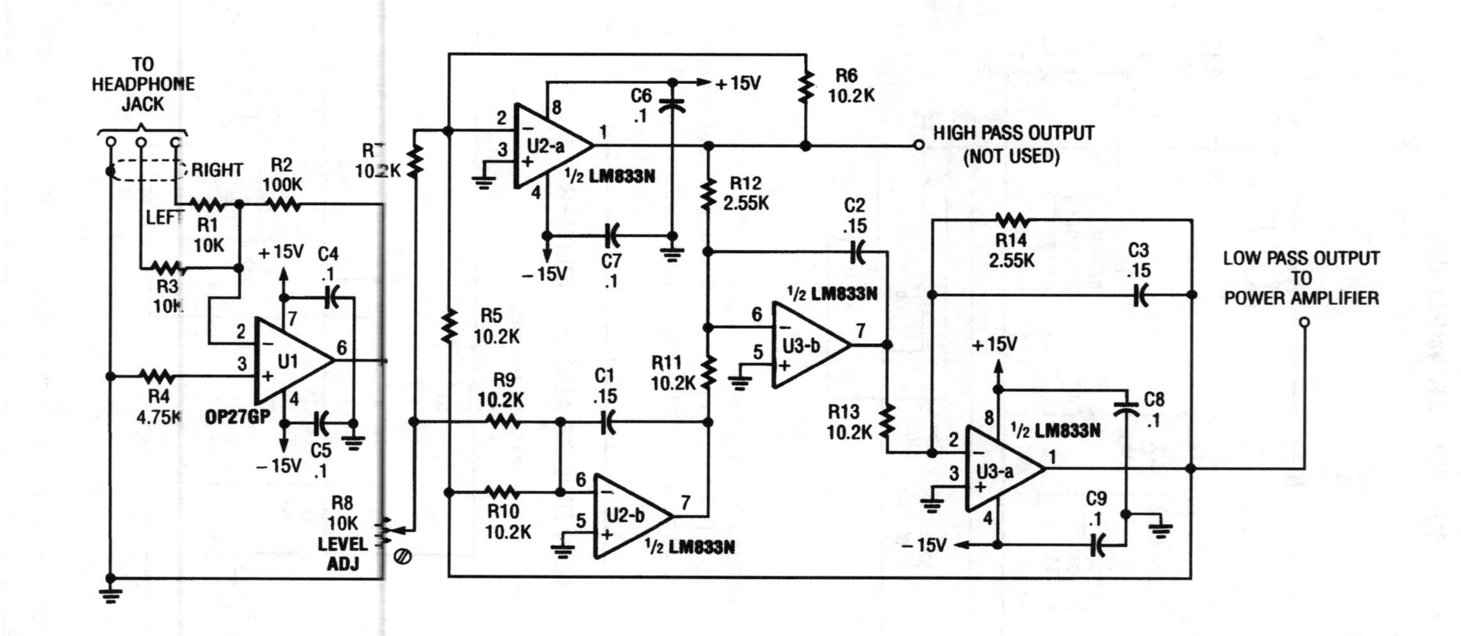

RADIO-ELECTRONICS

FIG. 5-13

The electronic-crossover circuit contains a summing amplifier that combines the left and right channels from a stereo's headphone jack. Originally used in a subwoofer system, the above circuit might be useful in similar audio applications.

AUDIO POWER AMPLIFIER

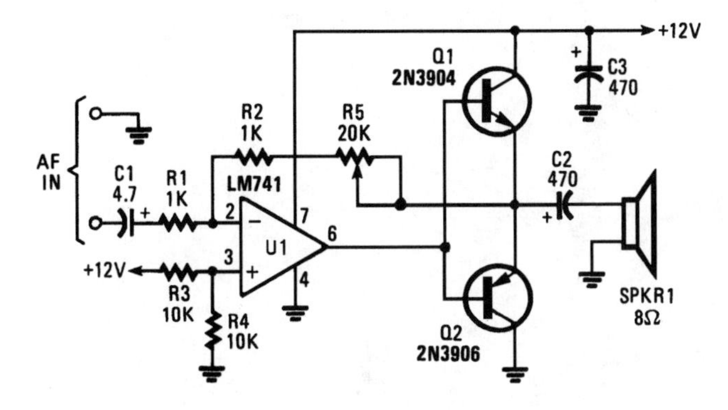

POPULAR ELECTRONICS *FIG. 5-14*

The circuit, built around an LM741 op amp configured as an inverting amplifier, is used to drive complementary transistors (Q1 and Q2). The op amp's feedback loop includes the base-emitter junctions of both transistors—an arrangement that helps to reduce crossover distortion that would normally occur as a result of the emitter-to-base junction voltage drop of about 0.6 V. Potentiometer R5 varies the amplifier's voltage gain from 1 to about 20. As much as 0.5 W can be obtained from the circuit if a heatsink is added to the transistors.

FAST HIGH-VOLTAGE LINEAR POWER AMP

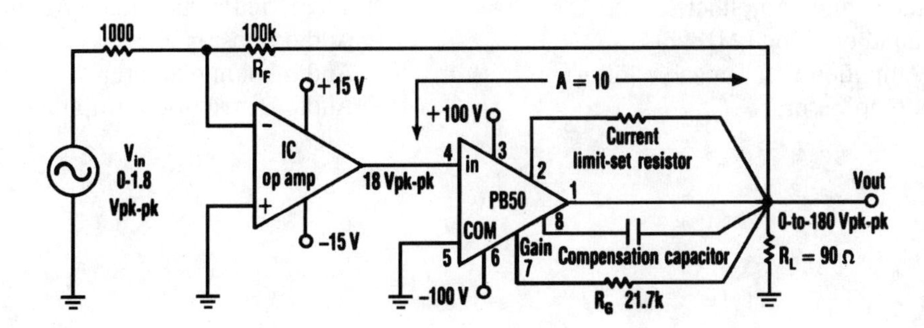

ELECTRONIC DESIGN *FIG. 5-15*

An Apex PB50 Booster Amplifier, plus an IC op amp, can be used in a high-voltage op amp that converts a small analog signal to a 180-V p-p signal.

Apex Microtechnology manufactures a number of power op amps. The above circuit uses a PB50 booster amplifier to deliver a 180-V p-p signal into a 90-Ω load, from a ±100-V supply.

6

Audio Signal Amplifier Circuits

The sources of the following circuits are contained in the Sources section, which begins on page 675. The figure number in the box of each circuit correlates to the entry in the Sources section.

HEADPHONE AMPLIFIER

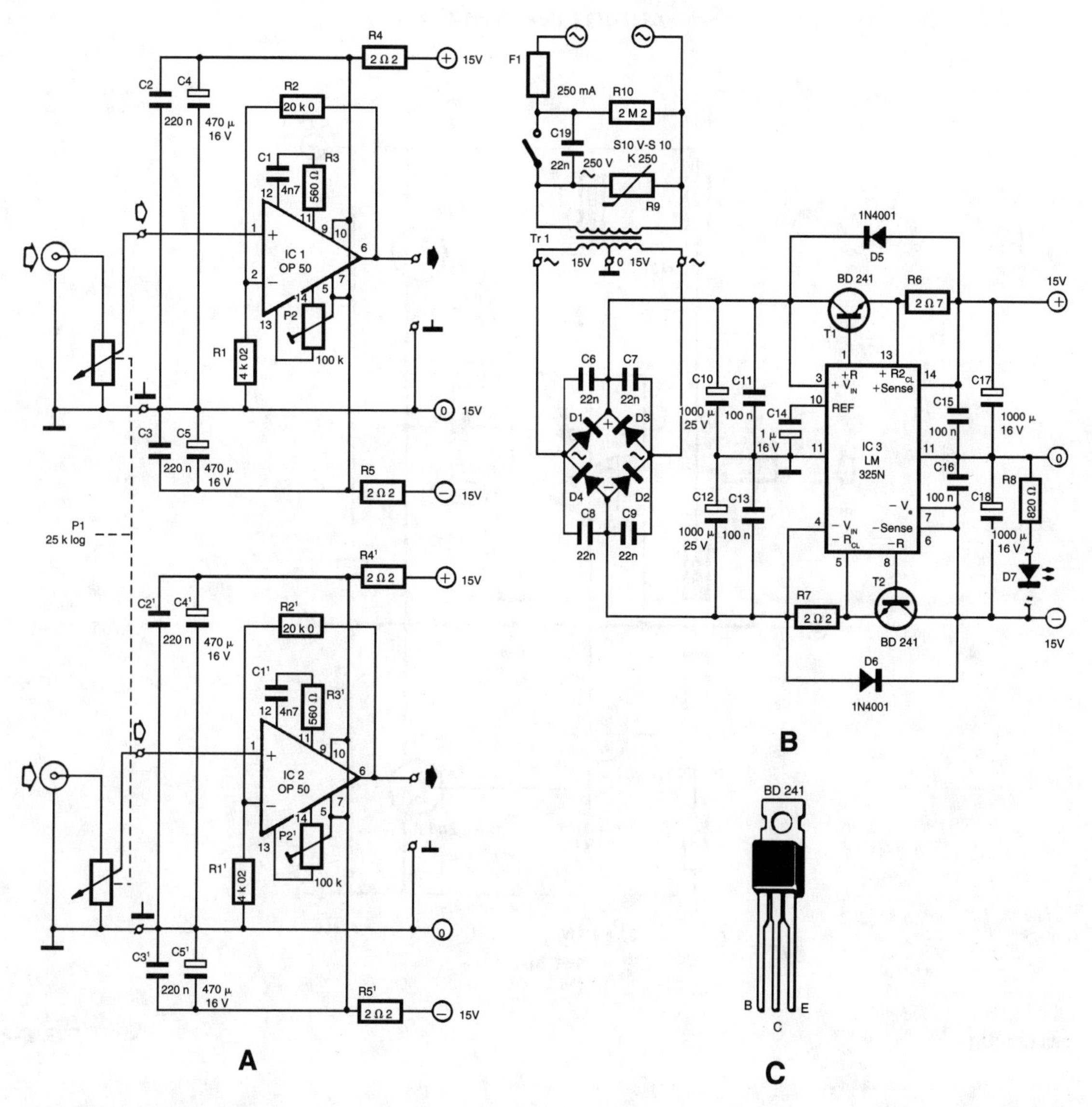

303 CIRCUITS *FIG. 6-1*

Built around Precision Monolithics Inc. OP-50 op amps, this amplifier will drive 100-Ω to 1-kΩ headphone, is flat within 0.4 dB from 10 Hz to 20 kHz, and has a THD of less than 0.01% over most of the audio range. Amplification factor is about 6X.

AUDIO LINE DRIVER

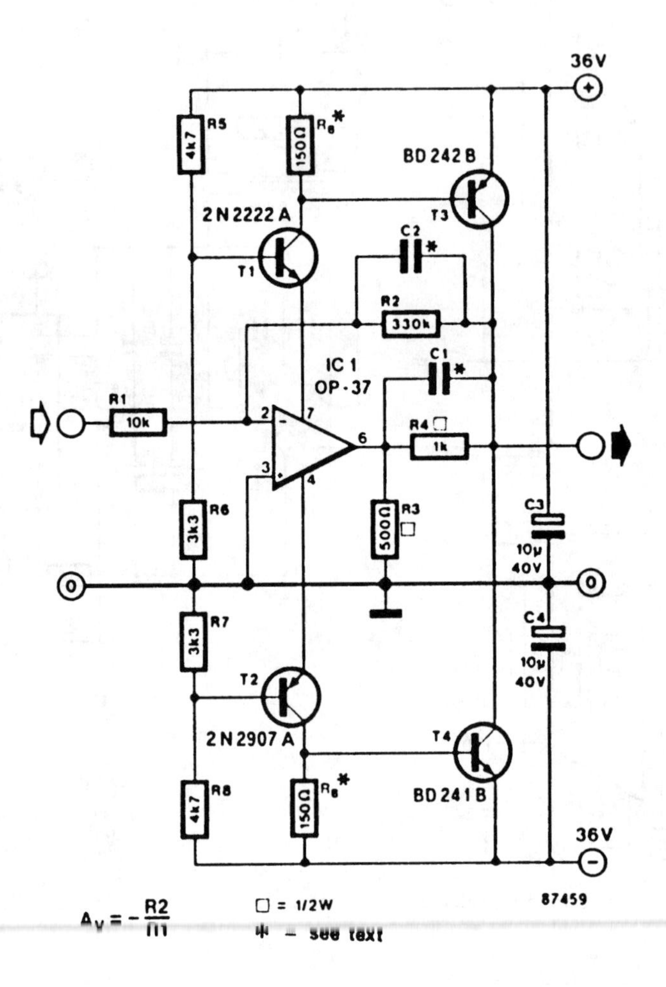

$A_V = -\frac{R2}{R1}$

303 CIRCUITS ***FIG. 6-2***

This line driver can drive low-impedance lines with up to 70 V p-p max. IC1 is a low-noise op amp suitable for ±15-V operation. T1 and T2 are regulators for the power supply for IC1. T3 and T4 form a complementary power output stage. Frequency response is flat up to 100 kHz.

CONSTANT-VOLUME AMPLIFIER

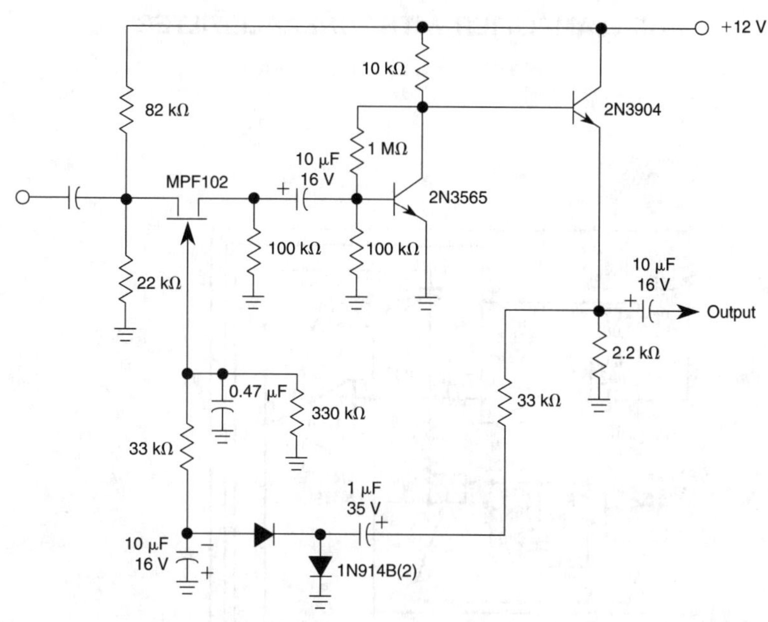

WILLIAM SHEETS

FIG. 6-3

The amplifier has an output level that shifts about 6 dB for a 40-dB input variation.

MINI AMPLIFIER USING LM1895N

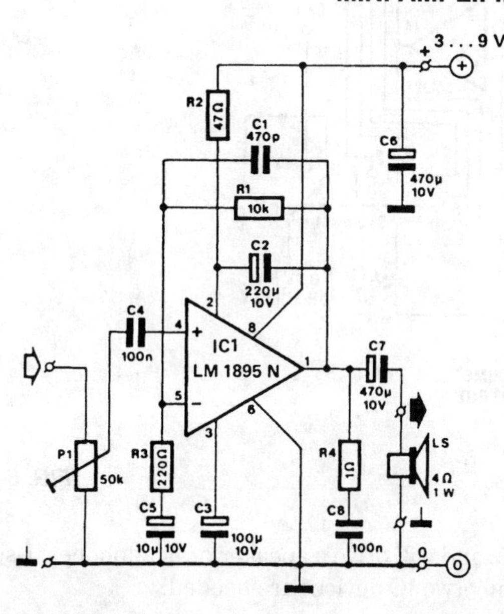

303 CIRCUITS

FIG. 6-4

With 3-V to 9-V supplies, this amplifier can provide from 100-mW to 1-W output into a 4 Ω and bandwidth is approximately 20 kHz @ 3 dB. This circuit is useful for low-power and battery applications. Drain is 80 mA @ 3 V or 270 mA @ 9 V at maximum signal conditions.

AUDIO AMPLIFIER WITH TUNEABLE FILTER

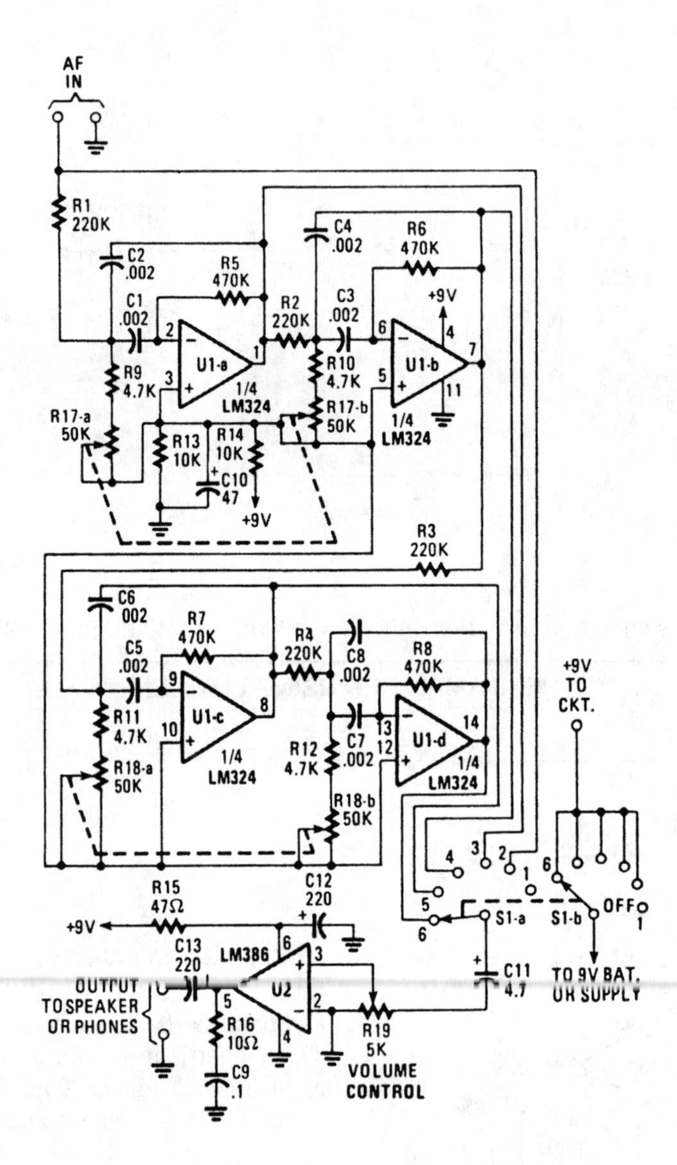

POPULAR ELECTRONICS **FIG. 6-5**

This audio amplifier can tune from 500 to 1500 Hz and will drive a speaker or headphones. Useful for CW reception or other receiver applications, only two IC devices are needed.

AUDIO COMPRESSOR

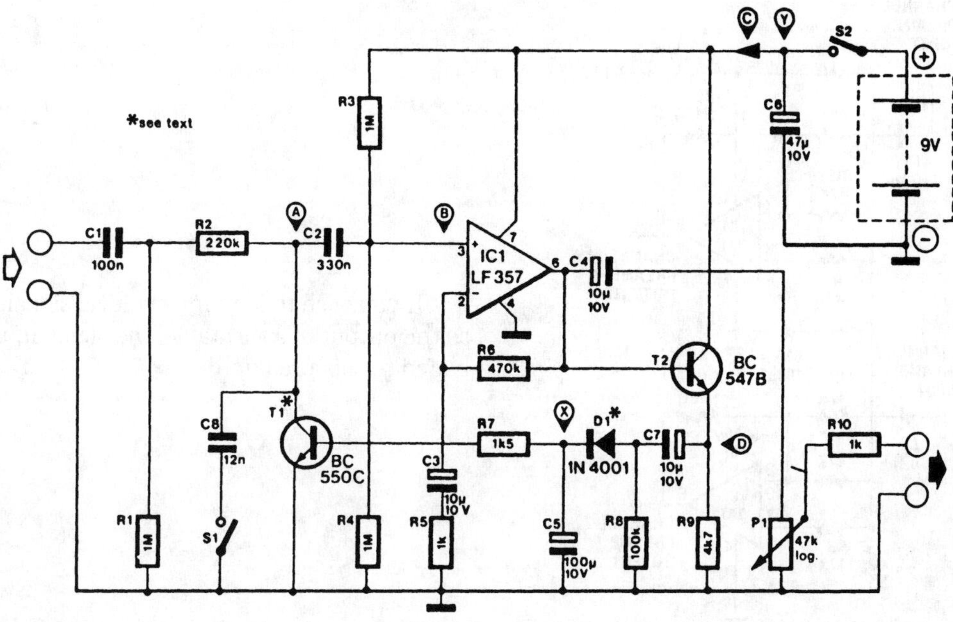

303 CIRCUITS **FIG. 6-6**

This compressor will compress a 25-mV p-p to 20-V p-p audio output to input levels remaining between 1.5 V p-p to 3.5 V p-p, and has a frequency response of 7 Hz to 67 kHz. It is suitable for audio and communications applications.

JFET HEADPHONE AMPLIFIER

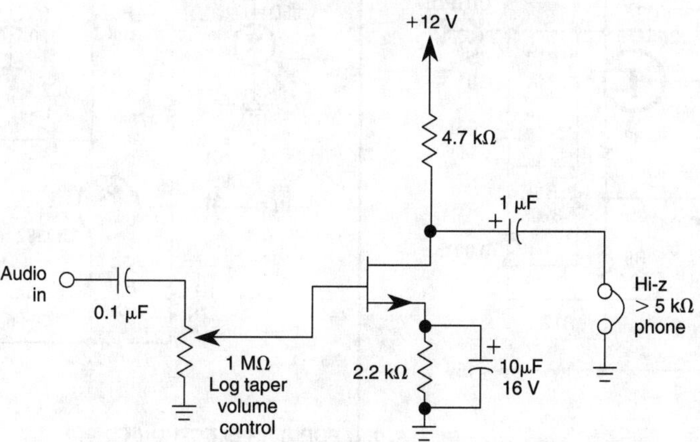

WILLIAM SHEETS **FIG. 6-7**

This circuit can drive high-impedance headphones from a low impedance low-level source. Gain is about 5X to 10X depending on headphone impedance. A volume control is included.

DUAL PREAMP

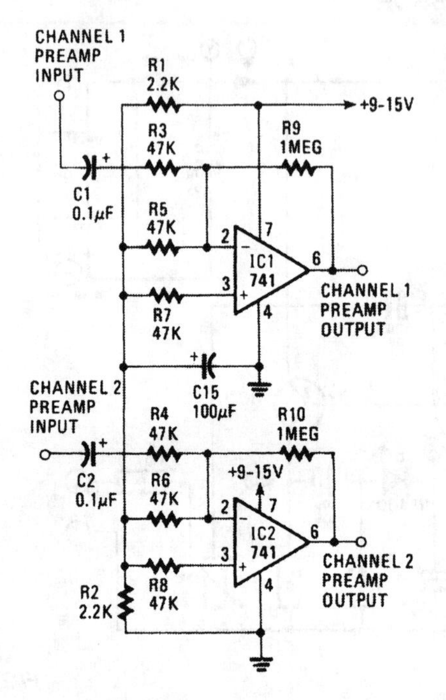

If you wish to amplify low-level signals, such as the output of a turntable, the signal must first be fed to this preamp.

1987 R-E EXPERIMENTERS HANDBOOK *FIG. 6-8*

MAGNETIC PICKUP PHONO AMPLIFIER

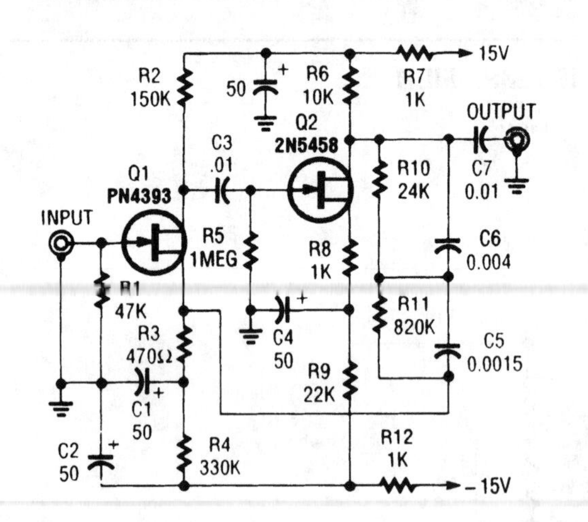

POPULAR ELECTRONICS *FIG. 6-9*

This preamp is RAA compensated for use with magnetic phone cartridges.

AUDIO BOOSTER

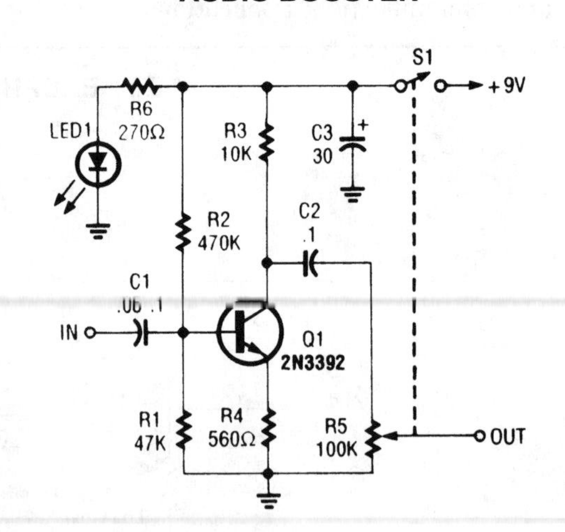

POPULAR ELECTRONICS *FIG. 6-10*

This circuit has a maximum gain of about 22 dB (voltage gain), and it can be used for miscellaneous audio circuits.

AUDIO VOLUME LIMITER

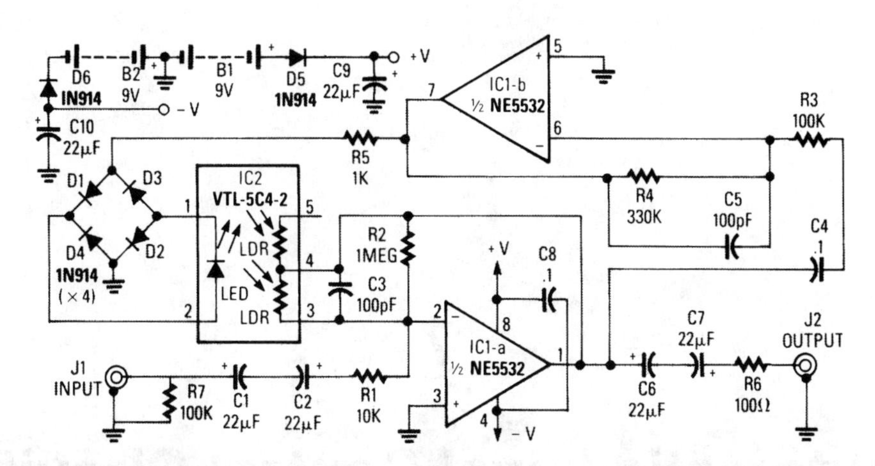

1992 R-E EXPERIMENTERS HANDBOOK ***FIG. 6-11***

IC1-a is connected as an inverting amplifier whose gain is controlled by the LDR portion of an optocoupler.

AUDIO DISTRIBUTION AMPLIFIER

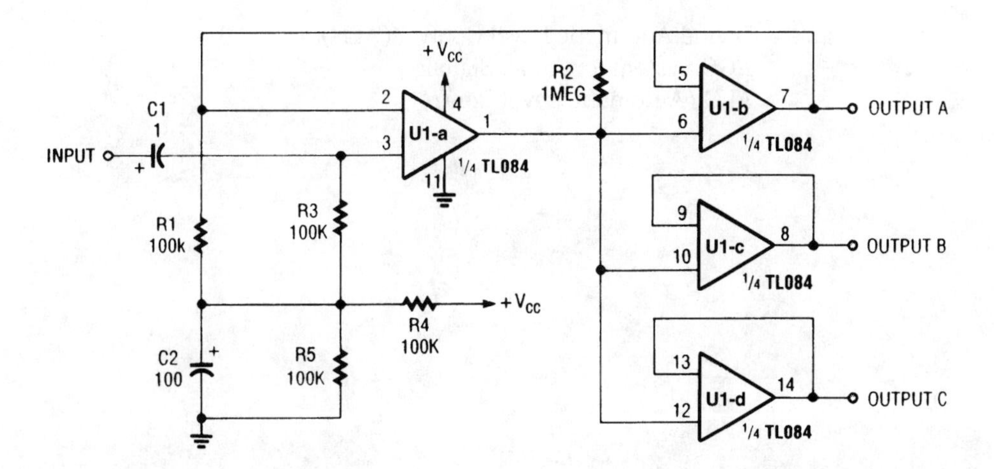

POPULAR ELECTRONICS ***FIG. 6-12***

Three low-*Z* audio outputs are available from this circuit, using a quad TL084 FET amplifier. The input is high impedance. V_{CC} can be 6 to 12 V for typical applications.

7

Automatic Level Control Circuits

The sources of the following circuits are contained in the Sources section, which begins on page 675. The figure number in the box of each circuit correlates to the entry in the Sources section.

DIGITAL AUTOMATIC LEVEL CONTROL (ALC)

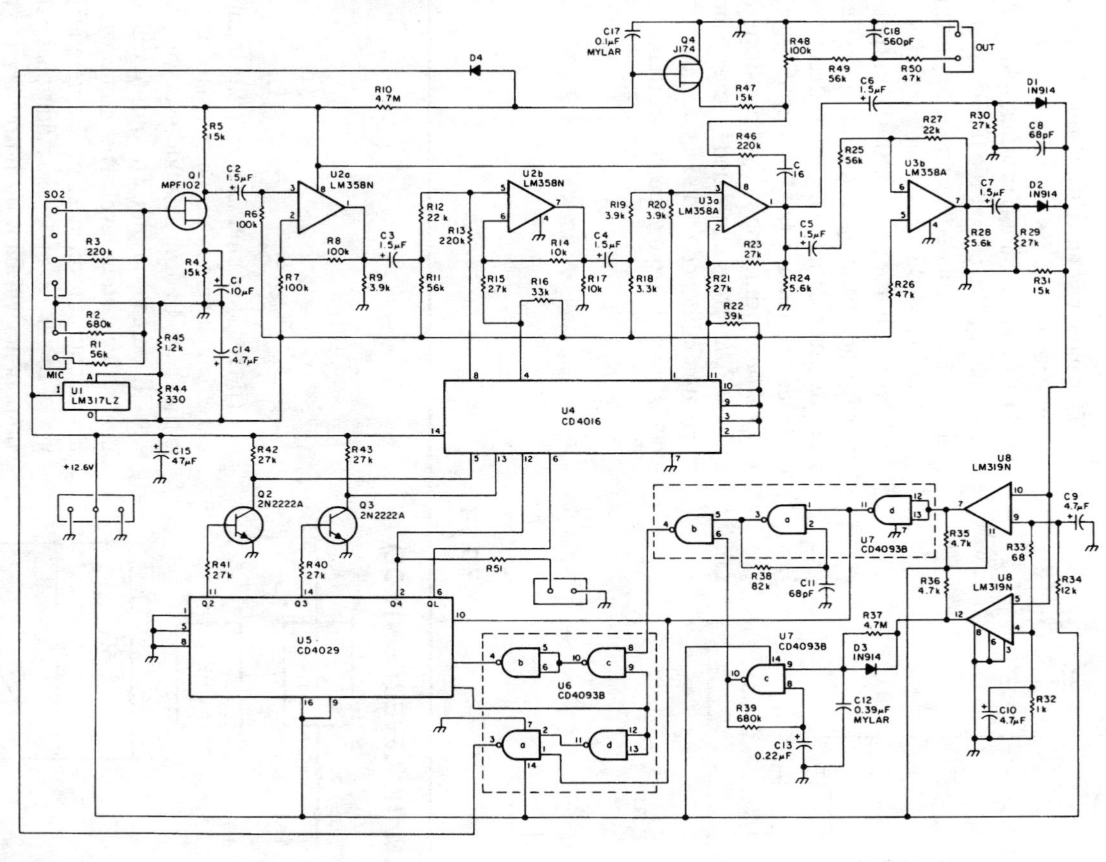

73 AMATEUR RADIO TODAY *FIG. 7-1*

This approach to automatic level control (ALC) makes use of digitally switched audio attenuators in the signal path. The output level of the system is sensed, compared to a reference, and audio pads are inserted via analog switches. This method is nearly instantaneous and eliminates the compromises necessary in conventional RC network ALC systems using fast attack, slow-decay approaches.

AGC SYSTEM FOR AUDIO SIGNALS

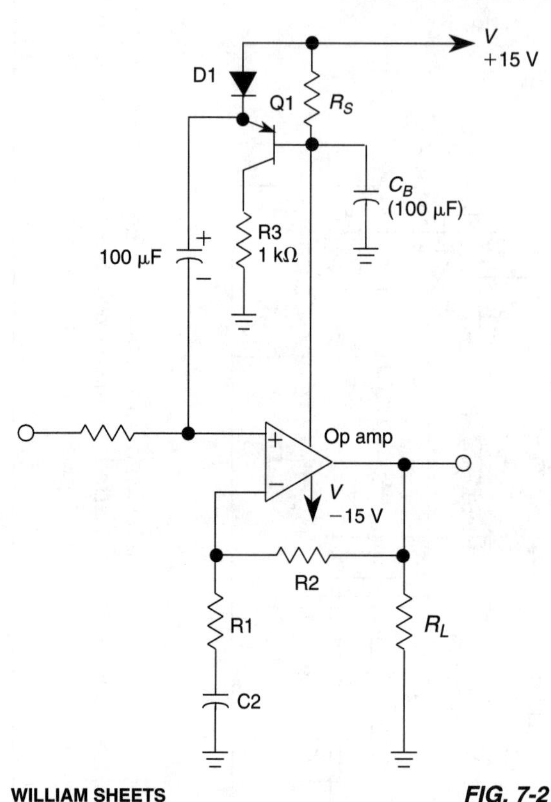

This circuit is an AGC system for audio-frequency signals. AGC systems usually consist of three parts: an amplifier, rectifier, and controlled impedance. In this circuit the functions of an amplifier and a rectifier are performed by a single op amp. This makes the system simple and cheap.

The rectifier is made with the output push-pull cascade of the op amp and R_s, R_L, and C_B. The transistor Q1 and D1 are used as a voltage-controlled resistance (Z). The input signal is $(Z + R_1)/Z$ times, diminished by the voltage divider and $1 + R_2/R_1$ times, amplified by the op amp. C2 eliminates influence of dc bias voltage. R3 protects Q1 and D1 from excessive current.

WILLIAM SHEETS

FIG. 7-2

ALC (AUTOMATIC LEVEL CONTROL)

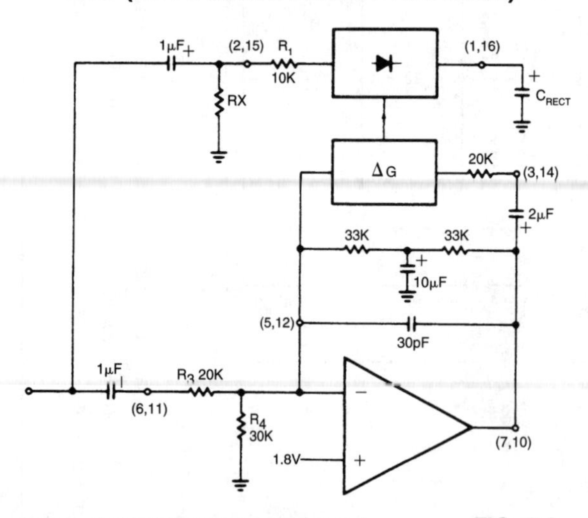

The rectifier input is tied to the input. This makes gain inversely proportional to input level so that a 20-dB drop in input level will produce a 20-dB increase in gain. The output will remain fixed at a constant level. The circuit will maintain an output level of ±1 dB for an input range of +14 to −43 dB at 1 kHz. Additional external components will allow the output level to be adjusted.

1989 RF COMMUNICATIONS HANDBOOK

FIG. 7-3

8

Automotive Circuits

The sources of the following circuits are contained in the Sources section, which begins on page 675. The figure number in the box of each circuit correlates to the entry in the Sources section.

CD IGNITION SYSTEM FOR AUTOS

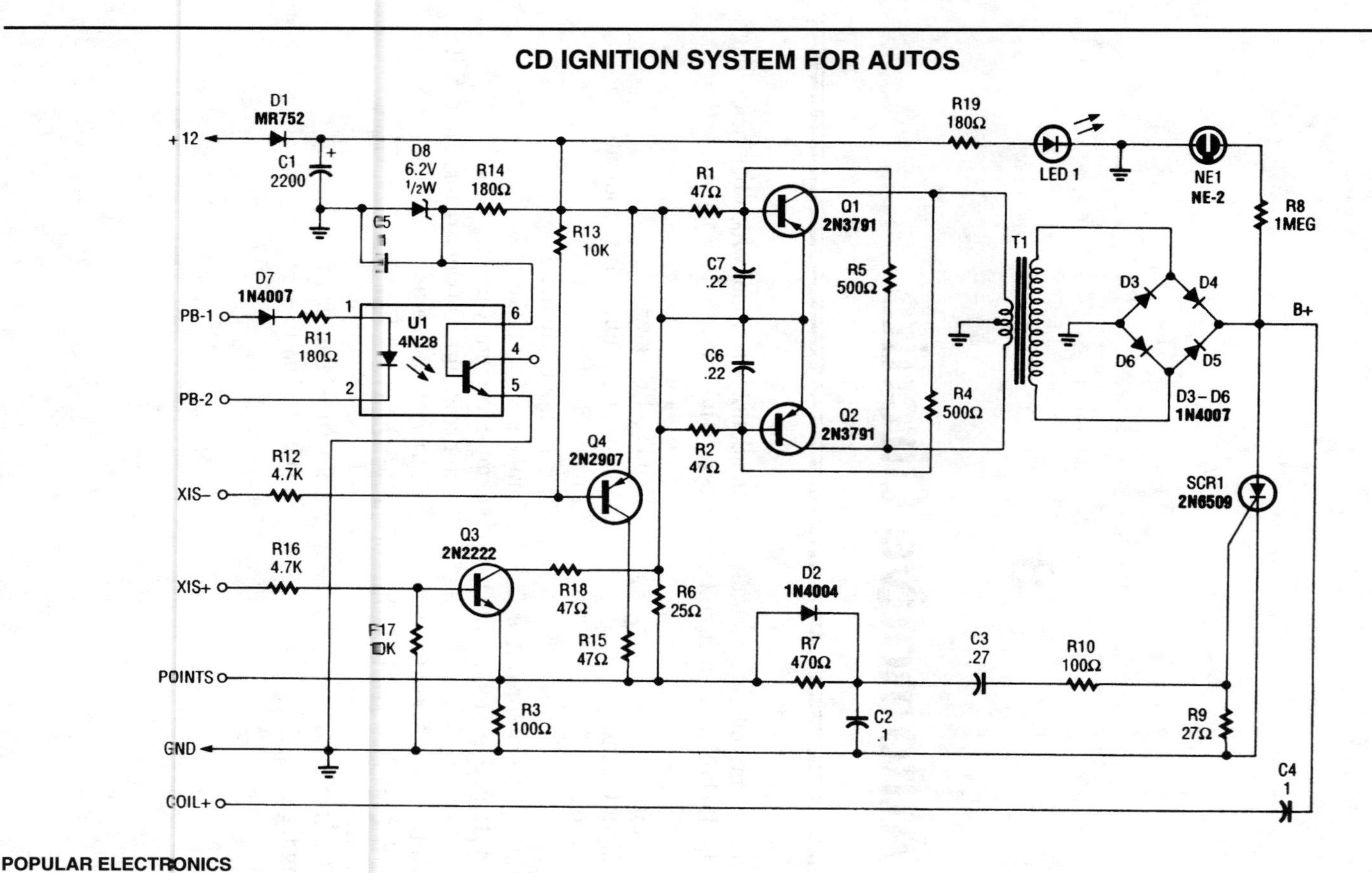

POPULAR ELECTRONICS **FIG. 8-1**

At the heart of the CD4-MX is an astable multivibrator, built around Q1 and Q2, that feeds step-up transformer T1. The output of T1 is rectified by D3 to D6 and used to charge capacitor C4. When the points close, a small voltage is fed to the gate of SCR1, causing it to fire, dumping the charge of C4 to the vehicle's ignition coil. The circuit also contains optional subcircuits to accommodate different types of auto ignitions.

X_{15} + and X_{15} – are alternative trigger configurations for nonpoint breaker ignition systems. R6 is not used for these systems and must be removed. Optocoupler U1 can be used (pin 4) in conjunction with X_{15} – or X_{15} + depending on polarity of sensor. Note that 60 to 70 kV is available from this system, so observe suitable safety precautions.

BRAKE AND TURN-SIGNAL LIGHT CIRCUIT

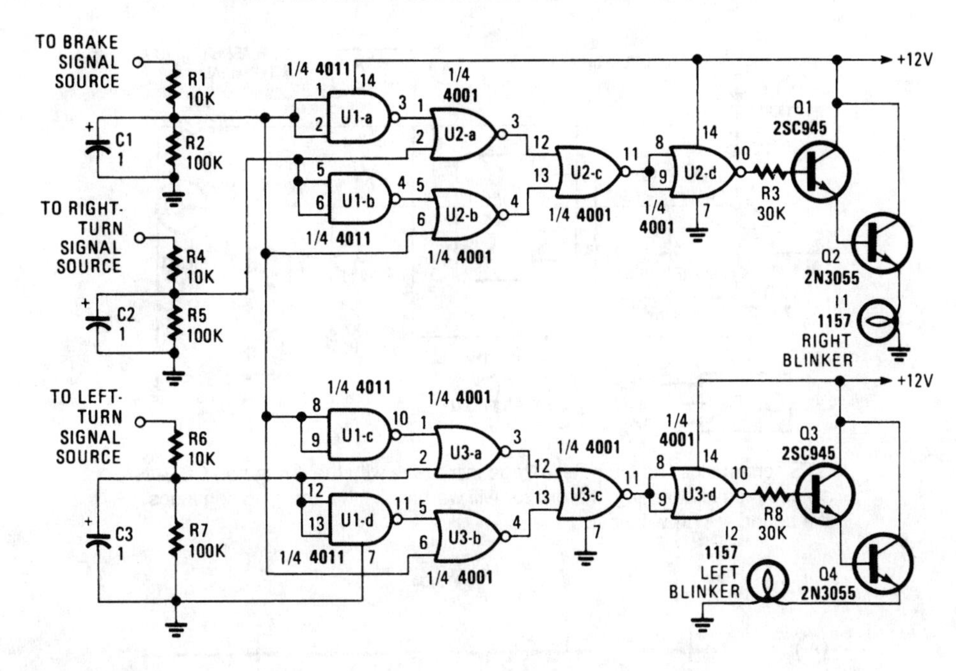

POPULAR ELECTRONICS ***FIG. 8-2***

This circuit enables single-filament tail lights to serve as combination brake lights and turn signals.

VEHICULAR TACHOMETER CIRCUIT

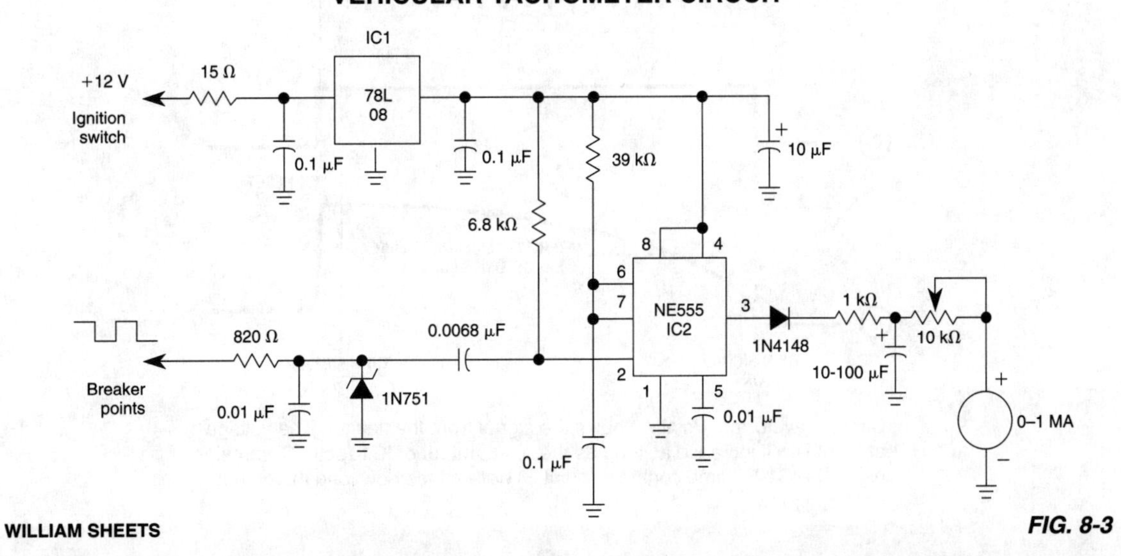

WILLIAM SHEETS ***FIG. 8-3***

In this automotive application, the 555 is a pulse counter. IC1 regulator provides proper operating voltage for IC2. This circuit is for vehicles with conventional breaker points.

SMART TURN SIGNAL

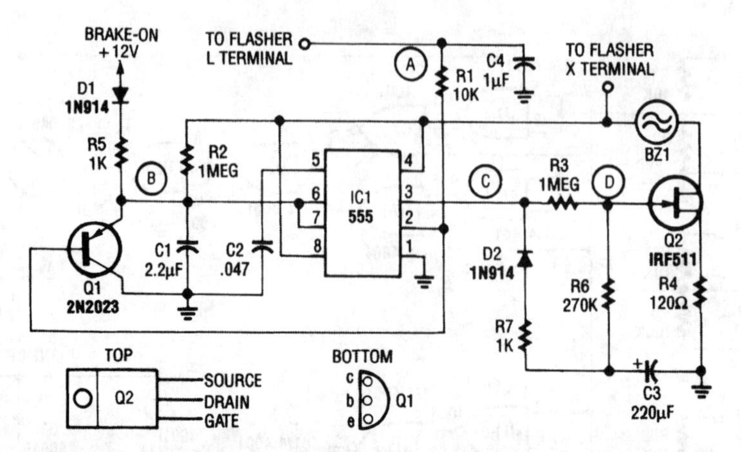

STS schematic. The Q2 gate voltage increases with the charge on C3. After 15 seconds of charging, the buzzer will warble. As the charging continues, the sound will grow louder.

A

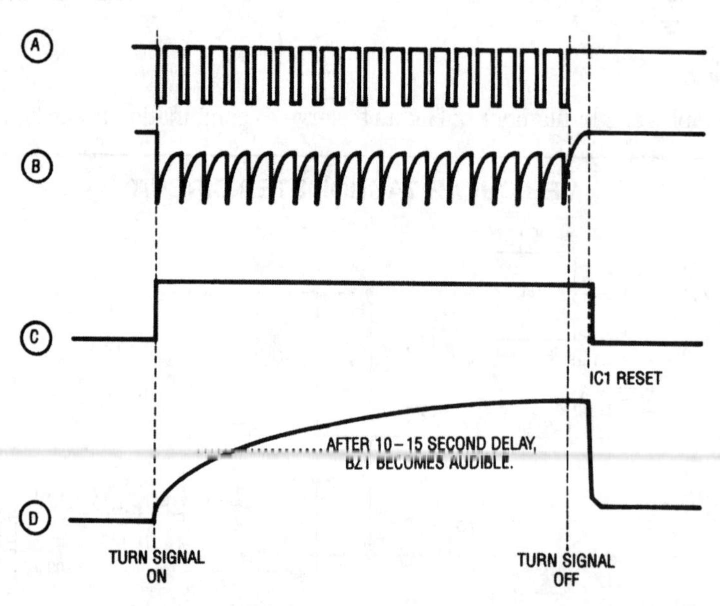

Circuit waveforms. Point A shows the signal from the flasher. The voltage at point D will increase as long as the pin-3 output of IC1 (point C) remains high. The C1-R2 time constant (point B) determines how long the output will be high.

B

ELECTRONICS NOW

FIG. 8-4

SMART TURN SIGNAL (*Cont.*)

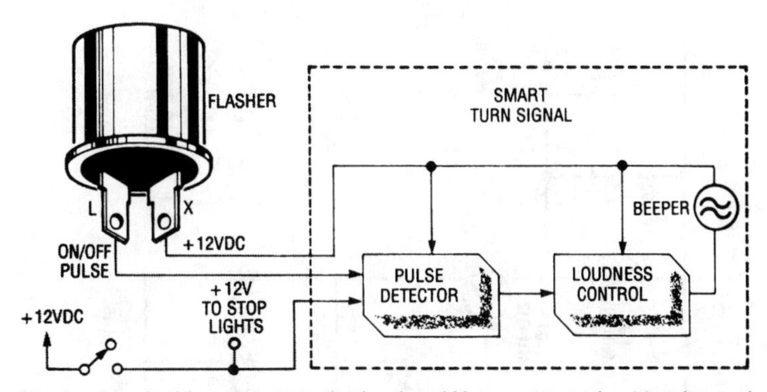

Flasher terminal L connects to the load and X connects to the 12-volt supply. When the driver engages the turn signal, the L terminal voltage varies with the blinking lights. The STS senses the changing voltage and, after 15 seconds, it applies power to a buzzer through a current-limiting device to control loudness.

C

This circuit reminds a driver that his turn signal has been left on for more than 15 seconds. When stopped for a light, the brake-on signal holds the warning off.

MANUAL HEADLIGHT/SPOTLIGHT CONTROL FOR AUTOS

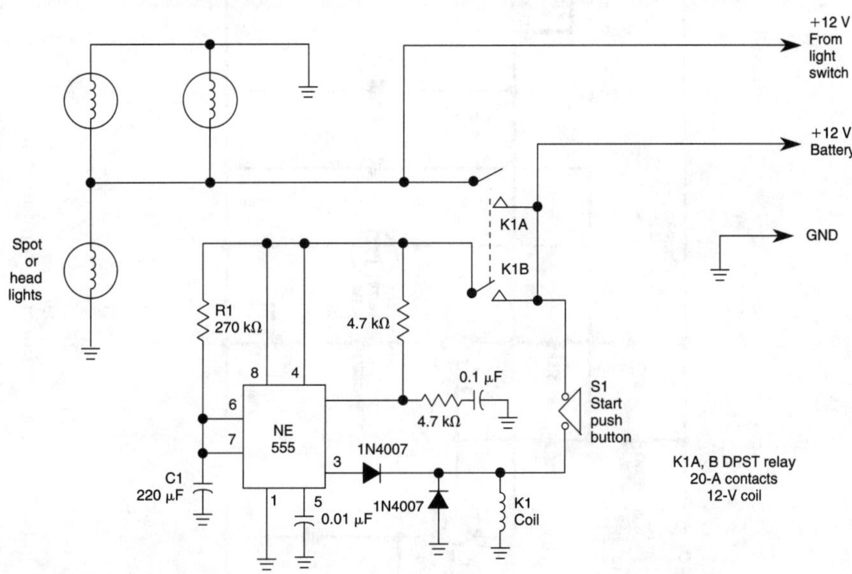

WILLIAM SHEETS ***FIG. 8-5***

Pressing the START pushbutton turns on either the headlights or spotlights for a predetermined time. After 1 minute (R1 and C1 determine this), the lights will shut off as the NE555 completes its cycle.

THERMOSTAT SWITCH FOR AUTOMOTIVE ELECTRIC FANS

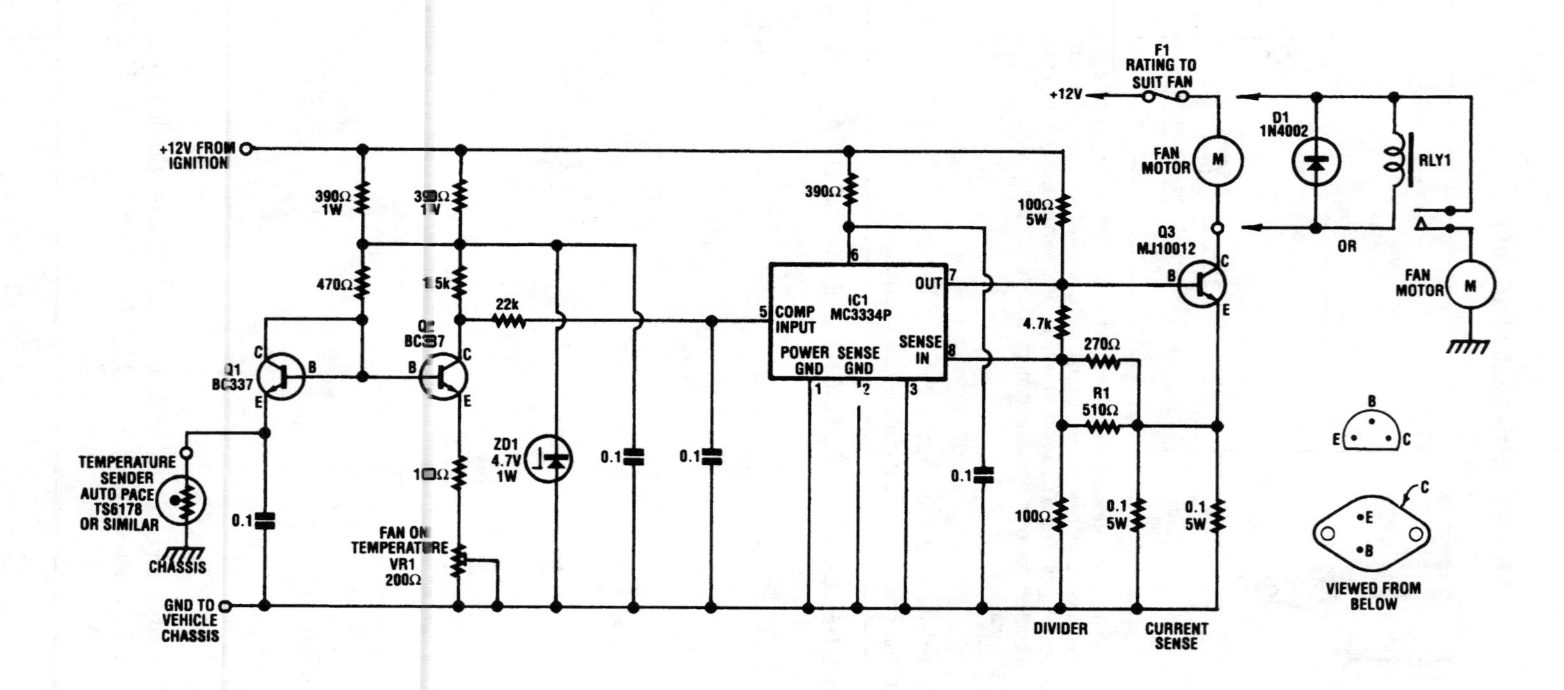

SILICON CHIP

FIG. 8-6

The circuit is based on a commercial temperature sensor (TS6178) and an MC3334P ignition chip. When the radiator temperature increases, the sensor pulls the base of Q2 low via Q1, which is wired as a diode. Q2's collector thus goes high and triggers IC1, which switches its pin 7 output high and turns on the fan motor via Q3.

FLASHING BRAKE LIGHT

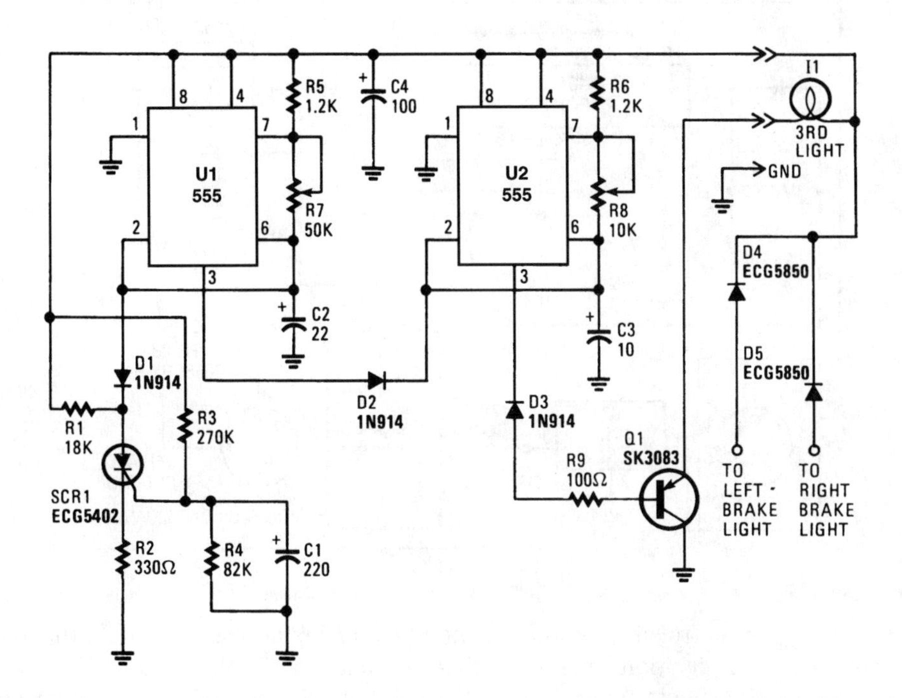

1990 PE HOBBYIST HANDBOOK ***FIG. 8-7***

When power is first applied, three things happen: the light-driving transistor (Q1) is switched on because of a low output from U2, pin 3; timer U1 begins its timing cycle, with the output (pin 3) going high, inhibiting U2's trigger (pin 2) via D2; and charge current begins to move through R3 and R4 to C1.

When U1's output goes low, the inhibiting bias on U2 pin 2 is removed, so U2 begins to oscillate, flashing the third light via Q1, at a rate determined by R8, R6, and C3. Oscillation continues until the gate-threshold voltage of SCR1 is reached, causing it to fire and pull U1's trigger (pin 2) low. With its trigger low, U1's output is forced high, disabling U2's triggering. With triggering inhibited, U2's output switches to a low state, which makes Q1 conduct, turning on I1 until the brakes are released. Removing power from the circuits resets SCR1, but the RC network consisting of R4 and C1 will not discharge immediately and will trigger SCR1 earlier. So, frequent brake use means fewer flashes.

Bear in mind that the collector/emitter voltage drop across Q1, along with the loss across the series-fed diodes, reduces the maximum available light output. If the electrical system is functioning properly (at 13 to 14 V for most vehicles), those losses will be negligible.

POWER CONTROLLER (FOR AUTOMOTIVE ACCESSORIES)

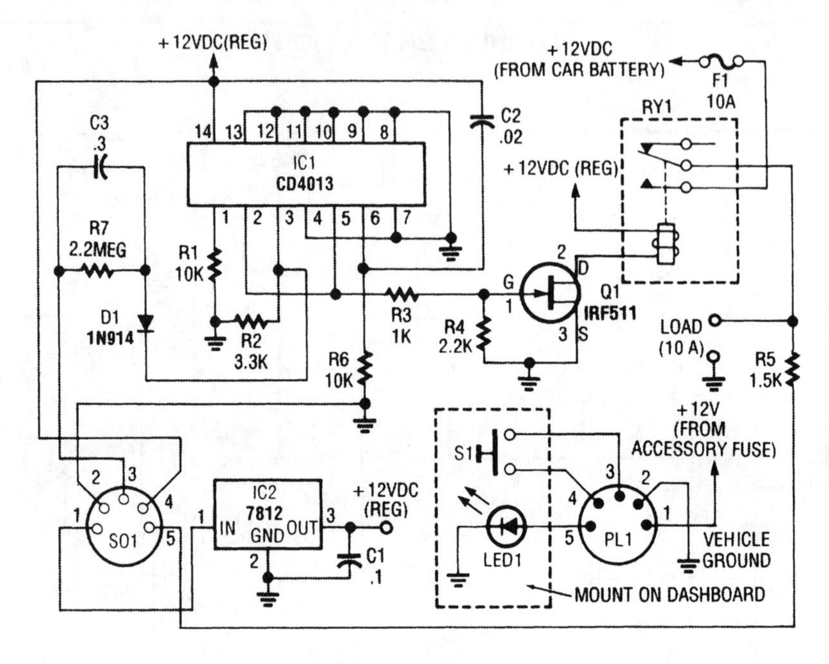

ELECTRONICS NOW *FIG. 8-8*

Because the power controller is powered from the vehicle's accessory switch, the load can receive power only when the ignition key is on. Using half of a dual flip-flop (CD4013), a load of up to 10 A is controlled by a momentary pushbutton. This circuit was originally intended for automotive power control, but could have other applications as well.

AUTOMOTIVE POWER ADAPTER FOR dc-OPERATED DEVICES

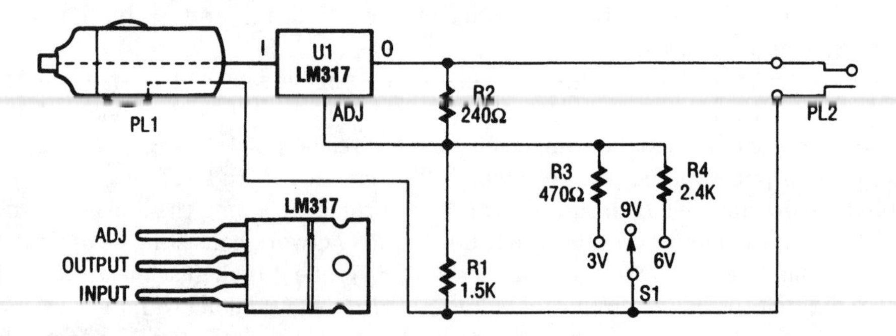

1993 ELECTRONICS HOBBYIST HANDBOOK *FIG. 8-9*

In the schematic diagram for the car-power adapter, note how the value of R_B (which is R1 and S1 in the center position) is changed by putting R3 or R4 in parallel with R1.

TIME-DELAY AUTO-KILL SWITCH

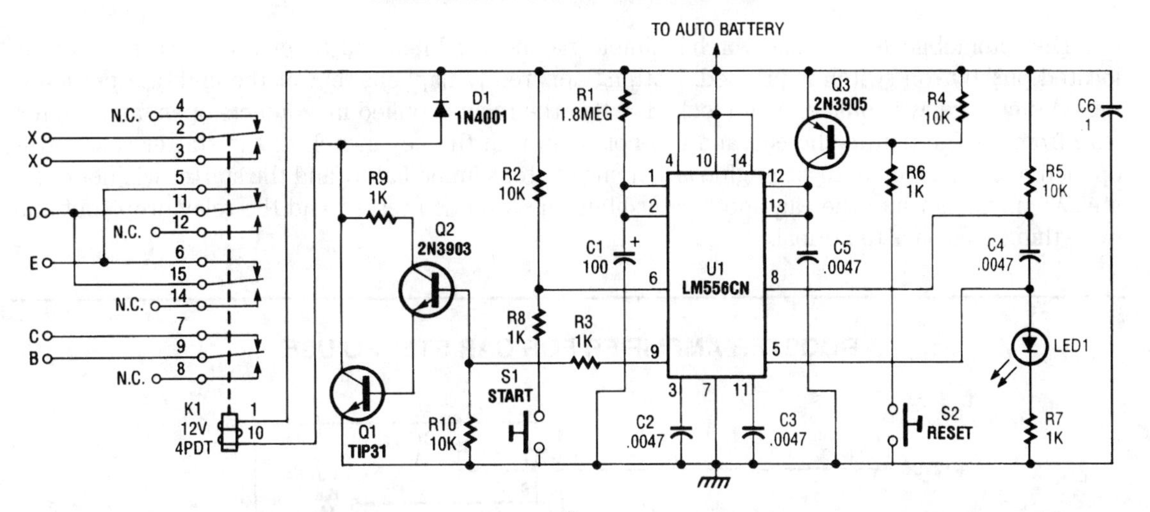

A

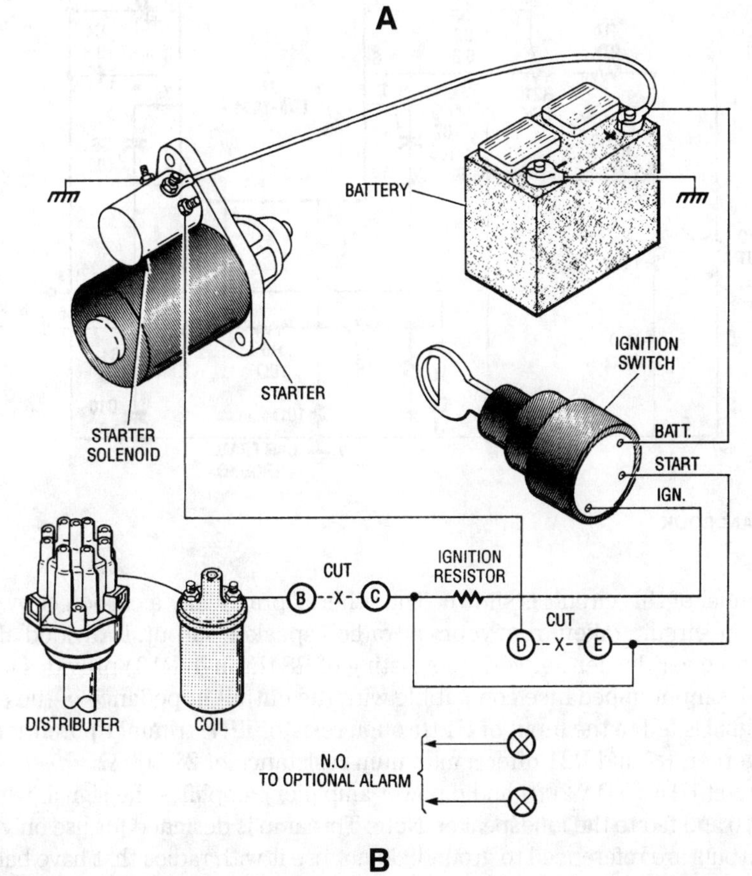

B

POPULAR ELECTRONICS

FIG. 8-10

The automobile delayed kill switch is simple in concept. When you get out of your car, a secretly located pushbutton switch is pressed. Nothing apparently happens, but at the end of a predetermined time, a relay is pulled in and locked. When the relay is pulled in, contacts open, and the hot lead from the ignition to the coil and the hot wire from the key switch to the starter solenoid is opened or disconnected. If the engine is running, it stops immediately and the starter will not operate. When you get into the car, another pushbutton switch is pressed and the relay drops out and everything goes back to normal.

BOOSTER AMPLIFIER FOR CAR STEREO USE

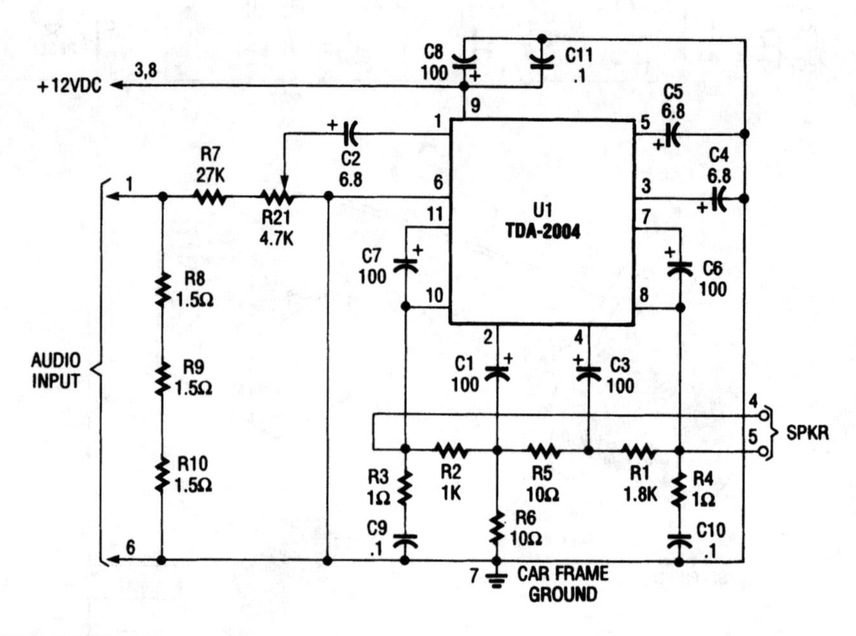

1990 PE HOBBYIST HANDBOOK ***FIG. 8-11***

Only one channel of this circuit is shown. The other is practically a carbon copy.

The input to the circuit, taken from your car radio's speaker output, is divided along two paths; in one path, a high-power divider network (consisting of R8 through R10) provides 4.5-Ω resistance to make the circuit's input impedance compatible with the output impedance of the car radio. In the other path, the signal is fed to the input of U1 through resistor LR7, trimmer potentiometer R21, and capacitor C2. Together, R7 and R21 offer a minimum resistance of 27,000 Ω.

Integrated circuit U1 (a TDA-2004 audio power amplifier) amplifies the signal, which is then output at pins 8 and 10 and fed to the loudspeaker. Note: This amp is designed for use only with car radios whose speaker outputs are referenced to ground: do not use it with radios that have balanced outputs.

AUTO TURN-SIGNAL REMINDER

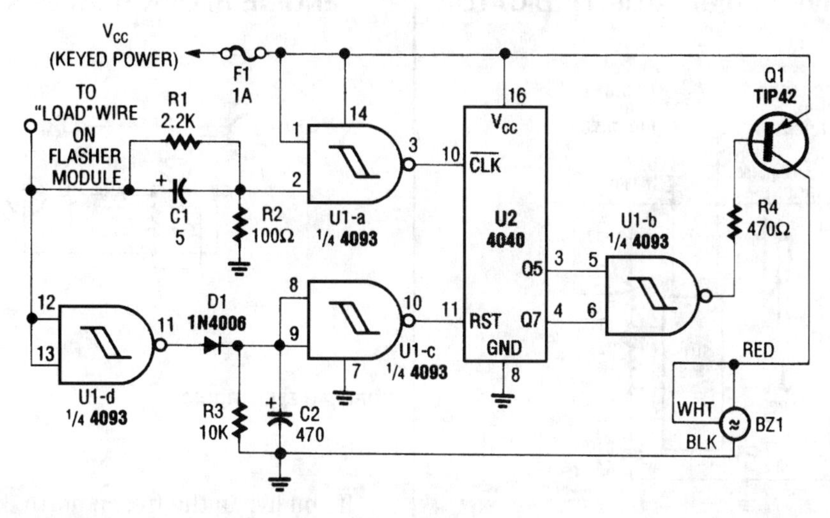

POPULAR ELECTRONICS *FIG. 8-12*

This circuit counts turn signal flashes. At the end of about 70 flashes, a chime sounds to remind the driver to turn off the turn signal. By using various taps on U2, the period can be changed if desired. BZ1 is a buzzer or chime module.

HEADLIGHT FLASHER

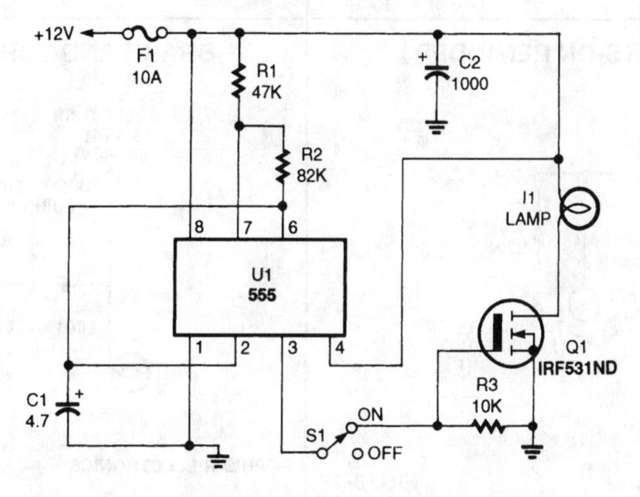

POPULAR ELECTRONICS *FIG. 8-13*

The headlight flasher is nothing more than a 555 oscillator/timer that's configured as an astable multivibrator (oscillator). Its input is used to drive the gate of an IRF53IND hexFET, which, in turn, acts like an on/off switch, turning the lamp on and off at the oscillating frequency (1 Hz).

AUTOMOTIVE AUDIBLE-TURN INDICATOR

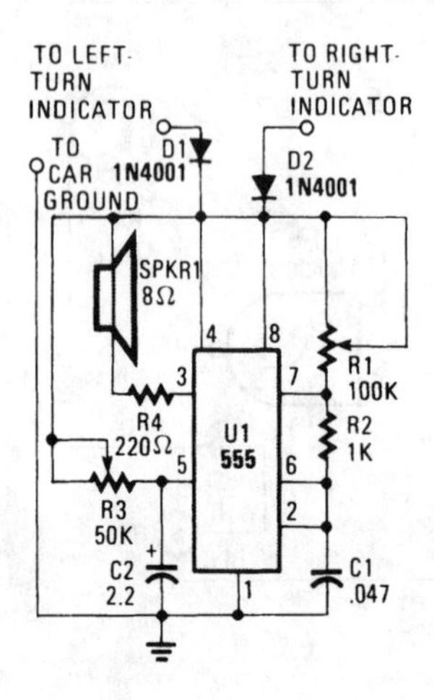

POPULAR ELECTRONICS *FIG. 8-14*

This little circuit should be useful to the hearing impaired. It produces a tone each time a dashboard turn indicator lights. The tone drops in frequency for as long as the indicator is lit.

ENGINE BLOCK HEATER MINDER

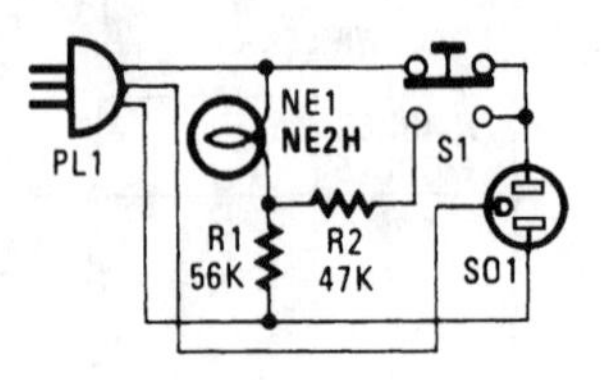

POPULAR ELECTRONICS *FIG. 8-15*

If you live in the frozen north, knowing your engine-block heater is working is a comfort. This device will let you know if yours is okay. Plug in PL1 to your power outlet. NE1 should light. Then, plug in the block heater. Depressing S1 should cause the indicator to get brighter. If not, your block heater might be open and inoperative.

HEADLIGHTS-ON REMINDER

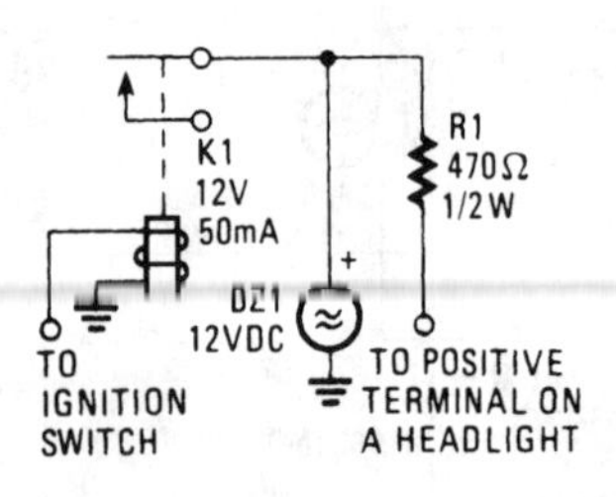

POPULAR ELECTRONICS *FIG. 8-16*

This circuit will sound alarm BZ1 if the ignition is turned off with the headlights on.

BRAKE AND TURN INDICATOR

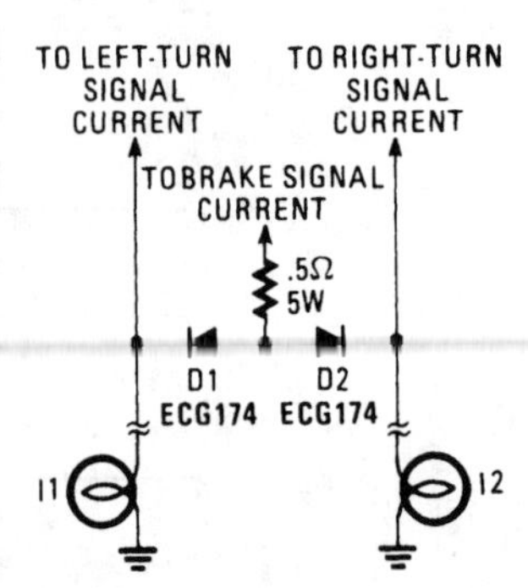

POPULAR ELECTRONICS *FIG. 8-17*

This might be a quick solution to getting the two-wire truck harness to support both turn and braking indications.

LAMP-SWITCHING CIRCUIT

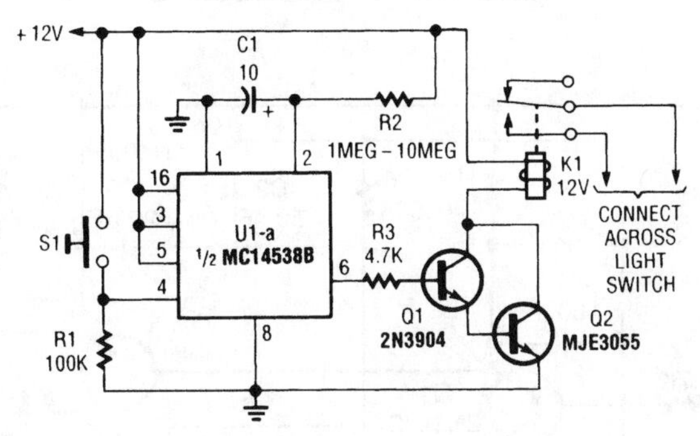

POPULAR ELECTRONICS ***FIG. 8-18***

A normally open pushbutton switch (S1) delivers a positive input pulse to pin 4 of U1, triggering the IC into action. The output of U1 at pin 6 supplies base-drive current to a Darlington pair comprised of Q1 and Q2, activating K1. A 10-µF capacitor and any resistor value of from 1 to 10 MΩ can be used as the timing components.

To use the circuit on an auto's headlights, connect the relay's normally open contacts across the car's headlight switch and press S1 to extend the on time. In connecting the circuit to control an ac-operated lamp, turn off the ac power and connect the relay contacts in parallel with the lamp's power switch contacts.

AUTOMATIC TURN-OFF CONTROL FOR AUTOMOBILES

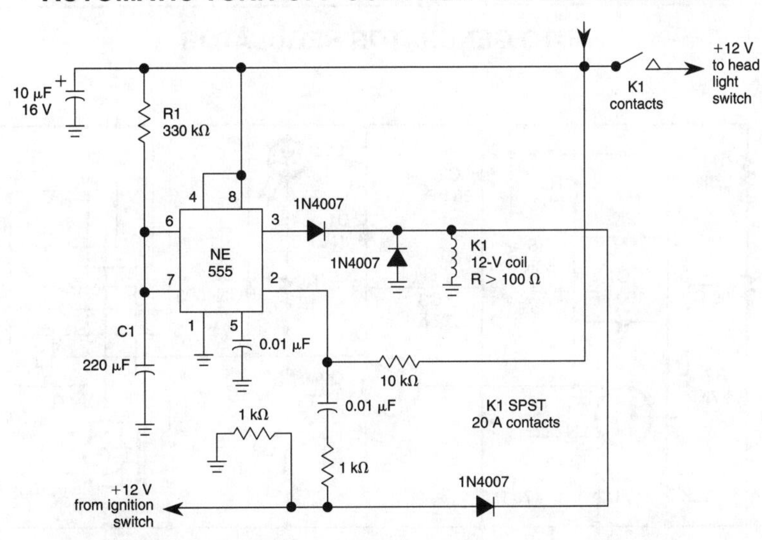

WILLIAM SHEETS ***FIG. 8-19***

When the ignition switch is on, relay K1 is energized continuously, and the headlights can be turned on. Turning off the ignition turns on timer IC1,which keeps IC1 energized for a time determined by R1 and C1. With the values shown approximately a 1 minute delay will result. The values of R1 or C1 can be changed to vary this delay time.

ALTERNATOR REGULATOR

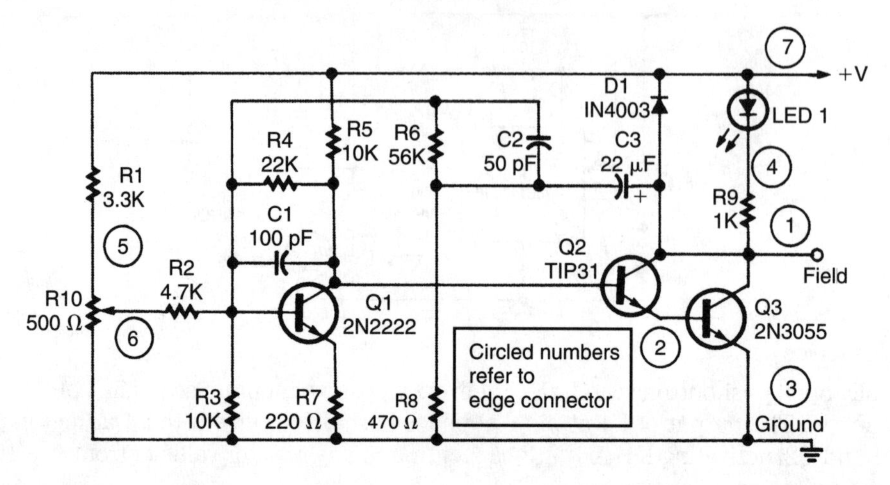

RADIO-ELECTRONICS *FIG. 8-20*

This alternator regulator uses a 3-transistor dc amplifier, and is designed for a "pulled up" field system, where one side of the alternate field returns to the +12-Vsupply, and the other end is pulled toward ground. The circuit monitors the state of the battery through a resistive divider and causes the voltage to change at the field terminal.

AUTO GENERATOR REGULATOR

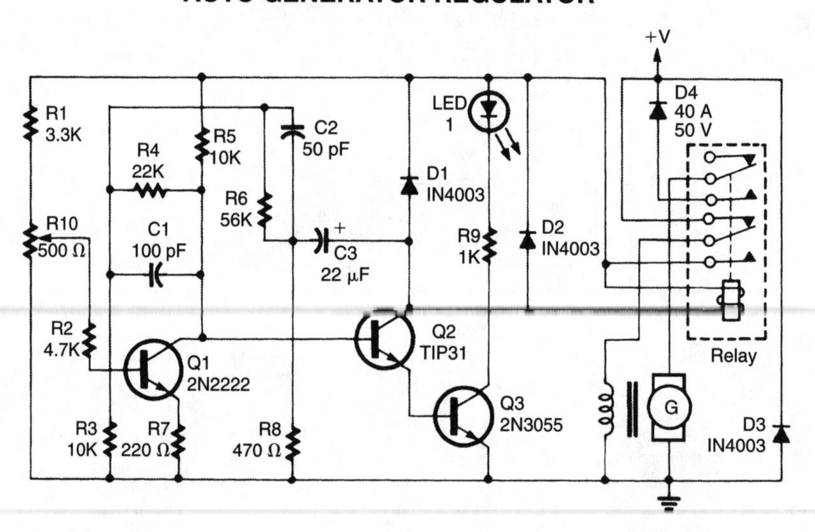

RADIO-ELECTRONICS *FIG. 8-21*

This regulator is for the purpose of controlling a dc generator. The field configuration is that one side of the field is grounded. D4 prevents the battery from discharging through the generator and takes the place of the mechanical cut-out relay. R10 adjusts the system voltage setting.

LIGHTS-ON REMINDER

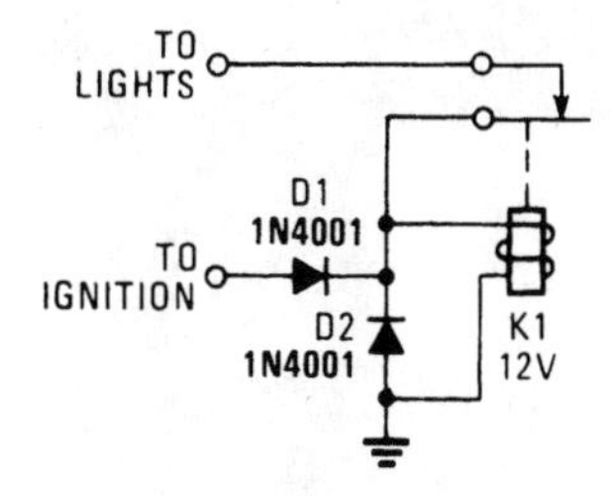

POPULAR ELECTRONICS *FIG. 8-22*

A relay and two diodes are all that is needed—the relay performs the job of a buzzer so no annunciator is required. When the lights are left on, but the ignition is off, the normally closed relay contacts are in series with the relay coil. That means the relay interrupts its own power each time it becomes active, so it chatters and acts like a buzzer. This is a real minimalistic headlight reminder. It doesn't even require an annunciator because the relay acts as buzzer.

AUTO FUSE MONITOR

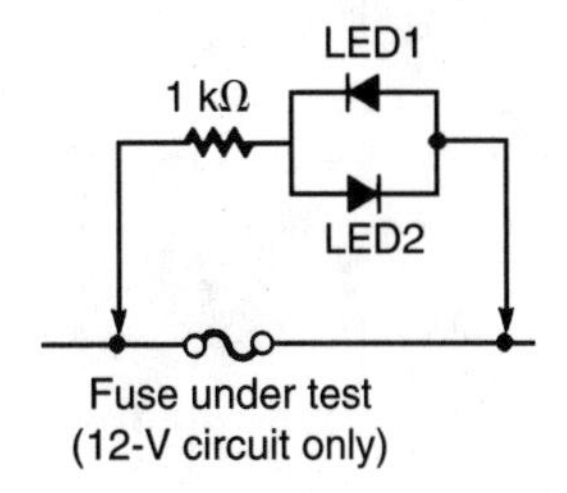

WILLIAM SHEETS *FIG. 8-23*

This circuit can quickly check a fuse in an automobile circuit. Connect across suspected fuse—either LED glows, fuse is blown. The circuit must be live for this test to work.

HEADLIGHT ALARM

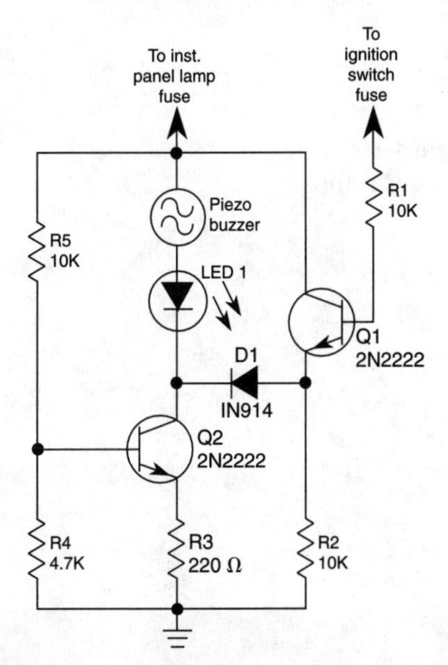

1989 R-E EXPERIMENTERS HANDBOOK *FIG. 8-24*

The base of Q1 is connected to the car's ignition circuit; the easiest point to make that connection is at the ignition switch fuse in the car's fuse panel. Also, one side of the piezoelectric buzzer is connected to the instrument-panel light fuse; when the headlights or parking lights are on, the instrument panel is lit, too. When the headlights are off, no current reaches the buzzer. Therefore, nothing happens. What happens when the headlights are on depends on the state of the ignition switch. When the ignition switch is on, transistors Q1 and Q2 are biased on, effectively removing the buzzer and the LED from the circuit.

When the ignition switch is turned off, but the headlight switch remains on, transistor Q1 is turned off, but transistor Q2 continues to be biased on. The result is that the voltage across the piezoelectric buzzer and the LED is sufficient to cause the buzzer to sound loudly and the LED to light.

9

Battery Charger Circuits

The sources of the following circuits are contained in the Sources section, which begins on page 675. The figure number in the box of each circuit correlates to the entry in the Sources section.

LEAD-ACID TRICKLE CHARGER

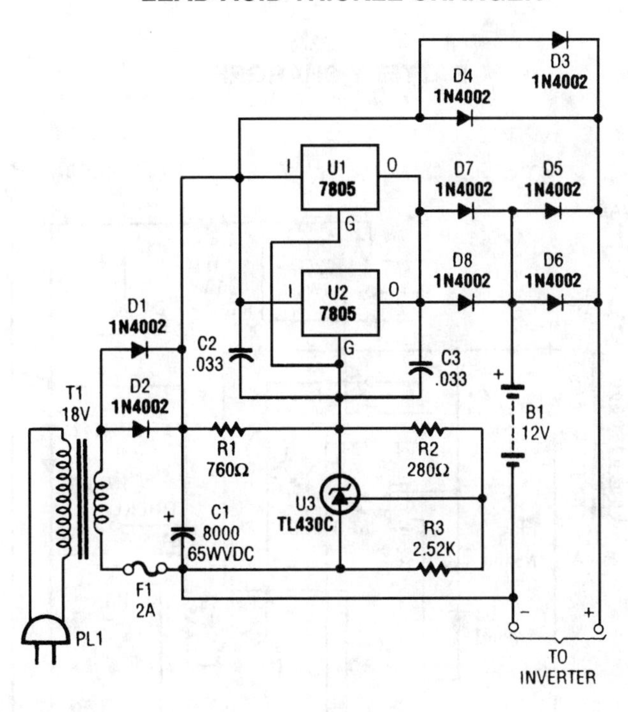

POPULAR ELECTRONICS **FIG. 9-1**

The charger can be used as a stand-alone charger or for emergency lighting and burglar alarm systems using lead-acid batteries.

RF-TYPE BATTERY CHARGER

This type of charger couples RF from L2 to an external pickup coil. The pickup coil connects to a rectifier and battery to be charged. This idea is handy because no wire or contacts are required. L2 is 10T #24 wire and L3 is 10T #30 wire. Both coils are mounted on a 1" × ¼" ferrite rod.

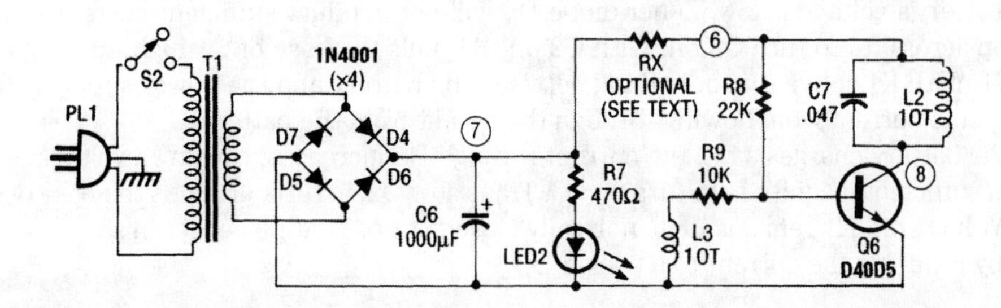

1992 R-E EXPERIMENTERS HANDBOOK **FIG. 9-2**

BATTERY CHARGER

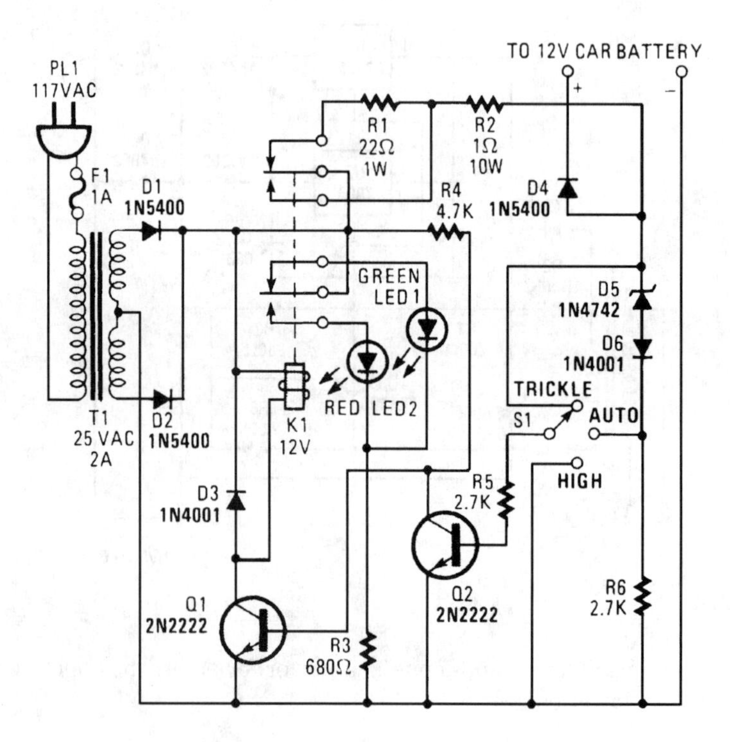

POPULAR ELECTRONICS *FIG. 9-3*

The circuit is capable of supplying either a trickle (50 mA) or high-current (1-A) charge. You can select either charging method or an automatic mode that will first trickle charge a battery if it is particularly low before switching to high current charging.

If the battery's voltage is low, Zener-diode D5 will not conduct sufficient current to produce a voltage drop across R6 to turn Q2 on. With Q2 off, R4 pulls the base of Q1 high, turning it on. That activates K1. With K1 active, the only thing between the battery and the power supply is R2 and D4 (which prevents current from flowing through the circuit from the battery).

Once the battery charges a bit, the current through D5 increases, causing a voltage drop across R6 that is of sufficient magnitude to turn on Q2. Transistor Q2, in turn, grounds the base of Q1, keeping it off. With Q1 off, K1 remains in its normally closed state. That places R1 in series with the battery, thereby reducing the current to a trickle.

SOLAR-POWERED BATTERY CHARGER

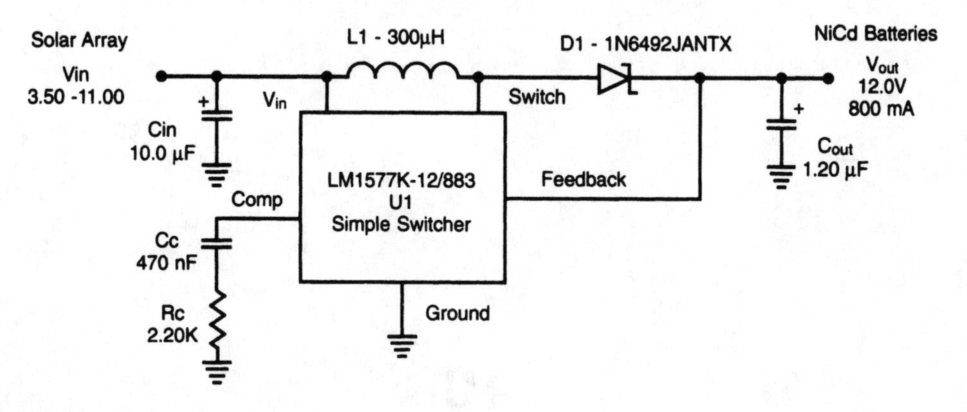

NATIONAL SEMICONDUCTOR *FIG. 9-4*

A National Semiconductor LM1577 IC is used in a step-up regulator to charge Nicad batteries from a solar panel.

INTELLIGENT BATTERY-CHARGING CIRCUIT

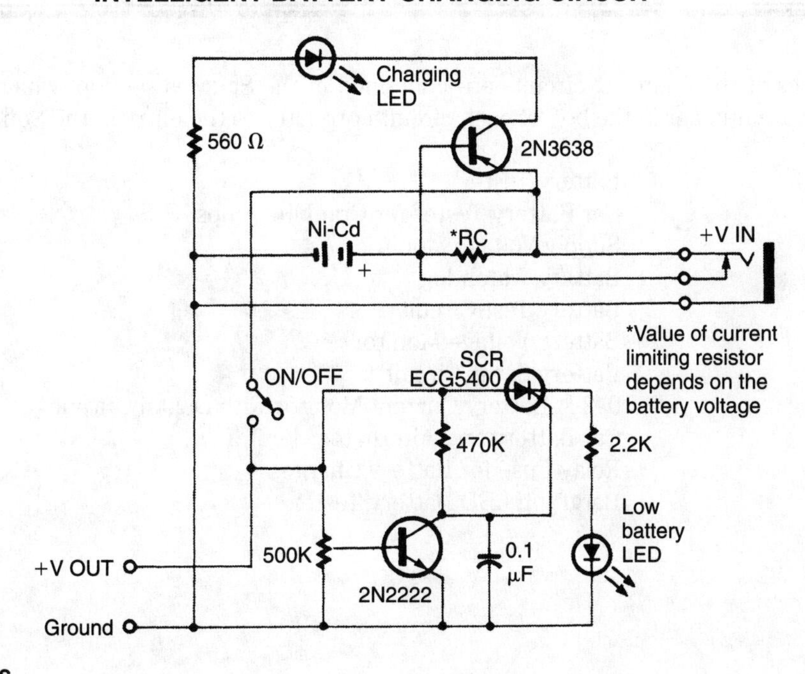

RADIO-ELECTRONICS *FIG. 9-5*

Intended for a Nicad application this charging circuit can be used with a wide range of batteries. A low-battery detector is intended. The trip voltage is set via the 500-kΩ pot. Select R_C for the battery you intend to use.

10

Battery Test and Monitor Circuits

The sources of the following circuits are contained in the Sources section, which begins on page 675. The figure number in the box of each circuit correlates to the entry in the Sources section.

BATTERY TESTER

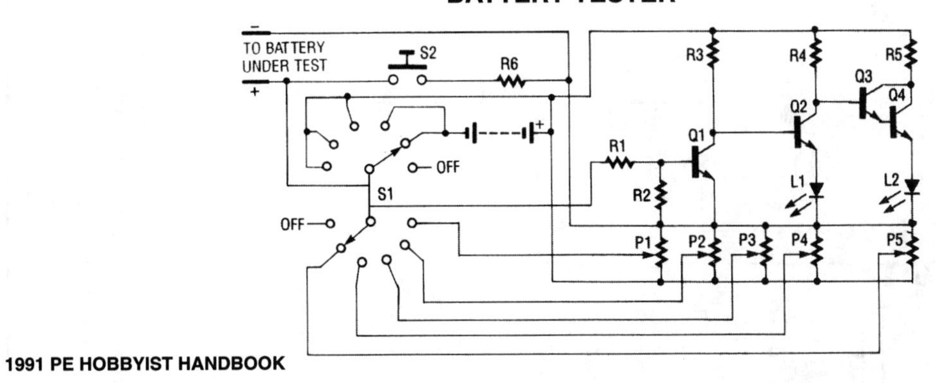

1991 PE HOBBYIST HANDBOOK *FIG. 10-1*

The battery tester uses four transistors and two LEDs to indicate the condition of any battery you want to test. Q3 and Q4 are connected in a Darlington configuration that has extremely high gain. LED L2 lights when a small positive potential appears on the base of Q3. Transistors Q1 and Q2 form a direct-coupled dc-amplifier circuit. The output of this stage drives the red LED L1. Rotary switch S1 is used to select different ranges (which have been previously set by adjusting trimmer resistors P1 through P5).

The positive (+) lead goes through the selected contacts of S1 to the biasing resistors R3, R4, and R5. The negative (–) lead of the battery under test goes to the ground or common lead of the circuit and the (+) side to one side of P1 through P5.

L1	Red LED
L2	Green LED
P1 through P5	5-kΩ trimmer resistor
R1	100 kΩ
R2, R3	33 kΩ
R4, R5	470 Ω
R6	12 Ω 1 W
S1	2 P6 position NS rotary switch
S2	NO pushbutton switch

Depending on the position of S1, a particular trimmer resistor (wiper lead) is selected. That lead goes through the contact on S1 to resistor R1 and into the base of npn transistor Q1. If the battery is good enough, (+) voltage goes to the base of Q1, turning it on. This turns Q2 off, which then allows Q3 to turn on. That causes Q4 to turn on and light green LED L2.

If the battery is weak, Q1 will not turn on, which will cause Q2 to be biased on by R3, which in turn lights red LED L1. When Q1 is on, it biases the base of Q3 negative, and causes Q3 to be turned off. That prevents L2 from turning on.

The circuit operates in the same manner for all ranges except the first two, where a 9-V battery has been added by S1 to be in series with the input voltage to allow for testing of very low voltage batteries. That is because at voltages below 2 Vdc, LEDs will not light and the circuit would be unable to set a low-voltage (<2-V) battery without the additional internal-battery voltage. A load resistor has also been included; it allows the battery under test to be connected to a load to give a better indication of its condition. That load resistor is connected across the battery when normally open (NO) switch S2 is depressed.

CAR BATTERY TESTER FOR CRANKING AMPS

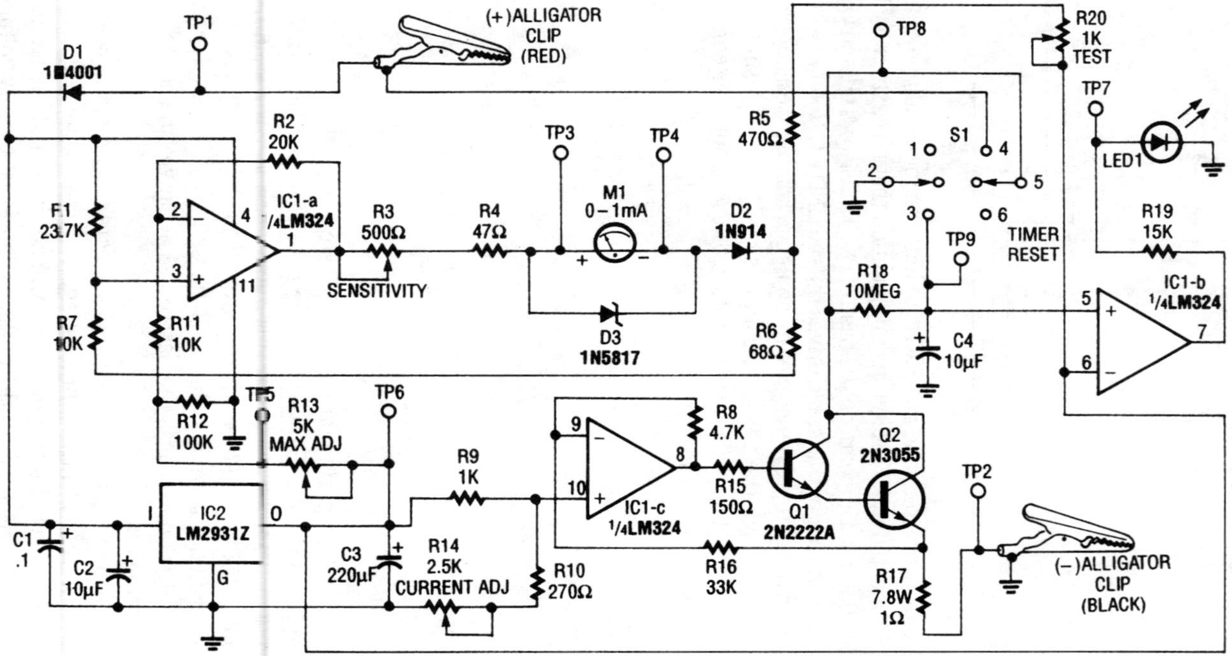

ELECTRONICS NOW

FIG. 10-2

This circuit determines the cold cranking amps of a battery by first discharging the surface charge, then checking the internal resistance. This gives a more realistic measurement than simply measuring the instantaneous drop in voltage with a load. A constant-current source draws 2.5 A. Then, after one minute, a voltage drop measurement is made under load.

SUPPLY VOLTAGE MONITOR

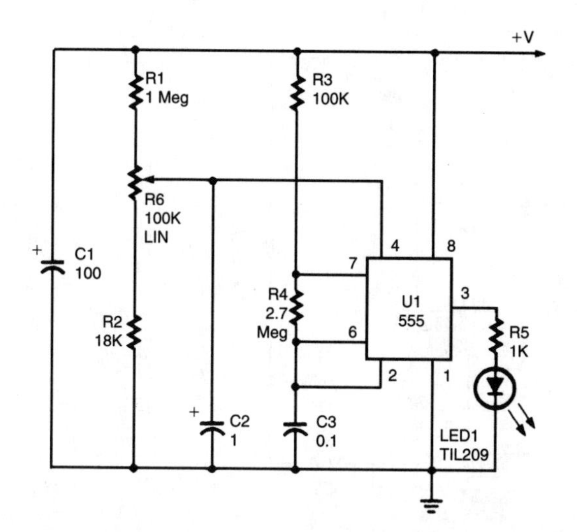

When supply voltage exceeds a preset level, the 555 oscillates, and flashes LED1. The flash rate is controlled by varying C3.

POPULAR ELECTRONICS

FIG. 10-3

BATTERY WATCHDOG

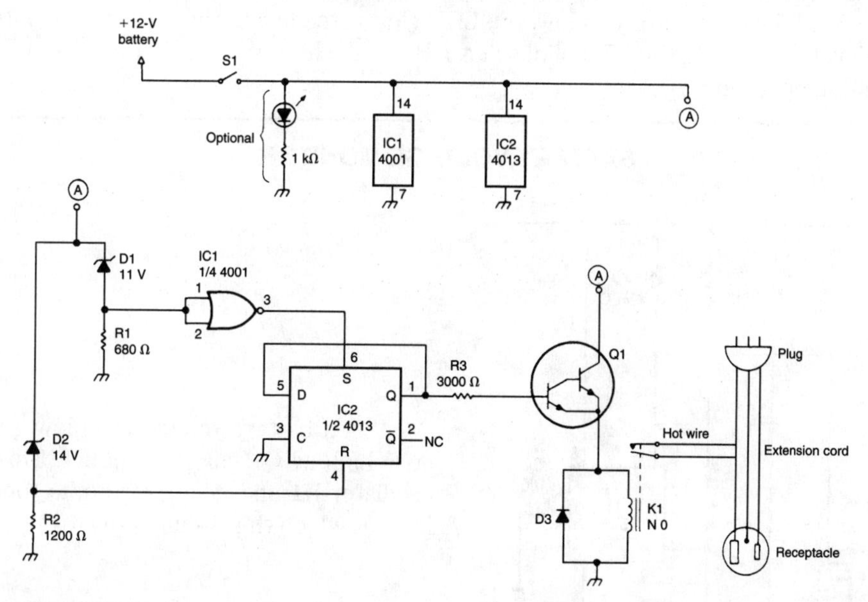

73 AMATEUR RADIO TODAY

FIG. 10-4

This circuit uses a pair of Zener diodes to monitor battery voltage of a 12-V battery. If below 11 V, D1 ceases to conduct, pin 3 of IC2 goes high, setting FF IC2 turning on Q1, K1, and the battery charger. At excess of 14-V battery voltage (full charge), D2 conducts, resetting FF IC2, and cutting off the battery charger.

BATTERY TEST CIRCUIT

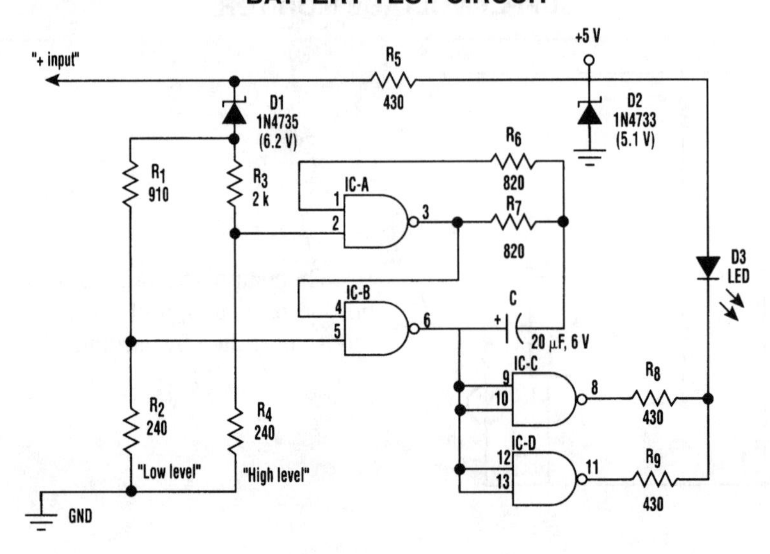

ELECTRONIC DESIGN

FIG. 10-5

Using this circuit, three levels of voltage can be displayed—normal (11 to 15 V), high (>15 V), and low (<11 V). When the voltage is low, the LED glows steadily. In the normal range, the LED is off. When the voltage is high, the LED blinks at a 1-Hz rate. This circuit is useful for assuring proper electrical system operation.

BATTERY VOLTAGE MONITOR

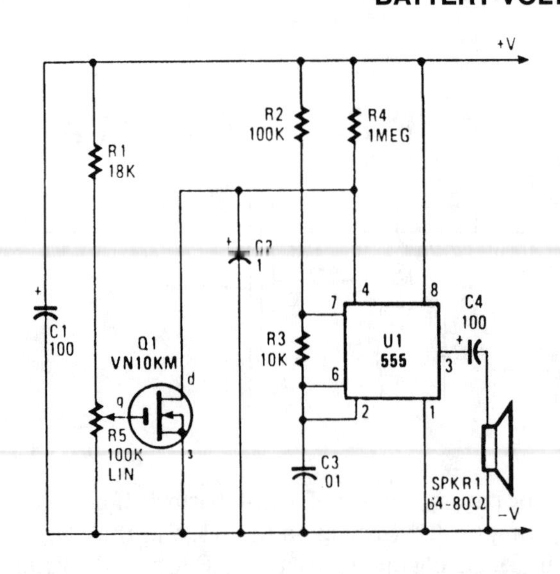

POPULAR ELECTRONICS

FIG. 10-6

When battery voltage goes low, pin 4 of U1 goes high as Q1 fails to conduct. This activates oscillator U1 and generates audio tone. R5 sets level at which the circuit activates.

BATTERY SAVER CIRCUIT

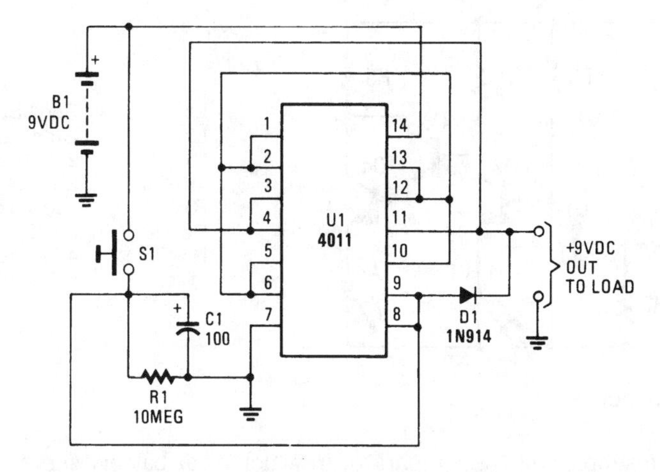

POPULAR ELECTRONICS ***FIG. 10-7***

This battery saver circuit can automatically turn off a small piece of test equipment after a desired period of time, allowing you to leave your shop worry free.

This circuit uses a CD4011 IC to act as a simple timer. One section acts as an RC discharge timer (pin 7). This causes its output to go low, holding the three other outputs high acting as a 9-V source. After C1/R1 discharges approximately 10 minutes, the output drops to zero. S1 resets the circuit.

0–2-A BATTERY CURRENT MONITOR WITH DIGITAL OUTPUT

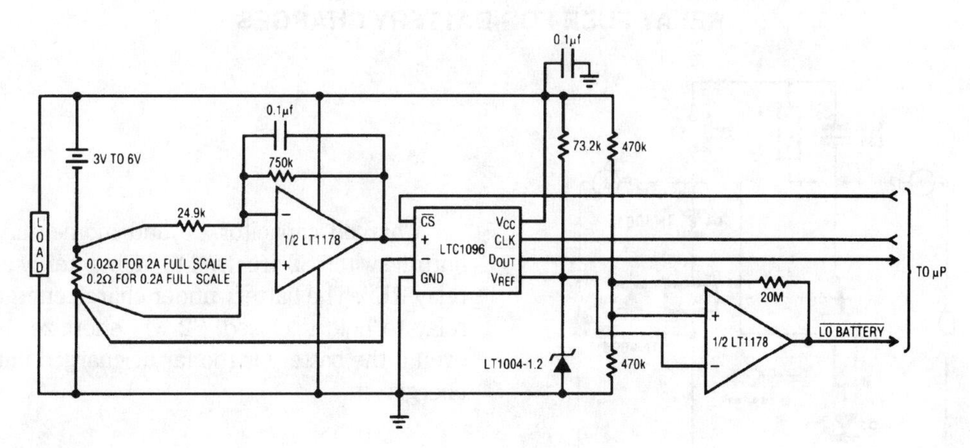

LINEAR TECHNOLOGY ***FIG. 10-8***

IC devices by Linear Technology make up this current monitor circuit. Drain is only 70µA from a 3- to 6-V battery.

CAR BATTERY AND ALTERNATOR MONITOR

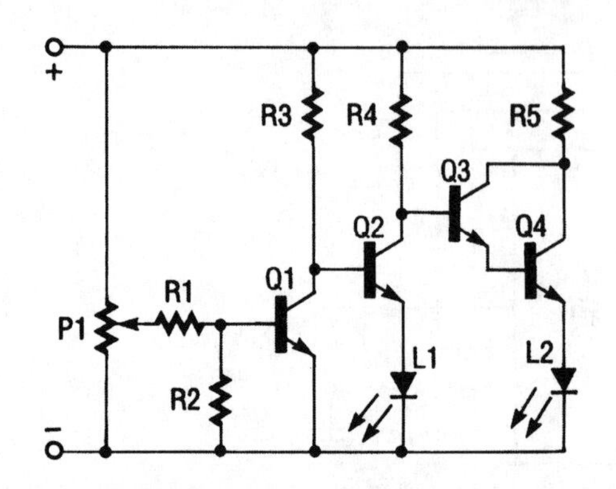

L1	Red LED
L2	Green LED
P1	2.5-kΩ trimmer resistor
Q1–Q4	2N3904 transistor
R1	100-kΩ resistor
R2, R3	33-kΩ resistor
R4, R5	470-Ω resistor
Misc.	PC board, wire

1991 PE HOBBYIST HANDBOOK *FIG. 10-9*

The monitor is a simple voltage comparator in which a car battery serves as the battery for operation. The input voltage to the comparator is set by adjustment potentiometer P1, which must be adjusted so that the green LED L2 is on when the alternator is operating properly and red LED1 is on when the alternator is inoperative.

The circuit operates as follows: When the alternator operates properly, the battery voltage is higher and P1 is set so that transistor Q1 causes Q2 to be off. That results in Q3 and Q4 being fully on, thus applying current to green LED L2. If the battery voltage is lowered (alternator inoperative), transistor Q1 is turned off. That allows transistor Q2 to turn fully on, applying current to red LED L1, indicating trouble. Once Q2 is on, it causes Q3 and Q4 to go out of conduction.

RELAY FUSE FOR BATTERY CHARGES

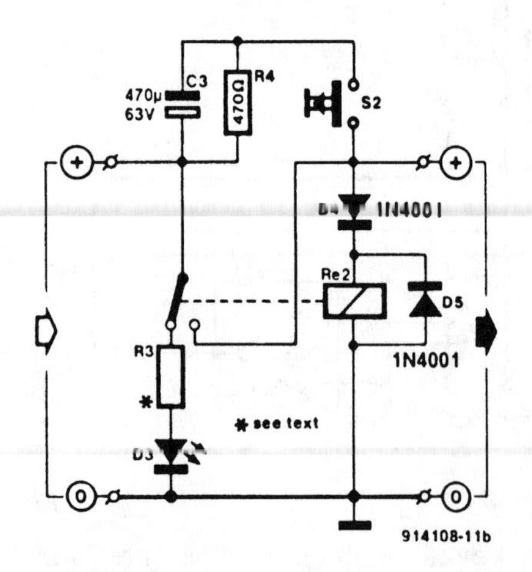

ELEKTOR ELECTRONICS *FIG. 10-10*

Charged capacitor C3 and momentary pushbutton switch S2 are used to momentarily energize relay RE2. The battery under charge energizes the relay to hold it closed. S2 will energize the relay even if the battery is too far discharged initially to energize it.

BARGRAPH LED BATTERY TESTER

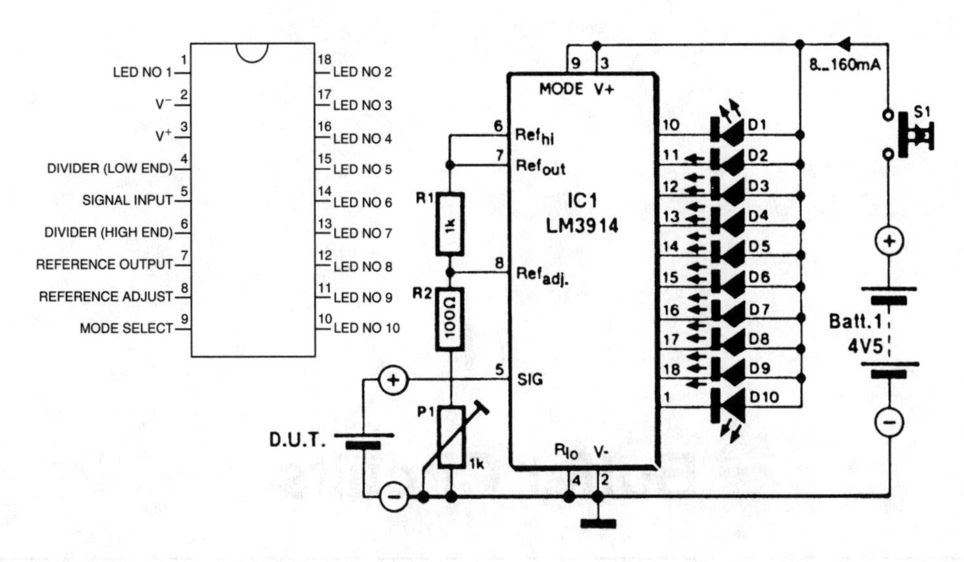

ELEKTOR ELECTRONICS USA *FIG. 10-11*

The LM3914A bargraph LED is used here as a voltmeter for battery testing. The circuit is powered by a 4.5-V battery and compares the battery under test with an internally derived reference, set by R1/R2/P1. Each LED of the 10 represent 10% of full scale. For best results, the battery (D.U.T.) should be loaded with an appropriate resistor.

11

Buffer Circuits

The sources of the following circuits are contained in the Sources section, which begins on page 675. The figure number in the box of each circuit correlates to the entry in the Sources section.

BUFFER/AMPLIFIERS

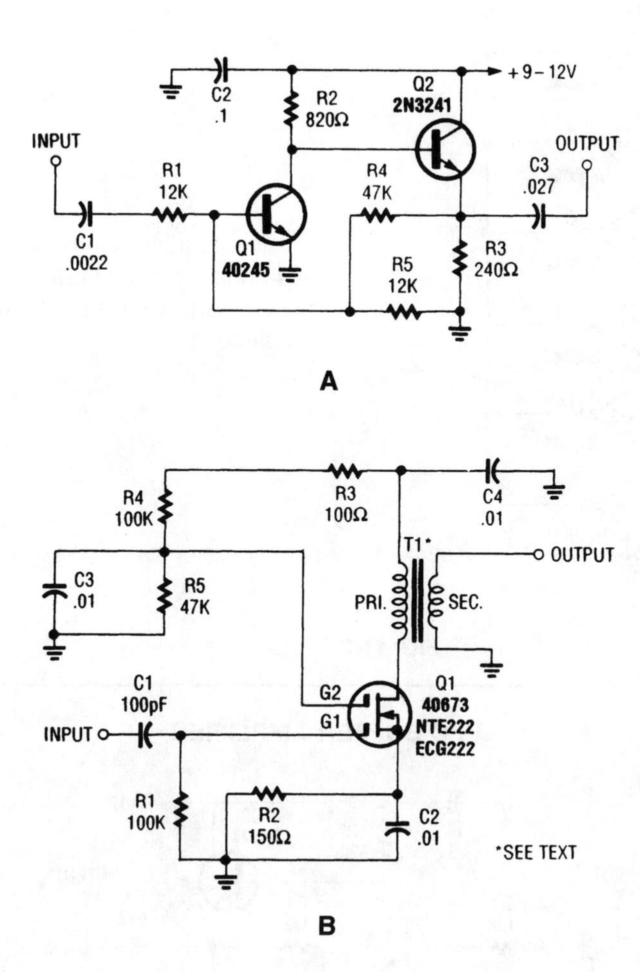

POPULAR ELECTRONICS ***FIG. 11-1***

These two buffer/amplifiers that have been successfully used with VFOs: one (shown in A) is based on a pair of bipolar npn transistors, and the other (shown in B) is built around a dual-gate MOSFET.

HIGH CURRENT BUFFER

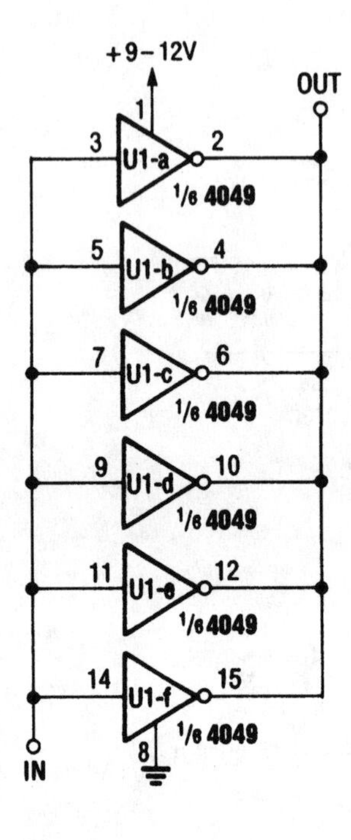

By parallel connecting all six gates of this 4049 hex inverting buffer, you can obtain a much higher output current than would otherwise be available.

POPULAR ELECTRONICS *FIG. 11-2*

VFO BUFFER AMPLIFIER

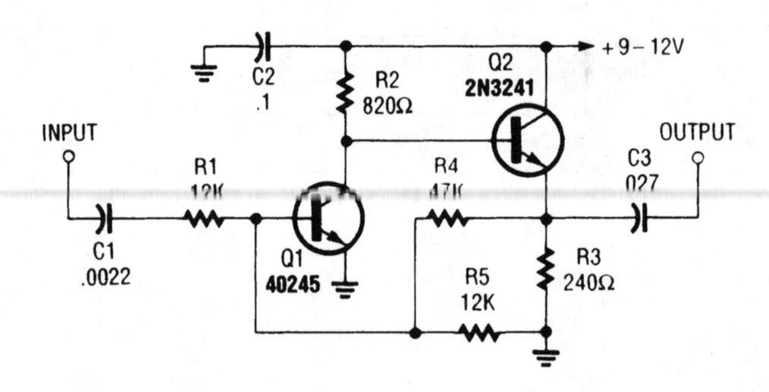

POPULAR ELECTRONICS *FIG. 11-3*

A two-transistor feedback pair provides broadband operation. The gain is approximately R_4/R_1.

MOSFET BUFFER AMPLIFIER

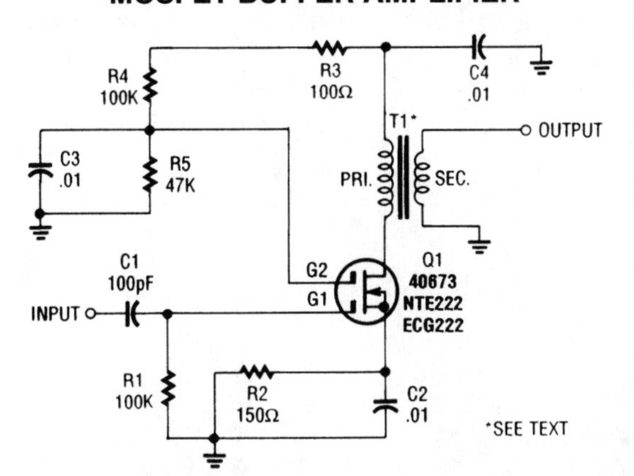

POPULAR ELECTRONICS *FIG. 11-4*

A MOSFET is used as a wideband buffer amplifier. T1 is wound on a toroid of approximately ½" diameter, with material suitable for frequency (usually 1- to 20-MHz range). The turns ratio should be about 4:1 depending on load impedance. Typically, at 4 MHz, there are 18 turns on the primary, 4 turns on the secondary, and the stage gain is about 14-dB voltage (Z_L = 50 Ω).

3-V RAIL-TO-RAIL SINGLE-SUPPLY BUFFER

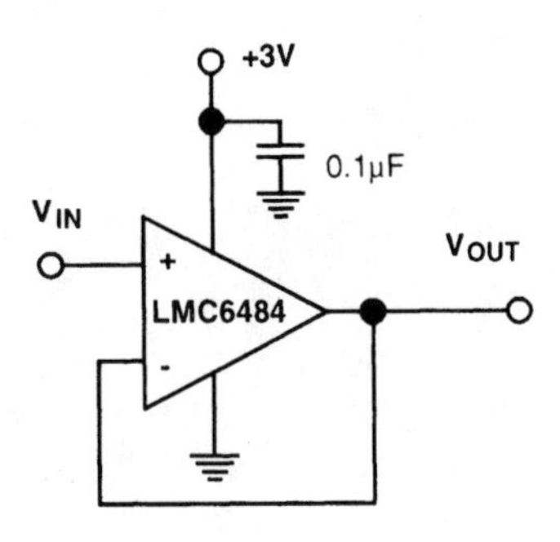

NATIONAL SEMICONDUCTOR *FIG. 11-5*

The LMC6484 provides a 3-V p-p rail-to-rail buffer with a +3-V supply commonly used for logic systems.

SIMPLE VIDEO BUFFER

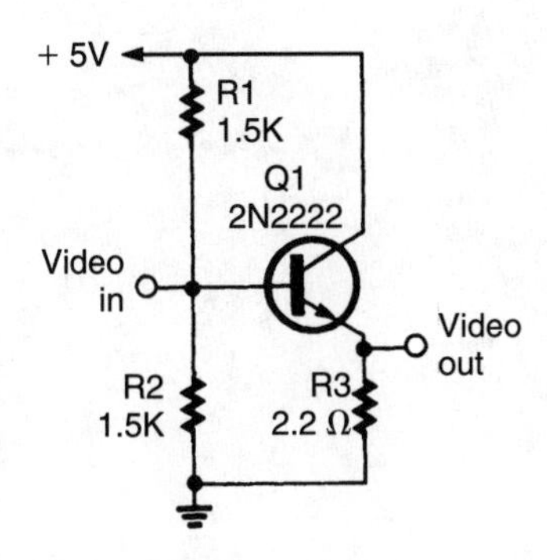

ELECTRONICS NOW *FIG. 11-6*

This simple emitter follower can be used as a video buffer.

LOW-OFFSET SIMPLE VIDEO BUFFER

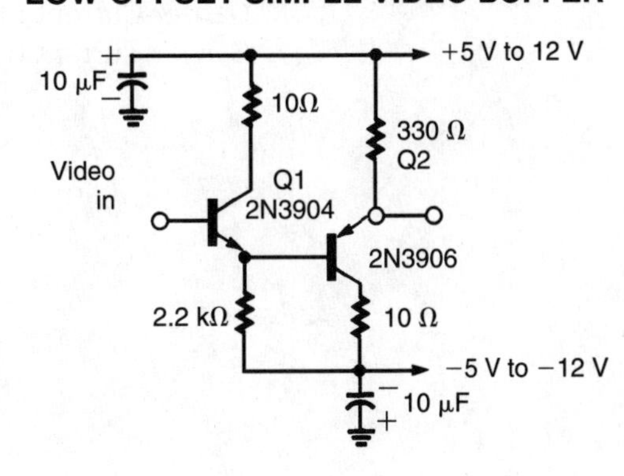

WILLIAM SHEETS *FIG. 11-7*

This circuit has proved to be an effective video buffer and will easily drive a 75-Ω load to 1.5-V p-p output. BW is better than 20 MHz and there is less than 0.05-V dc offset, which is the difference in V_{BE} of Q1 and Q2. The supply lines should be well bypassed, ± 5 V or more.

12

Carrier-Current Circuits

The sources of the following circuits are contained in the Sources section, which begins on page 675. The figure number in the box of each circuit correlates to the entry in the Sources section.

CARRIER-CURRENT BABY-ALERT TRANSMITTER

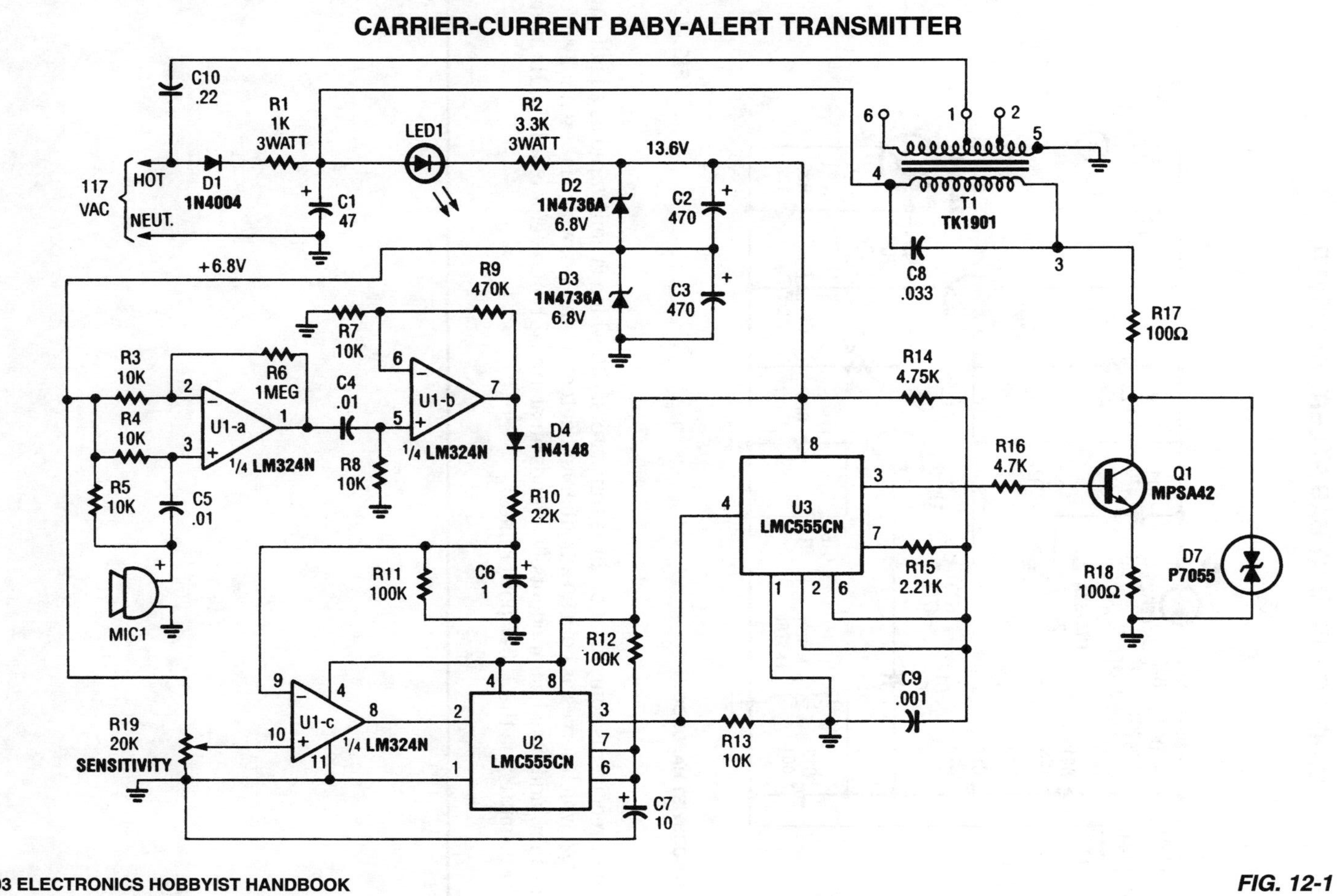

1993 ELECTRONICS HOBBYIST HANDBOOK ***FIG. 12-1***

The baby-alert transmitter is built around an LM324 quad op amp (U1), two LMC555CM CMOS oscillator/timers (U2 and U3), and a few support components. The transmitter sends a signal on receipt of a sound at MIC1. It has a frequency of around 125 kHz and can be used to trigger an alarm receiver.

CARRIER-CURRENT BABY-ALERT RECEIVER

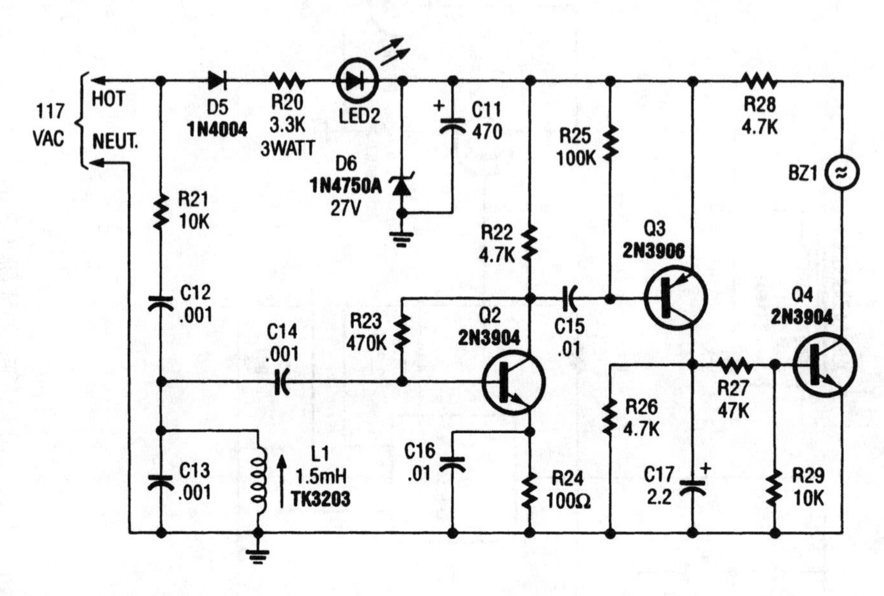

1993 ELECTRONICS HOBBYIST HANDBOOK *FIG. 12-2*

The baby-alert receiver is comprised of three transistors: Q2, which is configured as a high-gain linear amplifier; Q3, which serves as both an amplifier and detector; and Q4, which is essentially used as a switch; and a few additional components. It sounds an alarm BZ1 on receipt of a 125-kHz signal from an alarm transmitter via the 120-V power lines.

13

Clock Circuit

The source of the following circuit is contained in the Sources section, which begins on page 675. The figure number in the box of the circuit correlates to the entry in the Sources section.

Binary Clock

BINARY CLOCK

1992 PE HOBBYIST HANDBOOK

This circuit is an unusual clock in that the LEDs are bi-color red/green displays that indicate the time in binary coded decimal form.

LEDs 21 through 24 read out seconds

LEDs 5, 18, 19, and 20 read out 105 seconds

LEDs 14 through 17 read out in minutes

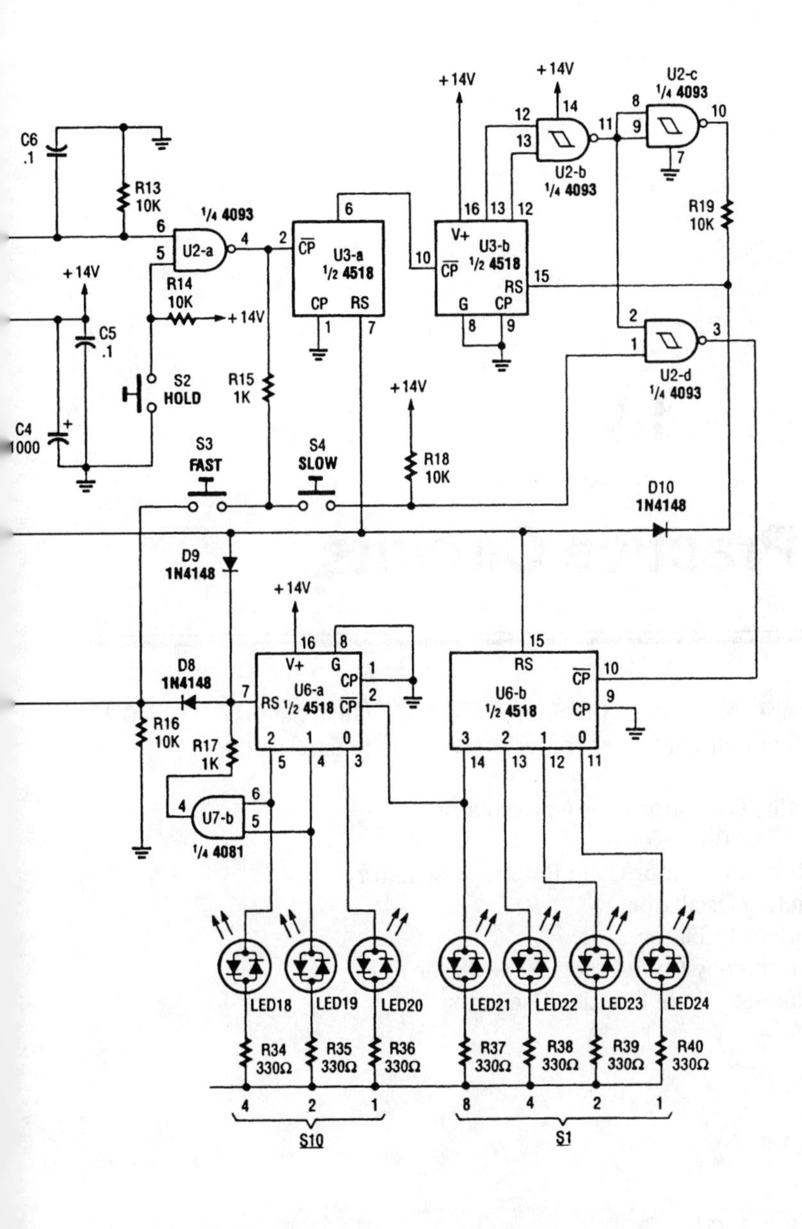

FIG. 13-1

LEDs 4, 11, 12, and 13 read out in 105 minutes
LEDs 7 through 10 read out the hours
LEDs 1, 2, 3, and 6 read out tens of hours
The 60-Hz line is used as a timebase.

14

Code Practice Circuits

The source of the following circuits are contained in the Sources section, which begins on page 675. The figure number in the box of each circuit correlates to the entry in the Sources section.

CODE PRACTICE OSCILLATOR USES OPTOISOLATOR

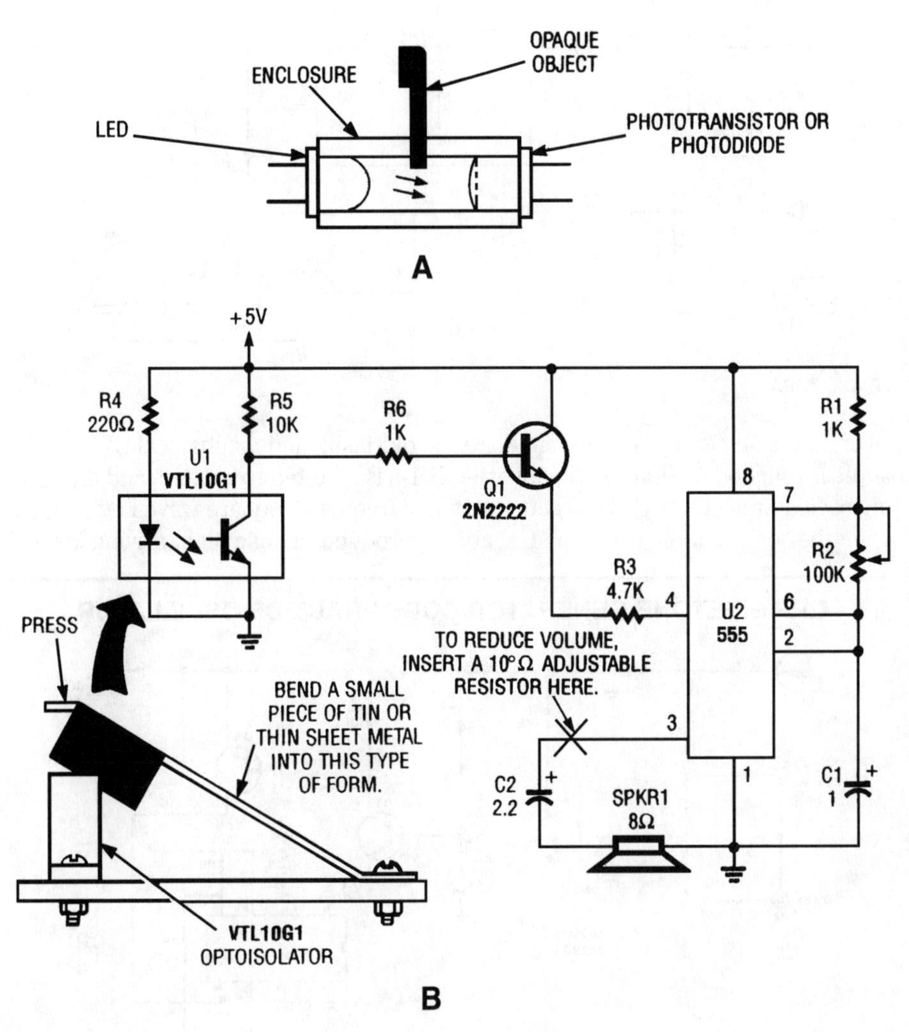

POPULAR ELECTRONICS ***FIG. 14-1***

A slotted-pair isolator (A) is effectively an enclosed-pair isolator with a slit that will allow an obstacle to interrupt the light path. That could be useful for building a code key (B).

ELECTRONIC CW "BUG" KEYER

73 AMATEUR RADIO TODAY ***FIG. 14-2***

This keyer uses skin conductivity to simulate the old-fashioned mechanical CW bug keyer. When the "dit" paddle is touched the bias on the inverter, IC1-a is shunted to ground, and it produces a logic high, causing oscillator sections C&D to generate a low-frequency square wave keying Q1 for a series of "dits." When the "dah" paddle is touched, section b produces a logic high, driving keyer Q1 on.

QRP SIDETONE GENERATOR/CODE PRACTICE OSCILLATOR

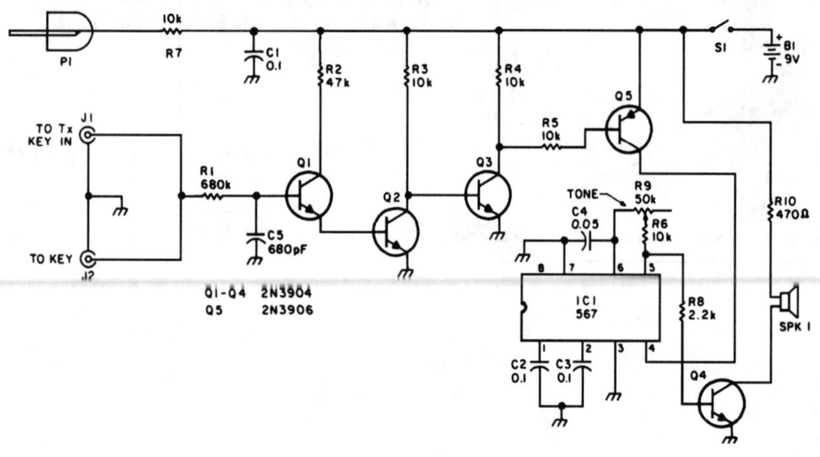

73 AMATEUR RADIO TODAY ***FIG. 14-3***

For use with low-power transmitters with a positive keying voltage. Q1/Q2/Q3 form a switching amplifier. When the key is pressed, the collector of Q3 goes to ground, turning on Q5 and activating IC1, an audio oscillator. Q4 drives the speaker. For use as a code practice oscillator, insert P1 and J1 and a key in J2.

MORSE PRACTICE OSCILLATOR

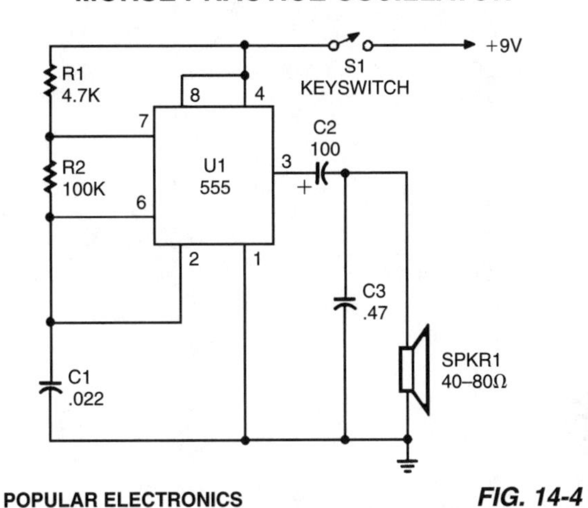

POPULAR ELECTRONICS *FIG. 14-4*

A 555 timer configured as an astable multivibrator is used in this circuit to generate an audio note. C1 can be changed to vary the audio note as desired.

CODE PRACTICE OSCILLATOR

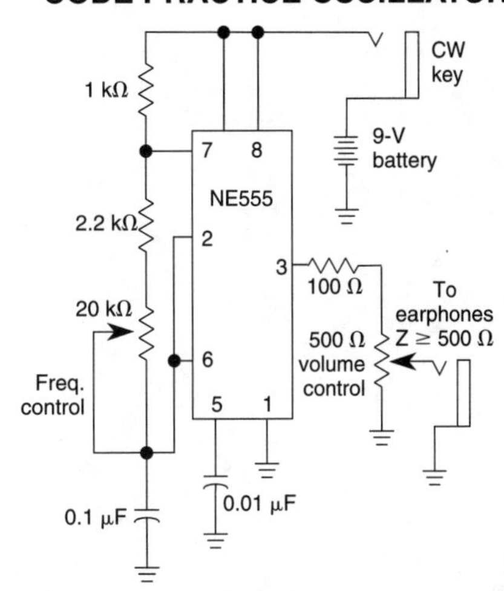

WILLIAM SHEETS *FIG. 14-5*

The tone and volume of the sound produced when the telegraph key is depressed can be varied in this code practice oscillator.

VARIABLE FREQUENCY CODE PRACTICE OSCILLATOR

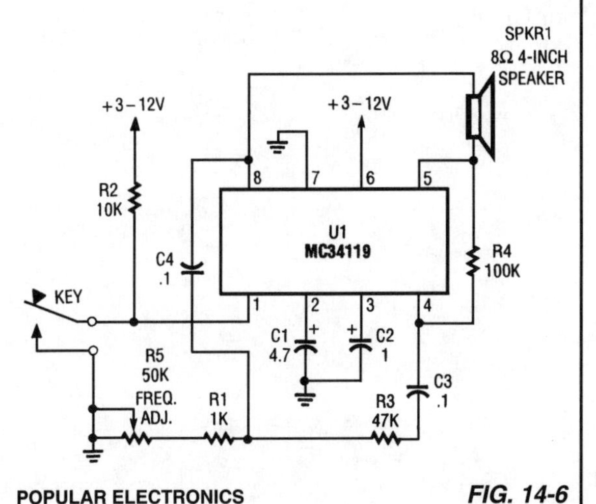

POPULAR ELECTRONICS *FIG. 14-6*

The variable frequency audio oscillator can be used as a low-level alarm sounder or a code-practice oscillator.

SINGLE-TRANSISTOR CODE PRACTICE OSCILLATOR

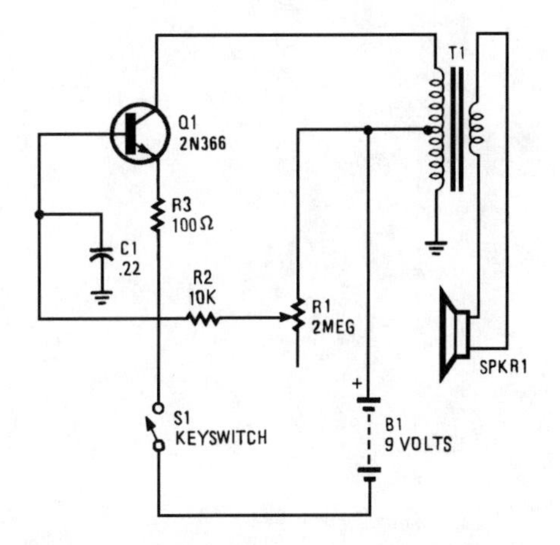

POPULAR ELECTRONICS *FIG. 14-7*

A 2N366 is configured as an audio feedback oscillator using an audio transformer is shown. Adjust R1 for proper operation and desired audio note.

15

Color Organ Circuit

The source of the following circuit is contained in the Sources section, which begins on page 675. The figure number in the box of each circuit correlates to the entry in the Sources section.

3-Channel Color Organ

3-CHANNEL COLOR ORGAN

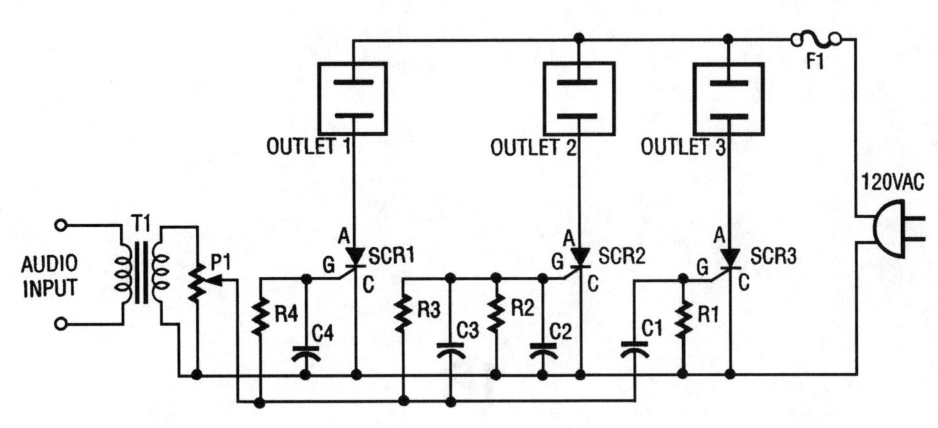

1991 PE HOBBYIST HANDBOOK ***FIG. 15-1***

The ac line power is brought back into the circuit through F1, a protective 5-A fuse. One side of the ac line is connected to one side of each ac outlet. The other side of the ac line is connected to each SCR or silicon-controlled rectifier. Each SCR is, in turn, connected to the other side of each ac outlet.

An audio signal is brought into the circuit from a stereo speaker by transformer T1. This transformer has 500-Ω impedance on the primary and 8-Ω impedance on its secondary. Connect T1 so that the 8-Ω side is connected to the speaker and the 500-Ω side is connected to potentiometer P1.

Potentiometer P1 is used as a level or sensitivity control. The signal from its wiper lead is applied to each RC filter stage. Because each SCR has a different RC (resistor/capacitor) filter on its gate lead, each will respond to different frequencies. The greater the capacitance in the filter, the lower the frequency that the SCR will respond to.

16

Computer Circuits

The sources of the following circuits are contained in the Sources section, which begins on page 675. The figure number in the box of each circuit correlates to the entry in the Sources section.

PRINTER SENTRY

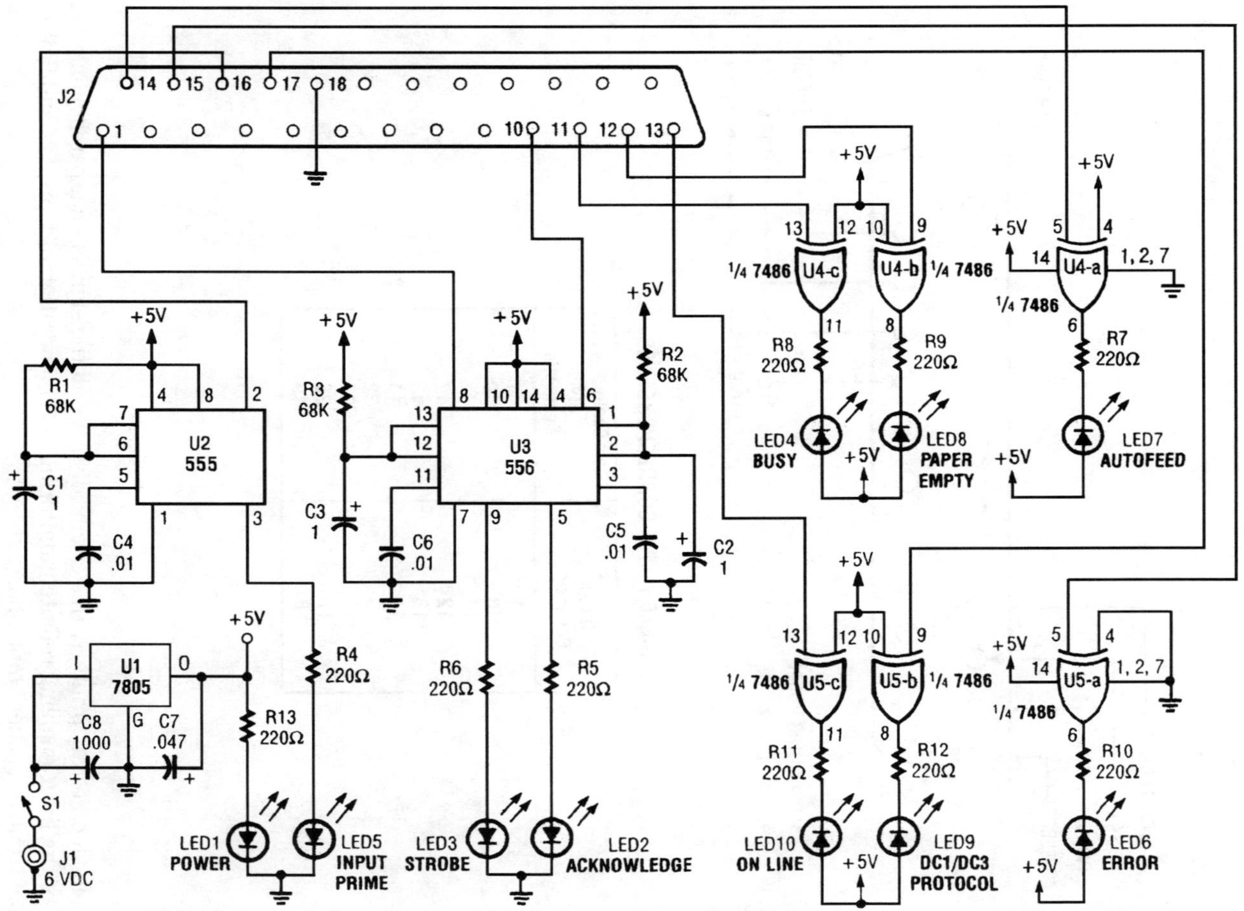

POPULAR ELECTRONICS

FIG. 16-1

PRINTER SENTRY (*Cont.*)

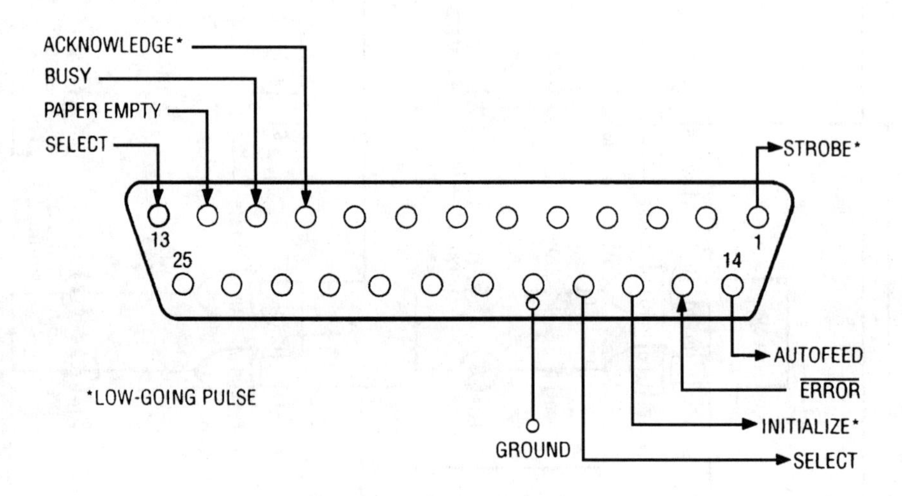

TABLE 1—PIN CORRESPONDENCE

DB-25 Connector	Centronics-Style Connector
1	1
10	10
11	11
12	12
13	13
14	14
15	32
16	31
17	36
18	19

Handy for monitoring printers, this circuit displays all the signals on a parallel link. It monitors the status of the lines, enabling remote monitoring of the operation of a printer, and it also gives an indication of troubles (paper empty, busy, etc.).

PC PASSWORD PROTECTION

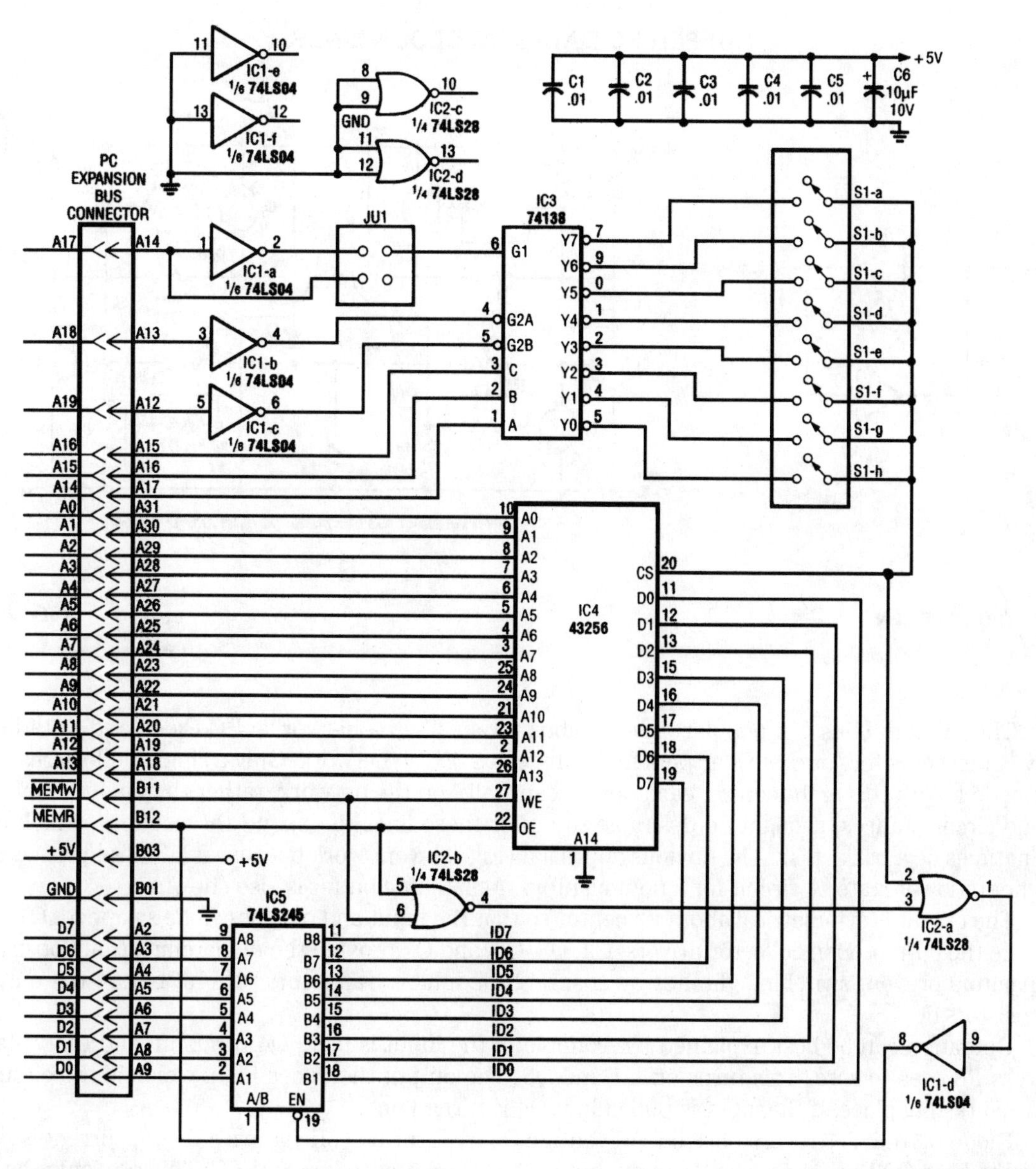

ELECTRONICS NOW

FIG. 16-2

With this circuit, a PC will be protected, requiring a password to boot. After three times, the computer will have to have a cold reboot and the password tried again. Software for this system is available—consult the reference for further details.

BUFFER I^2C DATA AND CLOCK LINES

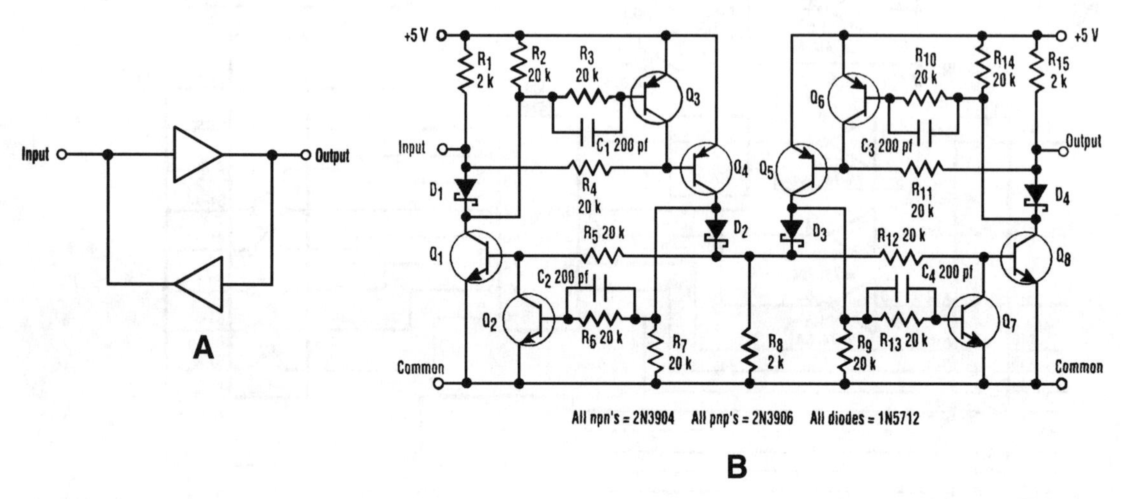

ELECTRONIC DESIGN *FIG. 16-3*

The I^2C serial bus is a popular two-wire bus for small-area networks. I^2C Clock and Data lines have open collector (or drain) outputs for each device on the network. Only a single pull-up resistor is needed. With this architecture, each device can "talk" on the network, rather than just "listen." In some circumstances, it might be desirable to buffer these lines to expand the network, which can sometimes be a tricky task. The obvious approach (Fig. 1) wont work because it latches in either the higher or lower state. A circuit for a noninventory nonlatching buffer is also shown.

The circuit is symmetrical about its center so that the input and output can be swapped. Q1 and Q8 are the output open collector drivers. Q2, Q3, Q6, and Q7 provide the nonlatching functions. The capacitors prevent switching glitches by ensuring the inhibit transistors turn off before the output transistors do.

Operation can be best explained by example: if the input is high, Q4 turns off, and the voltage across R8 goes to zero. This turns off Q1 and Q8. The output then goes high, which is the circuit's normal resting place. If the input is pulled low, Q4 is turned on.

Diode D1 remains reverse-biased, preventing Q3 from turning off Q4. With Q4 on, current is supplied to both Q2 and Q1 to turn them on, but Q2 turns on first to keep Q1 off. This prevents the input from latching. Q4 also turns on Q8. D4 is now forward-biased, so Q6 turns on, and thus turns off Q5. With Q5 off, Q7 will not turn on. The output remains low. Even with both the input and the output externally driven low, the circuit will not latch. The circuit, using the values shown in Fig. 2, reached a clock rate of 80 kHz with a VOH of 5.0 V and a VOL of 0.5 V.

17

Control Circuits

The sources of the following circuits are contained in the Sources section, which begins on page 675. The figure number in the box of each circuit correlates to the entry in the Sources section.

6-DIGIT CODED ac POWER SWITCH

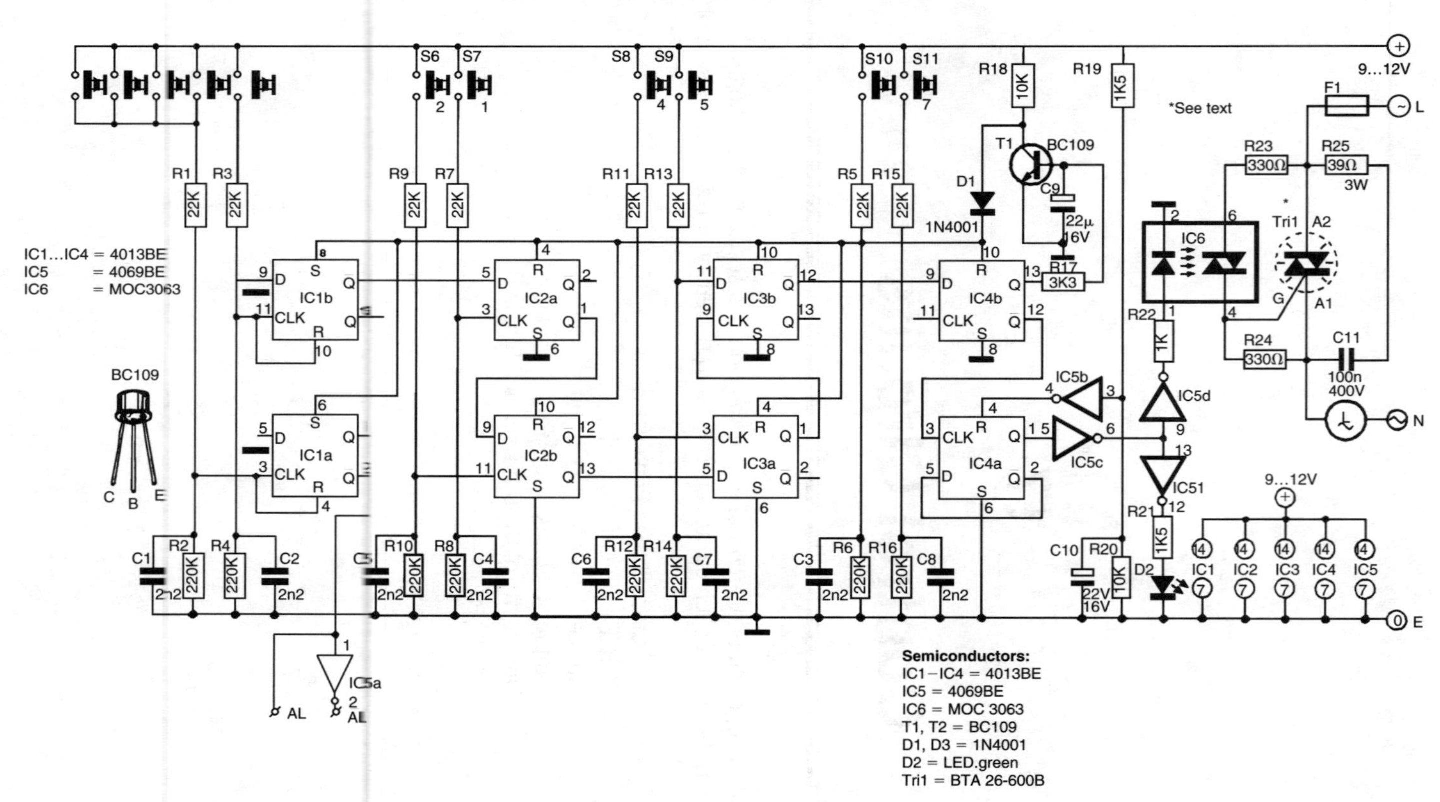

ELEKTOR ELECTRONICS

FIG. 17-1

This switch uses four CD4013 BE dual flip-flops, an inverter, and an optoisolator to drive a triac. The circuit can switch 25-A ac load current. A standard 4 × 3 telephone keyboard is used to enter a 6-digit code. In case of a wrong code, a signal is available to activate an alarm. The disarming method is a secret reset button that can be any number on the keyboard.

VCR TV ON/OFF CONTROL

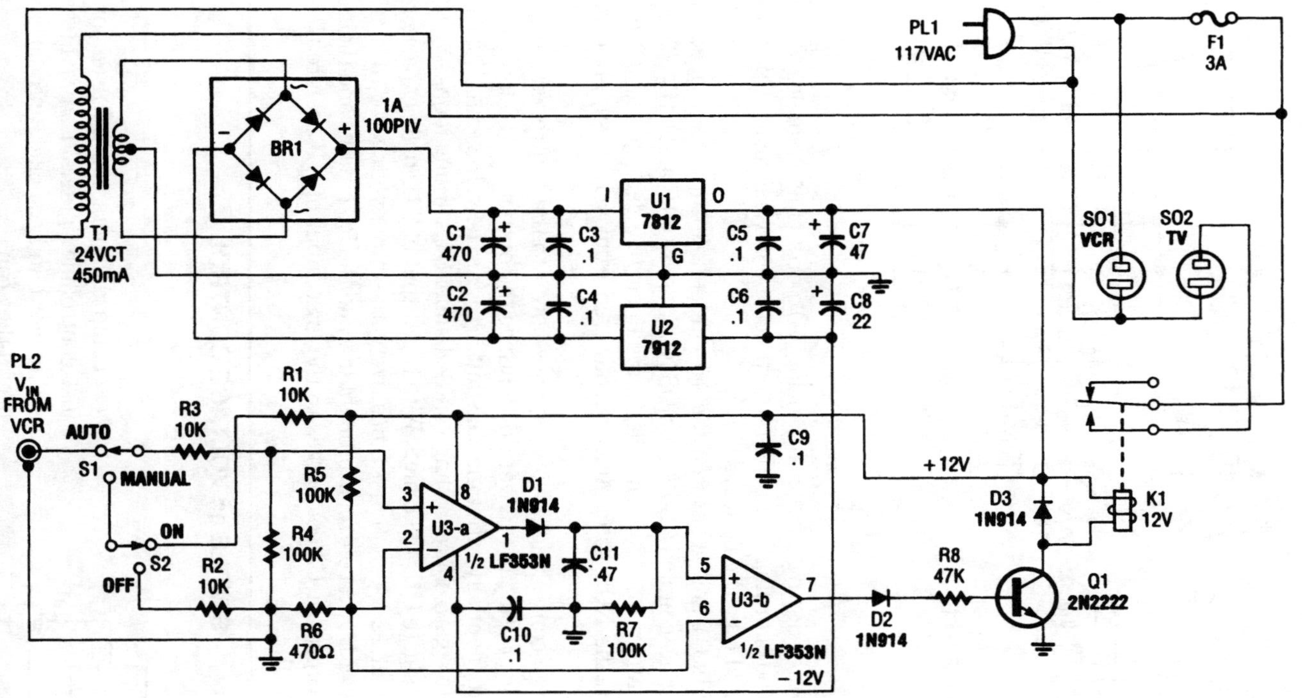

1993 ELECTRONICS HOBBYISTS HANDBOOK

FIG. 17-2

This circuit senses the video from the VCR. When the VCR is turned on, video signal is amplified by U3A and B to drive Q1, activating K1. In this manner, it is not necessary to turn on and off two video devices every time. In many cases, this avoids the use of a cable box, the cable-ready VCR performing this function.

SIMPLE POWER DOWN CIRCUIT

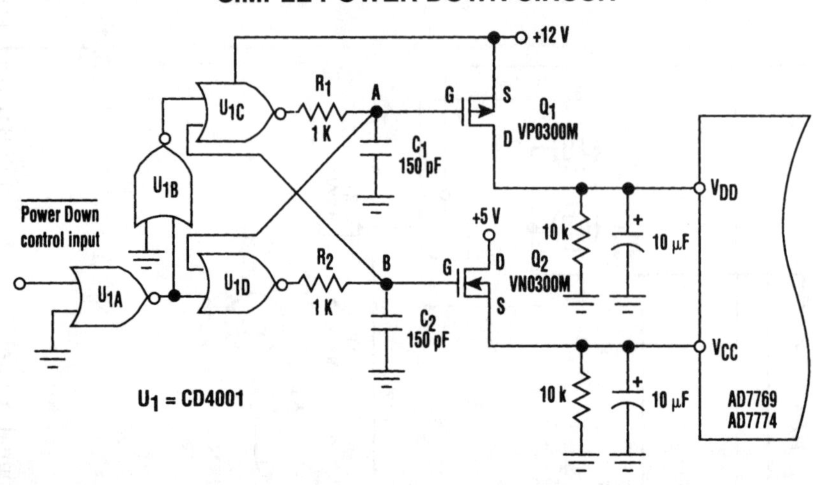

ELECTRONIC DESIGN

FIG. 17-3

This circuit adds a power-down function to analog I/O ports (for example, the AD7769 and AD7774). Moreover, the diodes ordinarily needed to protect the devices against power-supply mis-sequencing can be eliminated (see the figure).

In the circuit, MOSFETs Q1 and Q2 switch the +5- and +12-V supplies, respectively, in a sequence controlled by two cross-coupled CD4001 CMOS NOR gates (U1C and U1D). The sequence in which power is applied is important: The controlled circuits may be damaged anytime V_{CC} exceeds V_{DD} + 0.3 V. Consequently, the NOR gates must be powered from a 12-V supply throughout the power-down sequence.

Bringing the power down control high (+5 V) applies power to the controlled circuit by turning on all MOSFETs. Specifically, raising the power down brings the output of U1C low, causing capacitor C1 to discharge VOL exponentially with time constant R_1C_1. As the voltage on C1 falls, two events occur. First, it puts a negative gate-source voltage on P-channel Q1, turning it on.

Second, it causes output gate U1D to go high. With the output of U1D high, capacitor C2 charges exponentially to VOH—about 12-V—applying a positive gate-source voltage to turn on Q2. In the power down mode, the Power Down control is brought low and the RC circuits and their delays work in reverse. Consequently, capacitor C2 discharges to the logic input of U1C before C1 can charge. Hence, Q2 turns off before Q1.

SIMPLE ac VOLTAGE CONTROL

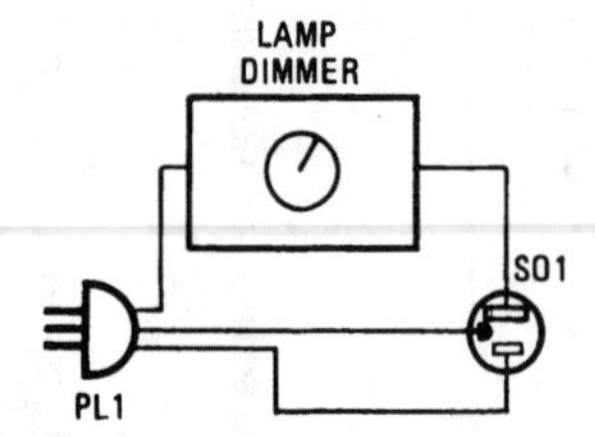

Lamp dimmers can be used for more than just controlling lights. Just provide one with an ac line cord and a socket, and discover just how useful they can be.

POPULAR ELECTRONICS

FIG. 17-4

DUAL-CONTROL SWITCH USES ac SIGNALS

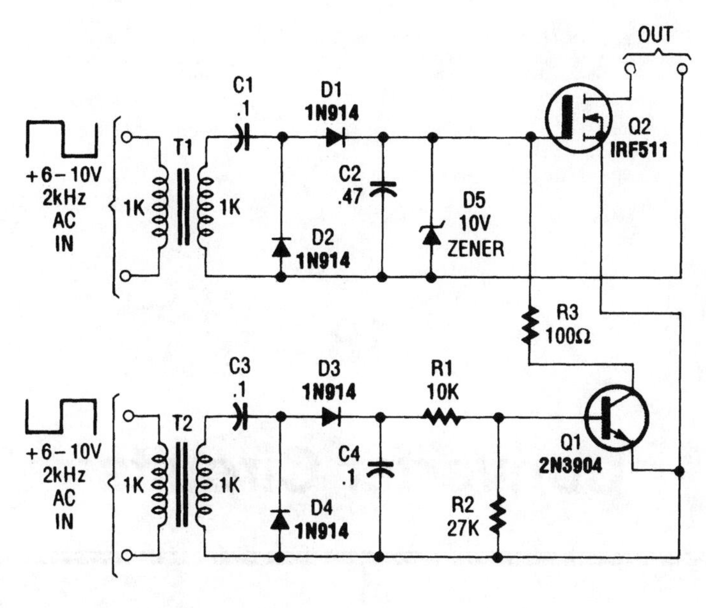

POPULAR ELECTRONICS *FIG. 17-5*

The Dual-Control Switch uses two 6–10-Vac sources to trigger the circuit on and off; one source for each function.

18

Converter Circuits

The sources of the following circuits are contained in the Sources section, which begins on page 675. The figure number in the box of each circuit correlates to the entry in the Sources section.

ONE-CHIP CRYSTAL-CONTROLLED CONVERTER

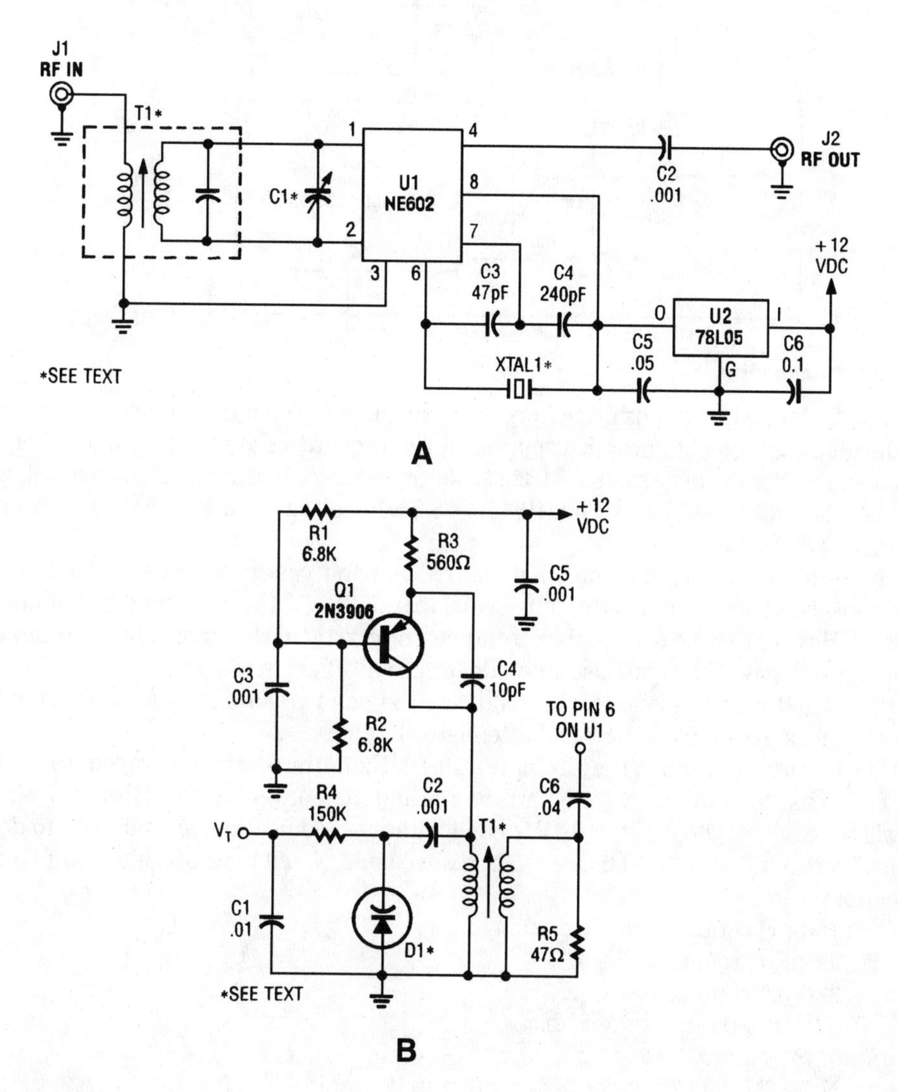

POPULAR ELECTRONICS *FIG. 18-1*

This circuit can work over a wide range of frequencies. XTAL 1 is a fundamental-frequency crystal. T1 and C1 are tuned to the input frequency. An application of this circuit is a simple shortwave converter for AM radios, etc. A tuneable oscillator can also be used, as shown.

DIGITAL RESISTANCE CONTROL

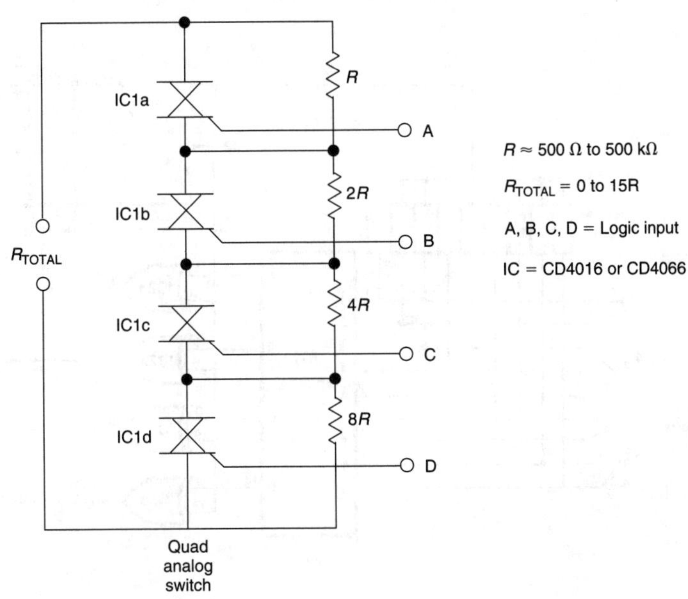

WILLIAM SHEETS

FIG. 25-4

Digital resistance control is possible with bilateral switches. Do not forget that analog switches have "on" resistance.

DIGITAL CAPACITANCE CONTROL

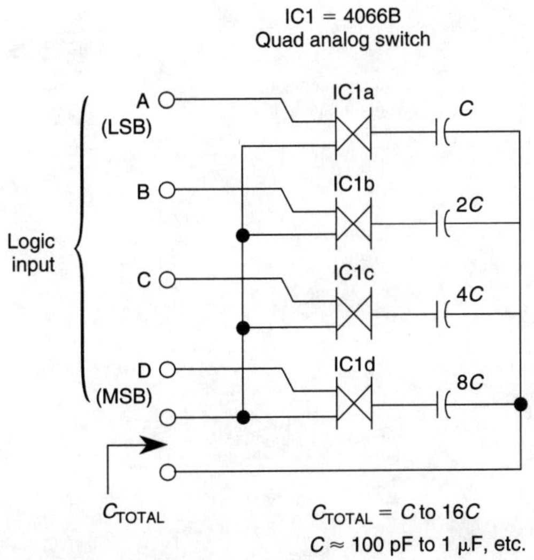

IC1 = 4066B
Quad analog switch

Logic input: A (LSB), B, C, D (MSB)

IC1a, IC1b, IC1c, IC1d

C, $2C$, $4C$, $8C$

C_{TOTAL}

$C_{TOTAL} = C$ to $16C$
$C \approx 100$ pF to 1 μF, etc.

WILLIAM SHEETS

FIG. 25-5

Digital capacitance control is possible with bilateral switches. Do not forget to consider "ON" resistance of the analog switches.

BCD ROTARY SWITCH

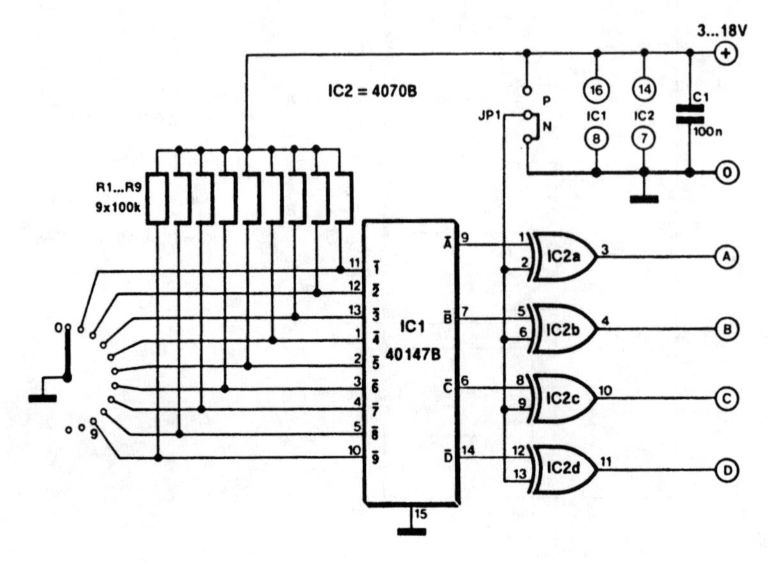

ELEKTOR ELECTRONICS USA

FIG. 25-6

This circuit allows a simple rotary switch to emulate a BCD switch. The circuit draws about 200 mA. A 10-position rotary switch is used.

26

Display Circuits

The sources of the following circuits are contained in the Sources section, which begins on page 675. The figure number in the box of each circuit correlates to the entry in the Sources section.

4033 DISPLAY CIRCUITRY COMMON CATHODE

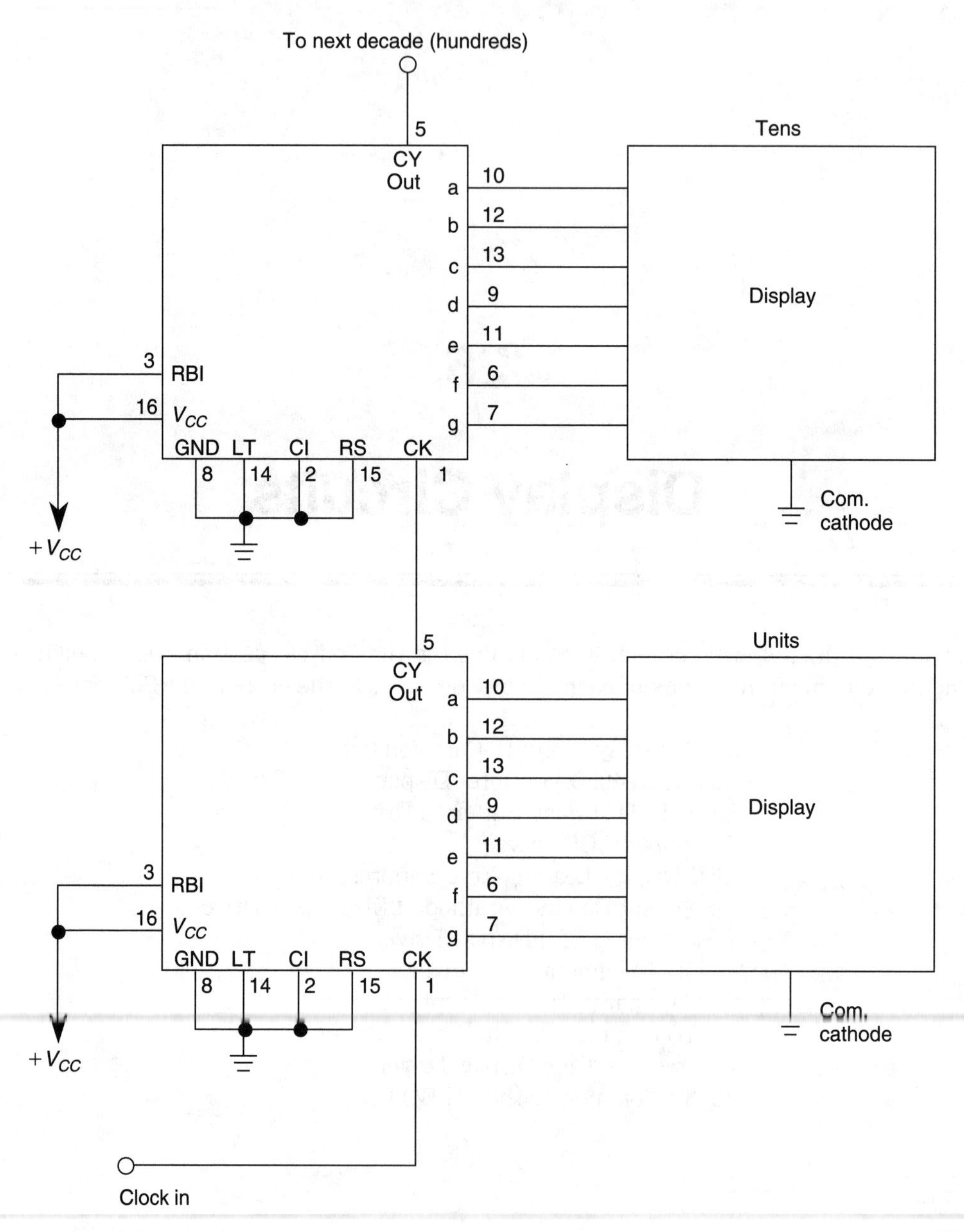

WILLIAM SHEETS ***FIG. 26-1***

To drive two or more common-cathode displays two or more 4033 decode counters can be cascaded.

CASCADED 4026B COUNTER/DISPLAY DRIVER CIRCUIT

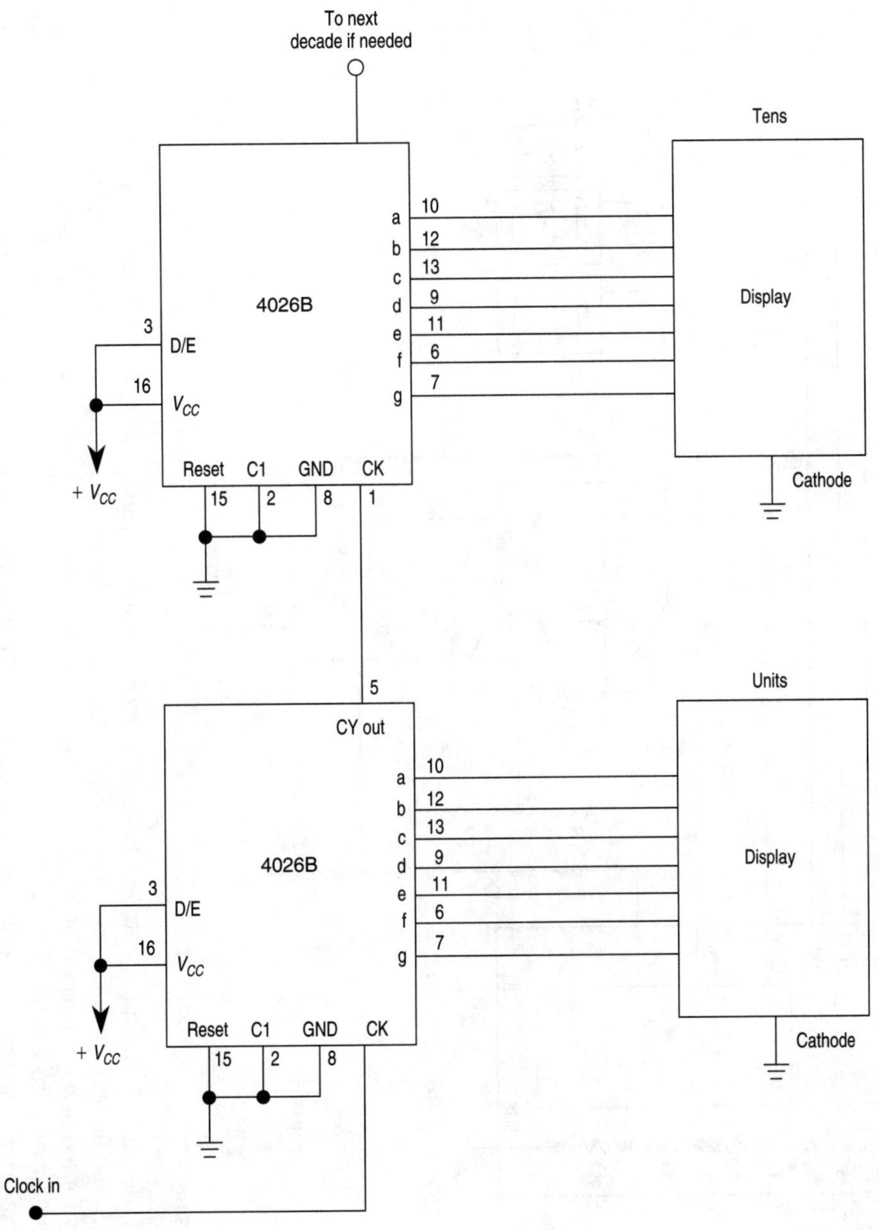

WILLIAM SHEETS ***FIG. 26-2***

Two or more 4026B counters can be cascaded as shown to give a multiple-digit display. Two, three or more displays can thus be connected.

LARGE LCD DISPLAY BUFFERING DRIVER

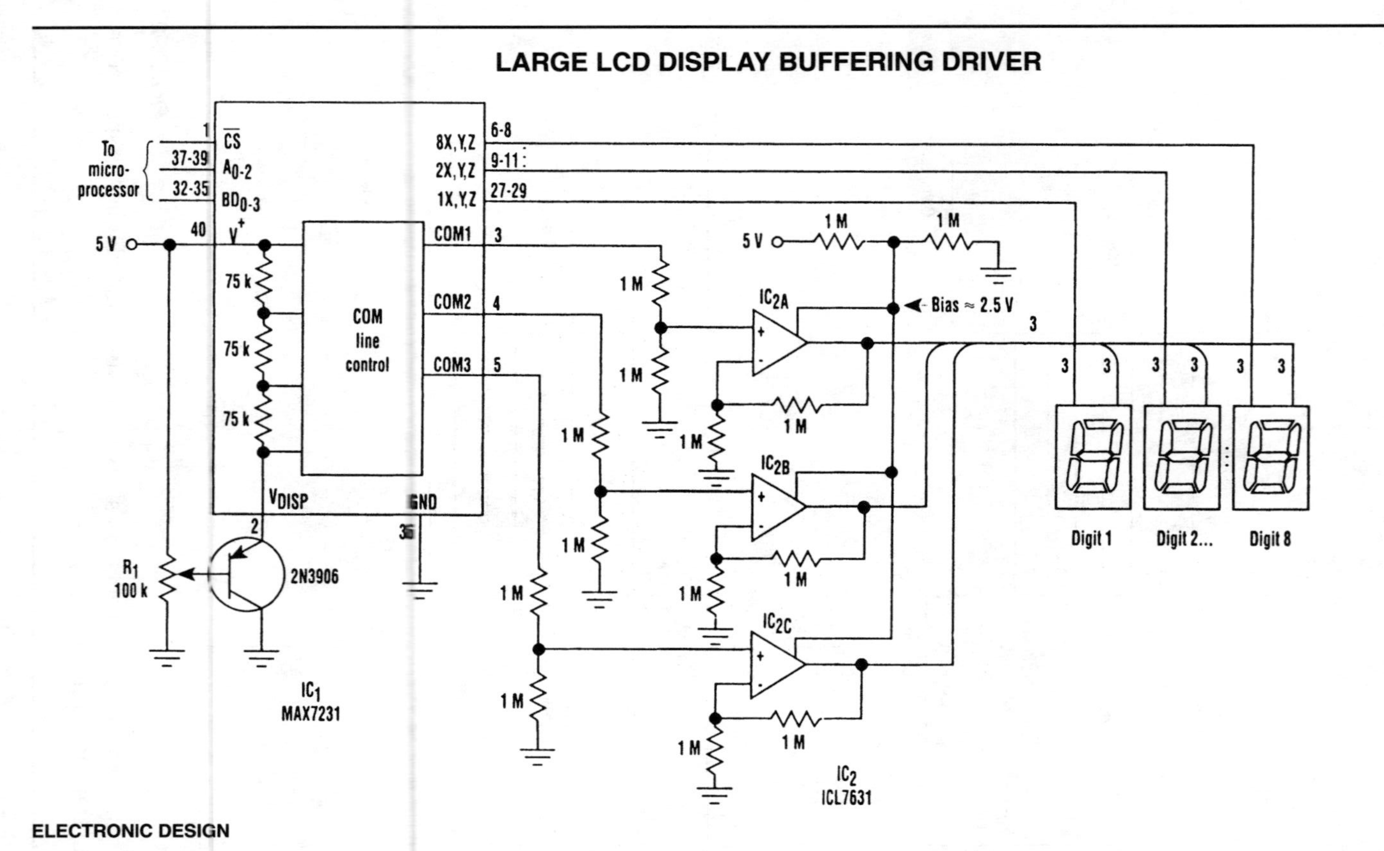

ELECTRONIC DESIGN

FIG. 26-3

Large LCD devices of 1" or more exhibit a large driving capacitance to the driver circuits. To solve this problem, the drive circuit shown (see the figure) introduces a buffer amplifier for each of the three common lines. Each amplifier can be programmed independently for a quiescent current of 10, 100, or 1000 μA. In this application, the bias network applies a voltage that sets the three quiescent currents to 100 μA.

The display driver and triple op amp operate between 5 V and ground, and the COM signals range from 5 V to ≈ 1 V. To ensure that these signals remain within the amplifiers' common-mode range, the signals are attenuated by one-half and the buffers operate at a gain of two. The circuit drives eight 1-inch displays, and is suitable for ambient temperature variations of 15°F or less. At the highest expected temperature, R1 should be adjusted so that no "off" segments are visible.

7-SEGMENT LCD DRIVER

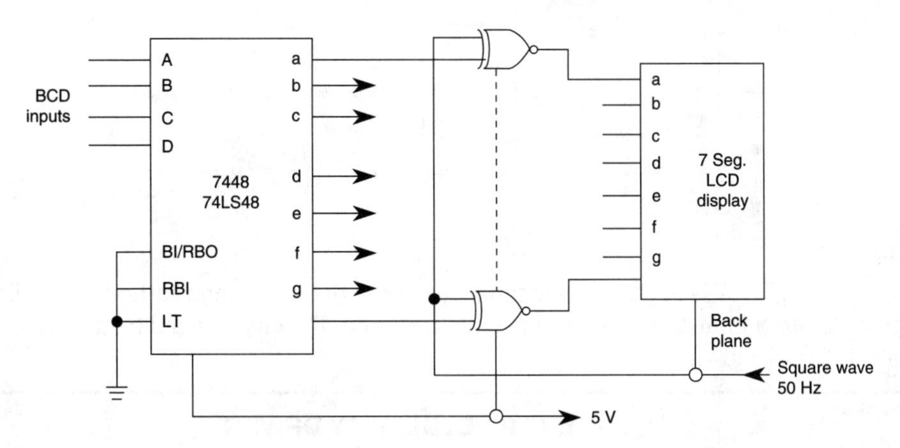

WILLIAM SHEETS **FIG. 26-4**

This circuit shows how a 7448 IC is used to drive a 7-segment LCD display. An external 50-Hz square wave supplies necessary phase signals to the back plane of the display.

LED DISPLAY LEADING-ZERO SUPPRESSOR

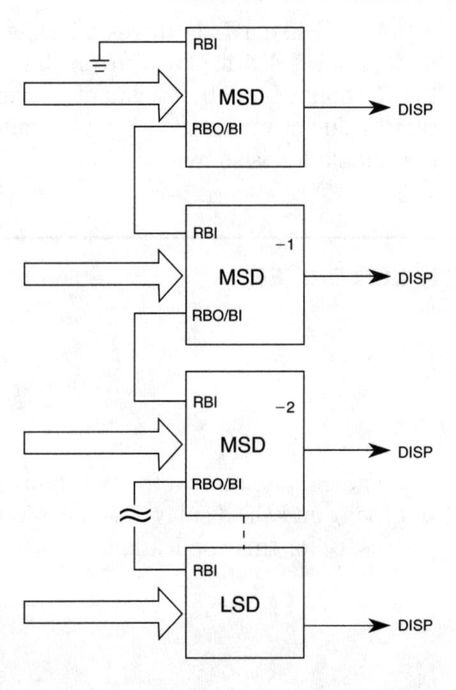

The diagram shows how to connect 7447-type IC devices for leading-zero suppression in an LED display.

WILLIAM SHEETS **FIG. 26-5**

7-SEGMENT COMMON-CATHODE LED DISPLAY DRIVER

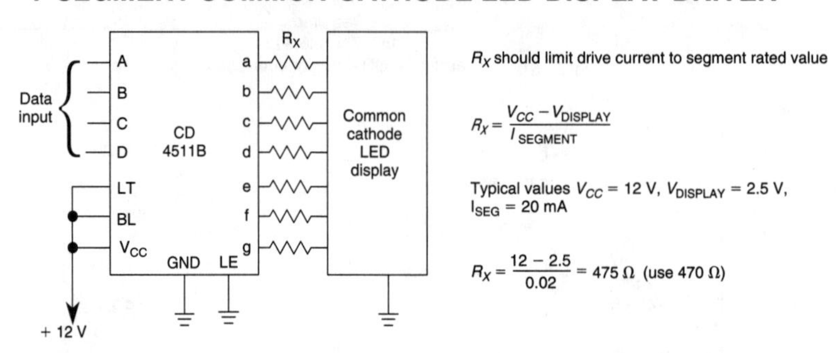

R_X should limit drive current to segment rated value

$$R_X = \frac{V_{CC} - V_{DISPLAY}}{I_{SEGMENT}}$$

Typical values V_{CC} = 12 V, $V_{DISPLAY}$ = 2.5 V, I_{SEG} = 20 mA

$$R_X = \frac{12 - 2.5}{0.02} = 475\ \Omega \text{ (use 470 } \Omega\text{)}$$

WILLIAM SHEETS ***FIG. 26-6***

A CD4511B CMOS LED display driver can be used to drive a common cathode LED display. Current limiting resistors limit the segment current to the rated value at maximum supply voltage. A sample calculation is shown.

7-SEGMENT (LED) DISPLAY DRIVER

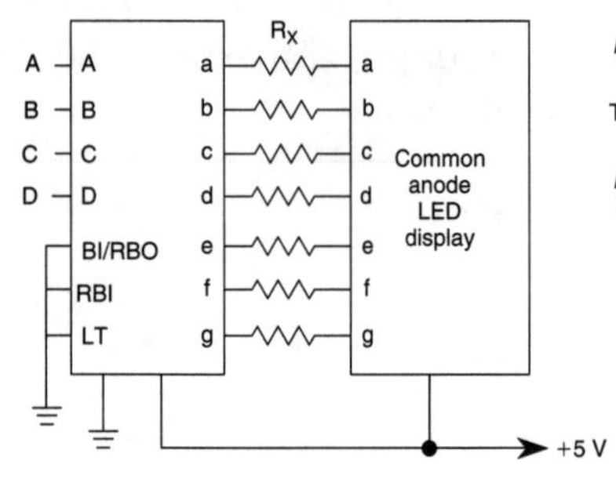

$$R_X = \frac{V_{OUT} - V_{SEGMENT}}{I_{SEGMENT}}$$

Typical : V_{OUT} = 4.3 V $V_{SEGMENT}$ = 2.5 V I_{SEG} = 20 mA

$$R_X = \frac{4.3 - 2.5}{0.02} = 180\ \Omega$$

An IC1 like a 7447 drives a 7-segment common anode LED display. Current limiting resistor R should limit the segment current to the rated value at maximum supply voltage. A sample calculation is shown.

WILLIAM SHEETS ***FIG. 26-7***

4543B 7-SEGMENT LCD DRIVER

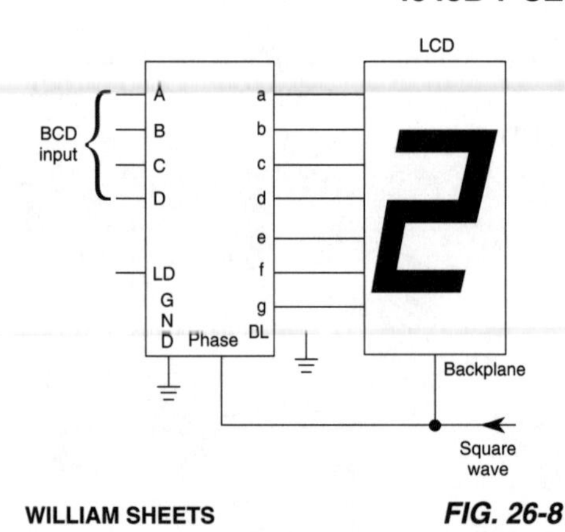

The circuit shows a frequently-used method of driving an LCD display. A square-wave drive is necessary for this application.

WILLIAM SHEETS ***FIG. 26-8***

GAS DISCHARGE TUBE OR DISPLAY DRIVER

WILLIAM SHEETS

FIG. 26-9

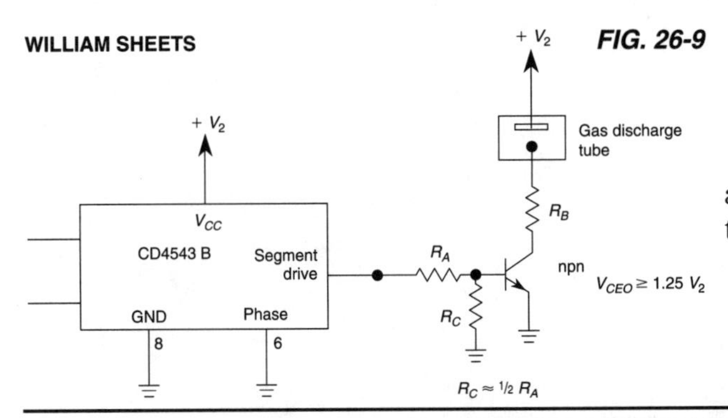

To drive the display, R_A should provide a drive of about 1 mA to the gas discharge tube. R_B is a current-limiting resistor.

4511B COMMON-ANODE DISPLAY DRIVER

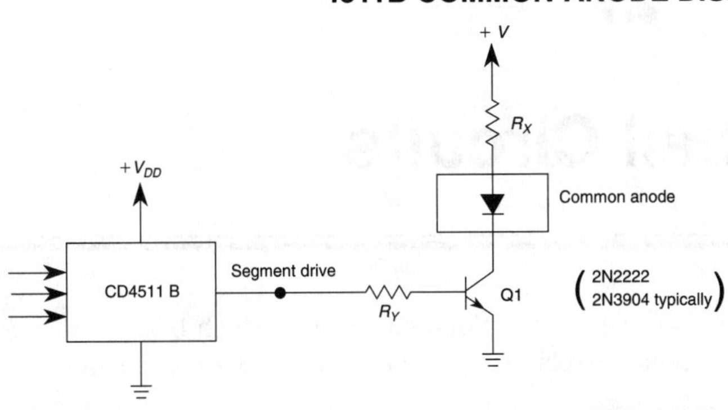

The use of a switching transistor (like a 2N2222 or 2N3904) allows use of the CD4511B with a common-anode display. R_y should be chosen to provide about 1 mA to drive Q1 and R_x should provide enough current to drive the display. For this circuit, the transistor gain (H_{FE}) should be at least the ratio of the segment drive current to the current through R_y.

WILLIAM SHEETS

FIG. 26-10

FLUORESCENT TUBE DISPLAY DRIVER

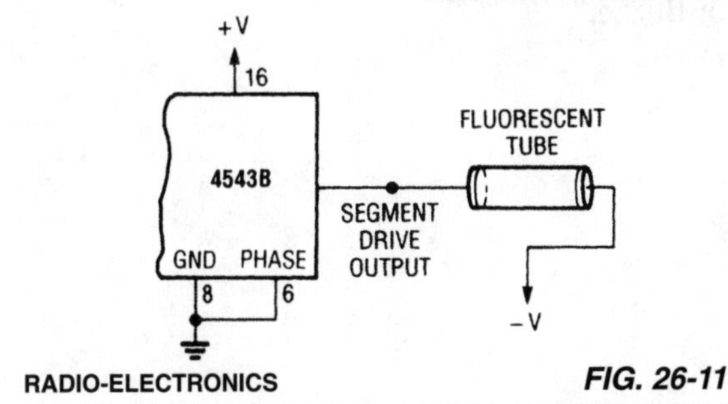

A fluorescent tube or display can be driven with a 4543B IC, as shown.

RADIO-ELECTRONICS

FIG. 26-11

4543B COMMON-CATHODE LED DRIVER

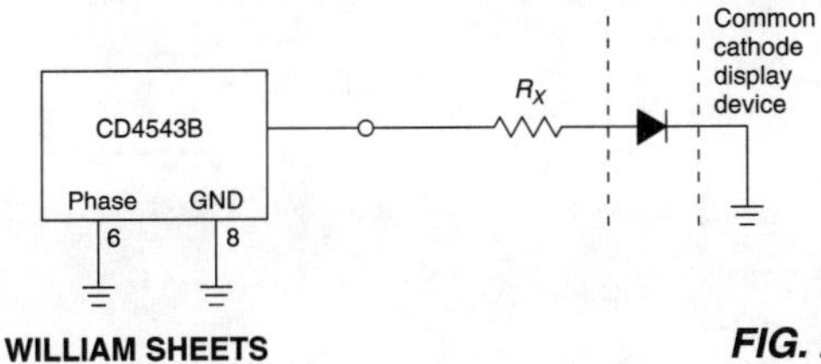

This circuit shows a way of driving a common-cathode display segment or an LED with a CD4543B.

WILLIAM SHEETS

FIG. 26-12

27

Doorbell Circuits

The sources of the following circuits are contained in the Sources section, which begins on page 675. The figure number in the box of each circuit correlates to the entry in the Sources section.

ELECTRONIC DOORBELL

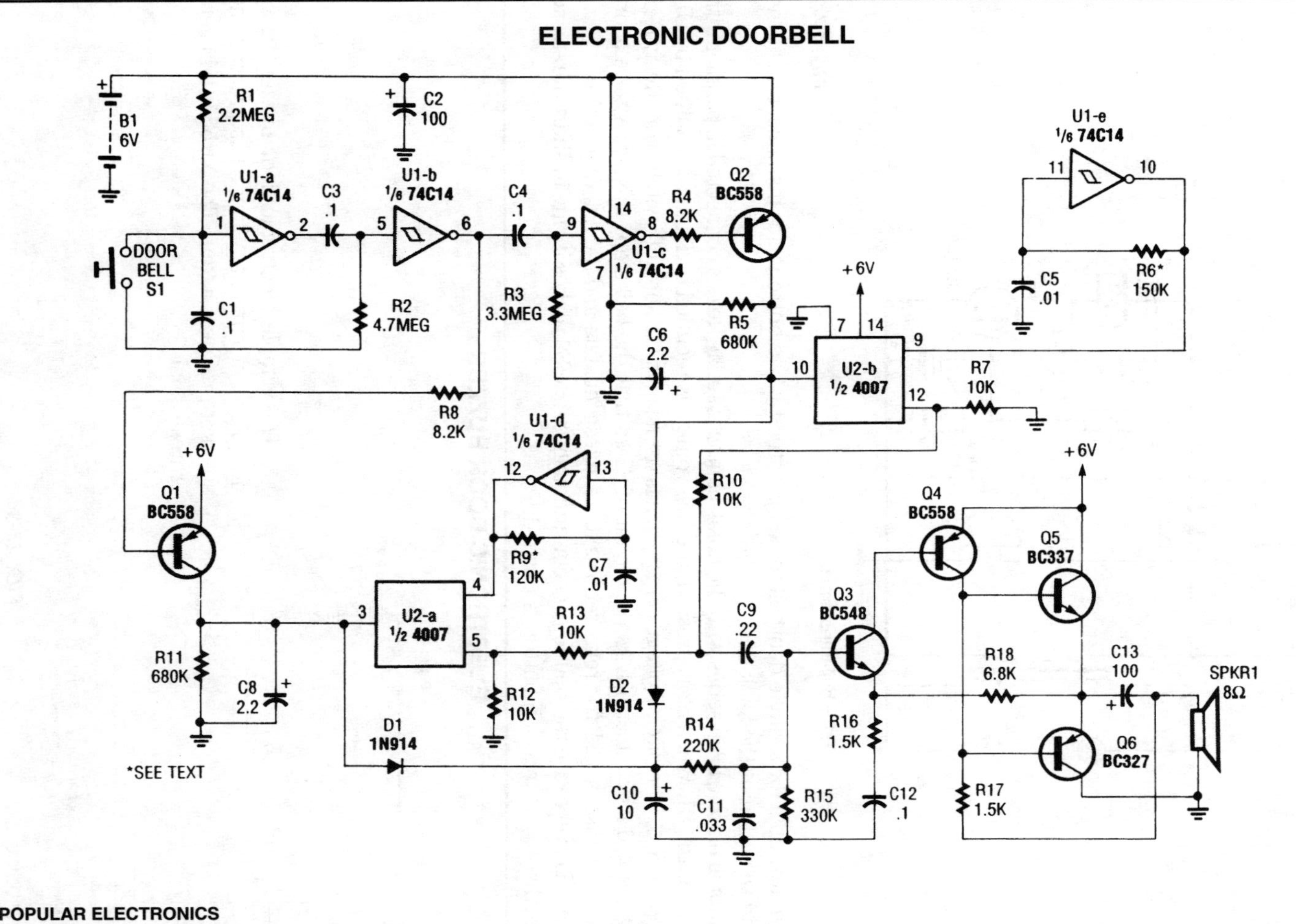

POPULAR ELECTRONICS

FIG. 27-1

When the doorbell switch is pressed, the two monostable stages are activated in sequence, applying bias to a pair of voltage-controlled resistor stages. These then modulate the outputs from a pair of tone generators. The resulting signals are fed to an audio amplifier, then to the speaker.

TWIN BELL CIRCUIT

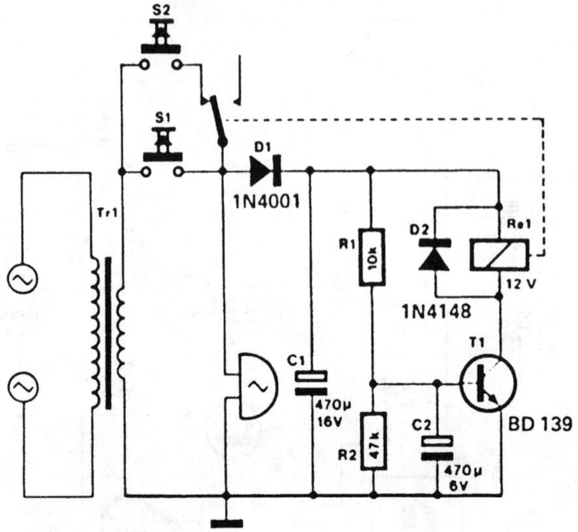

303 CIRCUITS

Tr1 = bell transformer

FIG. 27-2

It is often desirable for a single doorbell to be operated by two buttons, for instance, one at the front door and the other at the back door.

The additional button, S2 in series with the break contact of relay Re1, is connected in parallel with the original bell-push, S1. When S2 is pressed, the bell voltage is rectified by D1 and smoothed by C1. After a time, $t = R_1R_2C_2$, the direct voltage across C2 has risen to a level here T1 switches on. Relay Re1 is then energized and its contact breaks the circuit of S2 so that the bell stops ringing. After a short time, C1 and C2 are discharged, the relay returns to its quiescent state and the bell rings again.

In this way, S1 will cause the bell to ring continuously, while S2 makes it ring in short bursts, so that it is immediately clear which button is pushed.

ELECTRONIC DOOR BUZZER

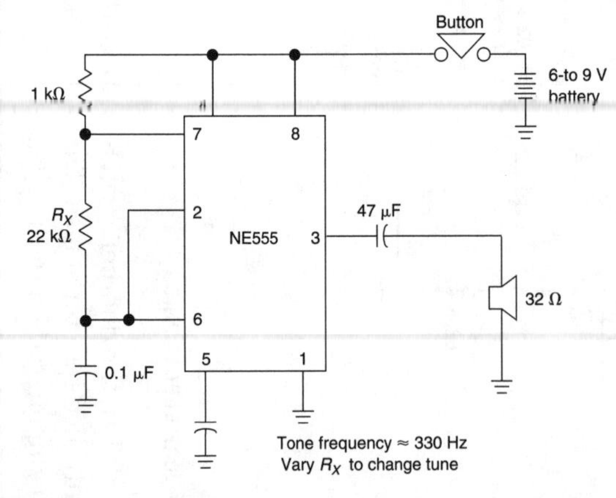

POPULAR ELECTRONICS

FIG. 27-3

This simple electronic door buzzer draws no quiescent current. When S1 is pressed the speaker produces a tone. The NE555 (U1) generates signal.

28

Fax Circuit

The source of the following circuit is contained in the Sources section, which begins on page 675. The figure number in the box of the circuit correlates to the entry in the Sources section.

Fax Mate

FAX MATE

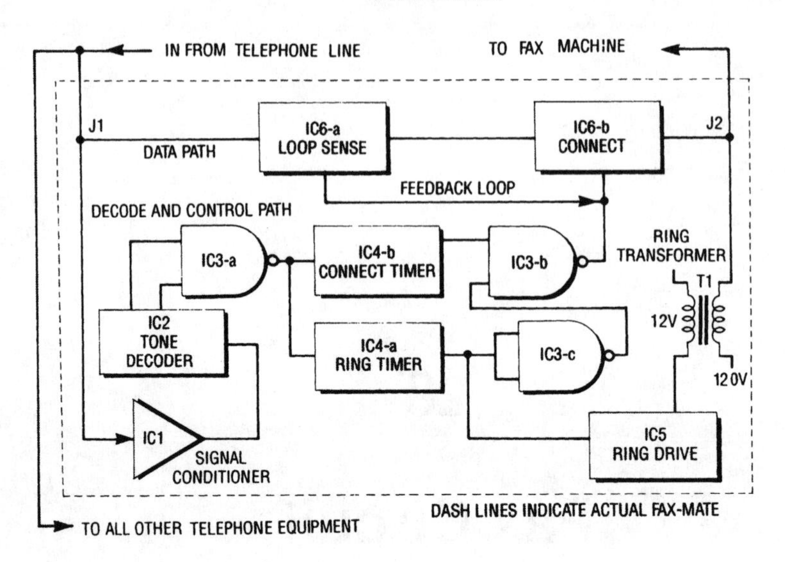

FIG. 1—BLOCK DIAGRAM for the Fax-Mate. The upper path is for data, and the lower one is the decode and control path.

1992 R-E EXPERIMENTERS HANDBOOK ***FIG. 28-1***

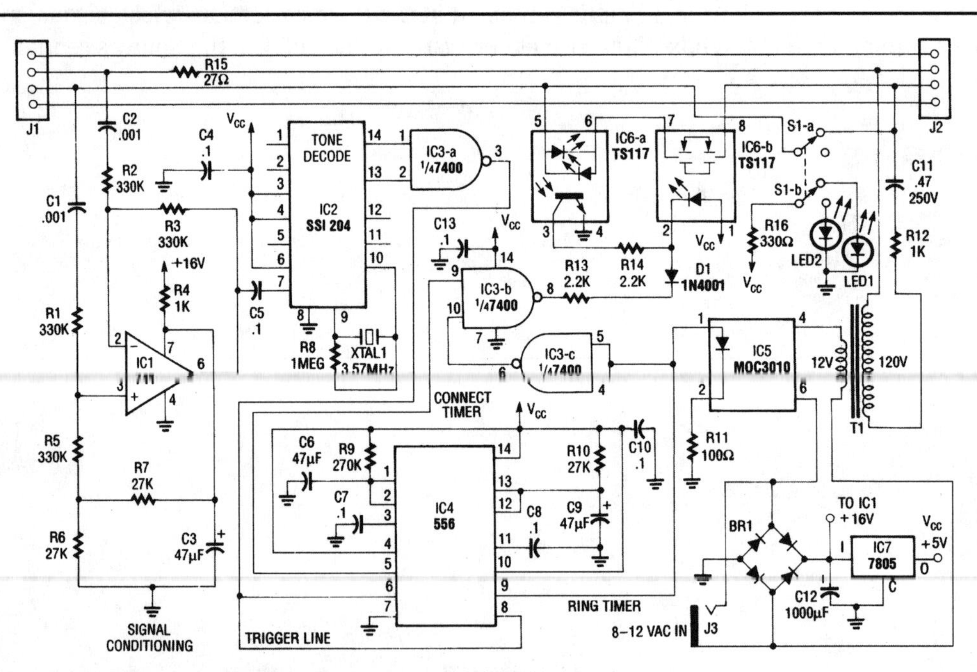

FIG. 2—SCHEMATIC for the Fax-Mate. Notice how it closely resembles the block diagram.

1992 R-E EXPERIMENTERS HANDBOOK ***FIG. 28-2***

FAX MATE(Cont.)

The fax mate separates the fax machine from the phone line, rings the fax machine on command, connects equipment to incoming lines, and senses the end of the message. When a touch tone pound signal (#) is detected, it actuates a ring greater and driver for the fax machine (the # signal is not used in ordinary dialing). The connect signal is inhibited for this time (ring cycle). 1C46 runs for 15 s and drives part of the connect IC. Then the fax or modem has fired up and is sending out a handshake tone. IC6 connects the equipment for initial hookup and keeps the connect section powered. When the fax machine hangs up, the loop current detector turns off, and resets the system.

29

Field-Strength Meter Circuits

The sources of the following circuits are contained in the Sources section, which begins on page 675. The figure number in the box of each circuit correlates to the entry in the Sources section.

REMOTE FIELD STRENGTH METER

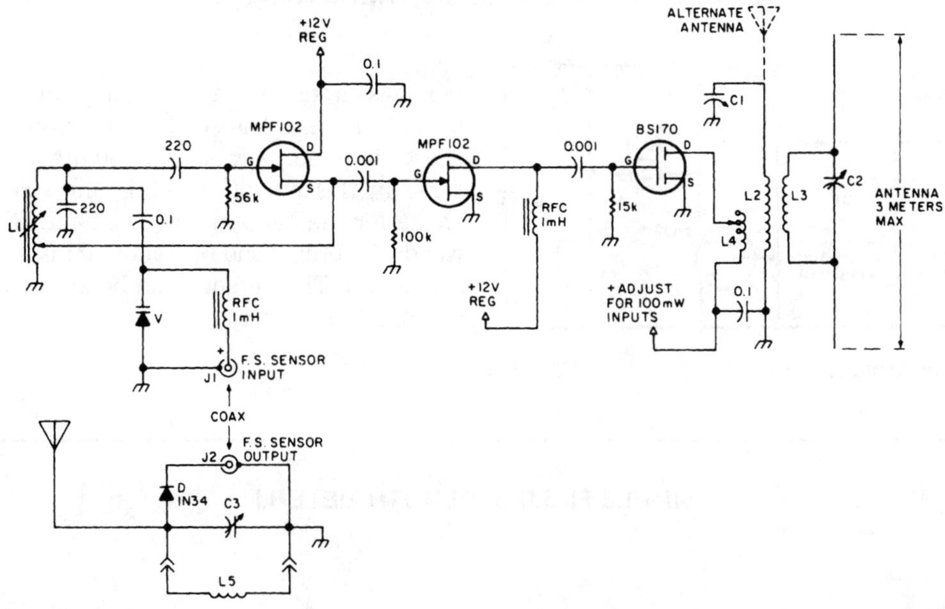

73 AMATEUR RADIO TODAY ***FIG. 29-1***

This field strength meter consists of a tuned crystal detector producing a dc output voltage from a transmitted signal. The dc voltage is used to shift the frequency of a transmitter of 100-mW power operating at 1650 kHz. The frequency shift is proportional to the received field strength. This unit has a range of several hundred feet and is operated under FCC part 15 rules (100-mW max power into a 2-m-long antenna between 510 and 1705 kHz).

AMPLIFIED FIELD STRENGTH METER

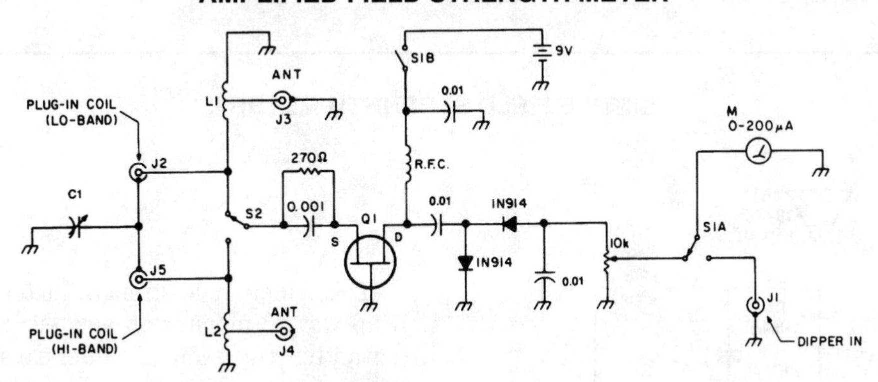

73 AMATEUR RADIO TODAY ***FIG. 29-2***

FET Q1 acts as an RF amplifier to boost sensitivity of the usual diode detector field strength meter.

SIMPLE AMPLIFIED FIELD STRENGTH METER

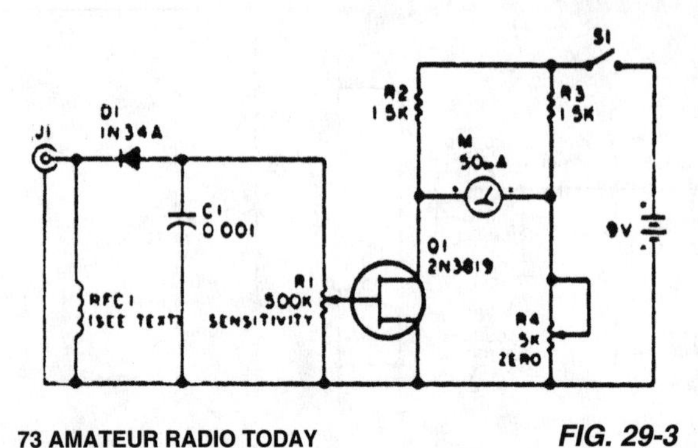

This circuit uses a FET as a dc amplifier in a bridge circuit. R4 is set for meter null with J1 short circuited. Any surplus 50-mA meter can serve in this circuit. RFC1 is any suitable RF choke for the band in use. A 2.5-mH RF choke will do for broadband operation. R1 is a sensitivity control. The antenna can be any small whip antenna (2 ft or less).

73 AMATEUR RADIO TODAY

FIG. 29-3

SIMPLE FIELD STRENGTH METER I

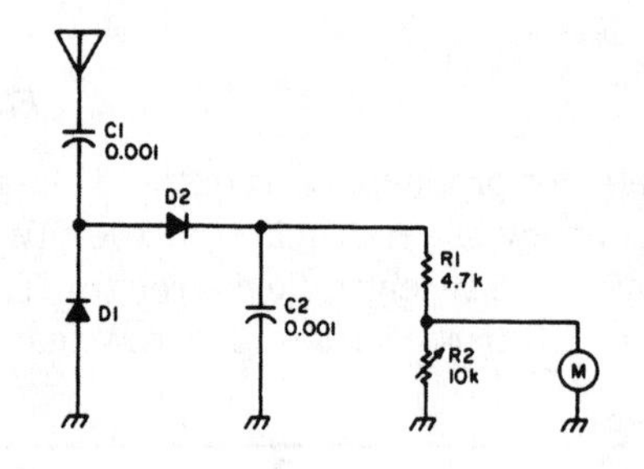

Useful for checking transmitters and antennas, this circuit uses a voltage-doubling detector D1 and D2 (HP 5082-2800 hot carrier types). D1 and D2 can also be type IN34 or IN82. M is a 100-mA meter movement.

73 AMATEUR RADIO TODAY

FIG. 29-4

SIMPLE FIELD STRENGTH METER II

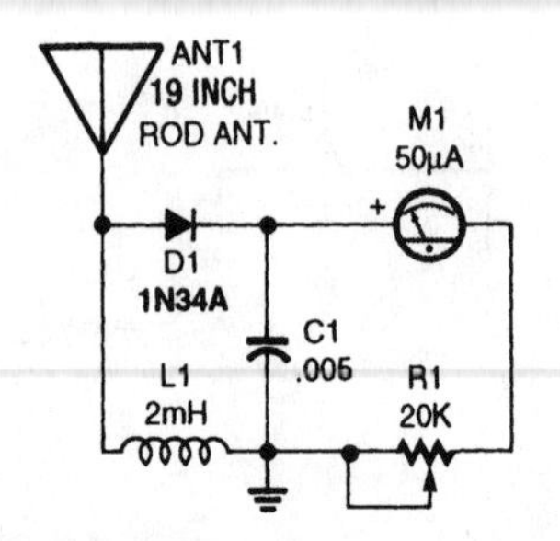

This simple field-strength meter provides a cheap way to monitor an amateur radio or CB transmitter (or even an antenna system) for maximum output.

POPULAR ELECTRONICS

FIG. 29-5

30

Filter Circuits

The sources of the following circuits are contained in the Sources section, which begins on page 675. The figure number in the box of each circuit correlates to the entry in the Sources section.

ACTIVE LOW-PASS FILTER

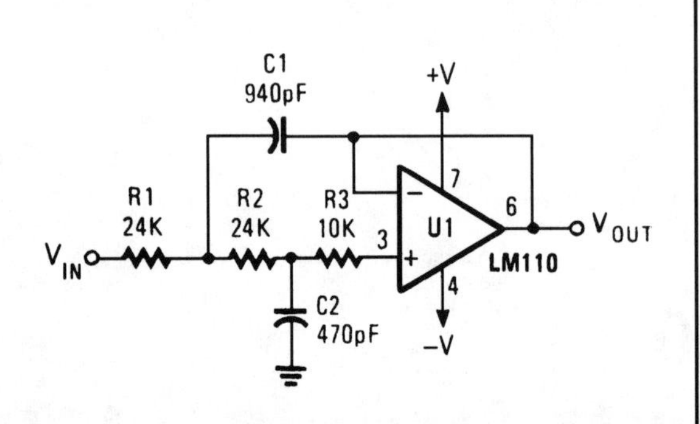

POPULAR ELECTRONICS

FIG. 30-1

HIGH Q NOTCH FILTER

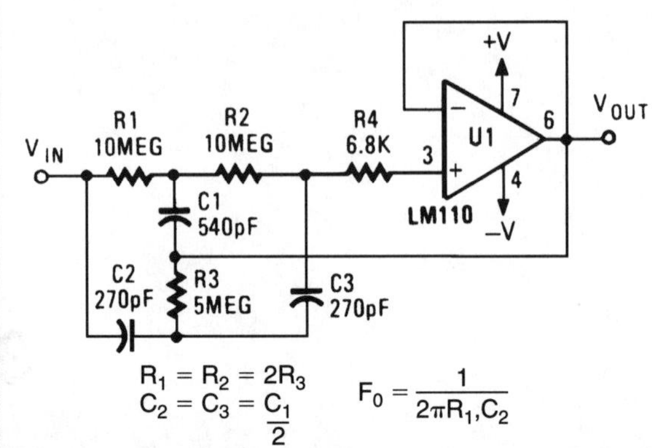

$$R_1 = R_2 = 2R_3$$
$$C_2 = C_3 = \frac{C_1}{2}$$
$$F_0 = \frac{1}{2\pi R_1, C_2}$$

POPULAR ELECTRONICS

FIG. 30-2

UNIVERSAL STALE VARIABLE FILTER

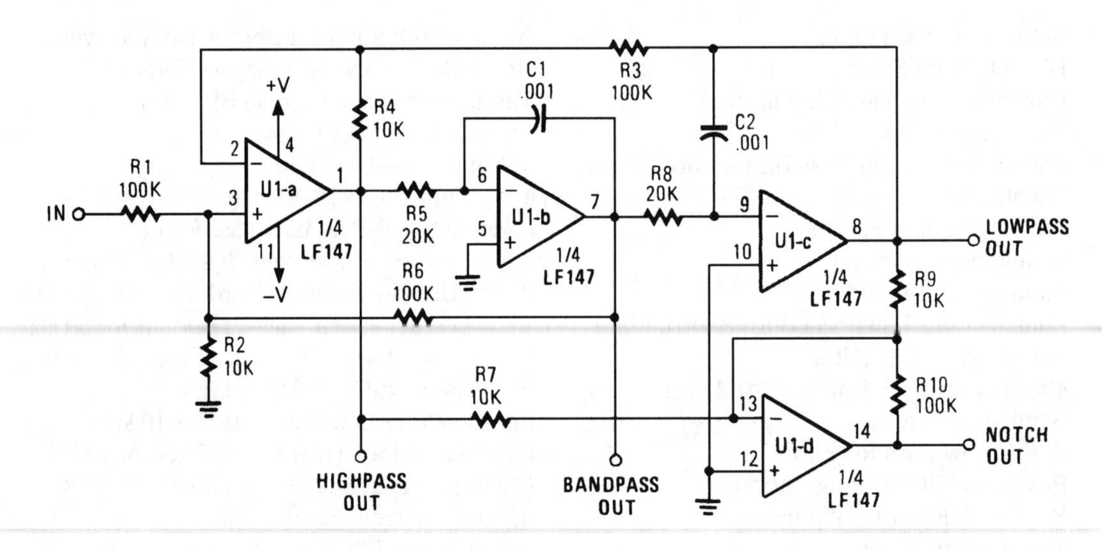

POPULAR ELECTRONICS

FIG. 30-3

ADJUSTABLE Q NOTCH FILTER

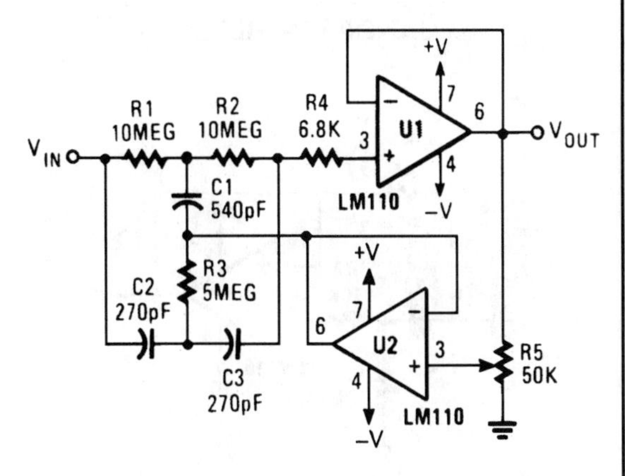

POPULAR ELECTRONICS

FIG. 30-4

FOURTH ORDER HIGH-PASS BUTTERWORTH FILTER

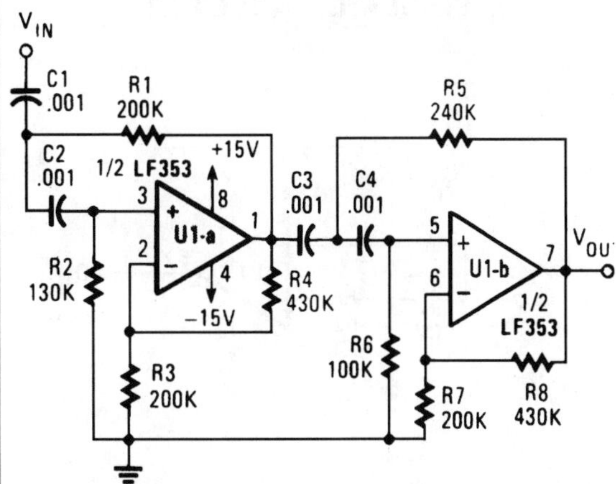

POPULAR ELECTRONICS

FIG. 30-5

TUNABLE NOTCH FILTER

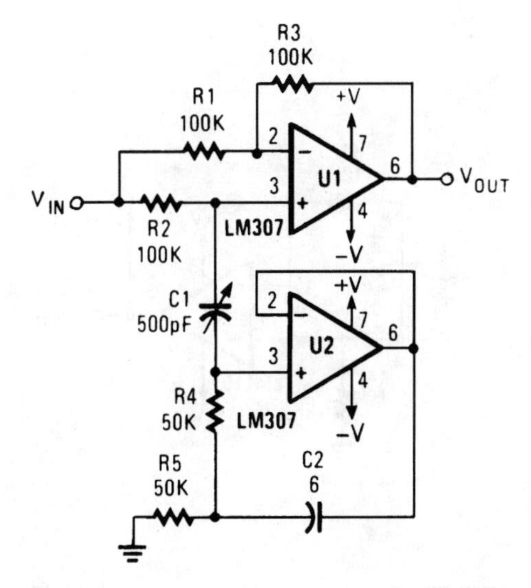

POPULAR ELECTRONICS

FIG. 30-6

HIGH Q BANDPASS FILTER

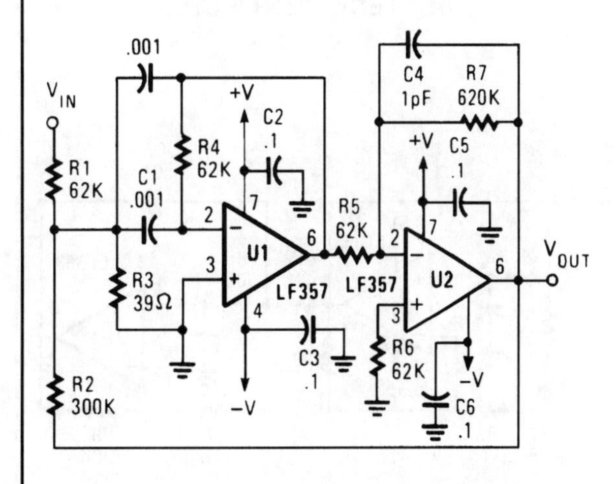

POPULAR ELECTRONICS

FIG. 30-7

SIMULATED INDUCTOR

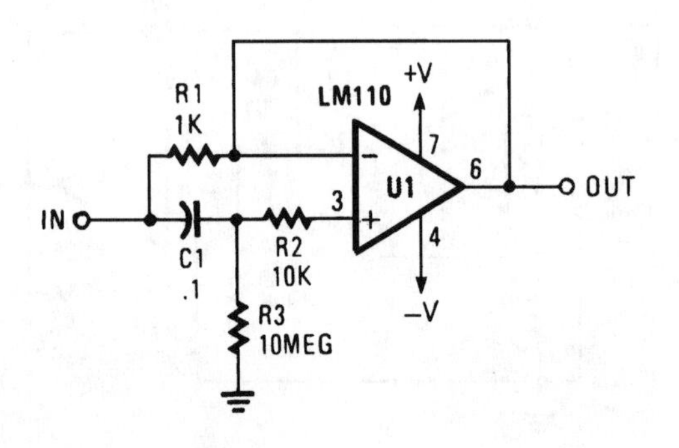

POPULAR ELECTRONICS *FIG. 30-8*

BANDPASS FILTER

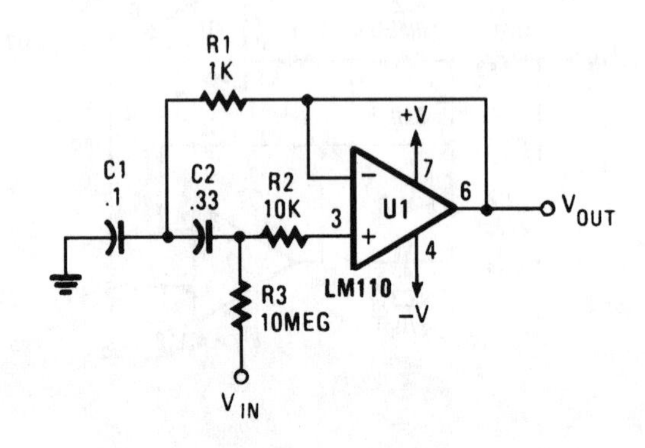

POPULAR ELECTRONICS *FIG. 30-9*

FOURTH ORDER LOW-PASS BUTTERWORTH FILTER

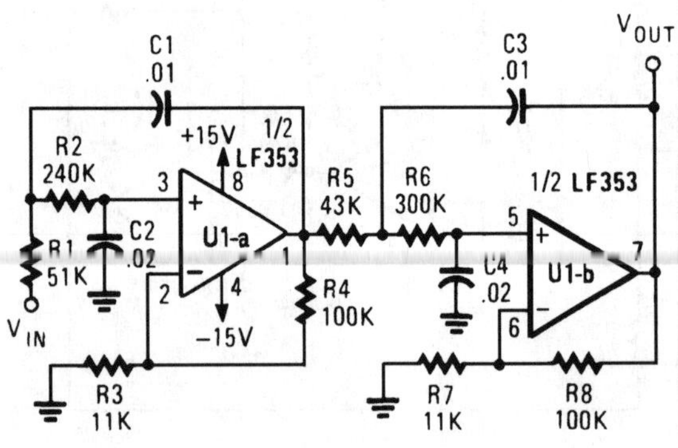

POPULAR ELECTRONICS *FIG. 30-10*

ACTIVE HIGH-PASS FILTER

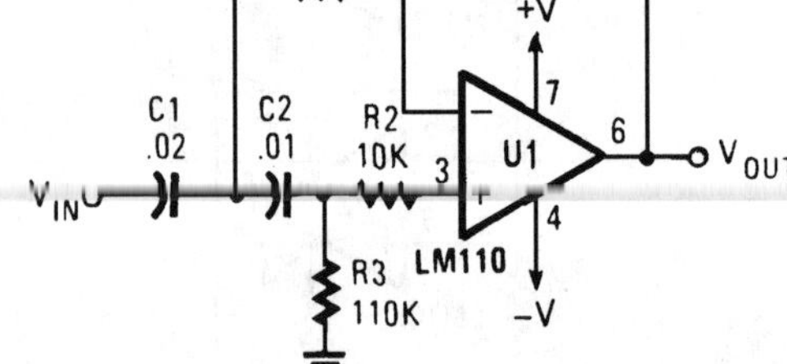

POPULAR ELECTRONICS *FIG. 30-11*

400-Hz LOW-PASS BUTTERWORTH FILTER

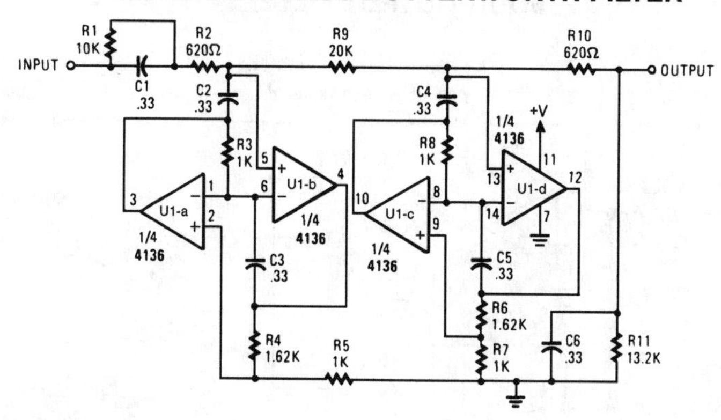

POPULAR ELECTRONICS

FIG. 30-12

Designed for a 400-Hz cutoff frequency, the cutoff can be scaled by varying the element values proportionally to frequency

BANDPASS FILTER

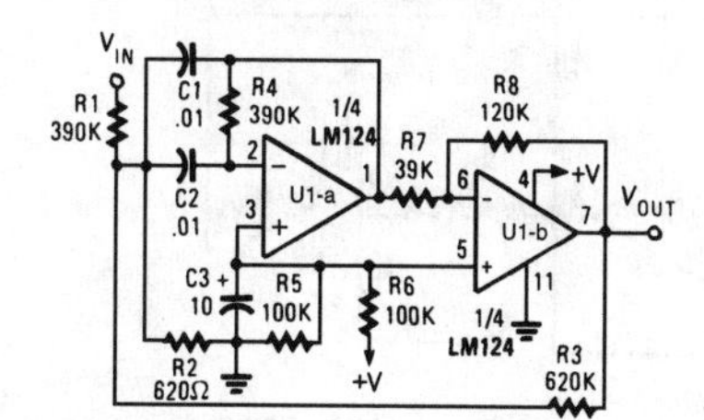

POPULAR ELECTRONICS

FIG. 30-13

Appropriate center frequency of this circuit is:

$$\frac{1}{R_4C_2}$$

$$C_1 = C2, R_1 = R_4$$

ACTIVE LOW-PASS RC FILTER

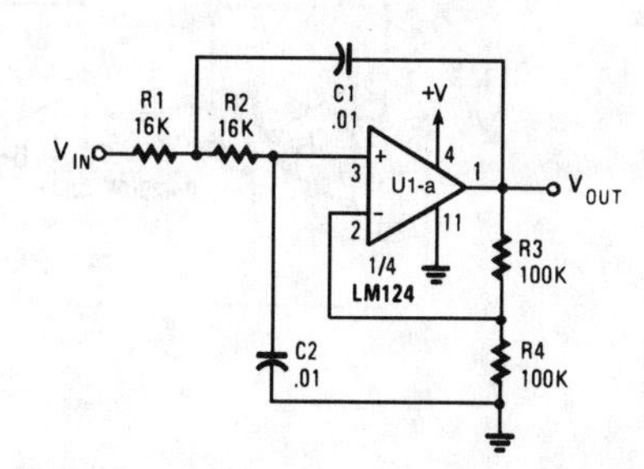

POPULAR ELECTRONICS

FIG. 30-14

The circuit shown has a cutoff frequency at about 1 kHz. R1, R2, C1, and C2 can be scaled to change this to any other desired frequency.

PASSIVE L FILTER CONFIGURATIONS

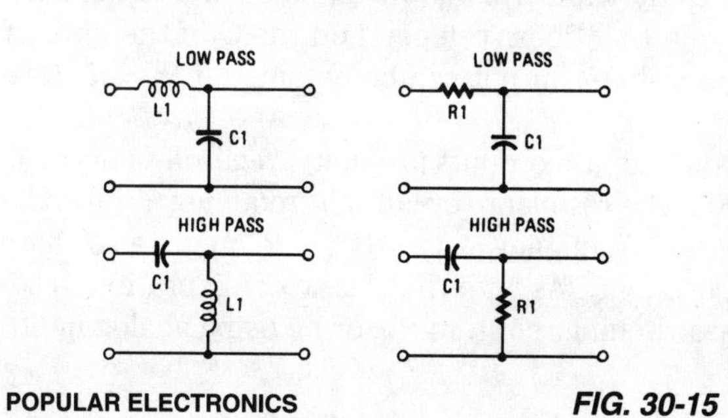

POPULAR ELECTRONICS

FIG. 30-15

PASSIVE PI FILTER CONFIGURATIONS

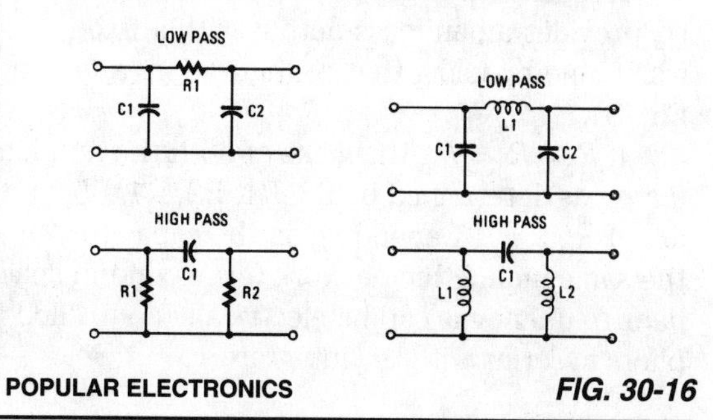

POPULAR ELECTRONICS

FIG. 30-16

FOUR-OUTPUT FILTER

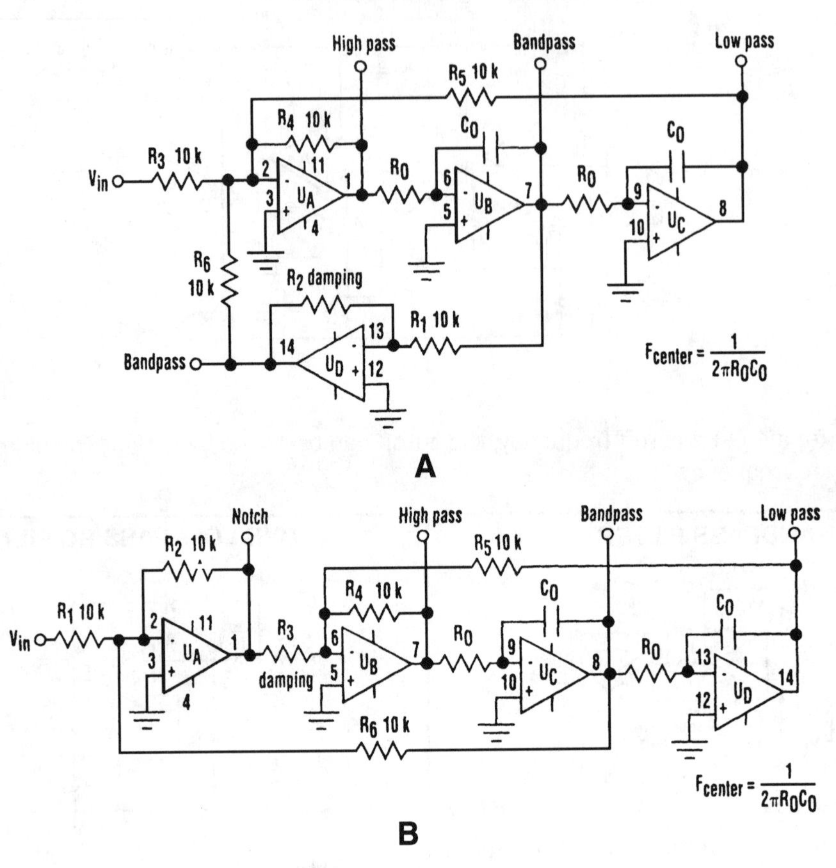

ELECTRONIC DESIGN

FIG. 30-17

The classic "state-variable" (two-integrator) filter (see Fig. A) is famous for its insensitivity to device parameter tolerances, as well as its ability to provide three simultaneous separate outputs: high pass, bandpass, and low pass. These advantages often offset the fact that a quad operational amplifier is needed to implement the circuit.

A modification of the classic scheme that applies the input voltage via amplifier U_D, rather than U_A provides a bandpass output with a fixed peak gain that doesn't depend on the *Q* of the filter. It was found by using that configuration, a fourth notch-filter output can be obtained if $R_1 = R_6$ (see Fig. B).

If $R_1 = R_6 = R_2$, the gains of both the notch and bandpass outputs are unity, regardless of the *Q* factor, as determined by R3, R1, R2, R4, R5, and R6. The resonant (or cutoff) frequency is given by ω, $- 1/R_O \times C_O$. Depending on the capacitor values and frequency ω, resistance R_O might also share the same monolithic network for maximum space economy. As with the classic configuration, resonant frequency ω can be electrically controlled by switching resistors R_O, or by using analog multipliers in series with the integrators.

VARIABLE Q FILTER FOR 400 Hz

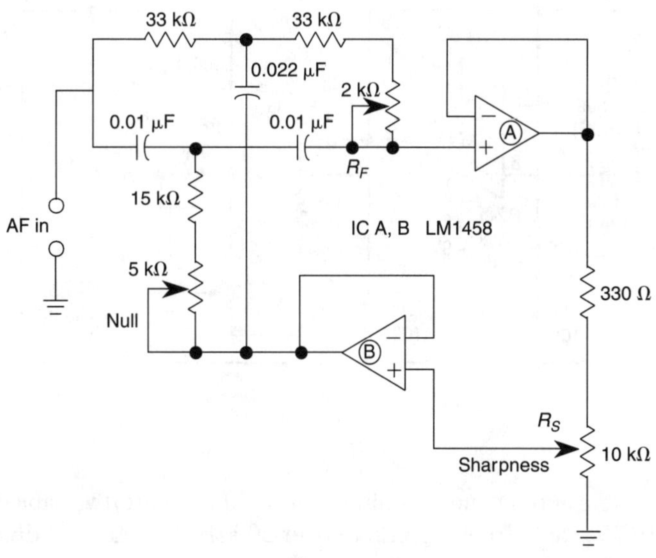

WILLIAM SHEETS **FIG. 30-18**

A bootstrapped twin T notch filter in this circuit can yield an effective Q of up to 10. R_S adjusts the feedback, hence the Q. Values of C_1 and C_2 can be changed to alter the frequency. R_F is a fine-tune null control.

TWIN T NOTCH FILTER FOR 1 kHz

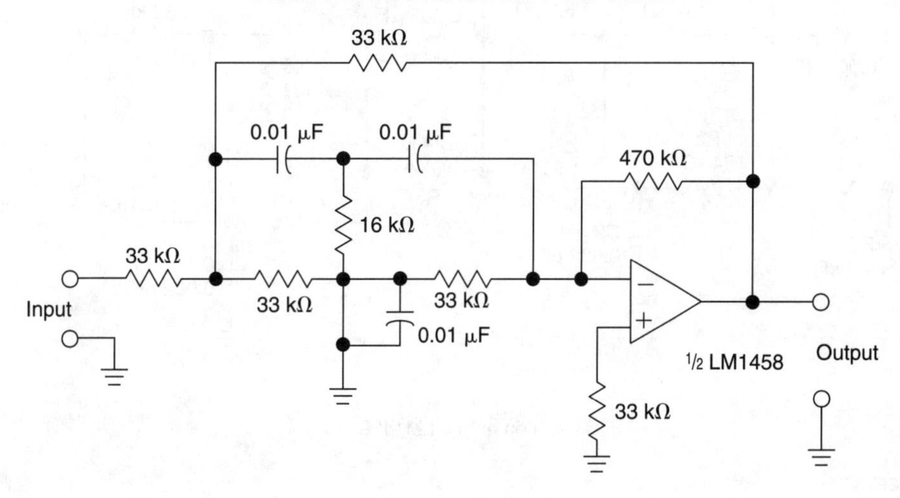

WILLIAM SHEETS **FIG. 30-19**

The circuit shown uses a twin T notch filter and an amplifier. Used to remove unwanted frequency.

VARIABLE BANDPASS AUDIO FILTER

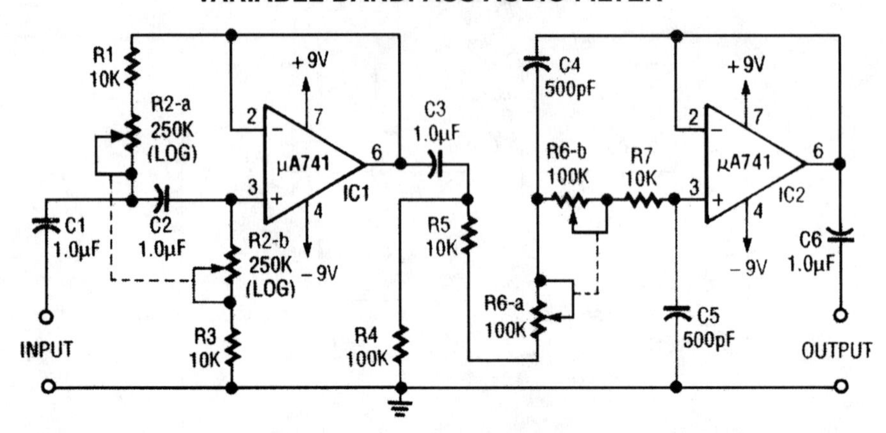

ELECTRONICS NOW *FIG. 30-20*

This circuit is a variable audio bandpass filter that has a low cutoff variable from about 25 Hz to 700 Hz and a high cutoff variable from 2.5 kHz to over 20 kHz. Rolloff is 12 dB/octave on both high and low ends. R2-a-b and R6-a-b are ganged potentiometers for setting lower and upper cutoff frequencies, respectively.

ACTIVE FOURTH-ORDER LOW-PASS FILTER

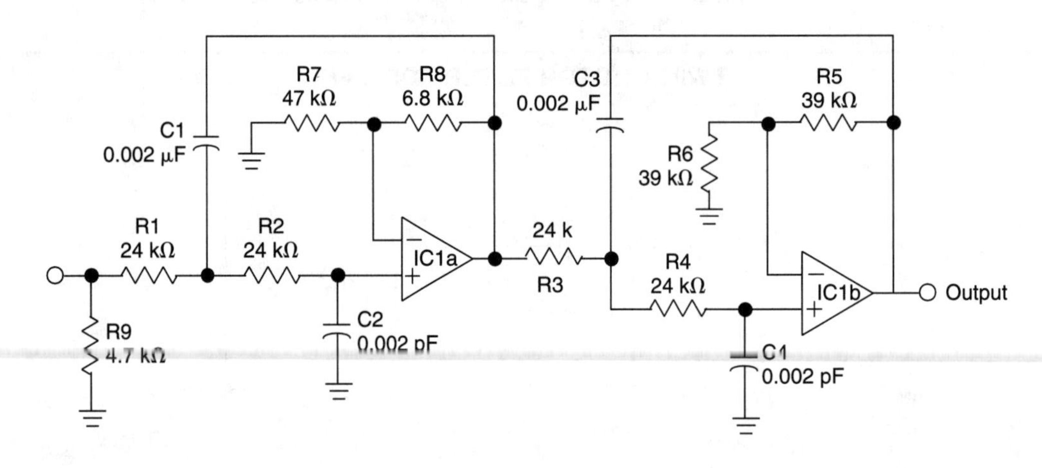

WILLIAM SHEETS *FIG. 30-21*

This circuit is a fourth-order low-pass filter with values for kHz. The values of R_1, R_2, C_1 and C_2, and R_3, R_4, C_3 and C_4 can be scaled for operation at other frequencies. Roll-off is 24 dB/octave.

AUDIO NOTCH FILTER FOR SHORTWAVE RECEIVERS

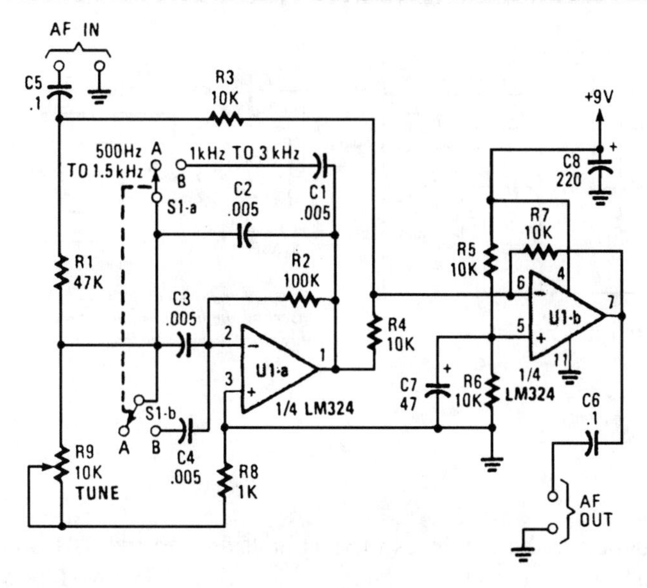

POPULAR ELECTRONICS *FIG. 30-22*

The notch filter can be added to just about any receiver to attenuate a single frequency by more than 30 dB. This filter should be handy for reducing heterodynes and whistles.

ACTIVE SECOND-ORDER BANDPASS FILTER FOR SPEECH RANGE

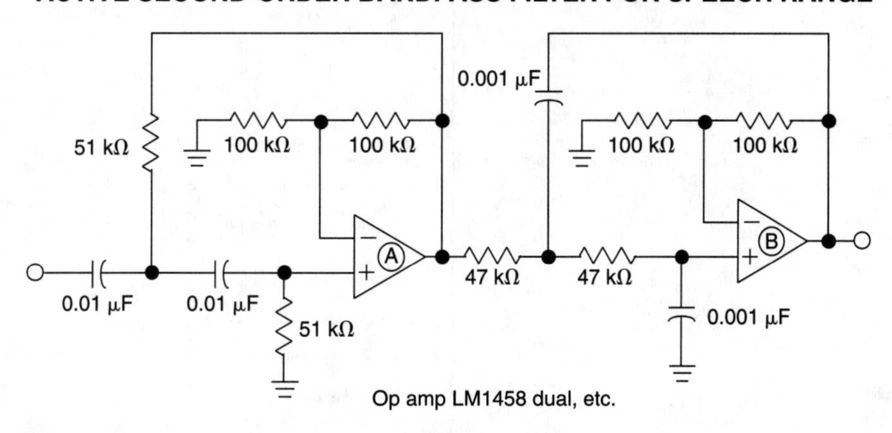

WILLIAM SHEETS *FIG. 30-23*

This filter circuit which uses LM1458 or similar op amp has a response of 300 Hz to 3.4 kHz with 12 dB/octave roll-off outside the pass band. Section A is the high-pass one, followed by low-pass section B. Values of either section can be scaled to alter the pass band.

VARIABLE-FREQUENCY AUDIO BP FILTER

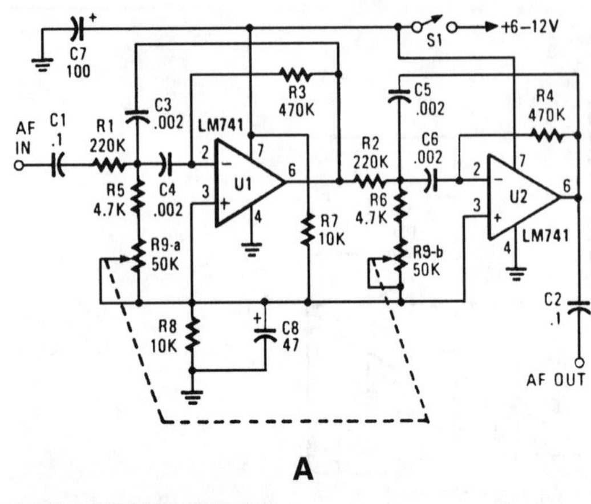

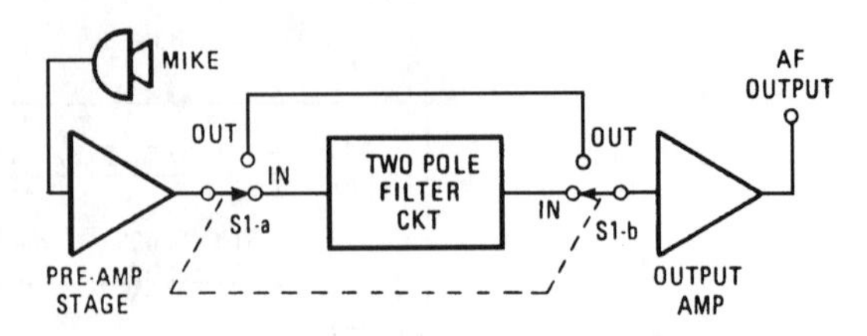

The filter can be wired into an existing amplifier by inserting the filter circuit between the amp's preamp and output stages as shown here.

POPULAR ELECTRONICS

FIG. 30-24

This variable-frequency, audio bandpass filter is built around two 741 op amps that are connected in cascade. Two 741 op amps are configured as identical RC active filters and are connected in cascade for better selectivity. The filter's tuning range is from 500 Hz to 1500 Hz. The overall voltage gain is slightly greater than 1 and the filter's is about 5. The circuit can handle input signals of 4 V peak-to-peak without being overdriven. The circuit's input impedance is over 200 kΩ and its output impedance is less than 1 kΩ .

VARIABLE LOW-PASS FILTER

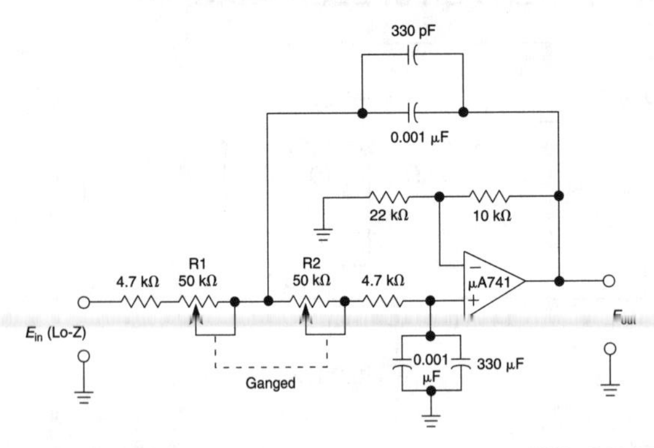

WILLIAM SHEETS

FIG. 30-25

This second-order low-pass filter uses a 741 op amp and is tuneable from 2.5 kHz to 25 kHz. This circuit is useful in audio and tone control applications. R1 and 2 are ganged potentiometers.

VARIABLE HIGH-PASS FILTER

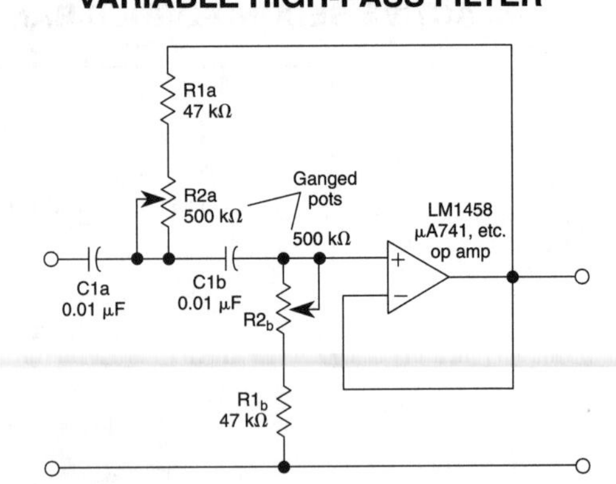

WILLIAM SHEETS

FIG. 30-26

This second order filter which should prove useful in audio applications uses an LM1458 or other similar of op amp. It is tuneable from 30 to 300 Hz cutoff. R2a, b are ganged log-taper potentiometers.

1-mV OFFSET, CLOCK-TUNABLE, MONOLITHIC 5-POLE LOW-PASS FILTER

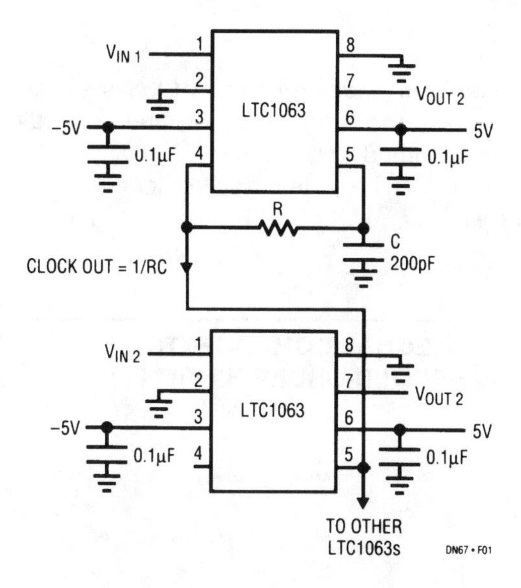

The LTC1063 is the first monolithic low-pass filter that simultaneously offers outstanding dc and ac performance. It features internal or external clock tunability, cutoff frequencies up to 50 kHz, 1-mV typical output dc offset, and a dynamic range in excess of 12 bits for over a decade of input voltage.

The LTC1063 approximates a 5-pole Butterworth low-pass filter. The unique internal architecture of the filter allows outstanding amplitude matching from device to device. Typical matching ranges from 0.01 dB at 25% of the filter passband to 0.05 dB at 50% of the filter passband.

An internal or external clock programs the filter's cutoff frequency. The clock-to-cutoff frequency ratio is 100:1. In the absence of an external clock, the LTC1063's internal precision oscillator can be used. An external resistor and capacitor set the device's internal clock frequency.

LINEAR TECHNOLOGY CORP. ***FIG. 30-27***

UNITY-GAIN SECOND-ORDER HIGH-PASS FILTER

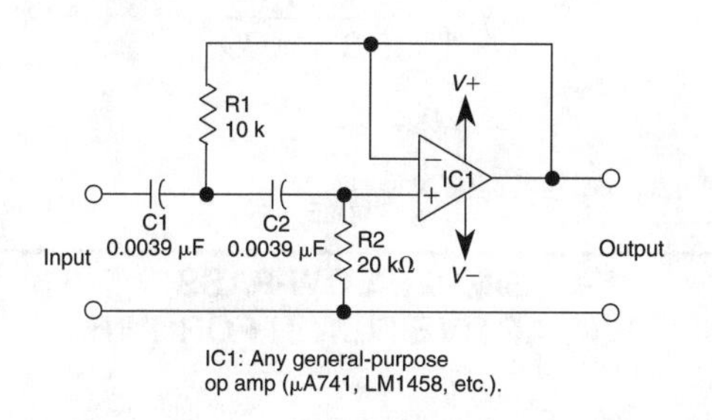

WILLIAM SHEETS ***FIG. 30-28***

This filter circuit has a cutoff frequency of 2900 Hz with the values shown.

$$f_{\text{cutoff}} = \frac{1}{2.83\pi RC}$$

$$R = R_1$$
$$R_2 = 2R_1$$
$$C = C_1 = C_2$$

ACTIVE UNITY-GAIN SECOND-ORDER LOW-PASS FILTER

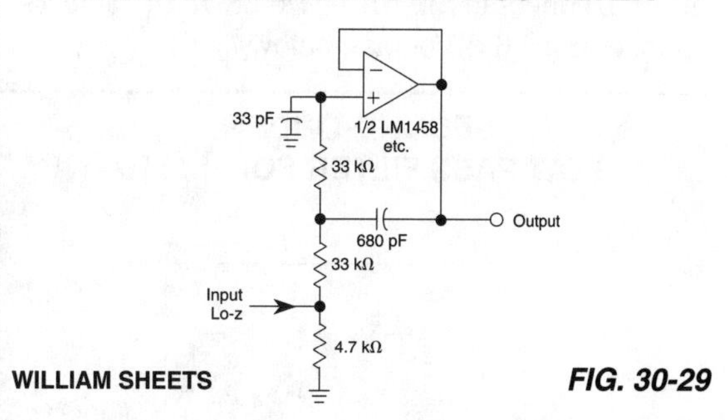

WILLIAM SHEETS ***FIG. 30-29***

This second-order Butterworth filter cuts off near 10 kHz. The values of C_1 and C_2 can be changed to alter the frequency, or else calculated from the formula.

$$f_{\text{cutoff}} = \frac{1}{2.83\pi RC}$$

$$C_1 = 2C_2$$
$$R_2 = R_3 = R$$

ACTIVE FOURTH-ORDER HIGH-PASS FILTER FOR 50 Hz

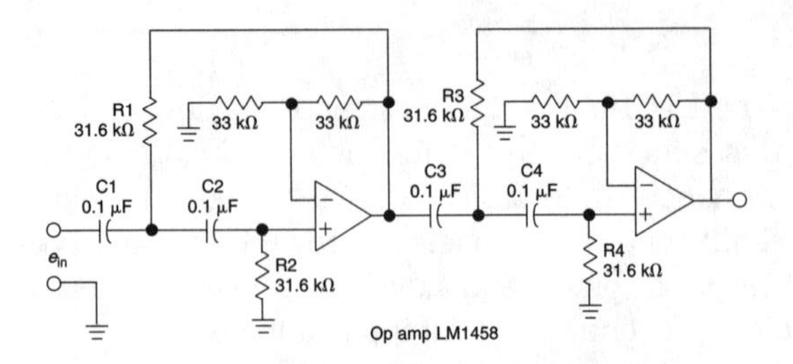

WILLIAM SHEETS *FIG. 30-30*

This circuit which uses an LM1458 or similar op amp is a fourth-order high-pass filter with a 24 dB/octave roll-off. The values of R_1/R_2, R_3/R_4, C_1/C_2, C_3/C_4 can be scaled to suit other cutoff frequencies.

SIMPLE HIGH-PASS (HP) ACTIVE FILTER FOR 1 kHz

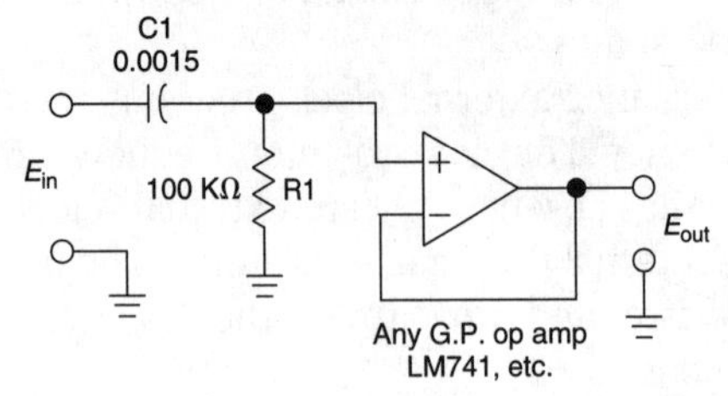

WILLIAM SHEETS *FIG. 30-31*

This simple 1 kHz filter uses a voltage follower and an RC section for a filter element. For other frequencies f_3 dB – 1/6.28 R_1C_1. The response drops 6 dB/octave below f_3 dB.

SECOND-ORDER LOW-PASS FILTER FOR 10 kHz

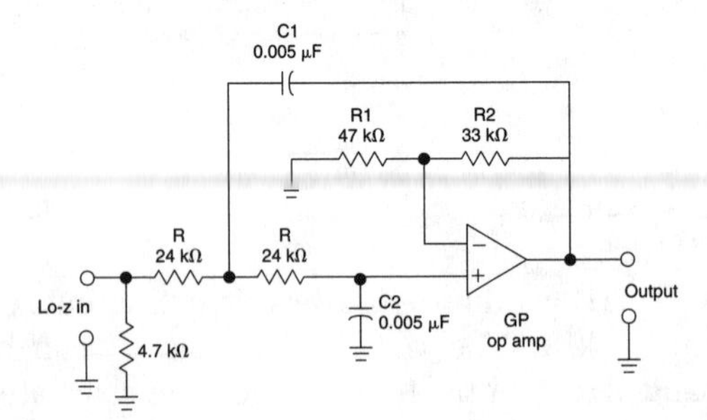

WILLIAM SHEETS *FIG. 30-33*

This circuit uses equal value capacitors. The cutoff frequency (f_c) is

$$f_c = \frac{1}{2.83\pi RC}$$

EQUAL COMPONENTS SECOND-ORDER HP FILTER

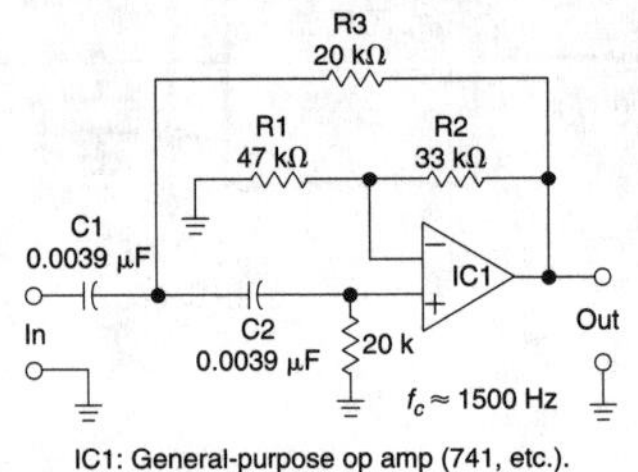

WILLIAM SHEETS *FIG. 30-32*

This filter circuit uses equal value components and is shown for 1500 Hz. The values can be scaled for other frequencies.

$$f_{\text{cutoff}} = \frac{1}{2.83\pi RC}$$

$$R = R_1$$
$$R_2 = 2R_1$$
$$C = C_1 = C_2$$

SIMPLE LOW-PASS (LP) ACTIVE FILTER FOR 1 kHz

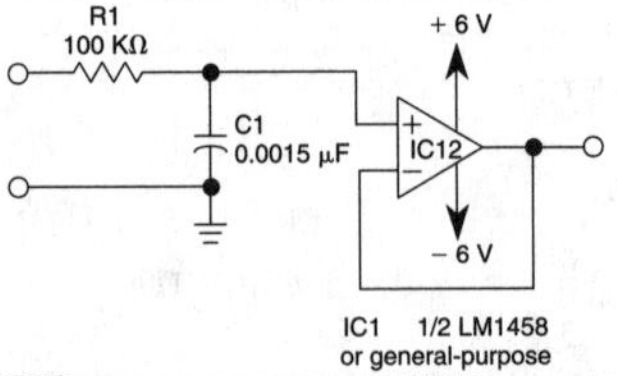

WILLIAM SHEETS *FIG. 30-34*

This simple filter uses an RC section for a filter element, with a voltage follower for other frequencies f_3 dB = 1/6.28 R_1C_1. Response drops 6 dB/octave above f_3 dB.

CURRENT-DRIVEN SALLEN KEY FILTER

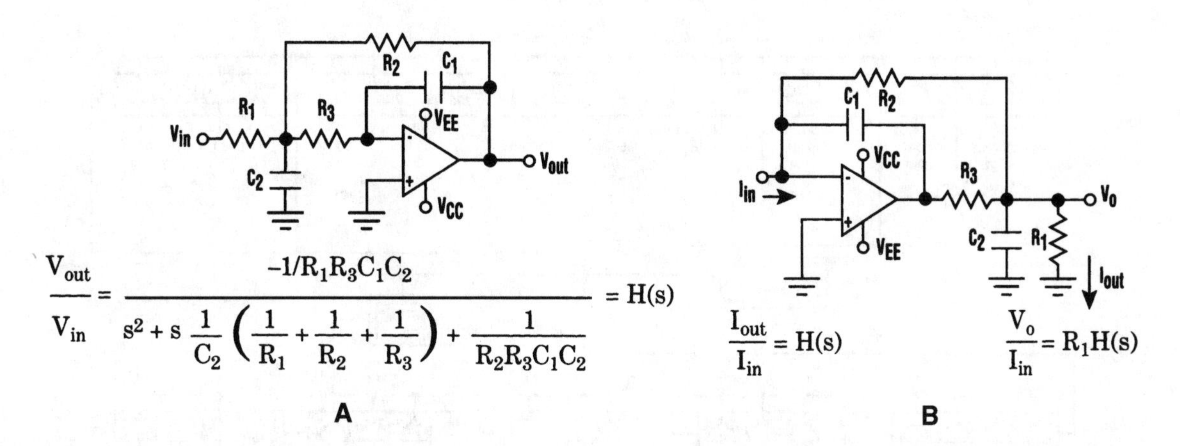

ELECTRONIC DESIGN *FIG. 30-35*

The low-pass Sallen-Key filter is staple for designers because it contains few components (A). By redesigning the filter, a current to voltage conversion can be avoided when the input signal to be filtered is in current form (B).

455-kHz NARROW-BAND IF FILTER

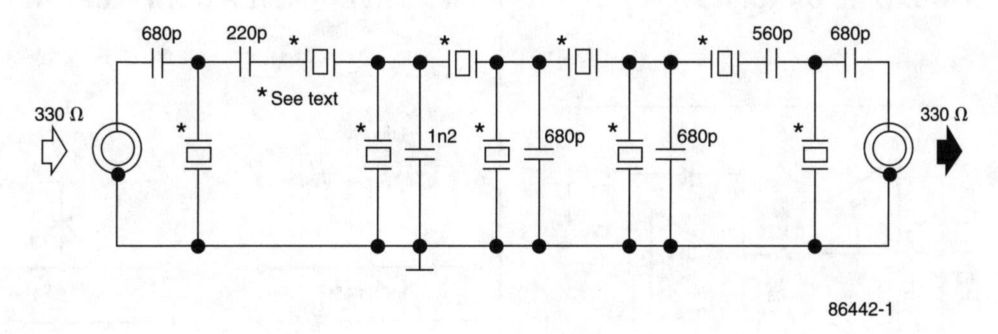

303 CIRCUITS *FIG. 30-36*

This filter uses five 455-kHz ceramic resonators. The impedance is 330 Ω, the bandwidth is 800 Hz, and the ultimate rejection ≥60 dB. The ceramic resonators could be replaced by crystals.

AUDIO-RANGE FILTER

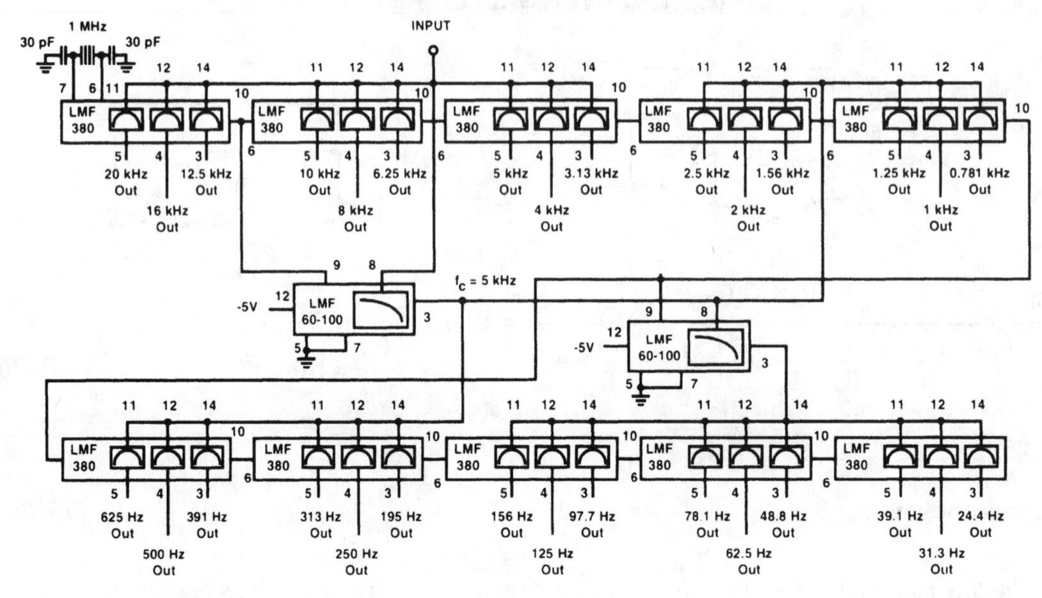

NATIONAL SEMICONDUCTOR

FIG. 30-37

The LMF380 switched audio filter by National Semiconductor is used here to obtain a third-octave filter set that covers the entire audio range.

BI-QUAD RC BANDPASS FILTER

PASSIVE T FILTER CONFIGURATIONS

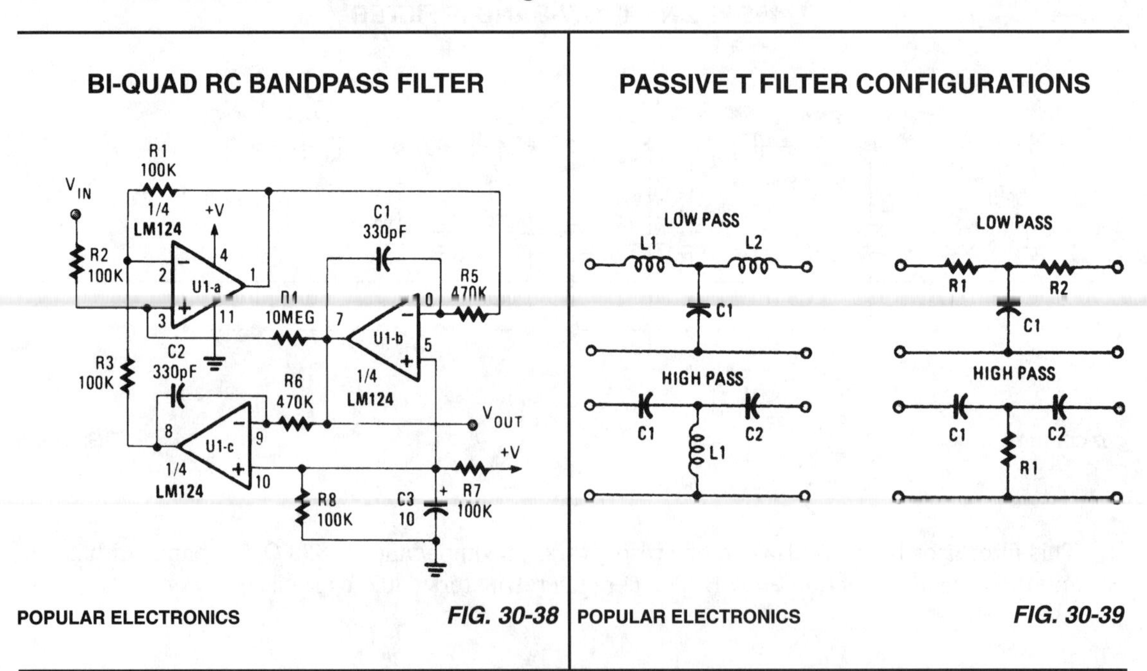

POPULAR ELECTRONICS

FIG. 30-38

POPULAR ELECTRONICS

FIG. 30-39

FULL-WAVE RECTIFIER/AVERAGING FILTER

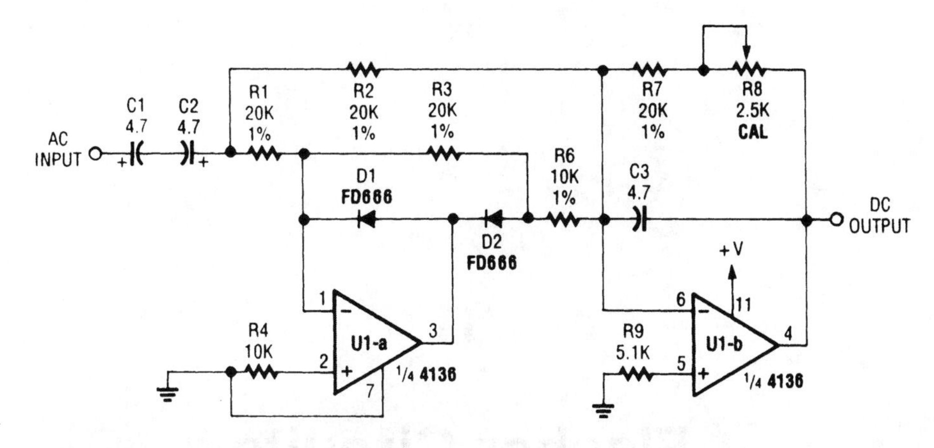

POPULAR ELECTRONICS **FIG. 30-40**

The input signal is rectified by D1 and D2 op amp U1-a, and fed to output amp U2. R8 is set for correct circuit calibration.

1-kHz TONE FILTER

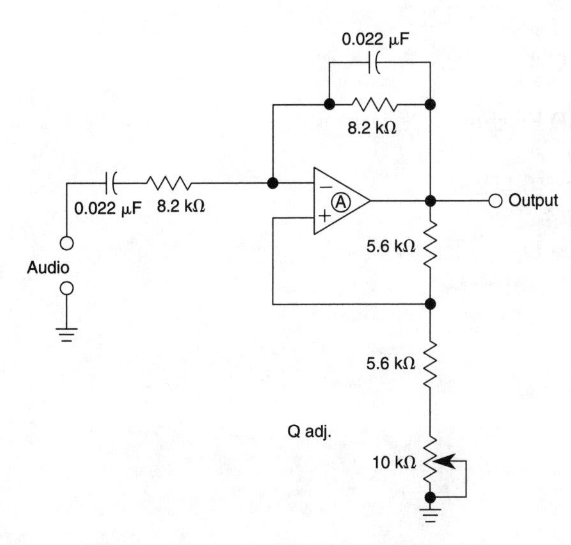

The Wien-bridge based filter has a variable bandwidth and a center frequency of 900 Hz. The circuit will oscillate if the 10-kΩ pot is set too low.

WILLIAM SHEETS **FIG. 30-41**

31

Flasher Circuits

The sources of the following circuits are contained in the Sources section, which begins on page 675. The figure number in the box of each circuit correlates to the entry in the Sources section.

SEQUENTIAL FLASHER

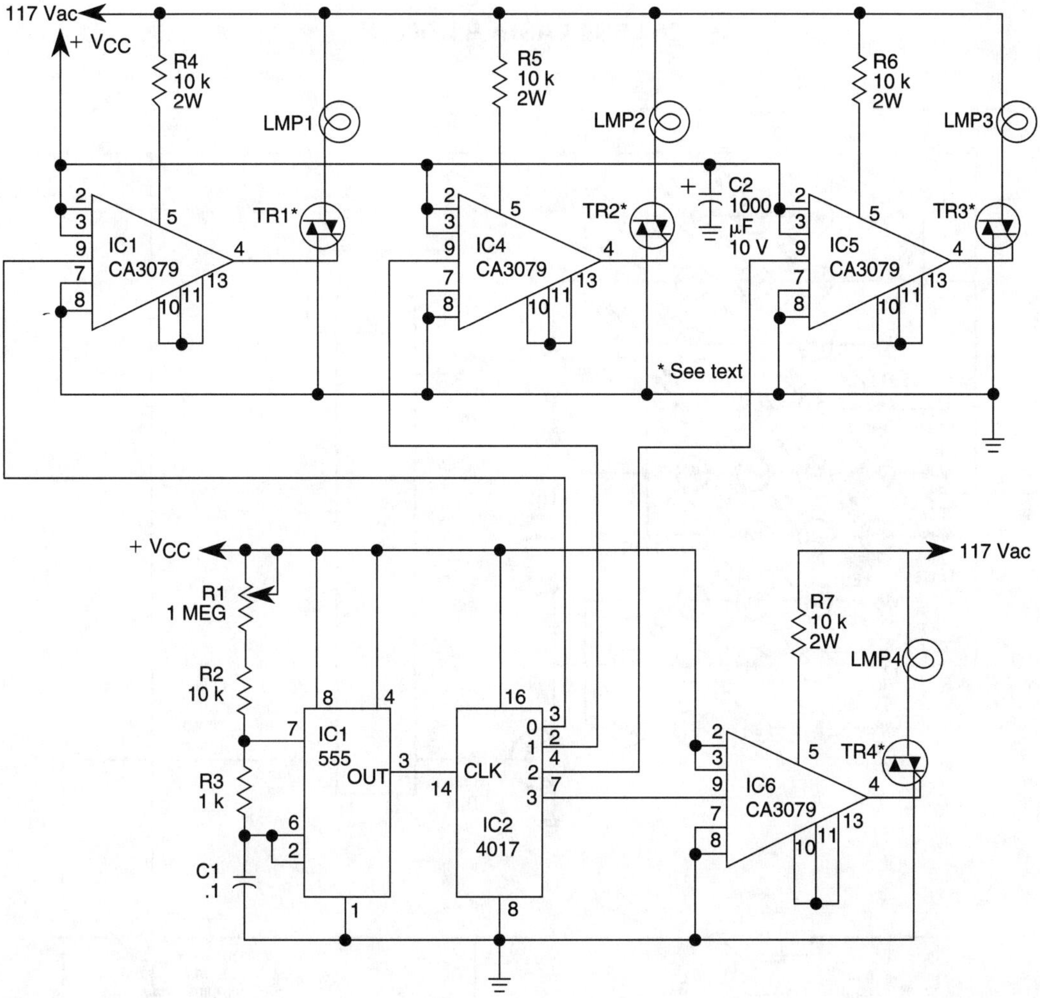

R-E EXPERIMENTERS HANDBOOK

FIG. 31-1

A 555 timer, IC1, drives a 4017 CMOS decade counter. Each of the 4017's first four outputs drives a CA3079 zero-voltage switch. Pin 9 of the CA3079 is used to inhibit output from pin 4, thereby disabling the string of pulses that the IC normally delivers. Those pulses occur every 8.3 ms, i.e., at a rate of 120 Hz. Each pulse has a width of 120 μs.

Because of the action of the CA3079, the lamps connected to the triacs turn on and off near the zero crossing of the ac waveform. Switching at that point increases lamp life by reducing an inrush of current that would happen if the lamp were turned on near the high point of the ac waveform. In addition, switching at the zero crossing reduces radio frequency interference (RFI) considerably. **Caution:** The CA3079s are driven directly from the 117-Vac power line, so use care.

36 LED FLASHER DRIVER

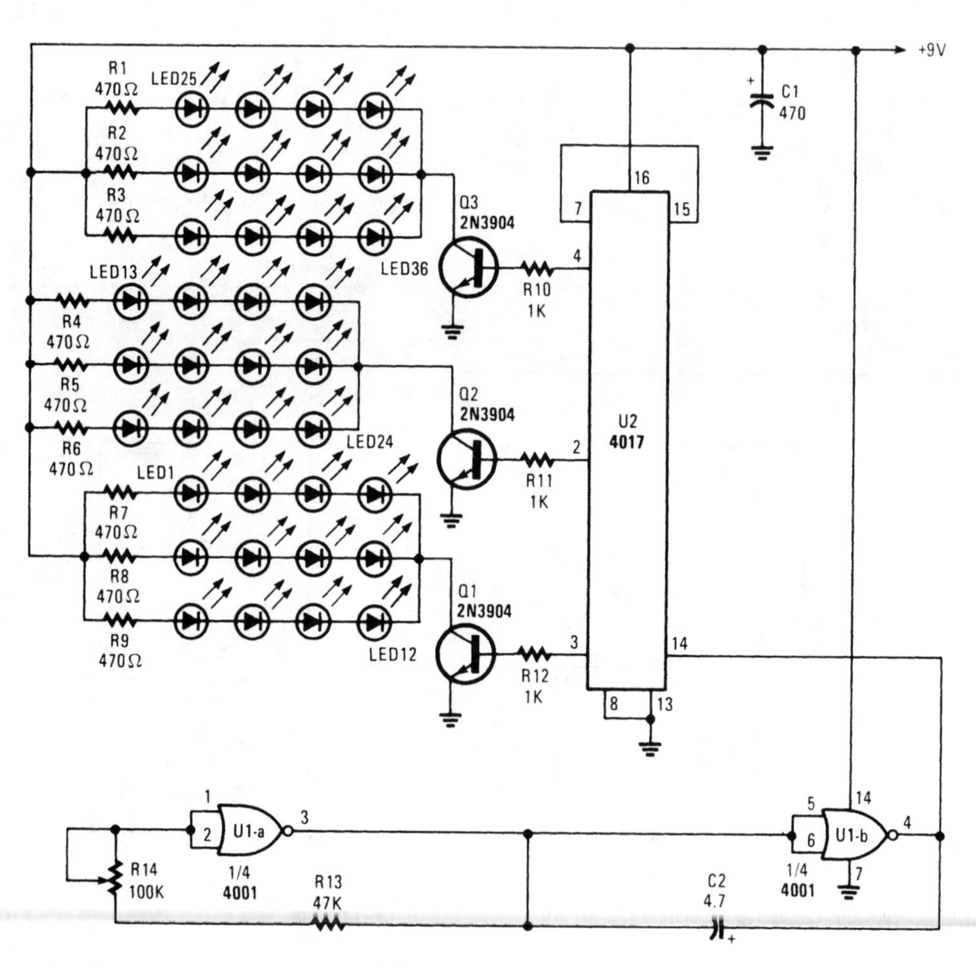

POPULAR ELECTRONICS *FIG. 31-2*

Originally intended as a 3-bell animation circuit for Christmas decorations, the circuit can be used for many other purposes that require a flasher of this kind. By re-connecting U2 (see the data manual), more than three outputs can be be obtained.

LED FLASHERS

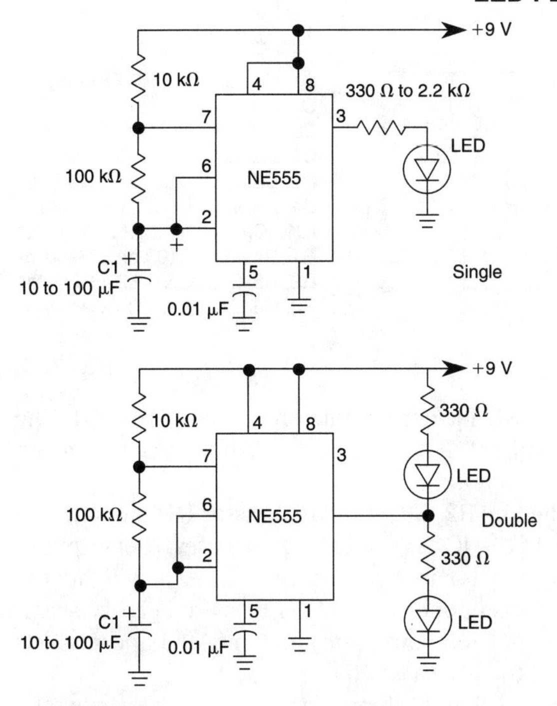

A 555 is used to switch an LED on and off. C1 determines the flash rate. Single ended (one LED) and double-ended (alternating) flashers are shown.

WILLIAM SHEETS

FIG. 31-3

DARK-ACTIVATED LED FLASHER

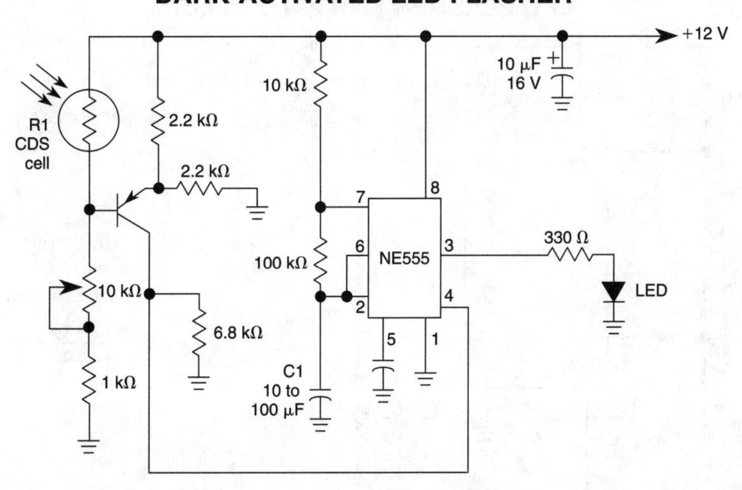

WILLIAM SHEETS

FIG. 31-4

This circuit can be used as a small beacon or marker light, and toys or novelty items. R1 is an LDR that has ≥10 kΩ dark-resistance, or a CDS photocell. C1 determines the flash rate.

SUPER LED FLASHER

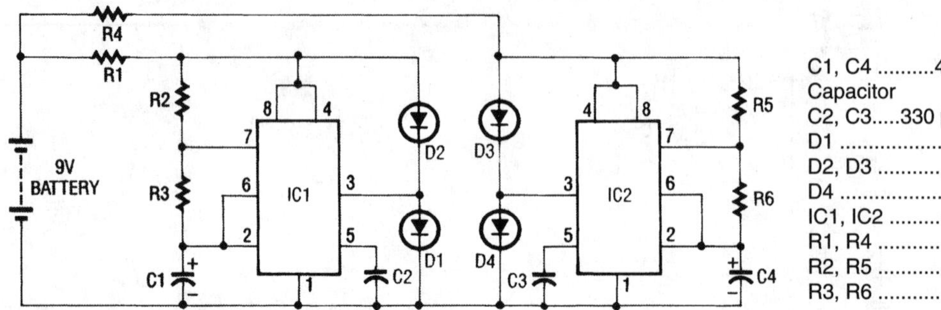

C1, C4 ....4.7 μF Electrolytic Capacitor
C2, C3.....330 pF Disc Capacitor
D1 .... Yellow LED
D2, D3 .... Red LED
D4 .... Green LED
IC1, IC2 .... 555 Timer IC
R1, R4 ....100 ohm Resistor
R2, R5 .... 82 k Resistor
R3, R6 .... 33 k Resistor

1991 PE HOBBYIST HANDBOOK ***FIG. 31-5***

The super LED flasher is actually two complete LED flasher circuits on one circuit board. The first LED flasher is made up of IC1 and LEDs D1 and D2. IC1 is a 555 timer IC configured as an astable (free-running) multivibrator with its output on pin 3.

The frequency of the 555's oscillation is controlled by R2, R3, and C1. Resistor R1 limits the input voltage to a low enough level to prevent damage to the IC. As the 555 IC oscillates, the output of pin 3 goes high (+) then low (–). When the output is high it supplies current to D1, which lights up. When it is low, pin 3 sinks current and D2 lights up. This happens because LEDs are polarity-sensitive (like all other diodes, they permit current flow in only one direction) and one lead of each LED has been connected to the respective polarity needed to light that LED.

The second LED flasher, made up of IC2 and LEDs D3 and D4, operates in the same way as the first LED flasher.

LED FLASHER FOR 2 TO 10 LEDs

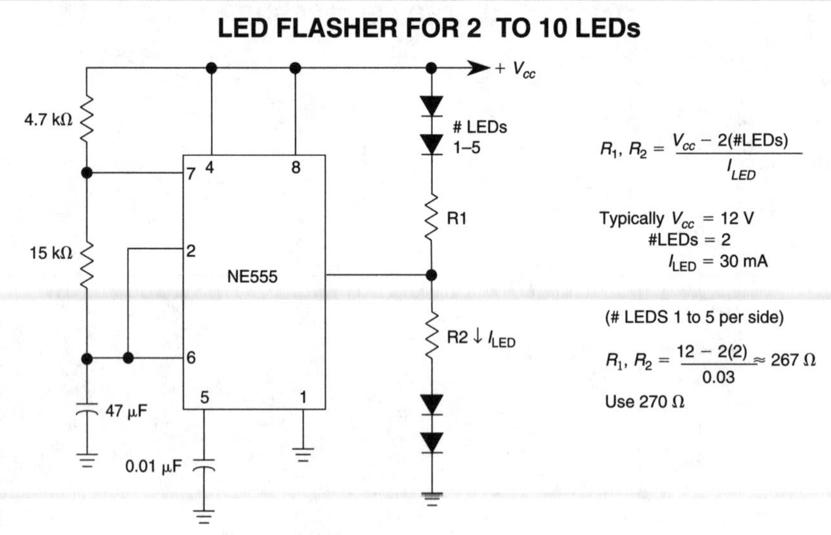

WILLIAM SHEETS ***FIG. 31-6***

This LED flasher has double-ended output connection. The circuit can be used with 1 to 5 LEDs on each side as indicated.

FLASH SIGNAL ALARM

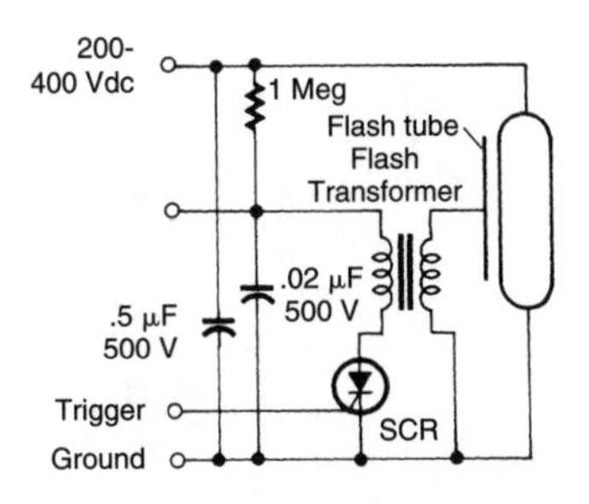

RADIO-ELECTRONICS

FIG 31-7

This circuit is useful if you need a low-energy flashing alarm. The 200 to 400-dc supply should have enough internal resistance to charge the 0.5 μF capacitor between flashes, about 2 or 3 time constants, which means about 500 kΩ to 1 MΩ for a 1-s rate. Use lower values for higher rates.

LED CHRISTMAS TREE LIGHT FLASHER

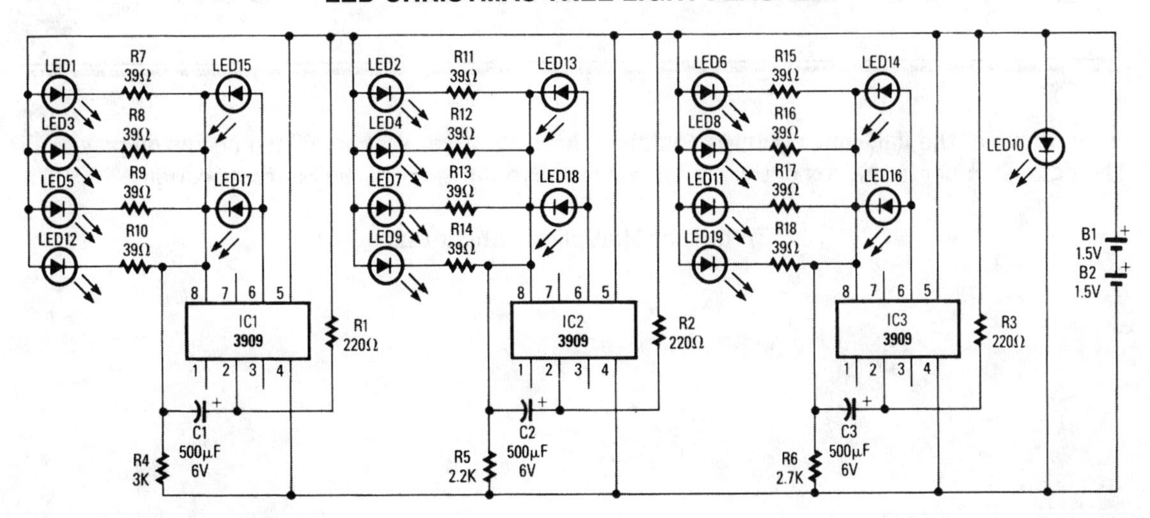

R-E EXPERIMENTERS HANDBOOK

FIG. 31-8

Three individual flashing circuits that use an LM3909 LED flasher/oscillator IC create the appearance of a pseudo-random firing order. The combination of C_1/R_4, C_2/R_5, and C_3/R_6 control the blink rate, which is between 0.3 and 0.8 s, and the inherent wide tolerance range (–20% to +80%) of standard electrolytic capacitors add to the irregularities of the blink cycles. The continuous current drain is about 10 mA; however, if you decrease the values of R4 through R6 or C1 through C3 in order to increase the blink rate, the current will then increase proportionally.

Note in particular that external current-limiting resistors aren't needed for LED13 through LED18; the resistors are built into the ICs. LED10, which serves as the tree's "star," is a special kind of flashing LED that blinks continuously at a fixed rate.

32

Frequency Multiplier Circuit

The source of the following circuit is contained in the Sources section, which begins on page 675. The figure number in the box of the circuit correlates to the entry in the Sources section.

Frequency Multiplier Without PLL

FREQUENCY MULTIPLIER WITHOUT PLL

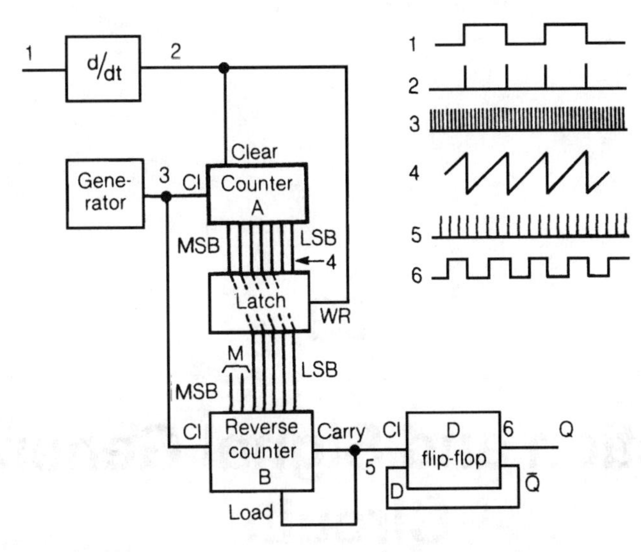

ELECTRONIC ENGINEERING

FIG. 32-1

An input rectangular signal is differentiated and short impulses are formed from its edges. These impulses write the content of counter A to a latch that clears the counter after a very short time. Counter A counts impulses of the frequency f_o that are much greater than that of the input signal. The pulses come from an impulse generator. Thus, the number, which is written to the latch, expresses the number of these impulses between the edges of the input signal. The impulses from the same generator pass to (reverse) counter B. The carry impulse loads the content of the latch to counter B. The latch is connected with the reverse counter such that the number written to this counter is $2M$ times smaller than the number introduced to the latch. This can be readily achieved by omitting M most significant bites of counter B. Because the number loaded to counter B is $2M$ times smaller than the number in the latch, the carry impulses of counter B have frequency $2M$ times greater than the frequency of the impulses at the output of the differentiator. The carry impulses are fed to a D flip-flop, which divides their frequency by two. In this way, the output frequency is $2M$ greater than input frequency f_o as long as the frequency of impulse generator f_g is much greater than $2Mf_o$.

33

Function and Signal Generator Circuits

The sources of the following circuits are contained in the Sources section, which begins on page 675. The figure number in the box of each circuit correlates to the entry in the Sources section.

Function Generator
100-dB Dynamic-Range Log Generator
Function Generator
Fast Logarithm Generator
Triangle-Wave Generator
555-Based Ramp Generator
Triggered Sawtooth Generator
Signal Generator
Transistorized Schmitt Trigger
Linear Sawtooth Generator
Capacitance Multiplier
Triangle-Wave Oscillator
Clock-Driven Triangle-Wave Generator
Triangle- and Square-Wave Generator
Root Extractor

FUNCTION GENERATOR

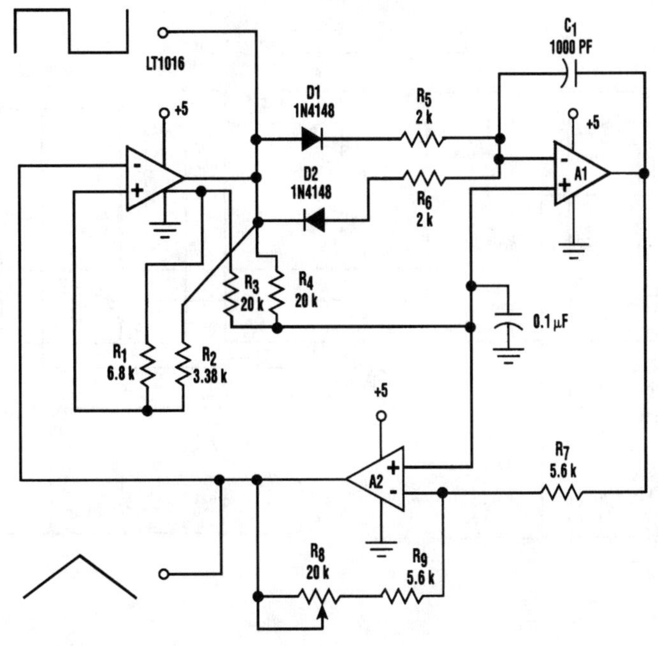

ELECTRONIC DESIGN

FIG. 33-1

This function generator, based on an LT1016 high-speed comparator, will generate from a single +5-V supply. The slow rate of the op amps used determines the maximum useable frequency of this circuit.

100-dB DYNAMIC-RANGE LOG GENERATOR

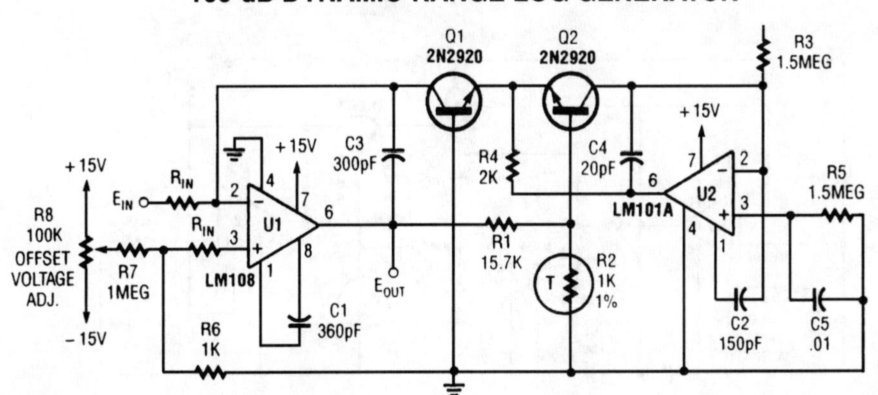

POPULAR ELECTRONICS

FIG. 33-2

E_{OUT} = constant × (Log E_{IN}). This circuit has 100-dB dynamic range, which is five decades of voltage change at the input.

FUNCTION GENERATOR

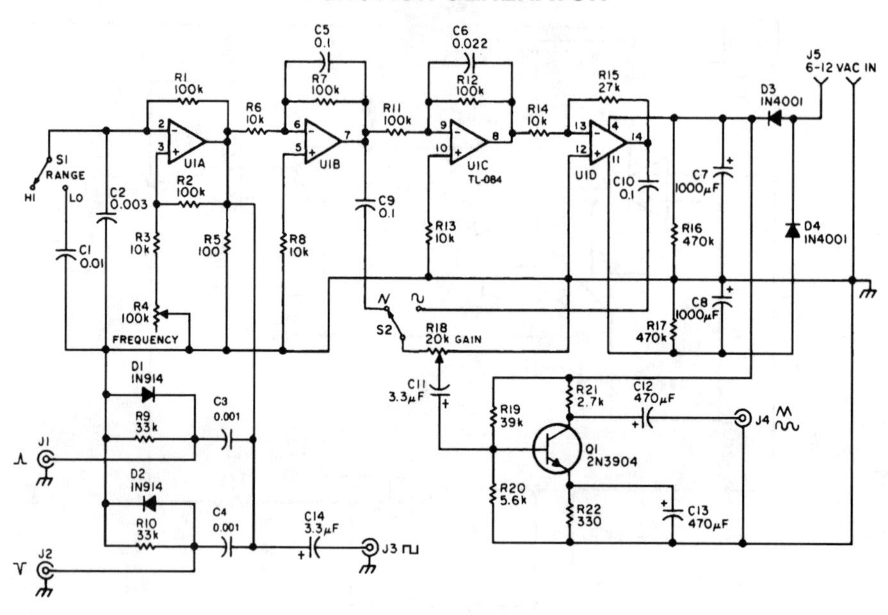

73 AMATEUR RADIO TODAY ***FIG. 33-3***

A quad op amp makes up the heart of this function generator. U1-a generates a square wave, and outputs this to J3. J1 and J2 are pulse outputs obtained by differentiating the square wave. Integrator U1-b generates a triangle-wave shaper to obtain a sine wave. Q1 is an output amplifier.

FAST LOGARITHM GENERATOR

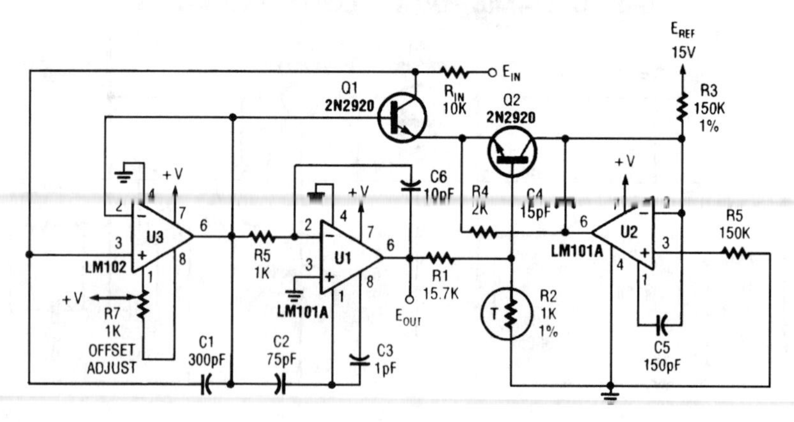

POPULAR ELECTRONICS ***FIG. 33-4***

In this circuit, E_{OUT} = (constant) × log E_{IN}. The circuit should be useable with op amps other than the ones illustrated.

TRIANGLE-WAVE GENERATOR

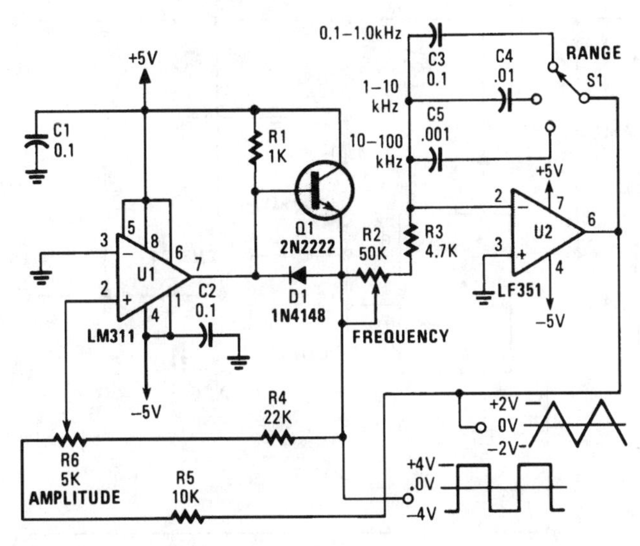

POPULAR ELECTRONICS

FIG. 33-5

This is a simple triangle-wave generator using two IC devices and a transistor. The triangle wave is used as feedback to the square-wave generator. S1 allows range switching in three ranges from 100 Hz to 100 kHz. Extra positions could be used to extend the range to lower frequencies, using larger values of capacitance.

555-BASED RAMP GENERATOR

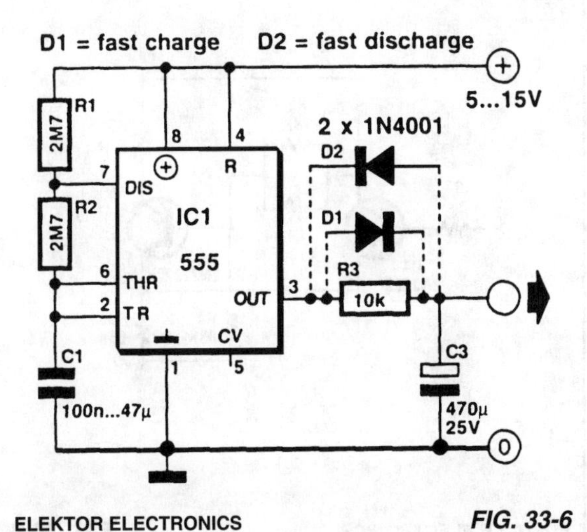

ELEKTOR ELECTRONICS

FIG. 33-6

This circuit is used to generate a ramp voltage for tuning a radio receiver. An NE555, running at about 0.1 Hz, is used as an astable multivibrator.

TRIGGERED SAWTOOTH GENERATOR

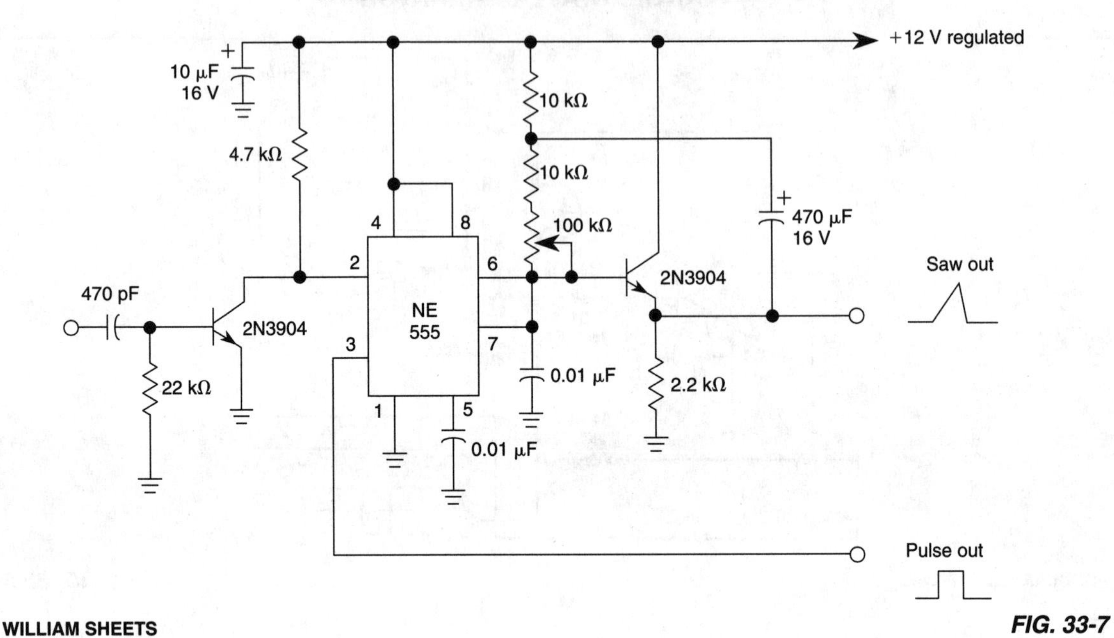

WILLIAM SHEETS

FIG. 33-7

Two 2N3904 transistors and a 555 form a triggered sawtooth generator. A sawtooth or other rising voltage input provides a pulse output when the trigger point is reached.

SIGNAL GENERATOR

WILLIAM SHEETS

FIG. 33-8

This simple oscillator is rich in harmonics which make this circuit useful for signal tracing applications.

TRANSISTORIZED SCHMITT TRIGGER

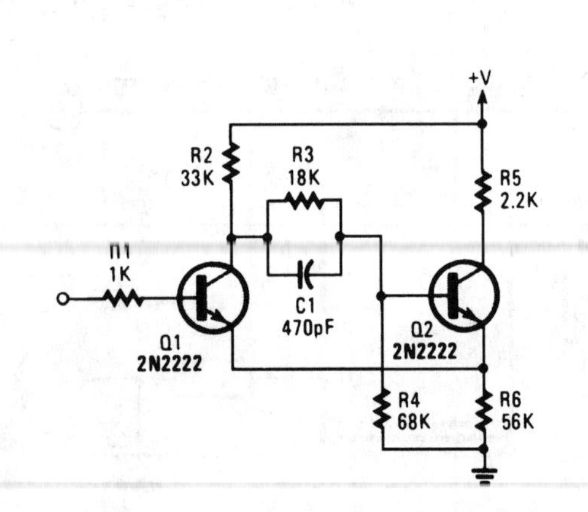

POPULAR ELECTRONICS

FIG. 33-9

LINEAR SAWTOOTH GENERATOR

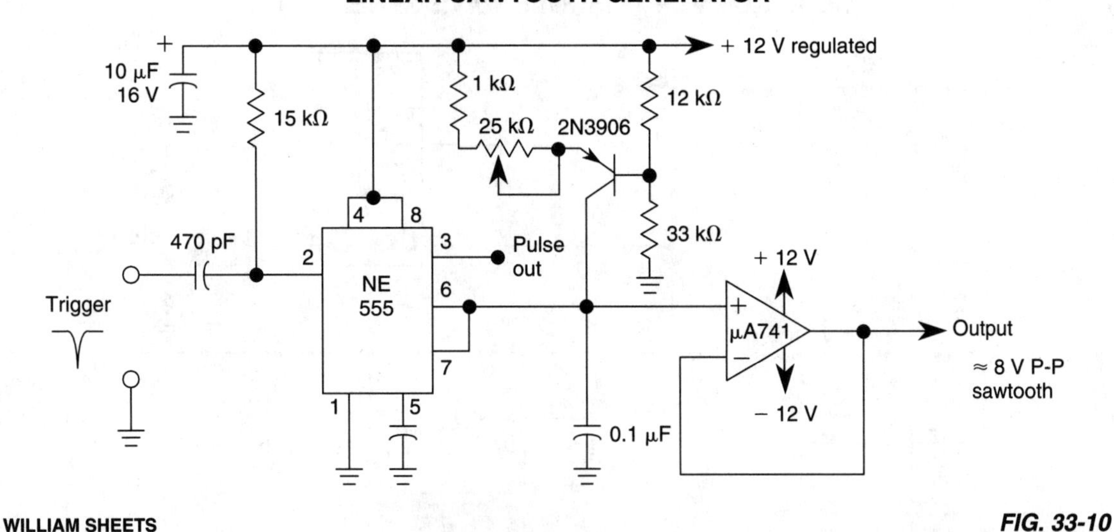

WILLIAM SHEETS ***FIG. 33-10***

The 2N3906 transistor is used as a constant-current source, to assure that the 555-based sawtooth generator generates a linear ramp waveform.

CAPACITANCE MULTIPLIER

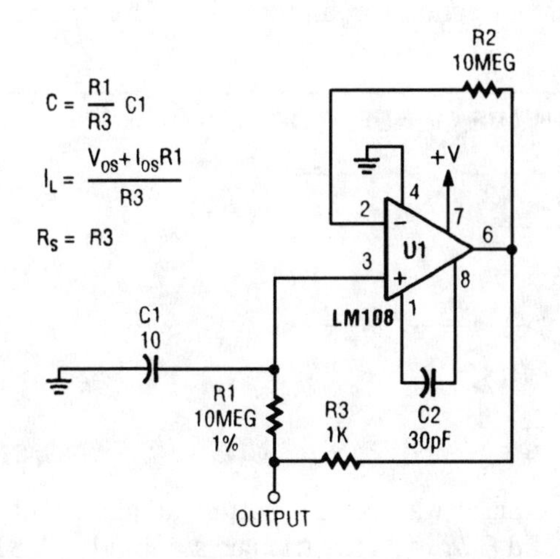

POPULAR ELECTRONICS ***FIG. 33-11***

Capacitance multiplier uses the gain of an op amp to produce an effective capacitance—in this case 100,000 μF.

TRIANGLE-WAVE OSCILLATOR

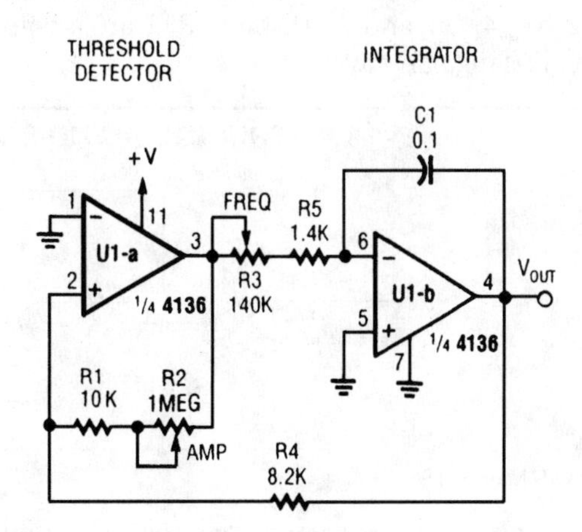

POPULAR ELECTRONICS ***FIG. 33-12***

U1-b acts as an integrator while U1-a is a threshold detector. R2 sets the trip level and therefore the amplitude. R3 controls charging current of C1 and the frequency.

CLOCK-DRIVEN TRIANGLE-WAVE GENERATOR

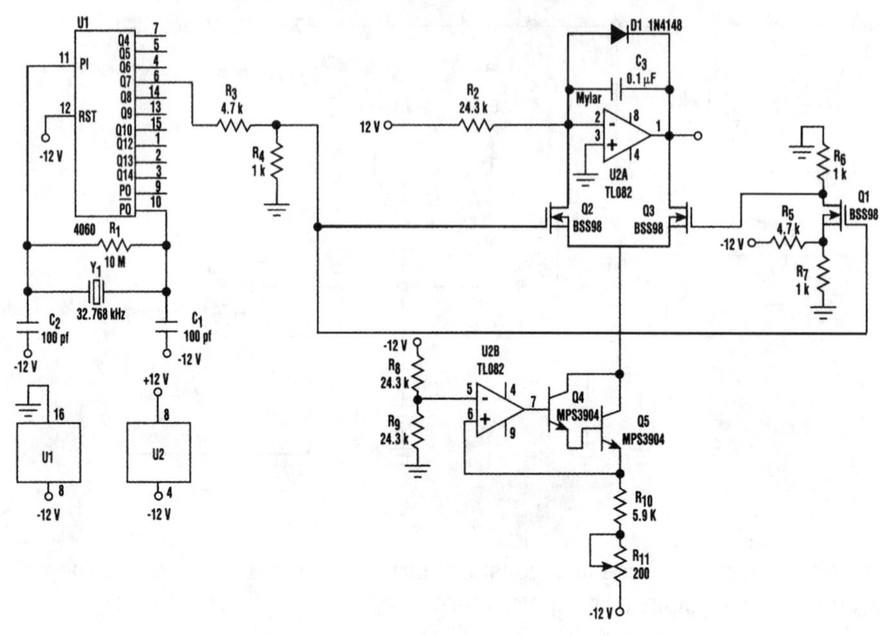

ELECTRONIC DESIGN *FIG. 33-13*

U2-a, C3 and R2 operate as an integrator. Q2 and Q3 are alternately switched at 256 cycles. U2-b, Q4, Q5, and R8 through R11 are a constant current generator, and R11 is set for a symmetrical triangular waveform.

TRIANGLE- AND SQUARE-WAVE GENERATOR

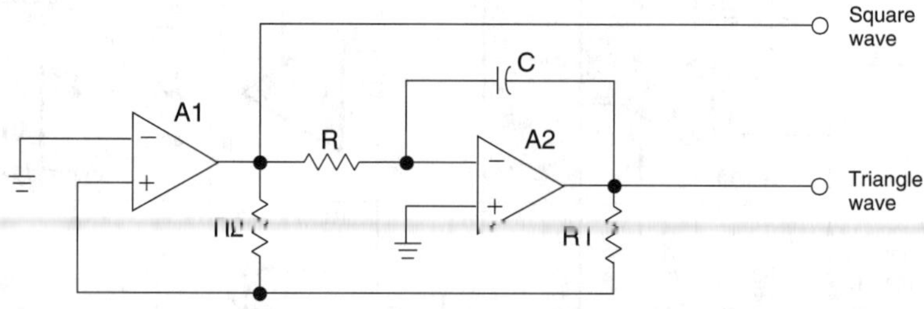

WILLIAM SHEETS *FIG. 33-14*

The circuit will generate precision triangle and square waves. The output amplitude of the square wave is set by the output swing of op amp A1, and R_1/R_2 sets the triangle amplitude. The frequency of oscillation in either case is approximately $1/0.69RC$.

The square wave will maintain 50% duty cycle—even if the amplitude of the oscillation is not symmetrical. The use of a fast op amp in this circuit will allow good square waves to be generated to quite high frequencies. Because the amplifier runs open-loop, compensation is not necessary. The triangle-generating amplifier should be a compensated type. A dual op amp, such as the MC1458, can be used for most applications.

ROOT EXTRACTOR

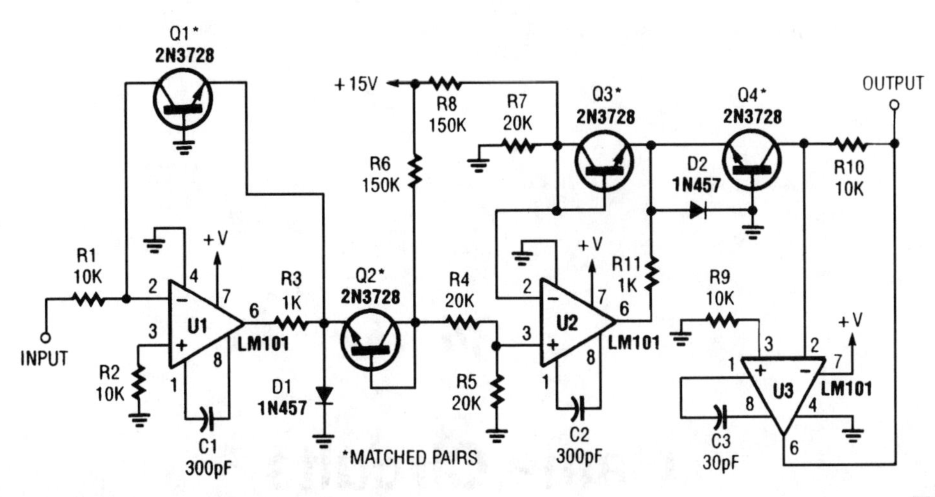

POPULAR ELECTRONICS

FIG. 33-15

This circuit produces a voltage that is proportional to the root of the input. This gives a logarithmic response, log $V_{IN}{}^{N} = N \log V_{IN}$.

34

Game Circuits

The sources of the following circuits are contained in the Sources section, which begins on page 675. The figure number in the box of each circuit correlates to the entry in the Sources section.

ELECTROMAGNETIC RING LAUNCHER

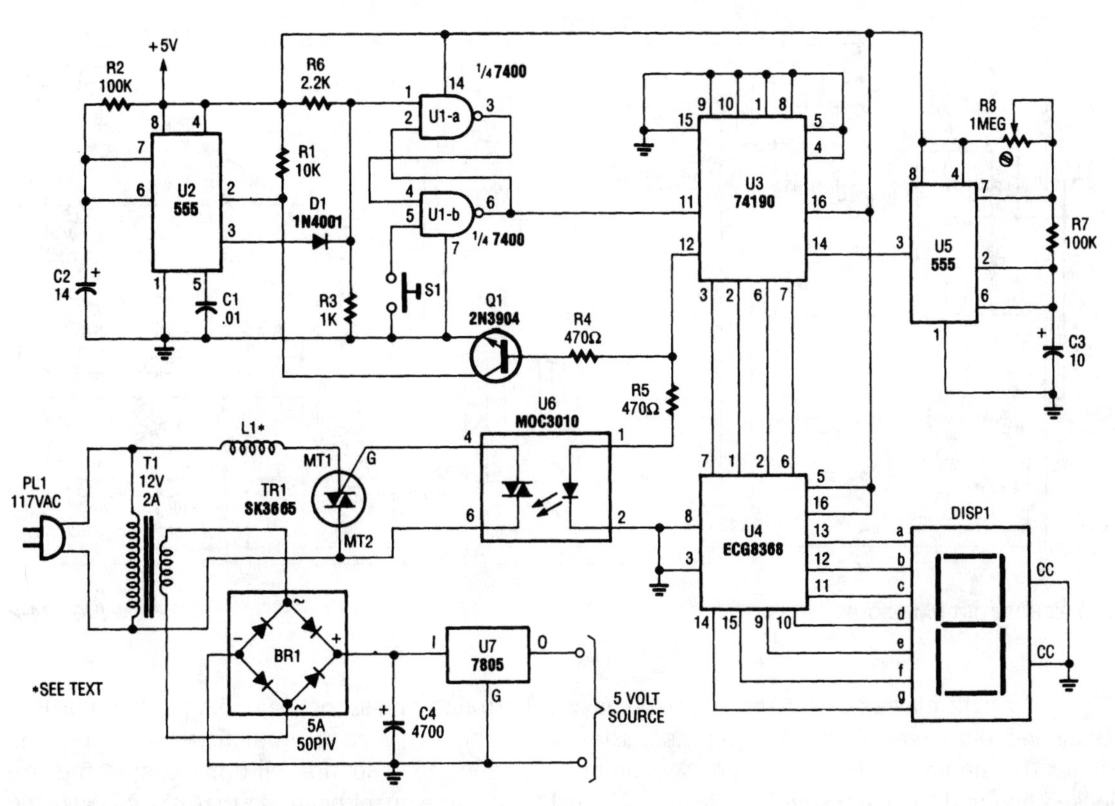

1993 ELECTRONICS HOBBYIST HANDBOOK **FIG. 34-1**

The electromagnetic ring launcher is comprised of four subcircuits: a clock circuit (built around U5, a 555 oscillator/timer configured for astable operation), a count-down/display circuit (built around U3), a 74190 synchronous up/down counter with BCD outputs that is configured for count-down operation; U4, a ECG8368 BCD-to-7-segment latch/decoder/display driver; and DISP1, a common-cathode seven-segment display), a trigger circuit (comprised of U6), an MOC3010 optoisolator/coupler with Triac-driver output; TR1, an SK3665 200-PIV, 4-A Triac; and a few support components), and a reset circuit (comprised of U1, a 7400 quad 2-input NAND gate; U2, a second 555 oscillator/timer configured for monostable operation; and a few support components).

This circuit is that of a repulsion coil (L1) used to demonstrate the principle of electromagnetic repulsion by propelling a metal ring around the core of L1 through the air. A countdown circuit is provided to count seconds before launch.

QUIZ MASTER

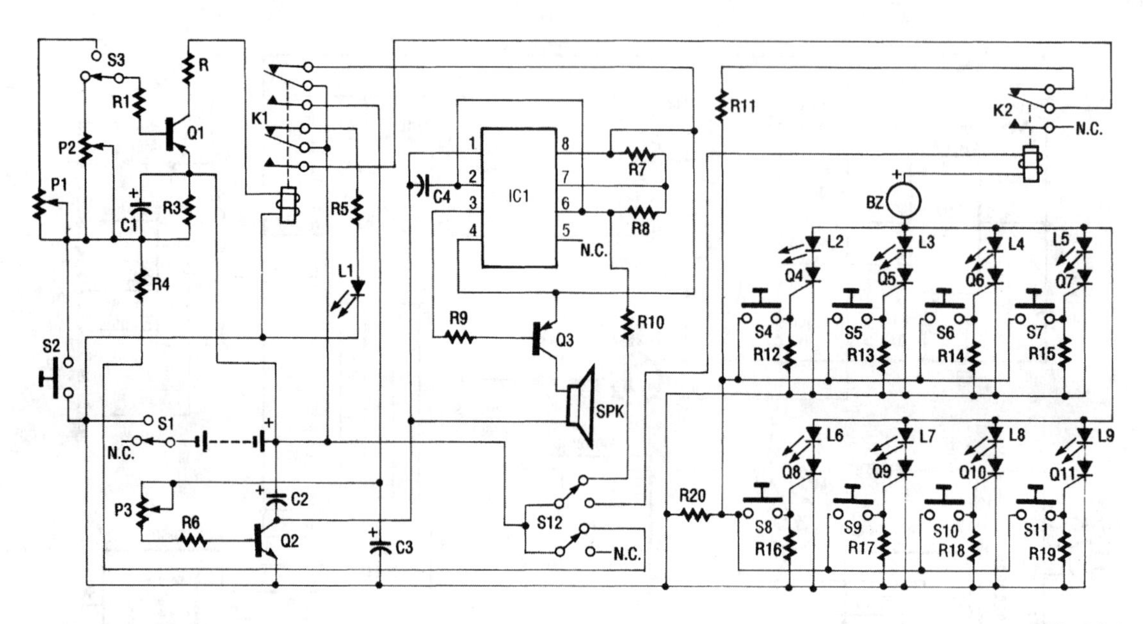

1991 PE HOBBYIST HANDBOOK **FIG. 34-2**

Up to eight players each have their own answer button to press, corresponding to the four Red Team and four Green Team LEDs on the master control board. As soon as the first contestant who thinks that he knows the answer presses the button, a loud tone sounds, all other contestants are locked out, and the contestant's indicator LED lights on the control board so that it's obvious who buzzed in first.

The control board also features two selectable "time out" periods—each adjustable from 3 to 15 seconds, setting specified time intervals in which the player must answer before the "time's up!" tone sounds. Eight SCRs form the heart of the circuit. The anode of each SCR has a positive (+) bias on it by way of an LED and a negative (–) bias on each cathode. As soon as a contestant depresses his or her switch button (S4 through S11), a positive bias is applied to the respective SCR gate. That bias latches the contestant's SCR on, which in turn lights up the appropriate LED on the master control board. At the same time, the activity of the SCR latching on turns on the answer buzzer (BZ) and locks out all other contestants. The lockout occurs because relay K2 contacts operate to remove the availability of a bias voltage to the gate of the other SCRs.

The other circuitry consists of a timer circuit and a "time's-up" tone-generating circuit. The timer circuit consists of transistor Q1, capacitor C1, resistors R1 through R3, and trimmer resistors P1 and P2. Depending on the adjustment of the trimmer resistors and selection switch S3, a specific time period can be set. The time's-up tone-generating circuit is made up of IC1, transistors Q2 and Q3, and the associated resistors and capacitors. The "on" time of the tone can be set by P3. Relay K1, which is operated by the timer circuit, serves to reset the entire unit for the next question.

ELECTRONIC SLOT MACHINE

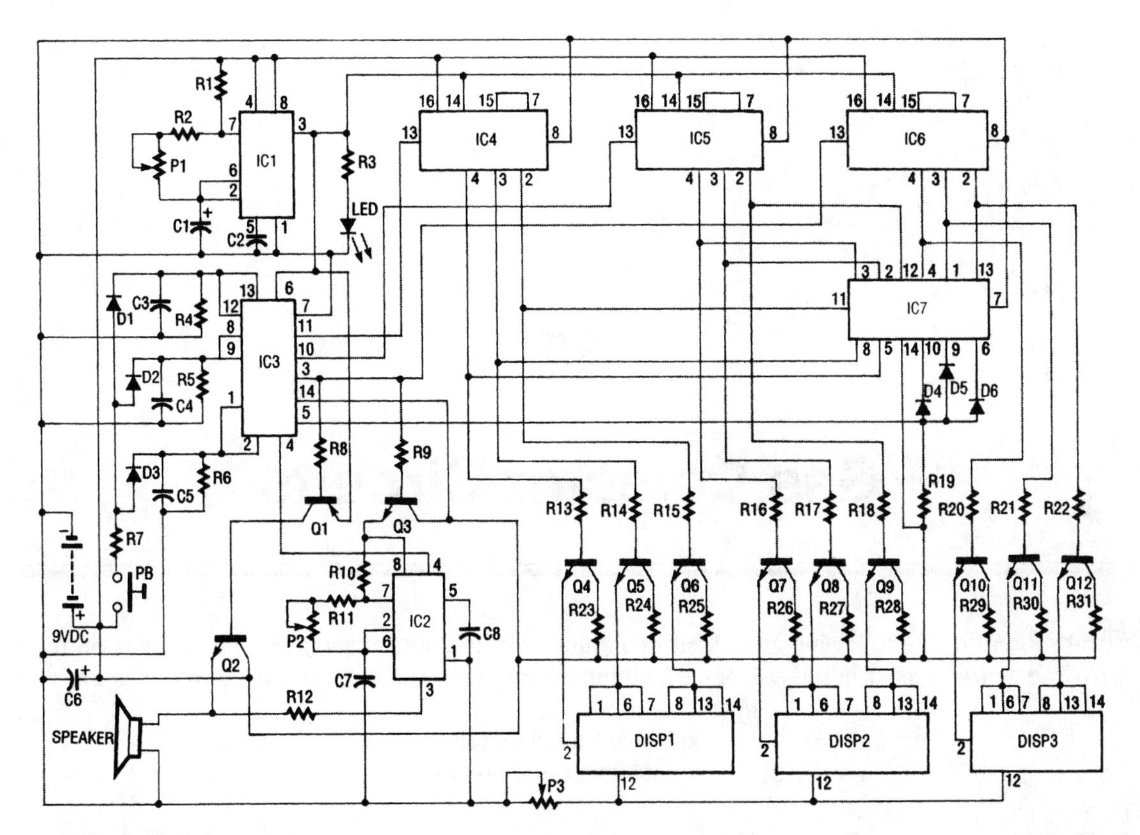

1991 PE HOBBYIST HANDBOOK ***FIG. 34-3***

The slot machine's realistic action is provided by seven ICs and three displays, as shown. Two 555 CMOS timer ICs generate pulses. IC1 is used to generate the clock pulses for the entire electronic slot machine. The pulses are coupled from the output (pin 3) to the clock inputs of IC4, IC5, and IC6, the display-driver ICs.

The displays are common-cathode 7-segment LED types. They are wired to display three different symbols, an "L," a "7," and "bar." When all three displays show the same symbols, IC7 (a 4023 triple 3-input NAND gate) decodes a winner and sends a signal to pin 5 of IC3. That IC is a 4001 CMOS NOR gate and it turns on IC2, a 555 timer IC. IC2 actually produces the winner tone on its output, pin 3.

Transistors Q4 through Q12 are used to drive the common-cathode displays. An LED is used to indicate the clock pulses, and a variable resistor is provided for each of these functions. Trimmer resistor P1 controls the overall clock rate, P2 controls the "winner" tone, and P3 controls the display brilliance.

35

Gas Detector Circuits

The sources of the following circuits are contained in the Sources section, which begins on page 675. The figure number in the box of each circuit correlates to the entry in the Sources section.

EXPLOSIVE GAS DETECTOR

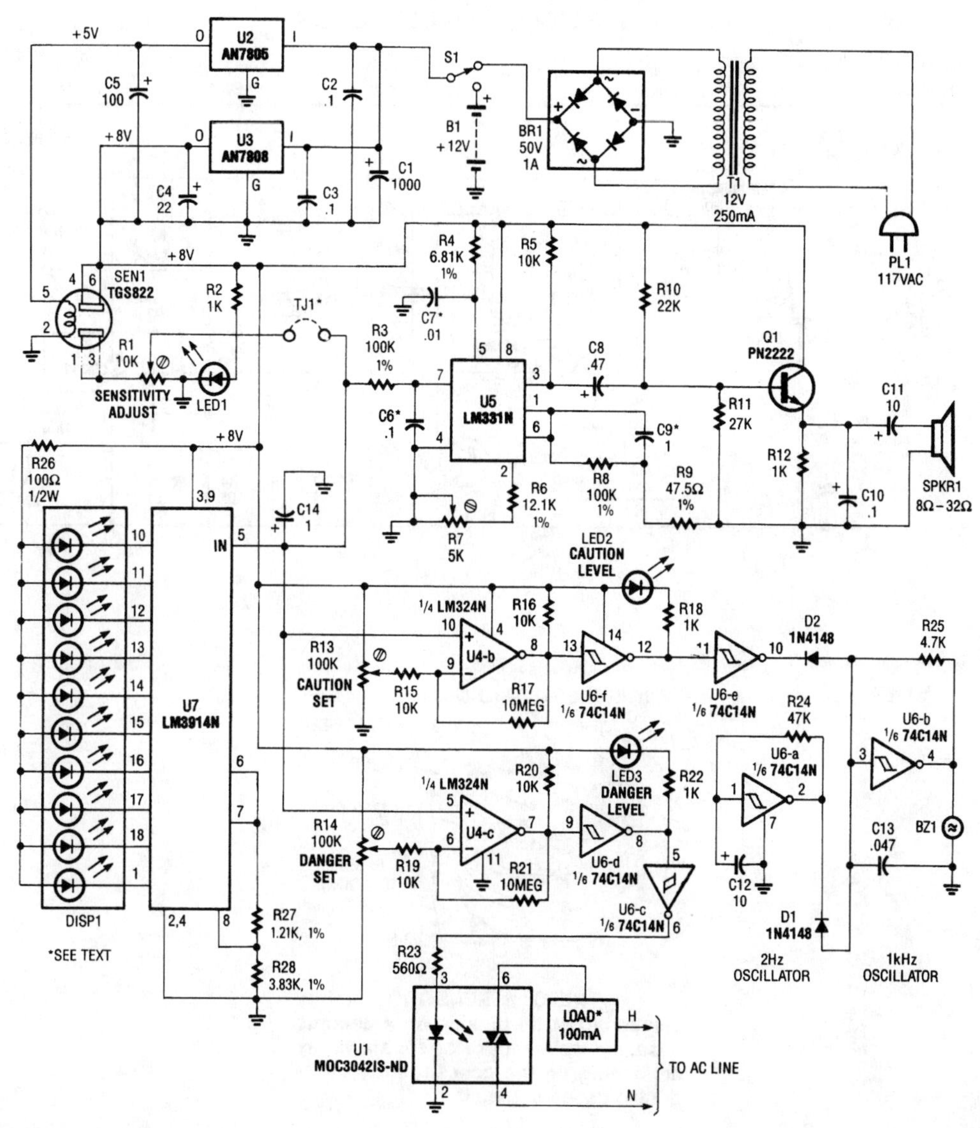

POPULAR ELECTRONICS ***FIG. 35-1***

A gas sensor (TGS823 from Allegro Electronics, Cornwall Bridge, CT 06754) conducts in the presence of explosive gases. U5 is a voltage-to-frequency converter that produces a frequency proportional to the sensor conductance. The output frequency ranges from 100 Hz in clean air to 8 kHz in a contaminated atmosphere. The dc voltage from the sensor also drives bar graph LED U7 and comparators U4-b and U4-c to sense present caution and danger levels. U1 drives an ac load up to 100 mA (relay, indicator, alarm, etc.).

COMBUSTIBLE GAS DETECTOR

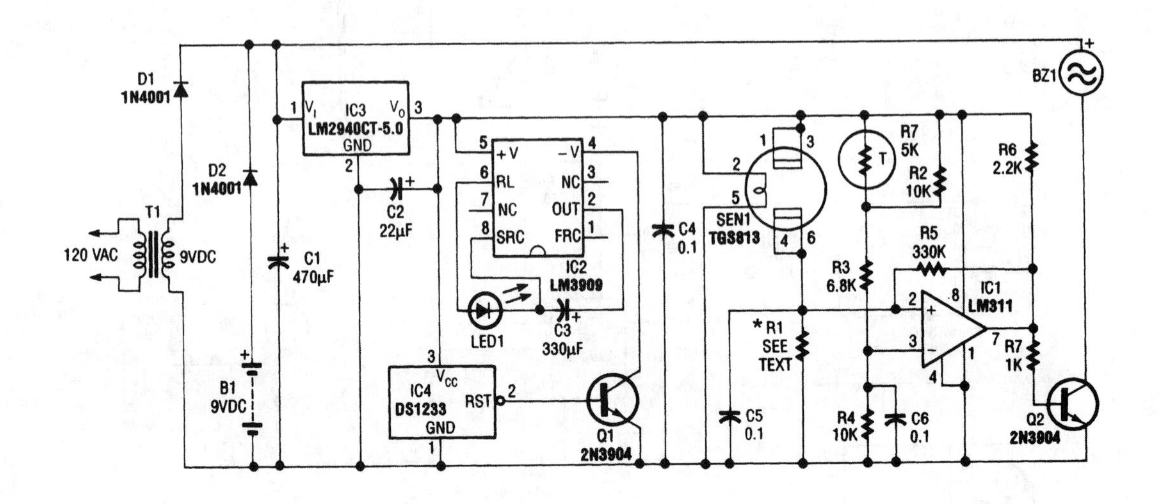

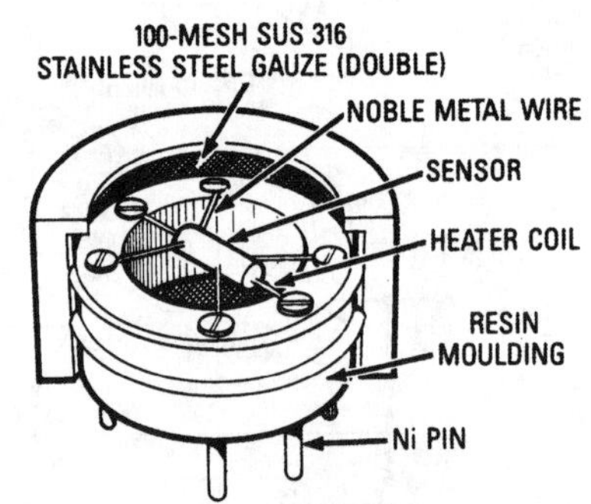

THE GAS SENSOR is mainly composed of tin dioxide on a ceramic base; the resistance of the sensor varies depending on the concentration of reducing gases in the air.

ELECTRONICS NOW

FIG. 35-2

The circuit shown is useful for the detection of dangerous levels of combustible fumes or gases. It uses a comparator circuit to trigger an alarm buzzer. The sensor's resistant element is connected in series with resistor R1 to form a voltage-divider circuit; R1 is specifically matched to each gas sensor by the manufacturer.

36

Gate Circuit

The source of the following circuit is contained in the Sources section, which begins on page 675. The figure number in the box of the circuit correlates to the entry in the Sources section.

AND Gate

AND GATE

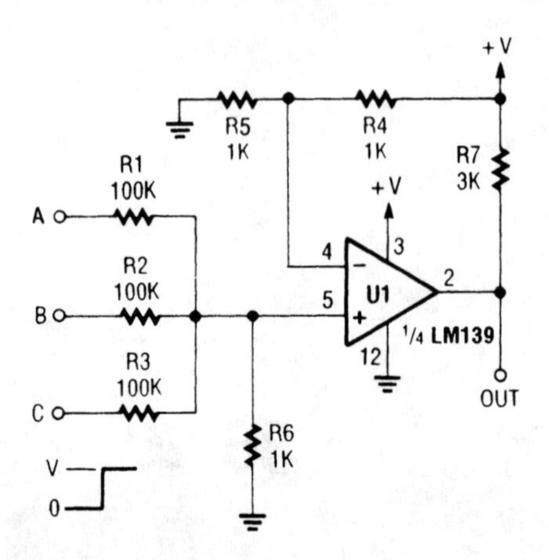

POPULAR ELECTRONICS *FIG. 36-1*

A left-over section of a quad op amp can be used to save cost and eliminate an extra logic chip for this AND gate.

37

Geiger Counter Circuits

The sources of the following circuits are contained in the Sources section, which begins on page 675. The figure number in the box of each circuit correlates to the entry in the Sources section.

GEIGER COUNTER I

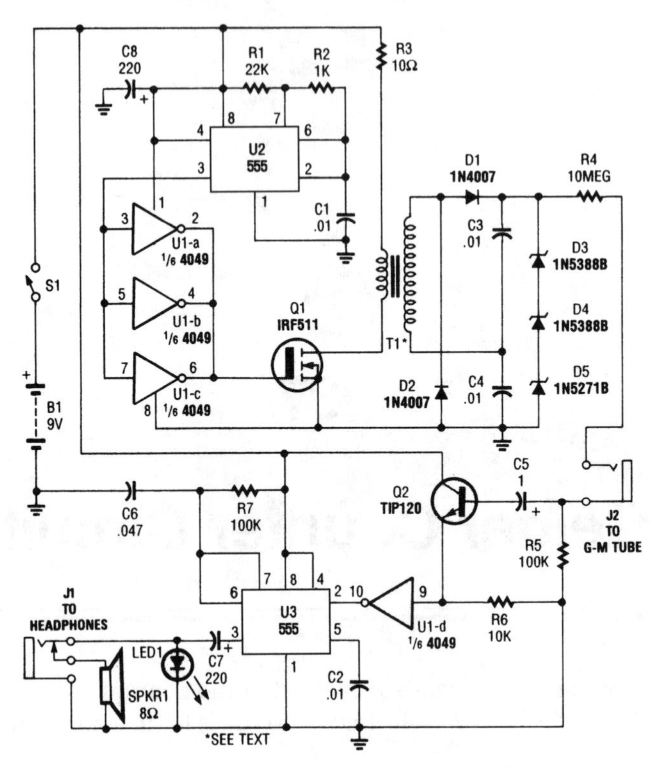

POPULAR ELECTRONICS *FIG. 37-1*

The circuit is built around a 4049 hex inverter (U1), a pair of 555 oscillator/timers (U2 and U3), two transistors, a Geiger-Muller tube, and a few additional support components. The first 555 (U2) is configured for astable operation. The output of U2 (a series of negative-going pulses) at pin 3 is fed to three parallel-connected inverters (U1-a, U1-b, and U1-c). The positive-going output pulses of the inverters are fed to the gate of Q1, causing it to toggle on and off.

The output of Q1, which is connected in series with the primary of step-up transformer T1, produces a stepped-up series of pulses in T1's secondary. The output of T1 (approximately 300 V) is fed through a voltage doubler (consisting of D1, D2, C3, and C4), producing a voltage of around 600 V. Three series-connected Zener diodes (D3, D4, and D5) are placed across the output of the voltage doubler to regulate the output to 500 V, fed through R4 (a 10-MΩ current-limiting resistor) and J2 to the anode of the GM tube. The limiting resistor also allows the detection ionization to be quenched.

The cathode side of the tube is connected to ground through a 100-kΩ resistor, R5. When a particle is detected by the GM tube, the gases within the tube ionize, producing a pulse across R5. That pulse is also fed through C5 and applied to the base of Q2 (a TIP120 npn transistor), where it is amplified and clamped to 9 V. The output of Q2 is inverted by gate U1-d, then it is used to trigger U3 (the second 555, which is configured for monostable operation). The output of U3 at pin 3 causes LED1 to flash, and produces a click that can be heard through speaker SPKR1 or headphones. The circuit is powered by a 9-V alkaline battery and draws about 28 mA when not detecting radiation.

GEIGER COUNTER II

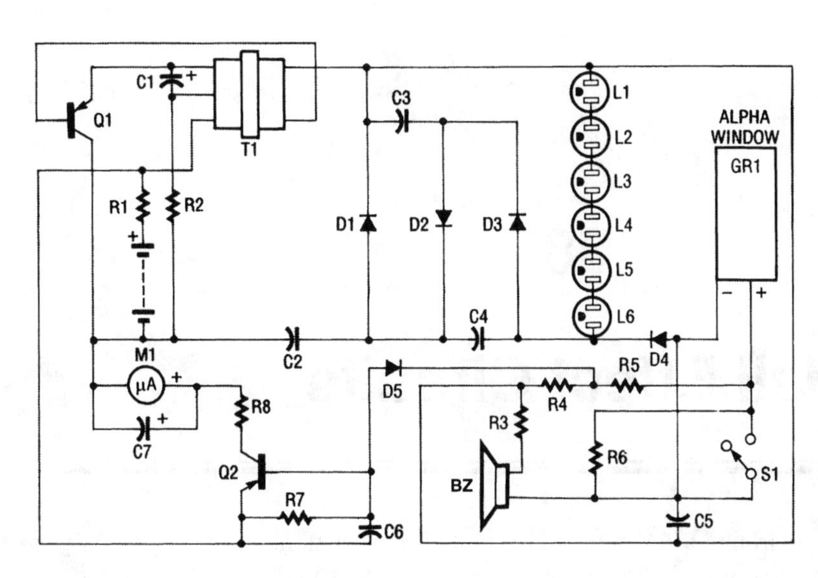

BZ ....Blue Piezo Buzzer
C1 .... 4.6-μF Electrolytic Capacitor
C2–C4 ....0.005-μF 1-kV Disc Capacitor
C5 .... .01-μF 1-kV Disc Capacitor (103 M)
C6 .... .1-μF 100-V Mylar Capacitor (104 k)
C7 .... 33-μF Electrolytic Capacitor
D1–D5 . . .1N4007 Diodes
GR1 .... Alpha Window Geiger Mueller Tube
L1–L6 .....Neon Lamps
M1 .... 0–200 Microamp Meter
Q1 .... 02-GE PNP Power Transistor
Q2 .... 2N3906 Transistor
R1 .... 47-ohm Resistor
R2, R3 ... 3.9-k Resistor
R4, R5 ... 4.7-Meg Resistor
R6 .... 220-k Resistor
R7 .... 27-k Resistor
R8 .... 18-kΩ Resistor
S1 .... SPDT Slide Switch
T1 .... Inverter Transformer

PE HOBBYIST HANDBOOK ***FIG. 37-2***

Q1 is a pnp power transistor used in conjunction with a ferrite transformer to form a blocking-type oscillator. This oscillator is a fixed-frequency type, and the feedback to sustain oscillations is from capacitor C1. Because of the turns ratio of T1, the small ac voltage produced on its primary is converted to a large ac voltage on its secondary. That high-voltage ac is applied to the voltage tripper stage, which consists of capacitors C2, C3, and C4 and diodes D1, D2, and D3. The resultant voltage is now over 800 V and it is regulated by neon lamps L1 through L6. Diode D4 rectifies the high voltage and applies it to the cathode lead of the GM tube. The positive (+) bias on the GM tube is applied to the anode by way of load resistors R4 and R5. Each time a radioactive particle strikes the GM tube, it causes the gas inside to ionize. This ionization of the gas creates a pulse, which drives the piezo speaker and is also coupled by diode D5 to the base of Q2. Transistor Q2 is a pnp type and is used to “integrate” the pulses in conjunction with capacitor C6. That produces a dc voltage level, which is in proportion to the quantity of pulses arriving at the base of Q2. The collector of Q2 is connected through resistor R8 to the (+) terminal of the meter. The other side of the meter goes directly to (–) of the battery.

38

Hall Effect Circuits

The sources of the following circuits are contained in the Sources section, which begins on page 675. The figure number in the box of each circuit correlates to the entry in the Sources section.

THE TALKING COMPASS

TABLE 1—74S188 TRUTH TABLE

Directory	Input					Output								Decimal Equivalent
	A4	A3	A2	A1	A0	B0	B1	B2	B3	B4	B5	B6	B7	
North	L	H	L	H	H	0	0	0	0	0	0	0	1	1
N.W.	L	L	L	H	H	0	0	0	1	0	1	0	0	20
West	L	L	H	H	H	0	0	1	0	1	0	0	0	40
S.W.	L	L	H	H	L	0	0	1	1	1	1	0	0	60
South	L	H	H	H	L	0	1	0	1	0	0	0	0	80
S.E.	L	H	H	L	L	0	1	1	0	0	1	0	0	100
East	L	H	H	L	H	0	1	1	1	1	0	0	0	120
N.E.	L	H	L	L	H	1	0	0	0	1	1	0	0	140

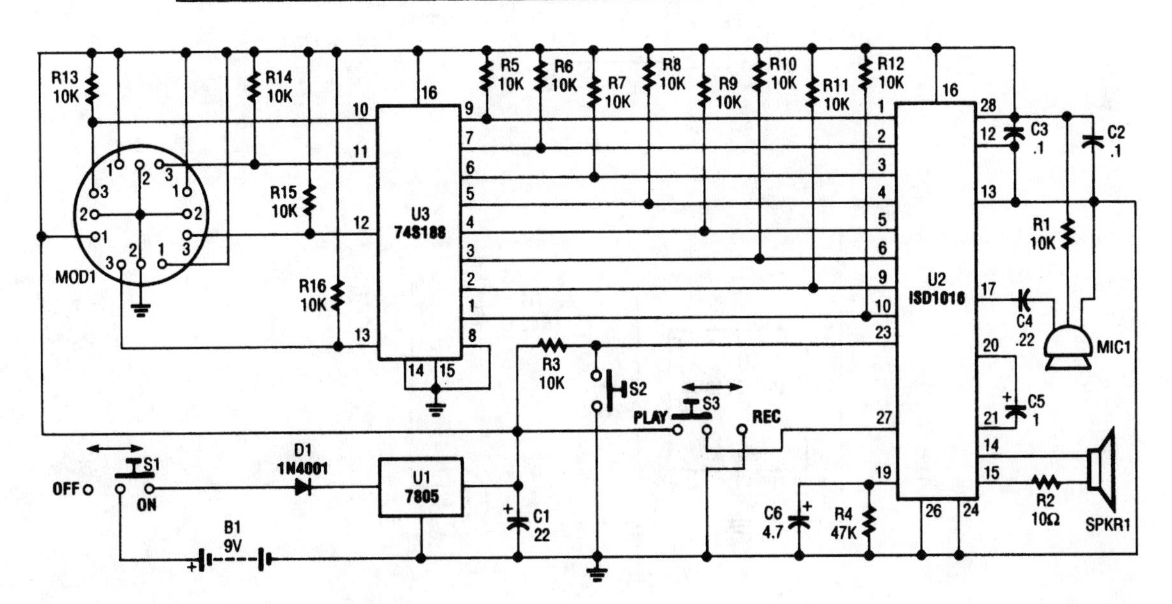

1993 ELECTRONICS HOBBYISTS HANDBOOK

FIG. 38-1

A talking compass is made up using a Hall-effect direction sensor (MOD1) and an ISD1016 analog audio storage device. It is possible to program eight two-second announcements, for each of the eight main compass directions.

The Talking Compass is comprised of a digital compass (MOD1), and ISD1016 analog storage device (U2), a 74S188 preprogrammed PROM (U3), and a handful of additional components.

UNUSUAL HALL-EFFECT OSCILLATORS

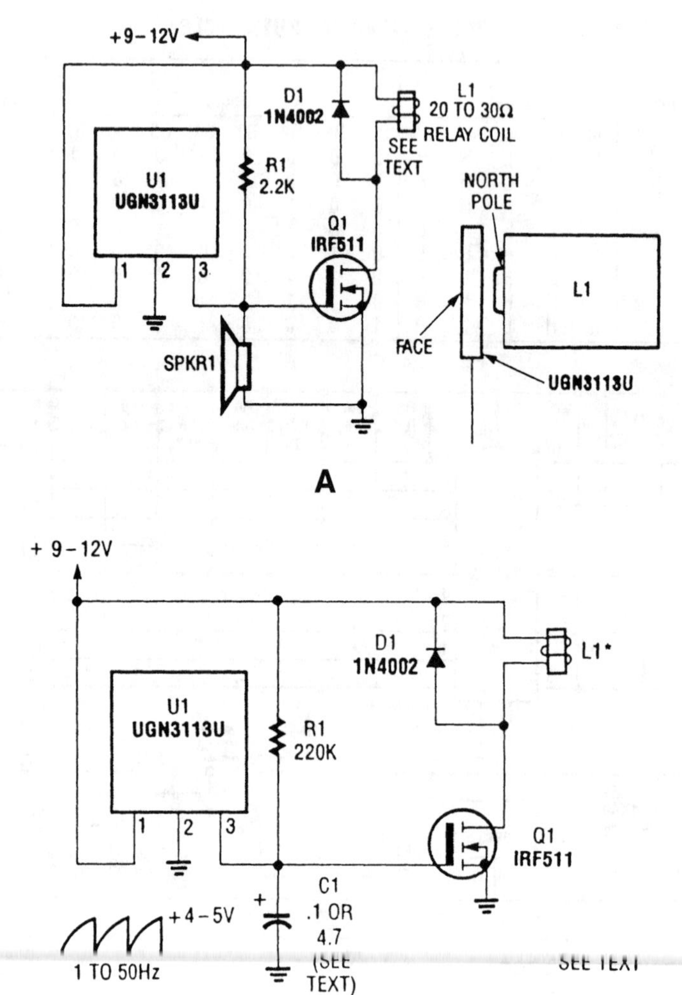

POPULAR ELECTRONICS *FIG. 38-2*

Although not intended for this application, Hall-effect switch can be used as the basis for a rather unusual oscillator. The oscillator can be reconfigured, as shown in Fig. B, to allow the circuit's oscillating frequency to be controlled via an RC network, comprised of R1 and C1.

39

Infrared Circuits

The sources of the following circuits are contained in the Sources section, which begins on page 675. The figure number in the box of each circuit correlates to the entry in the Sources section.

REMOTE-CONTROL ANALYZER

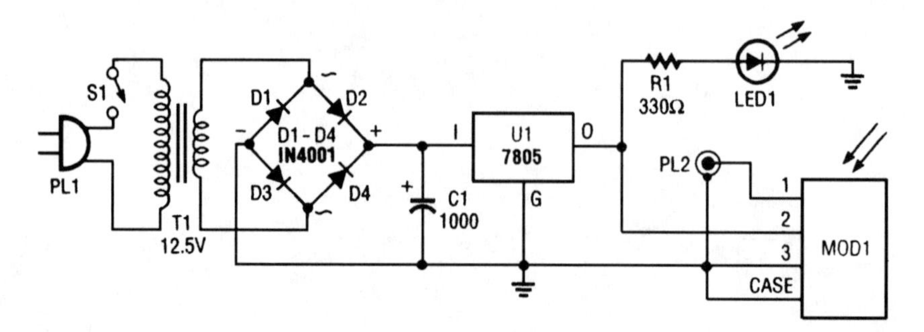

POPULAR ELECTRONICS *FIG. 39-1*

A schematic diagram for the remote analyzer is shown. The circuit is powered from a simple 5-V supply, consisting of PL1, S1, T1, a bridge rectifier (comprised of D1 through D4), capacitor C1, and a common 5-V regulator, U1. Switch S1 is the on/off control and is optional. The power-supply transformer used in the prototype is a 12.6-Vac unit, but any transformer that can supply at least 5.6-Vac will do. The 12.6-V unit was used solely because of its availability.

The output of T1 is full-wave rectified by diodes D1 through D4 and filtered by C1. The bumpy dc output from the capacitor is regulated down to 5 V by U1, a 7805 integrated regulator. LED1 acts as a power indicator to let you know that the circuit is active.

The 5-Vdc powers a GPIU52X infrared-detector module* (MOD1), which demodulates the 40-kHz carrier used by most infrared remotes. After demodulation, the resulting logic pulses are sent to an oscilloscope via PL2, a BNC connector.

*Radio Shack part #276-137

IR-PULSE-TO-AUDIO CONVERTER

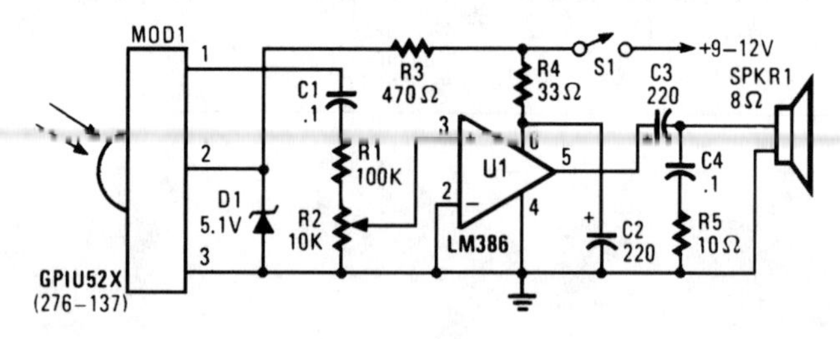

POPULAR ELECTRONICS *FIG. 39-2*

If your ear is good, you can use this IR-pulse-to-audio converter to troubleshoot infrared remote-controls. It is also a good project for detecting infrared-light sources. A photo cell module (Radio Shack P/N 276-137) detects IR radiation and drives audio IC U1. This circuit is useful for troubleshooting IR remote controls.

IR-CONTROLLED REMOTE A/B SWITCH

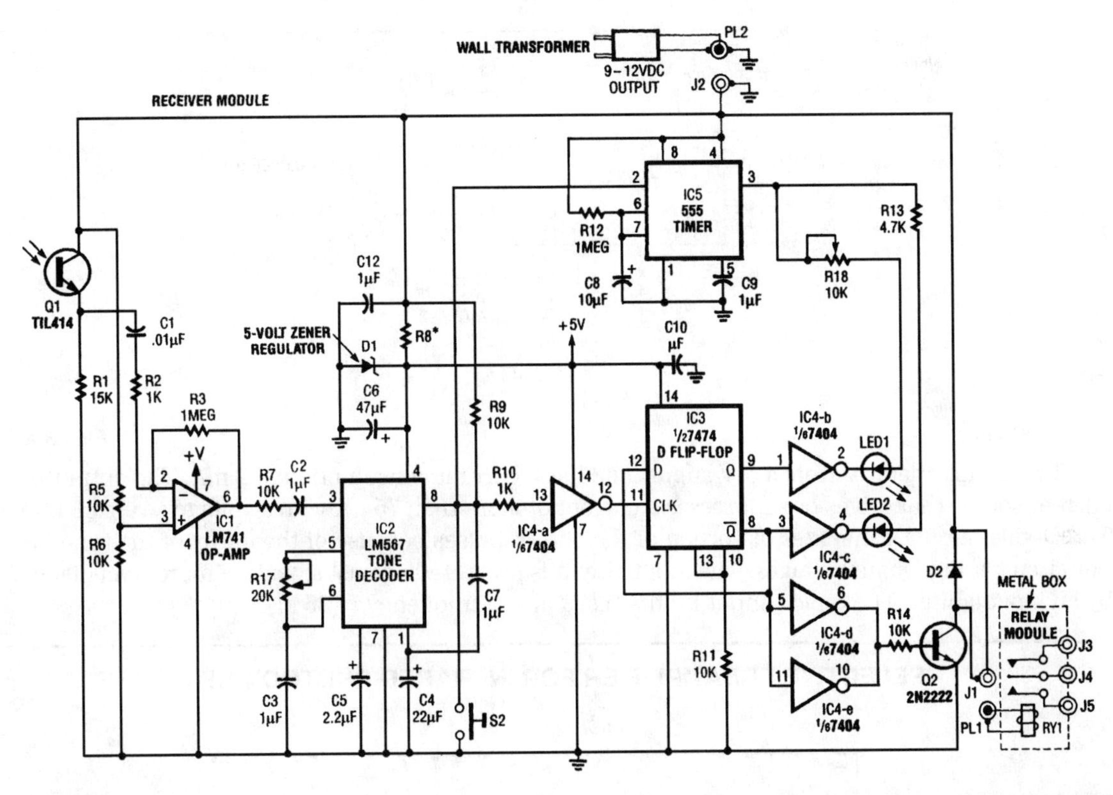

RADIO ELECTRONICS *FIG. 39-3*

Useful for A/B control, the IR receiver shown controls a relay from an infrared beam that has a pulsed tone-modulated signal. Q1 is the photo receptor feeding op amp IC1, tone decoder IC2, and flip-flop IC3. IC5 turns off the indicator LEDs after about 15 seconds.

SIMPLE IR DETECTOR

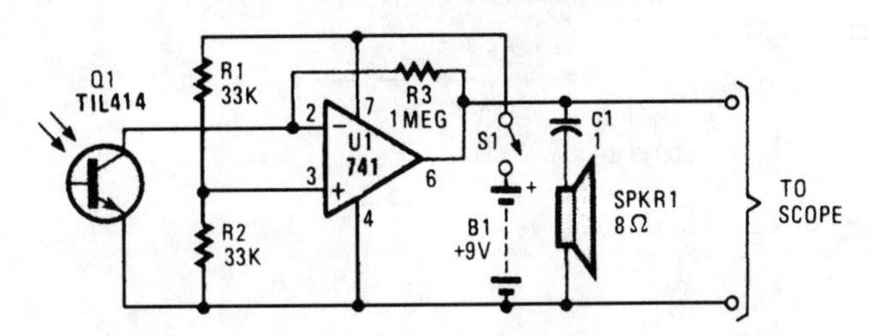

POPULAR ELECTRONICS *FIG. 39-4*

Useful for IR detection, this circuit uses an op amp of the 741 family (or similar) to detect and amplify IR pulses.

INFRARED RECEIVER

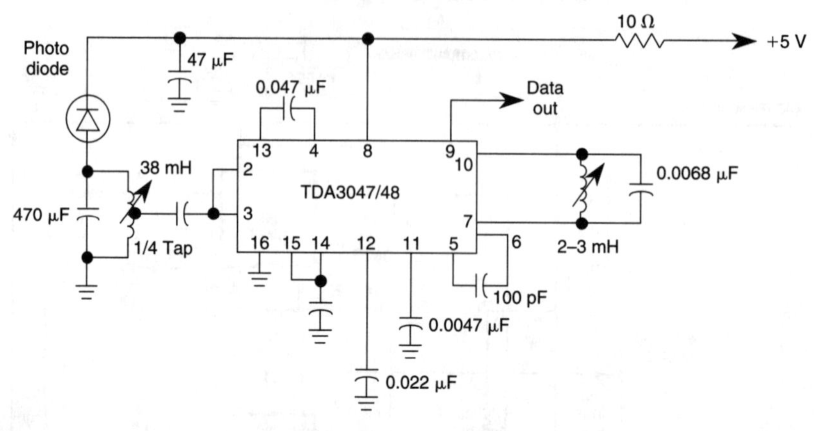

WILLIAM SHEETS

FIG. 39-5

The circuit operates from a 5-V supply and has a current consumption of 2 mA. The output is a current source that drives or suppresses a current of more than 75 µA with a voltage swing of 4.5 V. The Q-killer circuit eliminates distortion of the output pulses because of the decay of the tuned input circuit at high input voltages. The input circuit is protected against signals of more than 600 mV by an input limiter. The typical input is an AM signal at a frequency of 36 kHz.

SELECTIVE PREAMPLIFIER FOR INFRARED PHOTODIODE

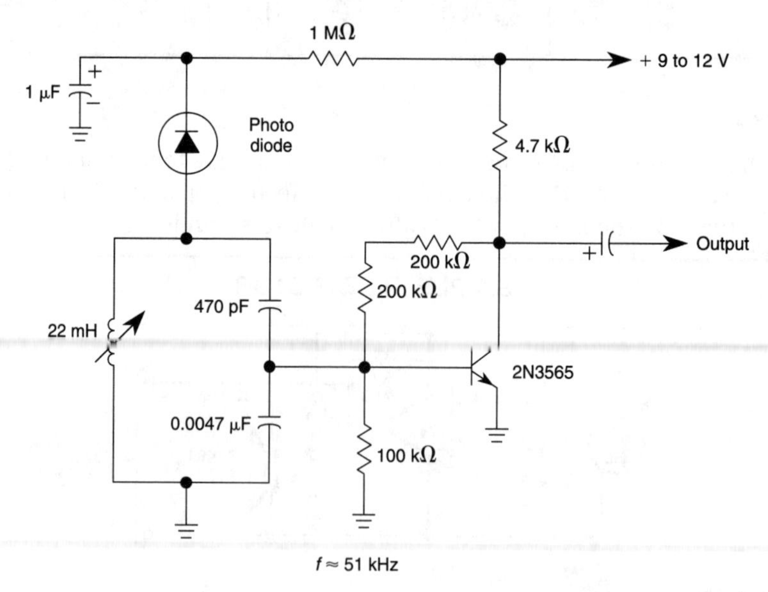

WILLIAM SHEETS

FIG. 39-6

The circuit uses a tuned circuit to achieve frequency selection. Values are for operation at about 51 kHz. The 2N3565 amplifies the output developed by the tuned circuit.

WIRELESS IR HEADPHONE TRANSMITTER

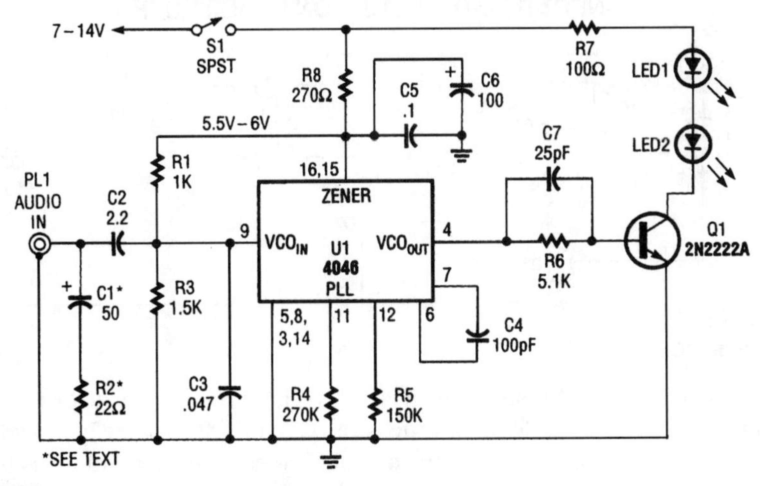

POPULAR ELECTRONICS *FIG. 39-7*

The transmitter for the wireless headphones is built around a CD4046 CMOS phase-locked loop, coupled with a driver transistor, and a pair of infrared LEDs. Although the CD4046 is comprised of two phase comparators, a voltage-controlled oscillator (or VCO), a source follower, and a zener reference, only its VCO is used in this application.

WIRELESS IR HEADPHONE RECEIVER

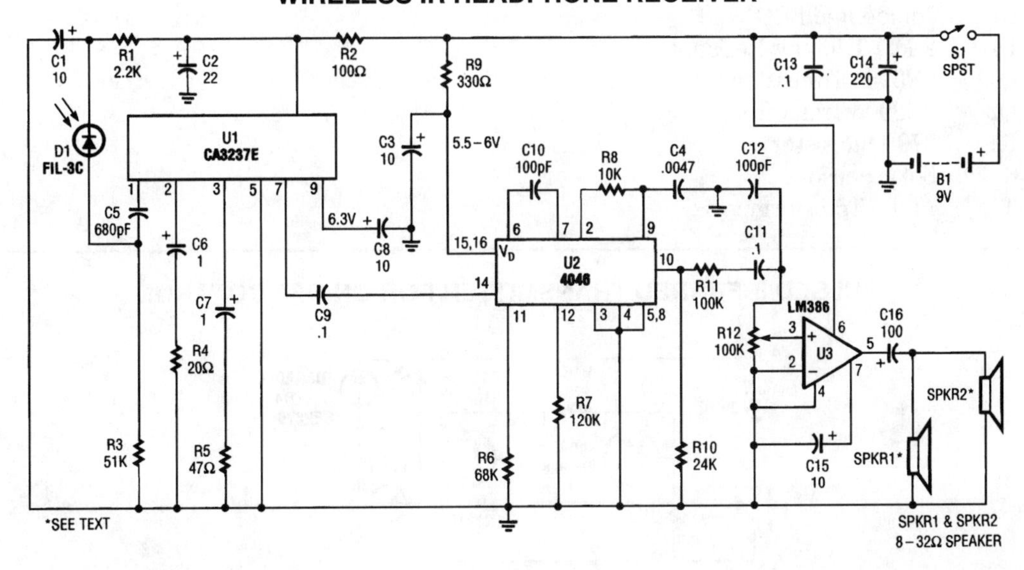

POPULAR ELECTRONICS *FIG. 39-8*

IR detector diode D1 intercepts the IR signal at around 40 kHz and feeds it from U1, a high-gain preamp, to PLL, U2, a 4046 configured to serve as an FM detector. U3 is an audio amplifier that feeds a pair of headphones or a speaker.

INFRARED REMOTE-CONTROL TESTER

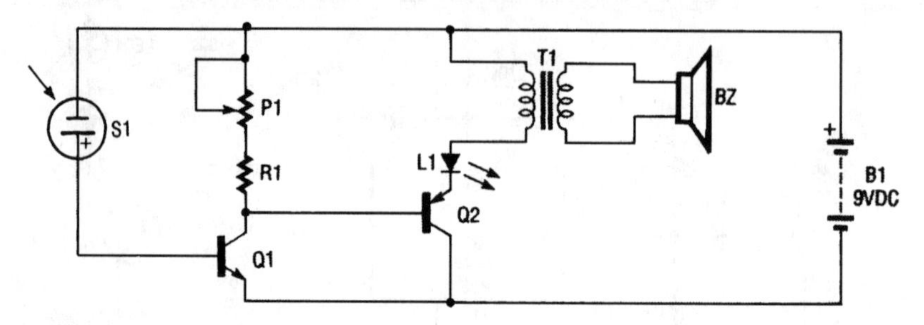

1991 PE HOBBYIST HANDBOOK **FIG. 39-9**

The infrared remote-control tester uses a sensitive PN-type solar sensor that is connected directly to a Darlington amplifier made up of transistors Q1 and Q2. Biasing is provided by R1 and P1, a variable resistor that serves as a sensitivity control. The collector lead of Q1 is the output lead of the Darlington amp, and it is connected to a red LED and the primary of transformer T1. The function of T1 is to convert the low-voltage output signal to a level high enough to drive a small piezo disc. That disc makes a clicking sound when the sensor picks up an infrared signal that is varying in frequency or amplitude. The infrared sensor will also pick up visible light. The use of an IR filter (Wratton #87) is recommended.

BZ	Piezo Disc
L1	Jumbo Red LED
P1	2-MΩ Trimmer Resistor
Q1	2N3904 Transistor
Q2	2N3906 Transistor
R1	270-Ω Resistor
S1	Solar Sensor
T1	Audio Transformer

PULSED INFRARED TRANSMITTER FOR ON/OFF CONTROL

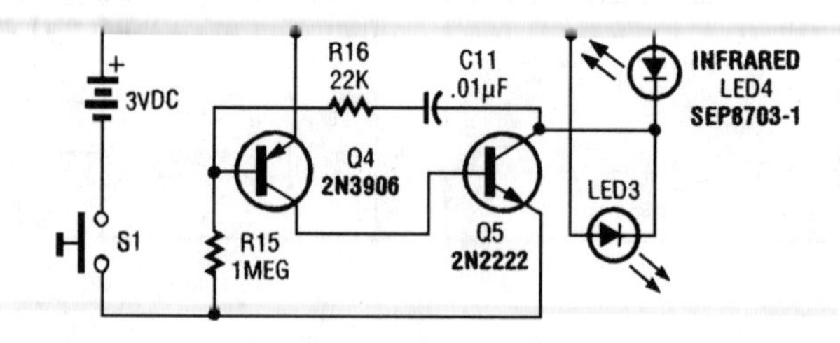

RADIO ELECTRONICS **FIG. 39-10**

This transmitter consists of an oscillator and LEDs. It generates a pulsed tone of around 850 Hz.

VERY SIMPLE IR REMOTE-CONTROL CIRCUIT

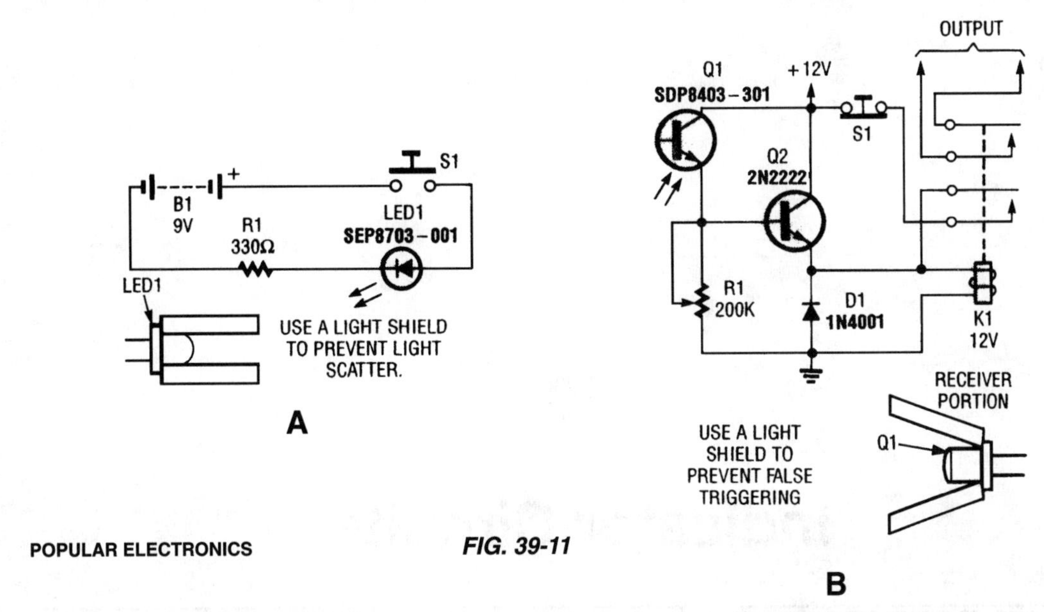

POPULAR ELECTRONICS *FIG. 39-11*

Here is a complete IR remote-control system that consists of a simple transmitter (A) and an equally simple receiver (B).

IR RECEIVER

9 Vdc
To LED
buzzer, etc.
4.7 k
1 Meg
1 k
741
Any
silicon
phototransistor

ELECTRONICS NOW *FIG. 39-12*

This circuit is just about the simplest IR receiver you can build. The parts are cheap, the layout is not critical, and a 9-V battery will last a long time.

REMOTE-CONTROL TESTER

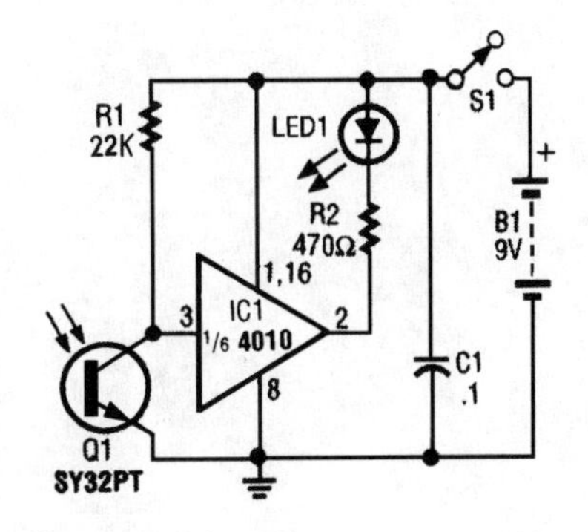

ELECTRONICS NOW *FIG. 39-13*

The IR Tester circuit lets you know if the button you press on a remote control is working. Q1 is a photo transistor that is activated by IR energy.

40

Indicator Circuits

The sources of the following circuits are contained in the Sources section, which begins on page 675. The figure number in the box of each circuit correlates to the entry in the Sources section.

POLARITY INDICATOR

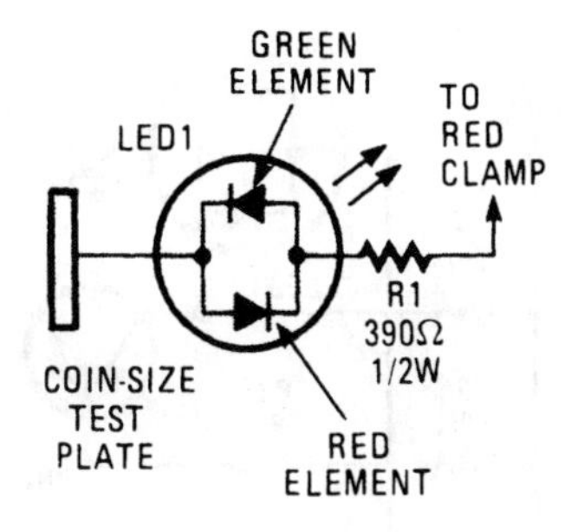

POPULAR ELECTRONICS ***FIG. 40-1***

This circuit consists of a tri-color LED, a resistor, wire, and a coin-size test plate. You will have to build two such circuits—one for each black clamp on a set of auto battery jumper cables. The author installed the circuits inside the black clamps themselves using lengths of wire to make the connections to the red clamps.

The first step is to connect one red clamp to what you believe is the positive post on the okay battery. Then, touch the test plate on the black clamp at the end of the cable to the negative terminal on the good battery. The LED will light red if the red clamp is on the wrong terminal. If so move the clamp to the other post and check again. If all is well, the LED will light green. Pick up the other black clamp and connect it to the remaining post on the good battery.

Connect the remaining red clamp to what you assume to be the positive terminal on the bad battery. Now, touch the test plate on the remaining clamp to the engine block or a bare area on the dead car's frame. If the LED appears or doesn't glow, switch the red clamp to the other terminal and test again. When the LED glows green, attach the black clamp to the car's frame (which will prevent any sparks from occurring near the battery). When you remove the clamps, take the clamps off in reverse order to avoid sparks.

BI-COLOR INDICATOR

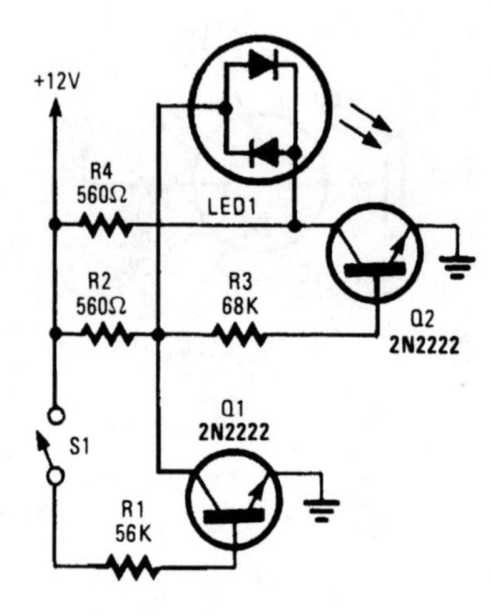

POPULAR ELECTRONICS

FIG. 40-2

With S1 open, base bias is supplied to Q2 through a voltage divider (formed by R2 and R3), thus turning on the green element in the LED. That indicates that power is being supplied to the project. If you close S1, current through R1 biases Q1 on, thereby grounding the voltage divider and turning off Q2. That reverses the flow of current through the LED, which causes its red element to light. That indicates that the circuit is under power and S1 (really a DPDT switch), whose remaining section controls another circuit, is active. In this circuit, a bi-color LED is used to indicate when a circuit is under power and the status of S1. In that way, the LED does the job of two indicators.

41

Instrumentation Amplifier Circuits

The sources of the following circuits are contained in the Sources section, which begins on page 675. The figure number in the box of each circuit correlates to the entry in the Sources section.

LMC6062 INSTRUMENTATION AMPLIFIER

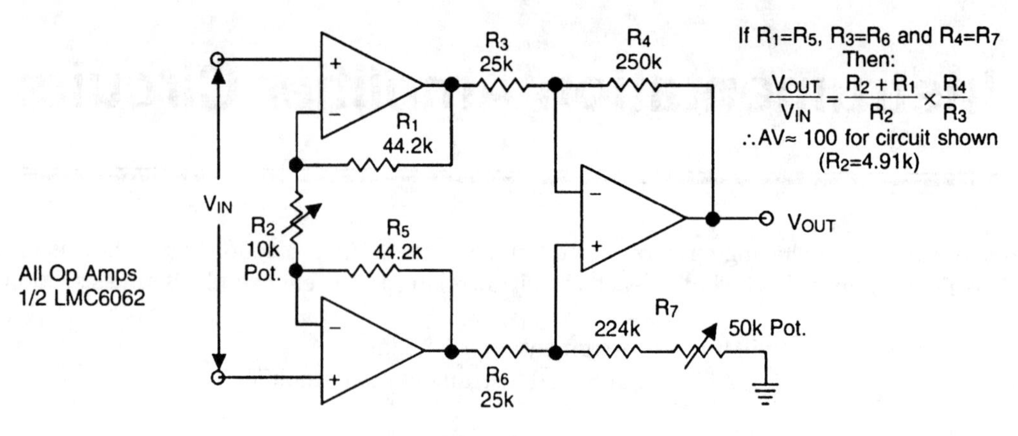

NATIONAL SEMICONDUCTOR

FIG. 41-1

Useful for +5-V single-supply applications, this op amp circuit features low drain (around 1 mA), high input resistance ($10^{14}\ \Omega$), and low bias current ($\approx 10^{-14}$A).

LM6218 HIGH-SPEED INSTRUMENTATION AMPLIFIER

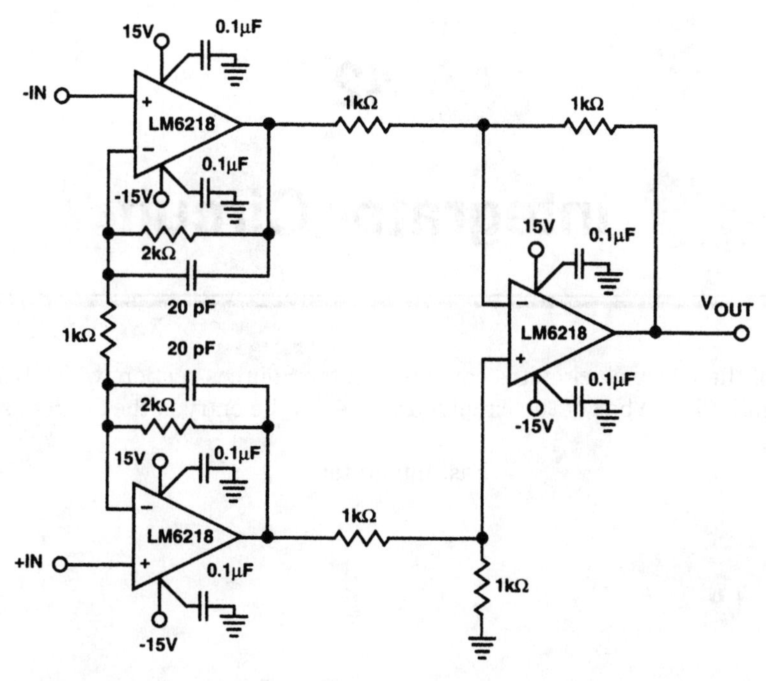

NATIONAL SEMICONDUCTOR *FIG. 41-2*

This amplifier features 400-μsec settling time (to 0.01%), 140-V/μsec slow rate, and 17-MHz gain-bandwidth product. The supply voltage can be ±5 to ±20 V.

42

Integrator Circuit

The source of the following circuit is contained in the Sources section, which begins on page 675. The figure number in the box of the circuit correlates to the entry in the Sources section.

Fast Integrator

FAST INTEGRATOR

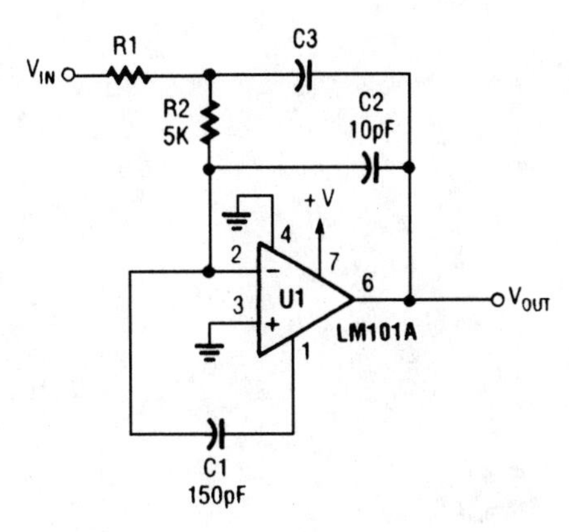

V_{OUT} is the integral of $V1$ in this circuit.

$$\frac{V_{OUT}}{V_{IN}} \approx \frac{1}{C_3} \frac{V_{IN}(A)}{R} \text{ dt.}$$

POPULAR ELECTRONICS

FIG. 42-1

43

Intercom Circuits

The sources of the following circuits are contained in the Sources section, which begins on page 675. The figure number in the box of each circuit correlates to the entry in the Sources section.

ONE-WAY VOICE-ACTIVATED INTERCOM

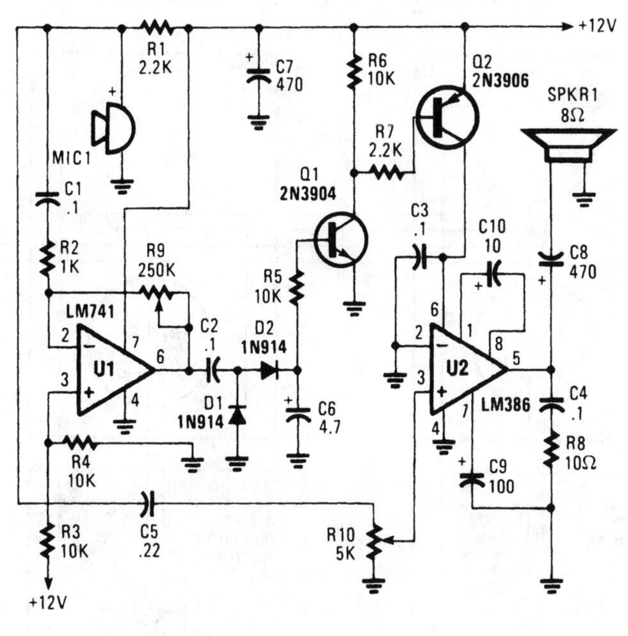

POPULAR ELECTRONICS *FIG. 43-1*

An omnidirectional electret microphone can be used to pick up the sound and convert it into an electrical signal. The output of the microphone is fed along two paths. In the first path, the signal is sent to the inverting input at pin 6. In the second path, the microphone signal is fed to the non-inverting input of U2, where it is amplified and output to the speaker, SPKR1.

VERY SIMPLE TELEPHONE INTERCOM CIRCUIT

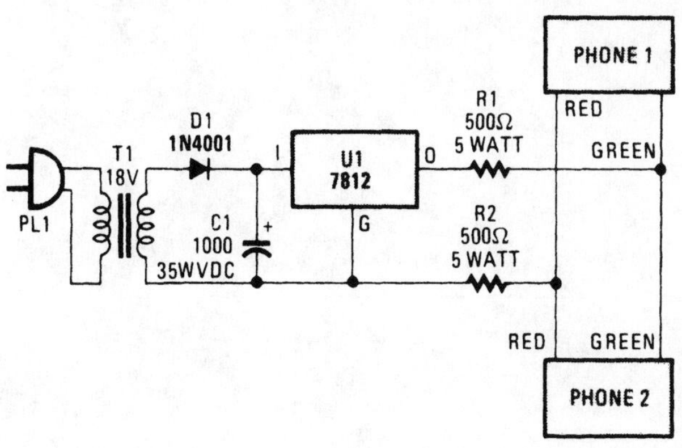

POPULAR ELECTRONICS *FIG. 43-2*

Two telephones can be used as an intercom by using this circuit. Older style rotary phones that are nonelectronic might work best in this application. Also, handsets only might be powered this way.

TELEPHONE INTERCOM

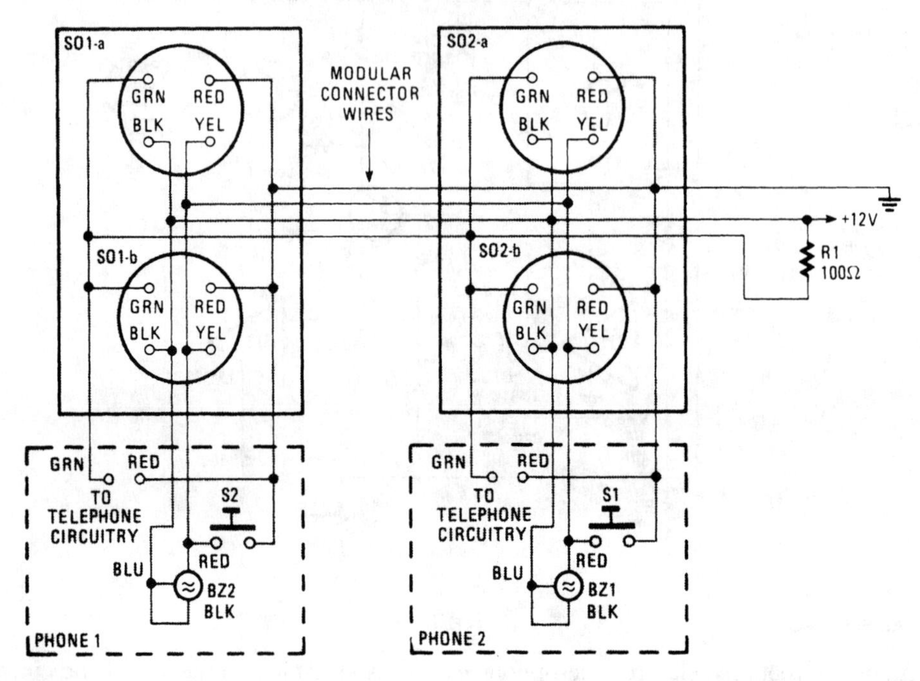

POPULAR ELECTRONICS ***FIG. 43-3***

An intercom using dual-modular wall jacks is shown in this circuit. If the wires are available in the home telephone cable, this system can be installed with little trouble.

44

Interface Circuits

The sources of the following circuits are contained in the Sources section, which begins on page 675. The figure number in the box of each circuit correlates to the entry in the Sources section.

AUDIO-TO-ADC INTERFACE

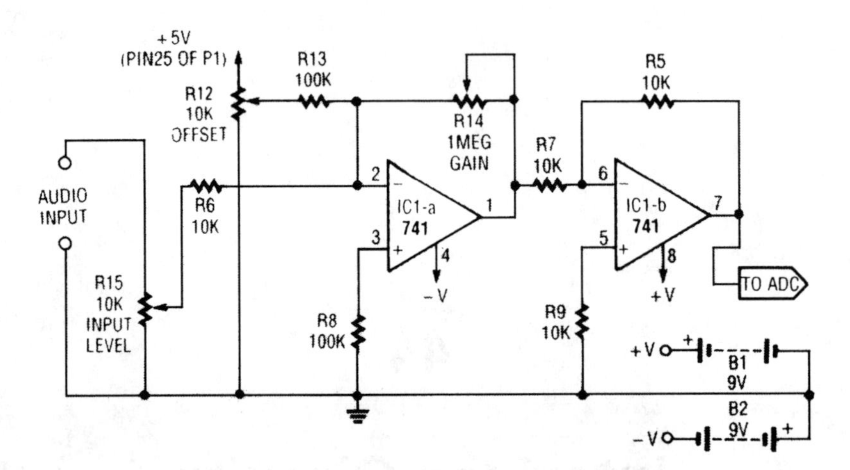

RADIO-ELECTRONICS

FIG. 44-1

This simple general-purpose driver for an analog/digital converter uses two 741 IC devices with adjustable gain and offset. Other op amps might be substituted, but some circuit adjustments might be needed.

PROCESS-CONTROL INTERFACE

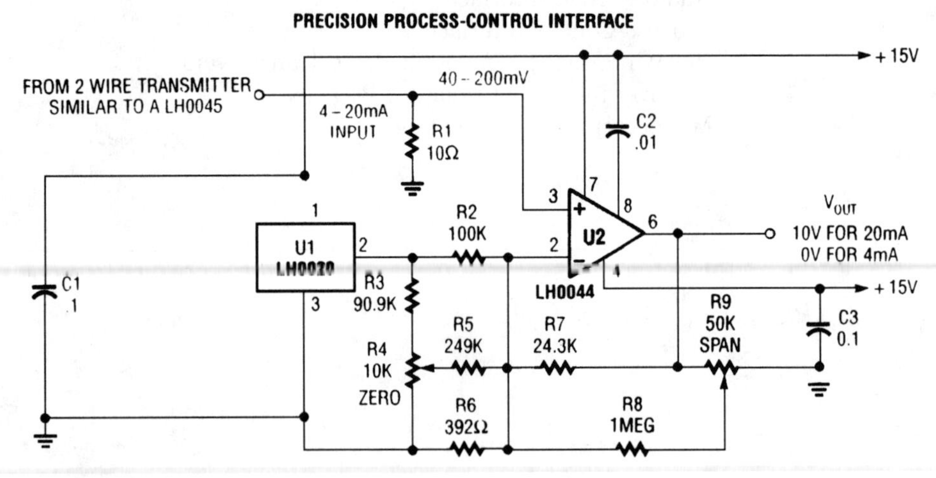

POPULAR ELECTRONICS

FIG. 44-2

This circuit can be used to interface a 2-wire transmitter/sensor combination to an external device or measurement setup.

RELAY INTERFACE FOR AMATEUR RADIO TRANSCEIVERS

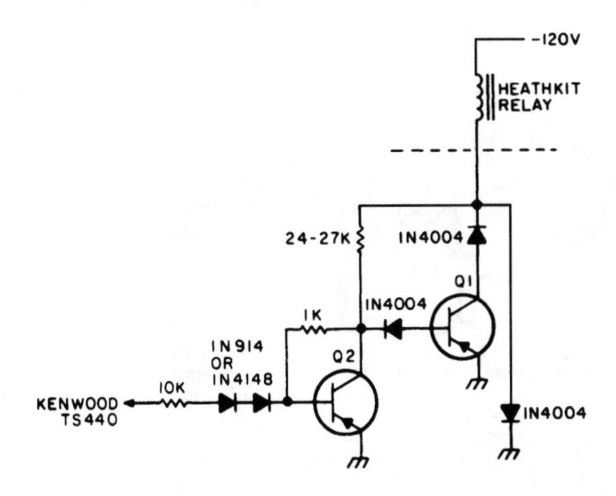

73 AMATEUR RADIO ***FIG. 44-3***

The relay power in the linear is obtained from the –120-V bias supply, and the transmit keying output from the Kenwood is +12 V at 10 mA maximum. The key ingredient in the circuit is the pnp driver transistor, which must be capable of handling at least 150 V at about 250 mA.

RECEIVER-INTERFACE CIRCUIT FOR PREAMPS

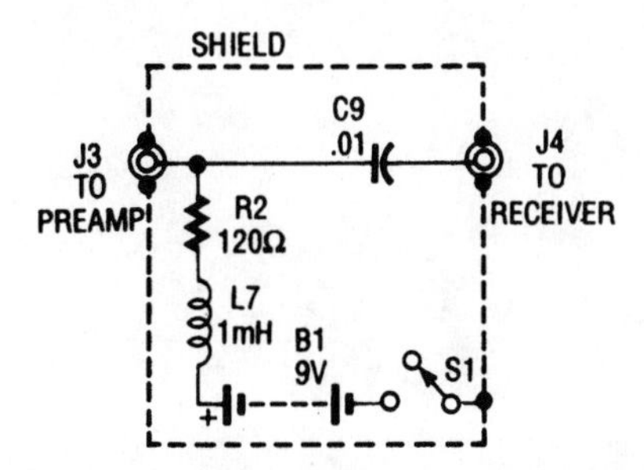

POPULAR ELECTRONICS ***FIG. 44-4***

The purpose of the receiver/interface circuit is to pass RF to the receiver through capacitor C9, while adding dc power to the feedline through R2 and RF choke L7.

MICROCOMPUTER-TO-TRIAC INTERFACE

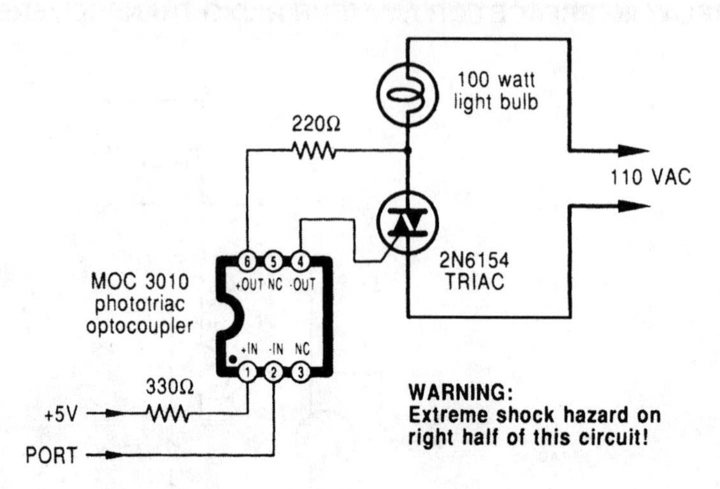

RADIO-ELECTRONICS *FIG. 44-5*

A microcomputer-to-triac interface uses a phototriac optoisolator to let safety-isolated logic signals directly control high-power loads. Depending on the input waveforms and the load, this circuit can be used in either an on/off switch or a proportional phase control. A low input powers the lamp.

45

Inverter Circuits

The sources of the following circuits are contained in the Sources section, which begins on page 675. The figure number in the box of each circuit correlates to the entry in the Sources section.

250-W INVERTER

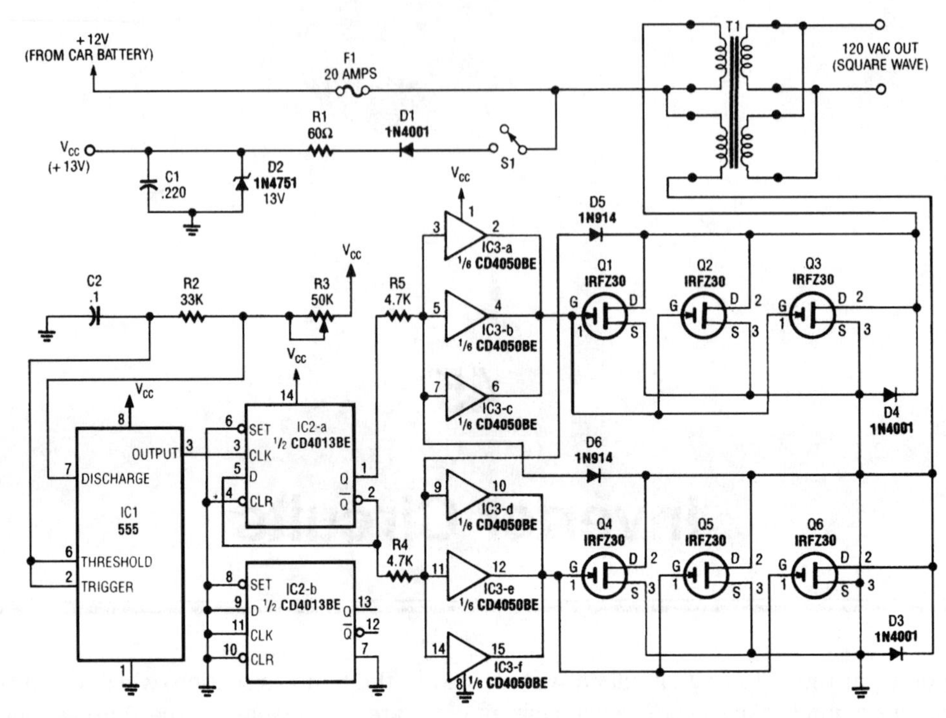

ELECTRONICS NOW *FIG. 45-1*

A 555 timer (IC1) generates a 120-Hz signal that is fed to a CD4013BE flip-flop (IC1-a), which divides the input frequency by two to generate a 60-Hz clocking frequency for the FET array (Q1 through Q6). Transformer T1 is a 12-/24-V center-tapped 60-Hz transformer of suitable size.

DIGITAL INVERTER

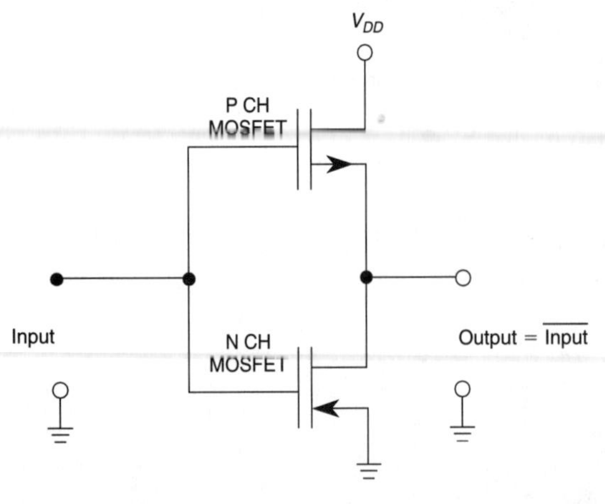

A CMOS digital inverter is formed by connecting two MOSFETS, as shown.

WILLIAM SHEETS *FIG. 45-2*

dc-to-ac INVERTER

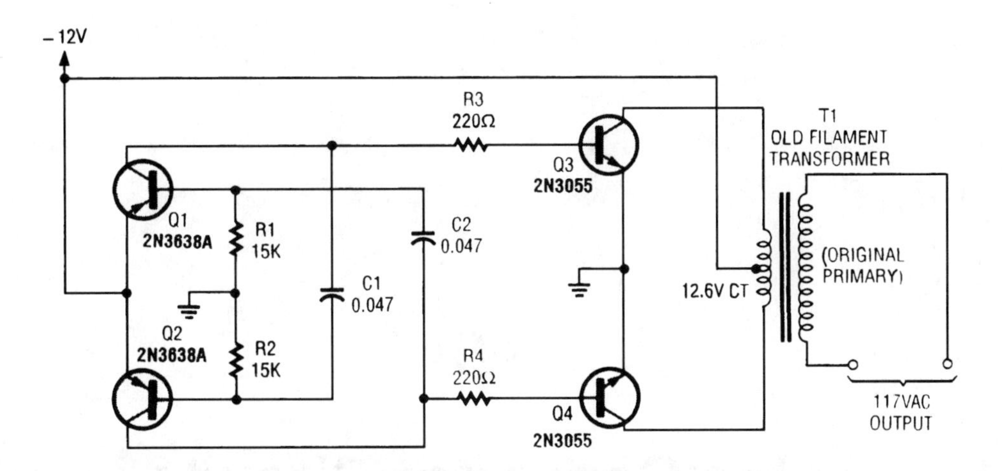

POPULAR ELECTRONICS *FIG. 45-3*

A multivibrator circuit drives a pair of 2N3055 power transistors. T1 is a 12.6-V CT filament transformer with a 120-V primary.

POWER MOSFET INVERTER

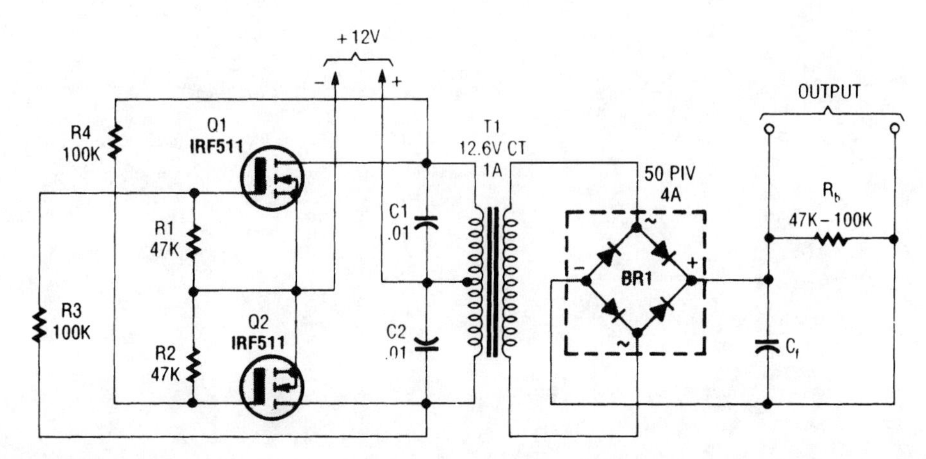

POPULAR ELECTRONICS *FIG. 45-4*

T1 is a suitable transformer for the voltage desired, with a 12.6-V CT winding.

46

Ion Generator Circuit

The source of the following circuit is contained in the Sources section, which begins on page 675. The figure number in the box of the circuit correlates to the entry in the Sources section.

Negative Ion Generator

NEGATIVE ION GENERATOR

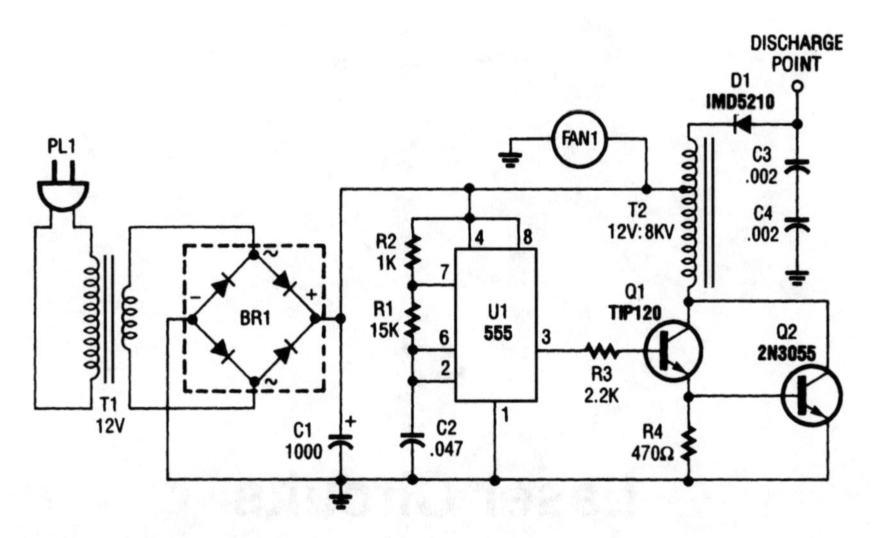

1993 ELECTRONICS HOBBYIST HANDBOOK *FIG. 46-1*

This oscillator-driver induces a high voltage in the windings of T2.

47

Laser Circuits

The sources of the following circuits are contained in the Sources section, which begins on page 675. The figure number in the box of each circuit correlates to the entry in the Sources section.

EFFICIENT LASER SUPPLY

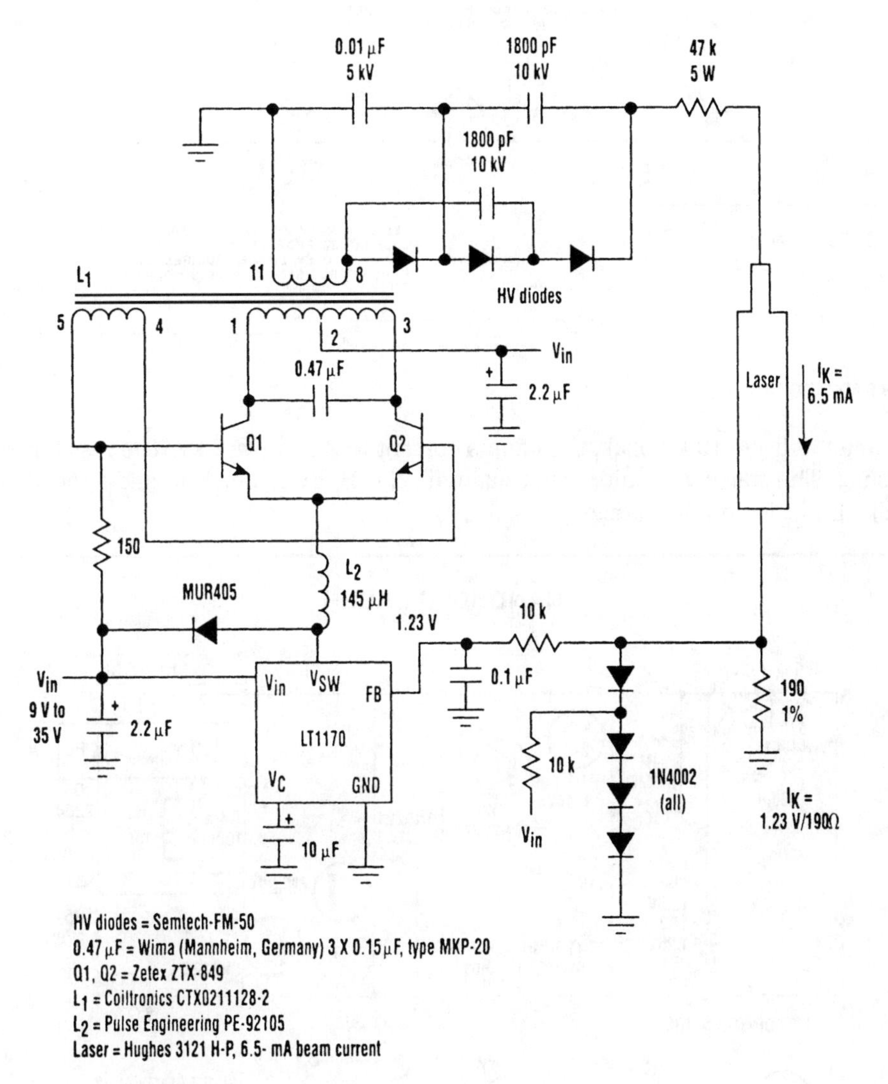

ELECTRONIC DESIGN

FIG. 47-1

Driving Helium-Neon Lasers can be simplified considerably using this power-supply configuration. When power is applied, the laser doesn't conduct and the voltage across the 190-Ω resistor is zero. However, a resonant circuit and a voltage tripler then produces over 10 kV to turn on the laser.

LASER POWER SUPPLY AND STARTING CIRCUIT

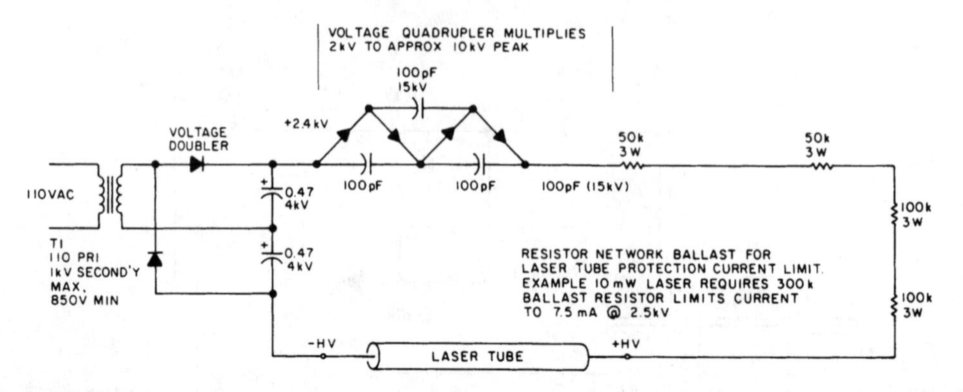

73 AMATEUR RADIO TODAY ***FIG. 47-2***

This circuit delivers 10 kV peak, then limits current to 7.5 mA @ 2 kV. The resistors shown provide ballasting. The starting circuit cannot maintain the 10 kV under load and appears as a series-pass circuit with little drop in voltage.

HANDHELD LASER

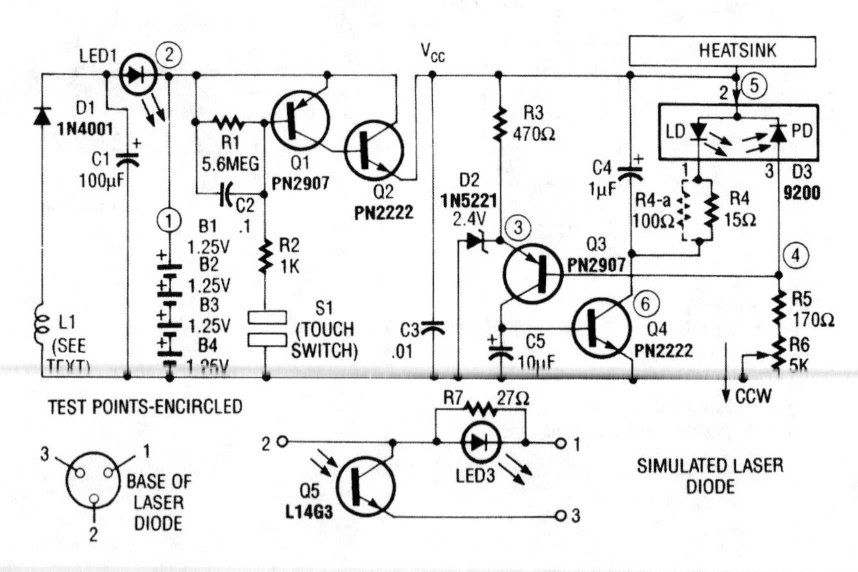

1992 R-E EXPERIMENTERS HANDBOOK ***FIG. 47-3***

A laser diode TOLD9200 (Toshiba) is used as a source of laser light. Q3, Q2, and S1 form a touch switch to control the laser. L1 is an RF pickup coil to pick up energy from an RF-type battery charger. It is 10 turns of #18 wire on a ½" diameter.

HIGH-VOLTAGE POWER SUPPLY

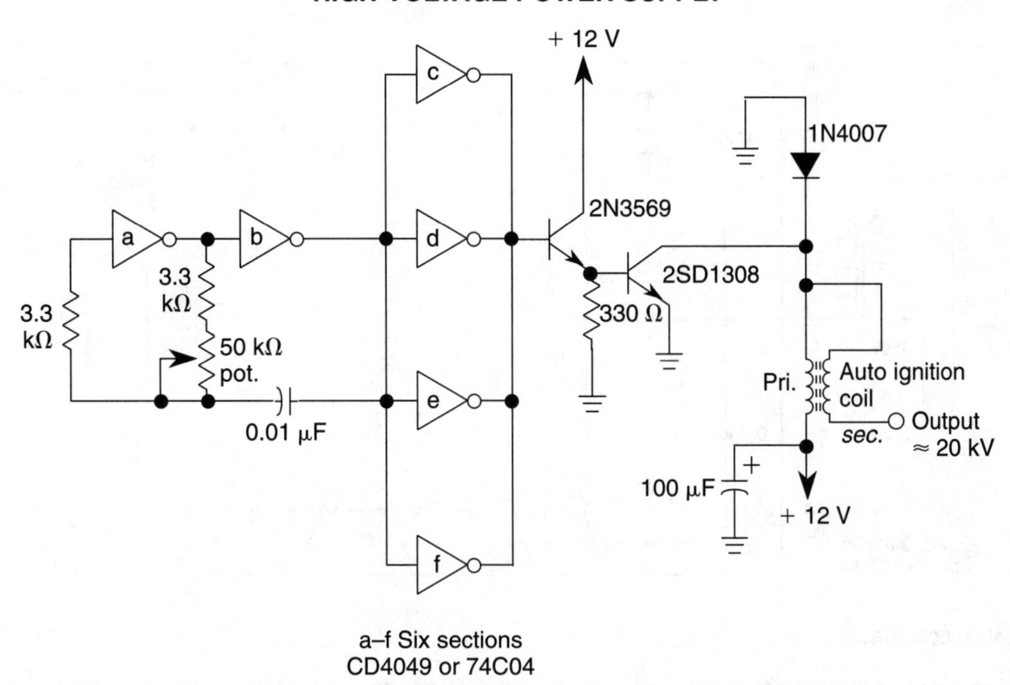

WILLIAM SHEETS ***FIG. 47-4***

The high-voltage power supply is a CMOS-based oscillator that pulses a high-voltage ignition transformer. The transformer output is around 20 kV.

FANTASTIC SIMULATED LASER

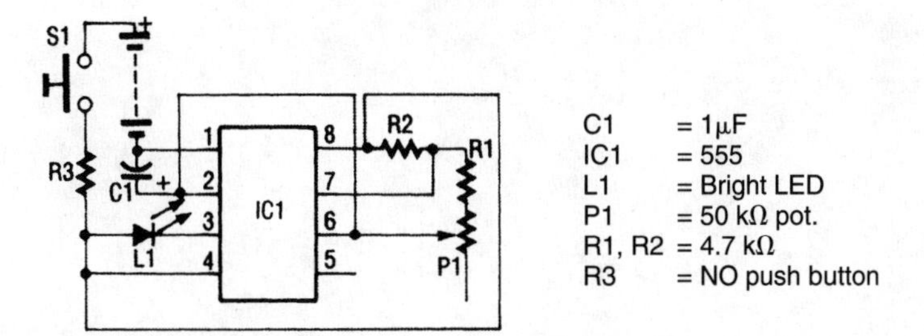

C1	= 1μF
IC1	= 555
L1	= Bright LED
P1	= 50 kΩ pot.
R1, R2	= 4.7 kΩ
R3	= NO push button

1991 PE HOBBYIST HANDBOOK ***FIG. 47-5***

The circuit uses a 555 timer IC to power an ultrabright LED. The output is a pulsing red light that can be projected using lenses. An ultrabright Stanley LED, capable of 300-millicandle output, is tied to pin 3 of the 555 timer IC. That IC has been configured as an astable multivibrator. The frequency of this multivibrator is controlled by R1, R2, C1, and P1. You can vary the frequency by adjusting P1, which changes the output from a slow blinking to a fast pulsating light. Resistor R3 is used to limit the current flowing into the circuit to a safe value, to prevent the LED and the IC from burning out. Switch S1 applies power to the circuit when its button is pressed.

LASER POWER SUPPLY

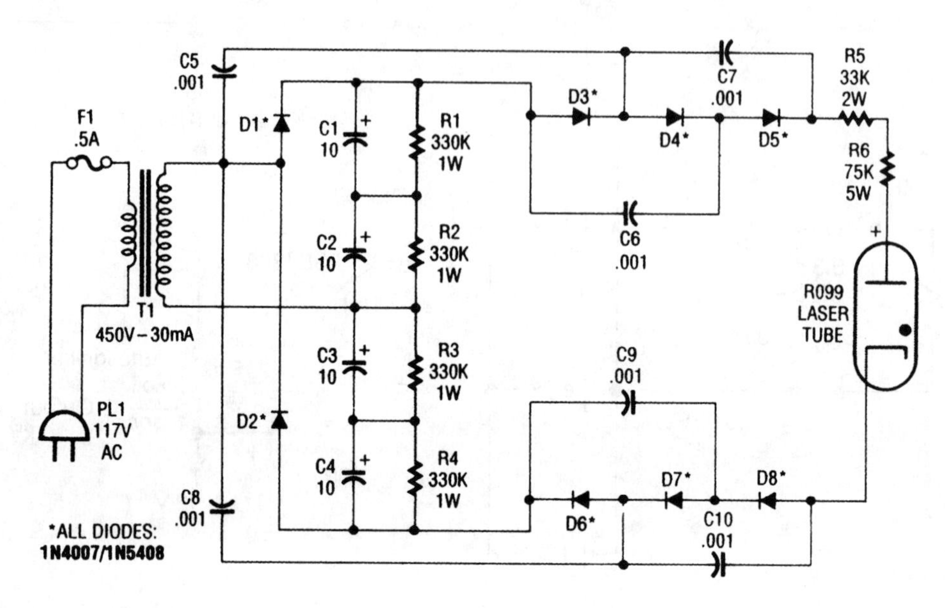

POPULAR ELECTRONICS ***FIG. 47-6***

This supply generates an initial high voltage for ignition purposes. After ignition, the supply generates about 1300 to 1500 V. If a higher ignition voltage (than the 6000 V supplied) is necessary, more multiplier stages can be added to D5 and D8.

48

Lie Detector Circuit

The source of the following circuit is contained in the Sources section, which begins on page 675. The figure number in the box of the circuit correlates to the entry in the Sources section.

Simple Lie Detector

SIMPLE LIE DETECTOR

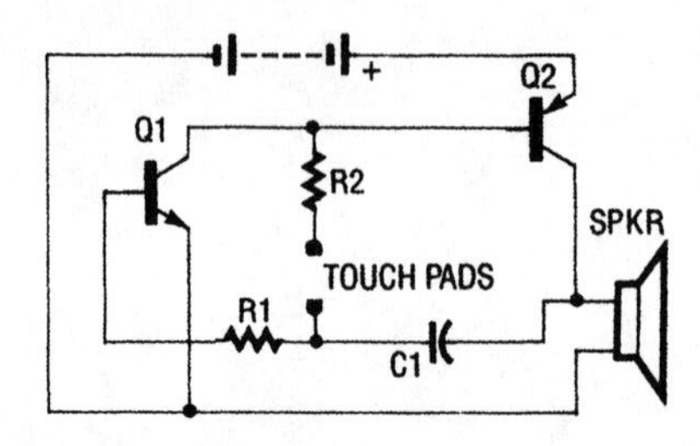

1991 PE HOBBYIST HANDBOOK ***FIG. 48-1***

The circuit uses a two-transistor direct-coupled oscillator that has a frequency determined by C1, R2, and the (skin) resistance across the touch pads. Since C1 and R2 are fixed values, only the skin resistance across the touch pads can vary the sound of the oscillator. To sustain oscillations, C1 feeds a portion of the output from Q2 back to the input of Q1 through resistor R1.

Transistor Q1 is an npn type and transistor Q2 is a pnp type. The output of Q2 is fed into a small speaker. The circuit relies on the fact that the human skin conducts electricity.

C1	0.01-µF Capacitor
Q1	2N3904 Transistor
Q2	2N3906 Transistor
R1	4.7 kΩ Resistor
R2	82 kΩ Resistor

49

Light Beam Communication Circuits

The sources of the following circuits are contained in the Sources section, which begins on page 675. The figure number in the box of each circuit correlates to the entry in the Sources section.

MODULATED LIGHT TRANSMITTER

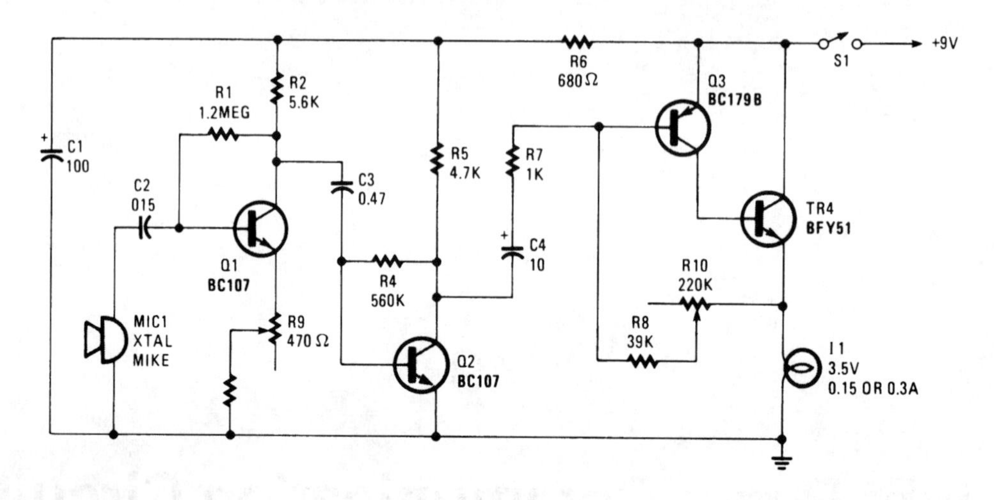

POPULAR ELECTRONICS

FIG. 49-1

A light-bulb filament can be modulated with audio as a method of optical transmission. Amplifier Q1/Q2/Q3 drives emitter-follower TR4. Adjust R10 for the *Q* point (light bulb) giving best results. It should have a filament with low thermal inertia for best audio responses.

MODULATED LIGHT RECEIVER

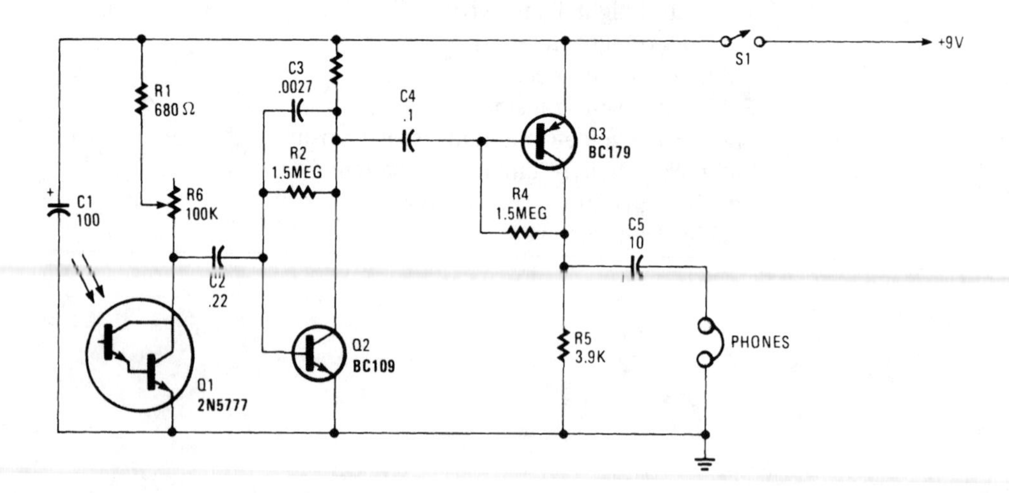

POPULAR ELECTRONICS

FIG. 49-2

Using a phototransistor, this receiver will detect and demodulate a modulated light beam. R6 affects sensitivity.

FM LIGHT BEAM RECEIVER

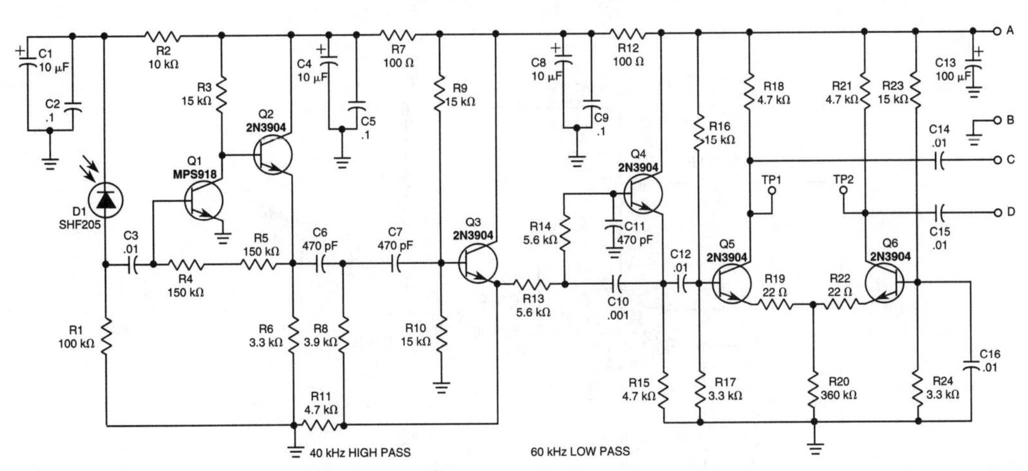

ELECTRONICS NOW ***FIG. 49-3***

This receiver will pick up IR or light beams that are frequency modulated on a 50-kHz carrier. Q2/Q1/Q3/Q4 from an active filter and amplifier and differential amp Q5/Q6 provide more gain.

FM LIGHT-BEAM TRANSMITTER

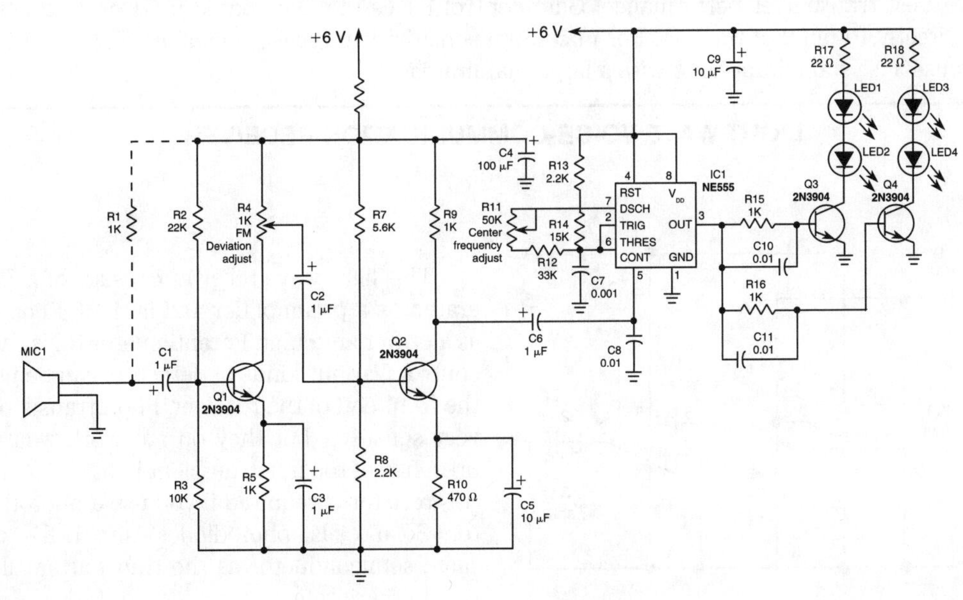

ELECTRONICS NOW ***FIG. 49-4***

This transmitter uses two-stage amplifier Q1/Q2 to frequency modulate an NE555 (configured as a VCO) operating at about 50 kHz. The resultant FM-modulated pulse train is converted to light pulses via LED1 through LED4, driven by Q3 and Q4.

LIGHT-WAVE VOICE-COMMUNICATION TRANSMITTER

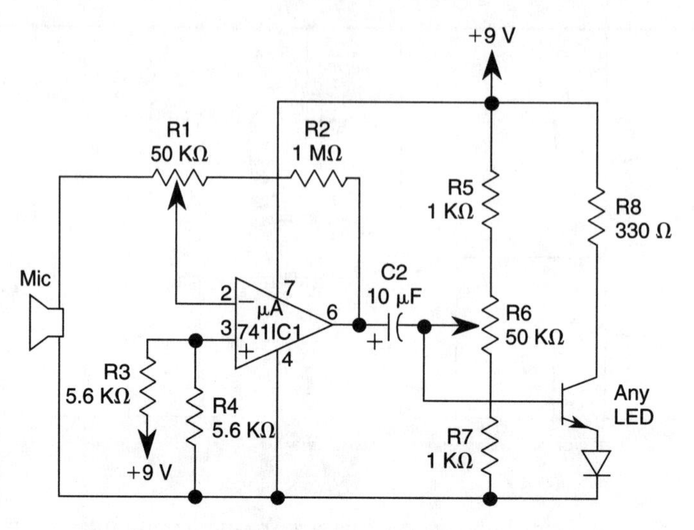

WILLIAM SHEETS

FIG. 49-5

This transmitter uses a 741 op amp as a high-gain audio amplifier, which is driven by a microphone. The output of the 741 is coupled to Q1, which serves as the driver for a LED. Potentiometer R1 is the amplifier's gain control. Miniature trimmer resistor R6 permits adjustment of the base bias of Q1 for best transmitter performance. Gain control R1 can be eliminated if C1 and R2 are connected directly to pin 2 of the 741. For maximum sensitivity, increase the value of R_2 from 1 to 10 MΩ and use a crystal microphone with a large diaphragm.

LIGHT-WAVE VOICE-COMMUNICATION RECEIVER

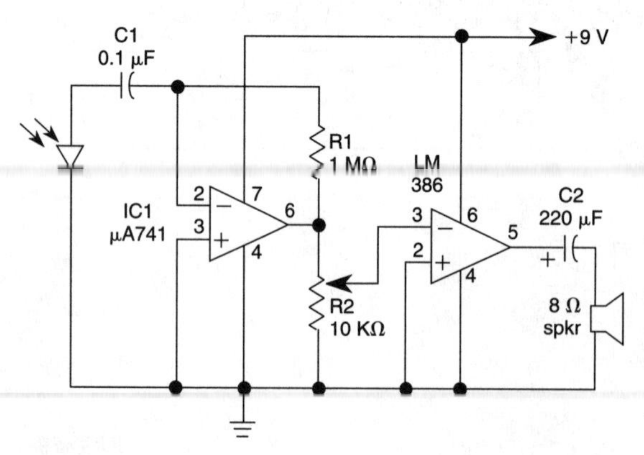

This light-wave receiver consists of a 741 operated as a preamplifier and an LM386 operated as a power amplifier. Potentiometer R2 is the gain control. Various kinds of detectors can be used as the front end of the receiver. Phototransistors are very sensitive, but they do not work well in the presence of too much ambient light. A 100-kΩ series resistor is required if you use a phototransistor. Solar cells, photodiodes, and LEDs of the same semiconductor as the transmitter all work well in this circuit.

WILLIAM SHEETS

FIG. 49-6

VISIBLE-LIGHT RECEIVER

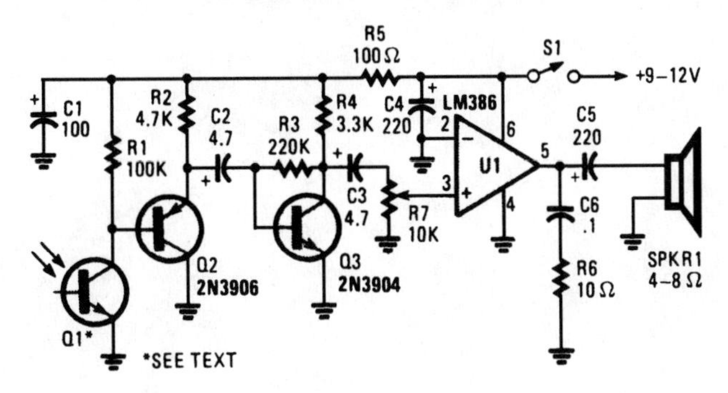

POPULAR ELECTRONICS *FIG. 49-7*

This receiver for amplitude-modulated light signals uses phototransistor Q1 mounted in a parabolic reflector (to increase range). Any npn phototransistor should work. Emitter-follower Q2 drives amplifier Q3. The output from Q3 feeds volume control R7 and audio amplifier U1. A 9- to 12-V supply is recommended for the receiver.

VISIBLE-LIGHT AUDIO TRANSMITTER

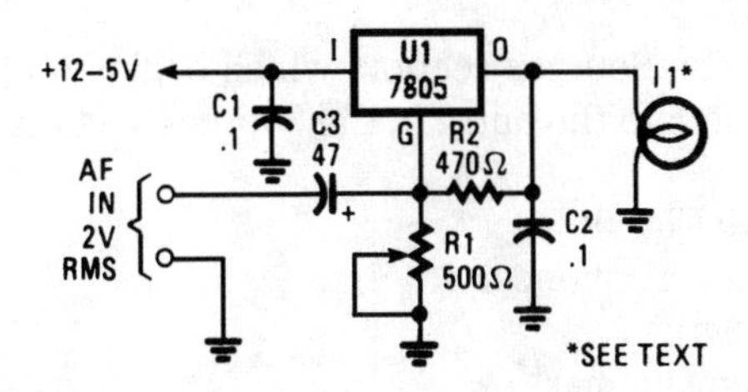

In the visible-light transmitter, a 7805 voltage regulator is connected in a variable-voltage configuration, and an audio signal is fed to the common input, to modulate the output voltage. The modulated output voltage is used to transmit intelligence via an incandescent lamp.

POPULAR ELECTRONICS *FIG. 49-8*

50

Light Control Circuits

The sources of the following circuits are contained in the Sources section, which begins on page 675. The figure number in the box of each circuit correlates to the entry in the Sources section.

LIGHT SEQUENCER

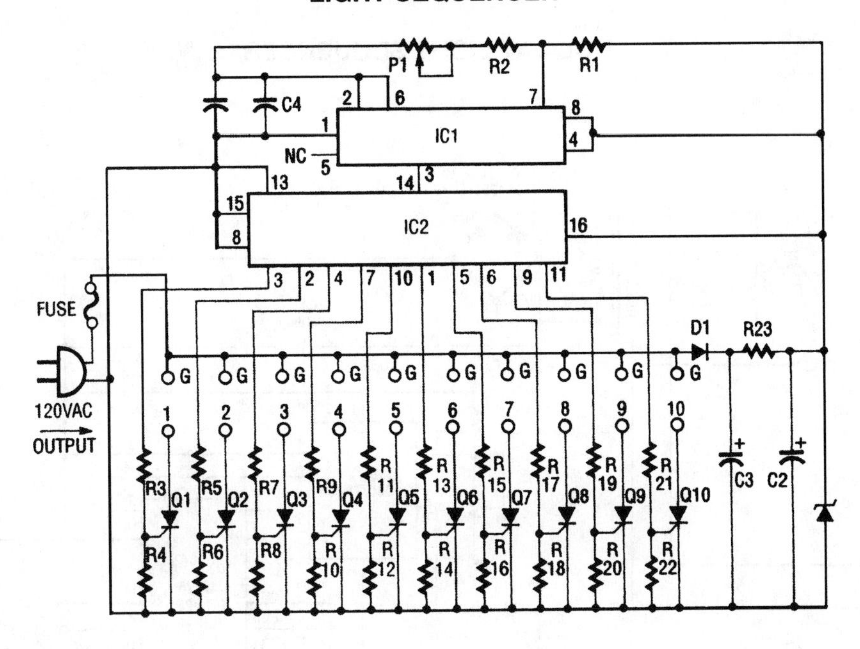

1991 PE HOBBYIST HANDBOOK

FIG. 50-1

The light sequencer uses two ICs and 10 SCRs to create an ac sequencer. The first IC, a 555 timer, is used to provide clock pulses for IC2. The IC is configured as an astable multivibrator, and its output is on pin 3.

Capacitors C1 and C4, along with resistor R2 and potentiometer P1, control the frequency of the pulses. IC2 is a 4017 Johnson counter, which shifts a high-signal level to each one of its 10 output pins in sequence. Each output pin is resistively coupled to the gate lead on an SCR. When the respective output pin on the 4017 is high and the positive half of the ac cycle is on the anode lead of the SCR, it turns on. The lamp that is connected to its anode lights.

Power is brought into the PC board by the line cord, then the circuit is fuse-protected. Diode LD1 changes the ac to pulsating, which is smoothed by C2 and C3. R23 limits the current, and zener diode D2 limits the dc voltage to 6 Vdc.

CI, C4	0.1-µF Capacitor
C2	100-µF Capacitor
C3	47-µF, 350-V Electrolytic Capacitor
D1	1N4007 Diode
D2	6-V Zener (M747814)
IC1	555 Timer IC
IC2	4017 CMOS IC
P1	500-kΩ Potentiometer
Q1–Q10	106 SCR
R1	560-Ω Resistor

R2, R4, R6, R8, R10, R12, R14, R16, R18 R20,R22	100-kΩ Resistor
R3, R5, R7 R9, R11, R13 R15, R17, R19 R21	2.2-kΩ Resistor
R23	15-kΩ 7-W Resistor

HOLIDAY LIGHT SEQUENCER

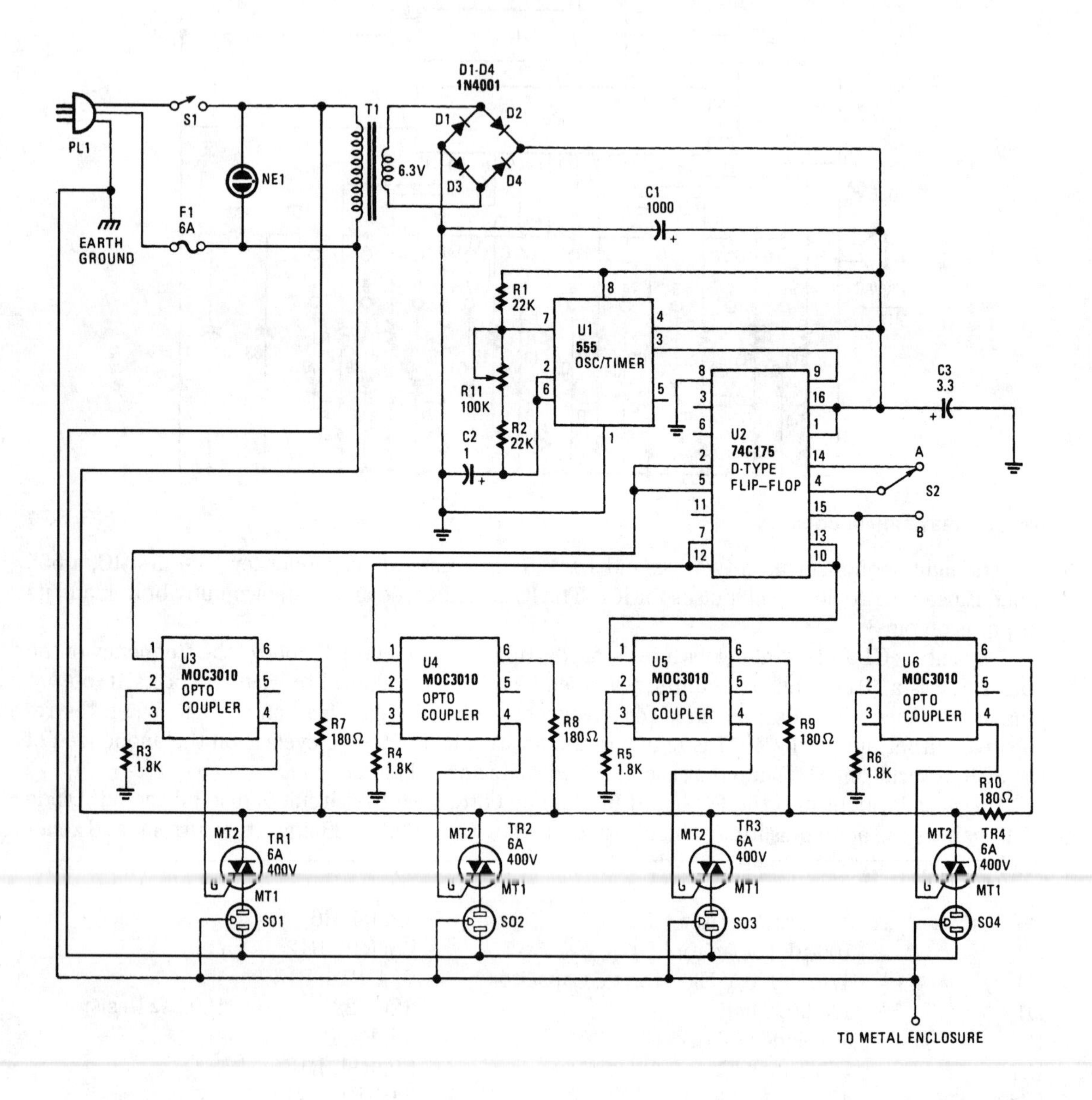

FIG. 50-2

HOLIDAY LIGHT SEQUENCER (*Cont.*)

Integrated circuit U1 (a 555 oscillator/timer) is wired as a conventional pulse generator. The frequency of the pulse generator is controlled by potentiometer R11. Resistor R2 puts a reasonable limit on the highest speed attainable.

The output of the pulse generator is fed to the common clock input of U2, a 74C175 quad D-type flip-flop. Each flip-flop is configured so that its Q output is coupled to the D input of the subsequent flip-flop.

Information on the D input of each flip-flop is transferred to the Q (and Q) outputs on the leading edge of each clock pulse. Switch S2 allows you to invert the information on the D input of the first flip-flop at any time during the cycle. This allows you to create a number of different sequences, which are determined by the state of the CQ output at the time of the switching.

Some of the possible sequences are:

- 1 through 4 on, 1 through 4 off;
- 1 of 4 on sequence;
- 1 of 4 off sequence;
- 2 of 4 on sequence;
- 1 and 3 on to 2 and 4 off;
- and other instances when the sequence of events is difficult to determine.

However, if S2 is switched to position B while all outputs are high or all are low (which seldom occurs), the sequence stops and the outputs remain either all on or all off. If that happens, you only need to switch back to position A for at least one pulse duration, then back to position B again.

Likewise, S2 should be in position A (pin 4 connected to pin 14) each time the power is turned on. This is because the data on pin 4 must be a logic 1 in order to start a sequence; otherwise all outputs remain at logic 0, regardless of the clock pulses.

Each output of the sequencing circuit is connected to an MOC3010 optoisolator/coupler (U3 through U6), which contains an infrared-emitting diode with an infrared-sensitive diac (triac driver or trigger) in close proximity. The diac triggers the triac, which carries the 117-volts ac.

Each time that the infrared-emitting diode receives a logic 1, it turns on and causes the diac to conduct. With the optoisolator/coupler's internal diac conducting, the triac turns on, and power is supplied to whatever load is plugged into the corresponding ac socket. So, the sequencing circuit and the 117-V ac outputs are "optically coupled" and are effectively isolated from each other.

Power for the sequencing circuit is provided by a 6.3-V miniature transformer. The output of the transformer is rectified by a four-diode bridge circuit, the output of which is filtered by C1 (1000-μF electrolytic capacitor). Capacitor C3 is added at the supply pin of U2 to suppress transients.

AUTOMATIC PORCH-LIGHT CONTROL

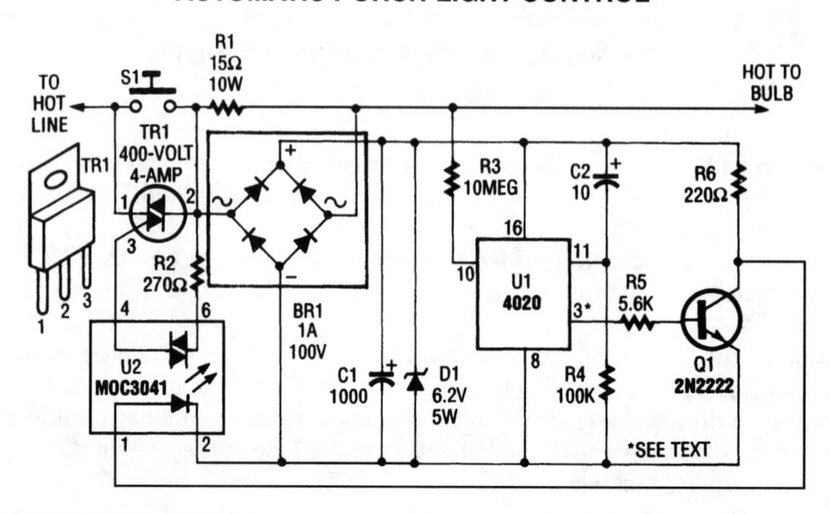

1993 ELECTRONICS HOBBYISTS HANDBOOK ***FIG. 50-3***

The automatic porch-light control circuit holds a triac on until a 4020 divider counts a number of 60-Hz powerline pulses. The circuit turns off a light after a predetermined time by using pins other than pin 3 of U1. Various times can be set. Consult the 4020 data sheet for information.

DIMMER FOR LOW VOLTAGE LOADS

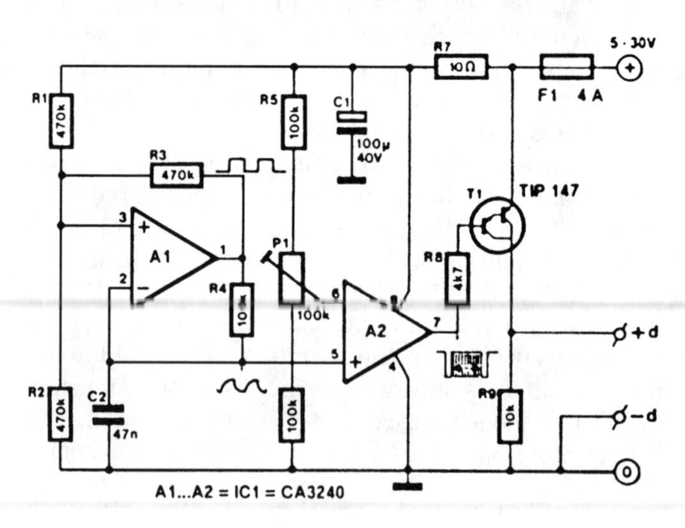

303 CIRCUITS ***FIG. 50-4***

This circuit controls a low voltage dc supply by pulse width modulation. The switching rate is 200 Hz. Input supply voltage should be +5 to +30 V. Up to 5 A can be controlled.

THREE-POWER-LEVEL TRIAC CONTROLLER

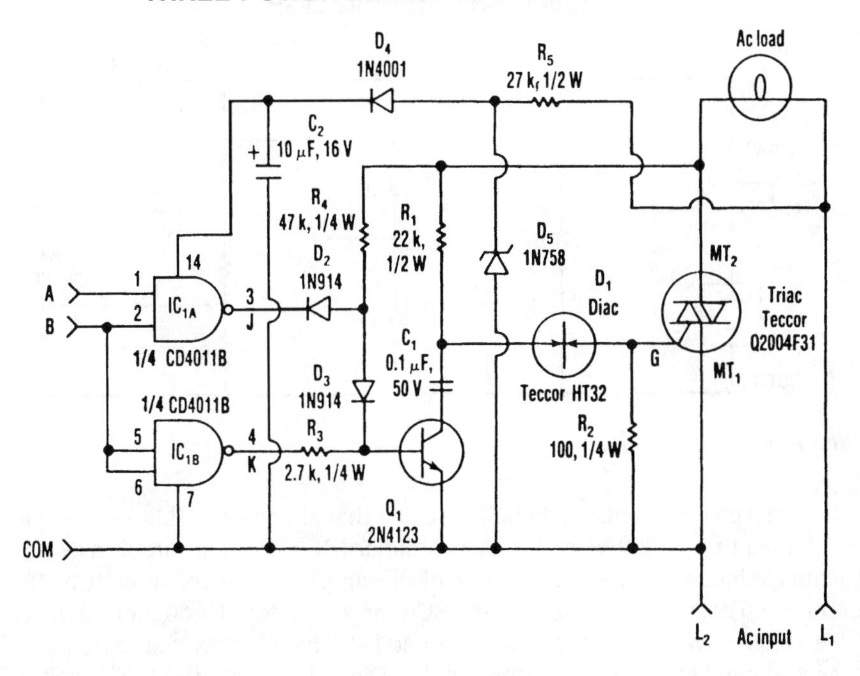

ELECTRONIC DESIGN

FIG. 50-5

Three power levels are supplied by the two logic inputs of this enhanced circuit. R5, D4, D5, and C2 form a power supply for the logic IC. They can be omitted if another source of low voltage is available.

PHASE-CONTROLLED DIMMER

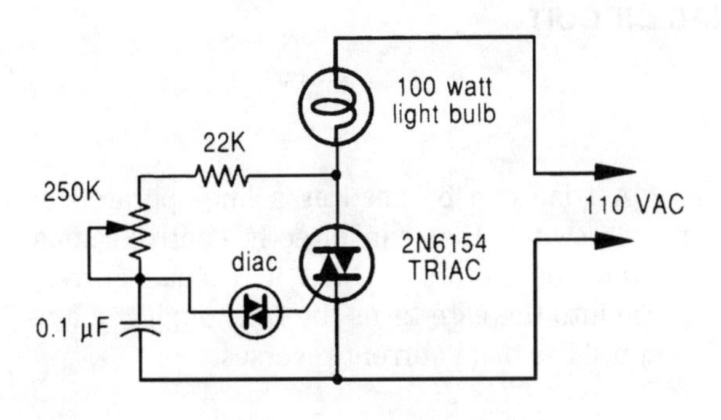

WARNING: Extreme shock hazard!

RADIO-ELECTRONICS

FIG. 50-6

A phase-controlled dimmer delays the triac turn-on to a selected point in each successive ac half cycle. Use this circuit only for incandescent lamps, heaters, soldering irons, or "universal" motors that have brushes.

120-ac SHIMMERING LIGHT

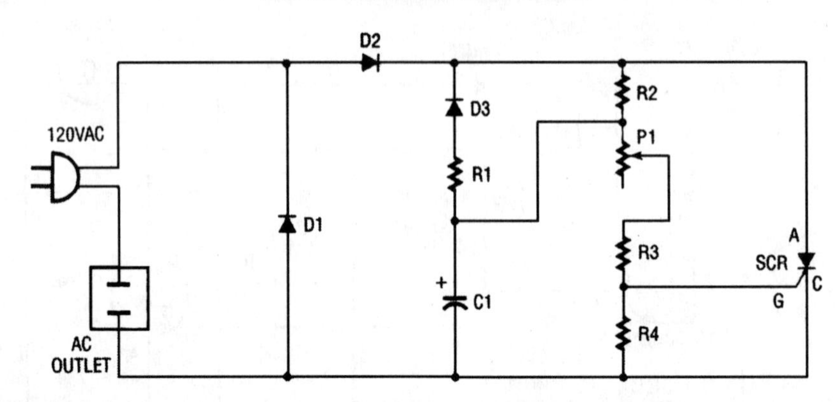

1991 PE HOBBYIST HANDBOOK

FIG. 50-7

You can turn any ordinary household bulb into one that shimmers or blinks. This circuit works on any incandescent light up to 200 W, and runs on standard 120 Vac. The circuit uses an SCR to cause an ordinary lamp to shimmer. Note that one side of the lamp is connected directly to 120 Vac, and the other side of the lamp goes to the cathode of the SCR. As ac voltage is brought into the circuit through the line cord, it is full-wave rectified by diodes D1 and D2. That changes the ac to dc, and a portion of that dc voltage is applied to capacitor C1 through R2. Diode D3 blocks the (+) dc voltage so that only the voltage from the path of R1 and D3 is clear. That forms an oscillator, which has a frequency determined by the setting of potentiometer P1 (because the other components have fixed values).

Remember to use **extreme caution** when using a device that connects to the ac line. **Never** use it outside or near water and always mount the entire kit inside a wooden or plastic (insulated) box to prevent any contact with the ac voltage.

SIMPLE TRIAC CIRCUIT

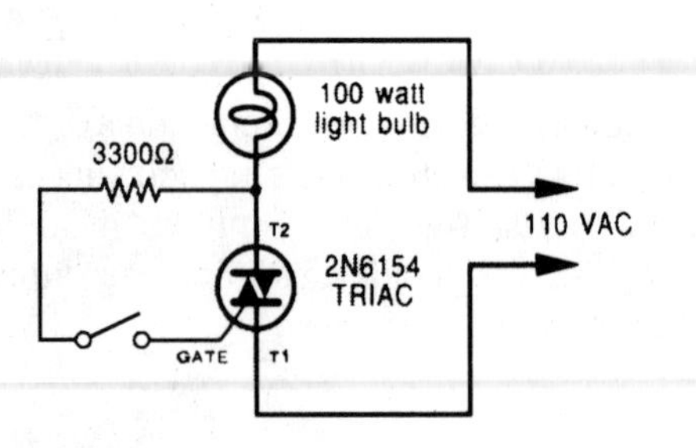

RADIO-ELECTRONICS

FIG. 50-8

A triac can be used as a line-operated ac power switch that can directly control lamps, heaters, or motors. A brief and small current pulse into the gate turns the triac on; it remains on until the main current reverses.

RUNNING LIGHT SEQUENCE

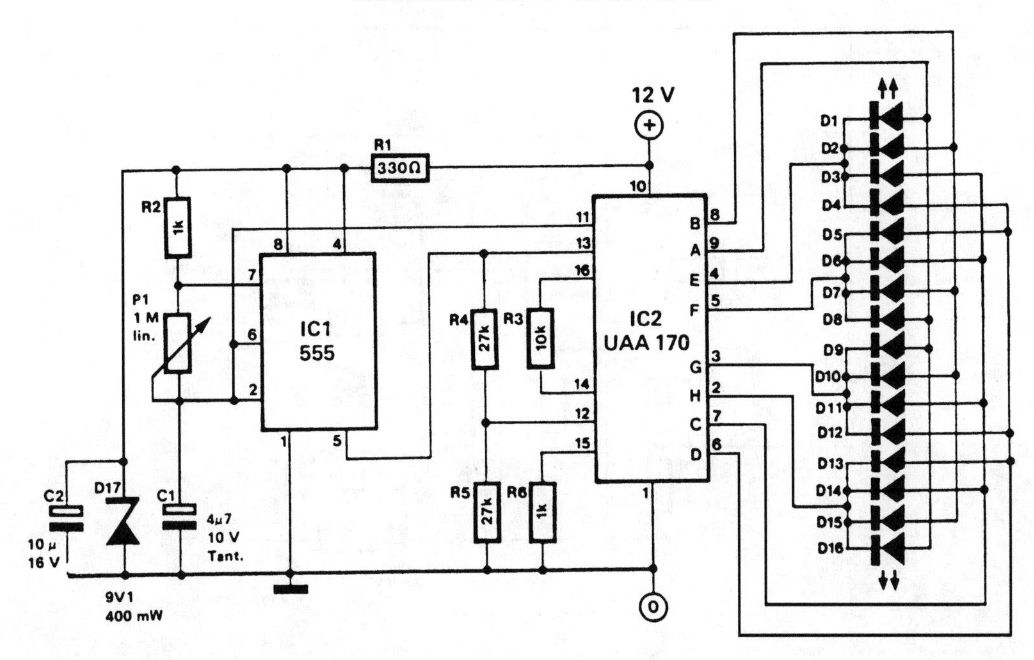

303 CIRCUITS *FIG. 50-9*

This running light sequencer drives 16 LEDs and runs from a 12-V supply. C1 can be varied to alter the rate of operation.

MOS LAMP DRIVER

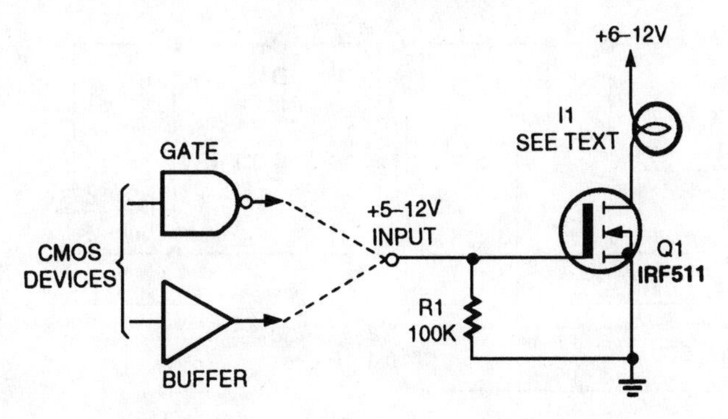

POPULAR ELECTRONICS *FIG. 50-10*

The circuit shows a way of using a MOSFET as a load driver. I1 can be a lamp, or any other load, that does not exceed the current rating of Q1.

CMOS TOUCH DIMMER

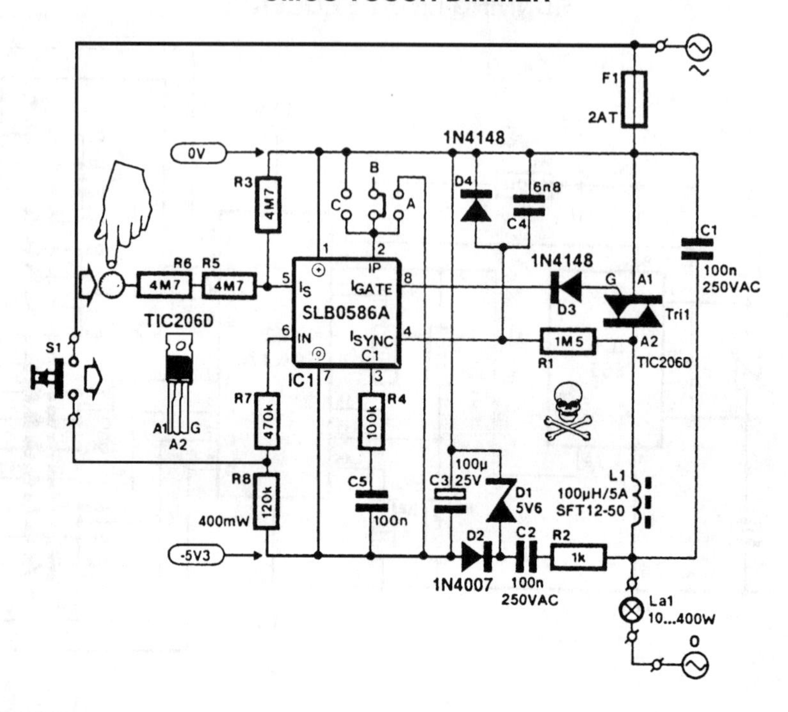

ELEKTOR ELECTRONICS *FIG. 50-11*

A Seimens SLB0586A IC allows the construction of a simple touch-controlled dimmer circuit. The circuit controls a triac ac switch, which allows control of loads from 10 to 400 W.

NEON LAMP DRIVER FOR 9-V SUPPLIES

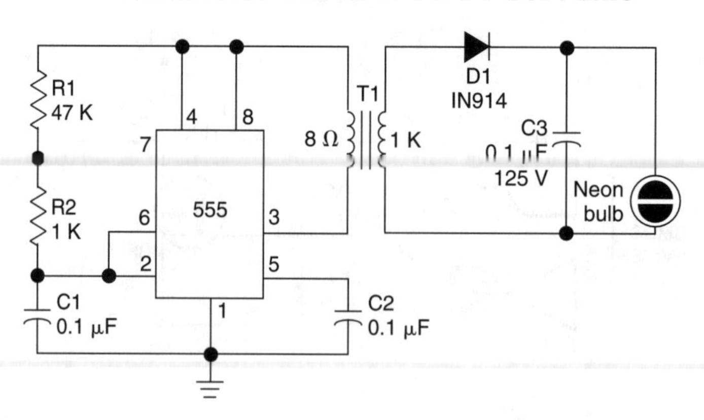

RADIO-ELECTRONICS *FIG. 50-12*

This circuit is for driving a neon lamp from a 9-V supply. The 555 generates an ac signal (stepped up by T1), and lights the neon bulb. T1 is any small audio output transformer.

SENSITIVE TRIAC CONTROLLER

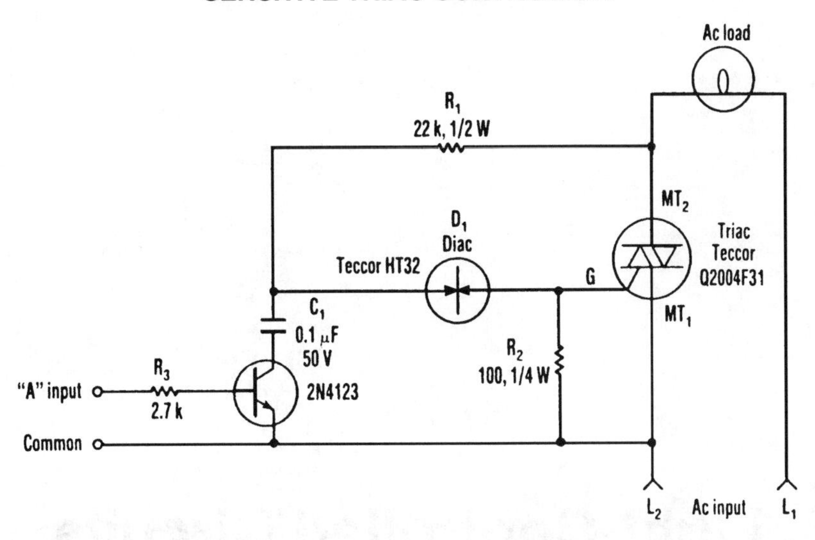

ELECTRONIC DESIGN

FIG. 50-13

The single transistor connected between the capacitor and the common side of the ac line allows a logic-level signal to control this triac power circuit. Resistor R2 prevents false triggering of the triac by the trickle current through the diac.

HALOGEN LAMP PROTECTOR

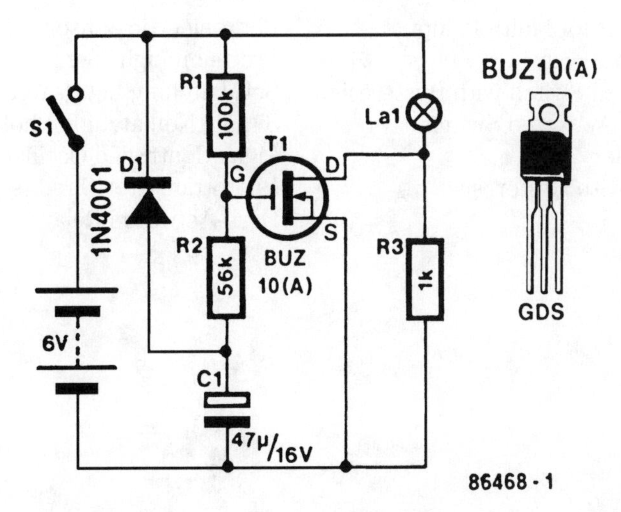

303 CIRCUITS

FIG. 50-14

This circuit produces a soft turn-on for halogen lamp filaments upon powering up. MOSFET used is a BUZ10, which has 0.2 Ω R_{DS} on. R1, R2, and C1 set the turn-on rate and D1 discharges C1 at turn-off.

51

Light-Controlled Circuits

The sources of the following circuits are contained in the Sources section, which begins on page 675. The figure number in the box of each circuit correlates to the entry in the Sources section.

LIGHT-DEPENDENT SENSOR FOR MULTIPLE INPUTS

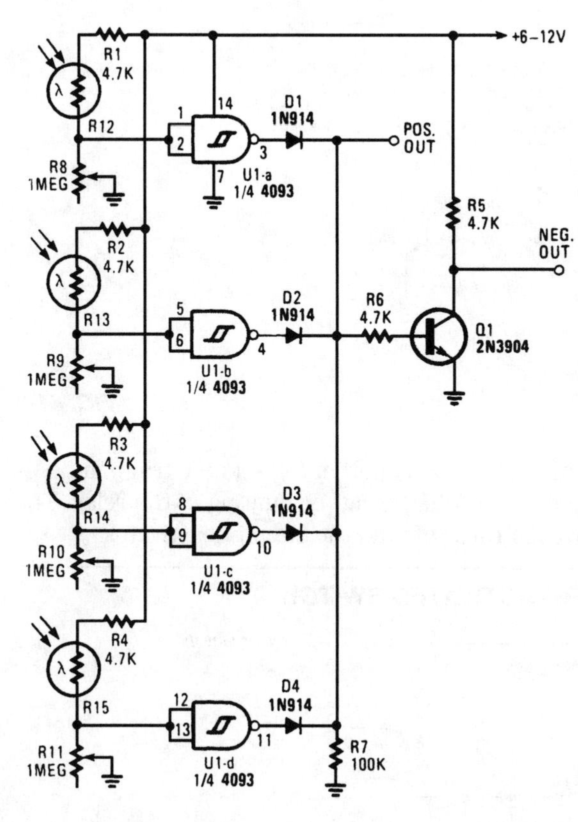

This light-dependent sensor uses LDRs to detect the presence or absence of light. As long as the light source striking the LDRs remains constant, the alarm does not sound. But when the light is interrupted, the alarm is triggered.

POPULAR ELECTRONICS *FIG. 51-1*

SIMPLE LIGHT-ACTIVATED ALARM

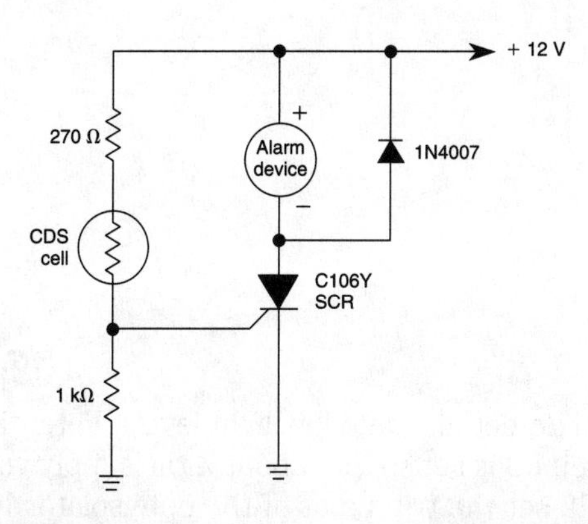

A cadmium-sulfide photocell conducts when a light beam strikes it. This triggers the SCR and activates the alarm device.

WILLIAM SHEETS *FIG. 51-2*

PRECISION DARK-ACTIVATED SWITCH WITH HYSTERESIS

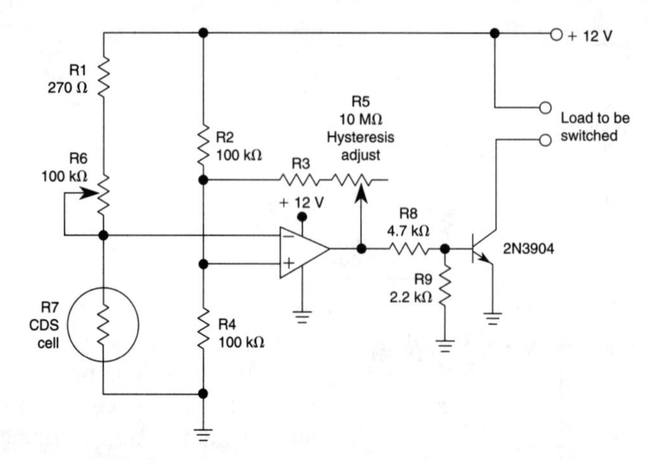

WILLIAM SHEETS

FIG. 51-3

A CdS cell is one leg of a bridge circuit. Potentiometer R6 in another leg sets the trip point. Potentiometer R5 provides hysteresis adjustment to prevent "chattering" or hunting of the relay. The light level has to increase noticeably before the 2N3904 turns off and the circuit deactivates.

COMBINED LIGHT-/DARK-ACTIVATED SWITCH

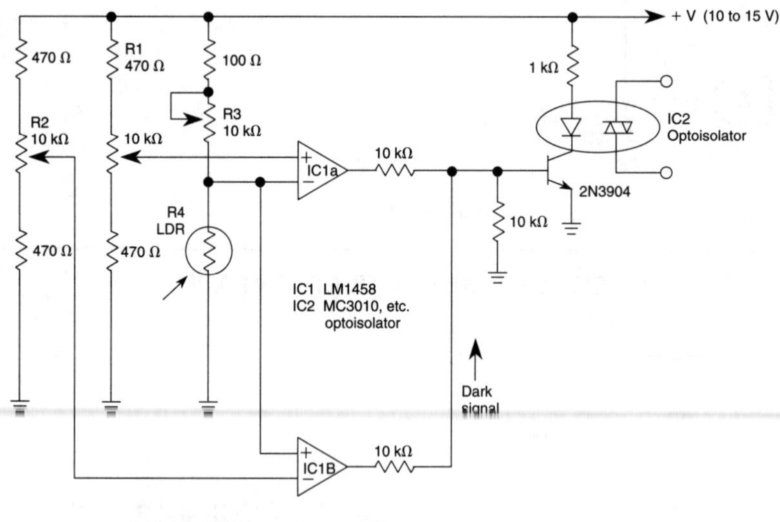

WILLIAM SHEETS

FIG. 51-4

Two op amps used in a bridge circuit configuration detect high and low light levels. Potentiometer R2 sets the dark level and R1 controls the light level. R3 is set so that about ½ the supply voltage appears across R4 at the desired light level. R1 and R2 set the trip point of the optoisolator IC2 at darker or lighter ambient levels, as required.

OUTDOOR LIGHT CONTROLLER

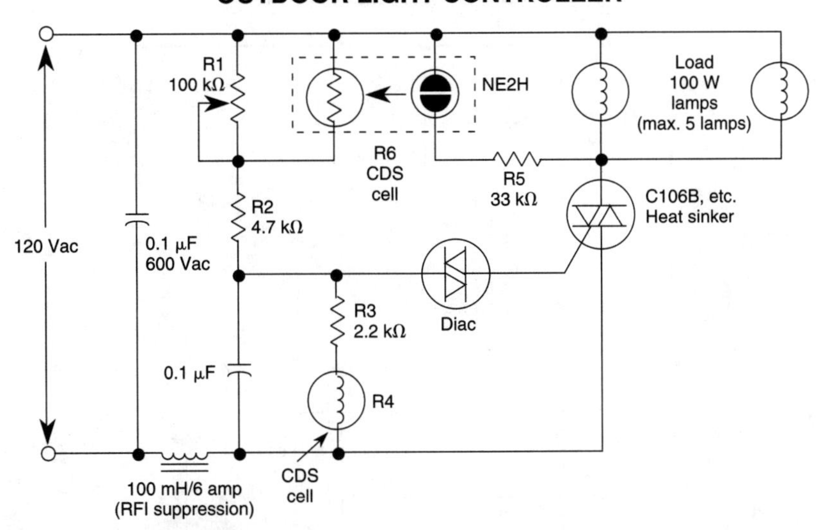

WILLIAM SHEETS *FIG. 51-5*

A neon bulb and a CdS photocell enclosed in a light-tight enclosure form an optocoupler. A diac/triac combination is used to provide the snap-switch effect. A second CdS photocell acts as the main sensor.

As darkness approaches, the resistance of R4 begins to increase. At a threshold level, the diac triggers the triac and causes the neon bulb to light. This reduces the resistance of R6, causing the diac to trigger the triac, which lights the neon bulb and provides power to the load.

As morning light comes up, the process is reversed. The neon bulb goes out and the SCR turns off.

DARK-ACTIVATED RELAY WITH HYSTERESIS

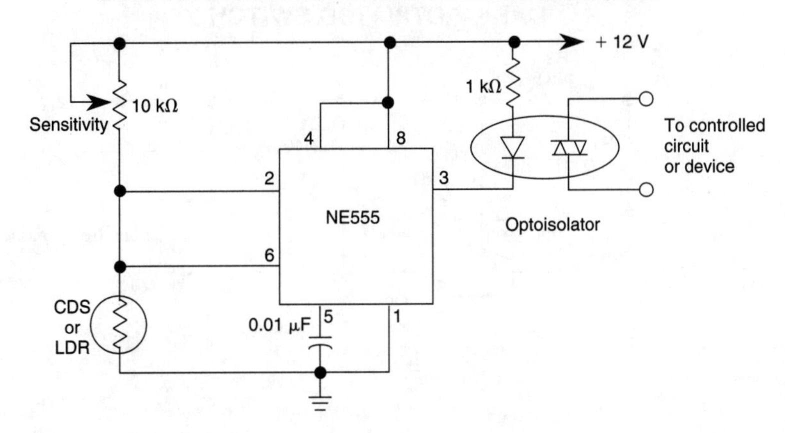

WILLIAM SHEETS *FIG. 51-6*

The hysteresis of a 555 IC can be used to advantage for sensing a drop in light. An LDR or CDS cell with about 2 to 8 k resistance at desired light level should be used.

PORCH LIGHT CONTROL

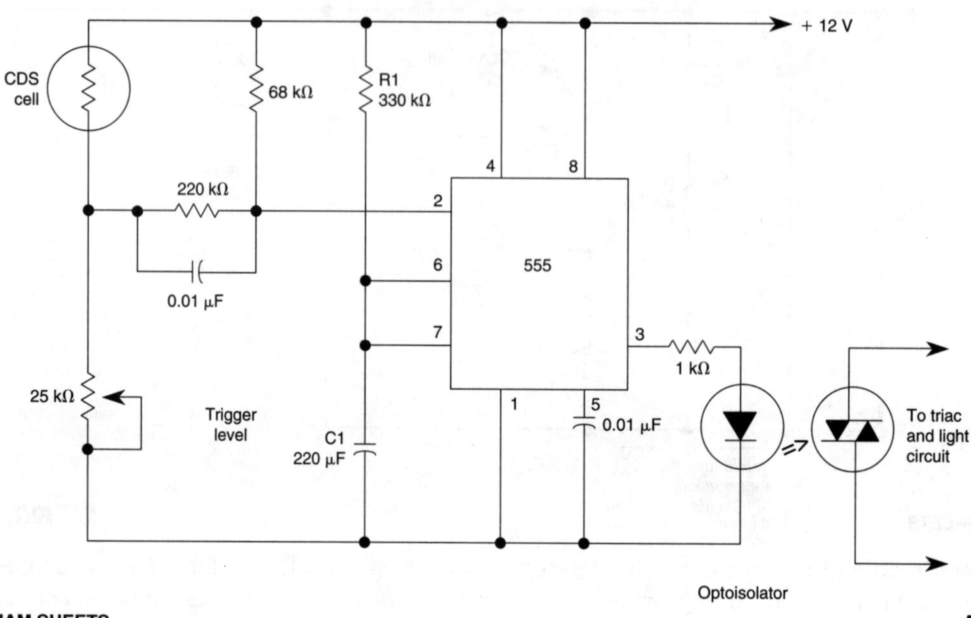

WILLIAM SHEETS

FIG. 51-7

This circuit can control the on/off cycle of a light via a CDS photocell, and turn it off after a preset period. The light can only be turned on when CDS cell is in darkness, and it stays on for a time determined by the 555 circuit. On time depends on R1 and C1 and is about 80 seconds with the values shown.

DARK-ACTIVATED SWITCH

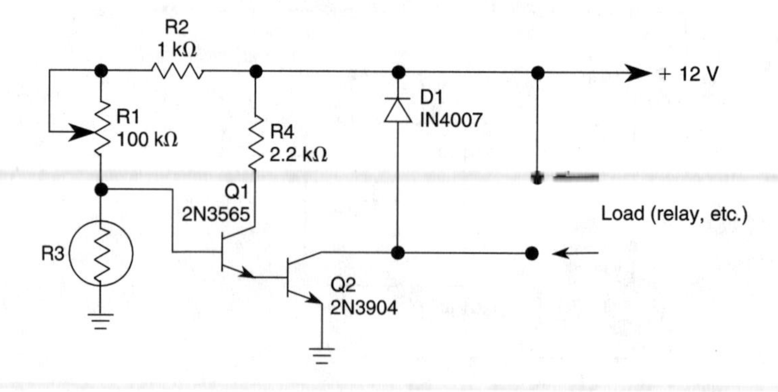

WILLIAM SHEETS

FIG. 51-8

In this circuit, lowering of the light level on the CDS cell turns on Q1 and Q2 which switches on the load which could be a relay, light, etc.

PHOTOELECTRIC SENSOR

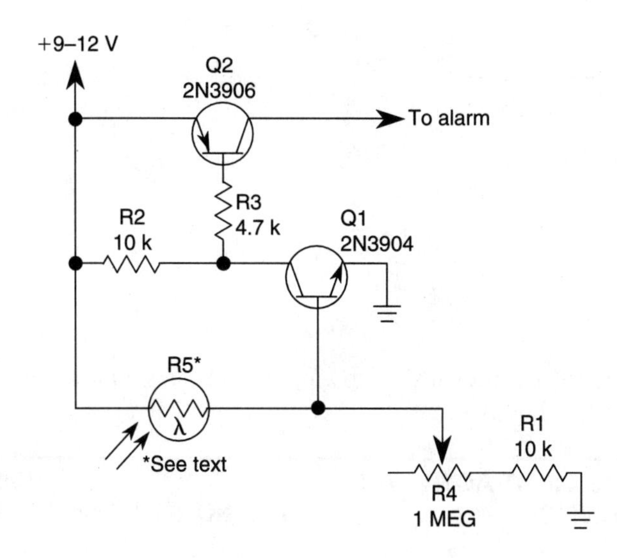

POPULAR ELECTRONICS ***FIG. 51-9***

The circuit can be used as a sensor that can trigger an alarm without direct contact being made by the intruder. In this circuit, a visible or invisible light source radiates on the sensor, keeping the detection loop in what could essentially be called a normally closed condition.

As long as the light source striking R5 remains uninterrupted, the switch remains closed. But if an intruder passes between the light source and the sensor, the circuit goes from closed to open, and triggers the alarm.

A light-dependent resistor (LDR), whose resistance varies inversely in with the amount of light hitting its sensitive surface, is used. A bright light aimed at R5 causes its internal resistance to drop as low as a few hundred ohms; in total darkness, the unit's resistance can rise to several megohms. The light-dependent resistor (R5) is connected between the $+V$ supply and the base of Q1. As long as R5 detects light, it supplies ample base current to cause Q1's collector to saturate to near ground level. That also pulls the base of Q2 (a 2N3906 general-purpose pnp transistor) to near ground level, turning it on and clamping its collector to the $+V$ rail.

PRECISION LIGHT-SENSITIVE RELAY SWITCH

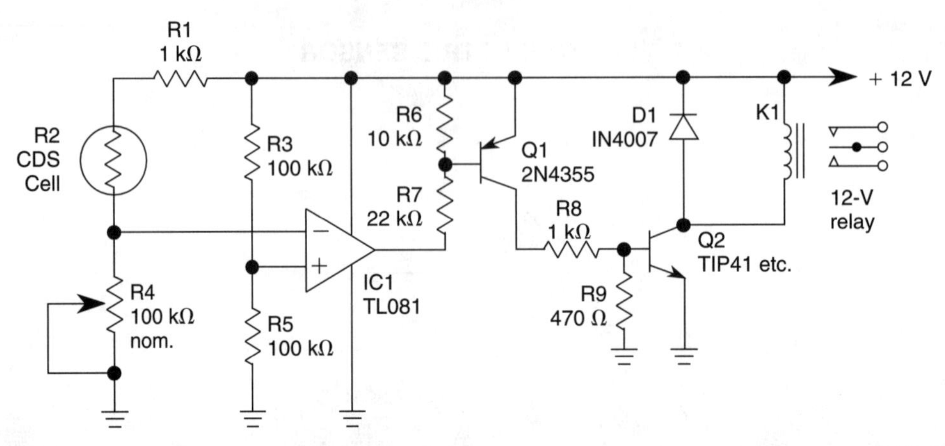

WILLIAM SHEETS **FIG. 51-10**

A CDS cell in a bridge circuit with an op amp provides a simple means of operating a relay at a predetermined light level. Potentiometer R4 sets the sensitivity.

SELF-LATCHING LIGHT-ACTIVATED SWITCH

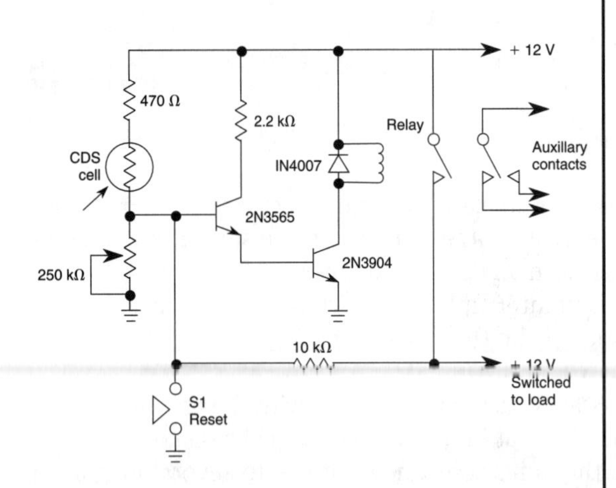

WILLIAM SHEETS **FIG. 51-11**

When light strikes the CDS cell it turns on the transistors which activates the relay which latches. Depressing S1 grounds the base of the 2N3565 and the relay resets. The 250 k potentiometer adjusts the sensitivity of the circuit.

SIMPLE NONLATCHING PHOTOCELL SWITCH

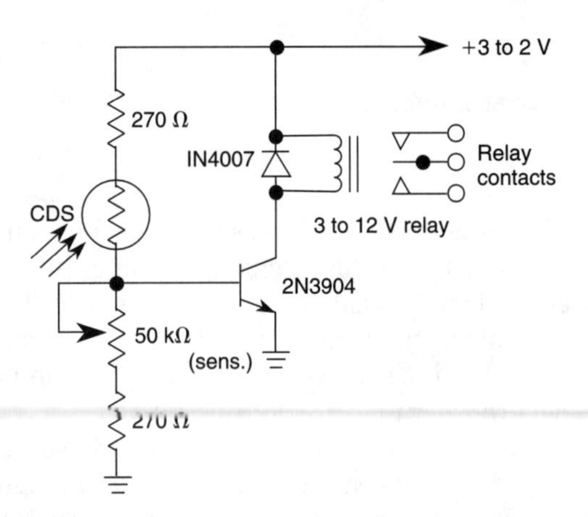

WILLIAM SHEETS **FIG. 51-12**

A CDS photocell is used to drive the relay. The circuit operates from a +12 V supply.

LIGHT-CONTROLLED OSCILLATOR

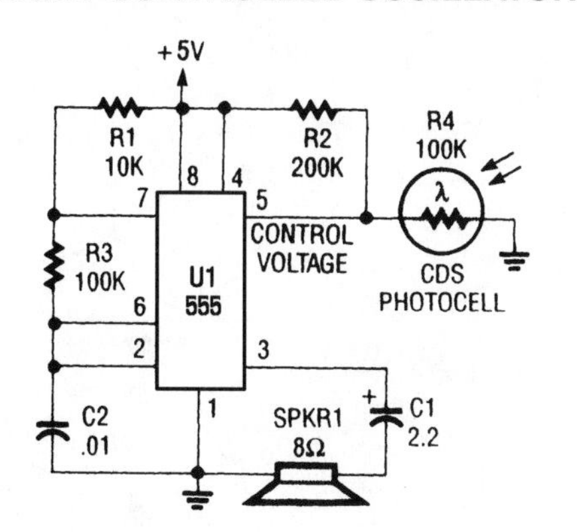

POPULAR ELECTRONICS ***FIG. 51-13***

This circuit can be used as a light detector and possibly as an aid for the visually handicapped. The frequency of the oscillator is determined by the amount of illumination striking LDR4.

PHOTOTRANSISTOR CIRCUITS

WILLIAM SHEETS ***FIG. 51-14***

Here are four ways to connect a phototransistor for general use in phototransistor circuits.

DARK-ACTIVATED RELAY

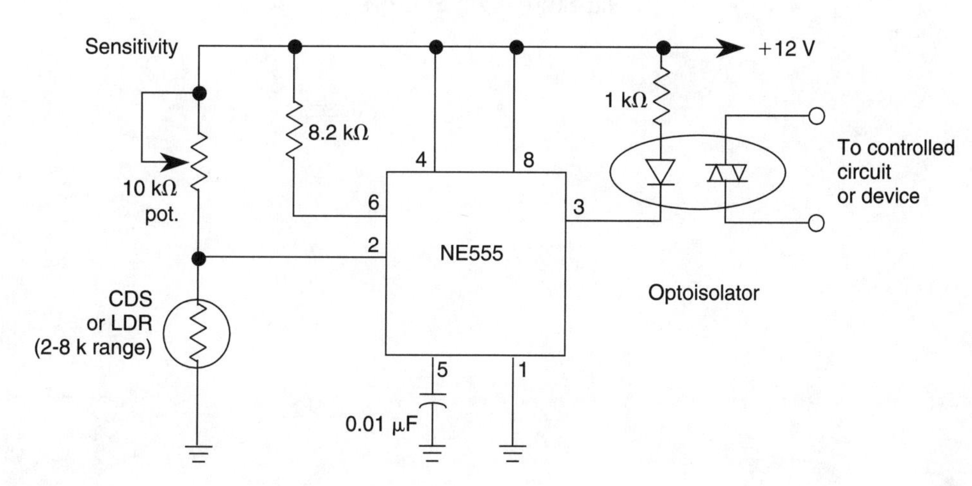

WILLIAM SHEETS ***FIG. 51-15***

Configuring a 555 IC as shown yields a dark-activated relay with low hysteresis. CDS or LDR should be in the 2 k to 8 k range at desired light level.

52

Light Sources

The sources of the following circuits are contained in the Sources section, which begins on page 675. The figure number in the box of each circuit correlates to the entry in the Sources section.

BATTERY-OPERATED BLACK LIGHT

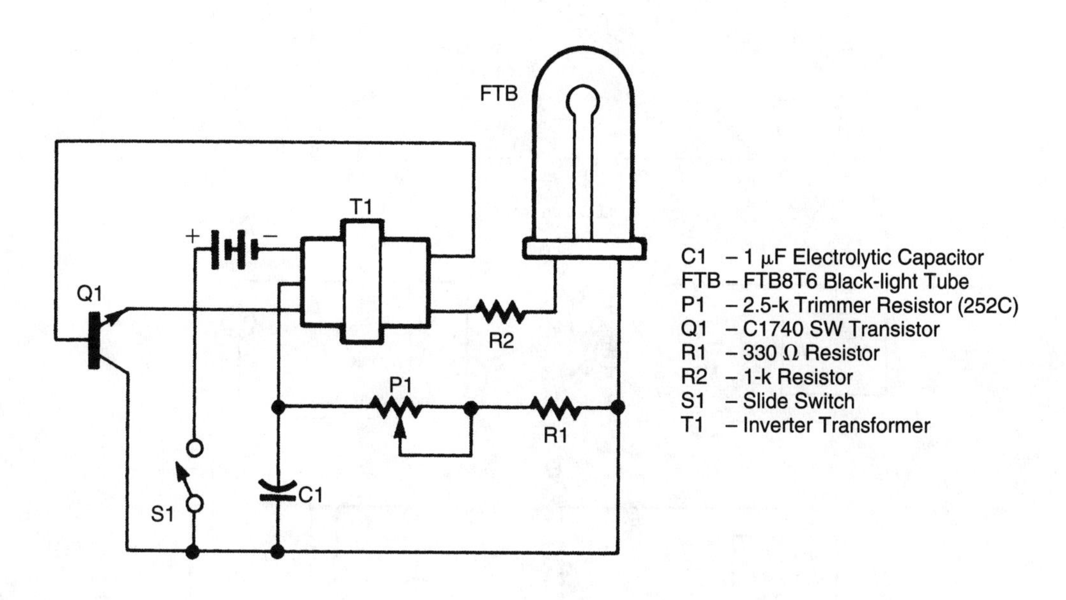

1989 PE HOBBYIST HANDBOOK ***FIG. 52-1***

The battery-operated black light uses a "U"-shaped, unfiltered, black-light tube, which requires approximately 250 Vac to operate. To create the 250-Vac 6-V battery, the circuit uses a one-transistor blocking oscillator that drives a ferrite inverter transformer. A blocking oscillator turns itself off after one or more cycles. In this circuit, it consists of C1, P1, Q1, R1, and T1. The oscillations are sustained because the base of Q1 is connected to one of the windings on T1.

Transformer T1 is a step-up transformer that consists of a ferrite core, which has a few turns on the primary and many turns on the secondary. The oscillating (ac) output of Q1 is fed to T1, which, because of its large turns ratio, converts the low-voltage signal into a high-voltage alternating current, which is coupled through resistor R2 to the black-light tube. Resistor R1 and trimmer resistor P1 limit the current flowing through the circuit. As the control on P1 is rotated, more current flows in the circuit, producing a brighter light output.

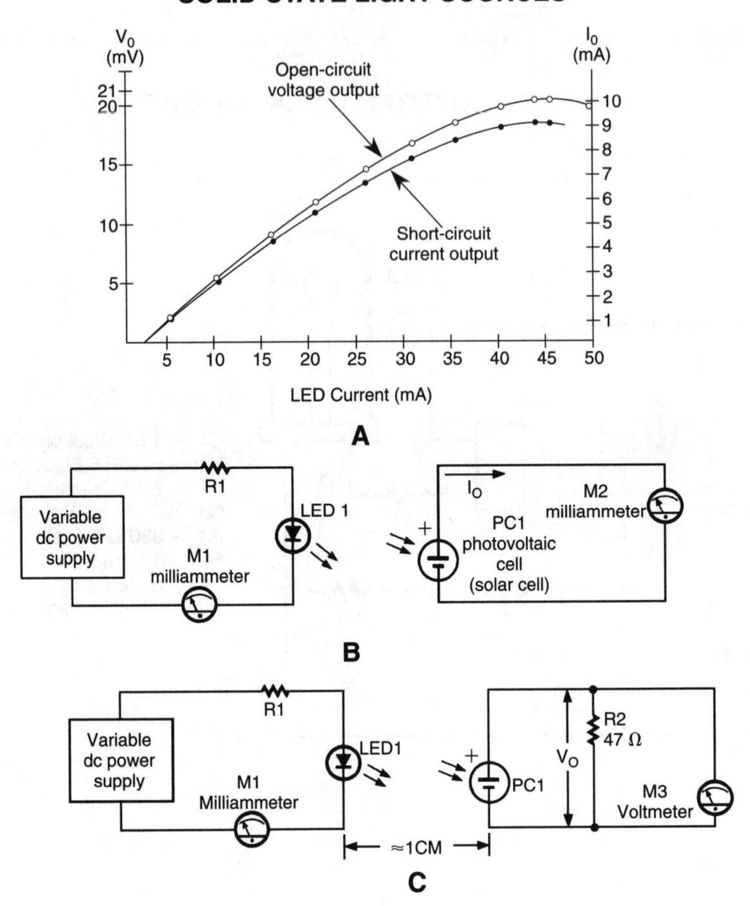

In A we show two LED output curves derived by experiment. The circuit in B was used to get the data for the short-circuit current plot, while the circuit in C yielded the data for the open-circuit voltage plot.

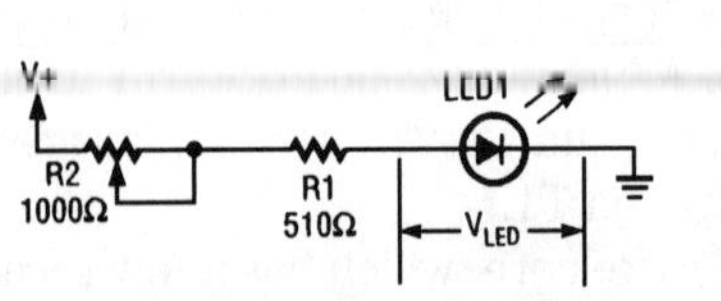

Since LED intensity is linearly related to the input current this circuit can be used to vary the LED's brightness via R2.

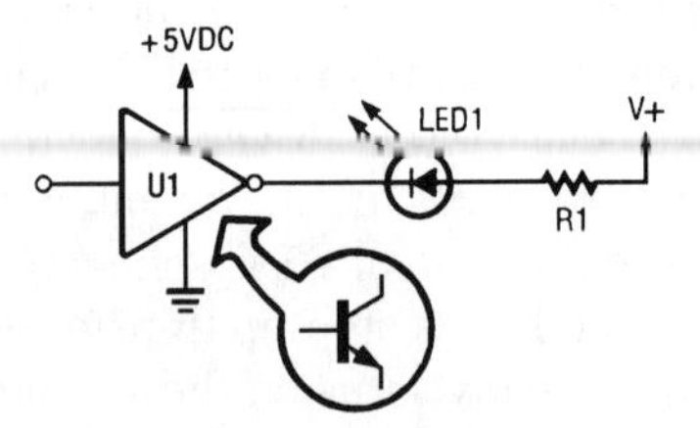

You can drive an LED with an open-collector TTL inverter. The inverter shown must ground the LED to turn it on.

POPULAR ELECTRONICS *FIG. 52-2*

The 12 LED circuits shown are useful for experiments and applications of LED devices. The captions are self-explanatory and illustrate many common LED applications.

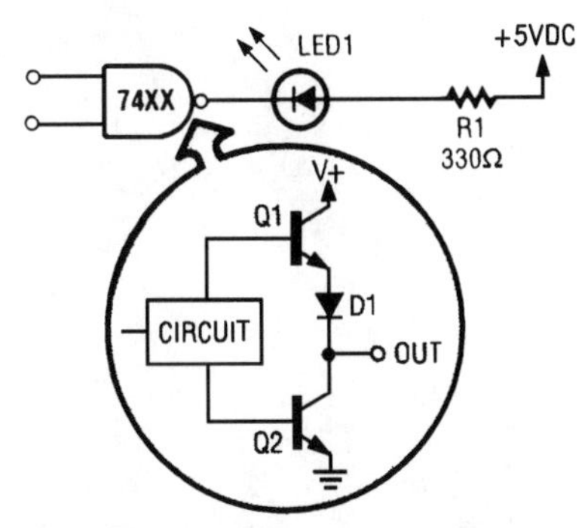

A totem-pole TTL output can drive an LED by grounding the LED's cathode, much like the open-collector driver.

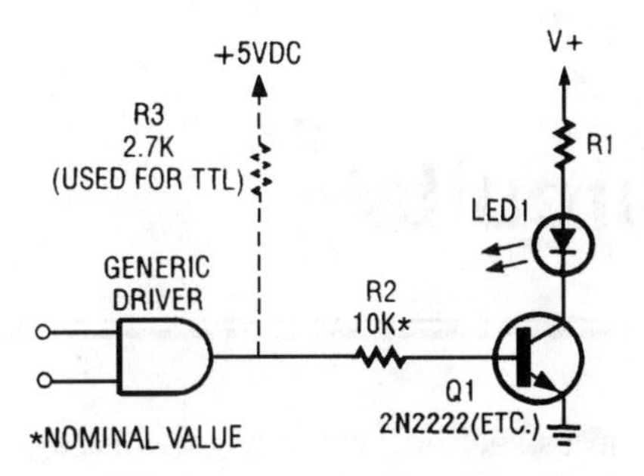

This driver circuit will work for either CMOS or TTL gates, but you don't need R3 in a CMOS-driven circuit.

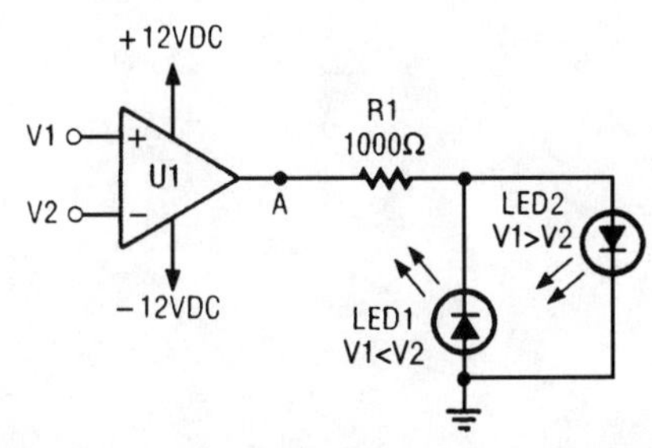

This is a bipolar output indicator that lets you know if one voltage is greater than, less than, or equal to another.

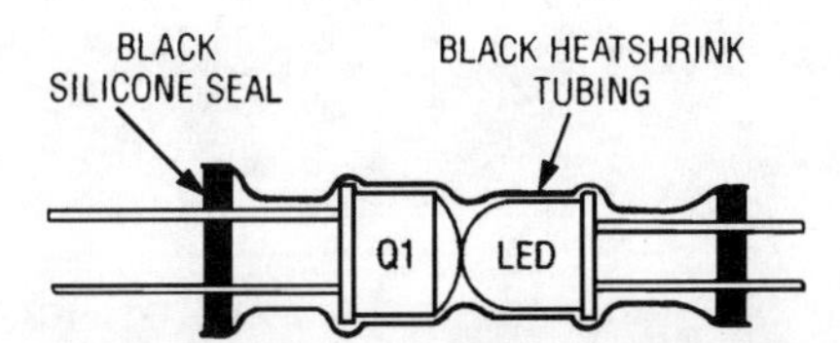

You can "roll your own" optocoupler by using some heat-shrink tubing, an LED, and optical transistor, and silicon sealant as shown here.

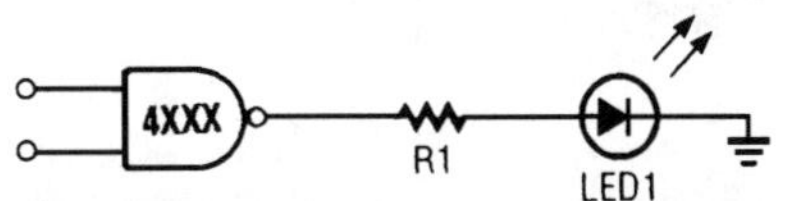

Unlike TTL devices, integrated circuits made with CMOS technology can source enough current to power an LED as shown here.

A CMOS-based gate can sink current much like a TTL gate in order to activate an LED.

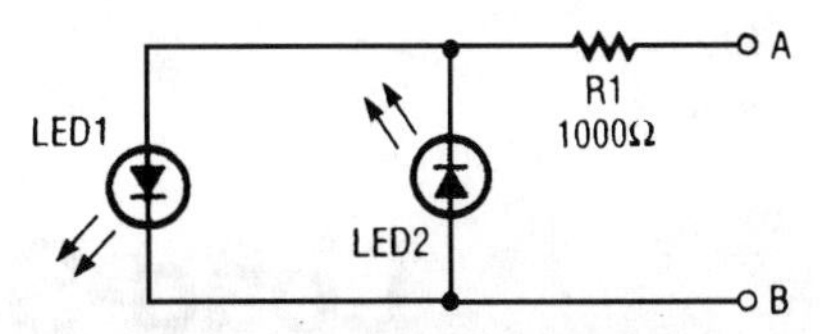

This simple polarity checker is easy to build and can be of help if you don't know much about a circuit's wiring or grounding convention.

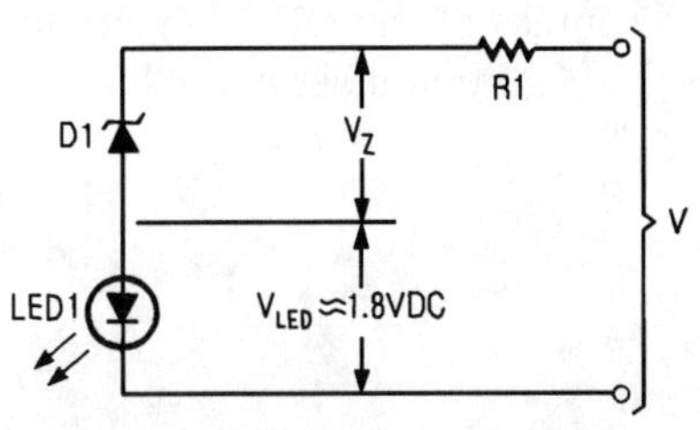

This is a simpler voltage-level sensor than that shown back in Fig. 9. To use it you have to know the polarity of the voltage it is to monitor.

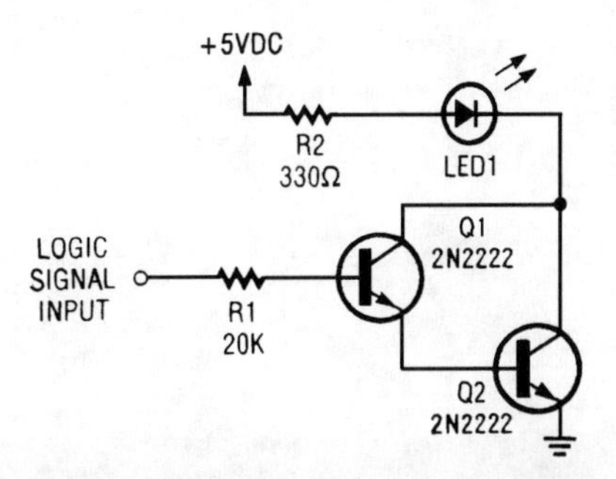

This high sensitivity Darlington LED driver circuit can be used as a simple logic probe. You may have to vary the value of R1 to suit the circuit under test.

53

Load-Sensing Circuits

The sources of the following circuits are contained in the Sources section, which begins on page 675. The figure number in the box of each circuit correlates to the entry in the Sources section.

LOAD-SENSING SOLID-STATE SWITCH

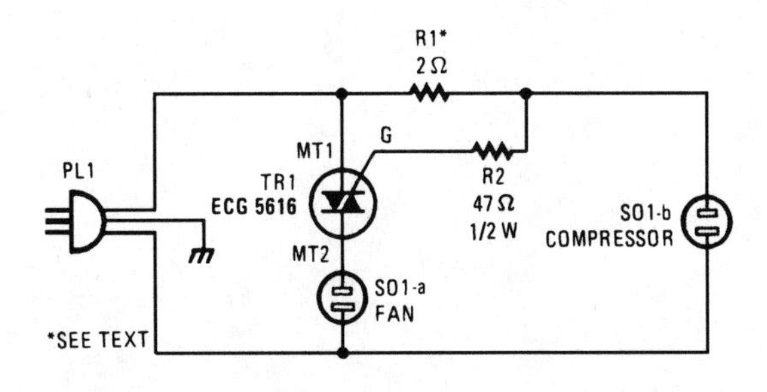

When this triac circuit senses current flow through SO1-a, it activates the device plugged into SO1-b. The values of the resistors must be chosen for the specific devices to be plugged in.

POPULAR ELECTRONICS

FIG. 53-1

LOAD-SENSING TRIGGER

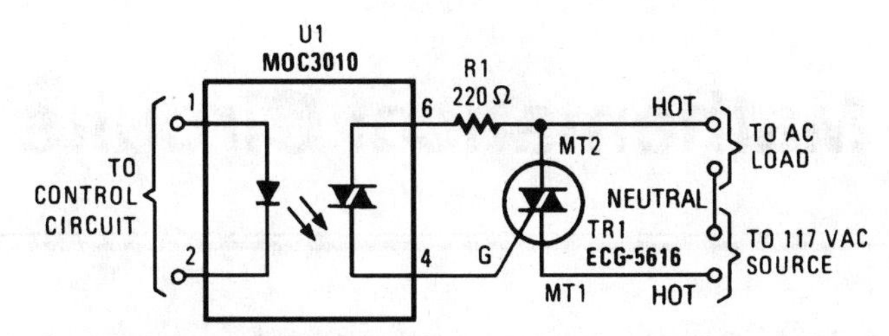

Triacs can be controlled by low-power circuits through Triac-driver optoisolators as shown here.

A

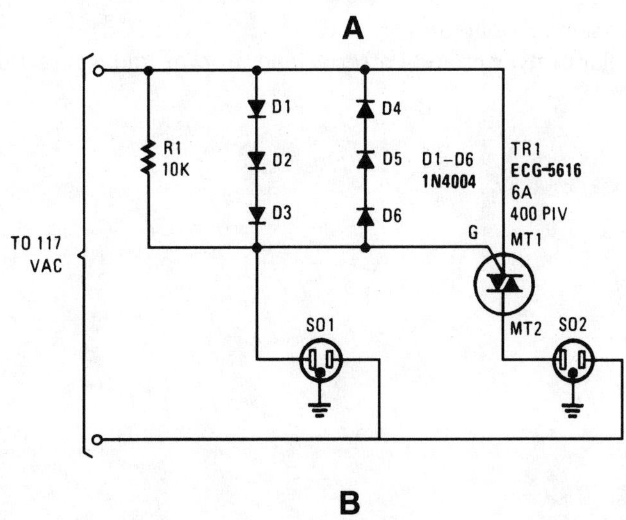

B

POPULAR ELECTRONICS

FIG. 53-2

A device plugged into SO1 causes a voltage-limited gate trigger for triac TR1, and causes power to be applied to SC2.

54

Mathematical Circuits

The sources of the following circuits are contained in the Sources section, which begins on page 675. The figure number in the box of each circuit correlates to the entry in the Sources section.

SECOND-ORDER POLYNOMIAL GENERATOR

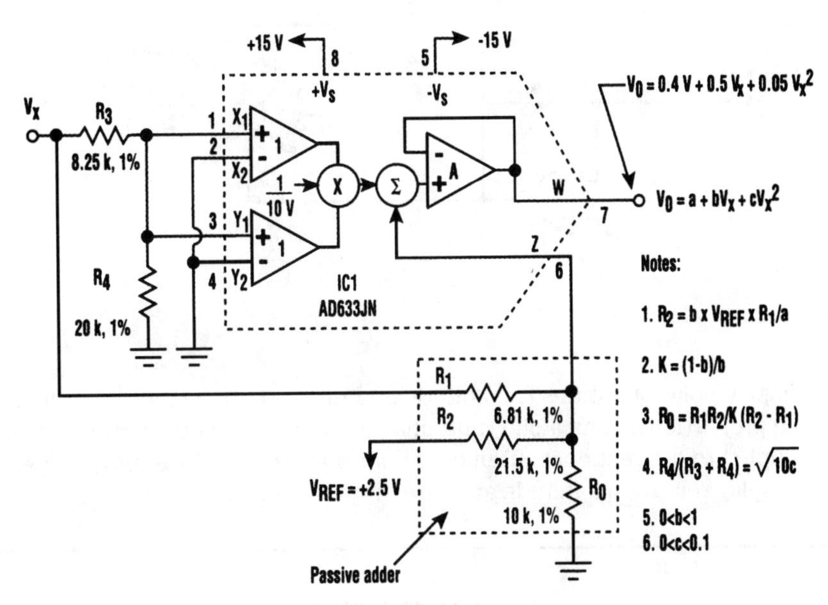

ELECTRONIC DESIGN

FIG. 54-1

By using a circuit built with a single analog multiplier and five precision resistors, an output voltage (V_o) can be made to create a second-order polynomial.

The circuit implements the following quadratic:

$$V_o = a + bV_x + cV_x^2$$

The input terminals of IC1 are connected to create a positive square term and present the V_x signal to the output with a 1-10-V scale factor. Incorporating the voltage-divider network (resistors R3 and R4) in the input signal path provides additional attenuation adjustment for the coefficient (c) of the square term in the quadratic. Then, the passive adder (resistors R1, R2, and R_o) is wired to IC1's internal summing circuit to generate the polynomial's other two terms; the offset term (a) and the linear coefficient (b).

POLAR-TO-RECTANGULAR CONVERTER AND PATTERN GENERATOR FOR RADIO DIRECTION FINDING

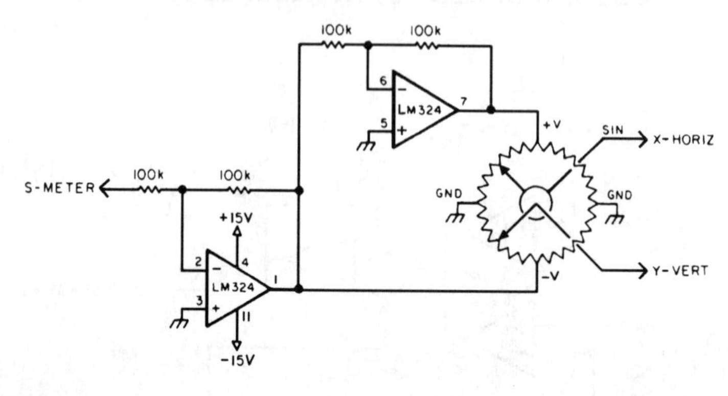

73 AMATEUR RADIO TODAY *FIG. 54-2*

In order to display polar quantities (magnitude and direction of a received radio signal), a sine and cosine voltage proportional to an angle (antenna direction) is needed. In this case, a sine-cosine potentiometer coupled to a directional antenna and a sample of a voltage proportional to received signal is used to display relative magnitude and direction of a received signal.

ROOT EXTRACTOR

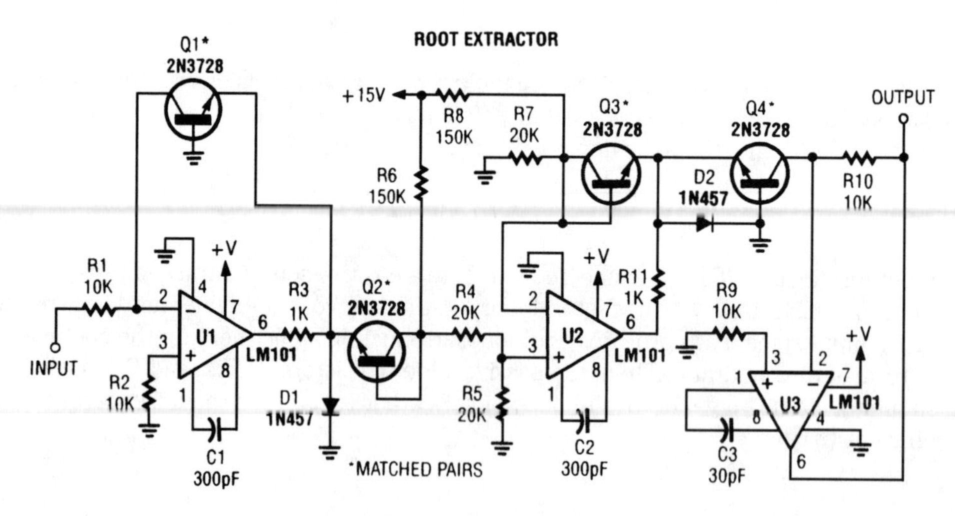

POPULAR ELECTRONICS *FIG. 54-3*

55

Measuring and Test Circuits

The sources of the following circuits are contained in the Sources section, which begins on page 675. The figure number in the box of each circuit correlates to the entry in the Sources section.

ENERGY CONSUMPTION MONITOR

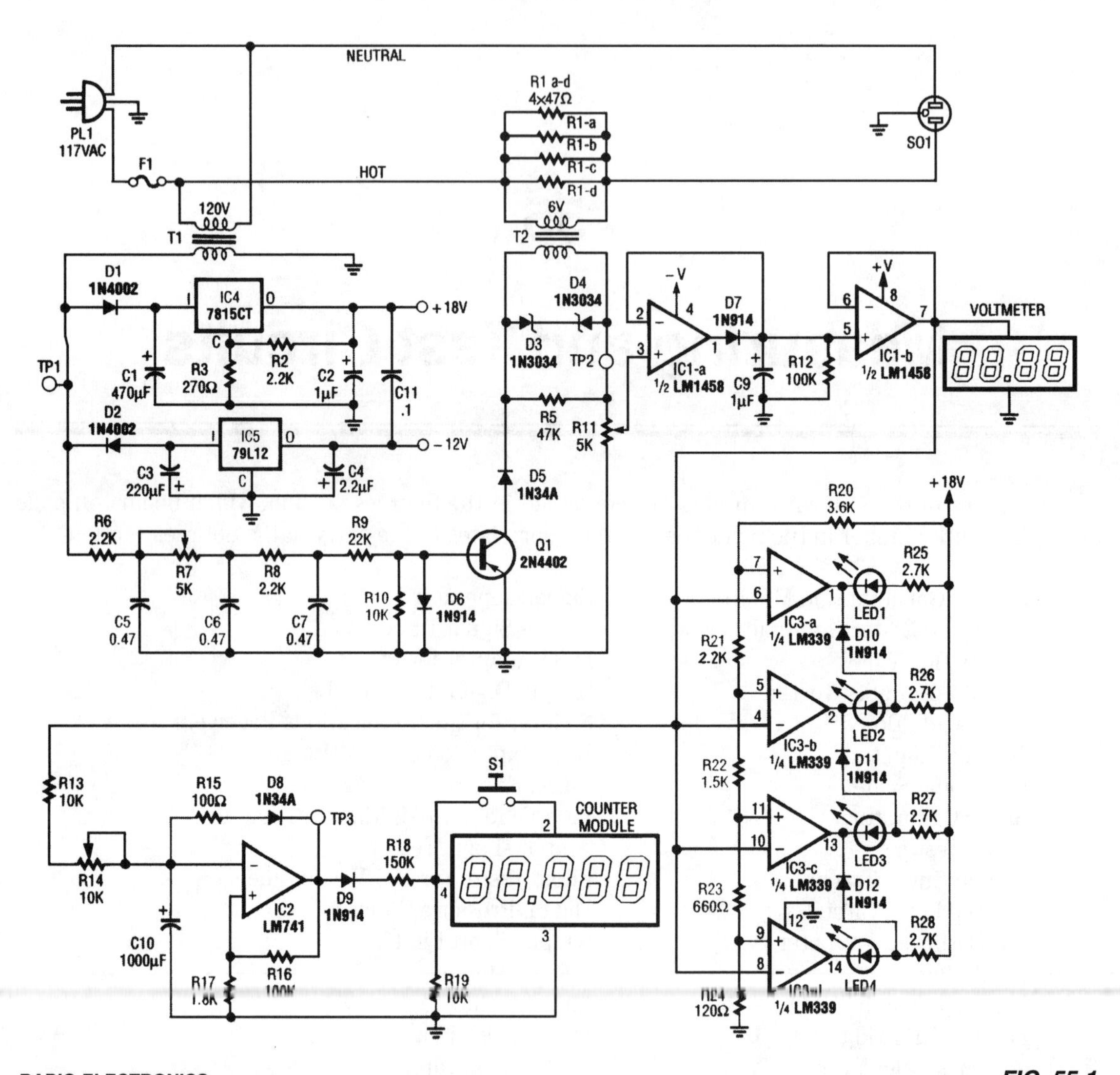

RADIO-ELECTRONICS

FIG. 55-1

The ECM circuit consists of four sections, as shown in the block diagram. A power converter generates a voltage that is proportional to the true of real power consumed by the load. That voltage feeds both a bargraph and a voltage-to-pulse converter. The bargraph gives an approximate indication of the amount of power used, and the voltage-to-pulse converter produces a pulse whose frequency is proportional to the power. The pulse triggers the counter module, which displays the cost of powering the monitored load.

HARMONIC DISTORTION ANALYZER

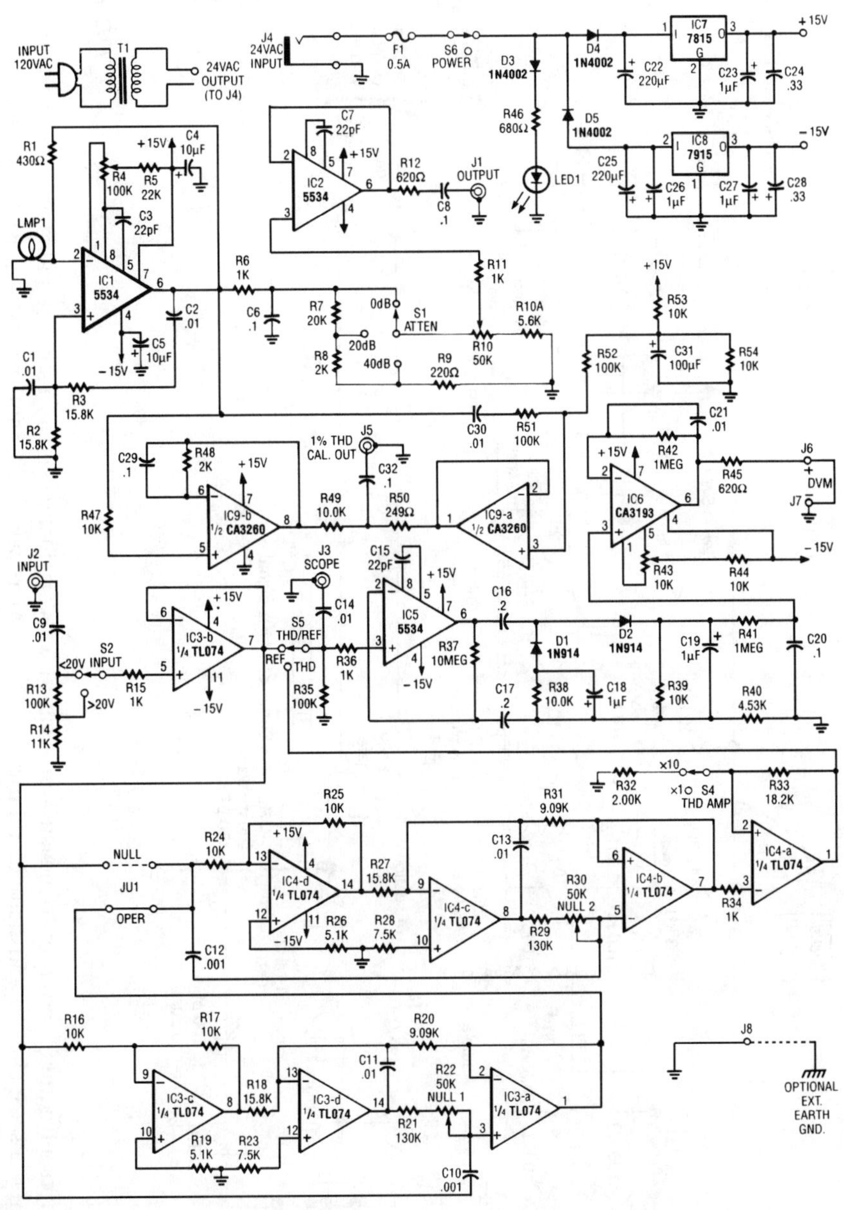

RADIO-ELECTRONICS

FIG. 55-2

The circuit includes a low-distortion, 1-kHz oscillator and will measure THD at a user selected voltage level for voltage amplifiers, or for checking amplifiers of power levels to 600 W. It will detect THD levels of .005% (–86 dB). A built-in one-percent THD calibrator is included. The output device is a digital multimeter (DMM).

WATCH TICK TIMER

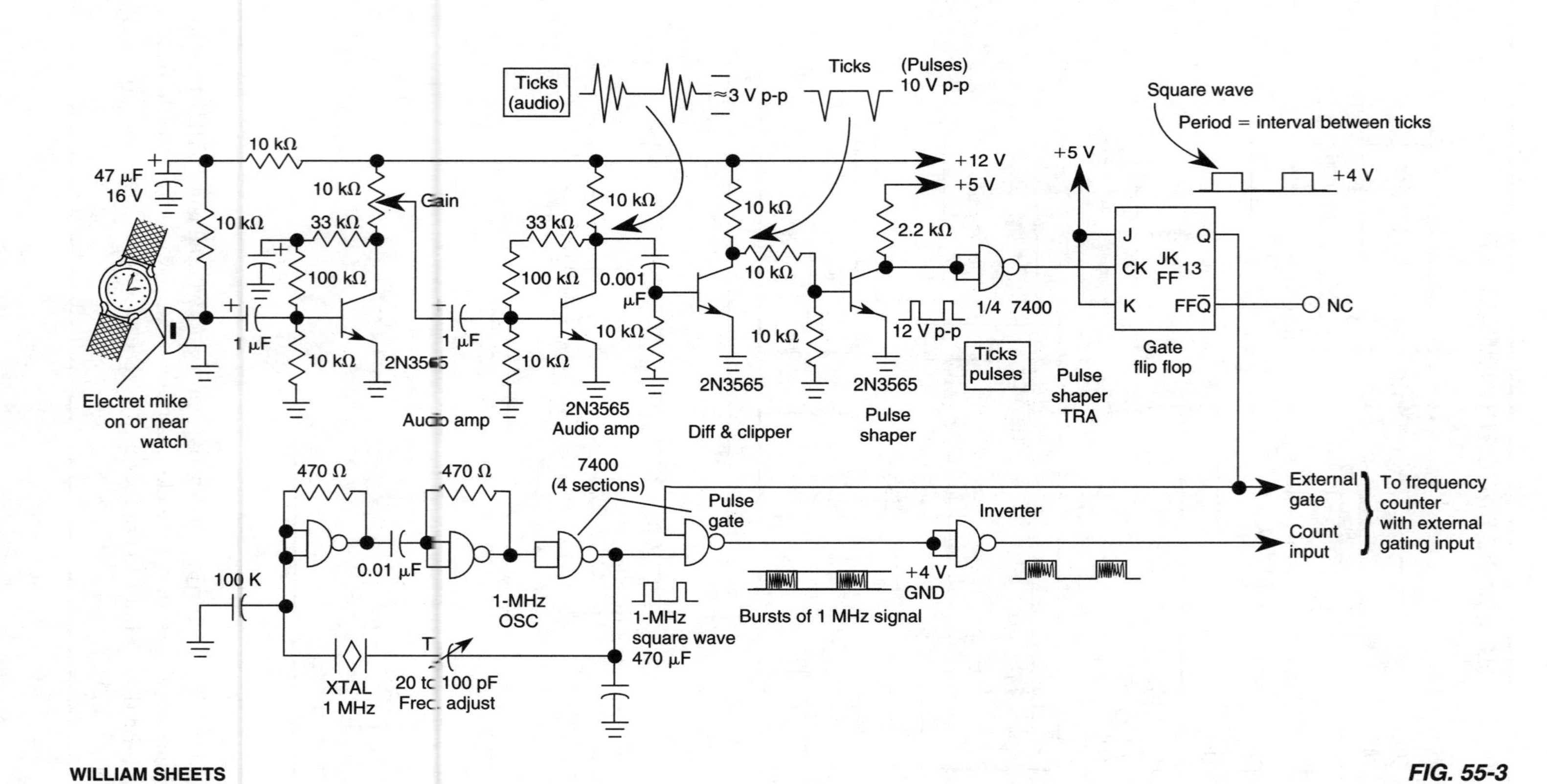

WILLIAM SHEETS

FIG. 55-3

This circuit adapts a frequency counter to measure intervals. It was originally used as a shutter speed checker for a photo application. The watch ticks are clipped and shaped and formed into a square wave. This square wave is used to gate an accurately known clock (1-MHz TTL XTAL OSC) and an external counter is used to directly count the clock pulses during the interval to be measured. A 1-MHz clock can be used to measure to a resolution of 1 µsec. Accuracy = ± time base ±1 µs ±1 count LSB.

VISUAL CONTINUITY TESTER

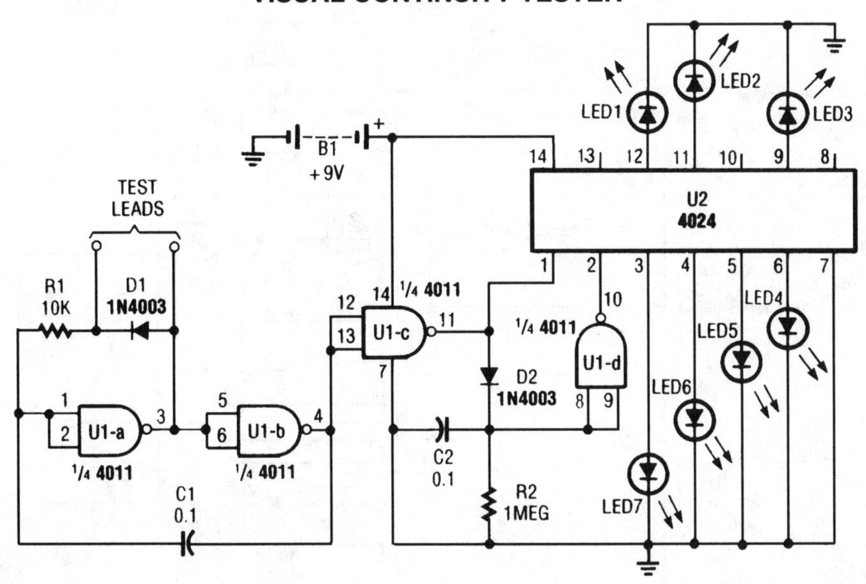

POPULAR ELECTRONICS *FIG. 55-4*

By judging the rate at which a particular LED flashes, you'll be able to estimate the resistance. The circuit consists of two IC's (1 4011 CMOS quad 2-input NAND gate, U1; and a 4024 binary counter, U2), seven LEDs, and a handful of additional components. All of the gates in U1 are wired as inverters.

Two of the inverters (U1-a and U1-b) comprise an astable-multivibrator (free-running oscillator) circuit, whose operating frequency depends on the amount of resistance detected between the test probes. Feedback from the output of the oscillator (at pin 4 of U1-b) back to the input of the circuit (at U1-a, pins 1 and 2) is provided via C1. Resistor, R1, along with the unknown resistance between the test probes, completes the RC timing circuit. The frequency of the oscillator decreases as the resistance between the test probes increases.

The output of the oscillator is fed to pin 12 and 13 of U1-c, the output of which then divides along two paths. In the first path, U1-c's output is applied to the clock input of U2 (a 4024 binary counter) at pin 1; in the other path, the signal is fed through D2 and across capacitor C2, causing it to begin charging. The charge on C2 is applied to U1-d at pins 8 and 9. The output of that inverter (U1-d) is fed to the reset terminal (pin 2) of U2. If there is continuity or a measurable resistance between the test probes, U2's reset terminal is pulled low, triggering the counter and allowing it to process the input pulses (count).

The rate of the count is proportional to the resistance between the test probes. If the resistance between the test probes is low, the counter advances slowly. The counter provides a 7-bit binary output that is wired to seven LEDs.

When the test probes are placed across a short circuit, LED7 flashes. If the tester is placed across a resistance of, for example, 2 MΩ, LED1 will flash. In either case, the LED whose assigned value most closely corresponds to the resistance connected between the two probes will flash continually at a steady pace, while the other LEDs will seem to flash intermittently.

RC DECADE BOX

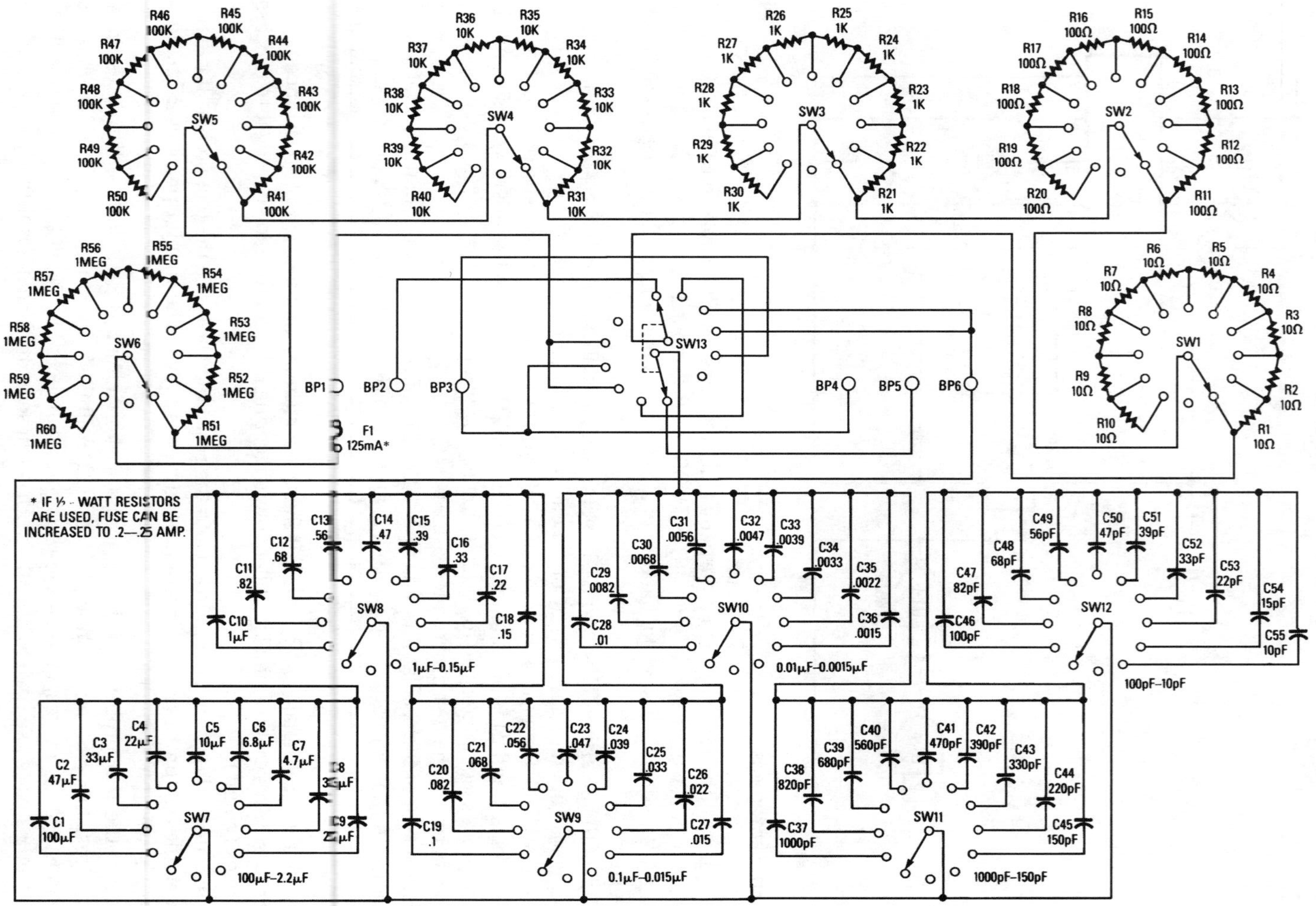

* IF ½-WATT RESISTORS ARE USED, FUSE CAN BE INCREASED TO .2—.25 AMP.

1989 R-E EXPERIMENTERS HANDBOOK

FIG. 55-5

RC DECADE BOX (*Cont.*)

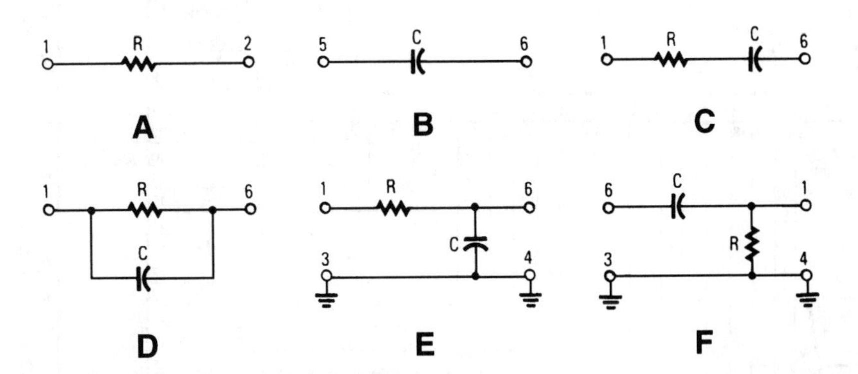

THE VARIOUS CONFIGURATIONS are set using S13: *(a)* resistor only and *(b)* capacitor only (both in position R/C); *(c)* series RC (position SER); *(d)* parallel RC (position PAR); *(e)* Low-Pass Filter (position LPF); and *(f)* High-Pass Filter (position HPF). The terminal numbers listed are those of binding-posts BP1–BP6.

TABLE 1—DECABOX TERMINAL CONNECTIONS

Configuration	S13 Position	IN/GND	OUT/GND
Resistance	R/C	IN: BP1	OUT: BP2
Capacitance	R/C	IN: BP5	OUT: BP6
Series RC	SER	IN: BP1	OUT: BP6
Parallel RC	PAR	IN: BP1	OUT: BP6
Low Pass Filter (Integrator)	LPF	IN: BP1 GND: BP3	OUT: BP6 GND: BP4
High Pass Filter (Differentiator)	HPF	IN: BP6 GND: BP3	OUT: BP1 GND: BP4

This decade box can be set for any resistance value between 10 Ω and 11.1 MΩ in 10-Ω stops. A switch can be used to configure several RC configurations. Use close tolerance components in the circuit. If possible, check components with an accurate bridge or other means to ensure accuracy.

DIGITAL ALTIMETER

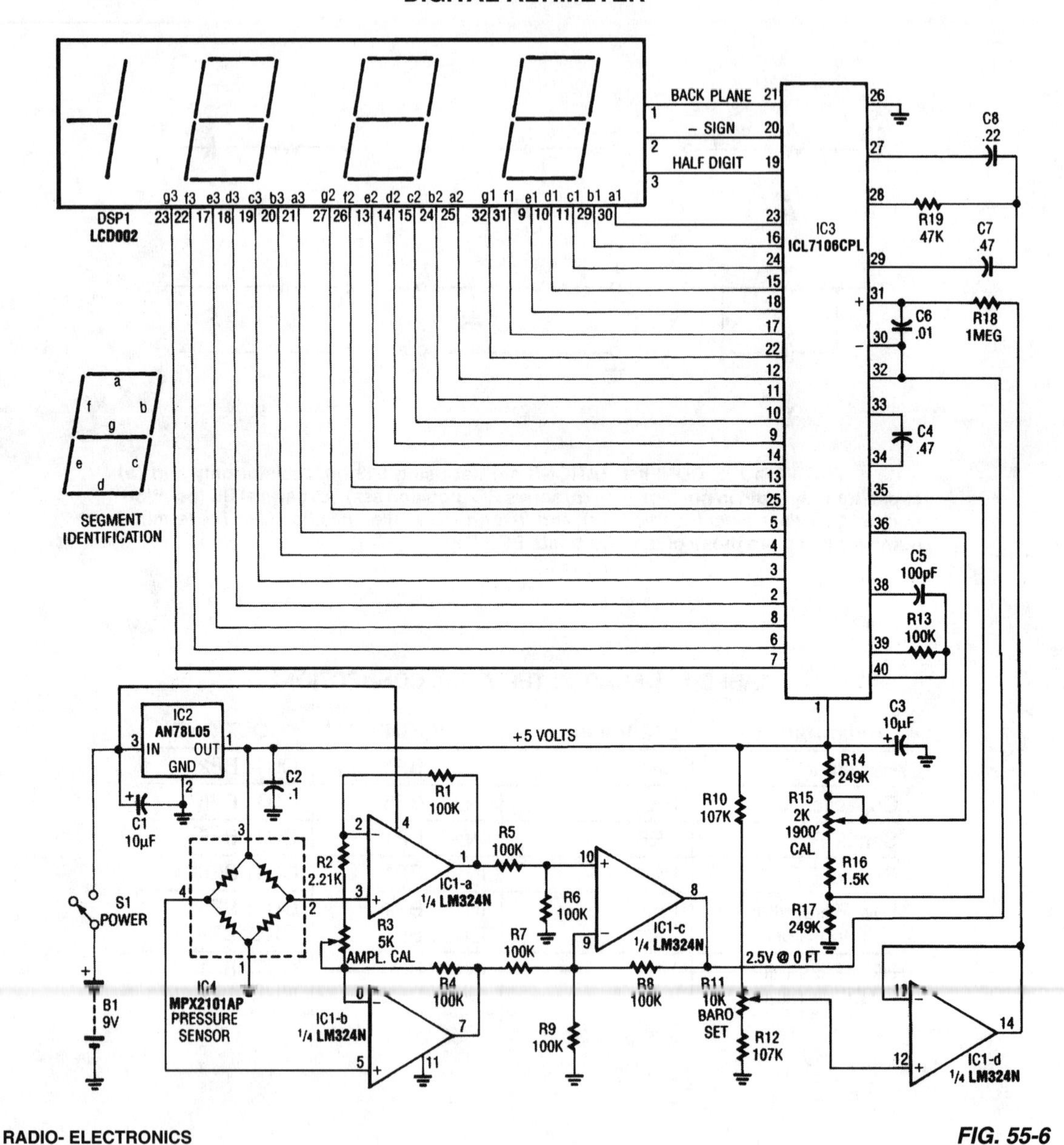

RADIO- ELECTRONICS

FIG. 55-6

A pressure sensor (IC4) is used with a dc amplifier to convert the bridge output (IC4) to a single-ended voltage. IC1d provides a reference voltage for setting barometric pressure. IC3 is an A/D converter manufactured by Intersil. This drives an LCD module. Calibration reads out in fact. A vacuum pump and a water-based manometer can be used for sensor calibration.

ELECTRONIC SCALE

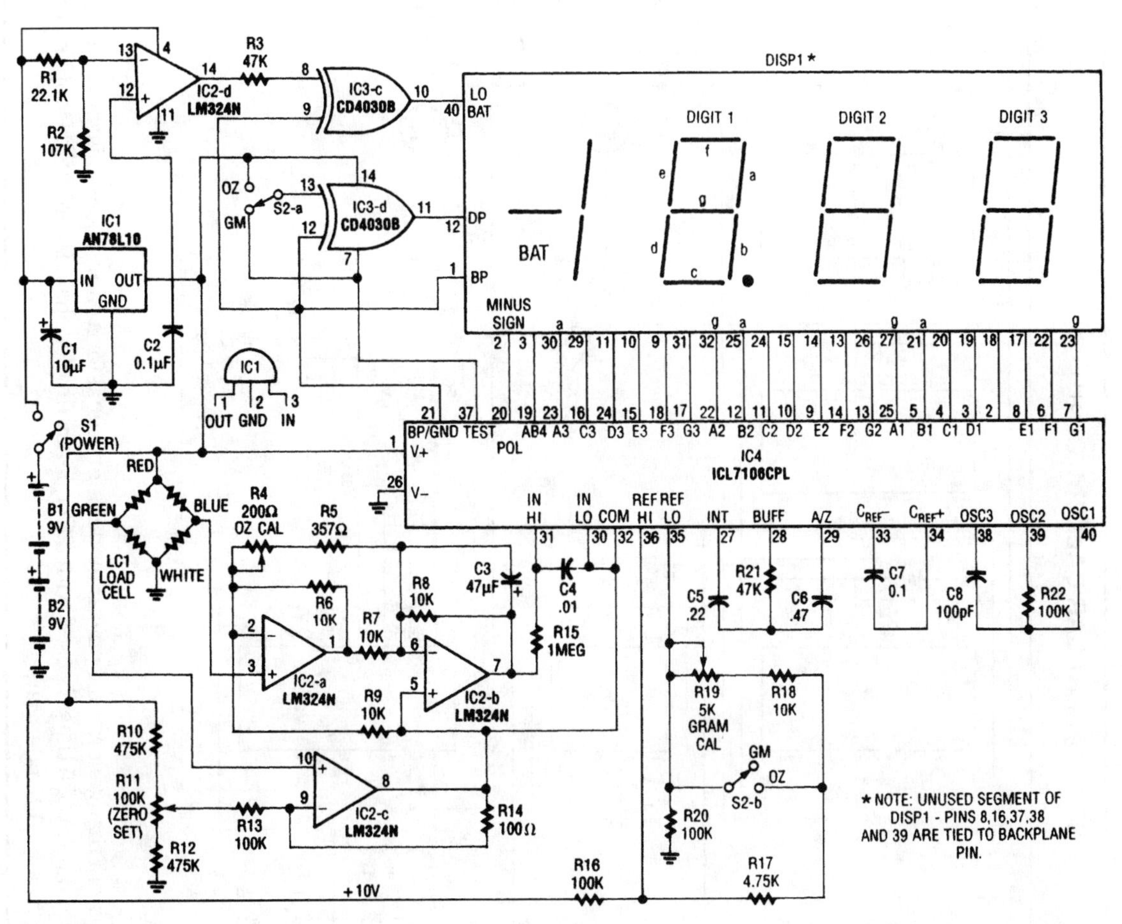

ELECTRONICS NOW ***FIG. 55-7***

An electronic scale using a pressure transducer (load cell) and an analog-digital (A/D) converter to drive a digital display is shown. The scale range depends on load cell. Display is calibrated in appropriate units. Components are on main circuit and display boards. The off-board controls are on the front panel and case. The cell in this scale is rated for 1.3 pounds (600 grams).

RADAR CALIBRATOR

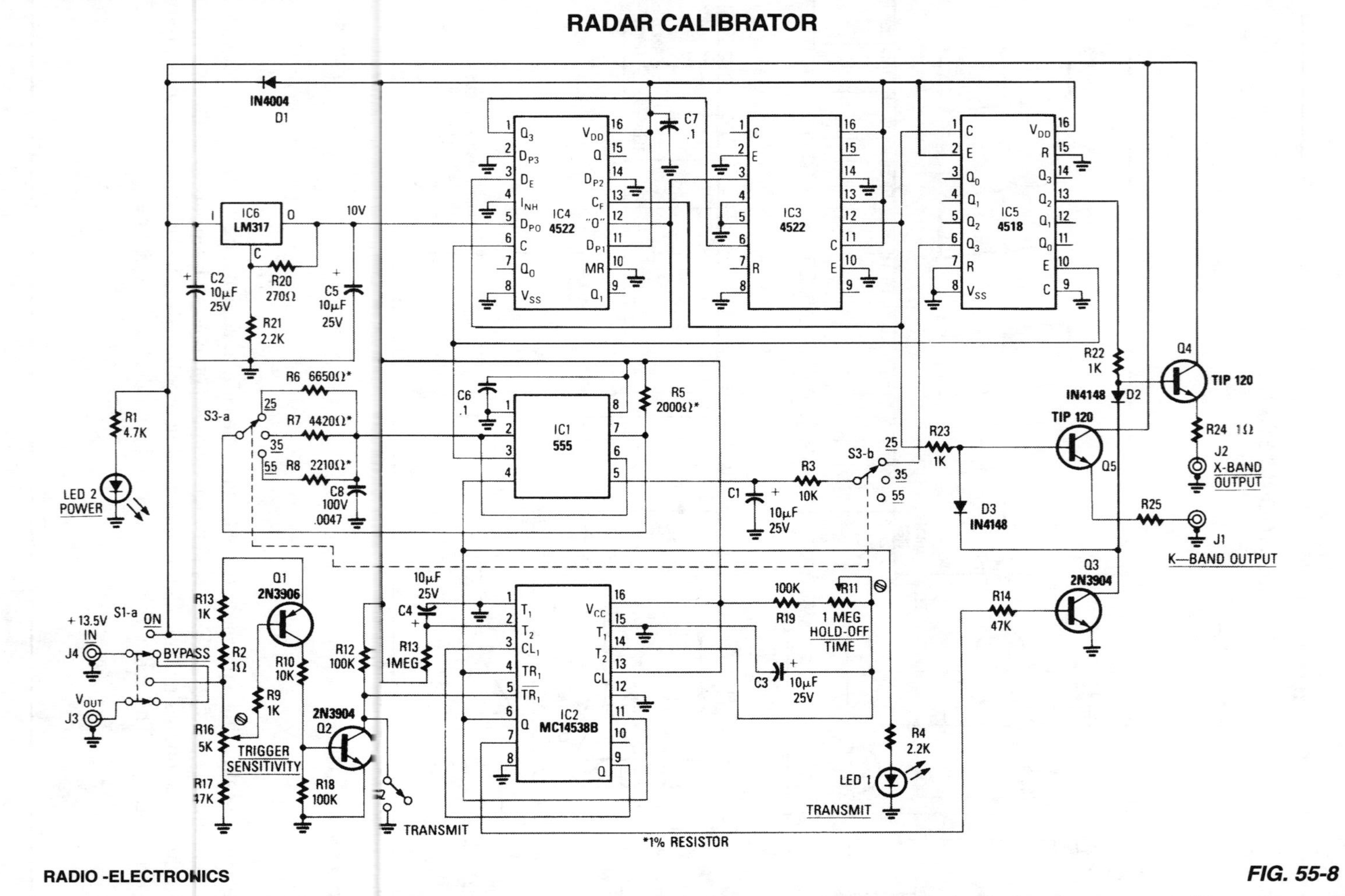

RADIO-ELECTRONICS

FIG. 55-8

This circuit is basically a system that generates a pulsed modulation signal for a Gunn diode microwave oscillator. Several speed settings are preset (S3 a and b). A 555 timer is used with a frequency divider chain to produce Doppler shift equivalents of 25, 35, and 55 mph, for both X- and D-band radars.

CABLE TESTER

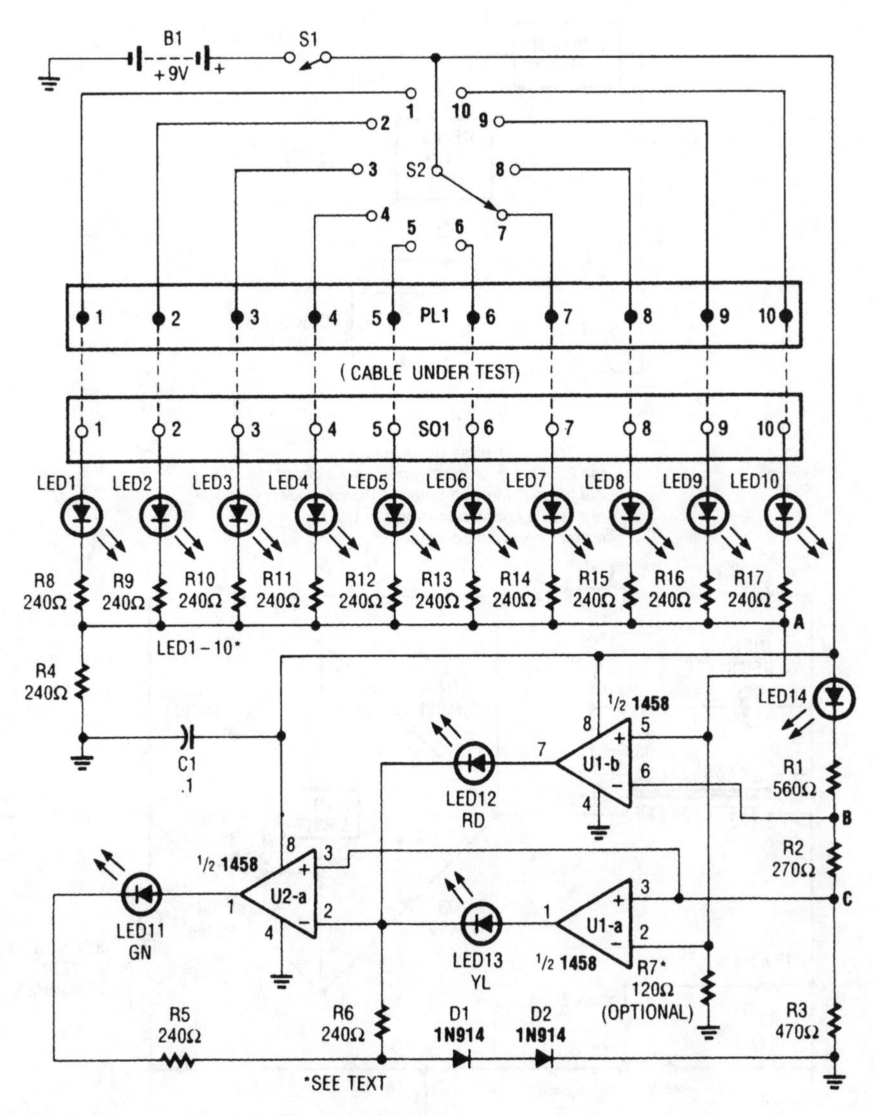

POPULAR ELECTRONICS ***FIG. 55-9***

At the heart of the cable tester are two op amps, which are used as a window comparator to indicate a short- or open-circuit condition. A third op-amp comparator is used to indicate a good circuit (i.e., neither open nor shorted). Colored LEDs are used to show the condition of individual conductors within the cable under test; a red one to indicate a short between conductors, a yellow one to identify an open conductor, and a green one to signify that the conductor is okay. Individual LEDs of a bar-graph display are used to show which conductor in the cable is being tested.

SIMPLE CURVE TRACER

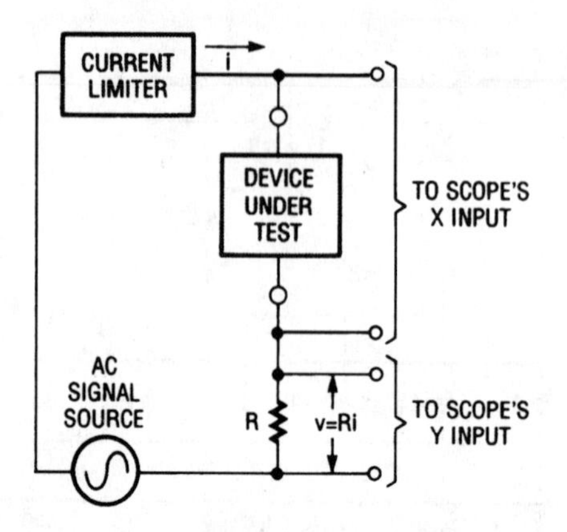

This is a simple block diagram of the EZ-Curve. Current-limited AC signals are passed through both the device under test and a precision resistor to yield current and voltage readings.

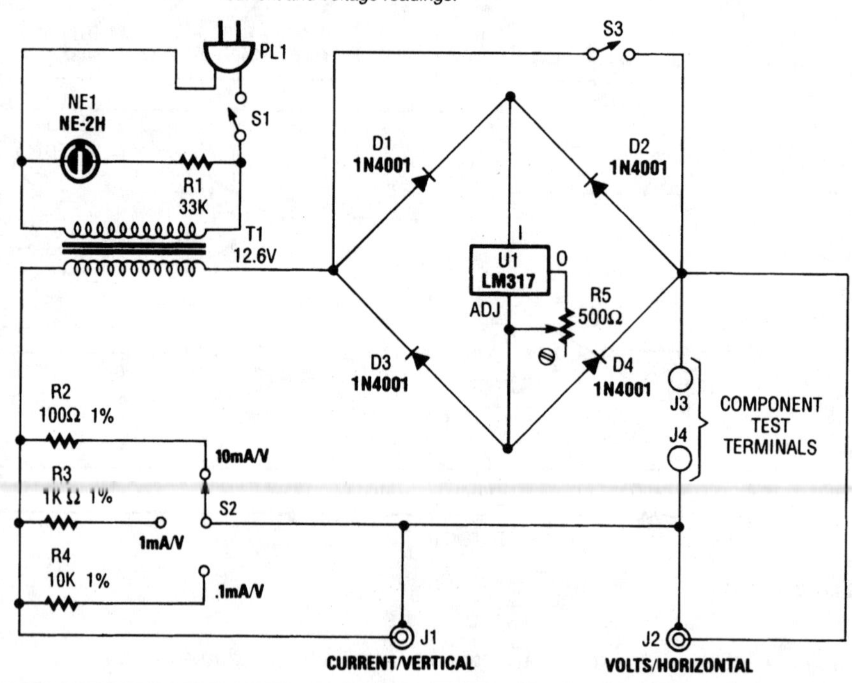

POPULAR ELECTRONICS

FIG. 55-10

Useful for checking diodes, transistors, triacs, SCRs, resistors, and LEDs, this curve tracer should prove useful in the experimenter's lab. It displays the volt-ampere characteristic of a two-terminal device on an oscilloscope.

VOLTAGE LEVEL CIRCUIT

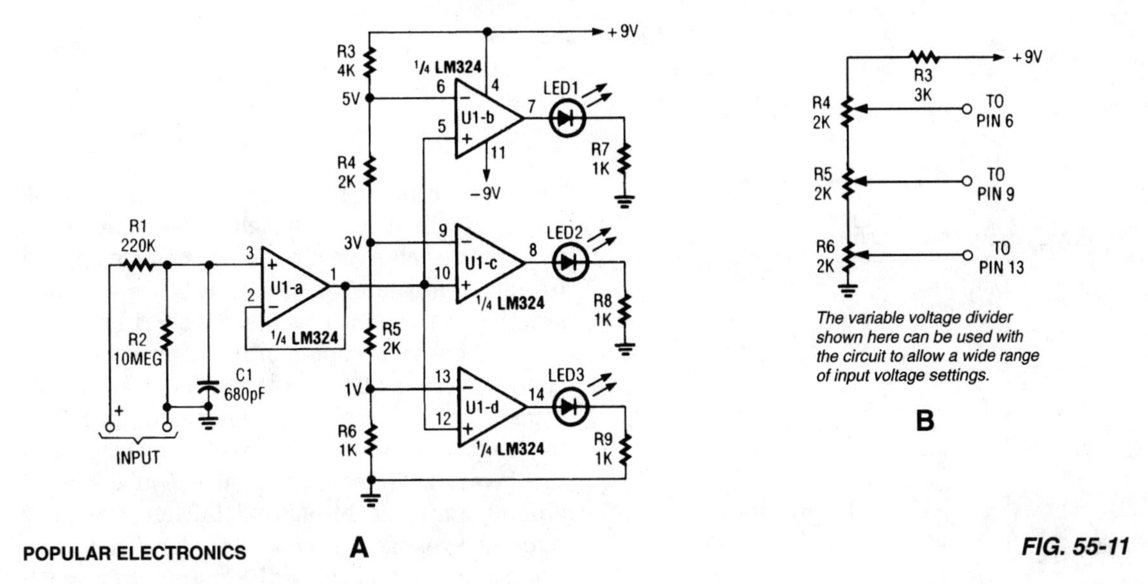

POPULAR ELECTRONICS **FIG. 55-11**

A DC op amp and a comparator with a ladder reference divider allow a dc input voltage to light one or more LEDs, depending on voltage levels.

LOW-DRIFT dc VOLTMETER

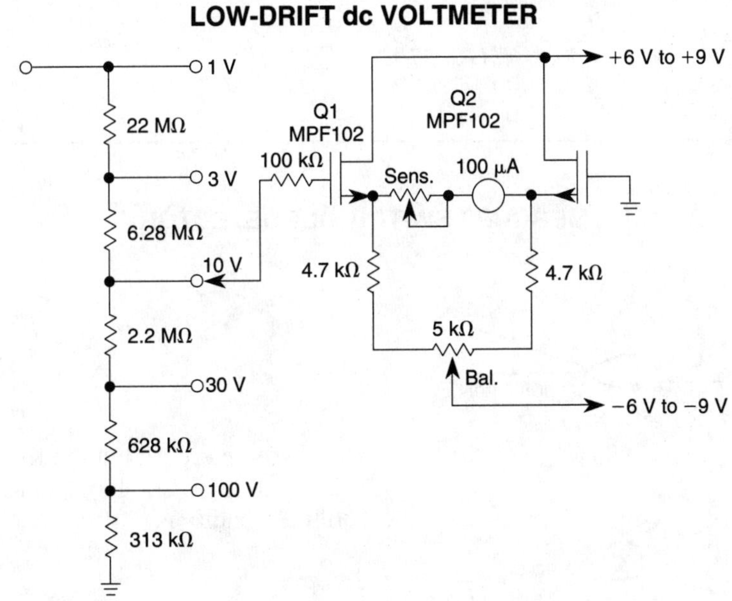

WILLIAM SHEETS **FIG. 55-12**

This voltmeter uses a pair of JFETs in a balanced-bridge source-follower amplifier circuit. Q1 and Q2 should be matched within 10% for I_{DSS}. This minimizes meter drift and maintains bridge balance over temperature.

LIGHT METER

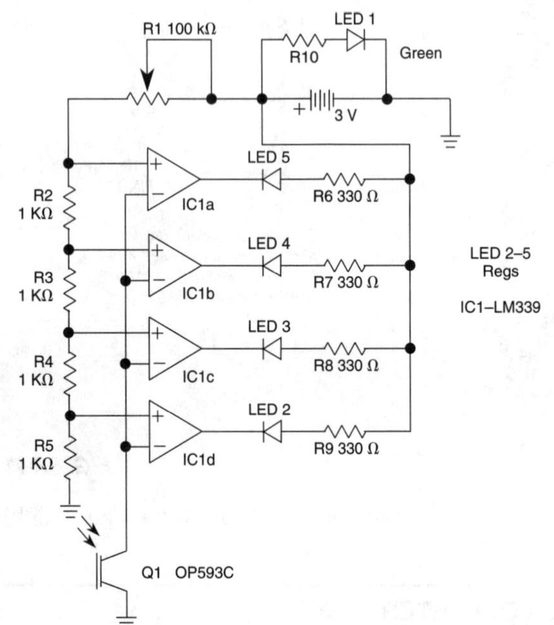

WILLIAM SHEETS

FIG. 55-13

The outputs from the comparators will swing, in sequence, from high to low as the input voltage rises above the reference voltage applied to each comparator. The output LEDs will then switch on in sequence as the voltage rises.

The inverting inputs of the comparators are connected in common to the collector of phototransistor Q1. When Q1 is illuminated, its collector-emitter junction conducts, thereby placing all the inverting inputs within a few millivolts of ground. For most settings of R1, each of the four reference voltages exceeds the value. Therefore, when Q1 is illuminated, the output from each comparator is high and its respective indicator LED is off.

MERCURY SWITCH TILT DETECTOR

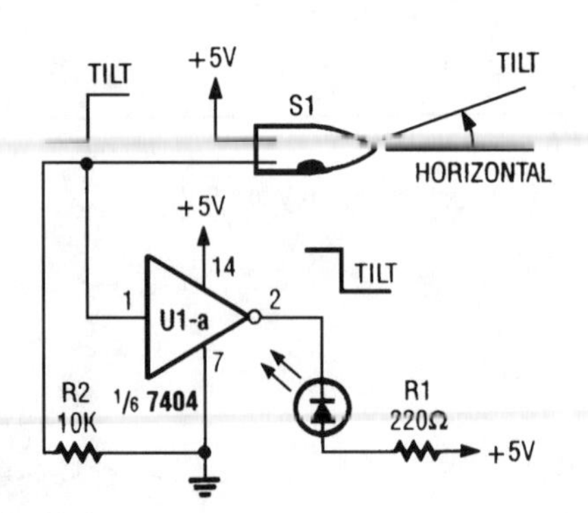

POPULAR ELECTRONICS

FIG. 55-14

If the mercury bulb in this circuit is tipped, U1-a will light LED1 by going low, indicating a “tilted” condition.

50-MHz RF BRIDGE

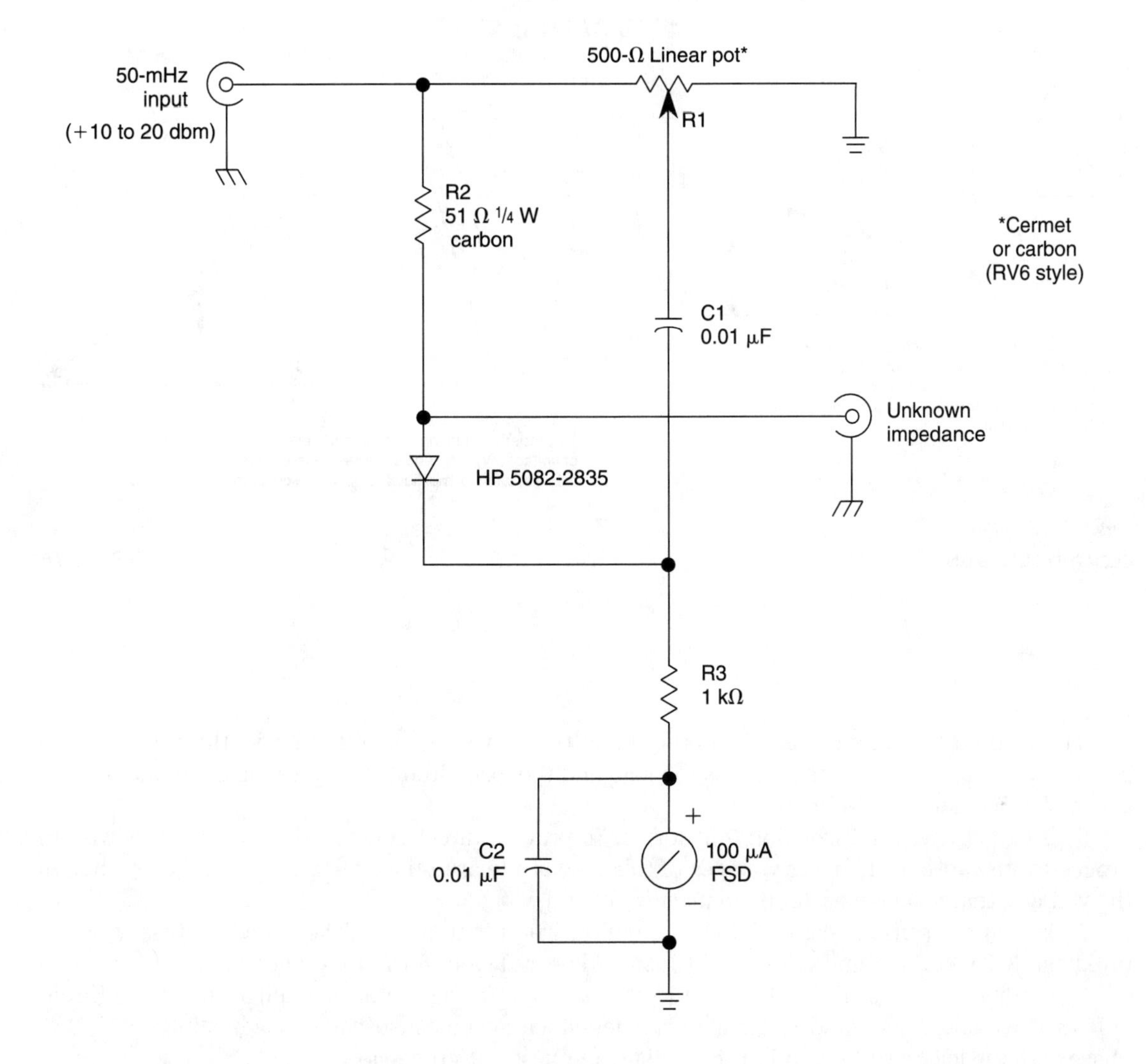

WILLIAM SHEETS *FIG. 55-15*

The bridge shown was used for measurements on 50-MHz amateur radio antennas. R1 is a miniature 500 Ω linear potentiometer. The unknown impedance is compared to R2, a 51-Ω resistor. An external signal source is required.

ac WATTS CALCULATOR

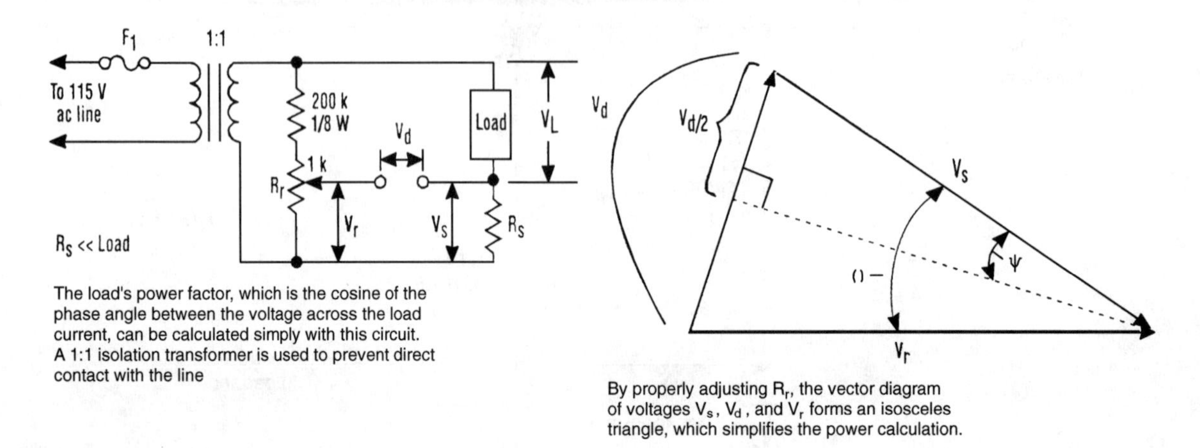

The load's power factor, which is the cosine of the phase angle between the voltage across the load current, can be calculated simply with this circuit. A 1:1 isolation transformer is used to prevent direct contact with the line

By properly adjusting R_r, the vector diagram of voltages V_s, V_d, and V_r forms an isosceles triangle, which simplifies the power calculation.

ELECTRONIC DESIGN ***FIG. 55-16***

The method basically consists of determining the power factor of the load—the cosine of the phase angle between the voltage across the load and the load circuit. Using a simple circuit, that angle can be calculated quite simply.

This circuit uses a 1:1 isolation transformer to prevent direct contact with the line. It is wise to proceed with caution whenever voltages of this magnitude are utilized in a test setup, even though the voltages that will be measured are usually below 1 V.

R_s is a circuit-sense resistor and R_r is a multi-turn potentiometer. The voltage across R_r is approximately 0.5% of the line voltage, which should be sufficient for most applications.

R_r is adjusted so that $|V_r| = |V_s|$; then V_d is measured. In the vector diagram according to Kirchhoff's voltage law, V_s, V_d, and V_r form a triangle, which becomes isosceles by adjusting R_r. V_s is in phase with the load current and V_r is essentially in phase with the load voltage.

The power delivered to the load can be calculated as follows:

$$P_L = V_L \times I_L \times \text{Cos}\ \theta$$

$$= V_L \times (V_s/R_s) \times \text{Cos}\ [2\ \text{Sin}-1\ (V_d/2V_s)]$$

$$[\theta\ 2\ \psi = 2\ \text{Sin}-1\ (V_d/2V_s)]$$

AUDIO-FREQUENCY METER

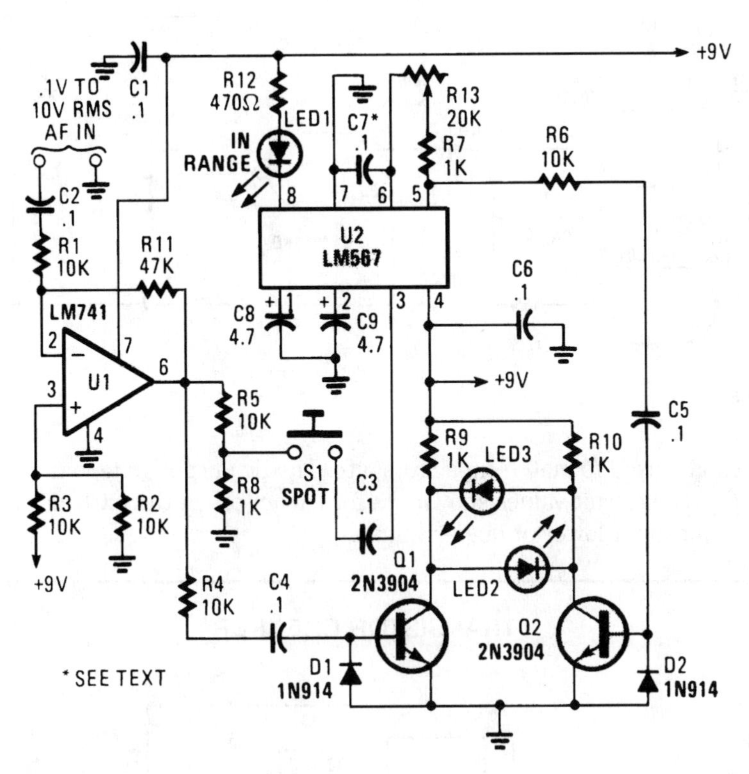

POPULAR ELECTRONICS

FIG. 55-17

This meter differs from the norm in that it does not use a D'Arsonval movement or digital display to give a reading of the input frequency. Instead, the measured frequency is read from a hand-calibrated dial.

Any audio signal applied to the circuit is amplified by U1 and the resulting output is divided along two paths. In one path, the output signal is applied to the mixer; in the other path, the signal is applied to the input of U2 through S1 (a normally open pushbutton switch).

The portion of the amplifier signal that is fed to the mixer is applied to the base of Q1, causing it to toggle on and off at the signal frequency. In the other path, when S1 is pressed, a portion of the op amp's output is applied to U2. If the signal is within the range of U2's internal oscillator's operating frequency, LED1 lights, and a signal is fed to the base of Q2. If the two signals arriving at the mixer do not match exactly, LED2 and LED3 light. That means that the circuit must be fine tuned, which is accomplished by releasing S1 and fine tuning R13 until LED2 and LED3 go out. The dial setting at that point gives the frequency of the input signal to within 1 Hz (or as close as the calibrated dial will allow).

ONE-IC CAPACITANCE TESTER

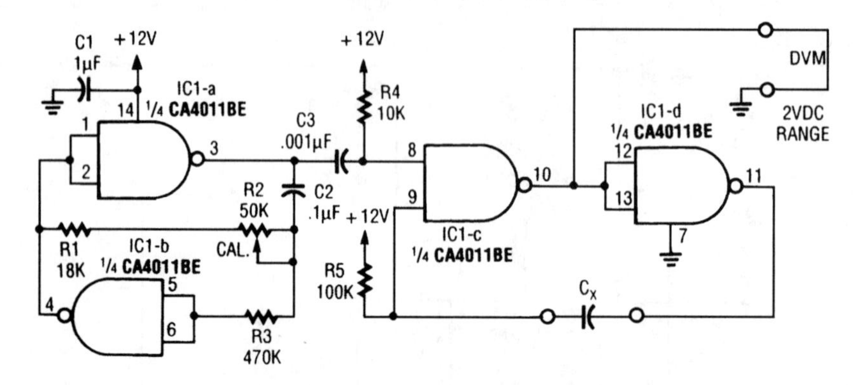

RADIO-ELECTRONICS **FIG. 55-18**

This circuit can be used to match capacitors, etc. The dc output voltage is related to the capacitance values of C_X. The circuit values shown are for capacitors in the 0.01-μF order of magnitude, but they can be changed for lower or higher values.

TRANSISTOR CHECKER

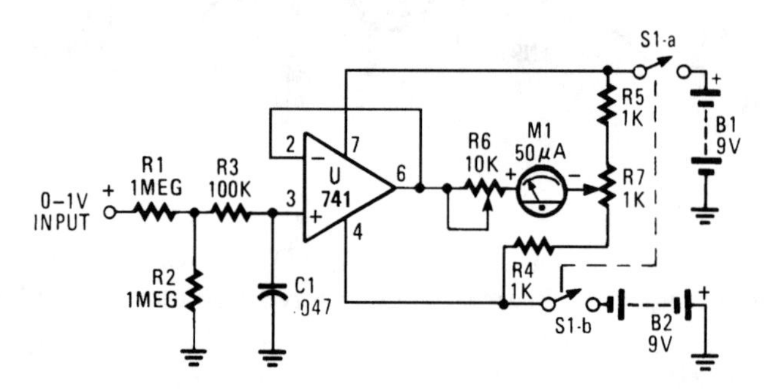

POPULAR ELECTRONICS **FIG. 55-19**

The circuit is built around a 741 general-purpose op amp that is configured as a voltage follower; with the components shown, the op amp has a voltage gain of one. The output of the 741 is used to drive a 50-μA meter movement. Potentiometer R7 is used to zero the meter and R6 sets the meter's full-scale reading.

Calibrating the meter is a snap. With no input applied to the circuit, set R6 to mid-position and adjust R7 to zero the meter. Once that is done, apply a positive 1-Vdc voltage to the input and adjust R6 for a full-scale reading. The voltmeter can be adjusted to read both positive and negative voltages by adjusting R7 for a center scale reading at the meter's zero position and a positive 1-V reading at the meter's full-scale position.

LOW-CURRENT AMMETER

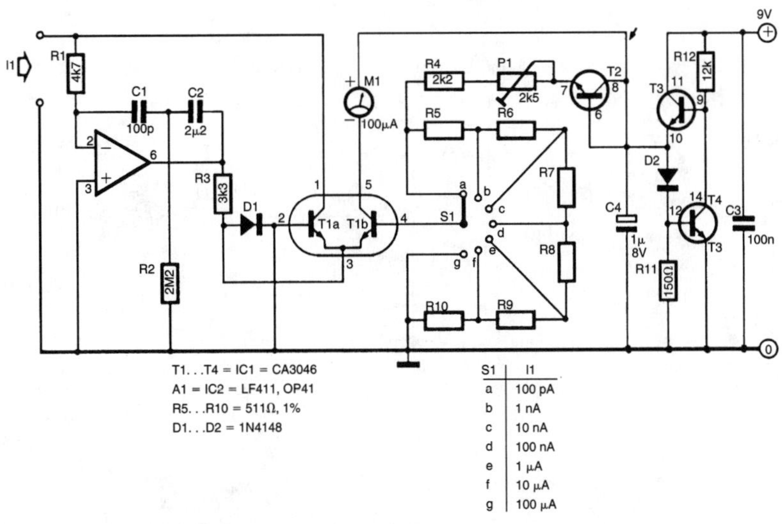

S1	I1
a	100 pA
b	1 nA
c	10 nA
d	100 nA
e	1 µA
f	10 µA
g	100 µA

303 CIRCUITS ***FIG. 55-20***

Without using high-value precision resistors, this circuit uses a current mirror, T1a/T1b. Currents of 100 pA can be measured with this circuit. M1 is a 100-mA meter. Make sure to use a high-quality PC board and low-leakage circuit construction.

ANALOG FREQUENCY METER

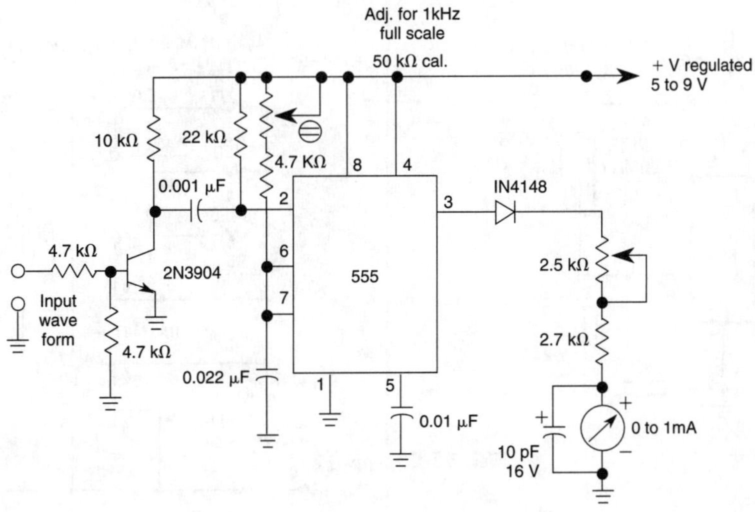

WILLIAM SHEETS ***FIG. 55-21***

This 1-kHz linear-scale analog frequency meter circuit uses the 555 as a pulse counter. Frequency is read on M1, (or 1 mA meter) which can be calibrated to read 0 to 1 kHz.

ELECTROMAGNETIC FIELD SENSOR

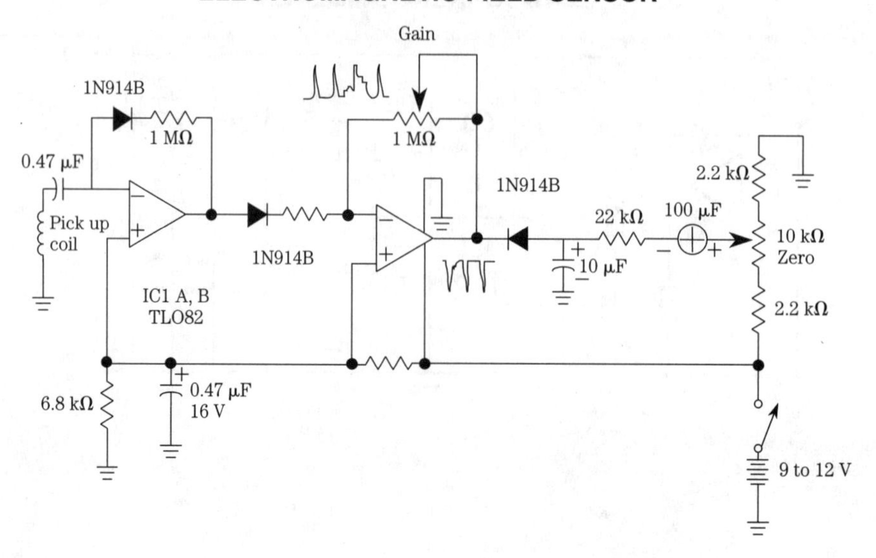

WILLIAM SHEETS *FIG. 55-22*

A telephone pick-up coil is used as a sensing coil. Any 60-Hz hum picked up by the sensing coil is rectified, amplified, and detected, and then drives a meter.

MAGNETIC PROXIMITY SENSOR

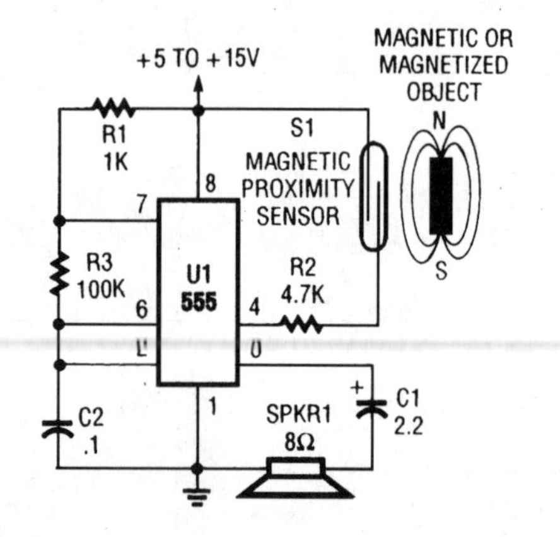

POPULAR ELECTRONICS *FIG. 55-23*

A magnetic need switch enables a 555 oscillator, which drives a speaker. C2 can be varied for different tone frequencies.

HIGH-IMPEDANCE VOLTMETER

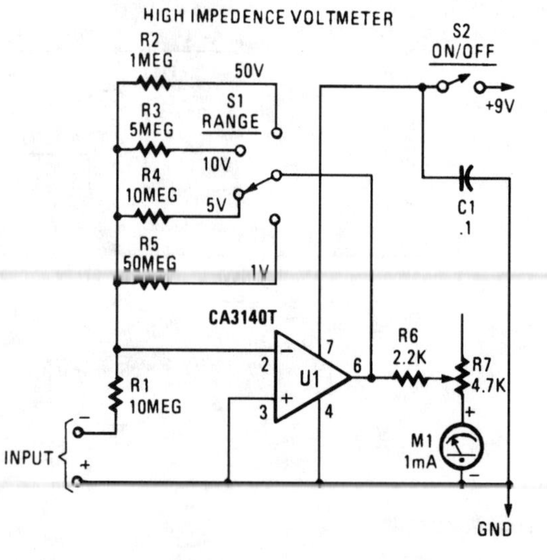

POPULAR ELECTRONICS *FIG. 55-24*

FAST VIDEO SIGNAL AMPLITUDE MEASURER

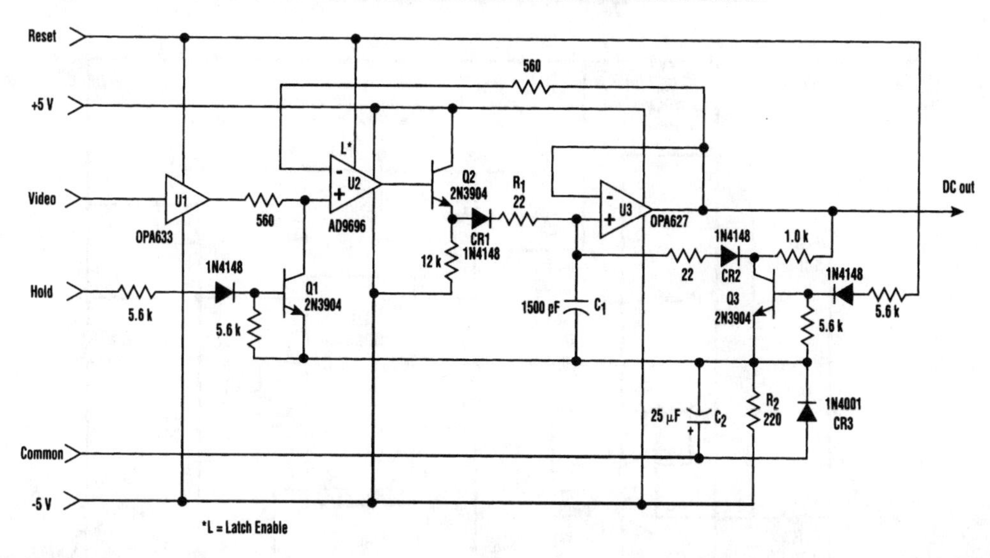

ELECTRONIC DESIGN

FIG. 55-25

Video-signal amplitude can be measured with this simple circuit, which is basically a modified standard peak detector. The device can verify RGB generated by video RAMDACs. U1 is a high-speed buffer and U2 is a latched comparator. C1 is a hold capacitor. Reset is performed by Q3. U2 has a latch that maintains the last comparator state. The reset holds the comparator output low during the reset operation. The dc output voltage is equal to the signal's maximum amplitude.

SIGNAL GENERATOR

OUTPUT
C1 .0022
R3 2K LEVEL
C2 .022
R1 5600Ω
Q2 2N464
R2 2MEG PITCH
Q1 2N366
S1
B1 +9V

POPULAR ELECTRONICS

FIG. 55-26

Useful for troubleshooting audio, video, and lower frequency RF amplifiers, this circuit generates a signal that is rich in harmonics.

SIMPLE SIGNAL TRACER

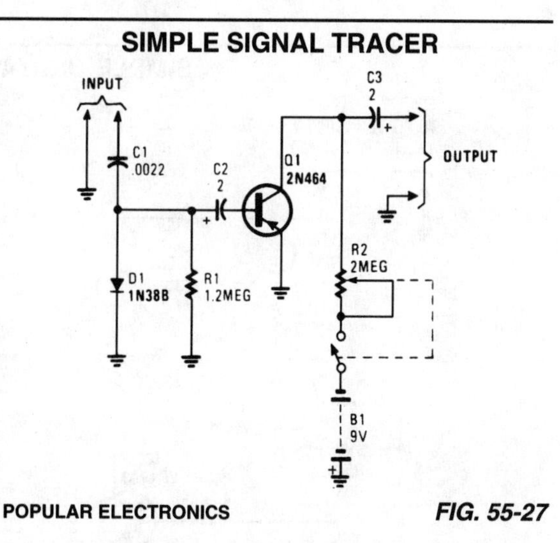

POPULAR ELECTRONICS

FIG. 55-27

In this circuit, C1/D1/R1 form an envelope detector. C2 couples audio to the base of Q1. R2 can be adjusted for the desired gain.

DVM ADAPTER FOR PC

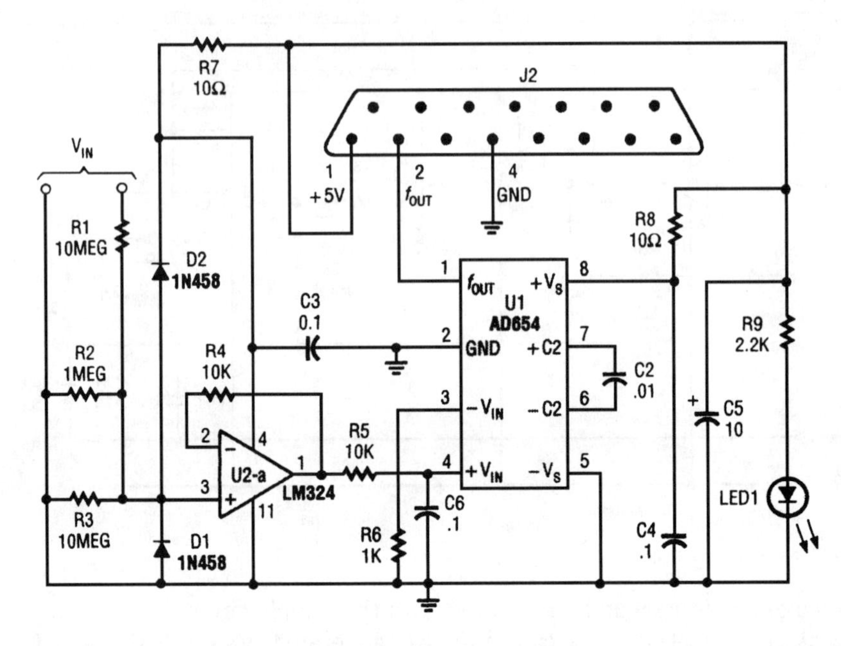

POPULAR ELECTRONICS ***FIG. 55-28***

The adapter consists of a voltage to frequency adapter with a signal conditioner and protection circuit. J2 connects to the game port of a PC. See reference listed for software for use with this circuit.

SIMPLE DIGITAL LOGIC PROBE

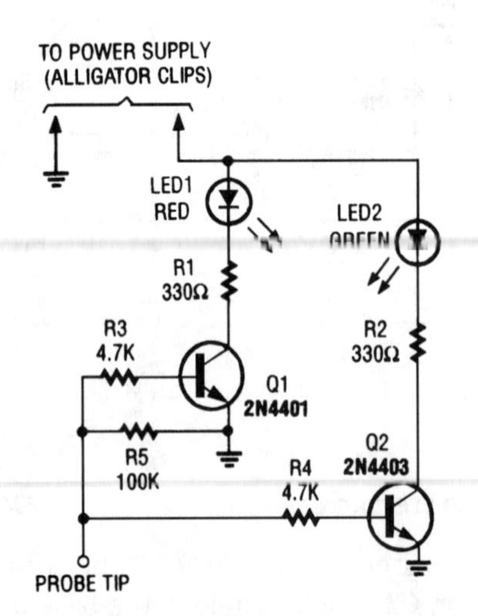

POPULAR ELECTRONICS ***FIG. 55-29***

The design of the digital logic probe centers around a pair of complementary bipolar transistors, which, in this application, are used as electronic switches.

S METER FOR COMMUNICATIONS RECEIVERS

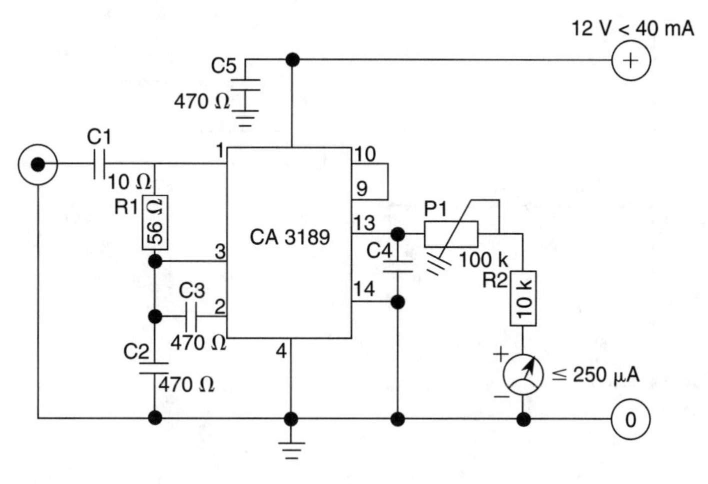

303 CIRCUITS ***FIG. 55-30***

Because many amateur receivers are fitted with an S meter that functions far from logarithmically, the proposed circuit should be a welcome extension of such receivers. Although ICs such as the CA3089 or the CA3189 are not in common use anymore, they serve a useful purpose in the meter circuit, because, apart from a symmetric limiter, a coincidence detector, and an AFC amplifier, they contain a very good logarithmic amplifier-detector.

As is seen, the circuit is fairly simple, but remember that these ICs operate up to about 30 MHz; the wiring of the meter and its connections in the receiver should be kept as short as possible.

LED EXPANDED SCALE VOLTMETER

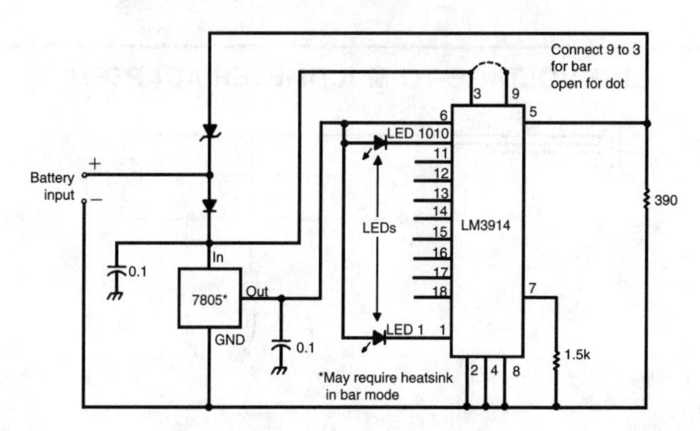

73 AMATEUR RADIO TODAY ***FIG. 55-31***

A 10-V zener diode is used to expand the scale of a 0- to 5-V voltmeter to a 10- to 15-V voltmeter. The LED bar graph lights one segment per 0.5-V input above 10 V. The 7805 IC provides a 5-V reference and 5 V for the bar graph LEDs.

1-KHz HARMONIC DISTORTION METER

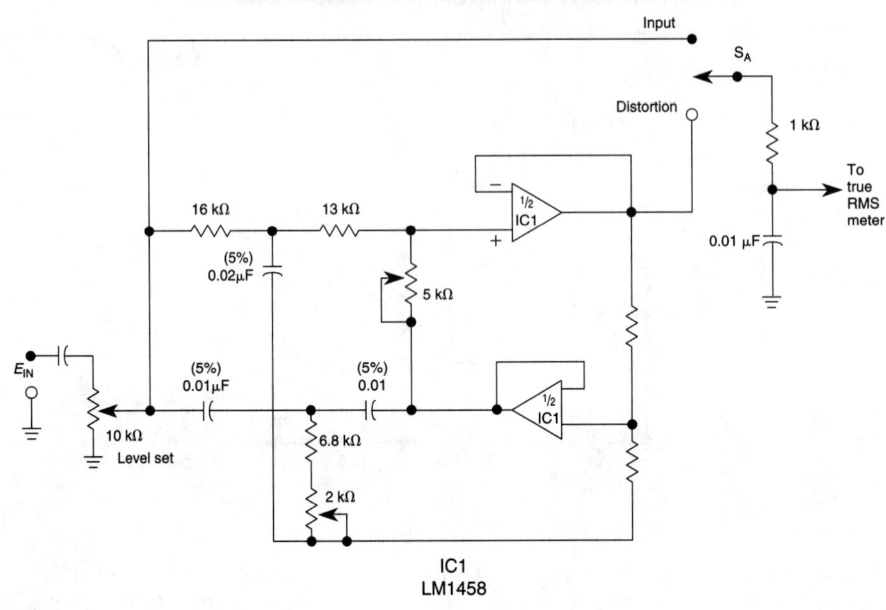

WILLIAM SHEETS

FIG. 55-32

The circuit useful for distortion measurements notches out the fundamental frequency of 1 kHz to allow measurement of the residual level of harmonics. First a true RMS meter is used to measure the 1-kHz input level E_{in} by setting S_A to the input position. Then, S_A is placed in the distortion position and the 2 k potentiometer is adjusted for a null. The residual reading is noted. The THD is then calculated based on the formula:

LINE VOLTAGE-TO-MULTIMETER ADAPTER

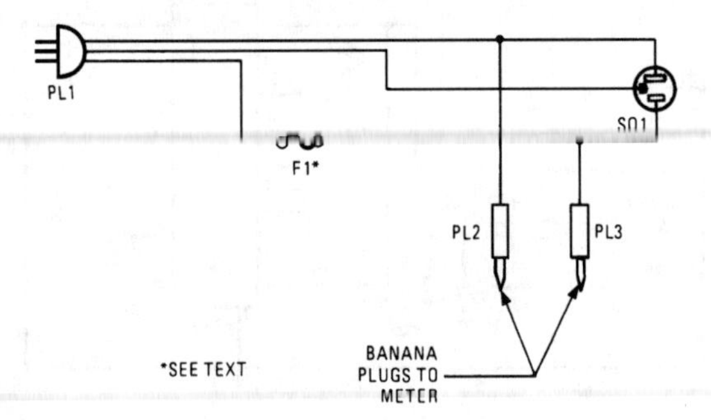

POPULAR ELECTRONICS

FIG. 55-33

This ac line-to-multimeter adapter can make checking line voltage safer. You can use it to find taxing loads on your household wiring.

AUDIBLE LOGIC TESTER

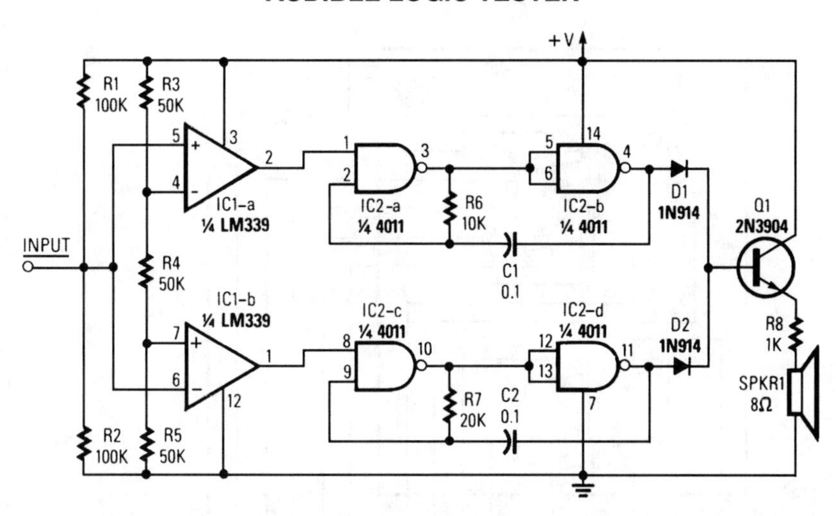

1989 R-E EXPERIMENTERS HANDBOOK **FIG. 55-34**

The tester provides an audible indication of the logic level of the signal presented to its input. A logic high is indicated by a high tone, a logic low is indicated by a low tone, and oscillation is indicated by an alternating tone. The input is high impedance, so it will not load down the circuit under test. It can be used to troubleshoot TTL or CMOS logic.

The input section determines whether the logic level is high or low, and enables the appropriate tone generator; it consists of two sections of an LM339 quad comparator. One of the comparators (IC1-a) goes high when the input voltage exceeds 67% of the supply voltage. The other comparator goes high when the input drops below 33% of the supply. Resistors R1 and R2 ensure that neither comparator goes high when the input is floating or between the threshold levels.

The tone generators consist of two gated astable multivibrators. The generator built around IC2-a and IC2-b produces the high tone. The one built around IC2-c and IC2-d produces the low tone. Two diodes, D1 and D2, isolate the tone-generator outputs. Transistor Q1 is used to drive a low-impedance speaker.

SHORT TESTER FOR 120-V EQUIPMENT

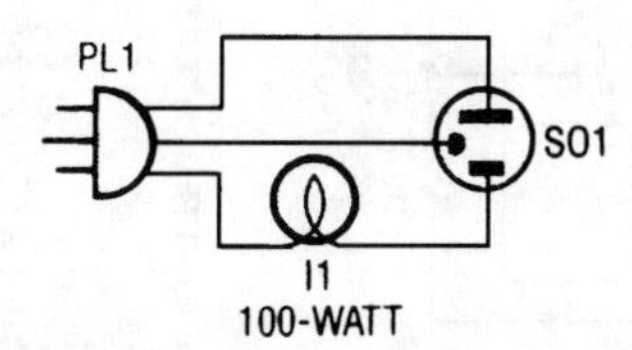

Do you deal with old equipment in unknown condition? If so, this little circuit could keep you from causing further harm to already shorted devices.

POPULAR ELECTRONICS **FIG. 55-35**

DIGITAL PRESSURE GAUGE

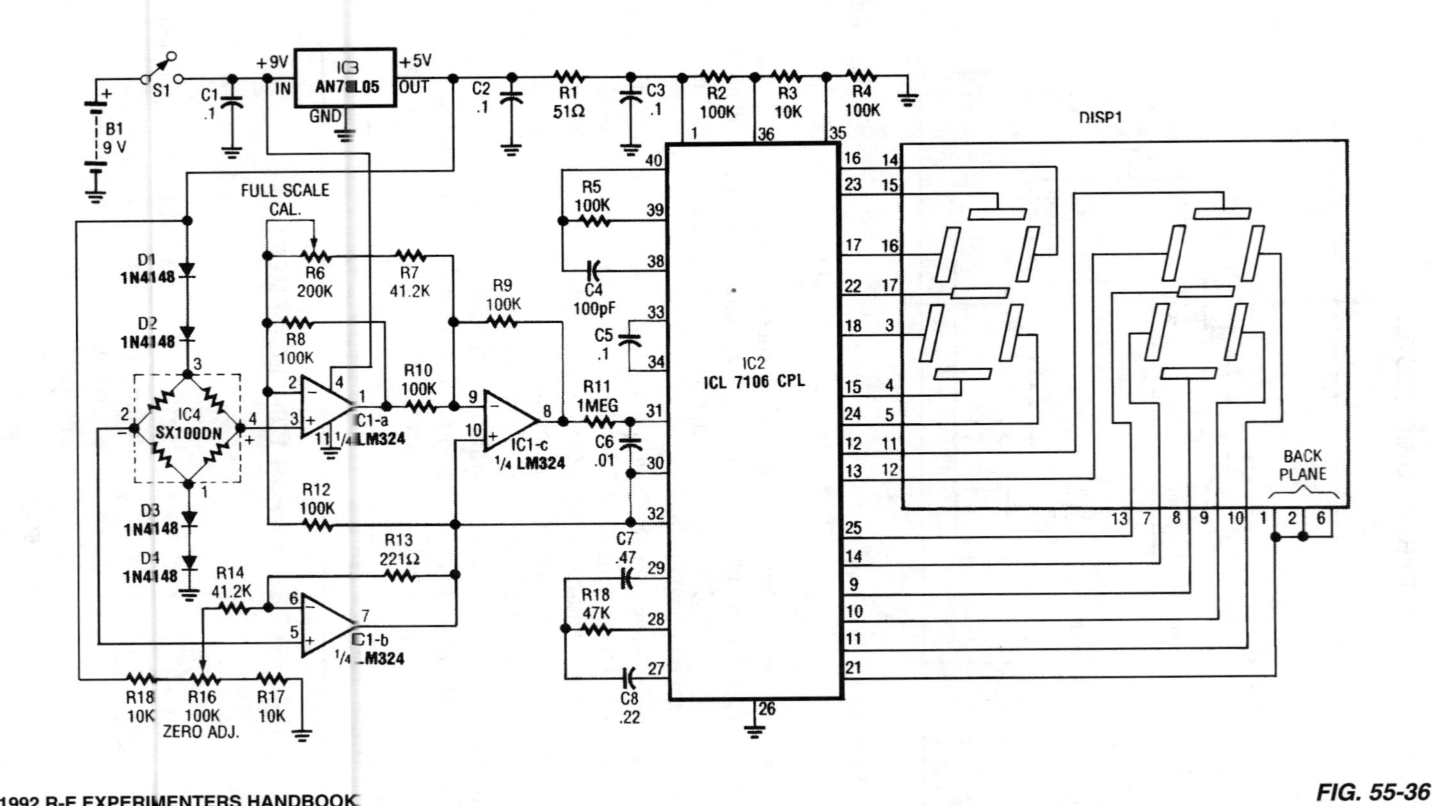

1992 R-E EXPERIMENTERS HANDBOOK

FIG. 55-36

This electronic pressure gauge uses a Wheatstone bridge-type pressure sensor to drive a 3½ digit A/D converter and a display. IC1 is a pump (quad) that interfaces the bridge sensor to the A/D converter. R16 provides zero adjustment and R6 provides full-scale calibration. D1 thru D4 provide temperature compensation.

SIMPLE SHORT FINDER

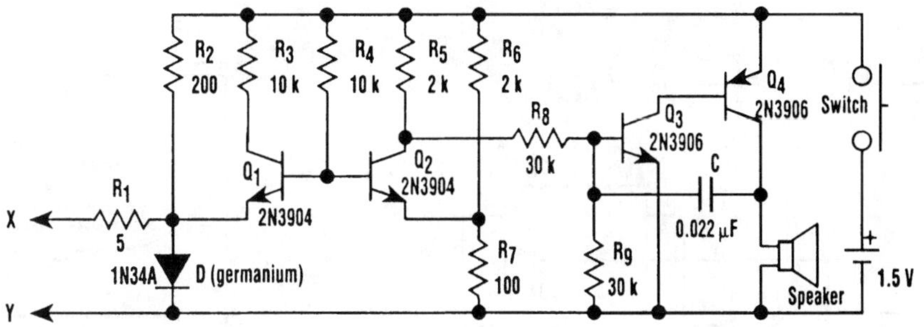

ELECTRONIC DESIGN

FIG. 55-37

Transistors Q1 and Q2, together with resistors R1 through R7, make up the input balancing stage, which senses the resistance between points X and Y. The input stage is essentially a bridge, consisting of R1, R2, R6, R7, and the resistance between points X and Y.

Transistors Q3 and Q4 and their associated passive components form a buzzer, which sounds when the tester detects a short. The buzzer is controlled by the output from Q2. When the input resistance is high (more than about 10 Ω), Q2 turns on, so its collector potential is close to ground, and the buzzer remains off. When the input resistance is sufficiently low, Q2 turns off, and the buzzer sounds. The frequency of the sound, which is about 1000 Hz, can be adjusted by varying the value of capacitor (C).

VOLTAGE MONITOR

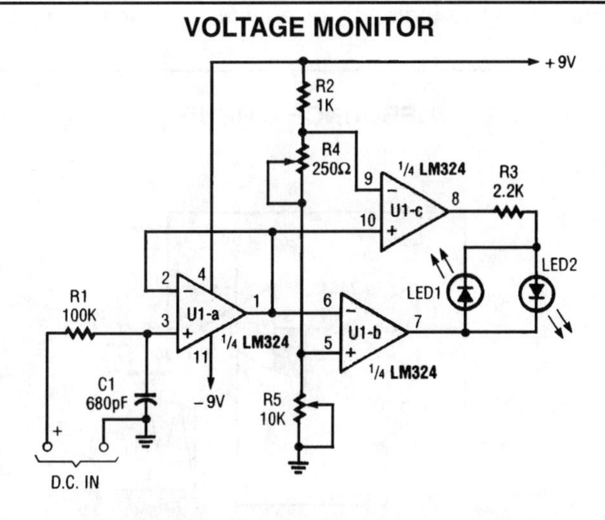

The adjustable voltage monitor can be used to check whether the voltage in a circuit remains within a given range.

POPULAR ELECTRONICS

FIG. 55-38

If the dc voltage is less than the voltage at pin 5 of U1-B, then LED 1 will light. If the voltage is over 5V, LED2 will light. If the voltage is within the window set by R4 and R5, neither LED will light. This circuit is useful as an under-or-over voltage monitor.

LINEAR INDUCTANCE METER

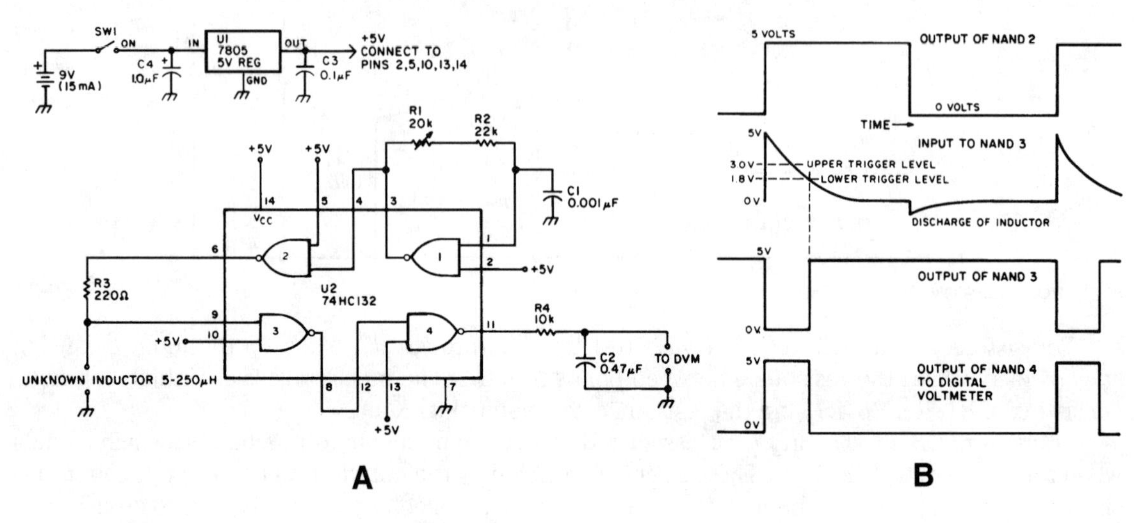

73 AMATEUR RADIO TODAY ***FIG. 55-39***

Using the fact that in an RL circuit, the pulse width seen across the inductor is proportional to the inductance, this circuit reads this indirectly on a DVM. The range is about 5 to 250 µH.

DEBOUNCE CIRCUIT

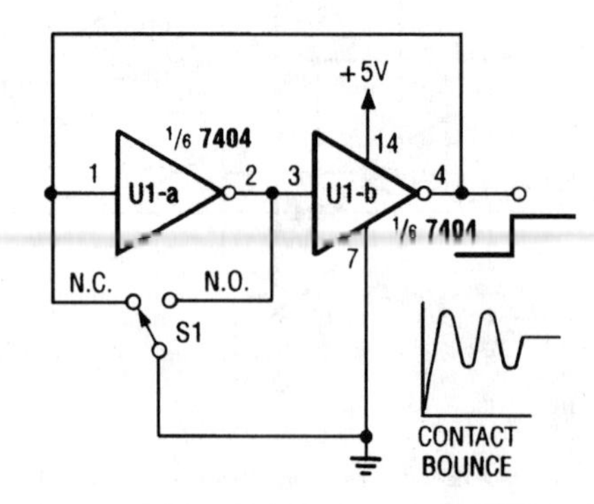

POPULAR ELECTRONICS ***FIG. 55-40***

This debounce circuit will keep the electrical noise generated by the mechanical switch (S1) from reaching the next circuit in line.

ac WIRING LOCATOR

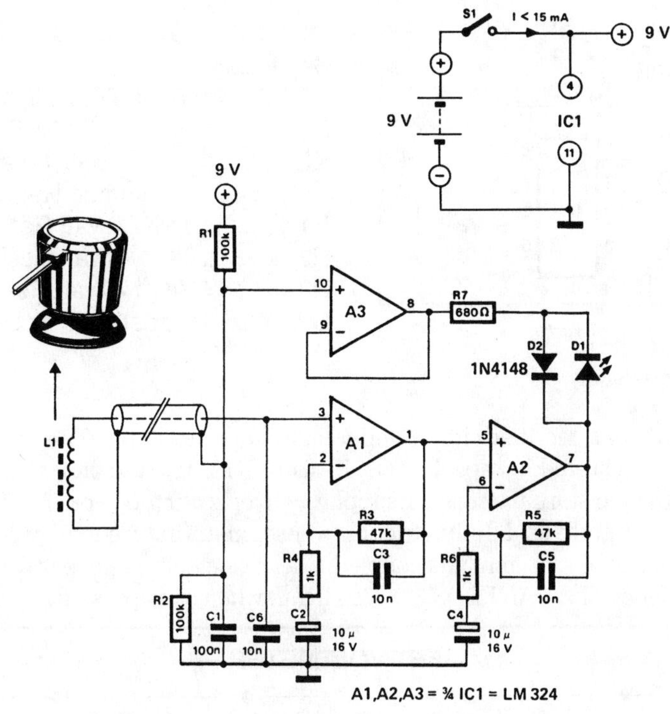

303 CIRCUITS *FIG. 55-41*

This circuit uses a pick-up coil to sense the 50- or 60-Hz field around wiring carrying ac. L1 is a telephone pick-up coil with a suction pad. D1 (LED) lights during positive half waves, indicating that ac current is present.

AUDIBLE CONTINUITY TESTER

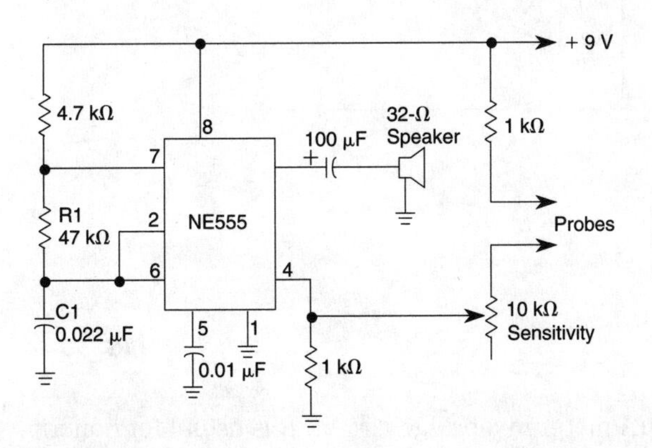

This 555 oscillator sounds a tone when continuity exists between the probes. Oscillator frequency is determined by the values of R1 and C1.

WILLIAM SHEETS *FIG. 55-42*

ac OUTLET TESTER

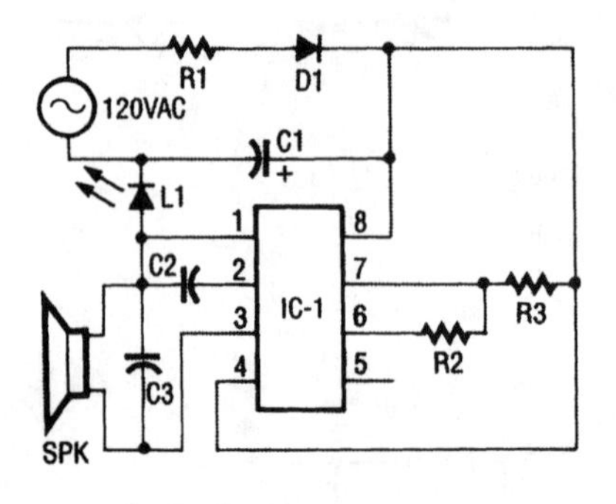

C1 .... 50 μF Electrolytic Capacitor
C2,C3 .. .047 μF Disc Capacitor
D1 .... 1N4003 Diode
IC1 .... 555 Timer IC
L1 .... Jumbo Red LED
R1 .... 3.9K, 1 watt Resistor
R2 .... 2K, 1/4 watt Resistor
R3 .... 4.7K, 1/4 watt Resistor
SPK .... Piezoelectric Speaker

1991 PE HOBBYIST HANDBOOK ***FIG. 55-43***

The tester consists of a rectifier circuit and a multivibrator circuit. The ac voltage is half-wave rectified by diode D1 and stored in capacitor C1. Resistor R1 is used to limit the current through D1 to a safe value. The voltage stored across C1 supplies IC1 operating power. The IC, the versatile 555 timer, is configured to operate as a multivibration whose operating frequency is determined by C2, R2, and R3. The output of IC1, on pin 3, is coupled to a piezoelectric speaker (SPK), which gives an indication of the presence of ac. An LED (L1) also lights when ac is present.

JFET VOLTMETER

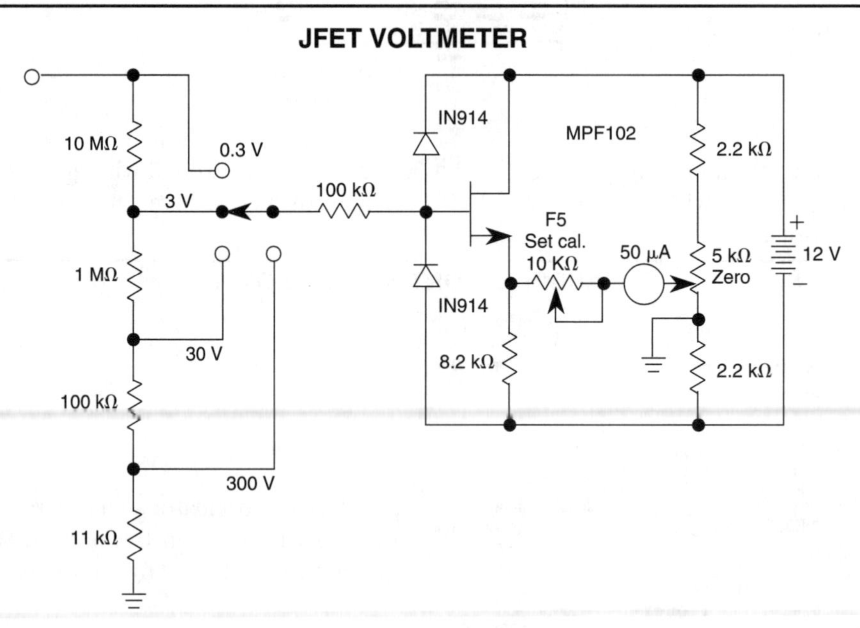

WILLIAM SHEETS ***FIG. 55-44***

This very simple voltmeter circuit uses a 50-μA meter in a bridge circuit. It is useful for noncritical applications.

CHECK FOR OP-AMP dc OFFSET SHIFT

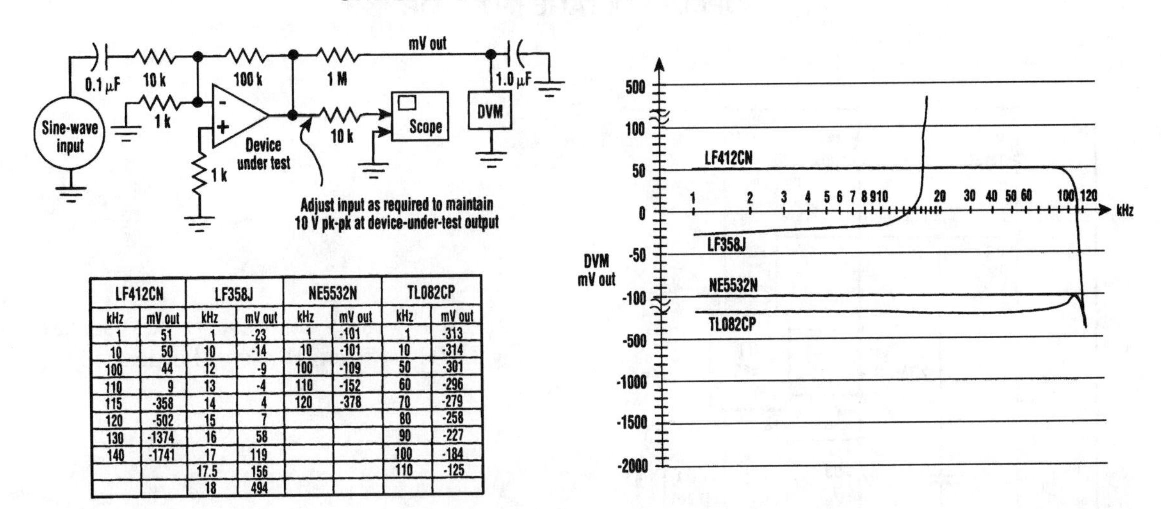

LF412CN		LF358J		NE5532N		TL082CP	
kHz	mV out	kHz	mV out	kHz	mV out	kHz	mV out
1	51	1	-23	1	-101	1	-313
10	50	10	-14	10	-101	10	-314
100	44	12	-9	100	-109	50	-301
110	9	13	-4	110	-152	60	-296
115	-358	14	4	120	-378	70	-279
120	-502	15	7			80	-258
130	-1374	16	58			90	-227
140	-1741	17	119			100	-184
		17.5	156			110	-125
		18	494				

ELECTRONIC DESIGN

FIG. 55-45

The dc values of op-amp offsets can't always be taken for granted when delivering ac outputs. No device is ever exactly symmetrical for maximum positive slew rate versus maximum negative slew rate. Consequently, there is always some range of output slew rates in which the device used limits in one direction more severely than in the other. What results in rectification of the ac signal and an apparent shift of the dc offset.

This test circuit can check for the shift phenomenon. The accompanying table and graph illustrate the results obtained for four devices, all of different types. As frequency and slew rate are increased, the effect can be either relatively abrupt (LF412CN and NE55532N) or relatively gradual (LF358J and TL082CP).

CONTINUITY TESTER FOR LOW-RESISTANCE CIRCUITS

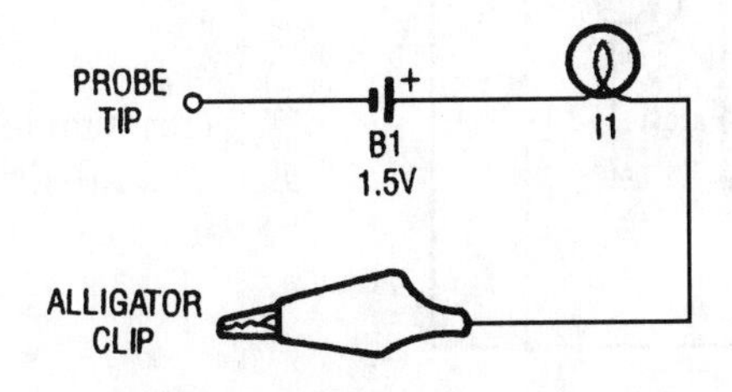

POPULAR ELECTRONICS

FIG. 55-46

The continuity tester is little more than a battery and a lamp connected in series, with one end of the string terminated in an alligator clip, and the other end connected to the probe tip.

SUPPLY VOLTAGE MONITOR

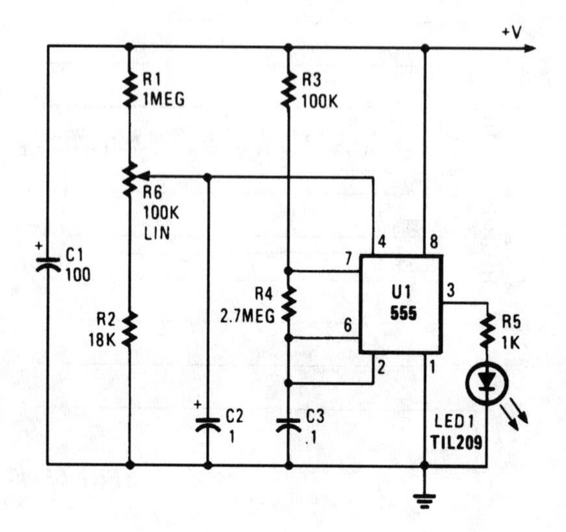

Excessive voltage causes U1 to oscillate, causing LED1 to flash. R6 sets the desired trip level.

POPULAR ELECTRONICS *FIG. 55-47*

AUDIO-FREQUENCY METER CIRCUIT

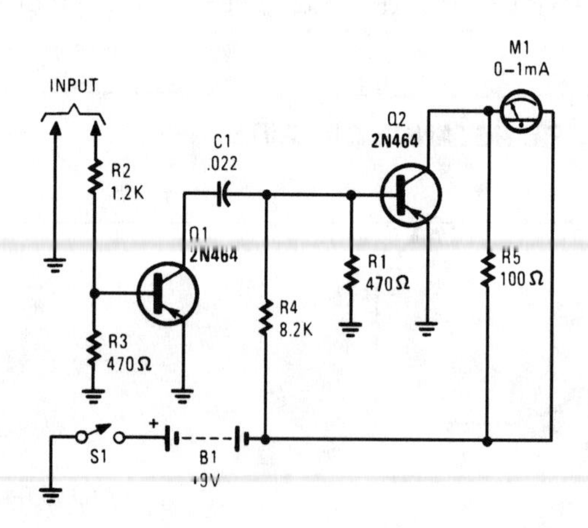

This simple tachometer circuit uses a pulse shaper Q1 to drive M1, a 0- to 1-μA meter. C1 can be varied to optimize operation.

POPULAR ELECTRONICS *FIG. 55-48*

ZENER DIODE TEST SET

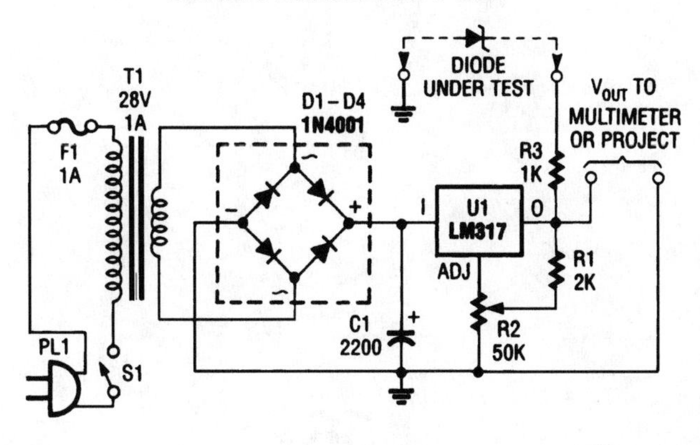

POPULAR ELECTRONICS ***FIG. 55-49***

This versatile circuit can be used to test zener diodes or act as a stand-alone power supply. It requires a voltmeter to work as a zener tester.

56

Metal-Detector Circuits

The sources of the following circuits are contained in the Sources section, which begins on page 675. The figure number in the box of each circuit correlates to the entry in the Sources section.

METAL PIPE DETECTOR

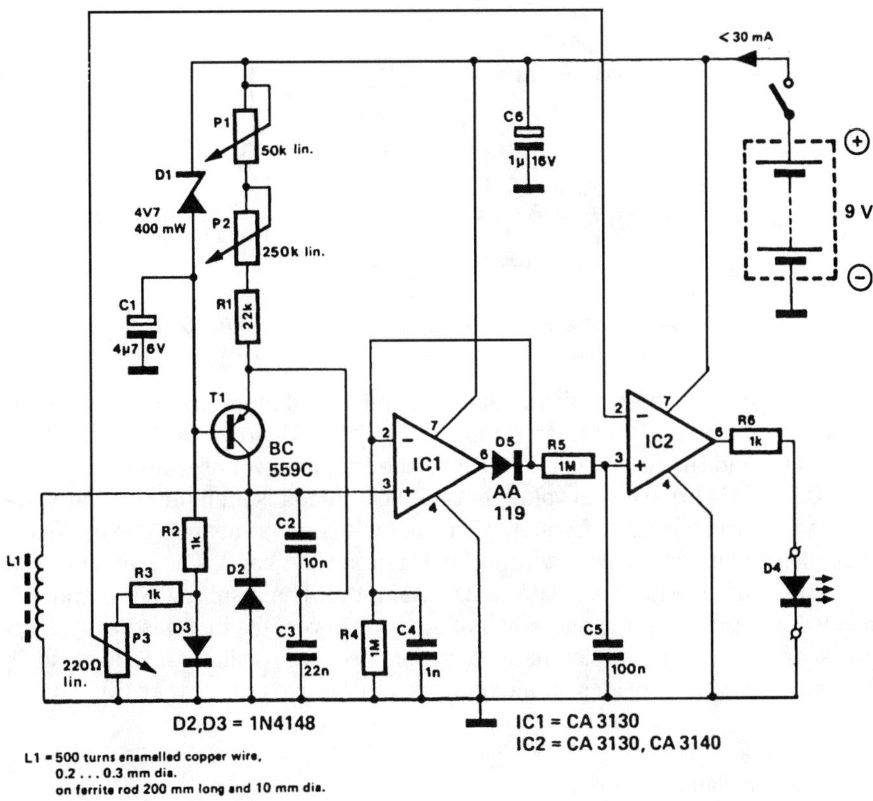

303 CIRCUITS

FIG. 56-1

This circuit uses a 15-kHz oscillator coil. When metal placed in the energy field is withdrawn, the oscillator voltage is rectified and compared to a reference. A drop in oscillator voltage therefore operates comparator IC2 and D4 (LED) extinguishes.

LOW-COST METAL DETECTOR FOR EXPERIMENTERS

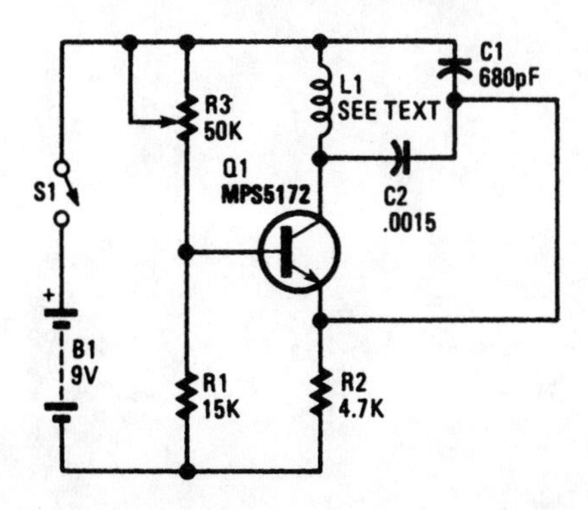

This circuit is on oscillator with L1 being a 4" diameter coil of 35 turns of #26 magnet wire. Metal in proximity to L1 will cause the oscillator to shift frequency. An AM transistor radio is used to detect the frequency shift.

POPULAR ELECTRONICS

FIG. 56-2

METAL LOCATOR

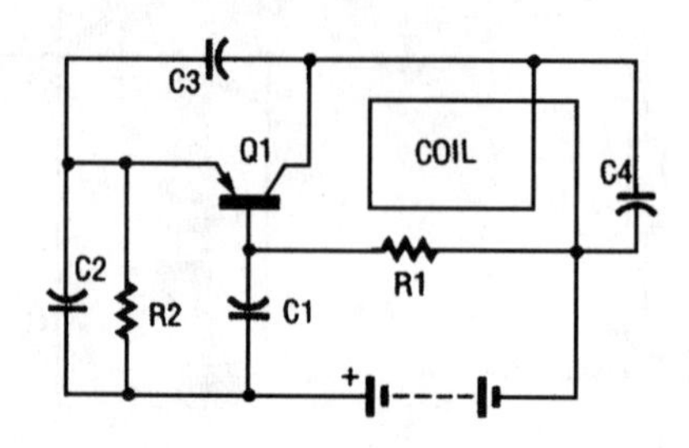

1991 PE HOBBYIST HANDBOOK *FIG. 56-3*

The metal locator uses a one-transistor oscillator and an AM radio to detect metal. Transistor Q1 is a pnp transistor that is connected to an oscillator. Resistor R1 provides the correct base bias and capacitors C3 and C4 and the search coil determine the frequency of oscillation.

Capacitors C3 and C4 are fixed in value, but the search coil is an inductor that varies in inductance (and thus varies the oscillator frequency) as metal is brought near it. The oscillator frequency is rich in harmonics and its output falls within the AM broadcast band. The metal detector works by combining its output with the local oscillator of the AM radio. The resulting net output of the radio is a low-frequency audio tone that changes—gets higher or lower—as metal is brought near or taken away from the search coil. Commercial metal detectors use two oscillators, so they don't require an AM radio. This metal locator provides an inexpensive alternative to an expensive commercial metal locator.

C1, C2	0.01-μF Capacitor (103)
C3, C4	0.001-μF Capacitor
Q1	2N3906 Transistor
R1	47-kΩ Resistor
R2	100-Ω Resistor

57

Miscellaneous Treasures

The sources of the following circuits are contained in the Sources section, which begins on page 675. The figure number in the box of each circuit correlates to the entry in the Sources section.

VOICE DISGUISER

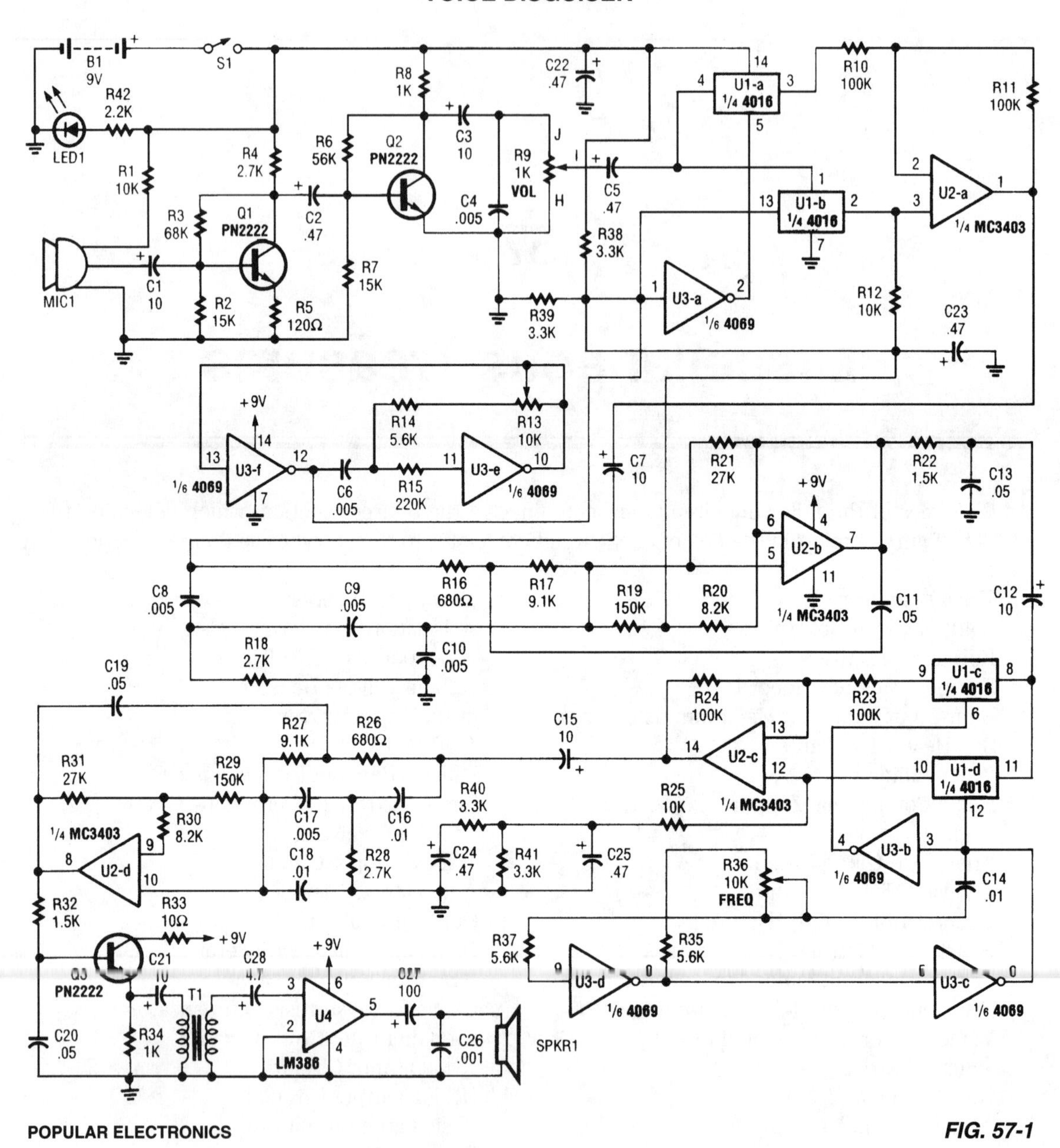

POPULAR ELECTRONICS ***FIG. 57-1***

A complete schematic diagram of the voice disguiser is shown. Microphone MIC1 picks up the voice signal and feeds it to an audio amplifier, consisting of Q1 and Q2, and a few support components. The amplifier has a low-pass gain response that limits the voice frequencies to 5 kHz or lower.

VOICE DISGUISER (*Cont.*)

The voice signal is then fed to the input of the first balanced modulator, which is comprised of U1-a, U1-b, U2-a, and U3-a. The output of the first 4-kHz oscillator, built around U3-f and U3-e, is fed to the carrier input of the first modulator. The frequency of the first oscillator is controlled by the setting of potentiometer R13. The modulator output—a double-sideband suppressed-carrier signal centered on 4 kHz—is then filtered by the first 5-kHz low-pass filter, formed by U2-b, which eliminates the upper-sideband signals.

At this point, the voice frequency spectrum is inverted (e.g., the frequencies that were low now become high, and vice versa), making the voice signal completely unintelligible. The output of the first low-pass filter is fed to a second modulator formed by U1-c, U1-d, and U3-b, where it is frequency modulated with the output of the second carrier oscillator, comprised of U3-c and U3-d; the frequency of the second oscillator is controlled by potentiometer R36.

The output of the second modulator is filtered by the second low-pass filter, which consists of U2-d and few support components, and amplified by Q3. The voice output signal from Q3 is fed to U4 (an LM386 low-voltage, audio-power amplifier) through an impedance-matching transformer, T1. The output of U4 is then used to drive SPKR1 (an 8-Ω speaker).

In operation, if both carrier oscillators are set to the same frequency, the voice signal from the speaker will be an exact duplicate of the input signal from the microphone. However, if the frequency of the second oscillator is varied (via R36), the output voice signal also shifts in frequency. That makes the voice reproduced by the speaker sound higher- or lower-pitched than normal.

SOLDERING IRON CONTROL

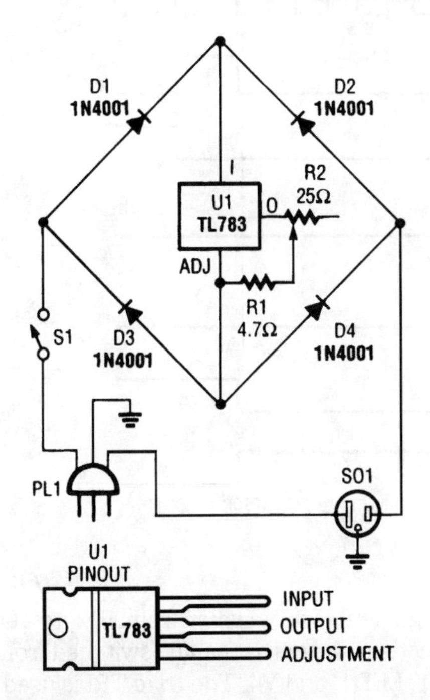

1992 ELECTRONICS HOBBYISTS HANDBOOK *FIG. 57-2*

A current control to temperature regulate a soldering iron uses a high-voltage integrated regulator, TL783 (U1). WIth the component values specified, the circuit should be used with a soldering iron of 25 W or less.

FURNACE FUEL MISER

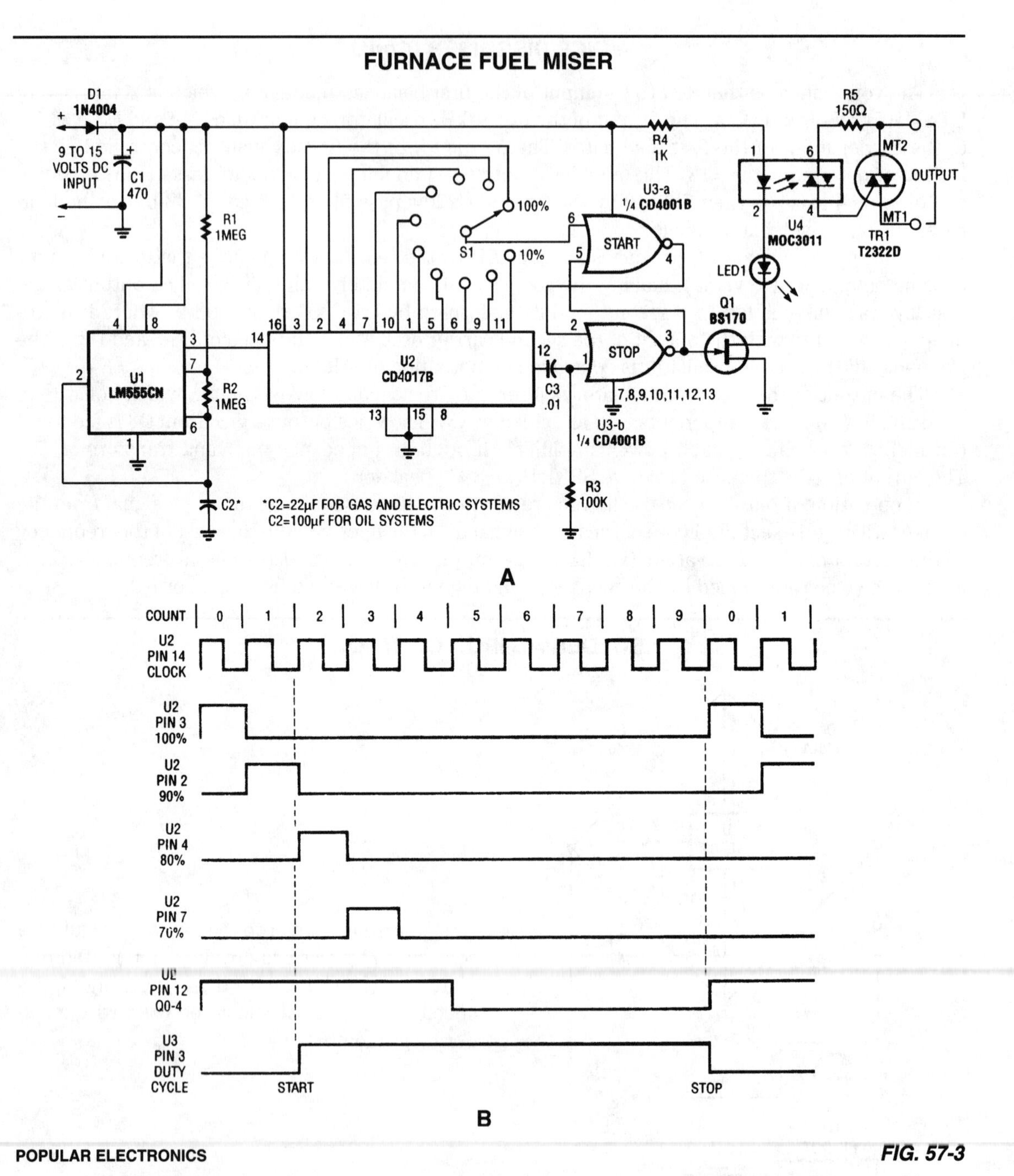

POPULAR ELECTRONICS ***FIG. 57-3***

A timer (LM555CN) and decode counter is used to generate duty cycles from 10% to 100% to control the time a heating system can operate. V2 is a decode counter that can be switched from 10% to 100% duty cycle. V3A and B form a latch that drive A1, LED1, and V4. The triac TRI is used as an ac switch, in series with the thermostat that controls the heating system.

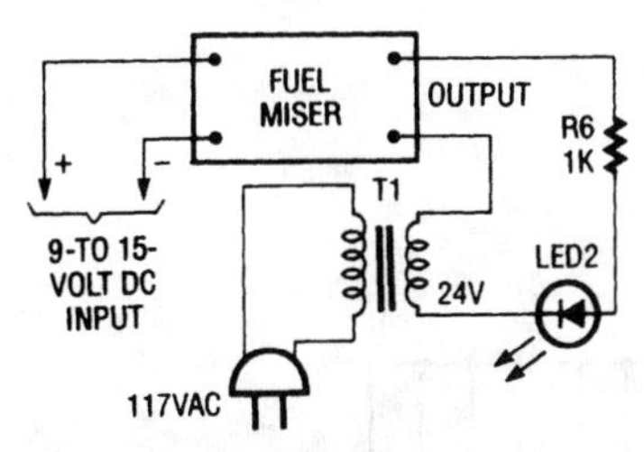

When the circuit is working properly, the output circuitry can be checked using a 24-volt step-down transformer, a 1k resistor, and an LED. Together those components simulate the load that the Fuel Miser sees during normal operation.

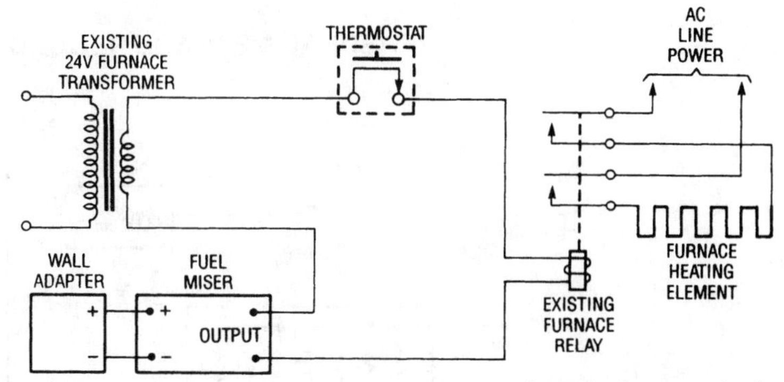

Electric-heating systems may or may not use a relay in the thermostat circuit. Those that do have a relay can be controlled by the Fuel Miser by wiring its output circuit in series with the relay coil connections as shown here.

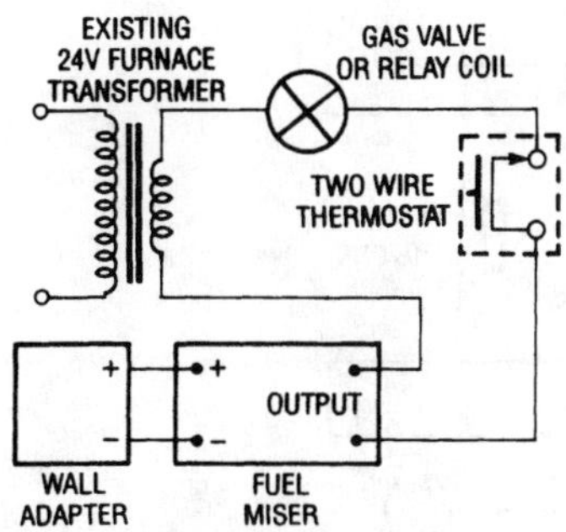

This drawing shows the Fuel Miser connected in series with the thermostat of a two-wire gas furnace that's powered by a 24-volt transformer.

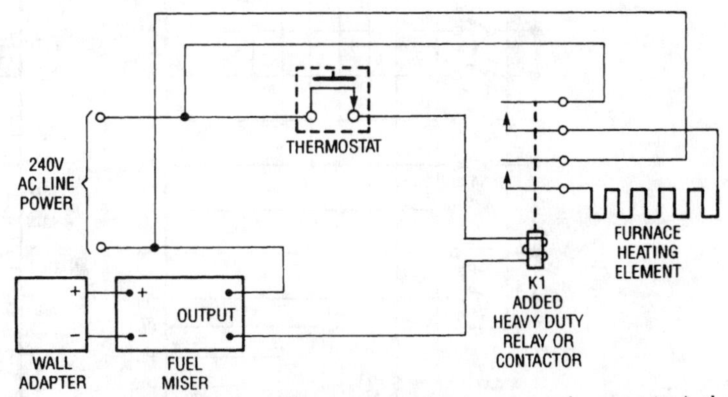

Electric-heating systems that do not contain a low-current thermostat (as in the previous installation), use a heavy-duty thermostat that directly feeds current to the heating element. For such systems, it will be necessary to install a heavy-duty relay (K1 in this example) to control the heavy heating-element current.

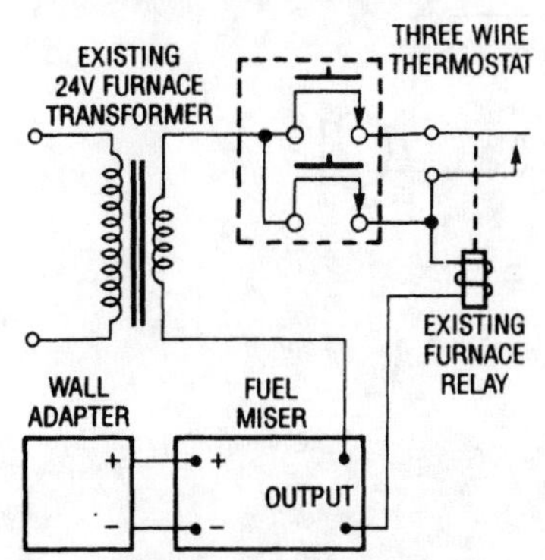

Some oil-fired systems use three-wire thermostats to control the operation of the burner motor and ignition system by activating a relay. This is a typical installation for such systems.

PERSONAL MESSAGE RECORDER

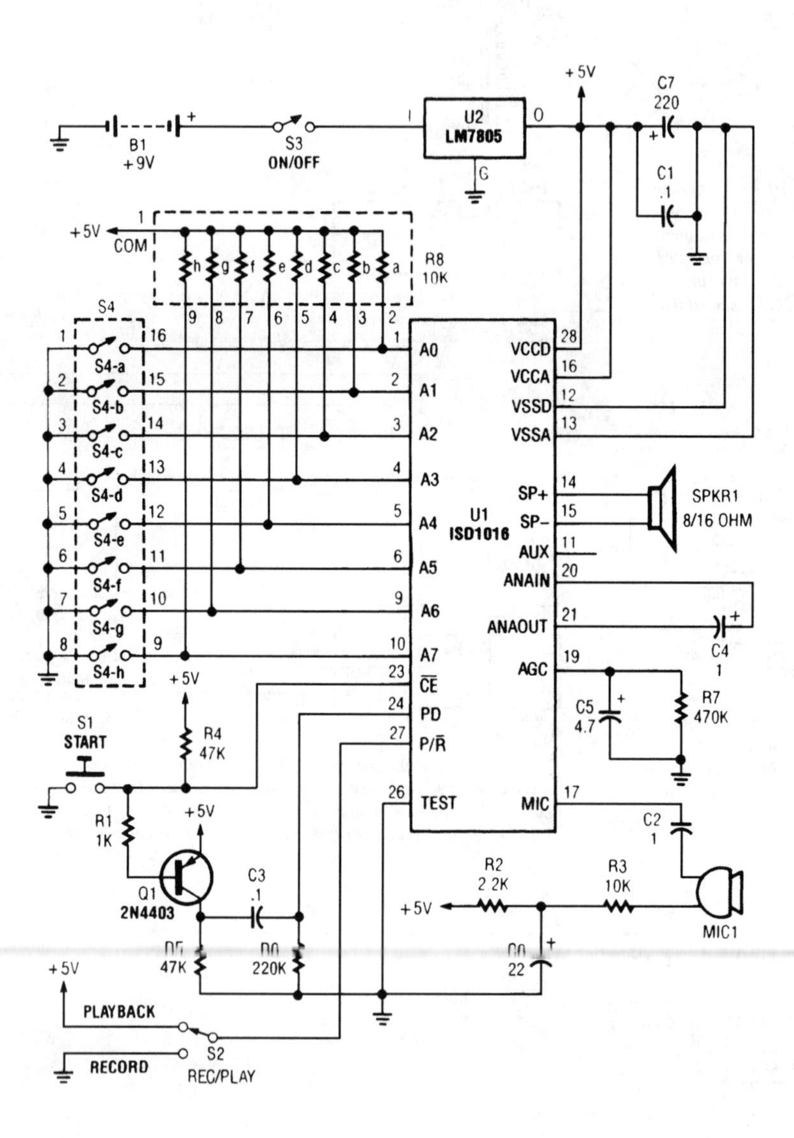

POPULAR ELECTRONICS

FIG. 57-4

PERSONAL MESSAGE RECORDER (*Cont.*)

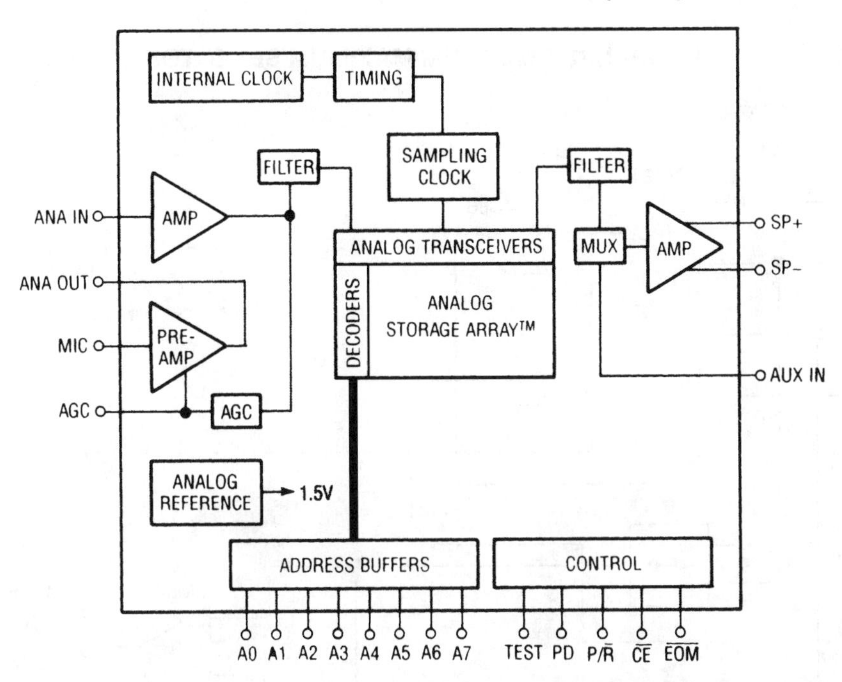

The personal message recorder is built around an ISD1016 CMOS voice messaging system, which does away with the cumbersome and expensive analog-to-digital and digital-to-analog conversion circuits.

A functional block diagram of the ISD1016 is shown. The ISD1016 contains all of the functions necessary for a complete message-storage system. The preamplifier stage accepts audio signals directly from an external microphone and routes the signals to the ANA OUT (analog out) terminal. An automatic-gain control (AGC) dynamically adjusts the preamplifier gain to extend the input signal range. Together, the preamp and AGC circuits provide a maximum gain of 24 dB. The internal clock samples the signal and, under the control of the address-decoding logic, writes the sampling to the analog-storage array. Eight external input lines allow the ISD1016's message space to be addressed in 160 equal segments, each with a 100-millisecond duration. When all address lines are held low, the storage array can hold a single, continuous, 16-second message.

However, there is a special addition to the POWER DOWN input (pin 24) of U1. If the internal memory becomes full during recording, an overflow condition is generated in order to trigger the next device. Once an overflow occurs, pin 24 must be taken high and then low again before a new playback of record operation can be started.

Transistor Q1, C3, R5, and R6 form a one-shot pulse generator that automatically clears any overflow condition each time that start switch (S1) is pressed. Switch S2 selects either the playback or the record mode. Switch S4—an 8-position (a–h) DIP switch—is included in the circuit to allow the circuit's record/playback time to be varied from 0 to 16 seconds. The maximum time available is when all 8 switch positions are closed (or set to the on position). Resistor network R8 (a–h) is included in the circuit to provide a pull-up function for the address lines, which thereby controls U1's record/playback time.

FOUR-INPUT MINIMUM/MAXIMUM SELECTOR

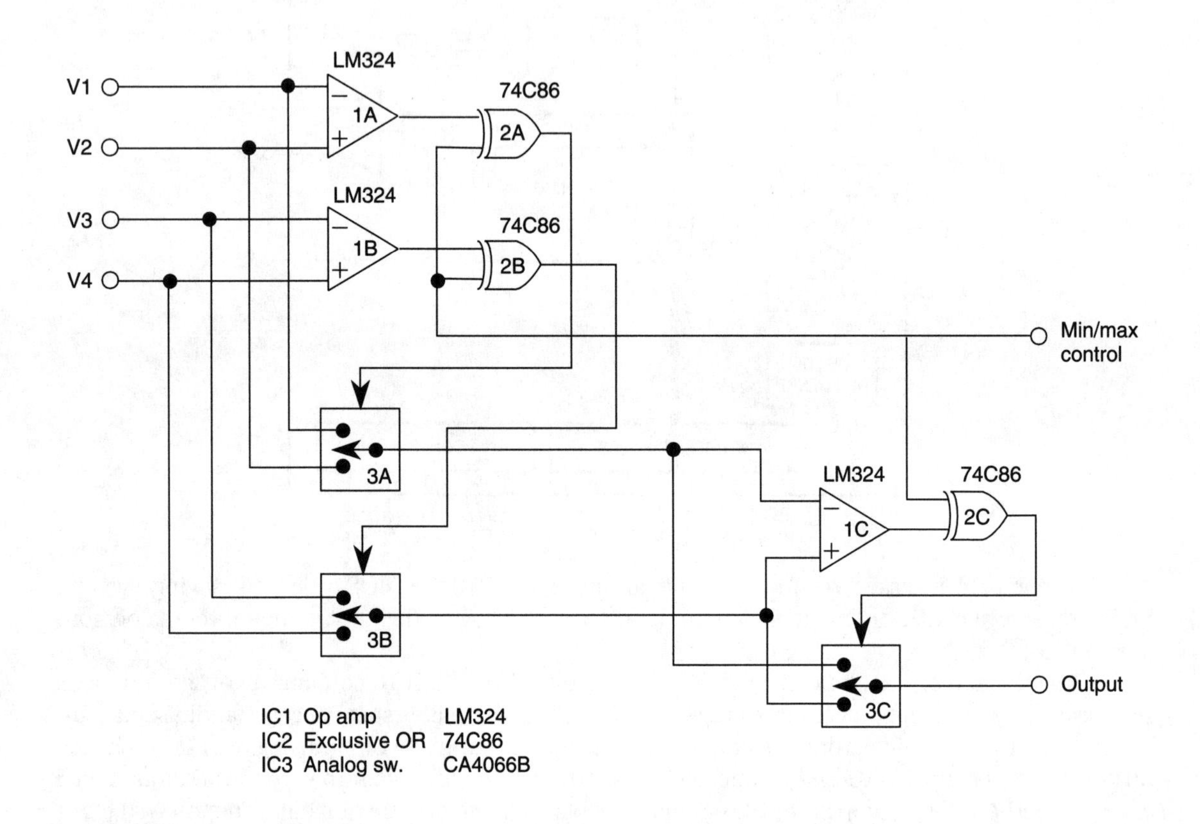

WILLIAM SHEETS

FIG. 57-5

This circuit outputs the maximum (or the minimum) of the four input voltages V_1, V_2, V_3, and V_4. Each of these input voltages is in the range 0 to 5 V.

The output of the unit is the maximum of V_1, V_2, V_3, and V_4 if the control voltage input is 5 V (i.e., logical 1). The output is the minimum of V_1, V_2, V_3, and V_4 if the control input is zero.

By cascading N such units, one can select the maximum (or the minimum) of $3N + 1$ input voltages.

Thus if k is the number of input voltages, we need $[(k+1)/3]$ units.

SOIL HEATER FOR PLANTS

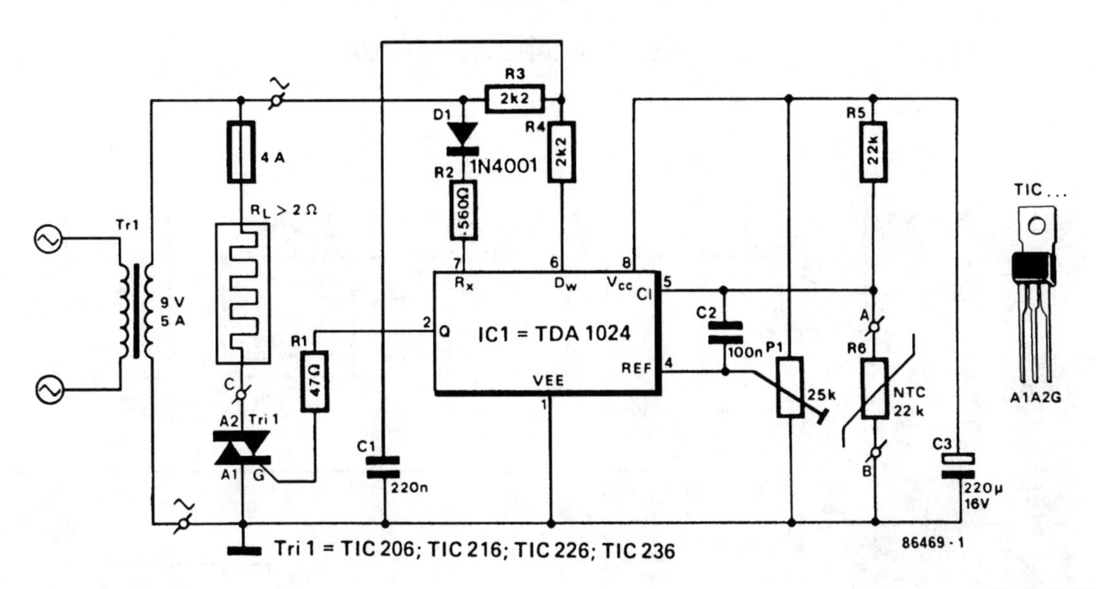

303 CIRCUITS *FIG. 57-6*

A TDA1024 electronic thermostat senses soil temperature via thermistor R6. The circuit uses zero-crossing switching of the heater. The heater is made of elastic-coated steel wire. P1 is used to set the temperature. The heater should have 2 Ω or more resistance and operate from the 9-V transformer. About 40 W of heat is available.

KEY ILLUMINATOR

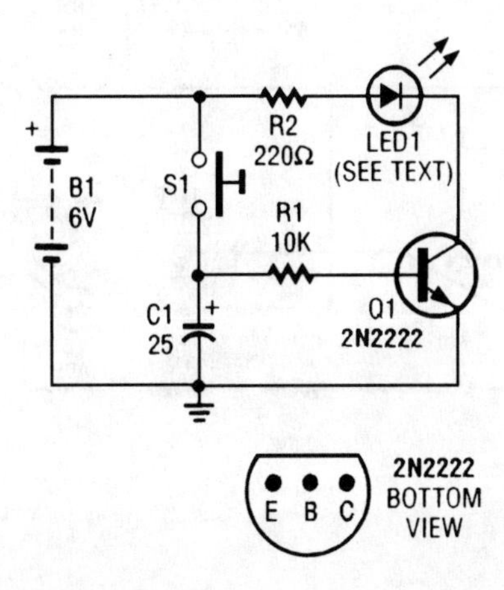

Used as a 10-second momentary illuminator, this circuit can be useful in other applications as well. Pressing S1 charges C1, which holds Q1 on and holds the LED lit for about 10 seconds.

POPULAR ELECTRONICS *FIG. 57-7*

RADIO COMMERCIAL ZAPPER

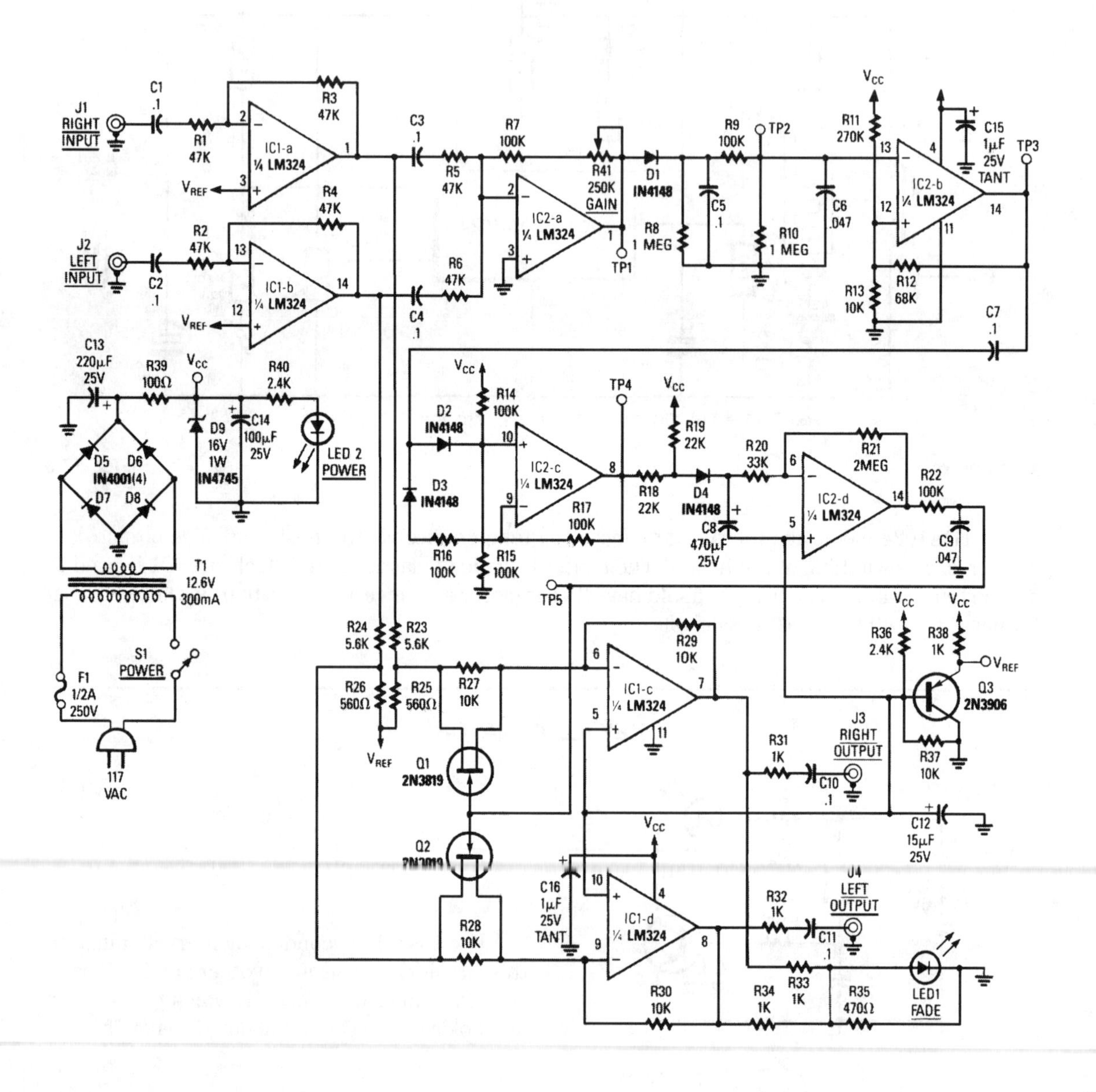

FIG. 57-8

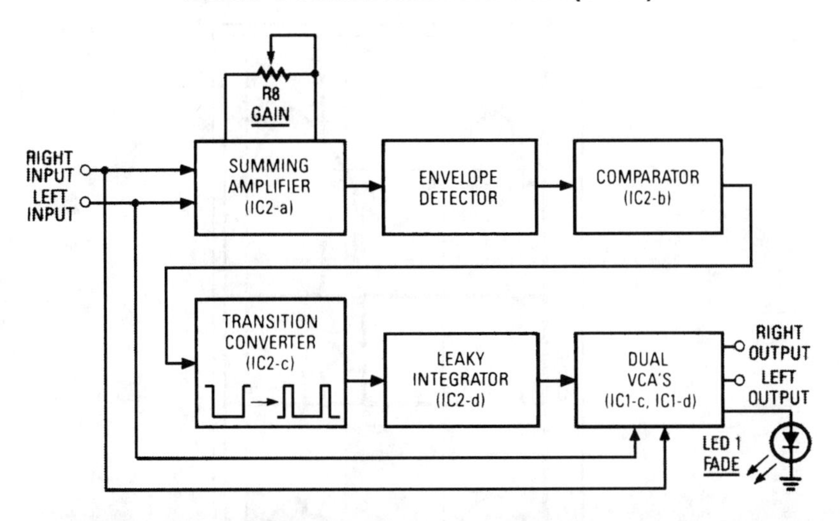

BLOCK DIAGRAM OF THE COMMERCIAL KILLER: The envelope of the signal is used to vary the pulse rate from IC2-c. The pulses are integrated; the resulting signal controls the gains of a pair of VCA's.

The L&R inputs are summed, dated and drive a comparator. The comparator senses level and generates a transition when audio inputs go above or below preset thresholds. The number of these transitions (corresponding to rapid volume changes) are integrated and feed voltage controlled amplifiers. This device actually senses dynamic range.

AUDIO LIMITER

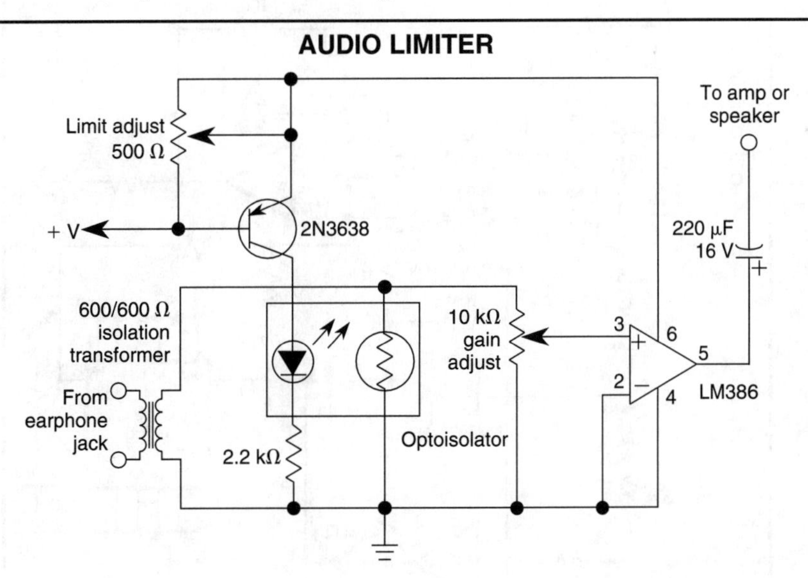

RADIO ELECTRONICS ***FIG. 57-9***

An optoisolator is used as an attenuator in this circuit. When the LM386 draws more current on audio signals, the 2N3638 turns on, which biases the optoisolator on, and reduces the volume.

ANALOG DE-GLITCH CIRCUIT

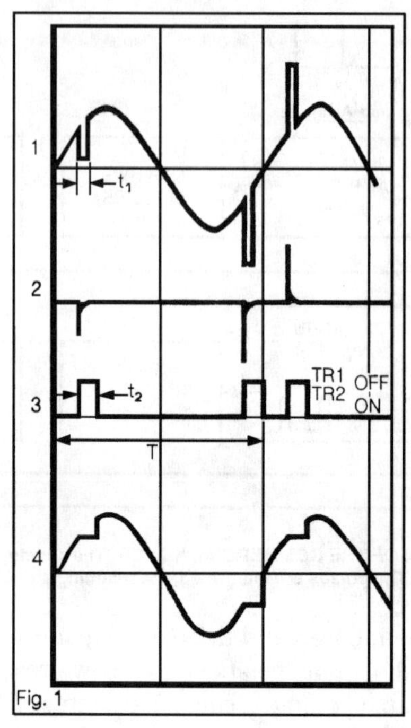

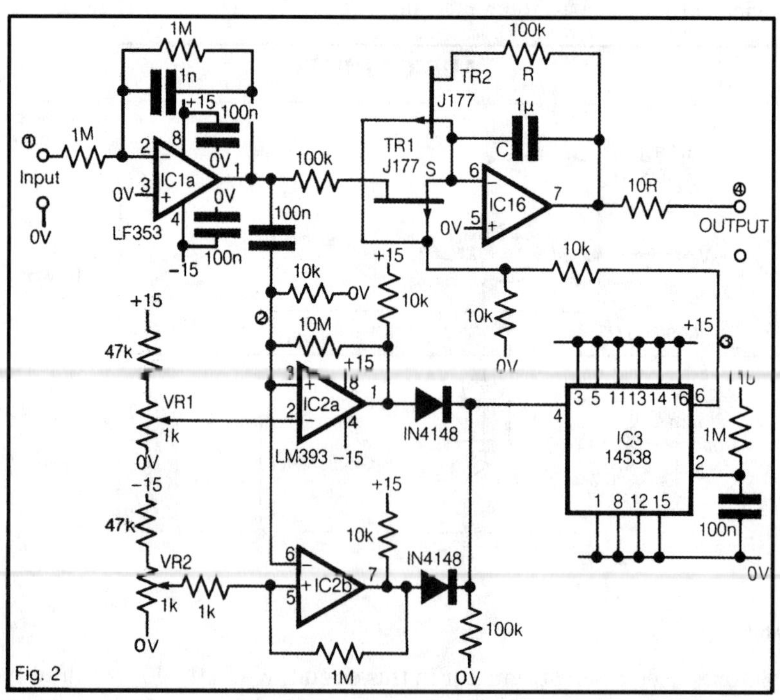

FIG. 57-10

ANALOG DE-GLITCH CIRCUIT (*Cont.*)

Low-frequency signals produced by transducers, measurement equipment, or data loggers often appear like the first waveform in the figure. The circuit shown operates as a tracking sample-hold, and the transients are replaced in the output by the stored value of the current signal at the instant of the transient.

The input signal is buffered and inverted by IC1a, and the differentiated result shown at 2 applied to the inputs of two comparators IC2-a and IC2-b. VR1 and VR2 set levels to prevent false or unnecessary operation. Either comparator output triggers the mono IC3 from positive or negative signal transients. When IC3 has not been triggered, TR1 and TR2 'p' channel JFETs are on, and IC1b operates as an integrator with a high leakage, and tracks the input signal. When the mono is triggered as at 3, TR1 and TR2 turn off and the previous signal value is held constant, as shown at 4. The resulting output waveform can then be easily filtered to remove the harmonics from the restoring step at the end of the mono period, if needed.

The criteria for successful operation are:

$t_2 > t_1$ (mono period longer than glitch)

t_2/T small (to optimize output waveform)

$$\text{Signal bandwidth } f_o = \frac{1}{2\pi CR}$$

$$\text{Signal phase } 0 = \tan^{-1} 2\pi fCR$$

The signal range is approximately ±5 V, depending on the transient amplitude and polarity. The mono period shown is 100 mS, but this can be optimized in practical applications. The shorter the mono period in relation to the signal waveform, the better the quality of the result.

ACOUSTIC FIELD GENERATOR

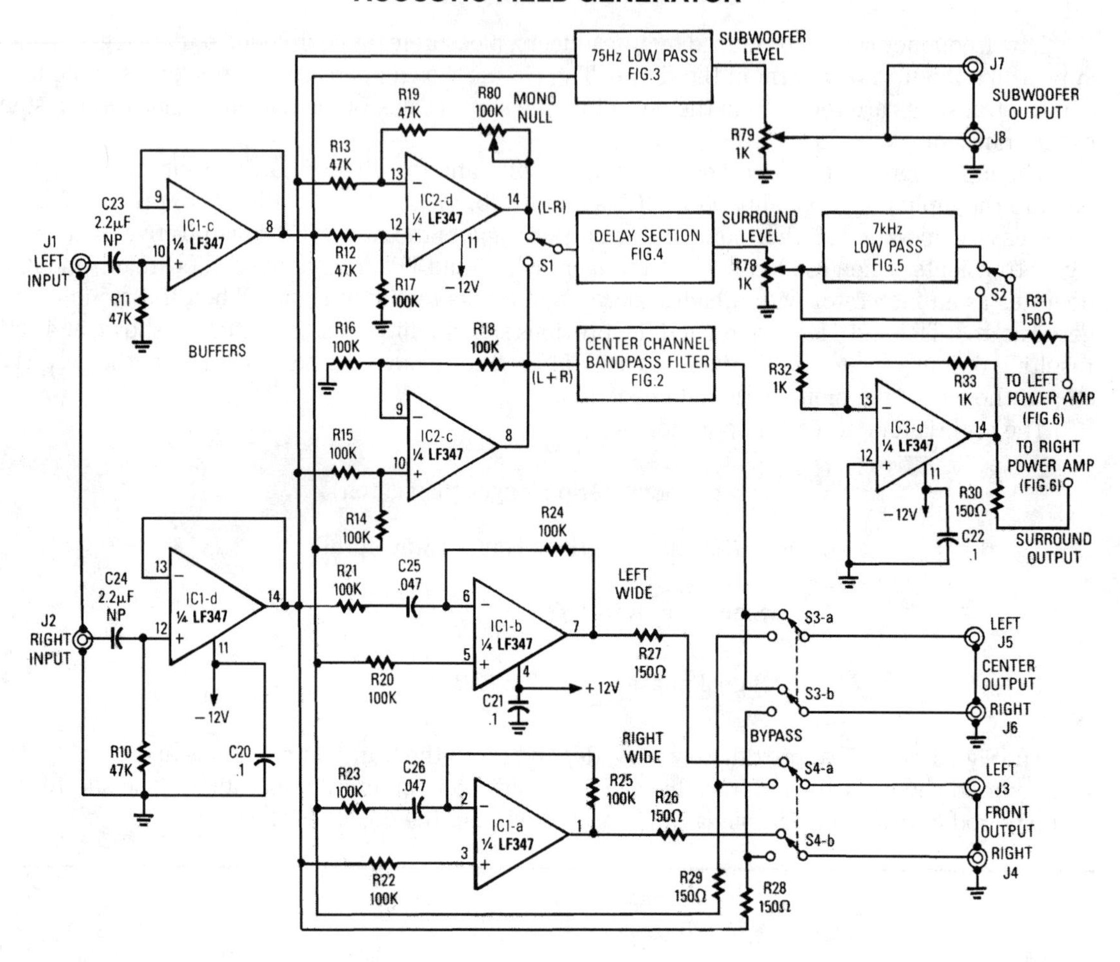

THE AFG IS MADE UP OF 10 relatively simple circuit elements.

A

R-E EXPERIMENTER'S HANDBOOK **FIG. 57-11**

Referring to the simplified schematic in A, the AFG is made up of 10 relatively simple circuit elements. IC1-c and IC1-d are configured as unity-gain noninverting buffer amplifiers.

The summing (*L*+*R*) amplifier, IC2-c, combines equal amounts of the left and right signals, via R14 and R15, to develop a total composite signal. Left- and right-channel signals are applied equally through R13 and R12 to IC2-d, the difference (*L*–*R*) decoder. Any common to both channels is canceled by IC2-d, which exactly balances the inverting and noninverting gains of the amplifier for a perfect null.

The stereo width-enhancement circuit made up from IC1-a and IC-b works similarly to the (*L*–*R*) decoder, except that C25 and C26 have been added in the inverting inputs of each op amp. IC1-b develops the "left wide" signal because its inverting and noninverting inputs are connected to the left

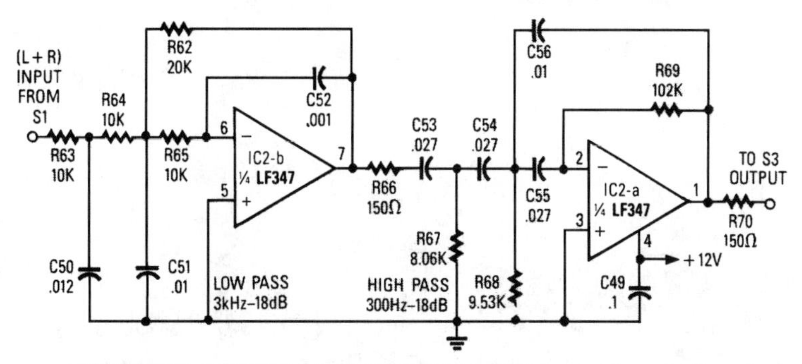

THE CENTER-CHANNEL SPEECH FILTER is built by cascading a 3-kHz low-pass filter with a 300-Hz high-pass filter to form a band-pass filter.

B

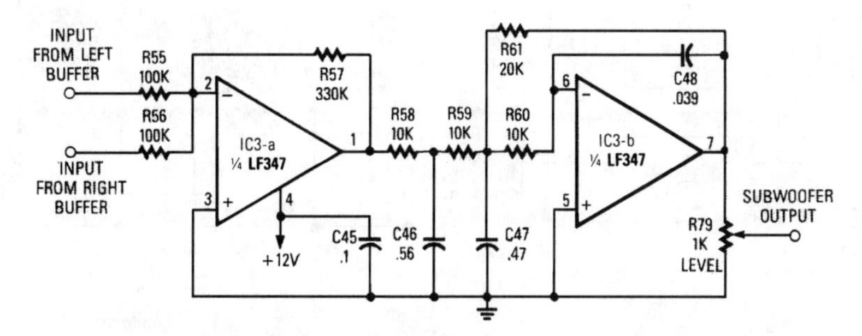

AN ACTIVE CROSSOVER NETWORK for driving a high-power subwoofer system is made from IC3-a and IC3-b.

C

and right channels opposite that of IC1-a. The output of the width-enhancement circuit is routed to S4, which selects either the "wide" or the bypass signal for feeding the front-channel amplifier.

The center-channel dialogue filter is built by cascading a 3-kHz low-pass filter with a 3-Hz high-pass filter to form a band-pass filter. It has a sharp –18 dB/octave cutoff, a flat voltage and power frequency response, and minimum phase change within the passband.

In C, IC3-a and IC3-b form an active crossover network for driving a subwoofer. IC3-a sums signals from the left- and right-channel buffer amps, it inverts the summed signal 180 degrees, and provides a low driving impedance for the following filter stage. IC3-b and its associated RC network form a 75-Hz, 3rd-order low-pass filter. The filter inverts the signal another 180 degrees, so the signal that appears across R79 (which is the output-level control) is back in phase with the original input signal.

The delay section of the AFG, shown in D, is built around the MN3008 bucket brigade device (BBD), and the MN3101 two-phase variable-frequency clock generator. The amount of delay required in this system varies between approximately 5 to 35 milliseconds. The delay time of a BBD is equal to the number of stages divided by twice the clock frequency. Values were chosen for R53, R54, R77, and C44, to produce a clock frequency, adjustable via R77, which varies from about 30 kHz to 130 kHz.

ACOUSTIC FIELD GENERATOR (*Cont.*)

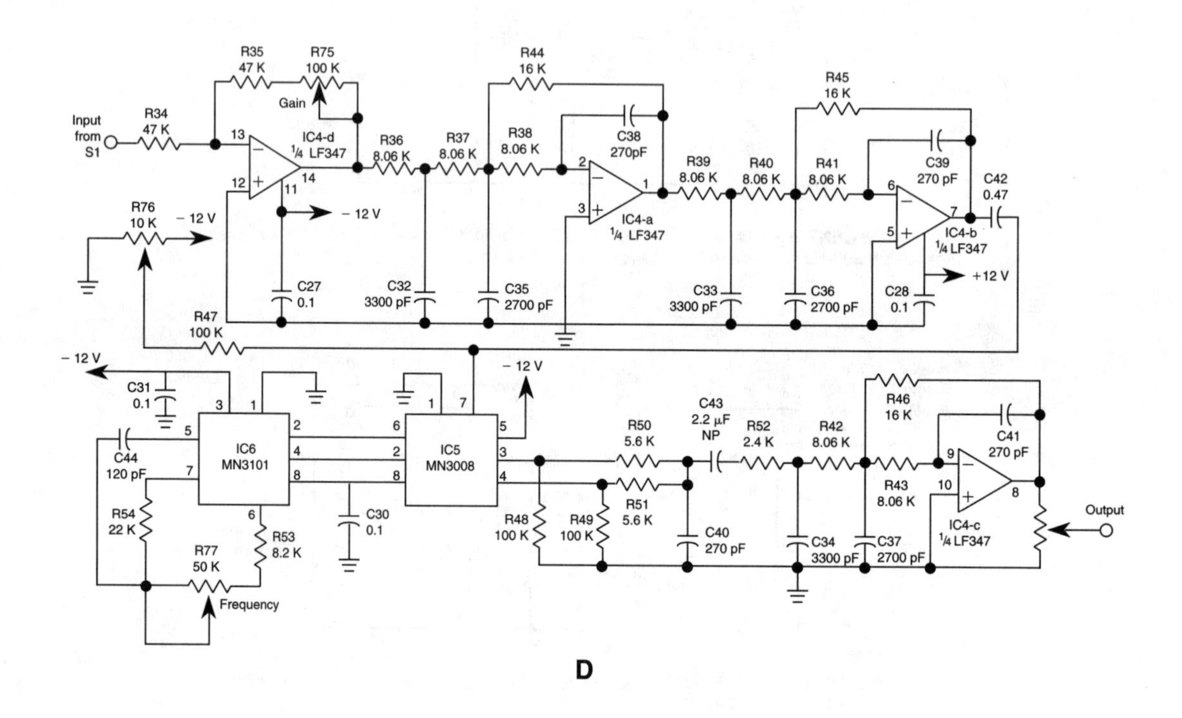

D

In A, S1 selects the signal to be delayed; either the difference signal (*L–R*) from IC2-d in the matrix mode or the sum signal (*L+R*) from IC2-c in the concert mode. The selected signal is fed from S1 to the delay section (D) where IC4-d is configured as an inverting amplifier; R75 adjusts the gain between unity and X3. Integrated circuits IC4-a and IC4-b, along with their associated RC networks, are identical 3rd-order 15-kHz low-pass filters. Cascading two filters produces a very sharp cut off (–36 dB per octave). Potentiometer R76 adjusts the bias voltage required by the BBD to exactly one half the supply voltage, as required.

The power supply of the AFG, shown in G, is of conventional design. A 25-V center-tapped transformer, along with diodes D1 and D2, produces about ±18-V unregulated dc. Two 2200-μF filter capacitors provide ample energy storage to meet the high-current demands of the audio output amplifier ICs during high output peaks.

ACOUSTIC FIELD GENERATOR (*Cont.*)

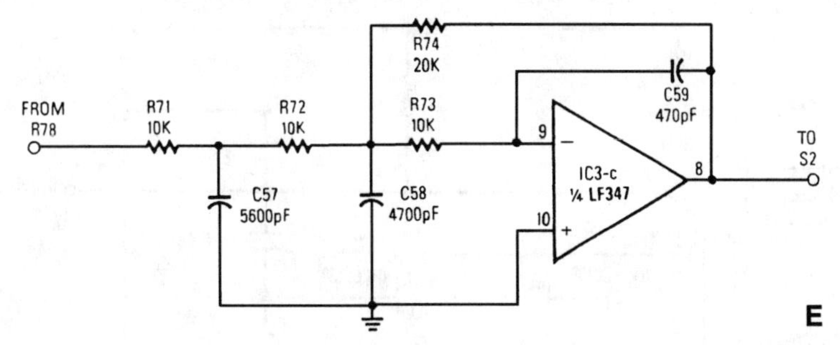

A 3rd-ORDER 7-kHz LOW-PASS FILTER is made from IC3-c and its associated RC network.

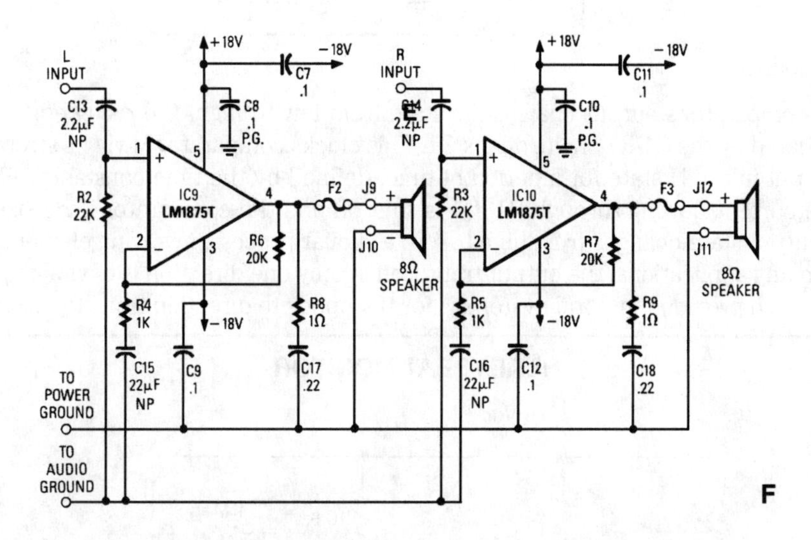

THE SURROUND CHANNEL POWER AMPLIFIERS are designed around a pair of LM1875 monolithic power-amplifier IC's.

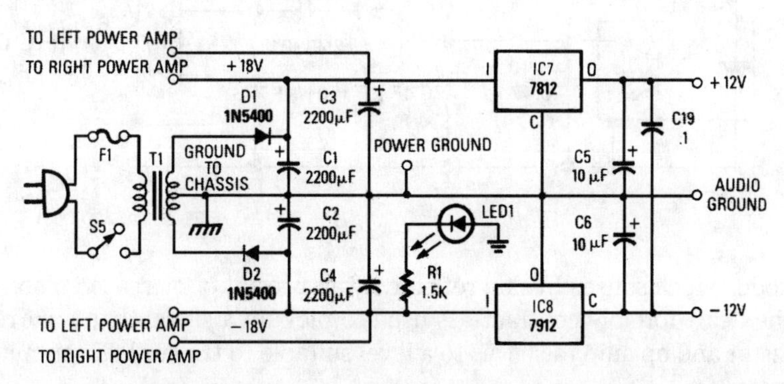

THE POWER SUPPLY produces about ±18-volts unregulated DC. **G**

SUPPRESS JITTER WITH HYSTERESIS

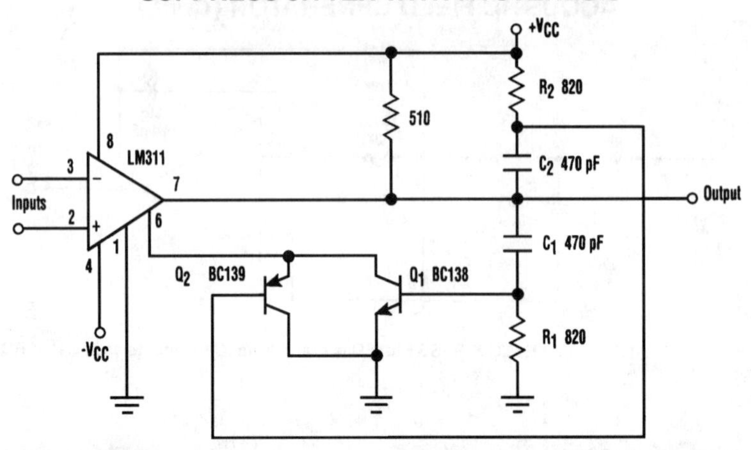

ELECTRONIC DESIGN **_FIG. 57-12_**

When the comparator's output changes its state from low to high, the rising edge of the output pulse, differentiated by the C1/R1 chain, opens Q1. This blocks comparator M via its strobing input and sustains its output in the H state for a period of time, defined by the time constant R_1C_1. After C1 is charged by the current flowing through R1, Q1 is shut off and the comparator is released. When the comparator's output state changes from high to low, a similar process, involving elements R2, C2, and Q2, occurs. In many applications, the output transition in only one direction is of vital importance, and the elements, which provide temporal hysteresis for the opposite direction transition, can be omitted.

HEARTBEAT MONITOR

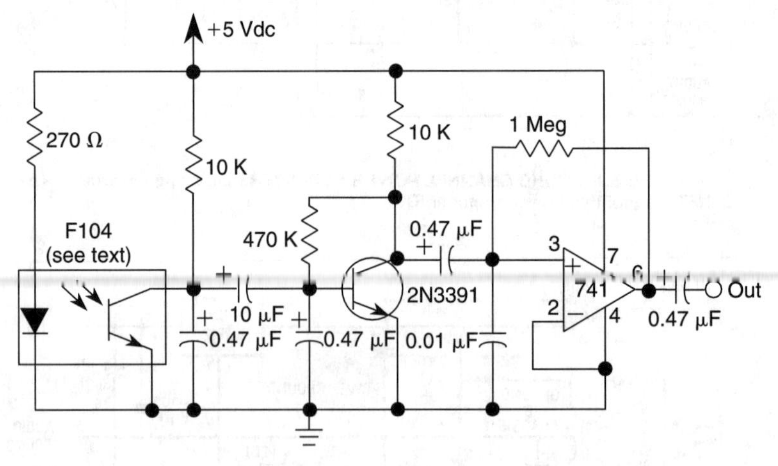

ELECTRONICS NOW **_FIG. 57-13_**

An IR photodiode, which senses IR skin reflectivity as a result of increased blood volume during the periods that the heart forcibly contracts, is used to pick up a signal that is correlated with the heartbeat. A transistor and op amp raise this to a level suitable to trigger logic circuitry or to be displayed on a scope.

SELF-RETRIGGERING TIMED-ON GENERATOR

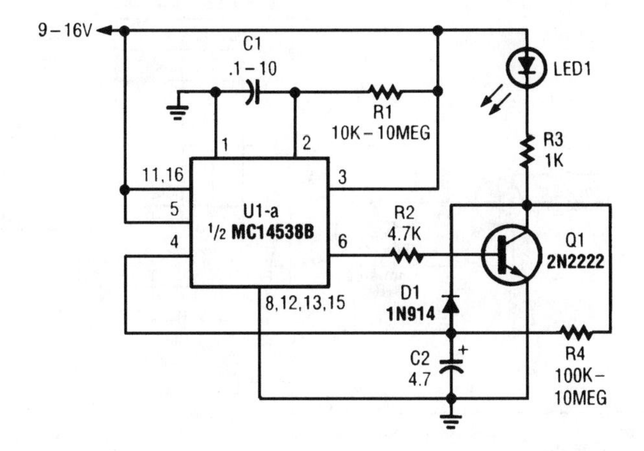

POPULAR ELECTRONICS

FIG. 57-14

When power is first applied to the circuit, C2 begins to charge via LED1, R3, and R4. When the voltage across C2 reaches U1's input trigger level, the output of U1 at pin 6 goes positive for a period that is determined by the values of C_1 and R_1. That turns Q1 on, discharging C2 through D1 and Q1.

At the end of the set period, the output of U1 at pin 6 goes low, turning Q1 off and allowing the current to begin flowing through LED1, R3, and R4 to gain charge C2, causing the cycle to repeat. The repeat time is determined by the values of R_3, R_4, and C_2. The previous formula won't be as accurate for this circuit, but it will at least get you close enough for the capacitor value; then R_4 can be fine-tuned to obtain the desired timing period.

FREQUENCY DIVIDER FOR MEASUREMENTS

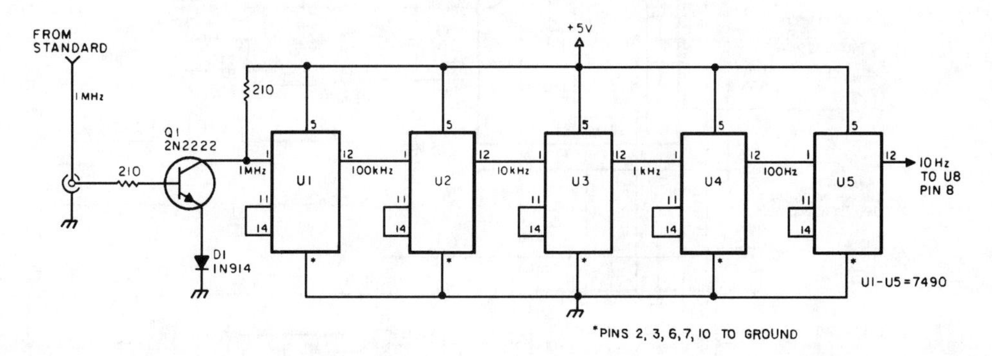

73 AMATEUR RADIO TODAY

FIG. 57-15

This circuit is meant to be driven by a 1-MHz standard signal of a few volts amplitude. U1 through U5 are 7490 decade counter/divider and produce a division ratio of 100,000:1. Successive divisions of 10 can be tapped off, if desired, between stages. One or more stages can be added for still lower frequencies.

VIDEO, POWER, AND CHANNEL-SELECT SIGNAL CARRIER

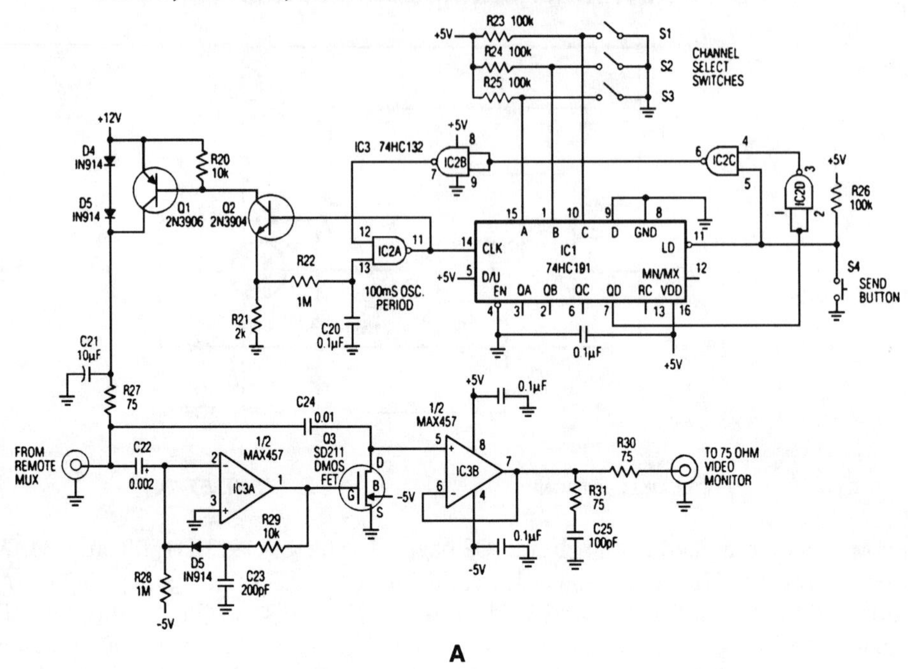

A

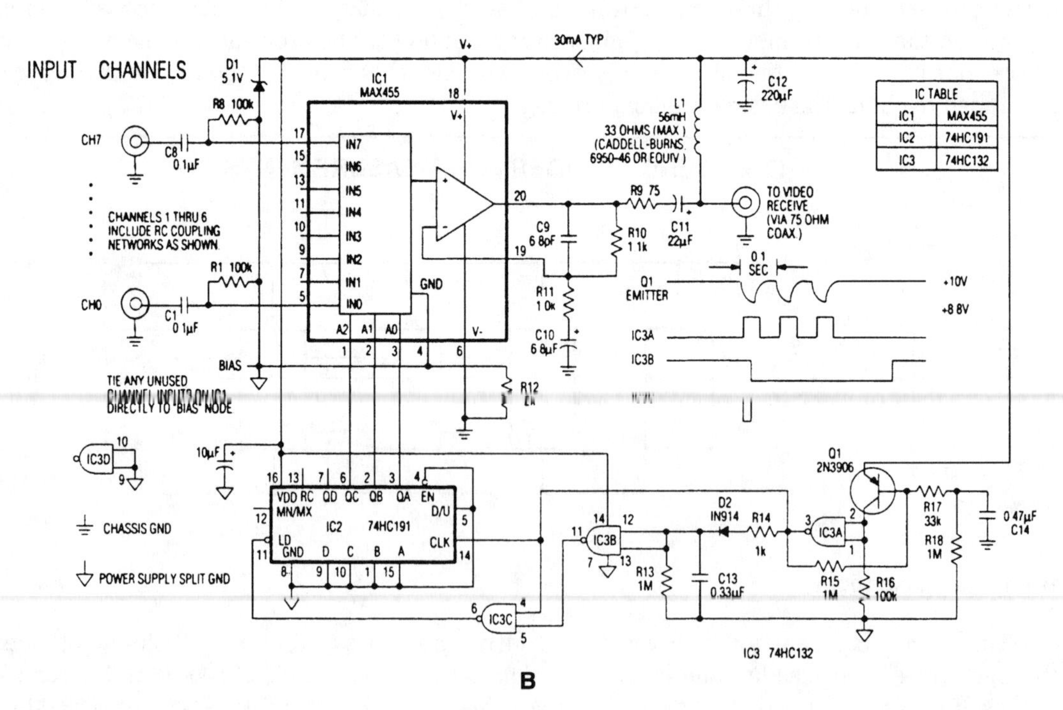

IC TABLE	
IC1	MAX455
IC2	74HC191
IC3	74HC132

B

MAXIM ENGINEERING JOURNAL

FIG. 57-16

In the video system of Figs. A and B, a single coaxial cable carries power to the remote location, selects one of eight video channels, and returns the selected signal. The system can choose one of several remote surveillance-camera signals, for example, and display the picture on a monitor near the interface box.

The heart of the multiplexer box (A) is a combination 8-channel multiplexer and amplifier (IC1). C11 couples the multiplexer's baseband video output to the coax, and L1 decouples the video from dc power arriving on the same line. This power—approximately 30 mA at 10 V—supplies all circuitry in the multiplexer box.

In interface box (B), a desired channel is encoded by three bits, set either by switches as shown or by an applied digital input. Momentary depression of the send button triggers downconverter IC1 and gated oscillator IC2A to initiate a channel-selection burst.

7805 TURN-ON CIRCUIT

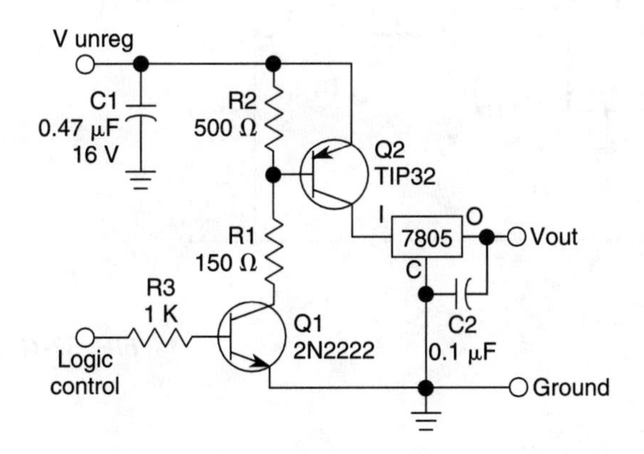

A logic level can control a 7805 regulator with this circuit. Q2 is a series switching transistor controlled by Q1. Q1 is turned on by a logic voltage to its base.

RADIO ELECTRONICS

FIG. 57-17

AF DRIVE INDICATOR

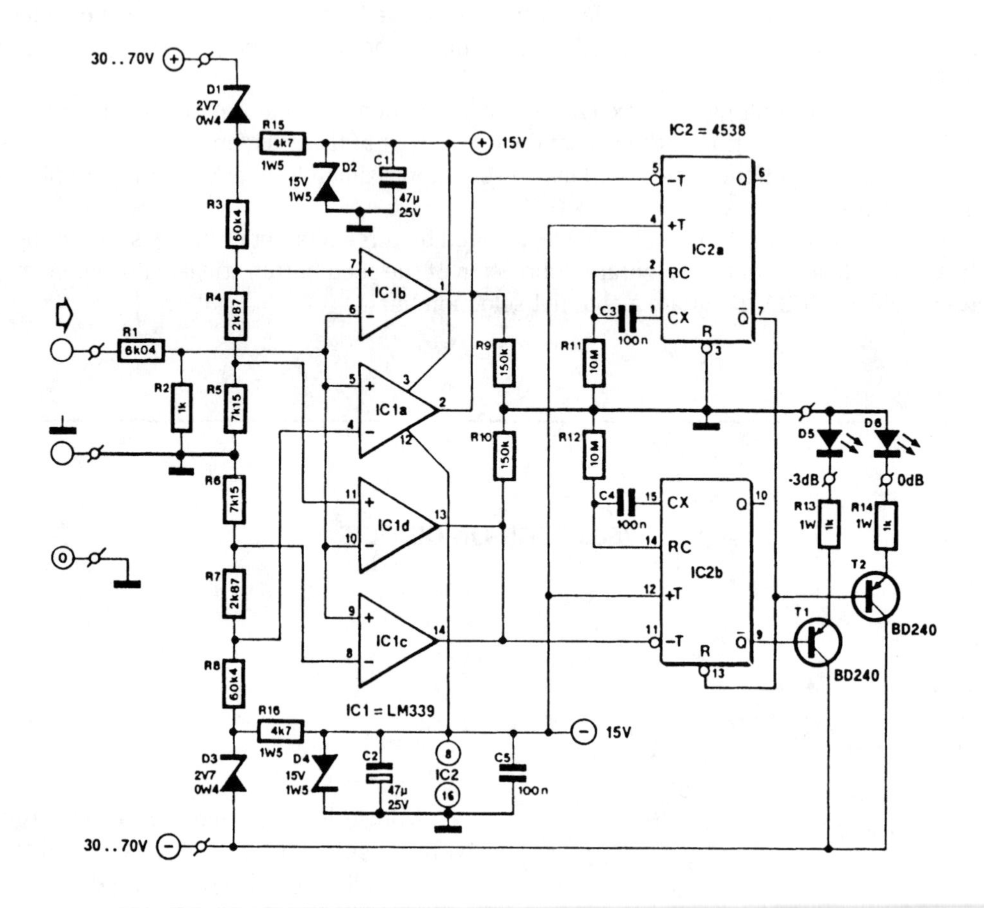

ELEKTOR ELECTRONICS

FIG. 57-18

This circuit was used with an audio power amplifier to detect the point at which output is –3 dB from maximum, indicated by LED D5, and at clipping, shown by LED D6. The indicator can be used with any amplifier operating from a ±30 to ±70 V symmetrical supply.

PHASE-LOCKED LOOP

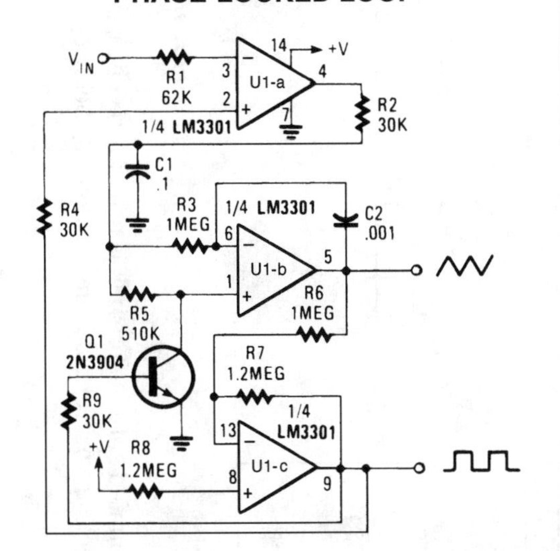

POPULAR ELECTRONICS *FIG. 57-19*

The PLL will lock onto an input signal. Both triangle- and square-wave outputs are available. A quad op amp can be used in this circuit, which should be useful in the audio and LF radio region.

CAPACITANCE MULTIPLIER

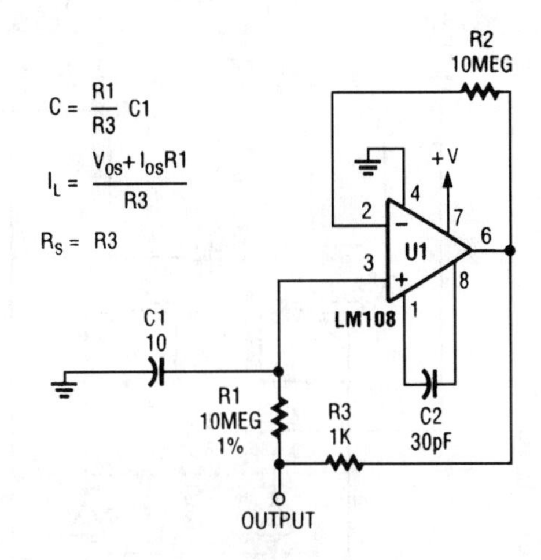

POPULAR ELECTRONICS *FIG. 57-20*

PRACTICAL DIFFERENTIATOR

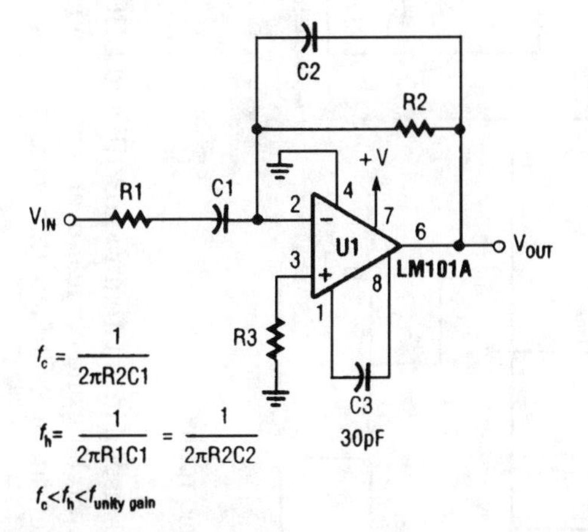

POPULAR ELECTRONICS *FIG. 57-21*

A differentiator has a high-pass characteristic. Components are chosen by using the design equations.

HUM REDUCER FOR DIRECT-CONVERSION RECEIVERS

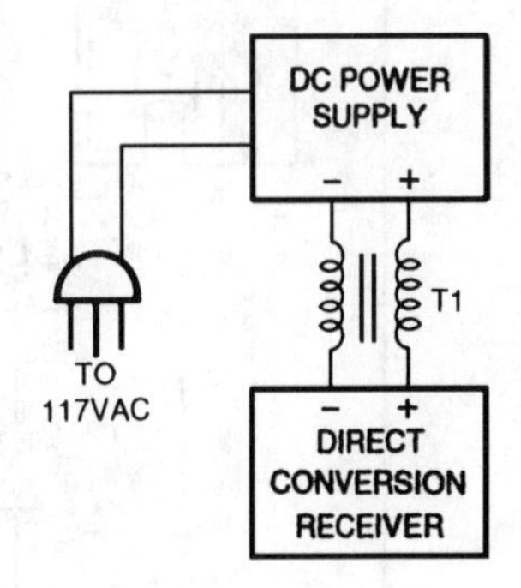

POPULAR ELECTRONICS *FIG. 57-22*

One cure for ac power line hum and ripple (caused by leakage current) is to use a well-regulated and filtered 9- to 18-Vdc power supply with a balancing choke (T1 in this illustration) between the power supply and the DCR.

PREAMP TRANSMIT-RECEIVE SEQUENCER

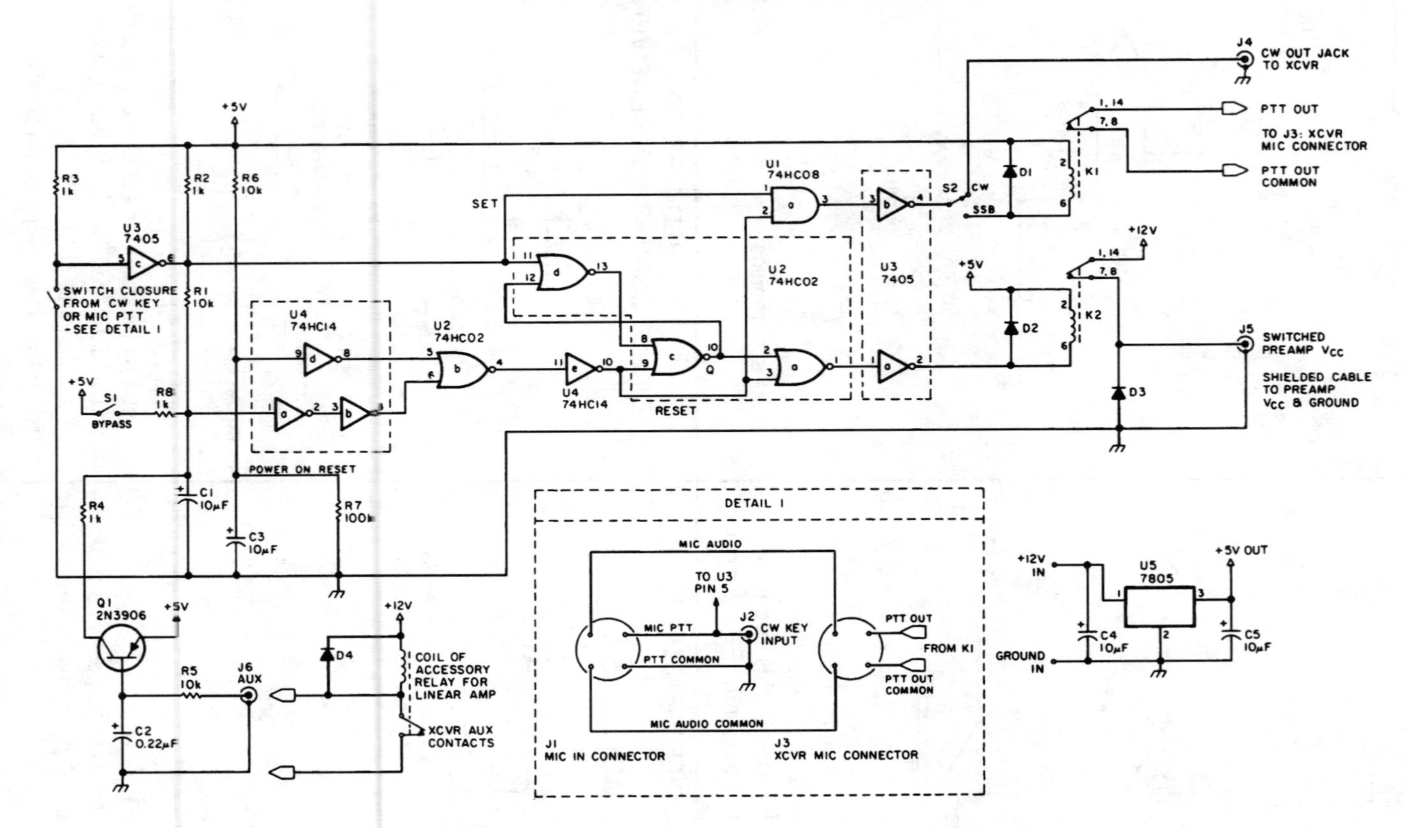

73 AMATEUR RADIO TODAY

FIG. 57-23

This circuit is useful in amateur radio VHF and UHF work where a mast-mounted antenna preamp is used for receiving. The kit controls T-R switching and change-over relay sequencing so that high RF levels are prevented from accidentally being applied to the preamplifier during switching intervals.

dc OUTPUT CHOPPER

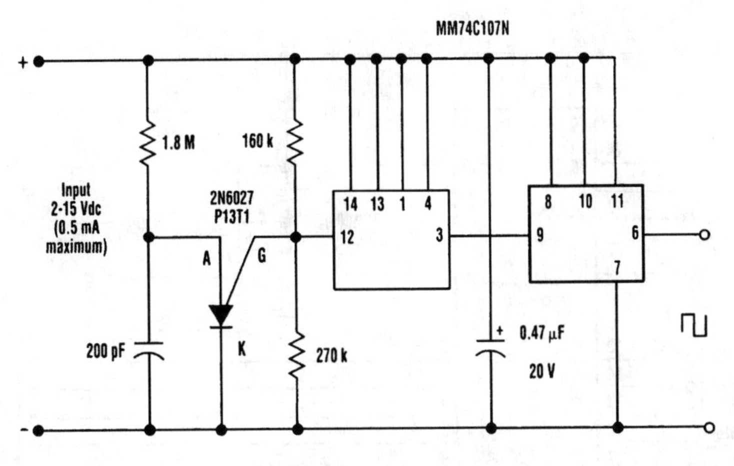

ELECTRONIC DESIGN

FIG. 57-24

Any dc voltage source in the 2- to 15-V range can be chopped into a unipolar square wave that has a peak amplitude nearly equal to the dc source voltage with circuit (lightly loaded CMOS will swing within a few millivolts of each rail at low frequencies). Depending on the actual voltage of the supply, the programmable-unijunction-transistor (PUT) relaxation oscillator produces 2000-Hz trigger pulses. These pulses operate the cascaded 74C107 flip-flop, producing a square wave.

ac ISOLATION TRANSFORMERS USE INEXPENSIVE 12-V TRANSFORMERS

"Safety first" is a good motto to follow when you play with electricity. You can follow that adage more closely with this homebrew isolation transformer.

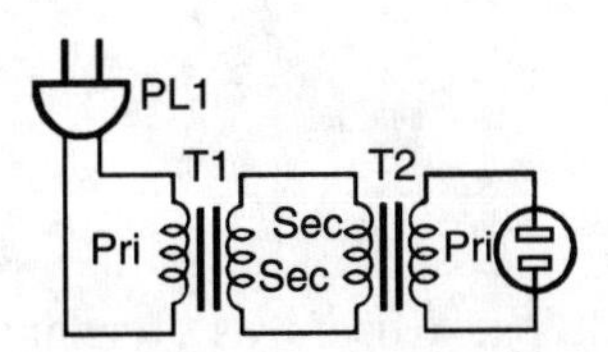

POPULAR ELECTRONICS

FIG. 57-25

ac LINE VOLTAGE BOOSTER

When incoming ac power drops, you can bring the voltage back up with this booster circuit. It adds the transformer's secondary voltage to the ac line voltage.

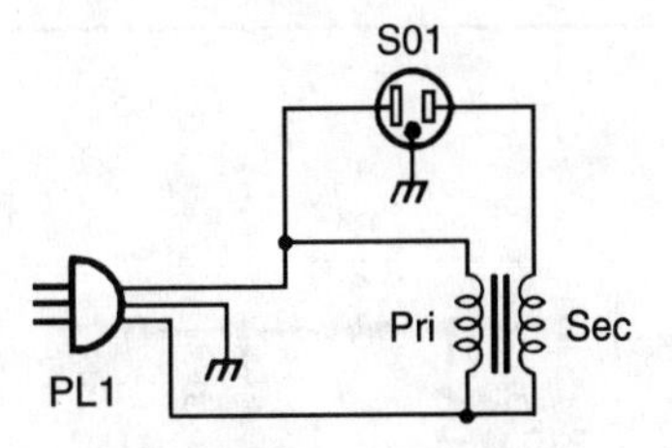

POPULAR ELECTRONICS

FIG. 57-26

OCTAL D/A CONVERTER

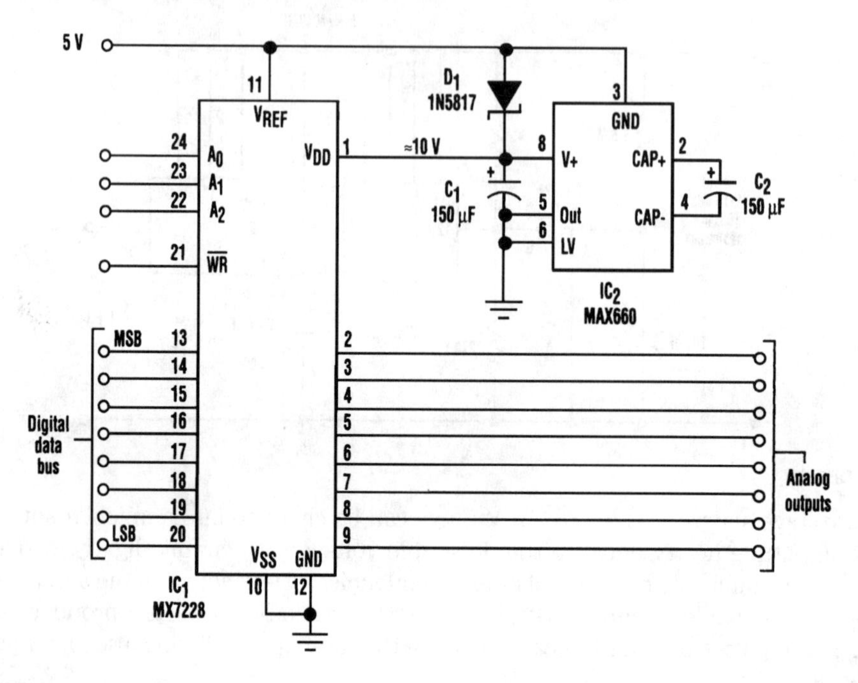

ELECTRONIC DESIGN

FIG. 57-27

This octal digital-to-analog converter operates on 5 V and provides eight output voltages, each digitally adjustable from supply rail to supply rail (0 to 5 V). Each output's resolution is 20 mV/LSB. The DAC chip (IC1) requires 3.5 V of "headroom" between its V_{DD} and reference voltages. However, a voltage-doubler charge pump (IC2) removes this limitation by generating an approximate 10-V supply for V_{DD}. All of the converter references are connected to the 5-V supply. IC2 doubles the 5-V input to an unregulated 10-V output that has an output impedance of less than 10 Ω. It can deliver 100 mA, which enables the eight DACs to issue their maximum output currents simultaneously (8 × 5 mA = 40 mA).

1-dB PAD

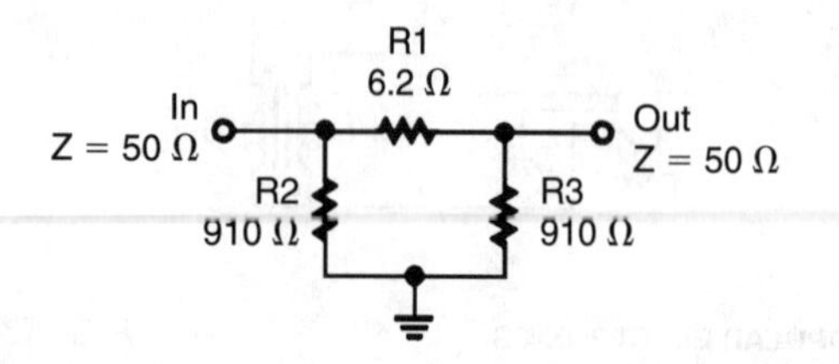

POPULAR ELECTRONICS

FIG. 57-28

The 1-dB pad is useful as a termination in RF work to limit possible mismatch range between system blocks, etc.

PSEUDO-RANDOM BIT SEQUENCE GENERATOR

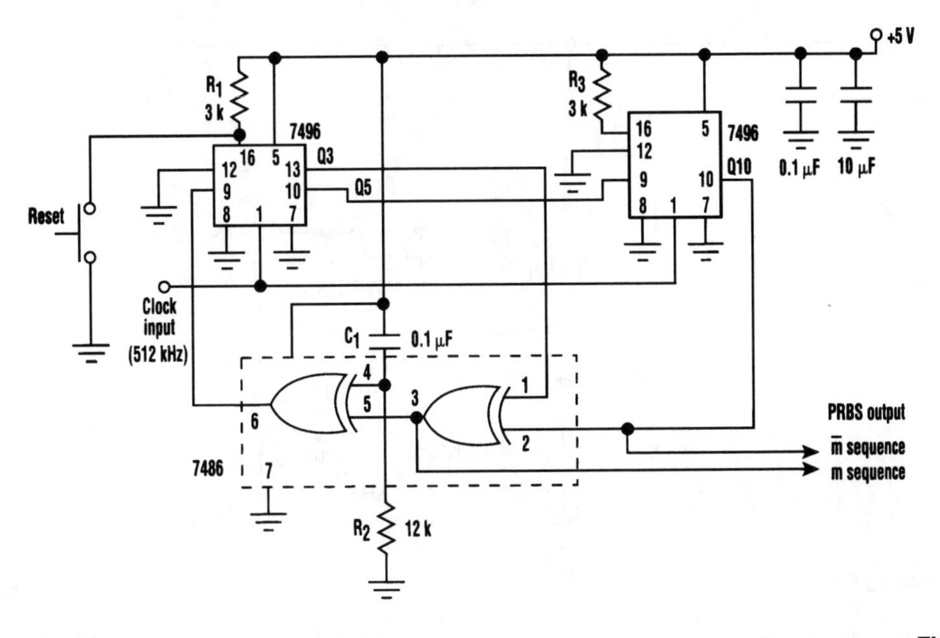

ELECTRONIC DESIGN *FIG. 57-29*

In this circuit, an additional exclusive-OR gate is connected after the modulo-2 feedback, with C1 and R2 applying the supply turn-on ramp into the feedback loop. This provides sufficient transient signal so that the PRBS generator can self-start a power-up. A shift-register length n of 10 is shown with feedback at stages 3 and 10, providing true and inverted maximal length sequence outputs.

This technique applies an input directly to the feedback loop. Therefore, it's considered more reliable than applying an RC configuration to the shift-register reset input to create a random turn-on state.

SIMPLE EXTERNAL MICROPHONE CIRCUIT FOR TRANSCEIVERS

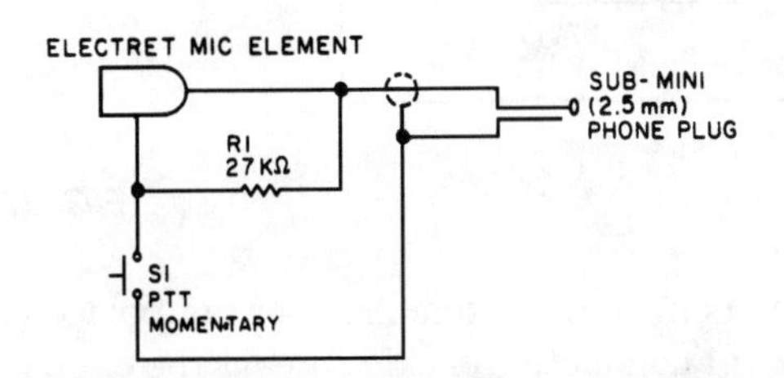

73 AMATEUR RADIO *FIG. 57-30*

Used originally for an Icom ICZAT handie talkie, this circuit might prove useful in other applications.

JFET CHOPPER CIRCUIT

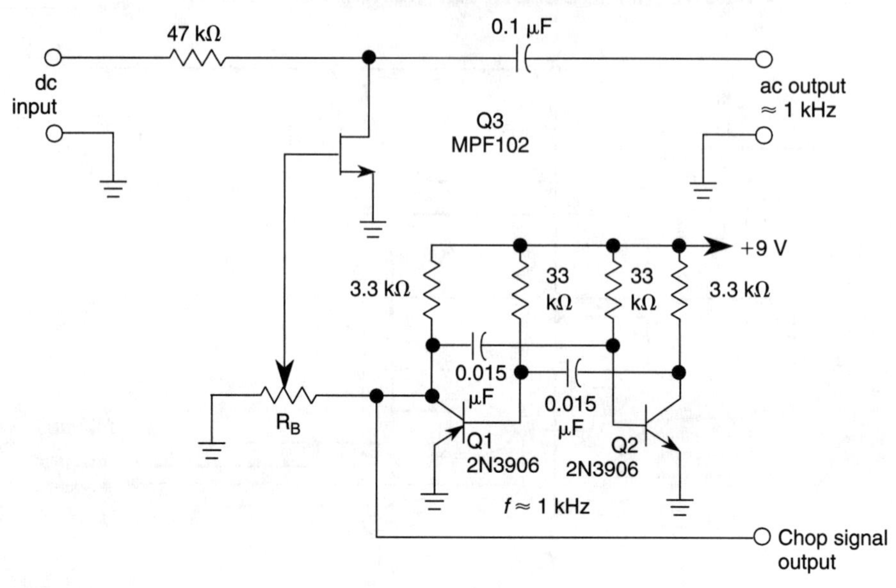

WILLIAM SHEETS *FIG. 57-31*

A JFET (MPF102) is used to chop a dc signal for amplification in an ac coupled amplifier. Q3 is the chopper element and Q1-Q2 forms the multivibrator to derive a chopping signal. R_B sets the bias on the FET to keep the drive to MPF102 as low as possible.

AUDIO MEMO ALERT

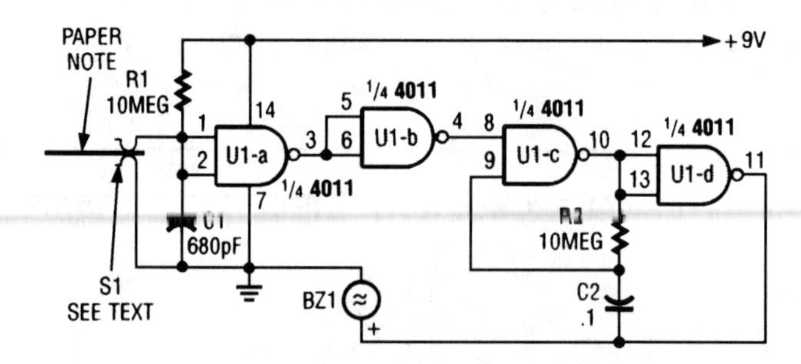

POPULAR ELECTRONICS *FIG. 57-32*

This device prevents paper notes and memos from being overlooked. A paper note placed between two fingers made of a conducting material (metal or conductive plastic) breaks the circuit, allowing pair 1 of U1-a to go high. This causes U1-c & U1-d to act as an oscillator, pulsing piezo buzzer BZ1.

OCTAVE EQUALIZER

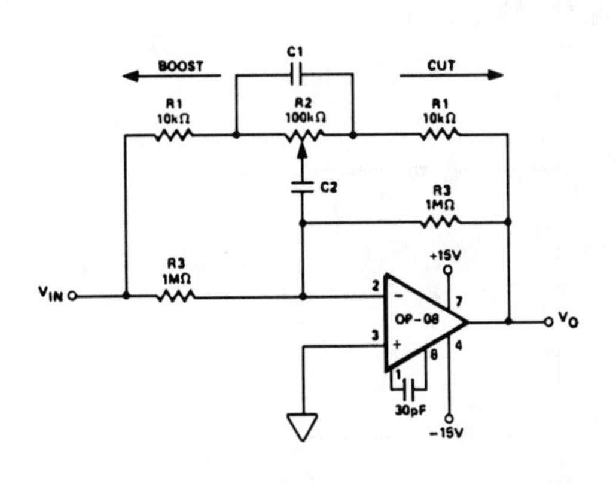

f_o (Hz)	C_1	C_2
32	0.18µF	0.018µF
64	0.1µF	0.01µF
125	0.047µF	0.0047µF
250	0.022µF	0.0022µF
500	0.012µF	0.0012µF
1k	0.0056µF	560pF
2k	0.0027µF	270pF
4k	0.0015µF	150pF
8k	680pF	68pF
16k	360pF	36pF

PRECISION MONOLITHICS INC.

FIG. 57-33

This circuit is one section of an octave equalizer used in audio systems. The table shows the values of C1 and C2 that are needed to achieve the given center frequencies. This circuit is capable of 12 dB boost or cut, as determined by the position of R2. Because of the low input bias current of the OP-08, the resistors could be scaled up by a factor of 10, and thereby reduce the values of C1 and C2 at the low-frequency end. In addition, 10 sections will only draw a combined supply current of 6 mA maximum.

COMPLEMENTARY OR BILATERAL ac EMITTER-FOLLOWER CIRCUIT

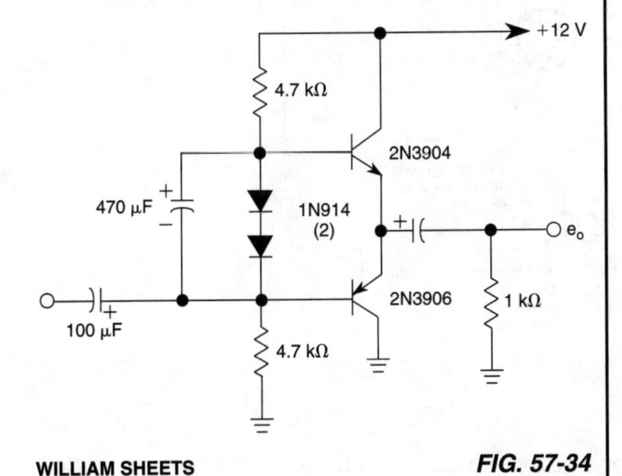

WILLIAM SHEETS

FIG. 57-34

This noninverting circuit uses a pair of complementary npn (2N3904) and pnp (2N3906) transistors.

CAPACITOR HYSTERESIS COMPENSATOR

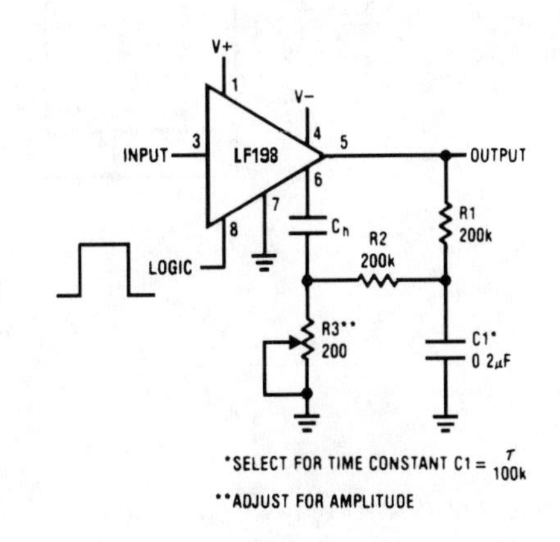

LINEAR DATABOOK

FIG. 57-35

AMPLIFIER COOL-DOWN CIRCUIT I

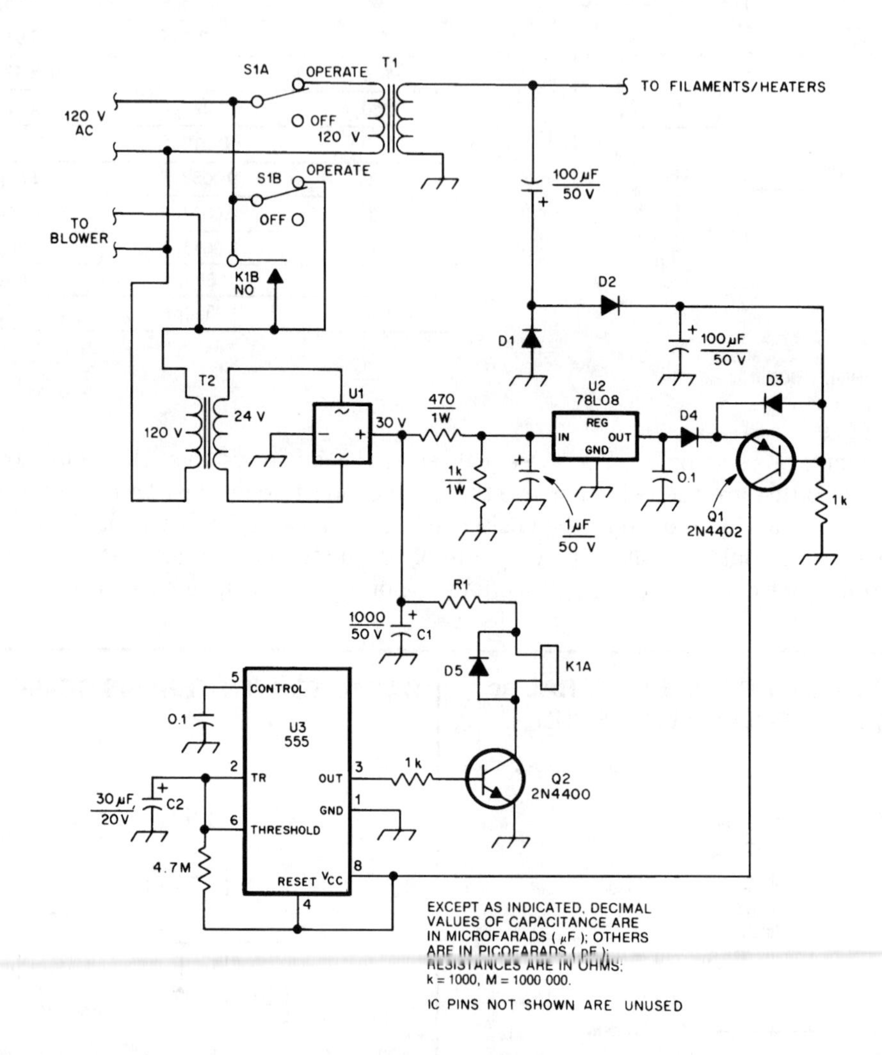

QST **FIG. 57-36**

This cool-down relay circuit uses an IC timer to drive a relay, which keeps the blower on for a time delay from timer U3. The value of C_2 can be changed to lengthen or shorten the time, as needed.

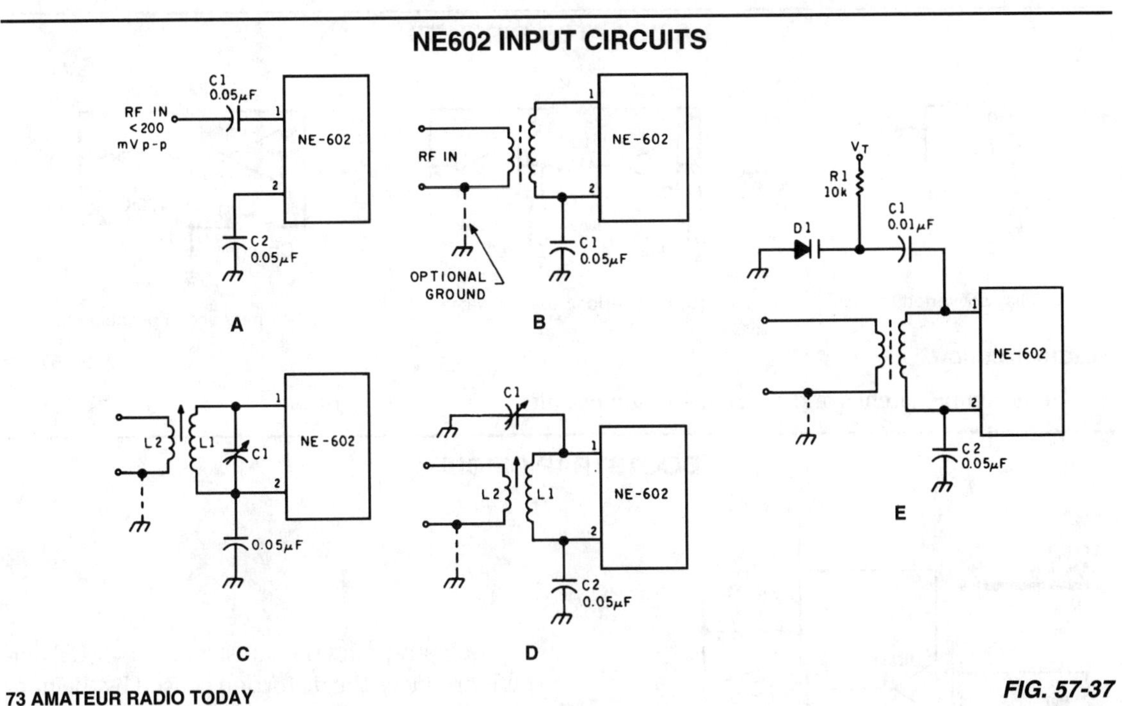

Input circuits for the NE-602.

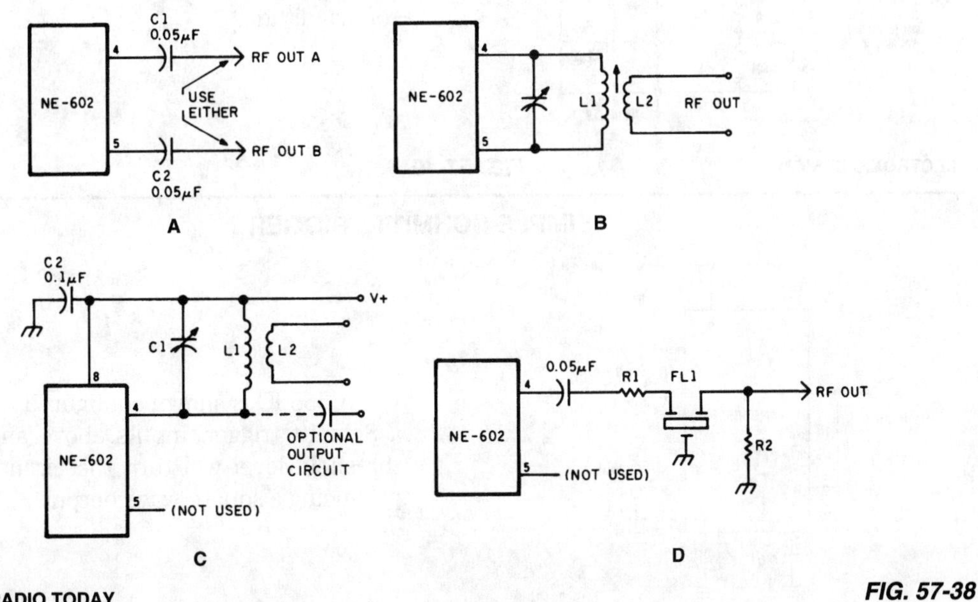

Output circuits for the NE-602.

BASIC LATCH CIRCUITS

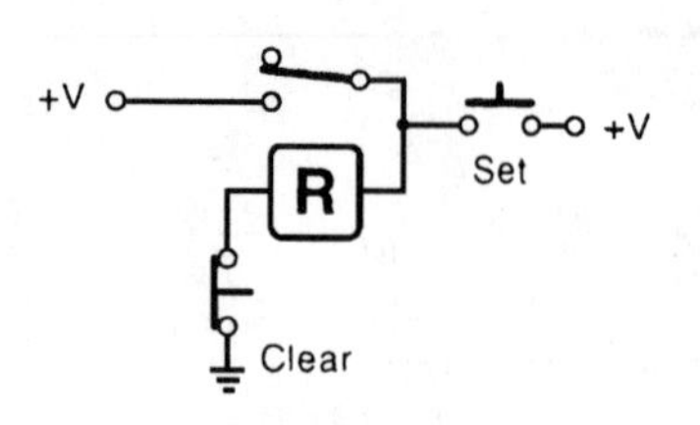

(A) Relay converted to latch.

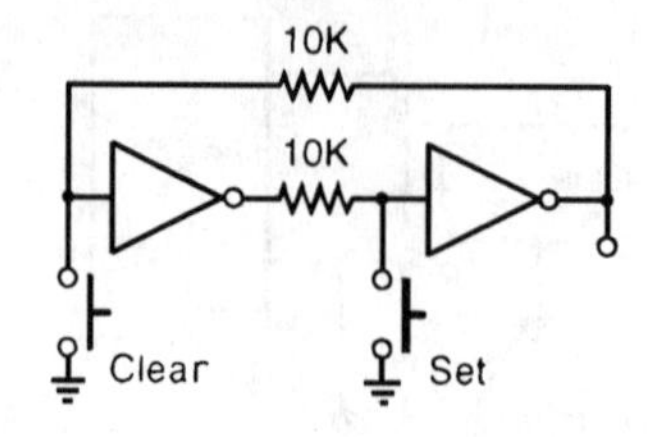

(B) Inverter pair used as latch.

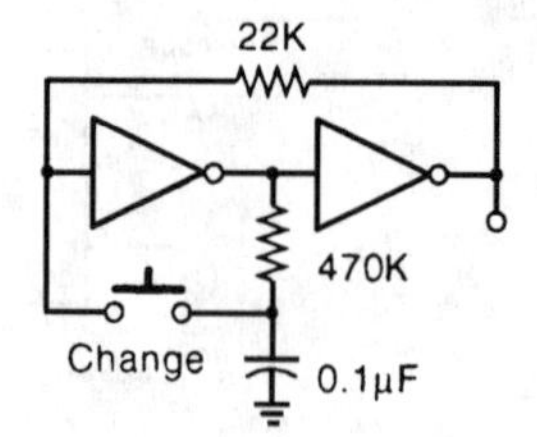

(C) Alternate action pushbutton.

ELECTRONICS NOW ***FIG. 57-39***

Some simple latches and alternate action circuits.

BOOTSTRAP CIRCUIT

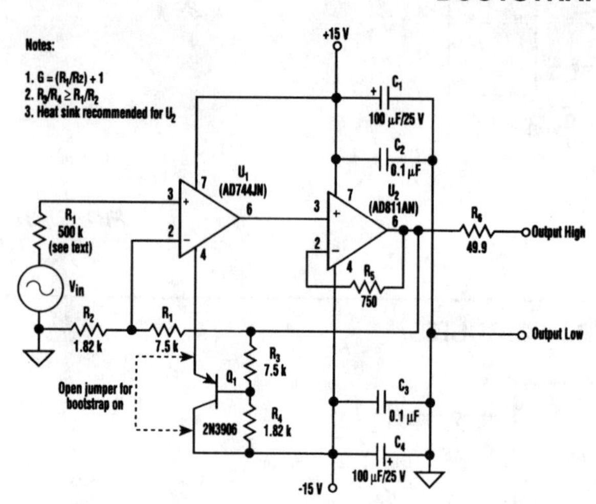

ELECTRONIC DESIGN ***FIG. 57-40***

Bootstrapping the substrate of a JFET amplifier reduces the distortion caused by the nonlinearity of the JFET input capacitance. In the figure, a second feedback divider bootstraps the substrate of U1. With R_1 = 500 kΩ (source impedance), THD at 10 kHz was reduced an order of magnitude.

SIMPLE SCHMITT TRIGGER

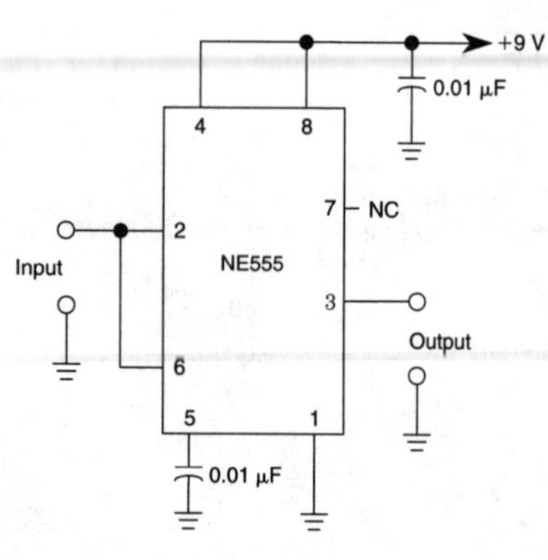

WILLIAM SHEETS ***FIG. 57-41***

A 555 IC is shown configured to function as a Schmitt trigger. Inputs above and below the threshold level will turn the circuit on and off producing a square wave output.

AMPLIFIER COOL-DOWN CIRCUIT II

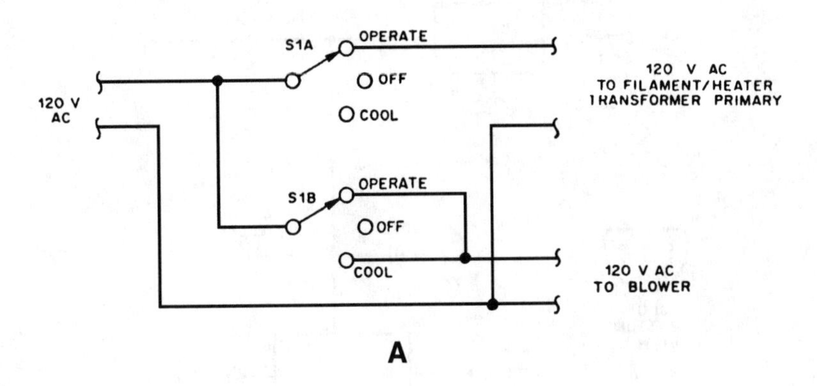

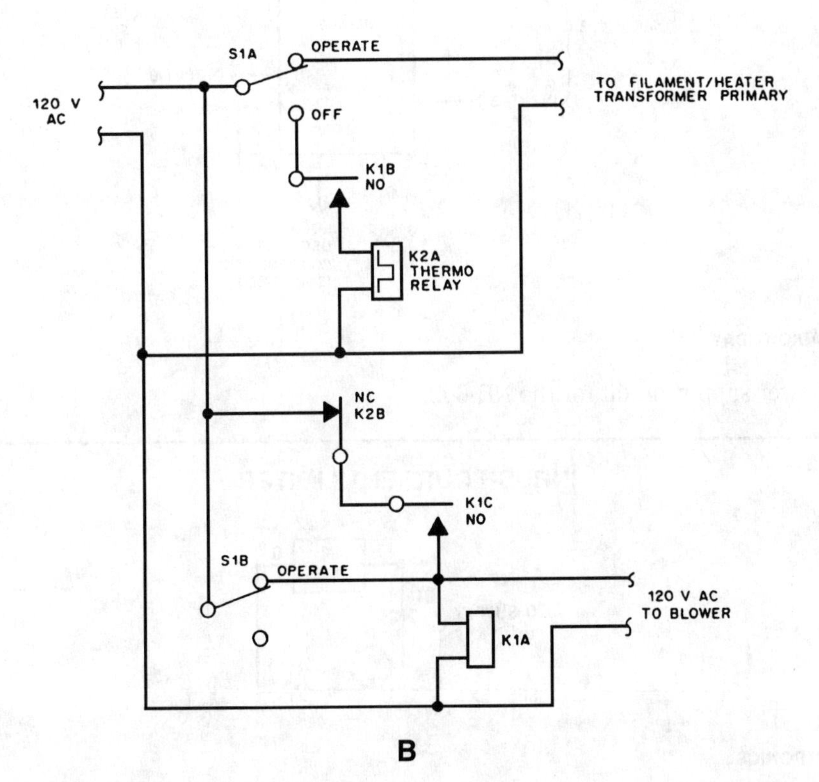

QST

FIG. 57-42

High-power amplifiers used in RF service, using vacuum tubes, often benefit from leaving the blower air flow on after removal of filament/heater voltage.

NE602 dc POWER CIRCUITS

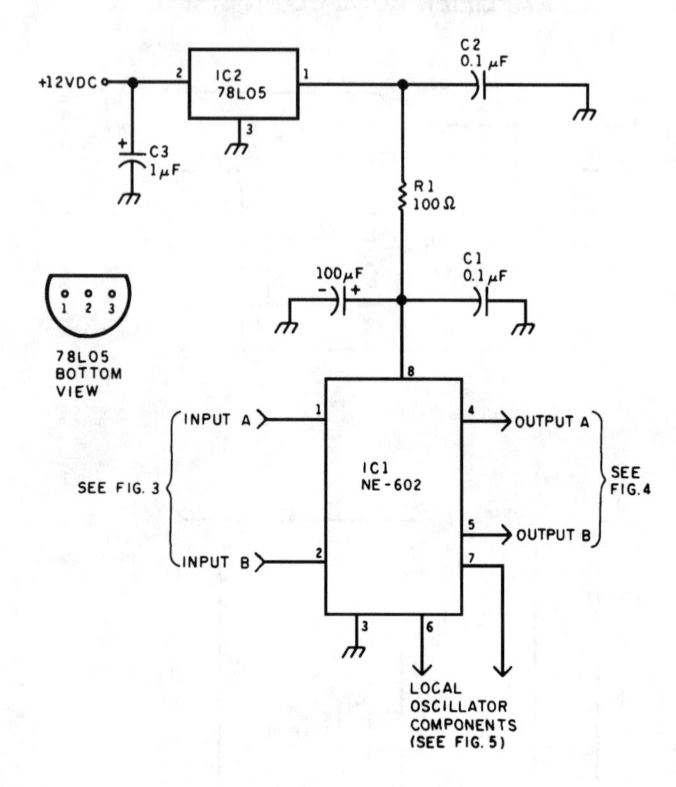

73 AMATEUR RADIO TODAY ***FIG. 57-43***

The dc power supply circuit for the NE-602.

INRUSH CURRENT LIMITER

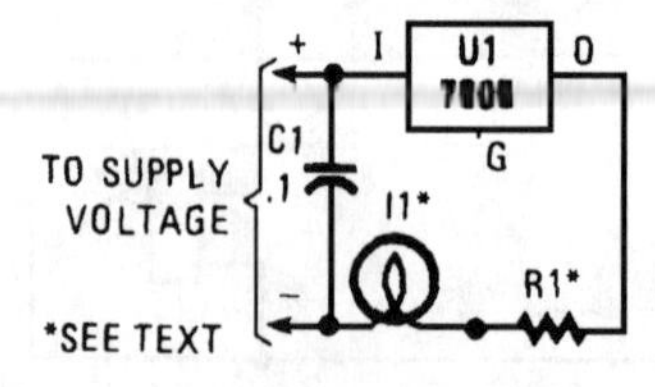

POPULAR ELECTRONICS ***FIG. 57-44***

A 7805 can be configured as a constant-current regulator, to serve as an inrush current limiter. R1 will have 5 V across it at all times so the total current through I1 will be 5 V/R_1 + 5 mA, the 5 mA being the regulator operating current. In this case, R_1 = 5 V/95 mA = 52.6 Ω for I1 current = 100 mA.

58

Mixer Circuits

The sources of the following circuits are contained in the Sources section, which begins on page 675. The figure number in the box of each circuit correlates to the entry in the Sources section.

LOW-NOISE 4-CHANNEL GUITAR MIXER

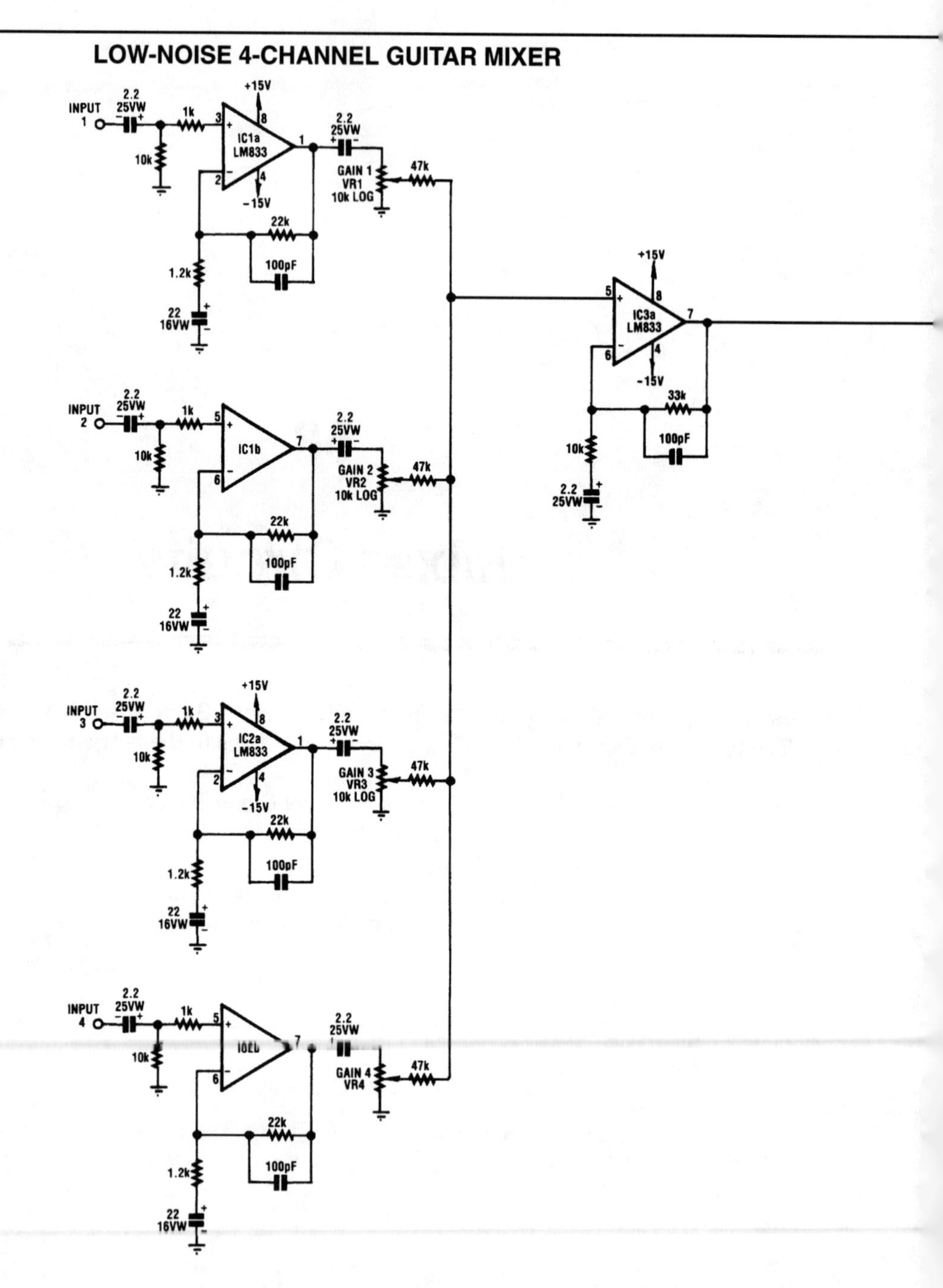

SILICON CHIP

IC1-a, IC1-b, IC2-a, and IC2-b all function with a gain of about 19. Their outputs are mixed via the level-control pots and the resulting signal amplified by IC3-a and fed to tone-control stage IC3-b. Finally, the output from IC3-b is fed to unity-gain buffer stage IC4-a via volume-control potentiometer VR8.

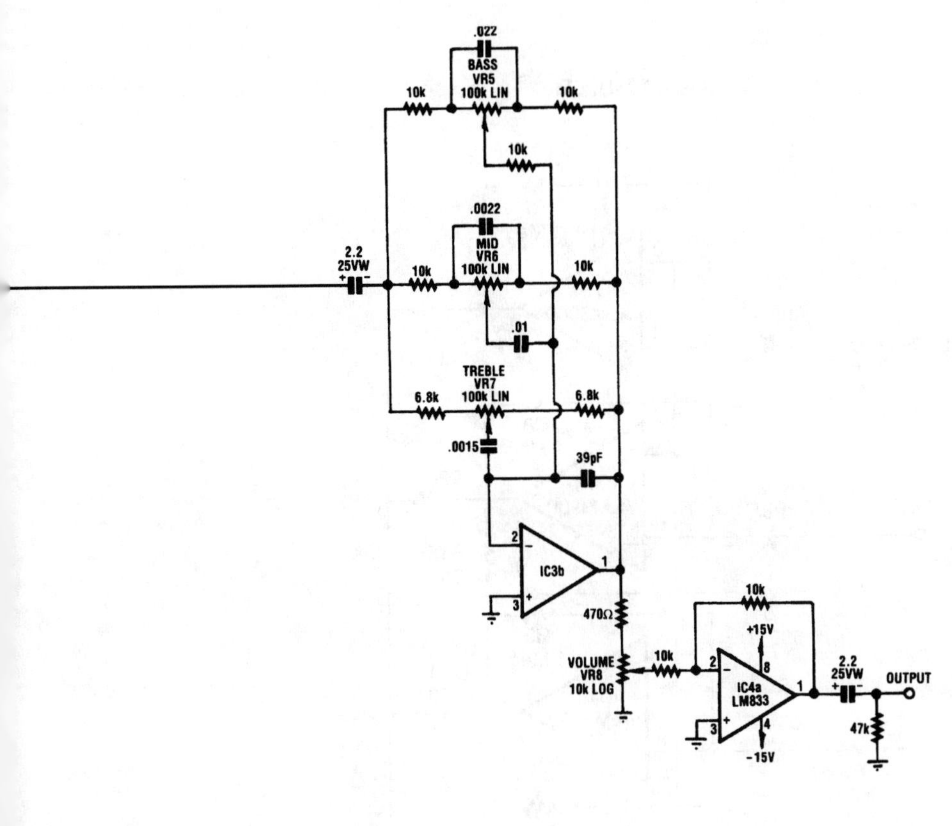
.022
BASS
VR5
100k LIN
10k
10k
10k
.0022
MID
VR6
100k LIN
2.2
25VW
10k
10k
.01
TREBLE
VR7
100k LIN
6.8k
6.8k
.0015
39pF
IC3b
470Ω
10k
+15V
VOLUME
VR8
10k LOG
10k
IC4a
LM833
-15V
2.2
25VW
OUTPUT
47k

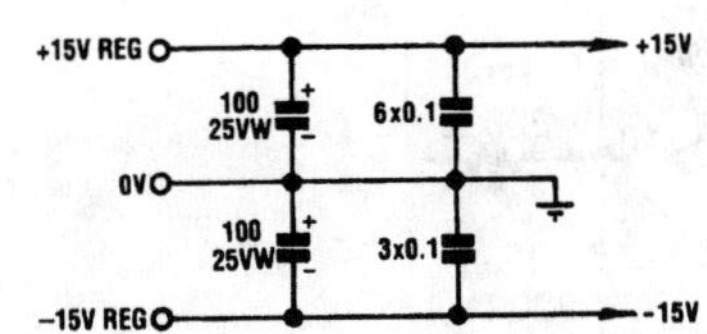
+15V REG
+15V
100
25VW
6x0.1
0V
100
25VW
3x0.1
-15V REG
-15V

FIG. 58-1

AUDIO MIXER

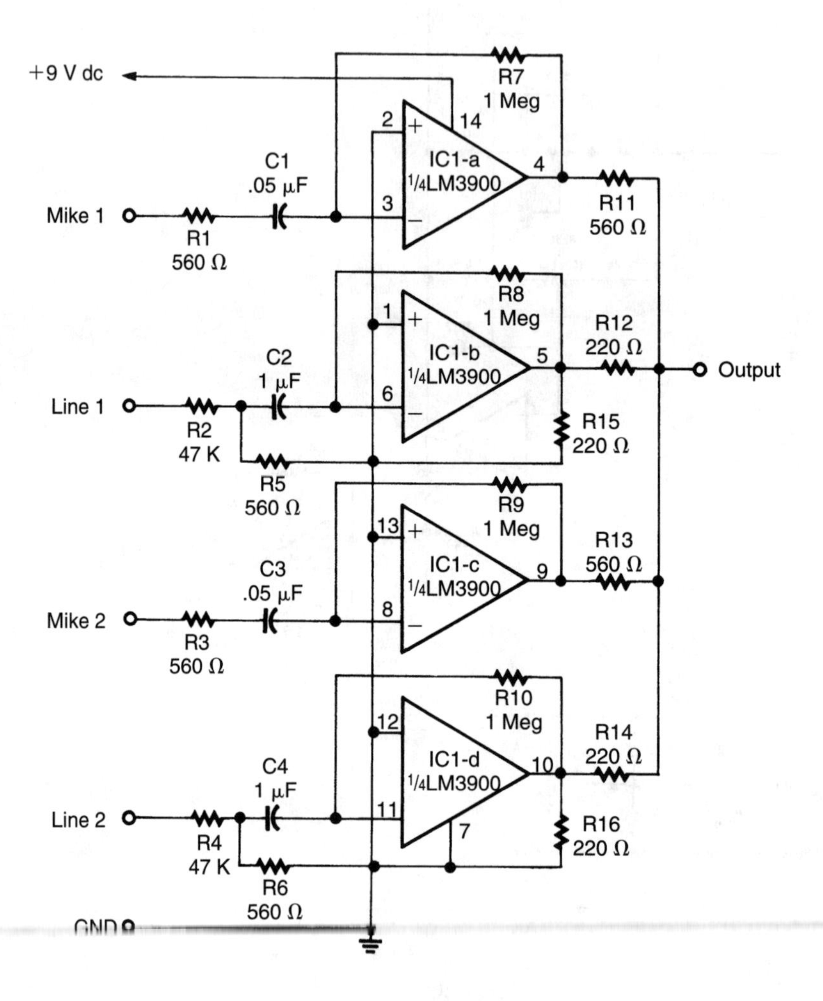

ELECTRONICS NOW *FIG. 58-2*

Designed around an LM3900 quad op amp, this mixer combines 2-line and 2-mike inputs and sums them at the output terminal. R7 through R10 can be changed to vary the gain (around +23 dB).

FET MICROPHONE MIXER

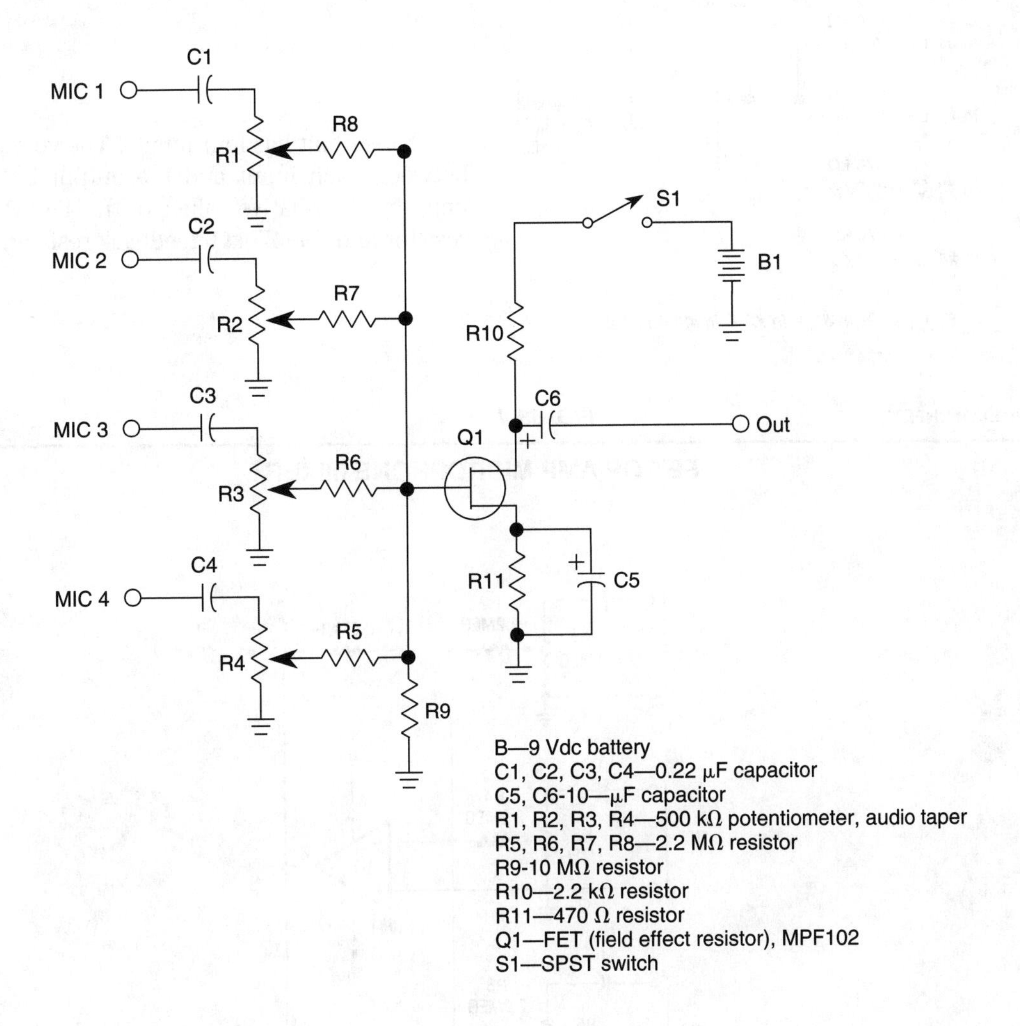

WILLIAM SHEETS ***FIG. 58-3***

A JFET transistor is used as a high-to-low impedance converter and signal mixer. Input impedance is approximately 500 kΩ but it can be increased by increasing R5 to R8 as high as 10 MΩ. Output Z is about 2 kΩ, but it can be increased or decreased by changing the value of R_{10}. Use 560 or 680 Ω to feed a 600-Ω input; use 100 kΩ to 1 MΩ for high impedance.

UNITY-GAIN FOUR-INPUT AUDIO MIXER

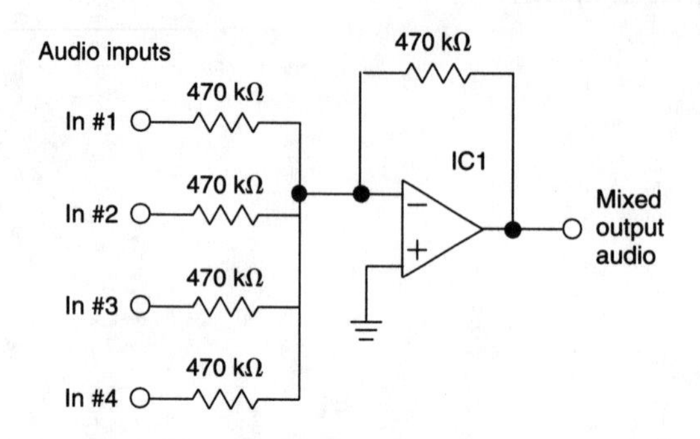

$E_{OUT} = -$ *(In #1 + In #2 + In #3 + In #4)*

IC1 = LM741, etc.

The circuit has four inputs. The voltage gain between each input and the output is held at unity by the relative values of the 470kΩ input resistor and the 470kΩ feedback resistor.

WILLIAM SHEETS

FIG. 58-4

FET OP AMP MICROPHONE MIXER

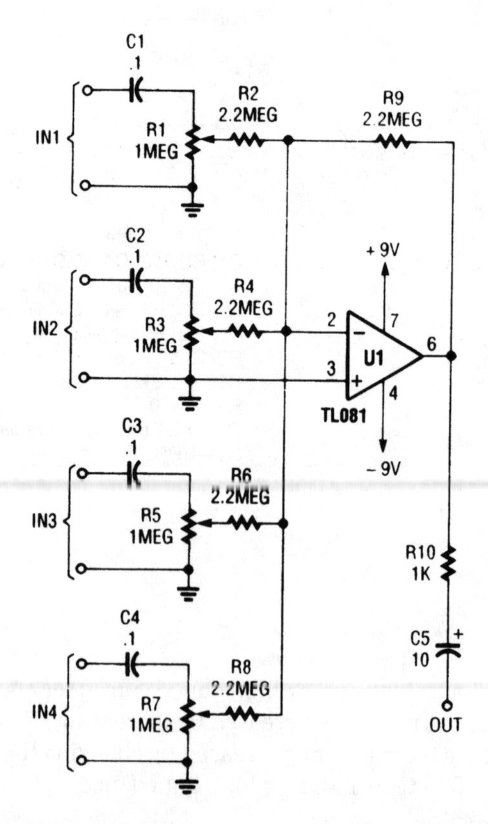

POPULAR ELECTRONICS

FIG. 58-5

59

Modulator Circuits

The sources of the following circuits are contained in the Sources section, which begins on page 675. The figure number in the box of each circuit correlates to the entry in the Sources section.

FM MODULATOR

IC-1 - Motorola MC-1648P
All resistors 5%, 0.25 W
Zener - 5.1 V, 0.5 W
All 0.1 and 0.01 uF capacitors ceramic, 16V
C4 - 100 uF, 16 V electrolytic
D1, D2 - Motorola MV-209
L1 - airwound, 6 turns, 3/16" dia., 5/16" long, 20 AWG
C3 - 500 pF, silver mica

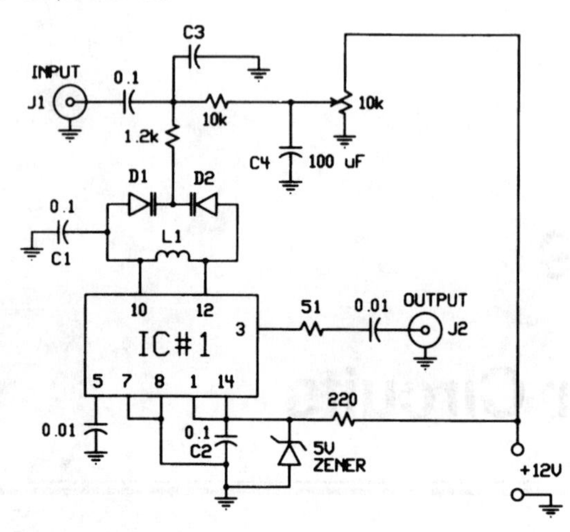

The FM modulator is built with a Motorola MC1648P oscillator. Two varactors, Motorola MV-209, are used to frequency modulate the oscillator. The 5000-Ω potentiometer is used to bias the varactors for best linearity. The output frequency of approximately 100 MHz can be adjusted by changing the value of the inductor. The output frequency can vary as much as 10 MHz on each side. The output level of the modulator is –5 dBm. In this prototype, the varactor bias was 7.5 V for best linearity; but this could be different with other varactors.

RF DESIGN ***FIG. 59-1***

455-kHz MODULATOR

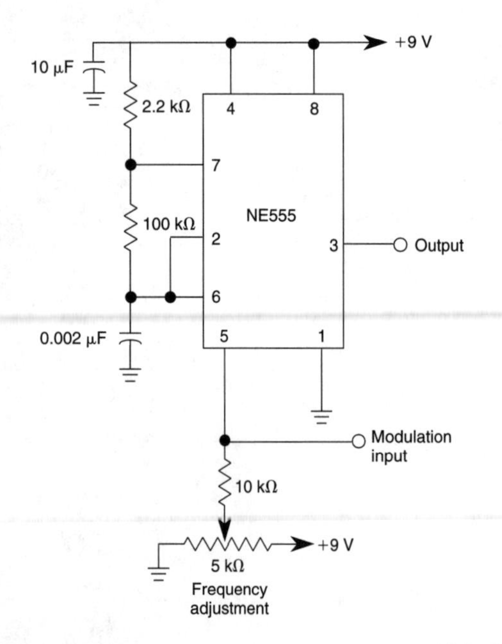

This circuit shows how to frequency-modulate the oscillator using a 555. Oscillator frequency is set with the 5-kΩ potentiometer and the modulation signal is dc-coupled.

WILLIAM SHEETS ***FIG. 59-2***

555 FM CIRCUIT

IC-1 - Motorola MC-1374P
IC-2 - National LH0002C
L1, L2 - Mouser Electronics #421IF200
C1, C2 - silver mica, 300 pF
All 0.1 uF cap., ceramic disc, 16V
C3 - 100 uF, 10 V, electrolytic
All resistors 5%, 0.25 W
ADJUSTMENT: Adjust R1 for minimum carrier; signal from function generator should generate 500 mVpp at pin 8 of IC-2 (suppressed carrier double sideband). Adjust R2 and function generator level to achieve 800 mVpp at pin 8 of IC-2 (standard AM with carrier). Adjust L2 for 455 kHz. Adjust L1 for maximum output.

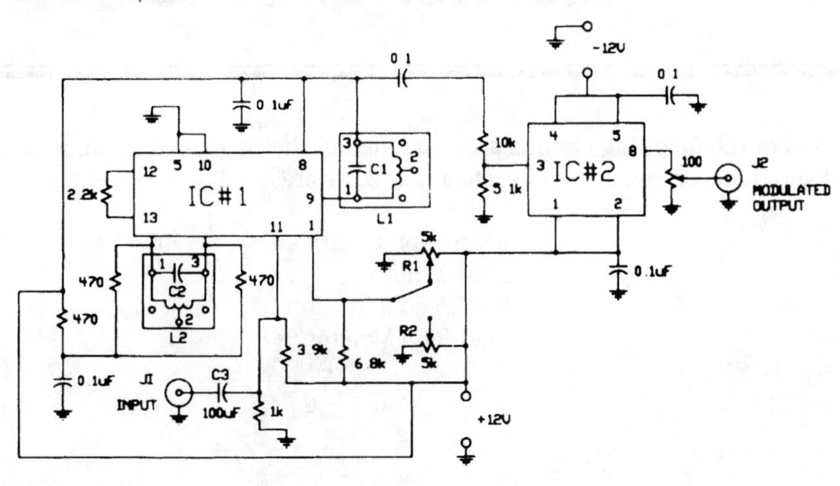

RF DESIGN ***FIG. 59-3***

Circuit for applying a dc-coupled FM or PPM to a 555 configured as an oscillator.

60

Monitor Circuits

The sources of the following circuits are contained in the Sources section, which begins on page 675. The figure number in the box of each circuit correlates to the entry in the Sources section.

ROOM MONITOR

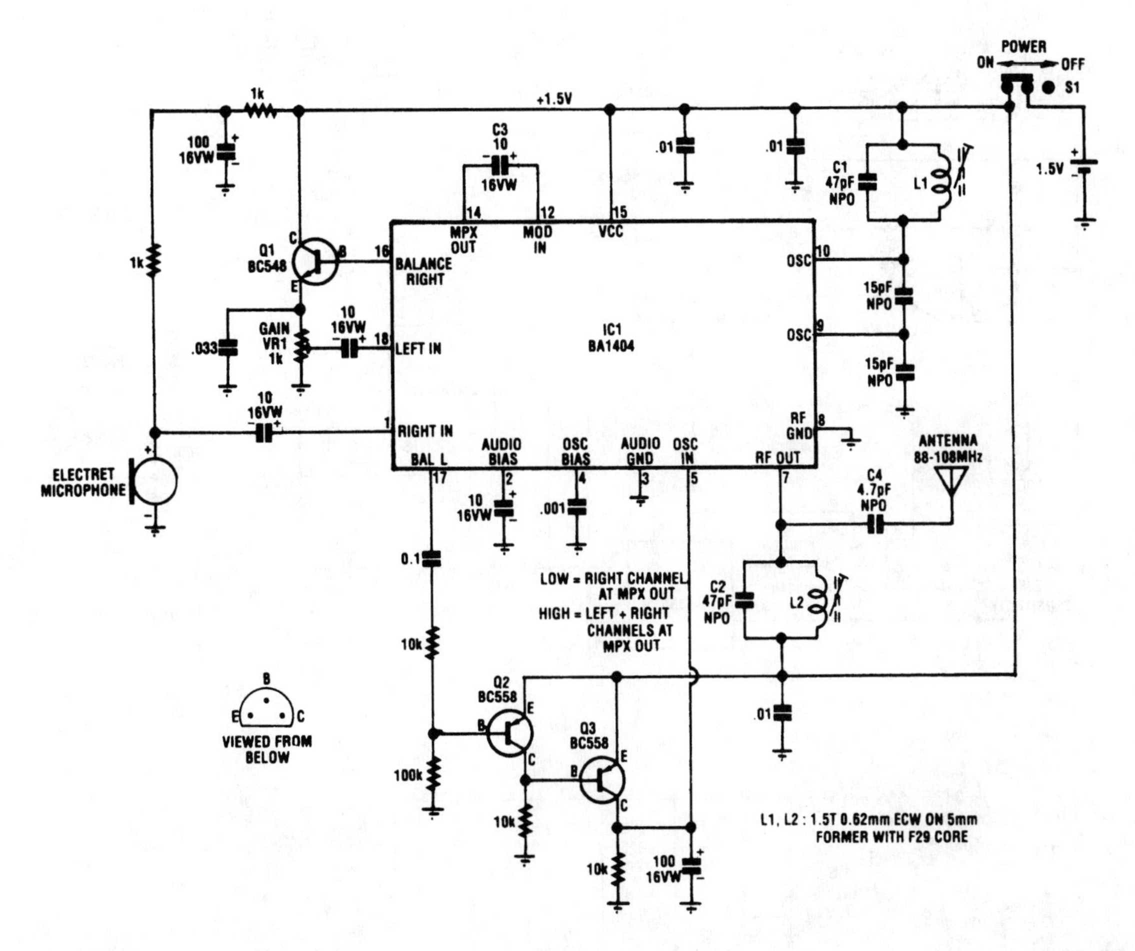

SILICON CHIP

FIG. 60-1

The circuit uses Q1 to buffer the right-channel balance output while Q2 and Q3 form a VOX circuit. When the signal level from the microphone goes high, the output of the VOX also goes high and the multiplexer inside IC1 switches the high-gain left-channel output through to a following buffer stage. This signal is then ac-coupled via C3 into an RF mixer stage and thence to an RF amplifier, which is tuned by C2 and L2.

BABY MONITOR

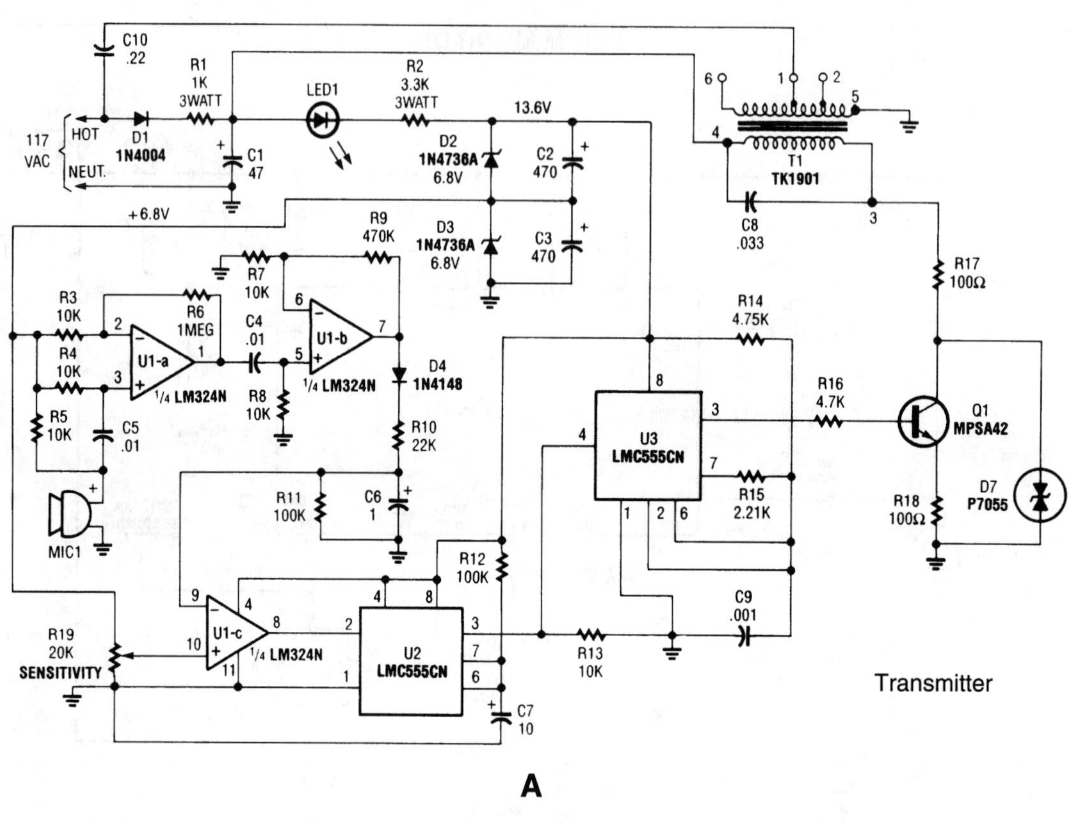

A

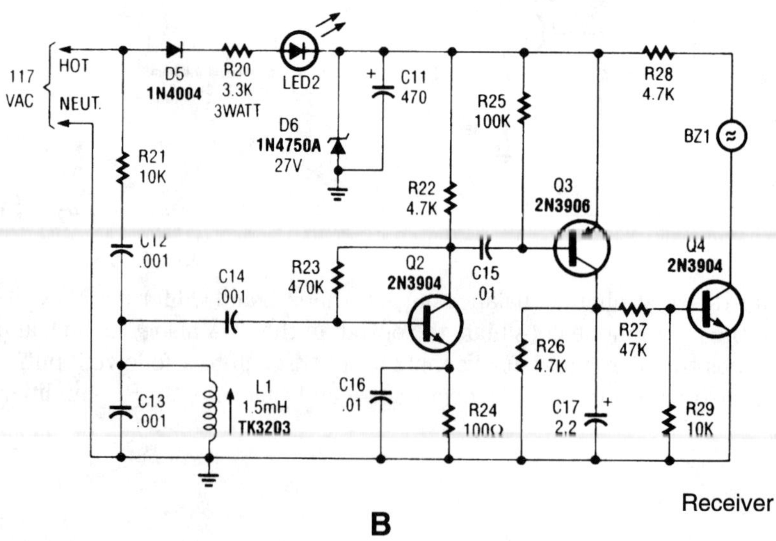

B

POPULAR ELECTRONICS *FIG. 60-2*

BABY MONITOR (*Cont.*)

Transmitter operation. Operating power for the transmitter circuit is derived directly from the ac line. The dc power to operate the circuit is generated in two stages, one for an RF power-amplifier stage, and the second for the remainder of the circuit.

The ac line voltage is applied to D1, which half-wave rectifies the ac input. The resulting dc voltage (approximately 30V under load) is fed across an RC filter (comprised of R1 and C1) and used to operate amplifier, Q1. The second stage of the power supply (composed of LED1, R2, D2, D3, C2, and C3, which forms a regulated +13.6-V, center-tapped supply) feeds the remainder of the circuit. LED1 is connected in series with R2 and is used as a visual power-on indicator for the transmitter.

An electret microphone element (MIC1) is used as the pick-up. The output of the microphone is ac coupled through C5 to U1-a (a noninverting op amp with a gain of about 100). The output of U1-a at pin 1 is ac coupled through C4 to the noninverting input of U1-b (which provides an additional gain of 48) at pin 5. The output of U1-b at pin 7 is then fed through D4 and R10, and across R11 and C6 to the inverting input of U1-c which is biased to a positive voltage that is set by SENSITIVITY-control R19. This represents a threshold voltage at which the output of U1-c switches from high to low.

During standby, the output of U1-c at pin 8 is held at about 12 V when the voltage developed across C6 is less than the bias-voltage setting at pin 10. When a sound of sufficient intensity and duration is detected, the voltage at pin 9 of U1-c exceeds the threshold level (set by R19), causing U1-c's output at pin 8 at go low. That low is applied to pin 2 of U2 (a 555 oscillator/timer configured as a monostable multivibrator). This causes the output of U2 to go high for about one second, as determined by the time constant of R12 and C7. The output of U2 at pin 3 is applied to pin 4 of U3 (a second 555 oscillator/timer that is configured for astable operation, with a frequency of about 125 kHz). That causes U3 to oscillate, producing a near square-wave output that is used to drive Q1 into conduction. The output of Q1 is applied across a parallel-tuned circuit composed a T1's primary and C8. The tuned circuit, in turn, reshapes the 125-kHz signal, causing a sine-wave-like signal to appear across both the primary and the secondary of T1.

The signal appearing at T1's secondary (about 1 or 2 V peak-to-peak) is impressed across the ac power line, and is then distributed throughout the building without affecting other electrical appliances connected to the line. Transient suppressor D7 is included in the circuit to help protect Q1 from voltage spikes that might appear across the power line and be coupled to the circuit through T1.

Receiver operation. Power for the receiver, as with the transmitter, is derived from a traditional half-wave rectifier (D5). The resulting dc voltage is regulated to 27 V by D6 and R20, and is then filtered by C11 to provide a relatively clean, dc power source for the circuit. A light-emitting diode, LED2, connected in series with R20 provides a visual indication that the circuit is powered and ready to receive a signal.

The 125-kHz signal is plucked from the ac line and coupled through R21 and C12 to a parallel-tuned LC circuit, consisting of C13 and L1. That LC circuit passes 125-kHz signals while attenuating all others. The 125-kHz signal is fed through C14 to the base of Q2 (which is configured as a high-gain linear amplifier), which boosts the relatively low amplitude of the 125-kHz signal. The RF output of Q2 is ac coupled to the base of Q3 through C15. Transistor Q3 acts as both an amplifier and detector. Because there is no bias voltage applied to the base of Q3, it remains cut off until driven by the amplified 125-kHz signal. When Q3 is forward biased, its collector voltage rises.

Capacitor C16, connected across Q3's collector resistor, filters the 125-kHz signal so that it is essentially dc. When the voltage at the collector of Q3 rises, Q4 is driven into conduction. That causes current to flow into piezo buzzer BZ1, producing a distinctive audio tone that alerts anyone within earshot that the baby needs attention.

BIRD FEEDER MONITOR

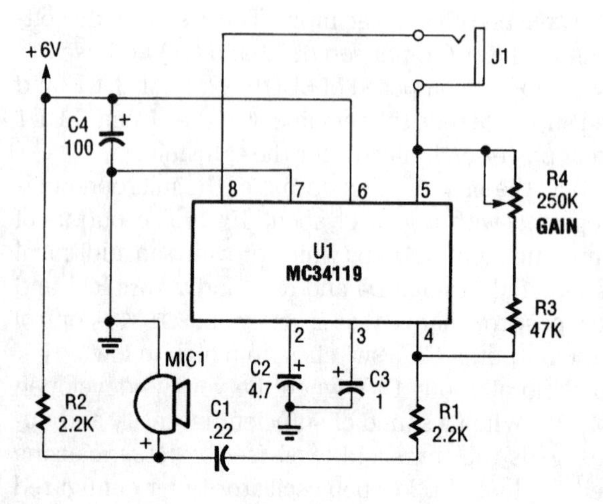

POPULAR ELECTRONICS *FIG. 60-3*

The first amplifier circuit is a bird phone. In this circuit, the electret mike (MIC1) is mounted in the neck of a large plastic funnel. The amplifier, built around an MC34119 (which is available from D.C. Electronics, P.O. Box 3203, Scottsdale, AZ 85271-3203; Tel. 800-467-7736, and elsewhere), is then placed outside of the funnel with the pick-up facing a nearby bird feeder. The output of the amplifier is then connected to a 16-Ω speaker.

The amplifier's voltage gain is determined by the values of the input resistor (R1) and the feed-back resistor (R3 and R4, respectively). The differential gain of the amplifier is given by: $R_3 + R_4/R_1 \times 2$. With the component values shown, the maximum voltage gain is about 270. This permits listening to the activity at the bird feeder.

ACID-RAIN MONITOR

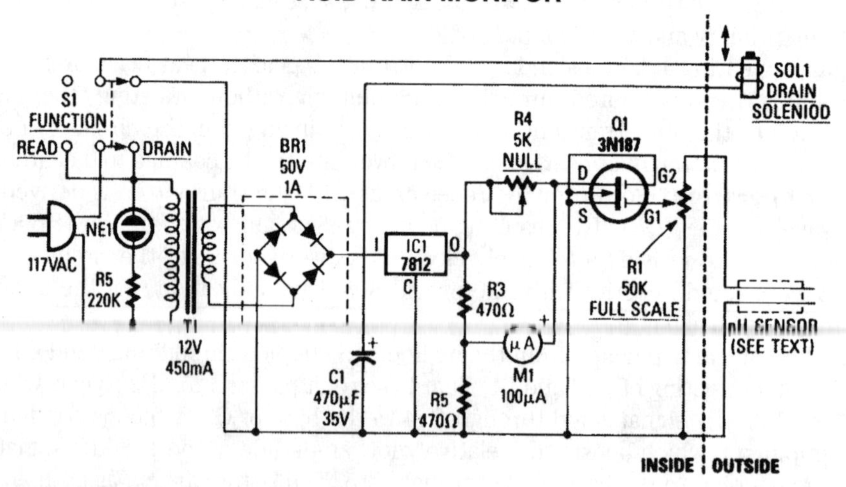

R-E EXPERIMENTERS HANDBOOK *FIG. 60-4*

The drain-to-source resistance of Q1 varies depending on the acidity of the sample presented to Q1's gate circuit. That variable resistance varies the current flowing through the bridge; that current is proportional to pH.

61

Moisture- and Fluid-Detector Circuits

The sources of the following circuits are contained in the Sources section, which begins on page 675. The figure number in the box of each circuit correlates to the entry in the Sources section.

WATER-ACTIVATED ALARM

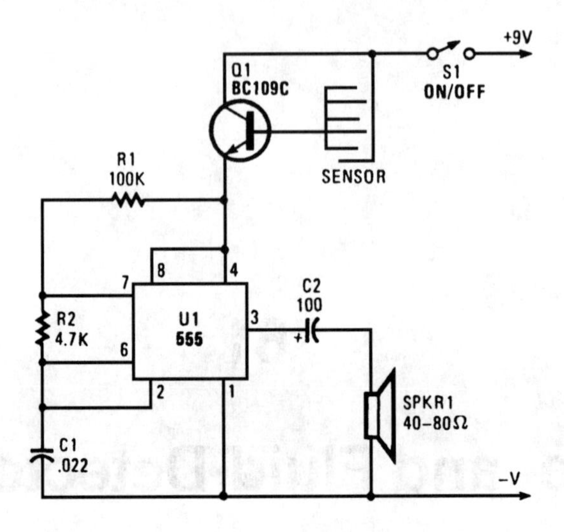

POPULAR ELECTRONICS *FIG. 61-1*

When sensor gets wet, it conducts, forward-biases Q1, and activates audio oscillator U1. A tone is heard from the speaker.

SIMPLE FLOOD ALARM

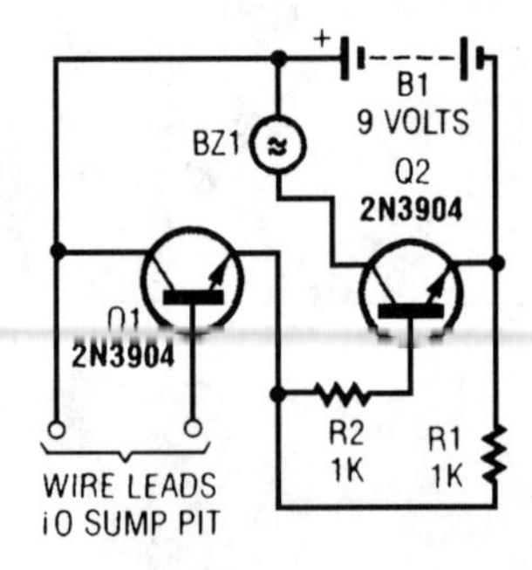

POPULAR ELECTRONICS *FIG. 61-2*

A common collector amplifier drives a 2N3904 switch to sound alarm BZ1. The wire leads to water sensor or sump pit, level switch, etc. and used to allow the alarm to operate and be mounted in a dry place.

MOISTURE DETECTOR

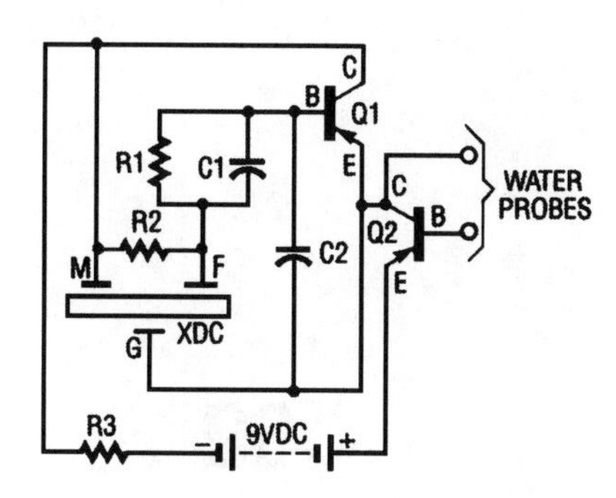

1991 PE HOBBYIST HANDBOOK ***FIG. 61-3***

The moisture detector uses two transistors and a piezoelectric transducer to sound an alarm tone when water is present. Transistor Q1 forms a crystal-controlled oscillator, using a portion of piezoelectric transducer XDC—which contains two piezoelectric crystal regions—as the crystal. The transducer has three separate leads. One lead goes to each of the crystals, and the third lead is common to both.

The smaller internal crystal region sets the frequency of operation and the larger element is driven by Q1 (when it is biased "on") to provide the loud tone output. To turn the pnp transistor Q1 (used as an oscillator) "on" pnp transistor Q2 (used here as a switch) must be on. To turn it "on" with the biasing that is normally connected, you would only need to connect a resistor from the collector of Q2 to the base, which gives the base a negative (–) bias. The resistor used is the water that is to be detected. That turns Q2 on, which, in turn, turns on Q1. The result when water touches the probe is that the transducer emits a loud sound.

C1, C2	0.1-µF Mylar Capacitor
Q1, Q2	2N3906 Transistor
R1	6.8-kΩ Resistor
R2	33-kΩ Resistor
R3	200-Ω Resistor
XDC	Piezoelectric Transducer

62

Motion Detector Circuit

The source of the following circuit is contained in the Sources section, which begins on page 675. The figure number in the box of the circuit correlates to the entry in the Sources section.

Microwave Motion Detector

MICROWAVE MOTION DETECTOR

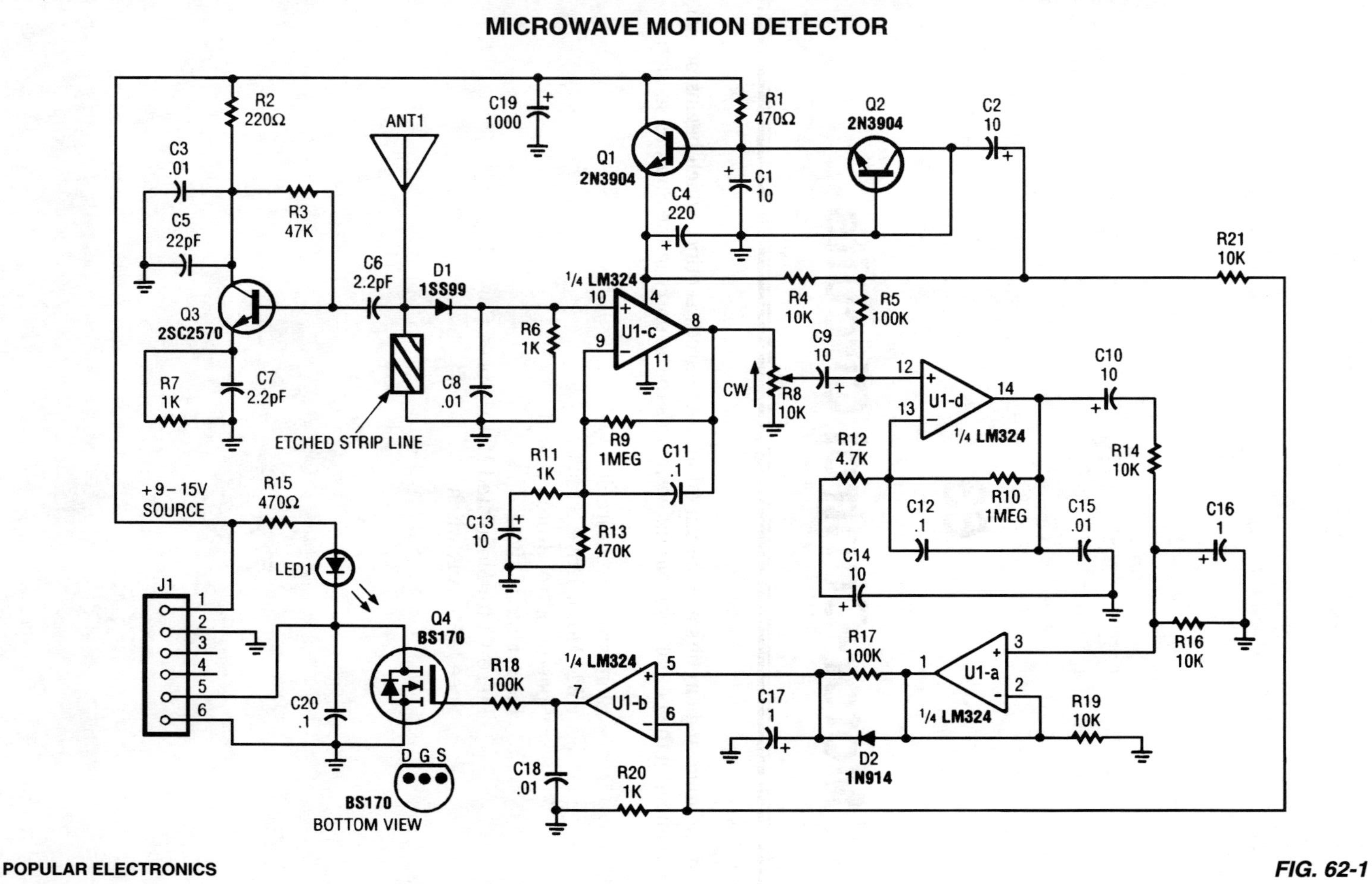

POPULAR ELECTRONICS

FIG. 62-1

Operating at around 1.1 GHz, the detector senses field disturbance in the neighborhood of the antenna. The Doppler signal from detector D1 is amplified and drives a power MOSFET switch. The antenna is a short (2 to 3") length of wire.

63

Motor-Control Circuits

The sources of the following circuits are contained in the Sources section, which begins on page 675. The figure number in the box of each circuit correlates to the entry in the Sources section.

BLENDER-CONTROL CIRCUIT

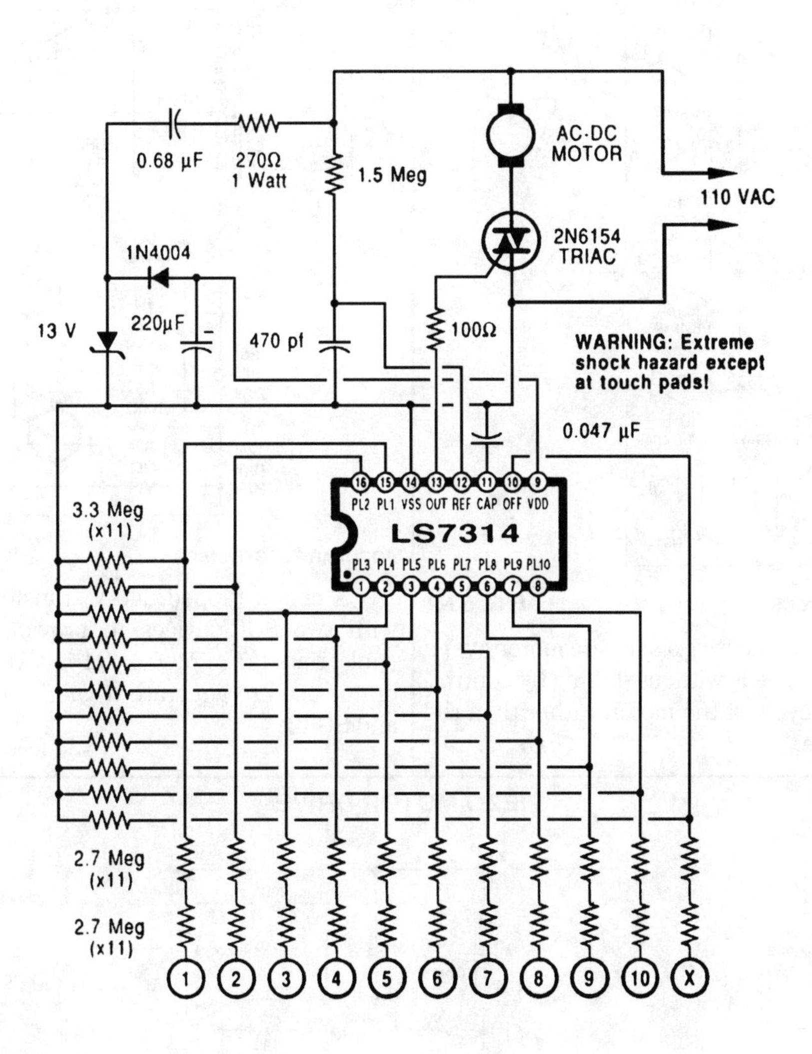

RADIO-ELECTRONICS

FIG. 63-1

A 10-speed touch-control blender circuit that uses the low-cost LS314 chip by LSI Systems. The 11th touch pad is for power off.

PWM MOTOR-DRIVE CIRCUIT

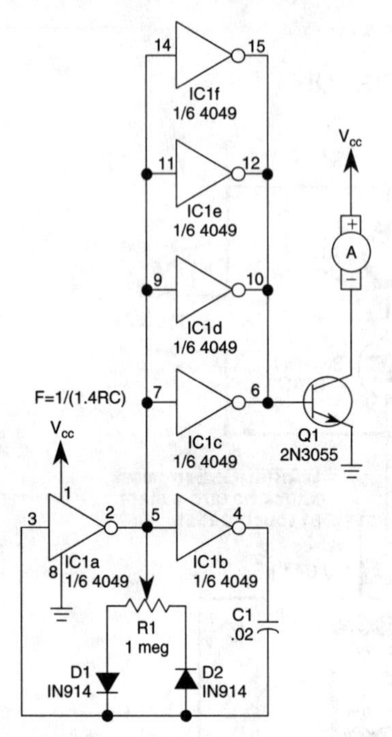

RADIO-ELECTRONICS *FIG. 63-2*

This circuit will drive a small dc motor over a wide range of speeds without stalling by controlling the duty cycle of the motor, rather than the supply voltage.

SPEED-CONTROL SWITCH CIRCUIT

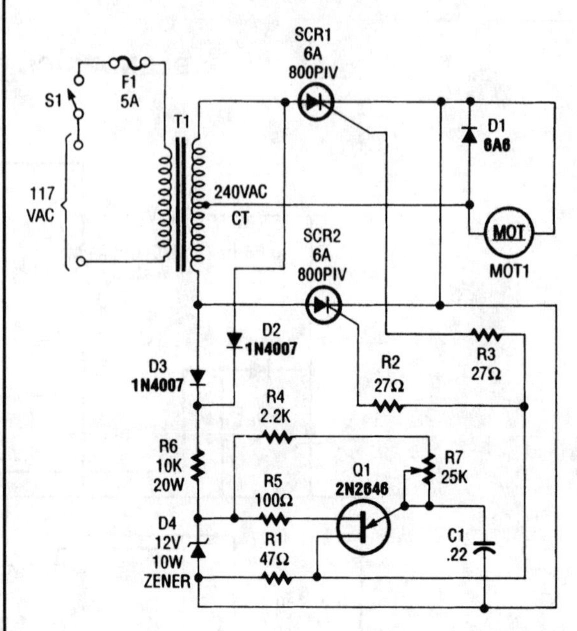

POPULAR ELECTRONICS *FIG. 63-3*

A center-tapped 240-V transformer is used with two SCR devices to provide rectified ac (pulsating dc) to MOT1. Q1 is a UJT ramp generator used to generate trigger pulses for SCR1 and SCR2.

PIEZO MOTOR DRIVE

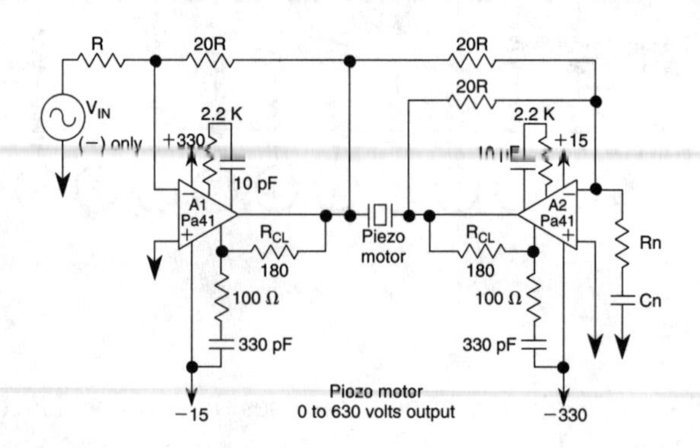

ELECTRONIC DESIGN *FIG. 63-4*

Using two Apex Microtechnology PA41 devices in a bridge circuit, this piezo motor driver delivers 0- to 630-V output.

PULSE-WIDTH-MODULATED MOTOR-SPEED CONTROL

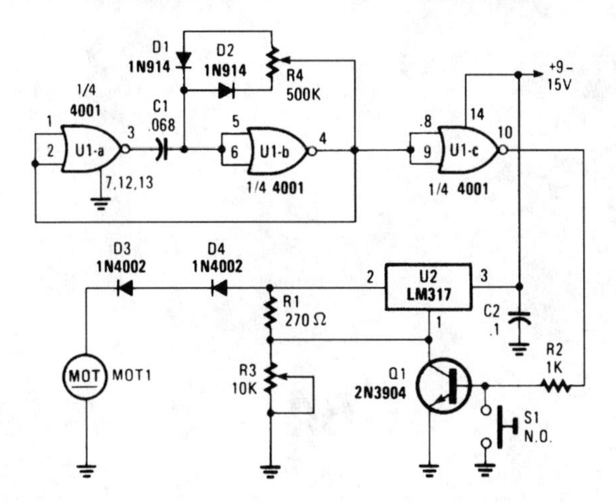

POPULAR ELECTRONICS *FIG. 63-5*

Connected in this manner, an LM317 1-A adjustable-voltage regulator can be used to control the speed of a miniature dc motor or vary the brilliance of a small lamp. The circuit does so by controlling the pulse width, and therefore the current, to the load device.

To set the desired maximum output voltage, momentarily close S1 and adjust R3. Connect either a lamp or small dc motor (as is shown in the schematic to the circuit's output) and adjust R4 for the desired results. Any device that is driven by this circuit should have a current requirement of 1 A or less. And you should be sure to use good-sized heatsink for the LM317 regulator IC.

SPEED-CONTROL SWITCH

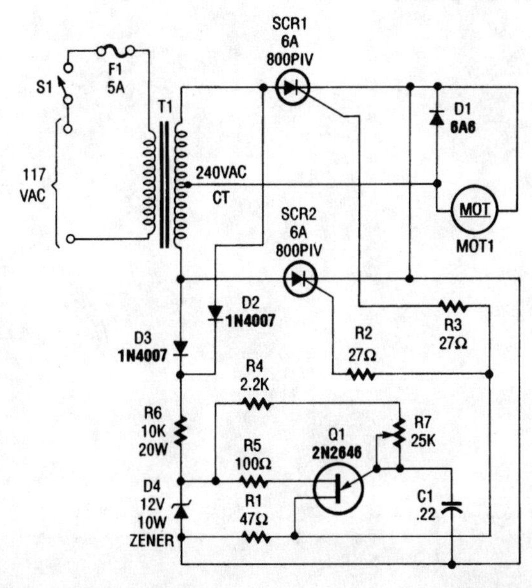

POPULAR ELECTRONICS *FIG. 63-6*

The speed-control switch offers reasonably good control and stability to both ends of its operating range. This circuit uses two SCR devices in a full-wave configuration to control the dc power to a motor. A center-tapped transformer is used to supply the SCRs.

64

Multiplexer Circuit

The source of the following circuit is contained in the Sources section, which begins on page 675. The figure number in the box of the circuit correlates to the entry in the Sources section.

32-Channel Analog Multiplexer

32-CHANNEL ANALOG MULTIPLEXER

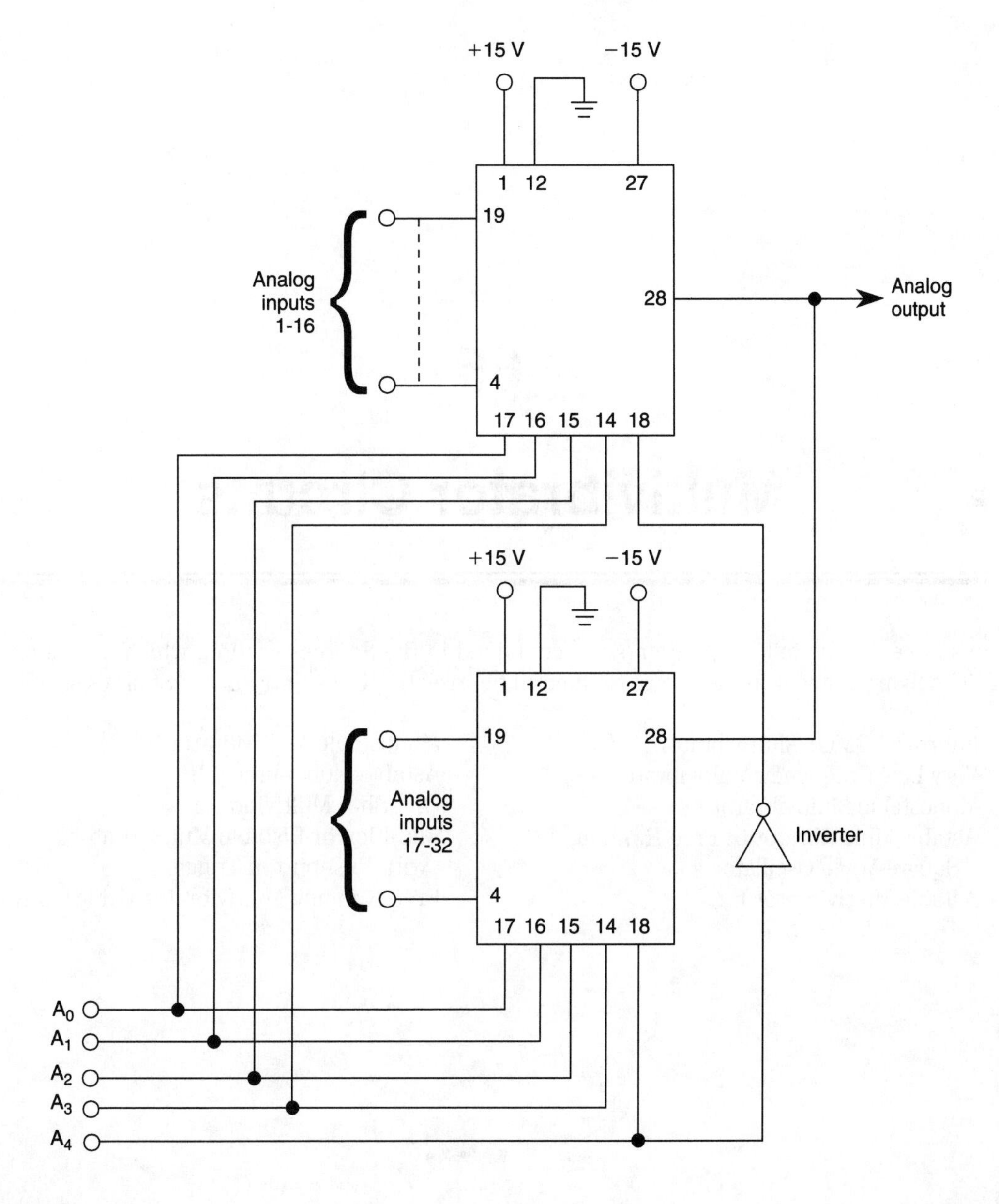

WILLIAM SHEETS ***FIG. 64-1***

Using two Siliconix DG506 multiplexer chips, this 32-channel analog multiplexer selects 1 of 32 channels, depending on the data inputs $A_0 - A_4$.

65

Multivibrator Circuits

The sources of the following circuits are contained in the Sources section, which begins on page 675. The figure number in the box of each circuit correlates to the entry in the Sources section.

- Improved CMOS Multivibrator
- Very Low Frequency Multivibrator
- Monostable Multivibrator I
- Astable Multivibrator or Free-Running Square-Wave Oscillator
- Astable Multivibrator I
- Monostable Multivibrator II
- Astable Multivibrator II
- One-Shot Multivibrator
- Flip-Flop or Bistable Multivibrator with Pushbutton Triggering
- Free-Running Multivibrator Using Op Amp

IMPROVED CMOS MULTIVIBRATOR

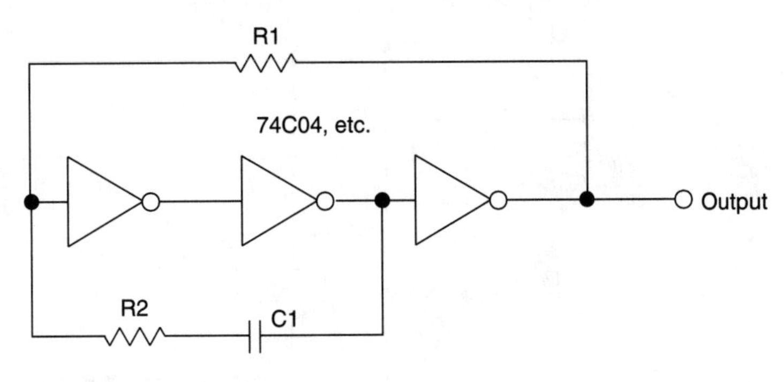

WILLIAM SHEETS

FIG. 65-1

This circuit uses a protective resistor $R2$ in conjunction with feedback resistor $R1$. Together, they form a voltage divider to reduce the input voltage amplitude for IC1-a so that the protective diodes never conduct. This improves temperature and voltage stability of the multivibrator.

VERY LOW FREQUENCY MULTIVIBRATOR

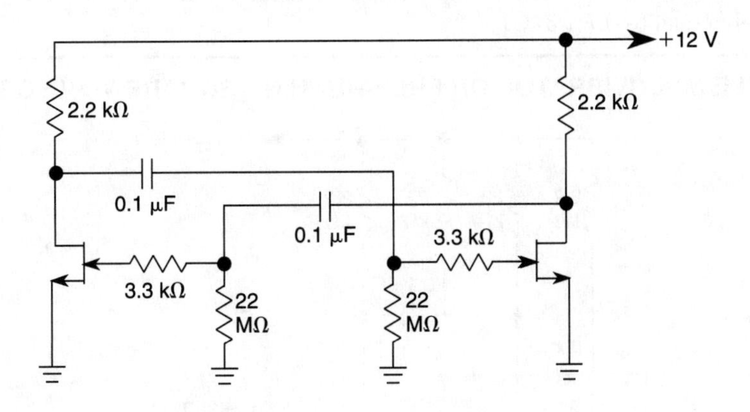

JFETs Transistor: N-channel (MPF102, etc.)

WILLIAM SHEETS

FIG. 65-2

The use of JFETs permits, high resistance and long time constants in this very low frequency multivibrator. The values shown are for 0.15 Hz operation.

MONOSTABLE MULTIVIBRATOR I

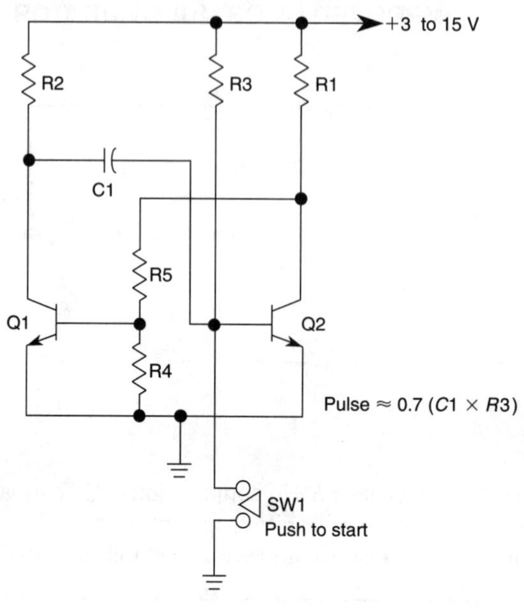

WILLIAM SHEETS

FIG. 65-3

This circuit is activated when SW1 is pushed to ground the base of transistor Q2. The pulse rate is approximately equal to 0.7(R3×C1).

ASTABLE MULTIVIBRATOR OR FREE-RUNNING SQUARE-WAVE OSCILLATOR

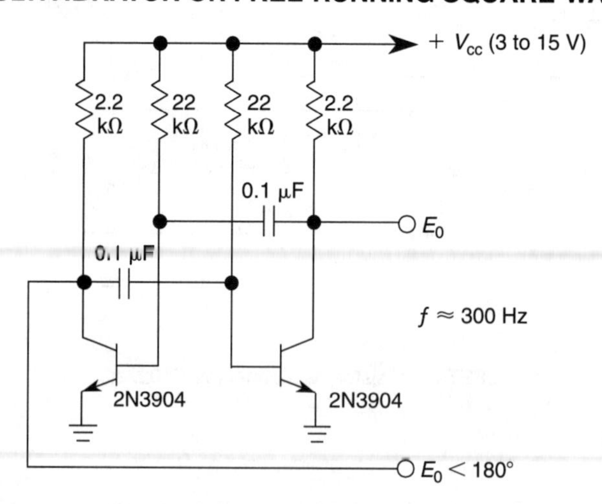

WILLIAM SHEETS

FIG. 65-4

This free-running square-wave oscillator uses two npn transistors. Output frequency is approximately 300 Hz with the values shown.

ASTABLE MULTIVIBRATOR I

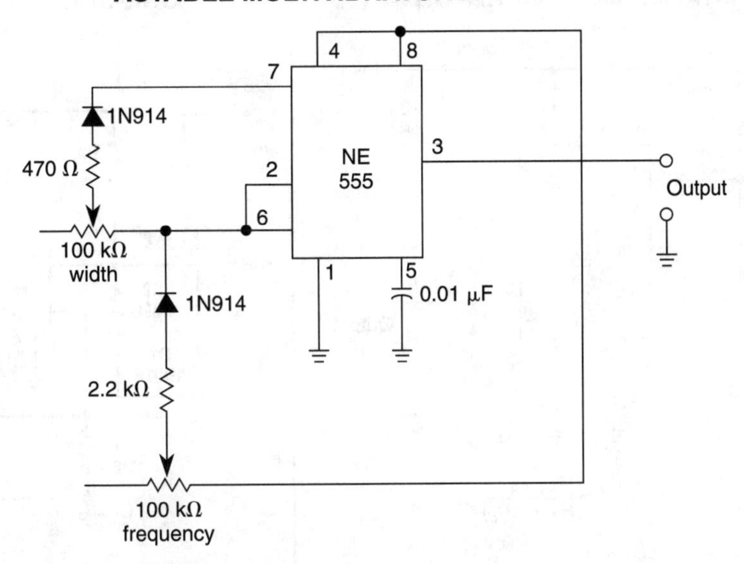

WILLIAM SHEETS ***FIG. 65-5***

In this multivibrator circuit frequency and pulse width can be separately controlled by using steering diodes (1N914) and two potentiometers.

MONOSTABLE MULTIVIBRATOR II

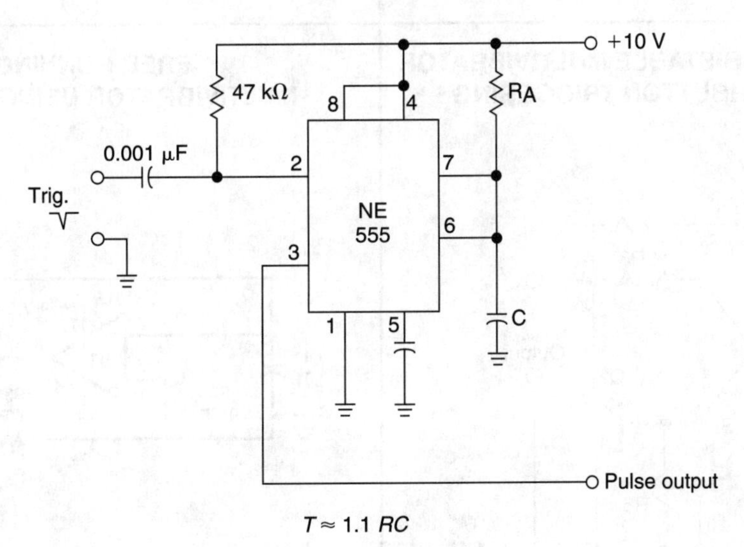

WILLIAM SHEETS ***FIG. 65-6***

The time constant of R_AXC determines the period of the monostable multivibrator. A negative pulse at pin 2 of the 555 starts the cycle.

ASTABLE MULTIVIBRATOR II

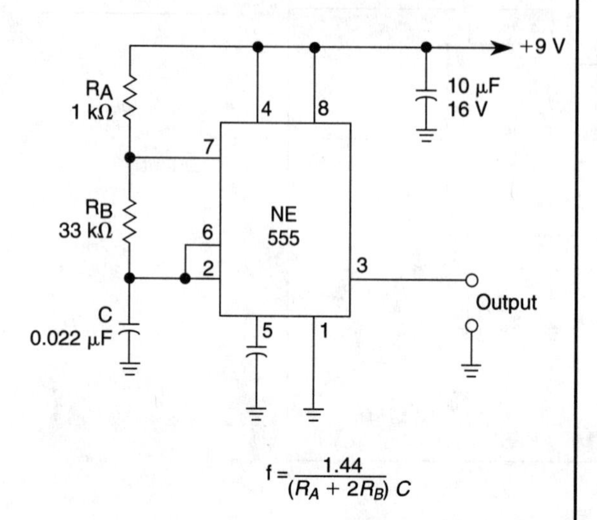

WILLIAM SHEETS *FIG. 65-7*

An astable multivibrator based on the 555 is shown. Freq is approximately 975 Hz as determined by the values of R_B and C.

ONE-SHOT MULTIVIBRATOR

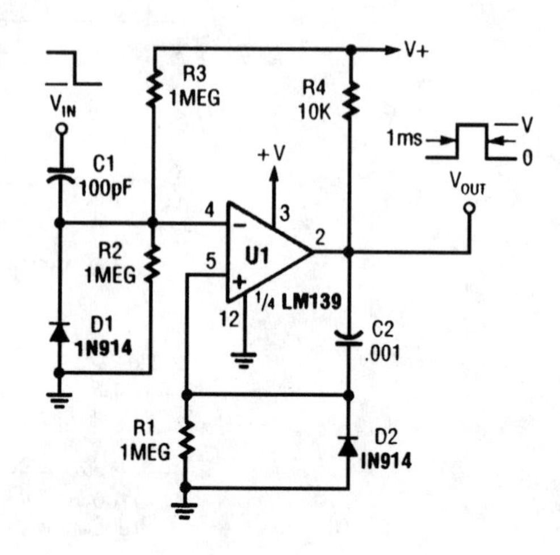

POPULAR ELECTRONICS *FIG. 65-8*

A section of a quad LM139 is used here as a one-shot pulse former.

FLIP-FLOP OR BISTABLE MULTIVIBRATOR WITH PUSHBUTTON TRIGGERING

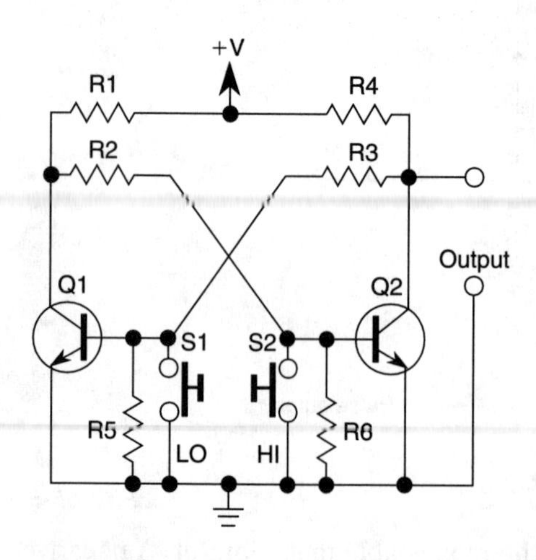

ELECTRONICS NOW *FIG. 65-9*

FREE-RUNNING MULTIVIBRATOR USING OP AMP

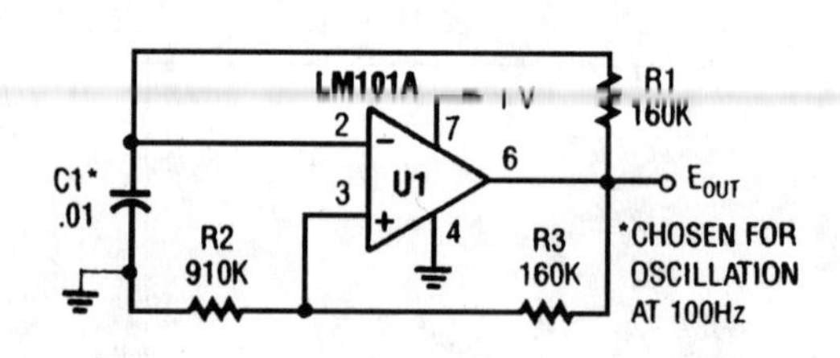

POPULAR ELECTRONICS *FIG. 65-10*

66

Musical Circuits

The sources of the following circuits are contained in the Sources section, which begins on page 575. The figure number in the box of each circuit correlates to the entry in the Sources section.

PRECISION AUDIO GENERATOR FOR MUSICAL INSTRUMENT TUNE-UP

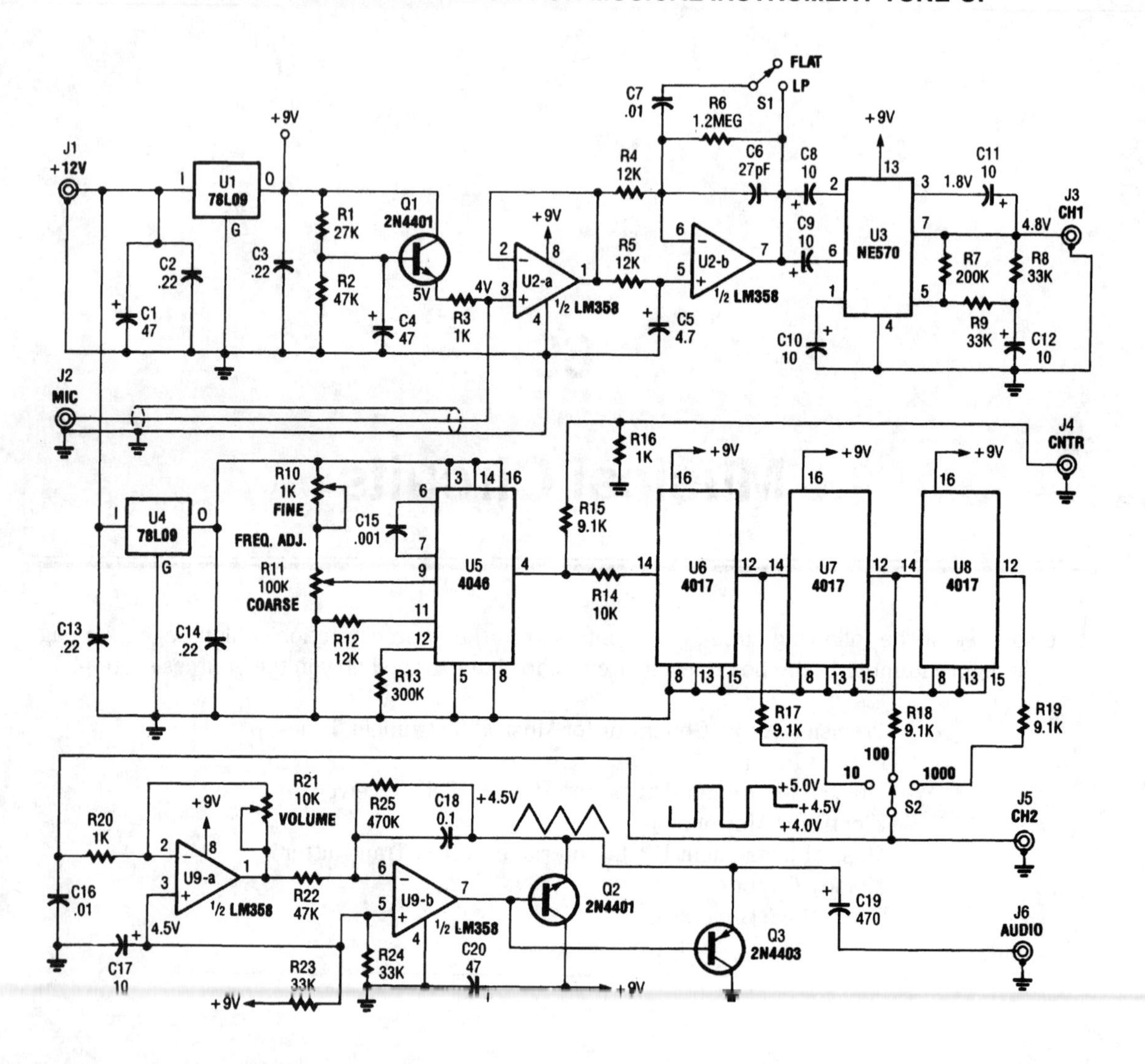

1993 ELECTRONICS HOBBYIST HANDBOOK *FIG. 66-1*

One section of the precision audio frequency generator uses an electret microphone element to pick up audio from the piano. That signal is then processed and sent to one channel of a dual-trace oscilloscope. The other section of the circuit is used to produce a variable-frequency signal that is fed to a digital frequency counter. After conditioning, the audio signal is presented to the second channel of the scope and output to a set of stereo headphones.

PERFECT PITCH

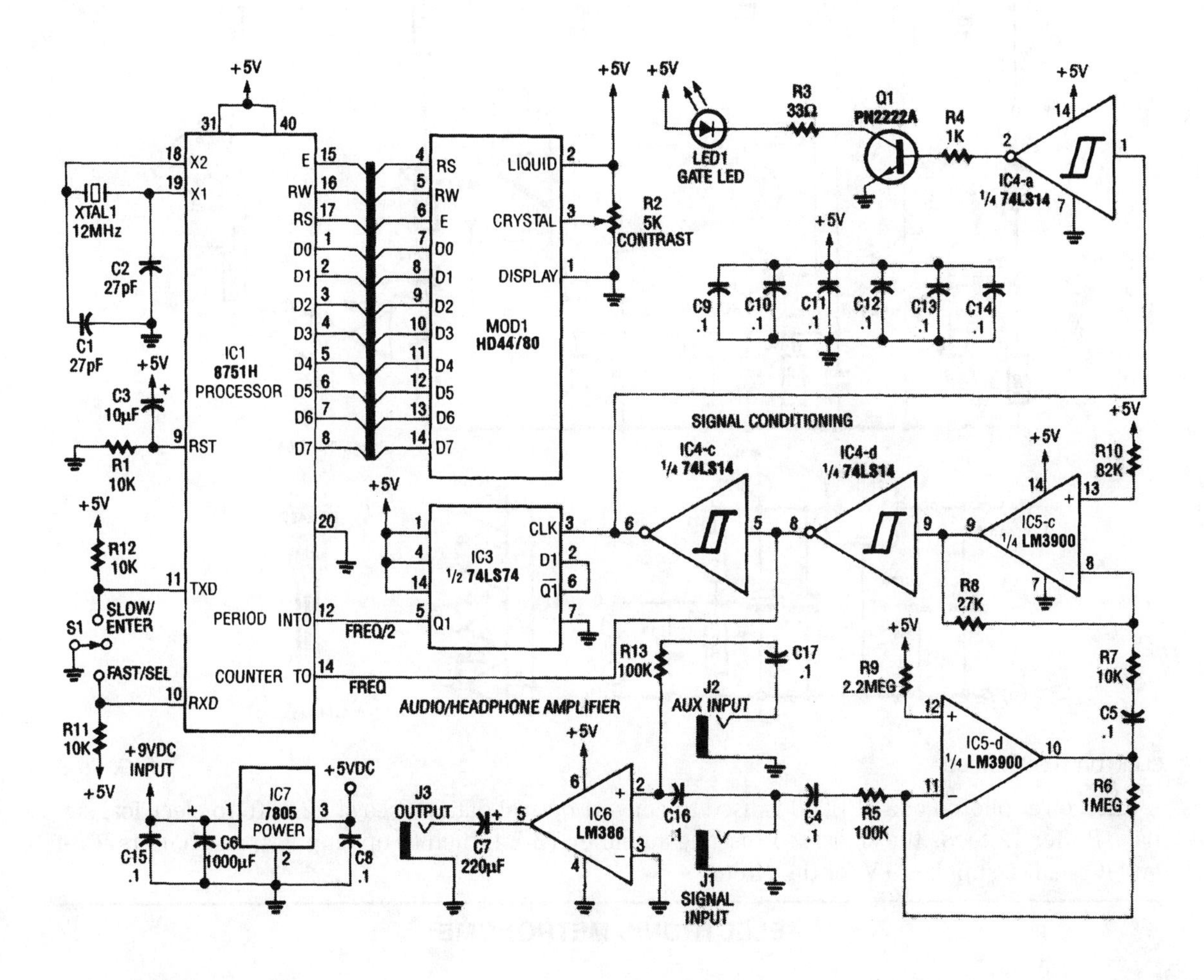

ELECTRONICS NOW

FIG. 66-2

Perfect pitch, which is based on the 8751 H microprocessor, is an inexpensive and easy-to-build instrument tuner/frequency counter with a built-in headphone amplifier and a visual metronome. Perfect pitch converts the audio signal from your instrument to a digital signal, and displays the musical note you are playing and its frequency in real time on a 16-character liquid-crystal display. It also has an auxiliary audio input for radio, tape, or CD players so that you can tune up and play along with your favorite artists.

MUSICAL INSTRUMENT DIGITAL INTERFACE (MIDI) RECEIVER

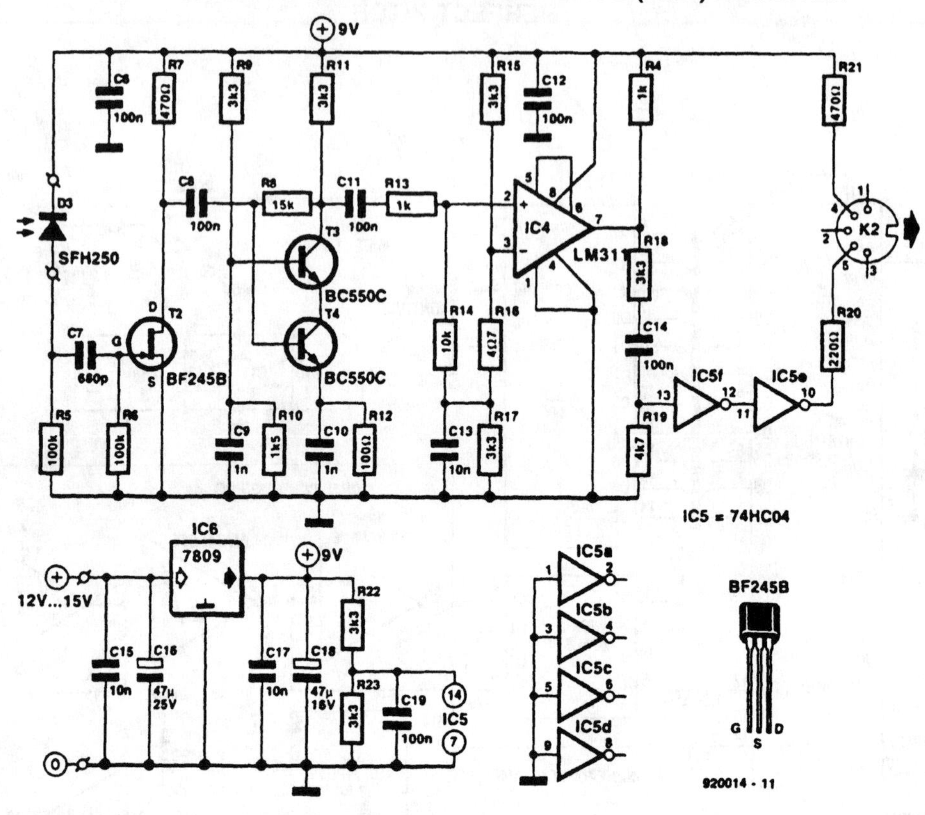

ELEKTOR ELECTRONICS ***FIG. 66-3***

Receiver photodiode SFH250 is used to convert optical data pulses at 32.5 Kb to electrical signals. Buffer T2 feeds the signals to cascade amplifier T3-T4, then to op amp IC4, and buffers IC5-f and IC5-e. IC6 supplies 9 V for the circuit.

ELECTRONIC METRONOME

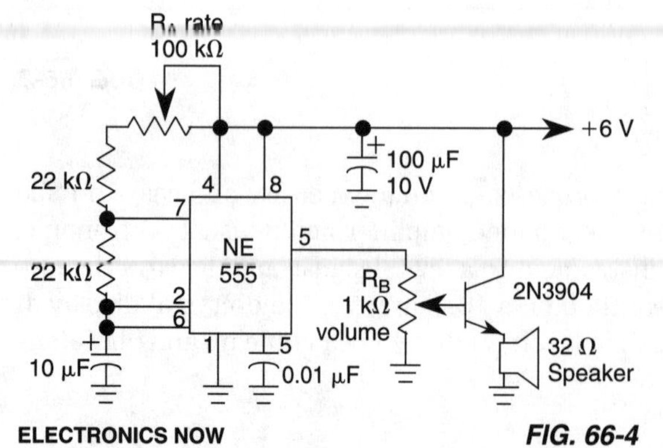

R_A sets the rate while R_B sets the volume of clocks in the speaker. The 555 is configured as a low frequency oscillator. The circuit is powered by a 6 V battery.

ELECTRONICS NOW ***FIG. 66-4***

MUSICAL INSTRUMENT DIGITAL INTERFACE (MIDI) TRANSMITTER

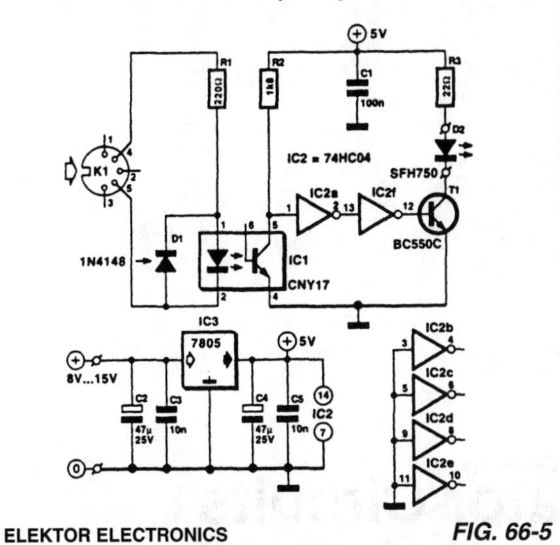

ELEKTOR ELECTRONICS *FIG. 66-5*

Used for digital control of musical instruments, this transmitter converts the digital data signals to equivalent optical signals for fiberoptic cable interface. Optocoupler IC1 provides isolation, and drives IC2-a and -b and T1, and finally provides a cable driver LED (SFH750).

MELODY CIRCUIT

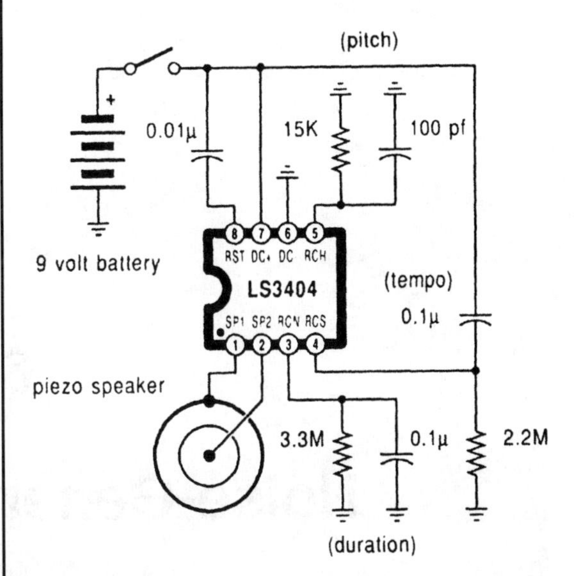

RADIO-ELECTRONICS *FIG. 66-6*

A high-quality melody circuit. The slow decay waveform produced will create chime-like notes. Pitch, tempo, and duration are all adjustable.

TOP OCTAVE GENERATOR

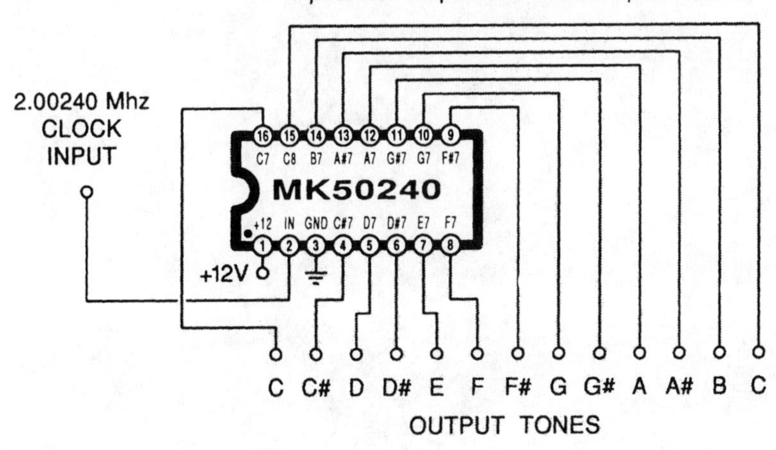

RADIO-ELECTRONICS *FIG. 66-7*

Using an MK50240, this circuit produces 12 top octave tones. The input and output lines can be divided using a binary divider IC to obtain the lower notes.

67

Noise-Generator Circuits

The source of the following circuit is contained in the Sources section, which begins on page 675. The figure number in the box of the circuit correlates to the entry in the Sources section.

Noise Generator

NOISE GENERATOR

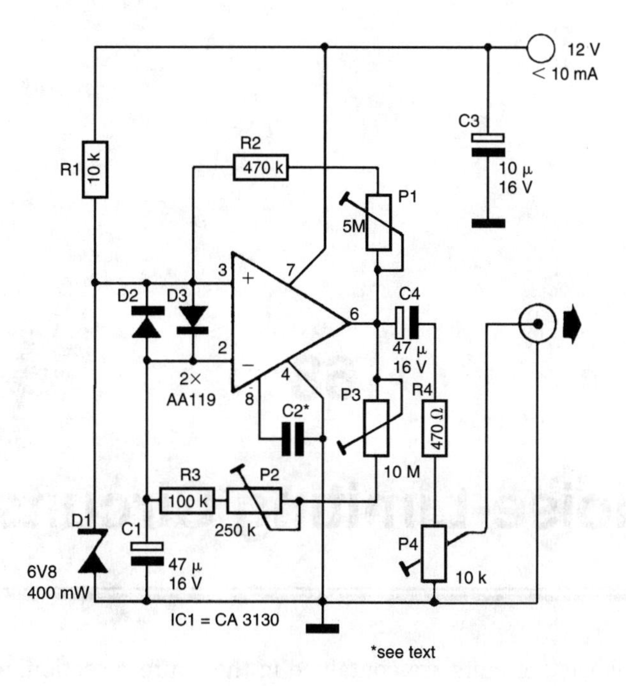

303 CIRCUITS **FIG. 67-1**

This circuit generates noise pulses that are suitable for test purposes, etc. A zener diode is used as a noise source. IC1 is a relaxation oscillator. P1 determines noise bandwidth, and P2 and P3 the noise amplification. Current consumption is 10 mA @ 12 Vdc.

68

Noise-Limiting Circuits

The sources of the following circuits are contained in the Sources section, which begins on page 675. The figure number in the box of each circuit correlates to the entry in the Sources section.

AUDIO DYNAMIC NOISE-REDUCTION SYSTEM

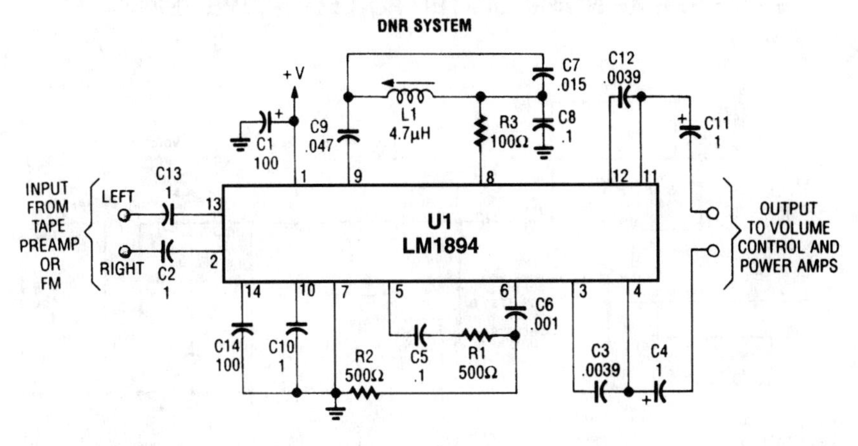

POPULAR ELECTRONICS

FIG. 68-1

U1 is a dedicated IC (National Semiconductor) that achieves up to 10 dB noise reduction by an adaptive bandwidth scheme and a psycho acoustic masking technique.

AMPLIFIED NOISE LIMITER FOR SW RECEIVERS

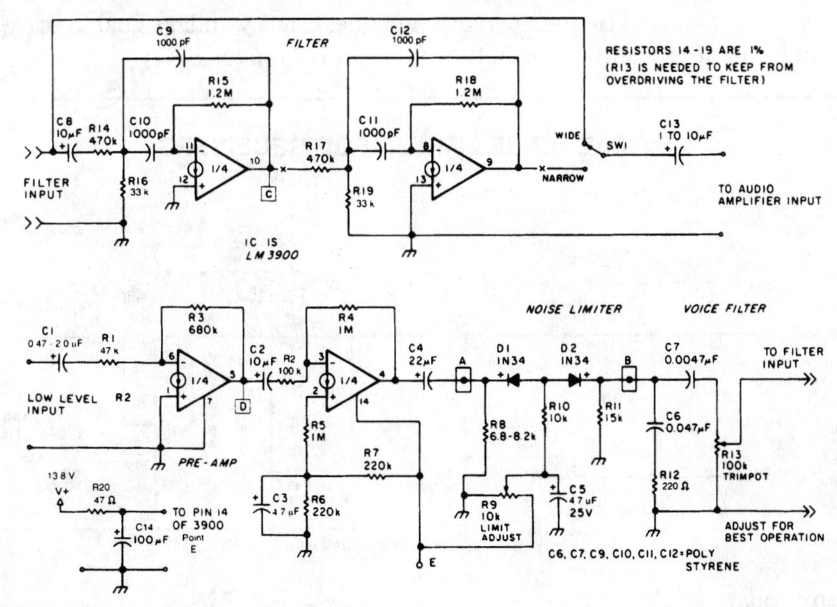

73 AMATEUR RADIO TODAY

FIG. 68-2

The noise limiter circuit has a preamplifier clipper, and a switchable audio bandpass filter. Audio levels in the 5- to 50-mV range are amplified in a preamp to several volts p-p, fed to a clipper, voice band filter, then to a narrow band active filter which can be switched in and out of the circuit.

RECEIVER AF NOISE LIMITER FOR LOW-LEVEL SIGNALS

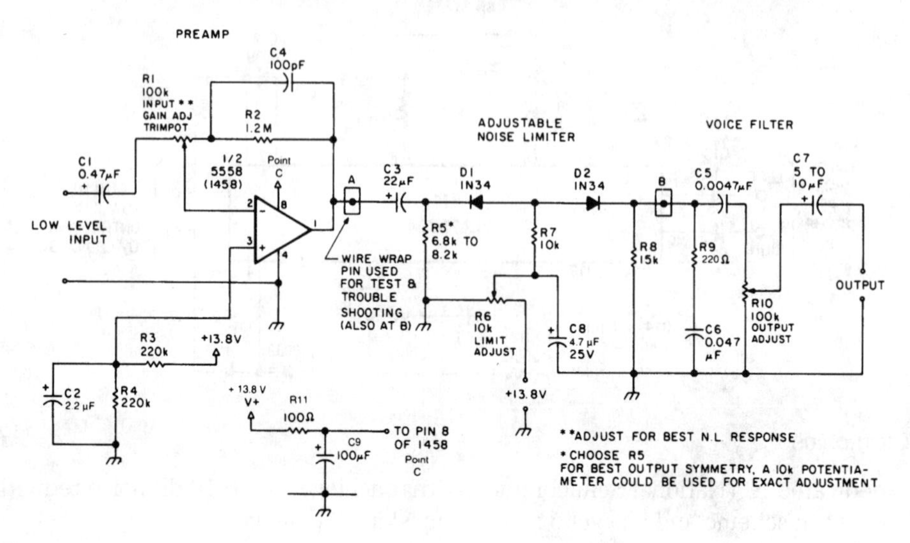

73 AMATEUR RADIO TODAY ***FIG. 68-3***

A preamplifier in the audio frequency range amplifies a noisy audio signal to drive a diode clipper. Suitable audio input levels would be in the 10-mV to 1-V range.

SIMPLE NOISE LIMITER FOR RECEIVERS

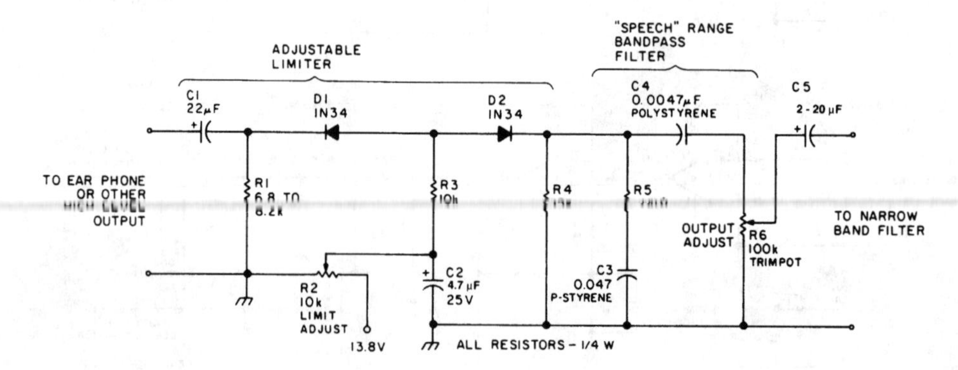

73 AMATEUR RADIO TODAY ***FIG. 68-4***

This circuit uses a diode series clipper to limit noise peaks on a received signal. It is best used where several volts p-p of audio signal are available.

69

Operational-Amplifier Circuits

The sources of the following circuits are contained in the Sources section, which begins on page 675. The figure number in the box of each circuit correlates to the entry in the Sources section.

POLARITY GAIN ADJUSTMENT

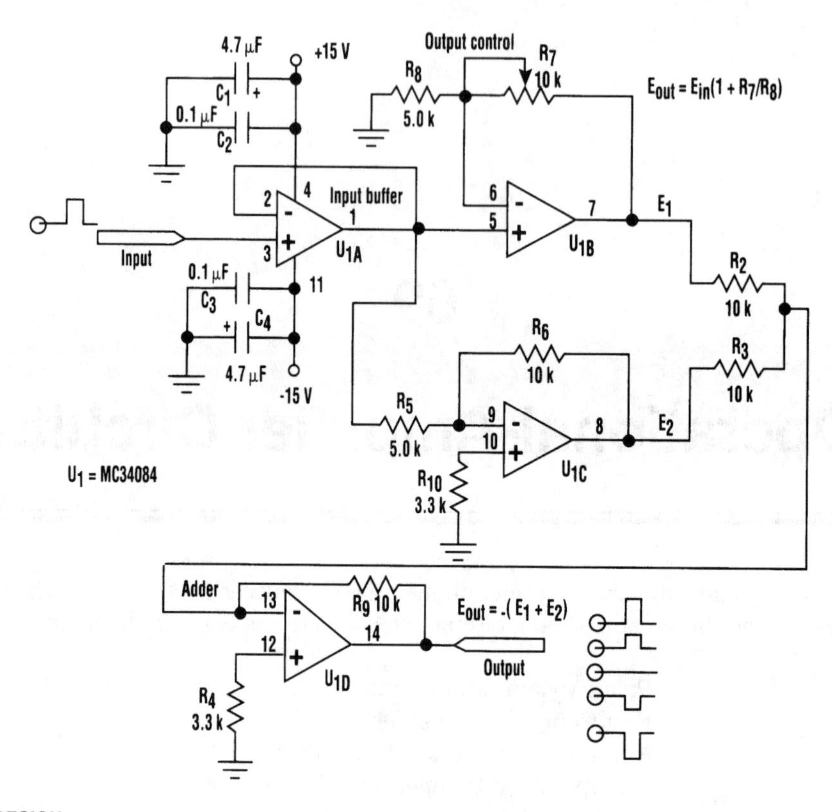

ELECTRONIC DESIGN

FIG. 69-1

By adjusting one potentiometer, this circuit's output can be varied from a positive-going version of the input signal, smoothly through zero output, then to a negative-going version of the input (see the figure). If the input signal is a positive pulse of, for example, +2-V peak, the output pulse amplitude can be smoothly varied from +2-V through ground (no output) to a –2-V peak.

Taking a closer look at the setup, assume that the signal has a +2-V peak input. The A section of the quad op amp is an input buffer, op amp C provides a fixed negative-going output of –4-V peak, and op amp B supplies a positive-going output that varies from +2-V to +6-V peak. The D section adds the B and C outputs. Thus, by varying the B output, the circuit output varies smoothly from –2-V to +2-V peak.

The circuit can, of course, also be used as a 0°/180° phase switcher. For instance, with a ground-centered sine-wave input of 4V p-p, the output varies from 4-V p-p in phase with the input, smoothly through 0 V, to 4V p-p 180° out of phase with the input.

FAST COMPOSITE AMPLIFIER

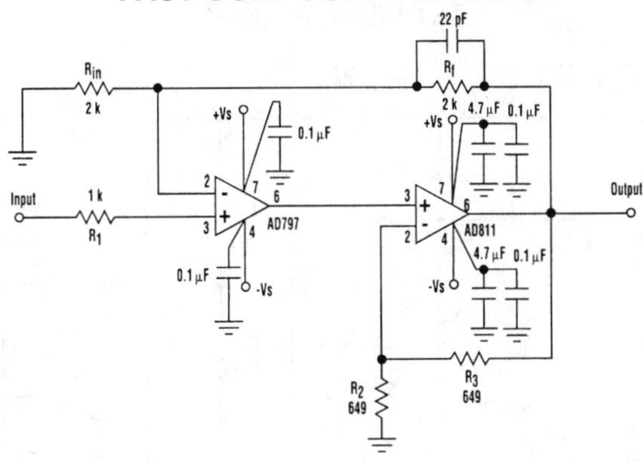

ELECTRONIC DESIGN

FIG. 69-2

An ultra-low-noise, low-distortion op amp—the AD797—is combined with the AD811 op amp, which offers a high bandwidth and a 100-mA output drive capability. The composite-amplifier circuit serves quite well when driving high resolution ADC's and ATE systems.

The fast AD811 operates at twice the gain of the AD797 so that the slower amplifier need only slew one-half of the total output swing. Using the component values shown, the circuit is capable of better than –90 dB THD with a ±5-V, 500-kHz output signal. If a 100-kHz sine-wave input is used, the circuit will drive a 600-Ω load to a level of 7 V rms with less than –109 dB THD, as well as a 10-kΩ load at less than –117 dB THD.

The device can be modified to supply an overall gain of 5 by changing both the R_f/R_{in} ratio and R_3/R_2 ratio to 4:1. This raises the gains of AD811 and the total circuit while maintaining the AD797 at unity gain. If only the R_f/R_{in} ratio is changed, the circuit might become unstable. In contrast, if only the R_3/R_2 ratio is varied, the AD797 will then operate at gain. Subsequently, the circuit will have a lower overall bandwidth. R_1 should be equal to the parallel combination of R_{in} and R_f.

NONLINEAR OPERATIONAL AMPLIFIER WITH TEMPERATURE-COMPENSATED BREAKPOINTS

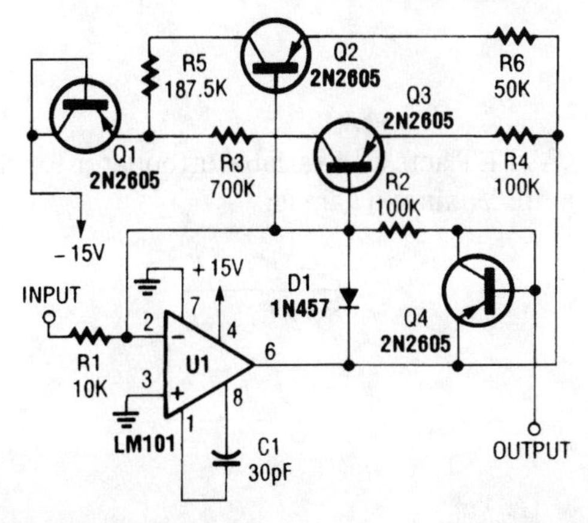

Using resistor and transistor feedback elements, this operational amplifier circuit can be used as a nonlinear amplifier. R4 and R6 can be varied to change breakpoints, as required.

POPULAR ELECTRONICS

FIG. 69-3

POWER OP AMP

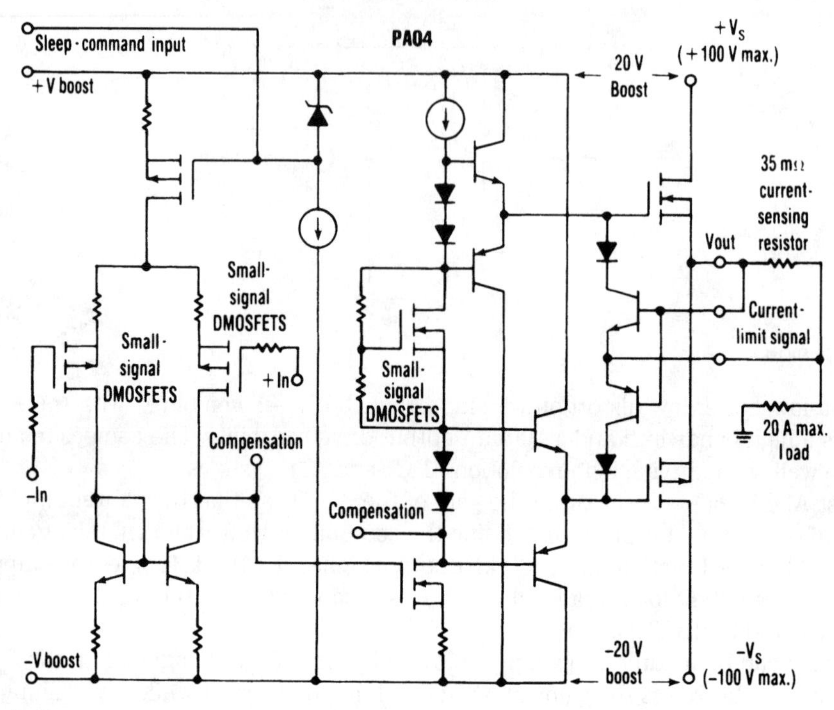

ELECTRONIC DESIGN

FIG. 69-4

This circuit from Apex Microtechnology can deliver 180 V p-p @ 90 kHz into a 4-Ω load. The PA04 can deliver 400-W RMS into an 8-Ω load with low THD at frequencies beyond 20 kHz.

VARIABLE GAIN OP-AMP CIRCUIT

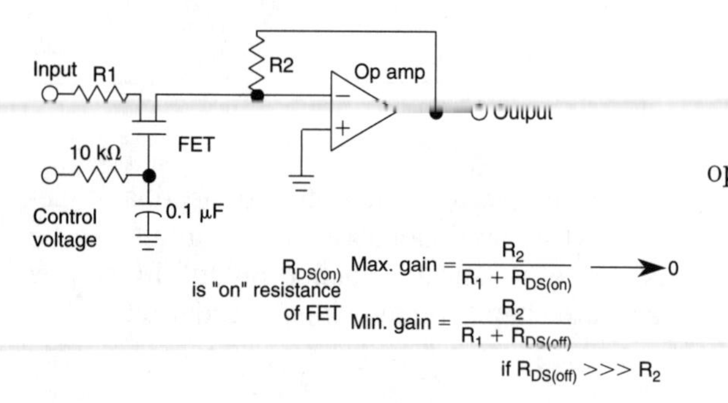

A JFET acts as a variable attenuator for this op amp. Maximum gain is:

$$\frac{R_2}{R_1 + R_{DS(ON)}}$$

ELECTRONICS NOW

FIG. 69-5

LOW NOISE AND DRIFT COMPOSITE AMP

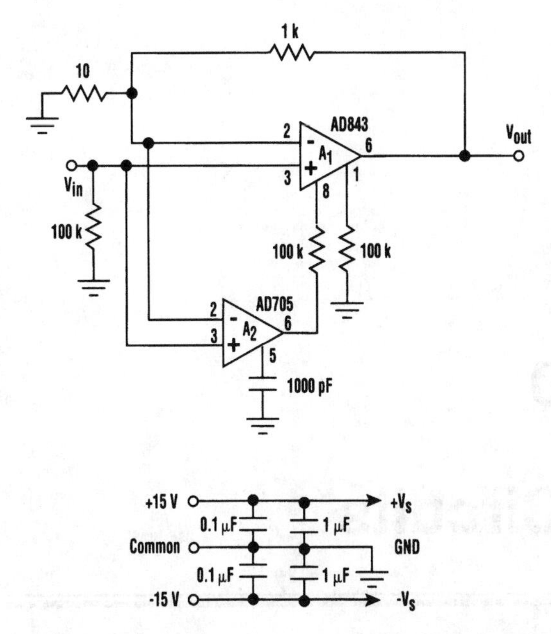

This circuit offers the best of both worlds. It can be combined with a low input offset voltage and drift without degrading the overall system's dynamic performance. Compared to a standalone FET input operational amplifier, the composite amplifier circuit exhibits a 20-fold improvement in voltage offset and drift.

In this circuit arrangement, A1 is a high-speed FET input op amp with a closed-loop gain of 100 (the source impedance was arbitrarily chosen to be 100 kΩ). A2 is a SuperBeta bipolar input op amp. It has good dc characteristics, biFET-level input bias current, and low noise. A2 monitors the voltage at the input of A1 and injects current to A1's null pins. This forces A1 to have the input properties of a bipolar amplifier while maintaining its bandwidth and low-input-bias-current noise.

ELECTRONIC DESIGN *FIG. 69-6*

HIGH-GBW OP AMP

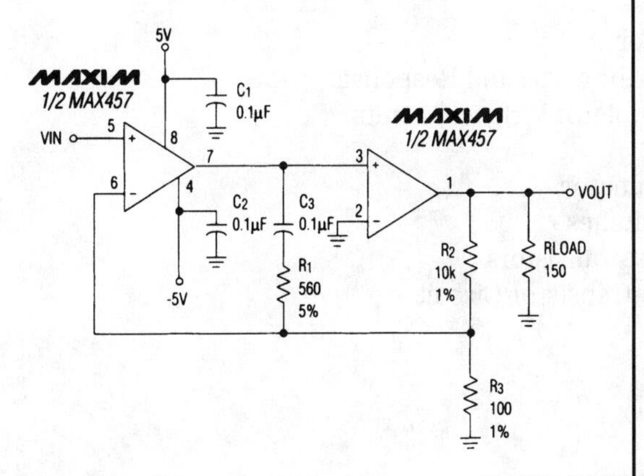

MAXIM ENGINEERING JOURNAL *FIG. 69-7*

You can build a composite amplifier featuring high gain, wide bandwidth, and good dc accuracy by cascading the sections of a dual video amplifier and adding two appropriate phase-compensation components. The op amp drives a 150-Ω load and provides a closed-loop gain of 40 dB.

SINGLE OP-AMP FULL-WAVE RECTIFIER

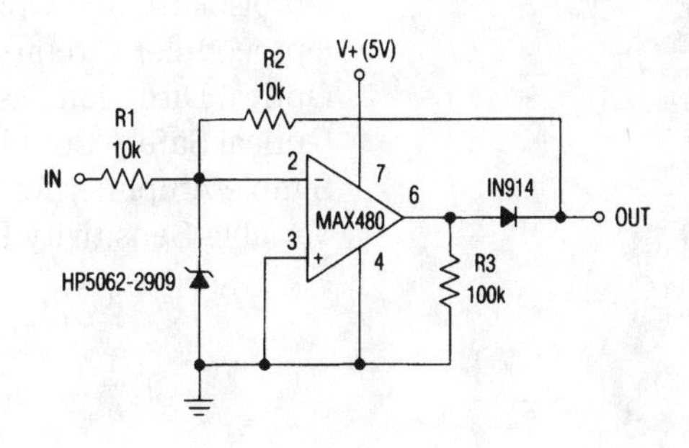

MAXIM ENGINEERING JOURNAL *FIG. 69-8*

This circuit operates from +5 V and uses a single op amp to deliver a full-wave rectified output of the input signal.

70

Optical Circuits

The sources of the following circuits are contained in the Sources section, which begins on page 675. The figure number in the box of each circuit correlates to the entry in the Sources section.

OPTICAL PROXIMITY DETECTOR

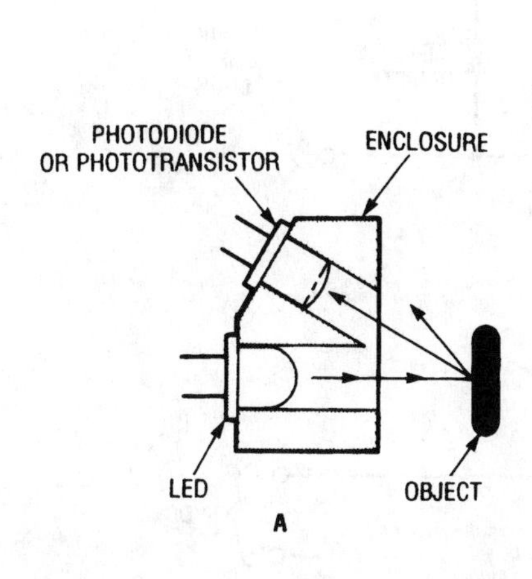

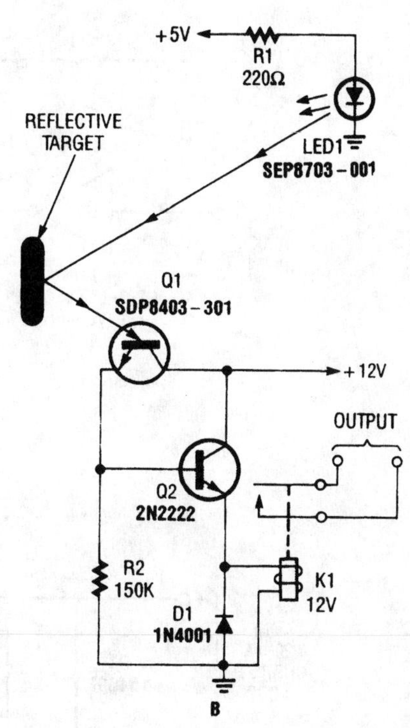

POPULAR ELECTRONICS

FIG. 70-1

A "reflector" isolator (A) detects the presence of an object by bouncing light off of it. This technique is useful in circuits that detect when an object is close enough to the sensor (B).

PHOTORECEIVER OPTIMIZED FOR NOISE AND RESPONSE

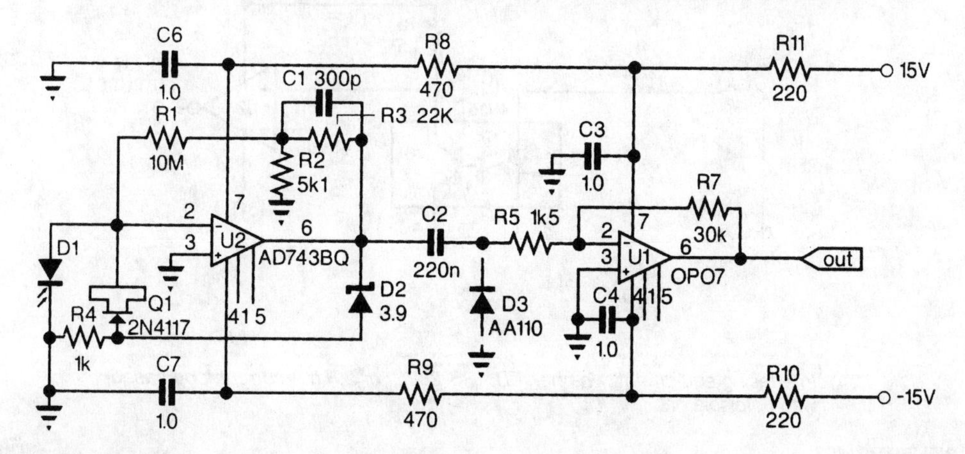

ELECTRONIC ENGINEERING

FIG. 70-2

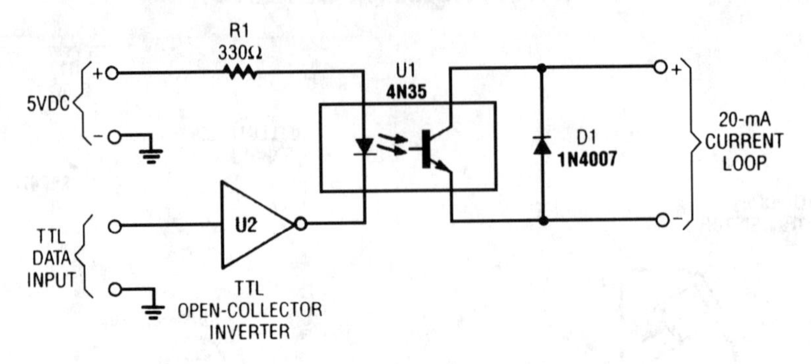

A

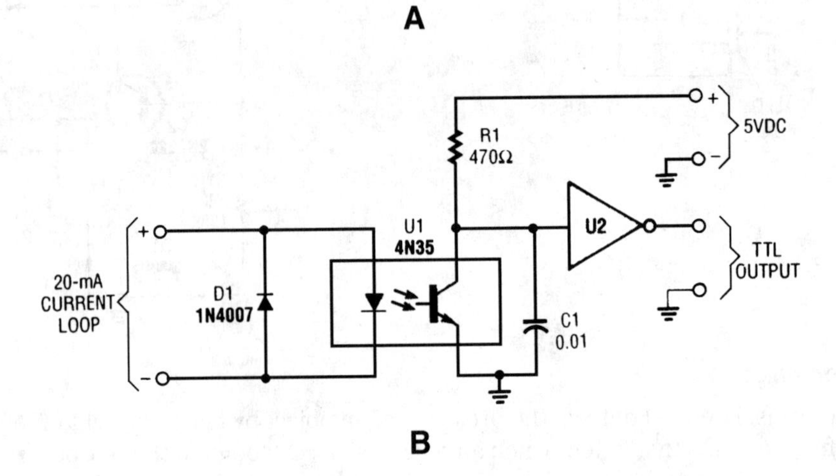

B

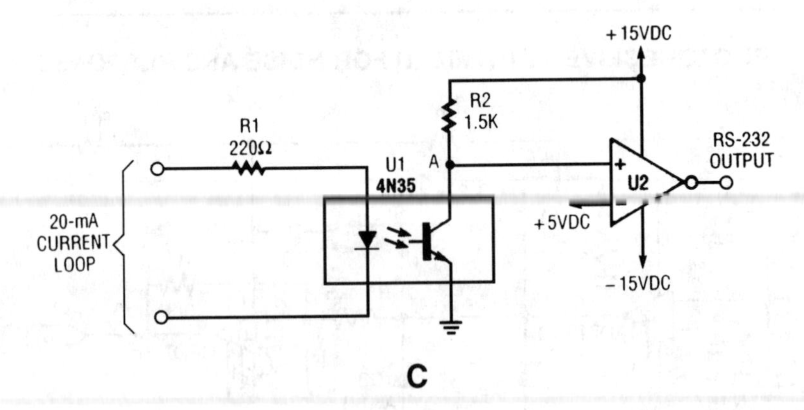

C

Interfacing equipment, whether TTL, RS-232C, or 20=mA current-loop based, with optoisolators.

POPULAR ELECTRONICS **FIG. 70-3**

OPTOISOLATOR AND OPTOCOUPLER INTERFACE CIRCUITS (*Cont.*)

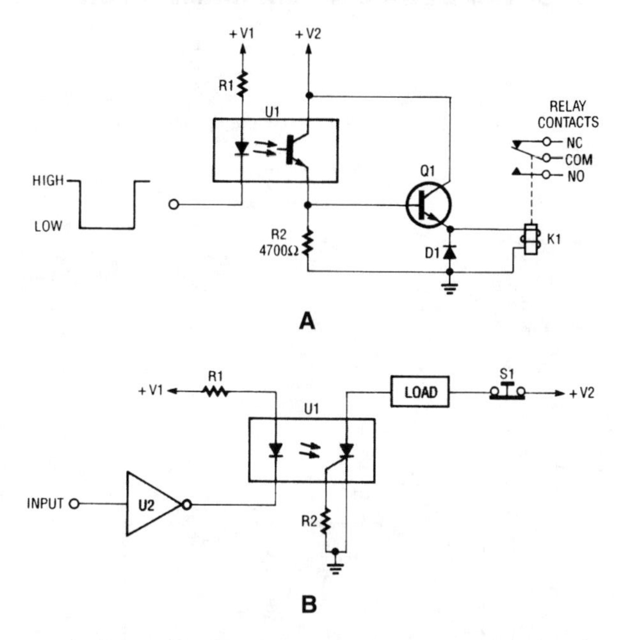

Very heavy loads, which can't be powered directly by an optoisolator, might require the use of a relay as shown in A. You can sometimes get away with using a circuit like that shown in B, but it won't turn itself off.

A circuit for isolating a variable resistor is shown. An optoisolator that has an LED and a photoconductive cell (or photoresistor) is used. The current through the LED controls its brightness, which in turn determines the resistance between terminals A and B. The LED current is set by the voltage of the dc power supply and the value of the two resistors (R1 and R2). The fixed resistor (R1) is used to limit the current to a maximum of 20 mA (when the resistance of the potentiometer, R_2, is set to zero ohms), otherwise, the LED might burn out.

OPTOCOUPLER CIRCUITS

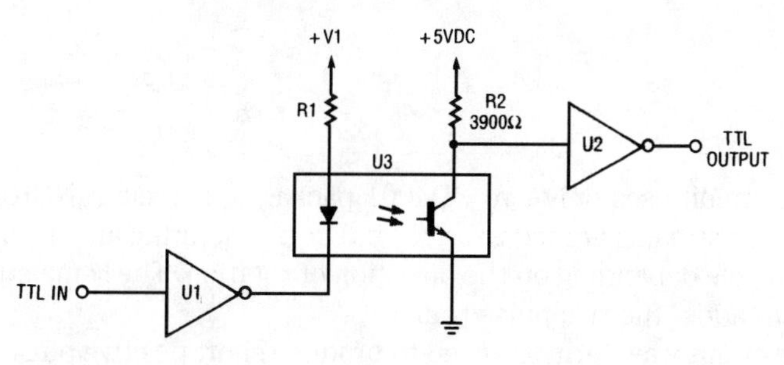

POPULAR ELECTRONICS ***FIG. 70-4***

This circuit is a TTL-to-TTL isolator circuit. The driver circuit is an open-collector TTL inverter (U1). When the input is high, then the output of the inverter is low. Thus, when the input is high, the output of U1 grounds the cathode end of the LED and causes the LED to turn on.

OPTICAL DIRECTION DISCRIMINATOR

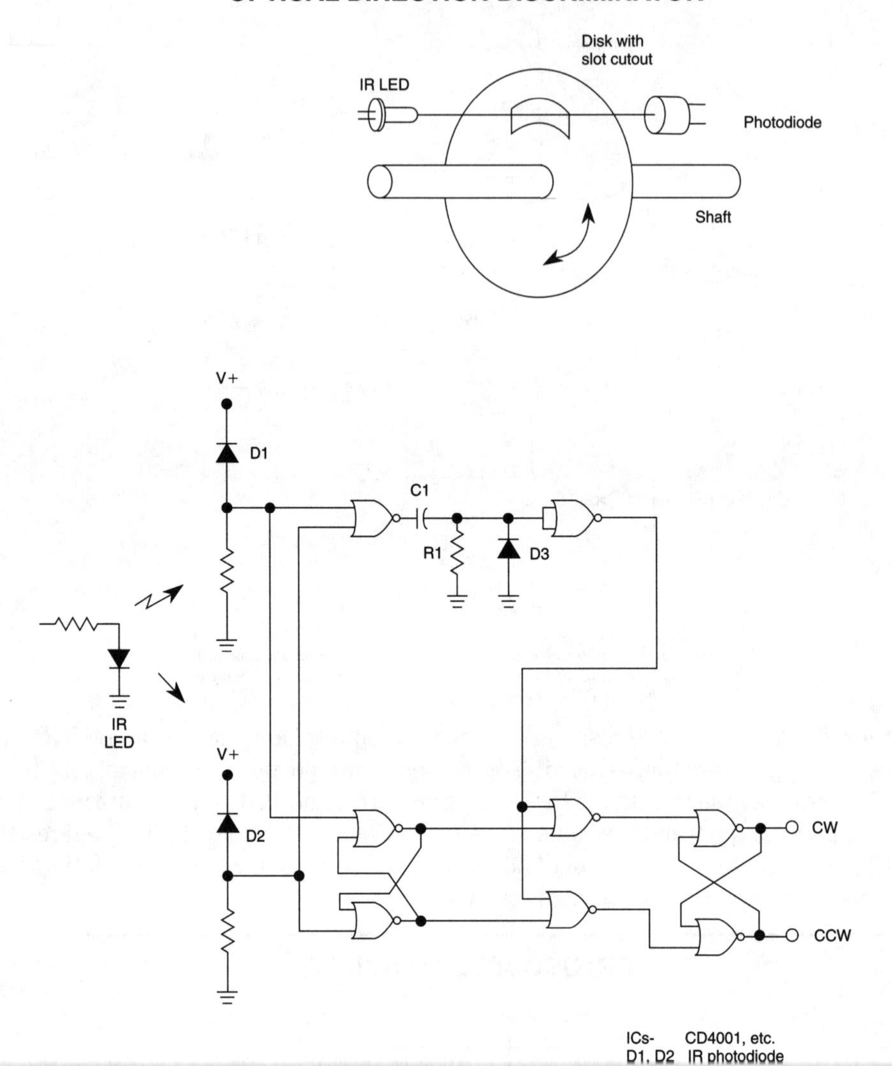

WILLIAM SHEETS

FIG. 70-5

The very simple circuit uses only two CD4001 packages, i.e., eight NOR gates and operates in the following way: Pulse streams are fed to an RS flip flop generating an output waveform which has a small or large duty cycle depending on the direction of rotation. The same input pulses are also fed to a NOR gate, which "adds" the two pulse trains.

The rising edges of this waveform are used to produce short positive pulses from the circuit consisting of R1, C1, D3, and a NOR gate used as an inverter. This is used to "sample" the outputs of the flip flop to detect the direction of rotation. The output, whose duty cycle is large, forces the sampling NOR gate to generate a pulse train which sets (or resets) the second RS flip-flop continuously giving a permanent indication of the direction of rotation.

OPTICAL SAFETY CIRCUIT SWITCHES

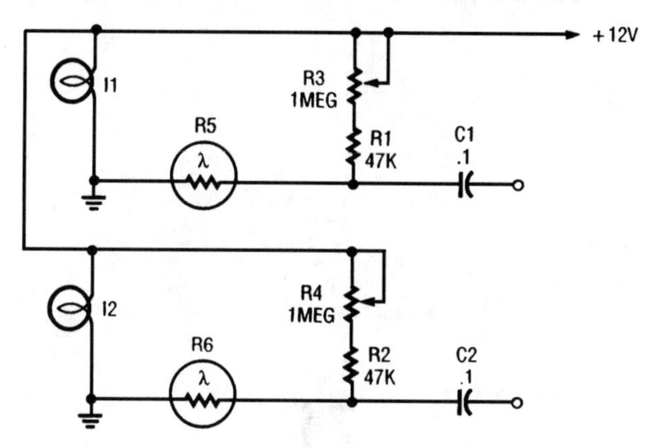

POPULAR ELECTRONICS *FIG. 70-6*

Use of two LDR devices replaces the two pushbuttons used in safety switches. The lamps provide light sources for the LDR devices.

SIMPLE AMPLIFIER FOR PHOTOTRANSISTORS

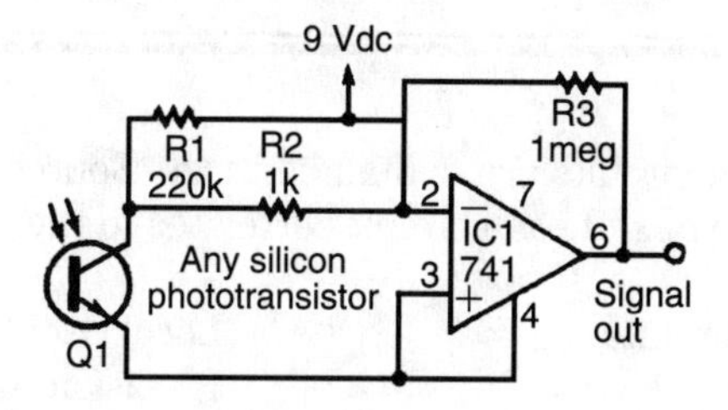

ELECTRONICS NOW *FIG. 70-7*

This simple amplifier will work well with just about any phototransistor. The 741, although designed to operate with a split supply, will work with a single-sided supply as well.

VARIABLE-SENSITIVITY PHOTOTRANSISTOR CIRCUIT

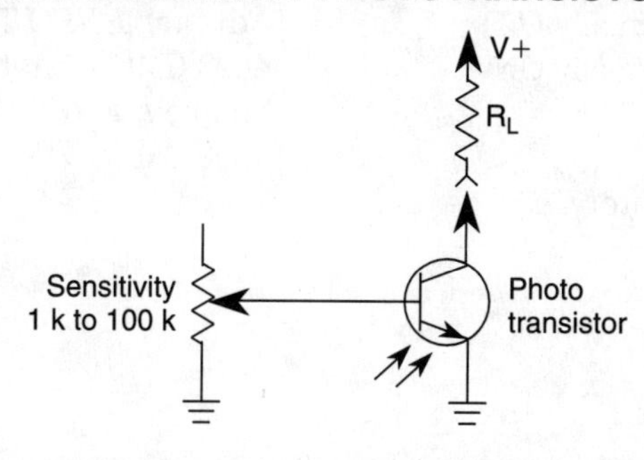

ELECTRONICS NOW *FIG. 70-8*

A variable resistor is used to vary the light-level response of a phototransistor. Phototransistors are more light sensitive than photodiodes, but they generally have poorer frequency response.

71

Oscillator Circuits

The sources of the following circuits are contained in the Sources section, which begins on page 675. The figure number in the box of each circuit correlates to the entry in the Sources section.

NE602 LOCAL OSCILLATOR CIRCUITS

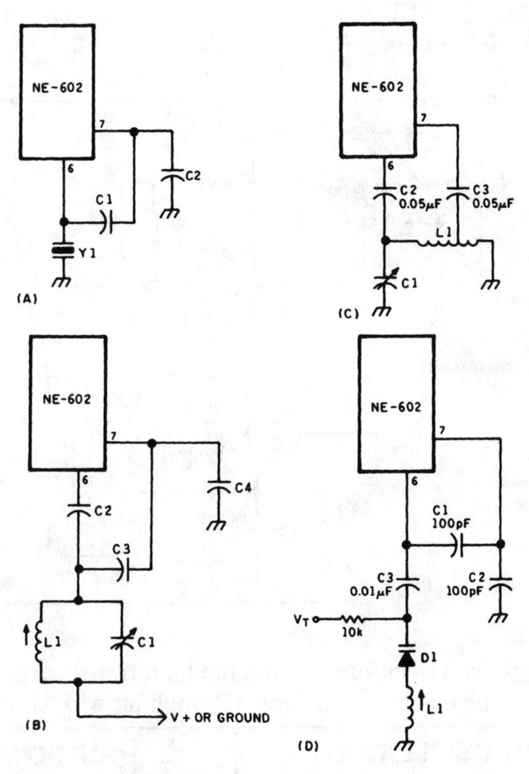

73 AMATEUR RADIO TODAY

FIG. 71-1

Local oscillator circuits for the NE602.

LC AUDIO OSCILLATOR

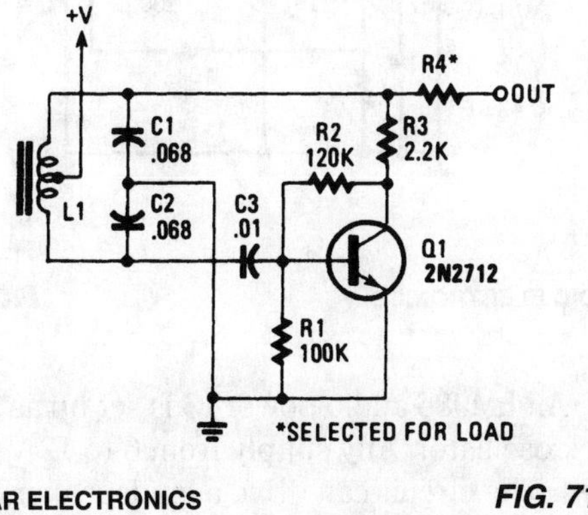

POPULAR ELECTRONICS

FIG. 71-2

COLPITTS OSCILLATOR

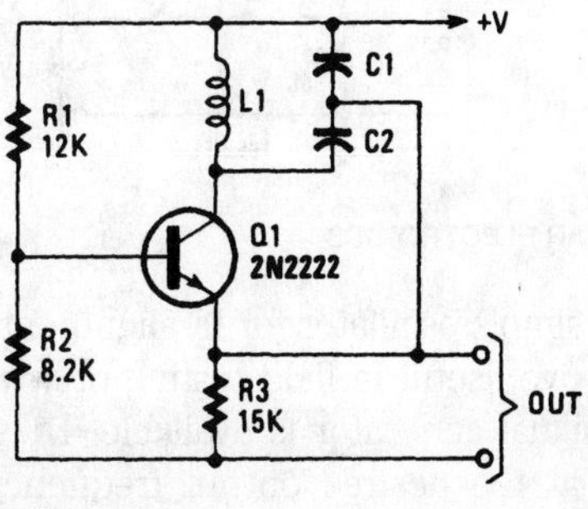

POPULAR ELECTRONICS

FIG. 71-3

MOSFET MIXER-OSCILLATOR CIRCUIT FOR AM RECEIVERS

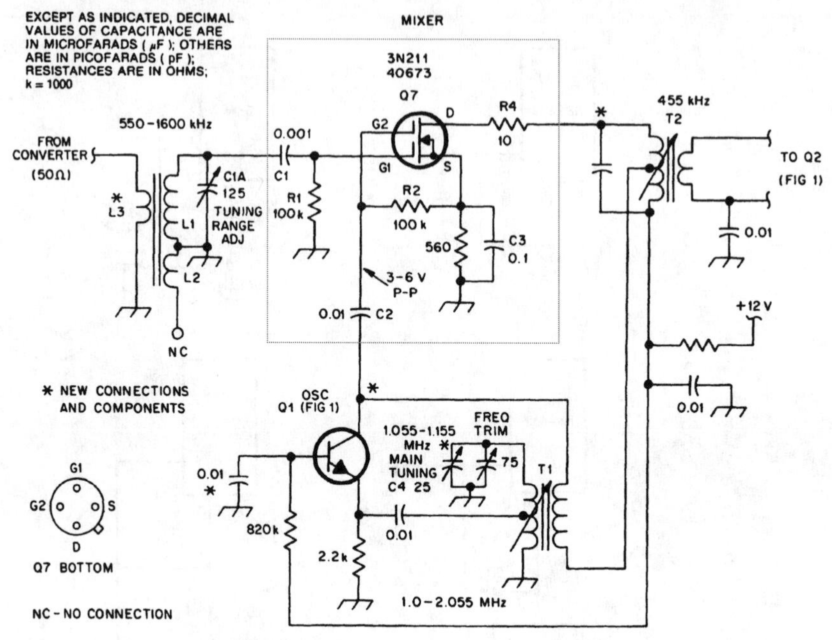

QST

FIG. 71-4

This circuit is an improved front end for upgrading a transistor AM receiver. This front end is useful when the radio is to be used as a tuneable IF amplifier with shortwave converters.

SIMPLE RF TEST OSCILLATOR

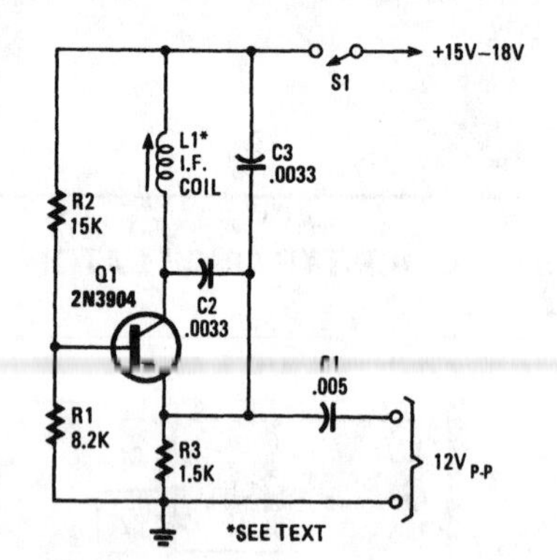

POPULAR ELECTRONICS

FIG. 71-5

A simple oscillator for IF alignment (455 kHz) can prove useful in field testing or where a standard signal generator is available. L1 should resonate at the desired output frequency with the series combination of C2 and C3.

AF POWER OSCILLATOR

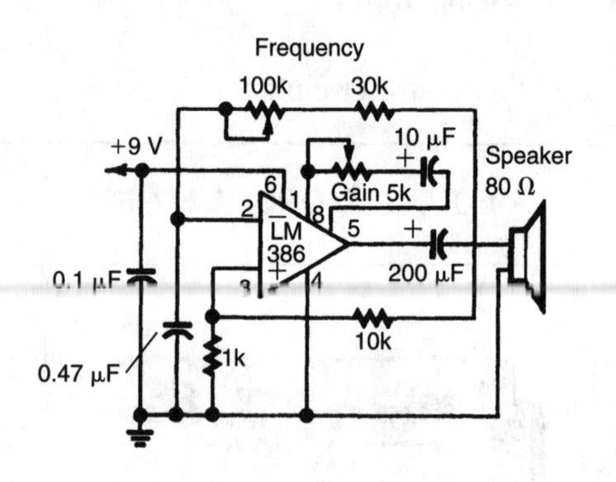

RADIO ELECTRONICS

FIG. 71-6

An LM386 audio power IC is set up as a feedback oscillator. Any supply from 6 to 12 V can be used. The circuit can drive a loudspeaker.

GATED 1-kHz OSCILLATOR (NORMALLY OFF)

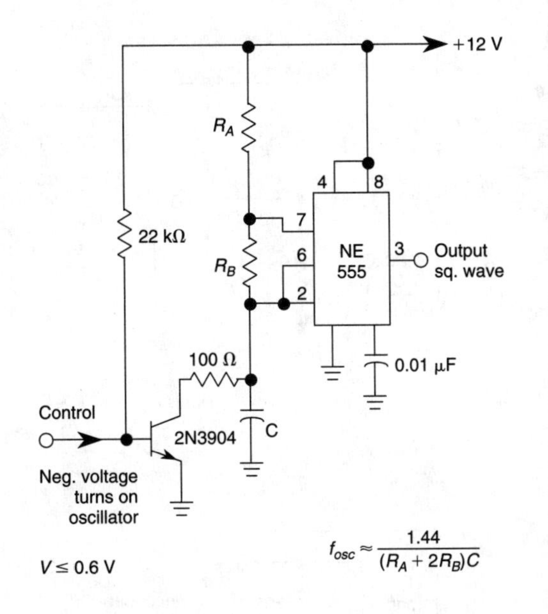

ELECTRONICS NOW

FIG. 71-7

This gated 1-kHz oscillator offers "press-to-turn-on" operation, A, and waveforms at the output of pin 3 and across C1, B.

GATED 1-kHz OSCILLATOR (NORMALLY ON)

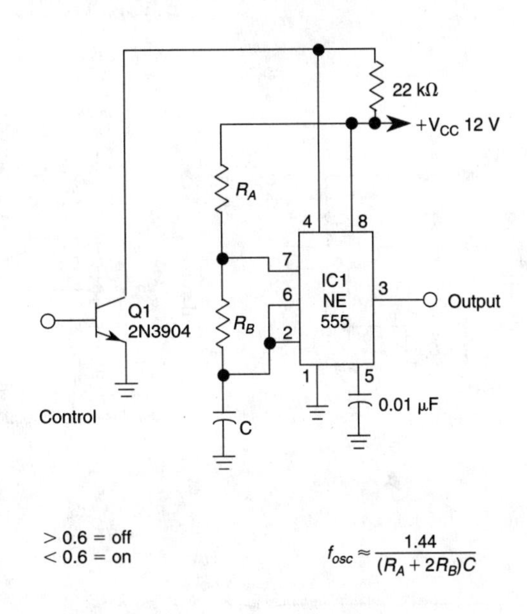

ELECTRONICS NOW

FIG. 71-8

This gated 1-kHz oscillator offers "press-to-turn-off" operation, A, and waveforms at the output of pin 3 and across C1, B.

PRECISION LF OSCILLATOR

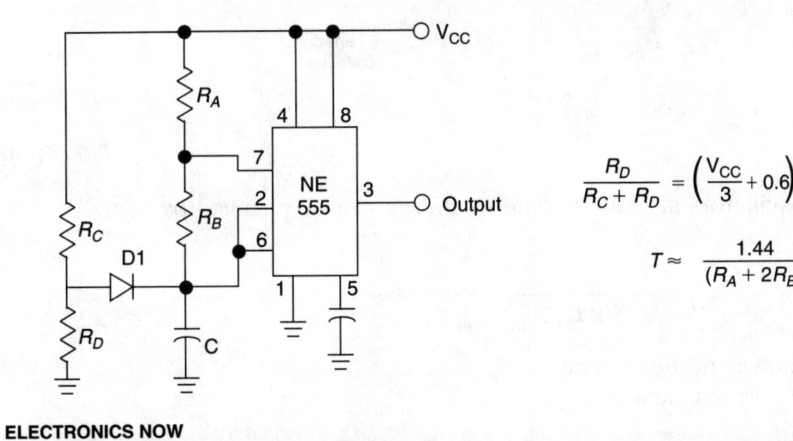

$$\frac{R_D}{R_C + R_D} = \left(\frac{V_{CC}}{3} + 0.6\right)$$

$$R_C + R_D << R_A + R_B$$

$$T \approx \frac{1.44}{(R_A + 2R_B)C}$$

ELECTRONICS NOW

FIG. 71-9

Using R1, R7, and D1 to preset C1 to one third of the supply voltage, this circuit avoids a longer first cycle period than subsequent cycles.

BASIC OSCILLATOR CIRCUITS

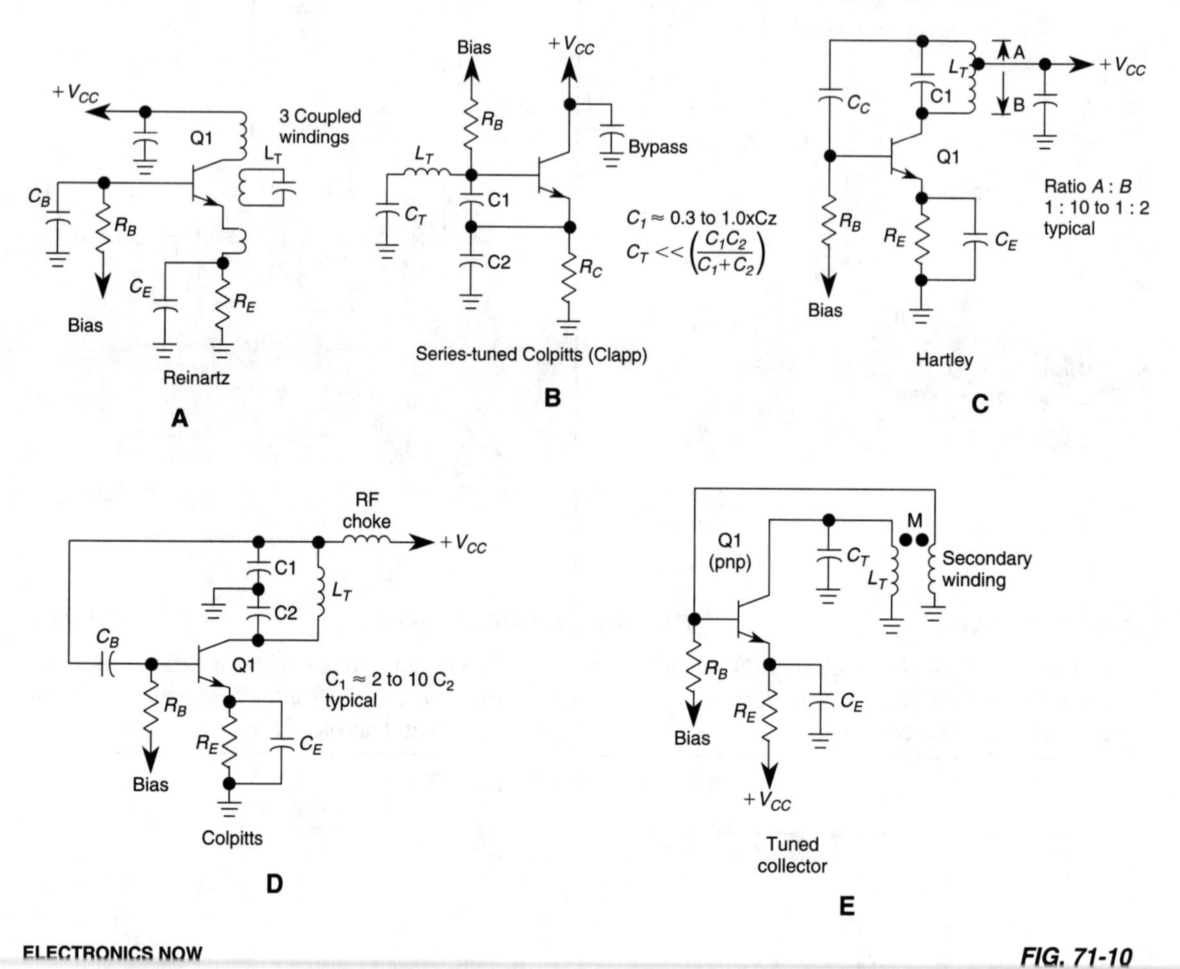

ELECTRONICS NOW

FIG. 71-10

Five basic types of LC oscillators are shown. The frequency can be changed by using the formula:

$$f = \frac{1}{2\pi L_{\text{effective}} C_{\text{effective}}}$$

where $L_{\text{effective}}$ = equivalent inductance
$C_{\text{effective}}$ = equivalent capacitance

VARIABLE WIEN-BRIDGE OSCILLATOR

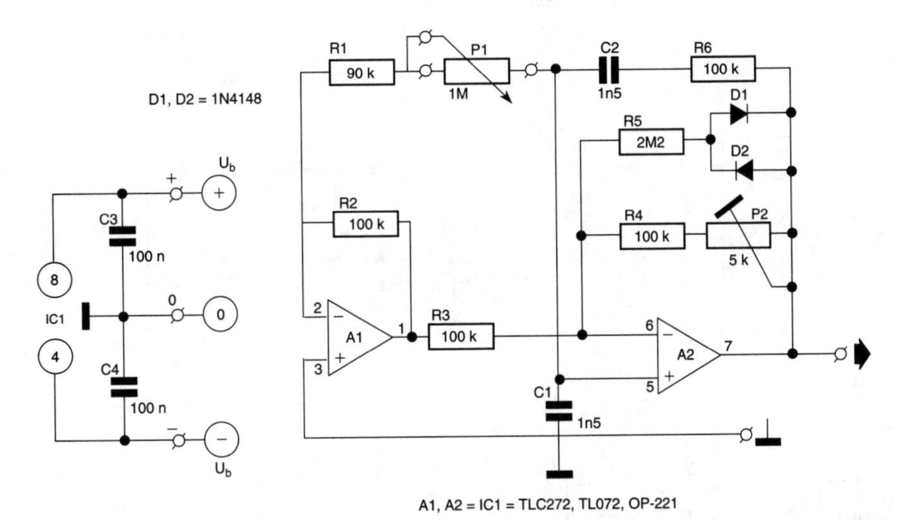

303 CIRCUITS ***FIG. 71-11***

This circuit uses a single potentiometer to tune a 300- to 3000-Hz range. A FET op amp is used at A1 and A2. The upper frequency limit is determined by the gain-bandwidth product of the op amps.

LOCAL OSCILLATOR FOR DOUBLE BALANCED MIXERS

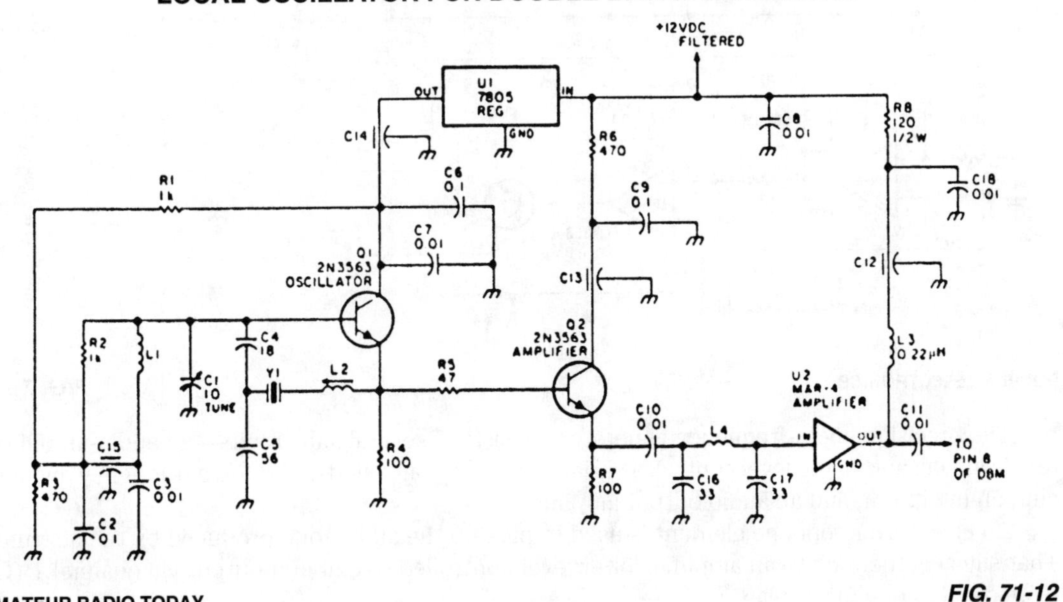

73 AMATEUR RADIO TODAY ***FIG. 71-12***

This circuit has an amplifier to supply +10 dBm to an SBL series (Mini-circuits) or similar type doubly-balanced mixer assembly. This circuit has values shown for ≈80- to 90-MHz crystals, although values of oscillator circuit constants can be scaled for higher or lower frequencies.

PRECISION AUDIO-FREQUENCY GENERATOR

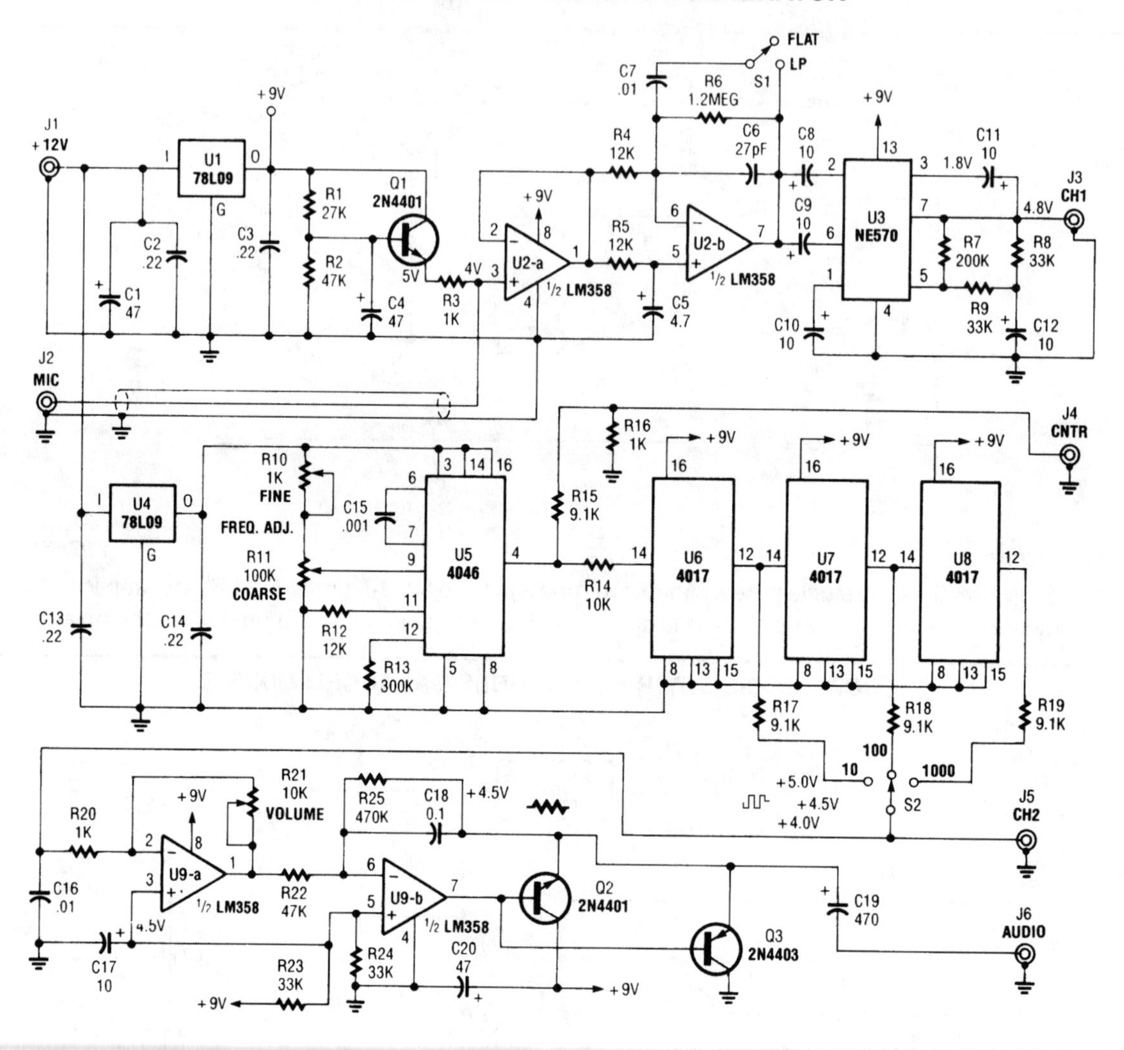

POPULAR ELECTRONICS

FIG. 71-13A

The precision audio-frequency generator consists of several subcircuits—an audio-amplifier/filter circuit, an automatic level control, a variable voltage-controlled oscillator, a frequency divider circuit, an integrator, and an audio output amplifier.

An electret microphone element is used to pick up the audio tone produced by the instrument. That signal is then fed to an amplifier/filter/level-controlled circuit and output via channel 1 (CH1) to an oscilloscope for display.

The variable voltage-controlled oscillator (VCO) is used to produce a signal of from less than 10 kHz to more than 99 kHz. The VCO output is fed to a digital frequency counter for display, and is also routed to a chain of frequency dividers, where the signal is divided by 10, 100, or 1,000, depending on the setting of a selector switch.

PRECISION AUDIO-FREQUENCY GENERATOR (*Cont.*)

Note/ Octave	Key#	Hertz	Stretch in Cents	Note/ Octave	Key#	Hertz	Stretch in Cents
A/0	1	27.184	−20	F/4	45	349.03	− 1
B♭/0	2	28.817	−19	G♭/4	46	369.78	− 1
B/0	3	30.548	−18	G/4	47	391.77	− 1
C/1	4	32.384	−17	A♭/4	48	415.07	− 1
D♭/1	5	34.329	−16	A/4	49	440.00	0
D/1	6	36.391	−15	B♭/4	50	466.16	0
E♭/1	7	38.578	−14	B/4	51	493.88	0
E/1	8	40.895	−13	C/5	52	523.25	0
F/1	9	43.352	−12	D♭/5	53	554.37	0
G♭/1	10	45.956	−11	D/5	54	587.33	0
G/1	11	48.717	−10	E♭/5	55	622.61	+ 1
A♭/1	12	51.644	− 9	E/5	56	659.64	+ 1
A/1	13	54.746	− 8	F/5	57	698.86	+ 1
B♭/1	14	58.035	− 7	G♭/5	58	740.42	+ 1
B/1	15	61.522	− 6	G/5	59	784.44	+ 1
C/2	16	65.180	− 6	A♭/5	60	831.57	+ 2
D♭/2	17	69.096	− 5	A/5	61	881.02	+ 2
D/2	18	73.204	− 5	B♭/5	62	933.41	+ 2
E♭/2	19	77.602	− 4	B/5	63	988.91	+ 2
E/2	20	82.217	− 4	C/6	64	1047.7	+ 2
F/2	21	87.106	− 4	D♭/6	65	1110.7	+ 3
G♭/2	22	92.285	− 4	D/6	66	1176.7	+ 3
G/2	23	97.773	− 4	E♭/6	67	1246.7	+ 3
A♭/2	24	103.65	− 3	E/6	68	1321.6	+ 4
A/2	25	109.81	− 3	F/6	69	1400.1	+ 4
B♭/2	26	116.34	− 3	G♭/6	70	1484.3	+ 5
B/2	27	123.26	− 3	G/6	71	1572.5	+ 5
C/3	28	130.59	− 3	A♭/6	72	1667.0	+ 6
D♭/3	29	138.35	− 3	A/6	73	1766.1	+ 6
D/3	30	146.58	− 3	B♭/6	74	1872.2	+ 7
E♭/3	31	155.29	− 3	B/6	75	1984.7	+ 8
E/3	32	164.53	− 3	C/7	76	2103.9	+ 9
F/3	33	174.31	− 3	D♭/7	77	2230.3	+10
G♭/3	34	184.73	− 2.5	D/7	78	2230.2	+10
G/3	35	195.71	− 2.5	E♭/7	79	2506.3	+12
A♭/3	36	207.41	− 2	E/7	80	2656.9	+13
A/3	37	219.75	− 2	F/7	81	2818.1	+15
B♭/3	38	232.81	− 2	G♭/7	82	2989.2	+17
B/3	39	246.66	− 2	G/7	83	3170.6	+19
C/4	40	261.32	− 2	A♭/7	84	3363.0	+21
D♭/4	41	276.86	− 2	A/7	85	3567.1	+23
D/4	42	293.33	− 2	B♭/7	86	3783.6	+25
E♭/4	43	310.86	− 1.5	B/7	87	4013.2	+27
E/4	44	329.44	− 1	C/8	88	4259.2	+30

•Standard pitch, A49= 440 Hz
Values shown are stretched for the average piano

POPULAR ELECTRONICS

FIG. 71- 13B

From there, the selected signal frequency divides along two paths; one going to CH2 (which feeds the oscilloscope's sweep synchronization input) and to an integrator that converts the square-wave output of the divider into a triangular waveform. The output of the integrator is then amplified and fed to a set of stereo headphones via an audio output jack.

One section of the precision audio-frequency generator uses an electret microphone element to pick up audio from the piano. That signal is then processed and sent to one channel of a dual-trace oscilloscope. The other section of the circuit is used to produce a variable-frequency signal that is fed to a digital frequency counter and, after conditioning, is presented to the second channel of the scope and output to a set of stereo headphones.

CMOS VFO

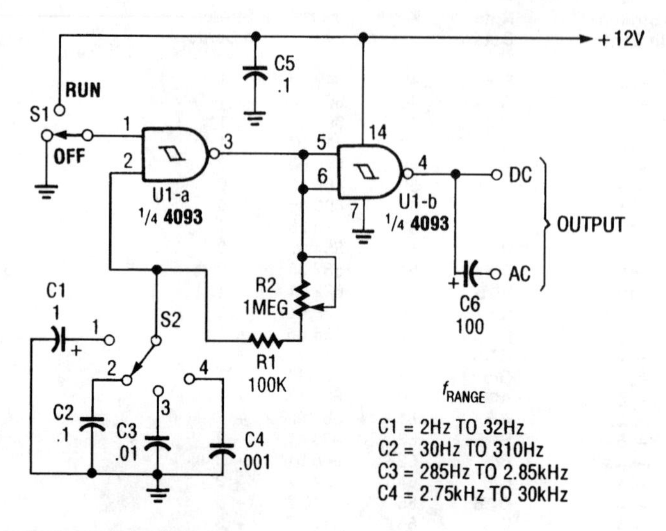

POPULAR ELECTRONICS

FIG. 71-14

The circuit shown has a frequency range of 2 Hz to 30 kHz. R2 is a linear or log potentiometer.

FREQUENCY SWITCHER

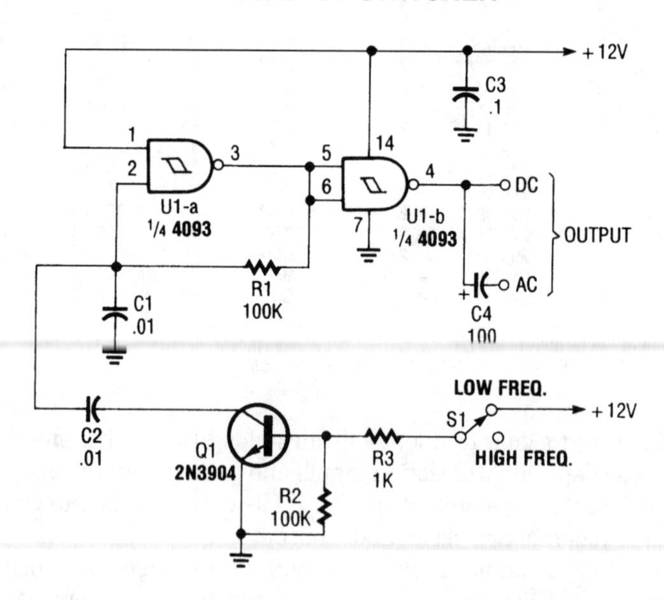

POPULAR ELECTRONICS

FIG. 71-15

This transistor can achieve frequency switching in this CMOS astable oscillator.

PRECISION GATED OSCILLATOR

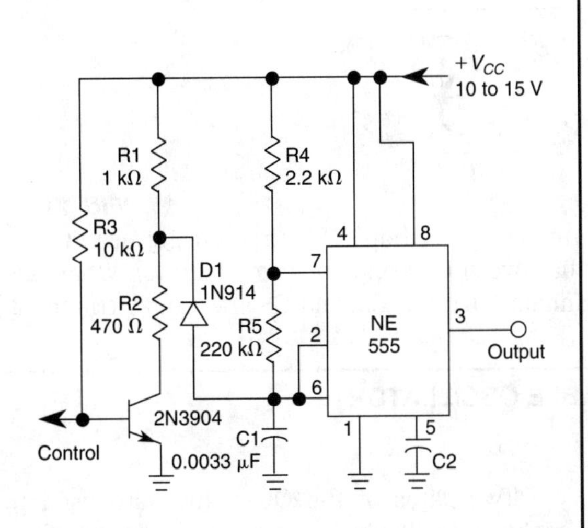

ELECTRONICS NOW *FIG. 71-16*

A 1-kHz gated oscillator with no long "turn-on" cycle is shown. R2, R3, and D1 preset the voltage on tuning capacitor C1 to ⅓ of the supply voltage.

WIEN-BRIDGE AUDIO OSCILLATOR

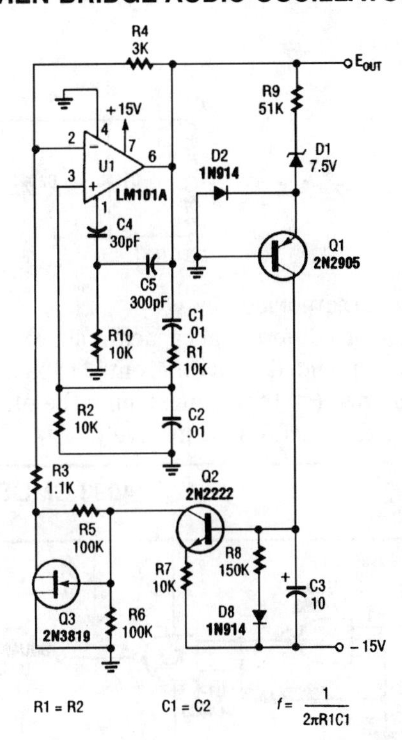

POPULAR ELECTRONICS *FIG. 71-17*

For variable-frequency operation, R1 and R2 can be replaced by a dual potentiometer.

VARIABLE DUTY-CYCLE OSCILLATOR

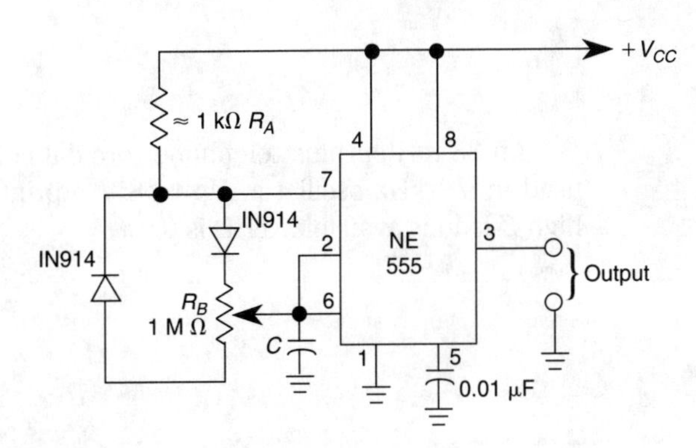

$$T \approx \frac{1.44}{(R_A + 2R_B)C}$$

NOTE: Diodes have the effect of slightly reducing the observed frequency—especially if $V_{CC} < 10$ V as a result of 0.6 V offset.

ELECTRONICS NOW *FIG. 71-18*

Using a potentiometer and steering diodes, this 1.2-kHz oscillator will provide 1 to 99% duty cycle. Vary C1 to change frequency.

ADJUSTABLE VFO TEMPERATURE COMPENSATOR

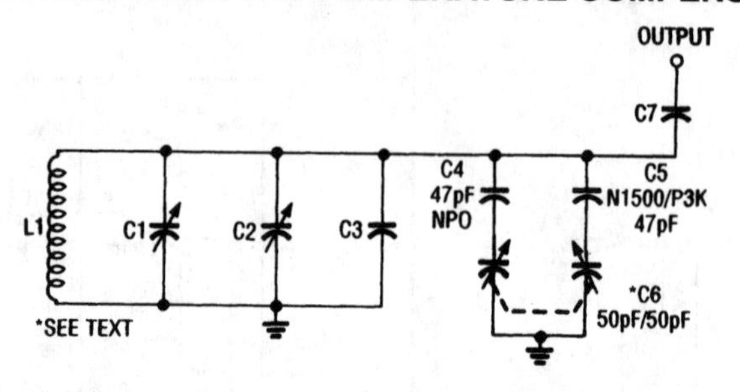

POPULAR ELECTRONICS — *FIG. 71-19*

Use of a differential capacitor allows temperature compensation of LC circuit using an NPO and N1500 ceramic. C6 is a differential capacitor that has two stators and one common rotor. When one capacitance (stator) is maximum, the other is minimum. L1, C1, C2, and C3 are tuning, trimming, and fixed capacitors, respectively.

4093 CMOS ASTABLE OSCILLATOR

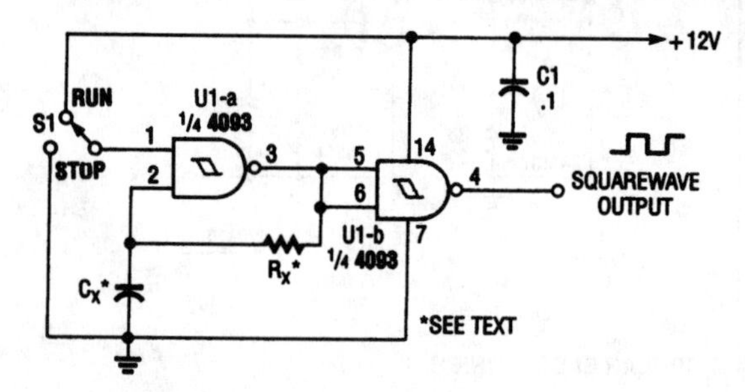

Two gates of the Quad 4093 are used to make an oscillator. R_x can be from about 5 kΩ to around 10 MΩ. C_x can be from about 10 pF to many μF, the limit being set by the leakage of the capacitor. Frequency is approximately $2.8/R_xC_x$ (R MΩ, Cmfd).

POPULAR ELECTRONICS — *FIG. 71-20*

SIMPLE AUDIO TEST OSCILLATOR

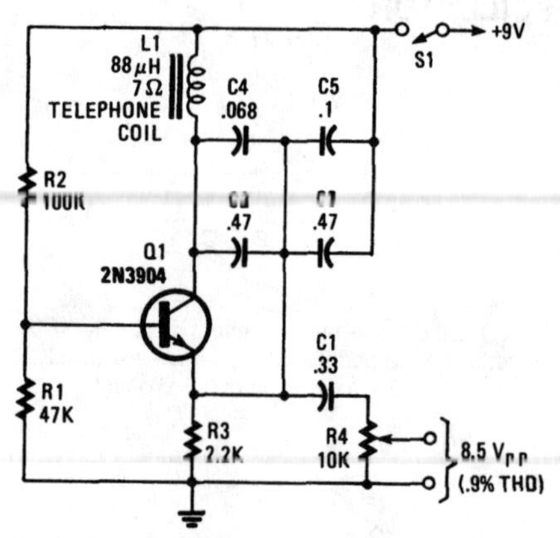

An 88-mH surplus telephone toroidal coil is used in a 1-kHz oscillator. Up to 8 V p-p into a high-Z load is available. THD is 0.9%.

POPULAR ELECTRONICS — *FIG. 71-21*

4093 CMOS VFO

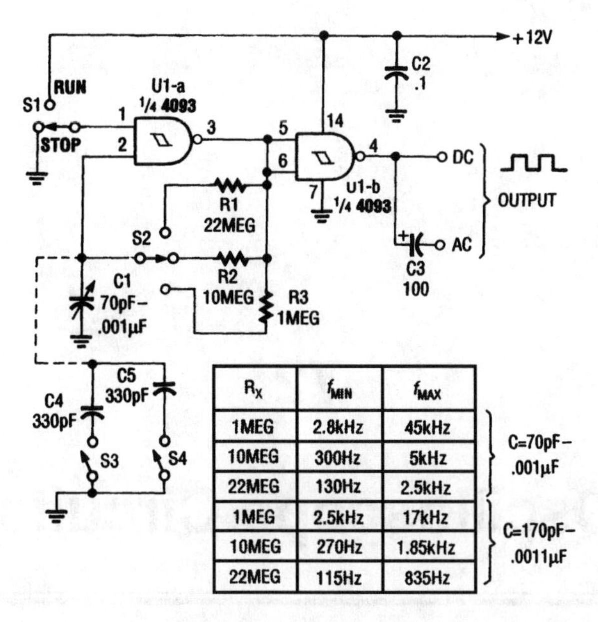

R_X	f_{MIN}	f_{MAX}	
1MEG	2.8kHz	45kHz	C=70pF–.001μF
10MEG	300Hz	5kHz	
22MEG	130Hz	2.5kHz	
1MEG	2.5kHz	17kHz	C=170pF–.0011μF
10MEG	270Hz	1.85kHz	
22MEG	115Hz	835Hz	

POPULAR ELECTRONICS ***FIG. 71-22***

Two gates of a Quad 4093 are used in an astable multivibrator. C1 is a three-gang 365 pF variable capacitor with sections paralleled. S3 and S4 switch in optional extra capacitors.

72

Oscilloscope Circuits

The sources of the following circuits are contained in the Sources section, which begins on page 675. The figure number in the box of each circuit correlates to the entry in the Sources section.

OSCILLOSCOPE PREAMPLIFIER

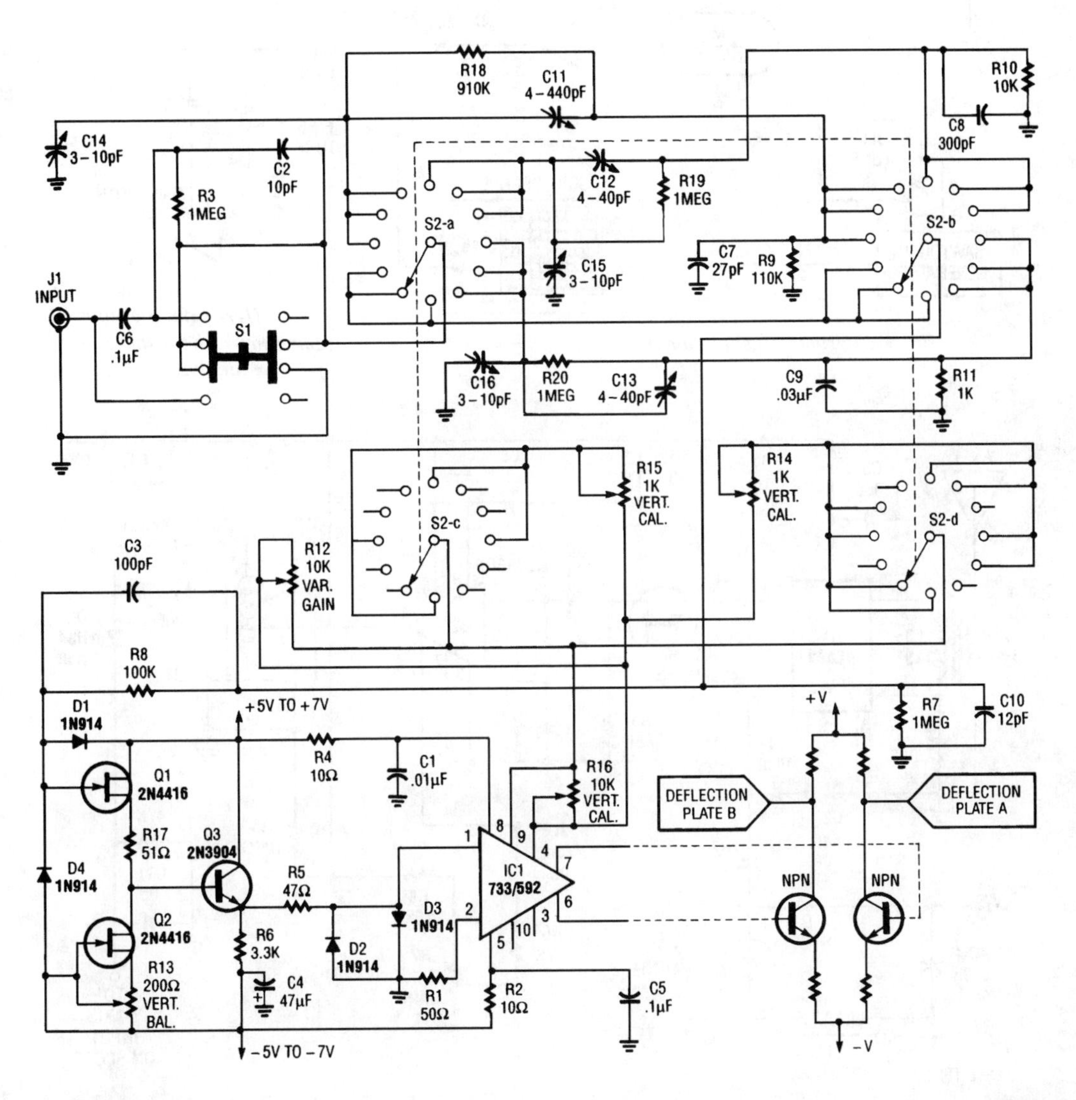

RADIO ELECTRONICS

FIG. 72-1

An oscilloscope front-end amplifier can be built with low-cost transistor and video amp ICs. This preamp uses a FET input and compensated attenuators, and has approximately 100-MHz bandwidth, which is adequate for most general-purpose oscilloscopes.

SIMPLE SPECTRUM ANALYZER ADAPTOR FOR SCOPES

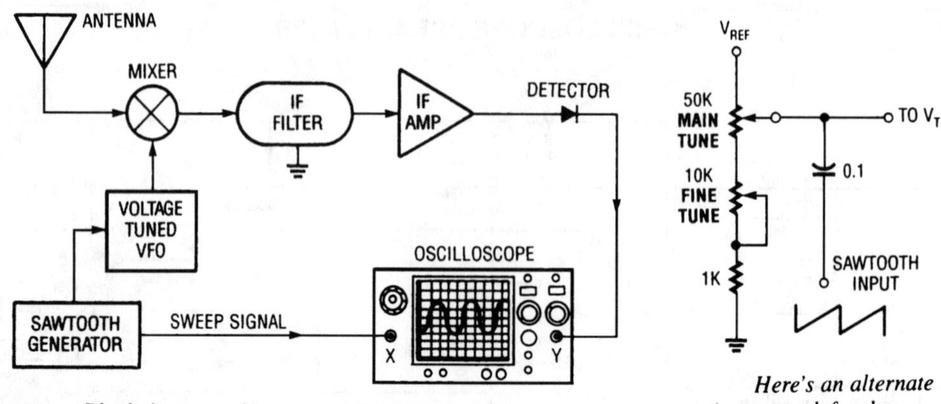

Block diagram of a spectrum analyzer.

Here's an alternate tuning network for the spectrum analyzer.

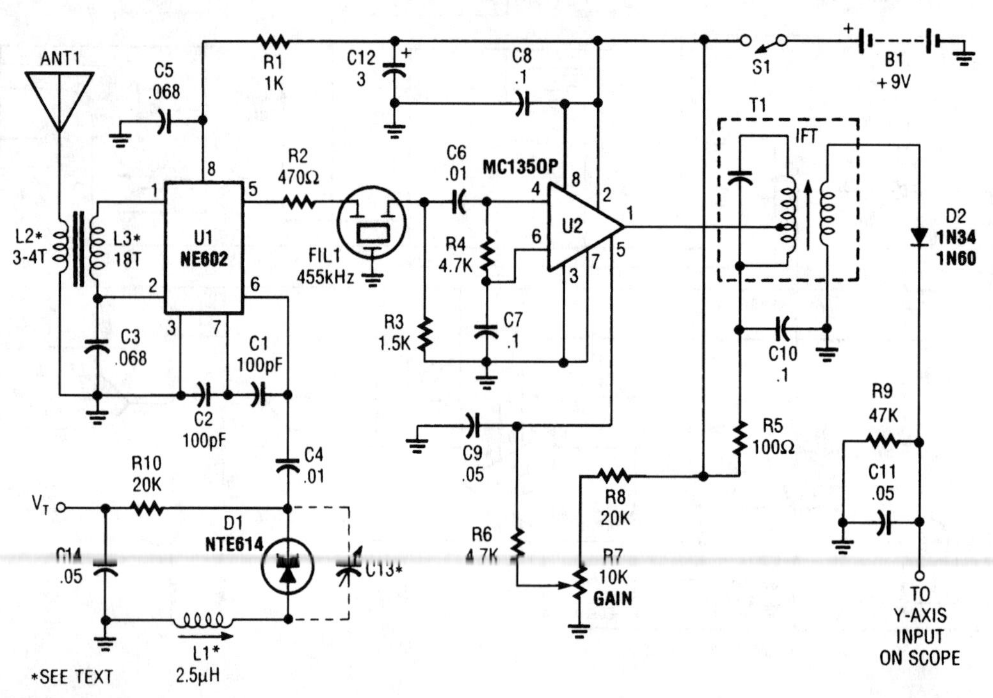

POPULAR ELECTRONICS

FIG. 72-2

Suitable for monitoring an amateur band or a segment of the radio spectrum, this simple adaptor uses an NE602 mixer-oscillator chip to produce a 455-kHz IF signal, which U2 amplifies, then feeds to detector D2 and the Y axis of an oscilloscope. V_T is used to drive the horizontal axis input of a scope. L2 and L3 are coils suitable for the frequency range in use. For this circuit, coils are shown for the 10- to 15-MHz range. L2 and L3 are wound on Amidon Associates, T-37 or T-50 toroidal cores, and L1 is a commercial or homemade variable inductor, etc.

SIMPLE OSCILLOSCOPE TIMEBASE GENERATOR

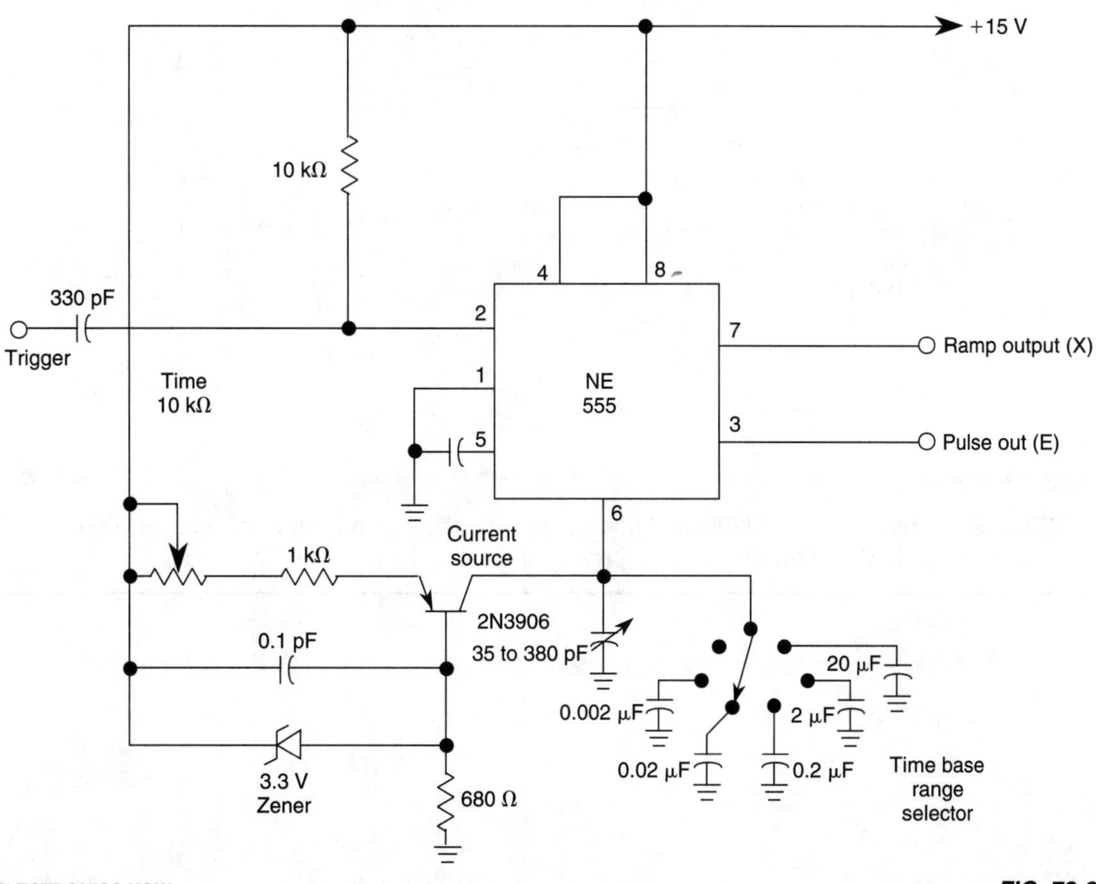

ELECTRONICS NOW ***FIG. 72-3***

The 555 timer generates both a linear ramp and an output for Z-axis modulations of the CRT electron beam.

TRIGGER SELECTION CIRCUIT FOR OSCILLOSCOPE TIMEBASE

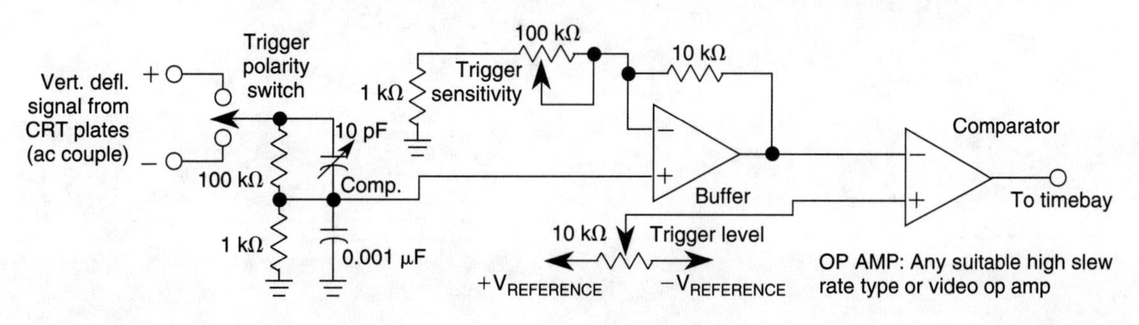

ELECTRONICS NOW ***FIG. 72-4***

VARIABLE GAIN AMPLIFIER

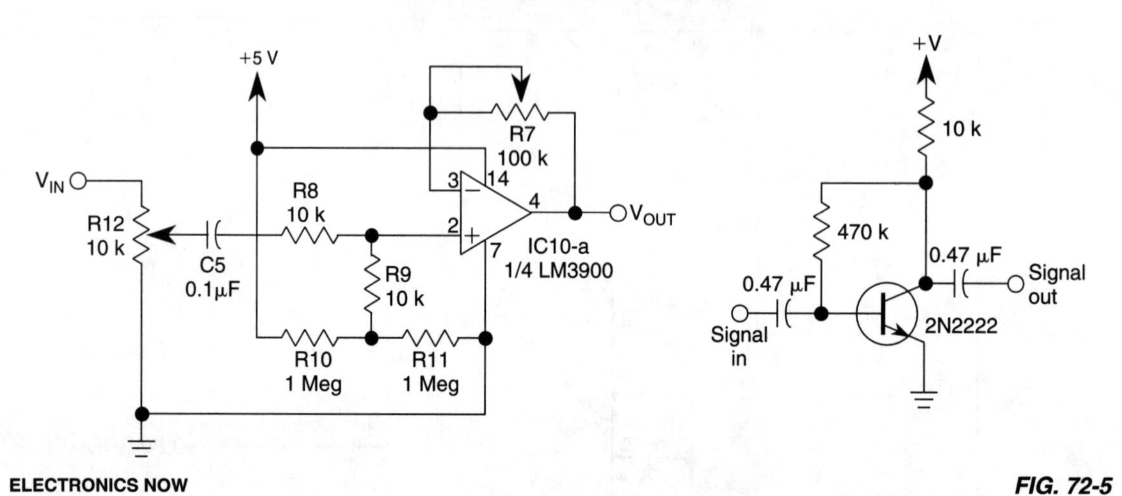

ELECTRONICS NOW *FIG. 72-5*

This circuit uses ¼ of an LM3900 to build a simple variable-gain front end for an oscilloscope. R7 is the gain control. Also shown is a simple preamp if you need more than 10X of gain.

73

Pest-Control Circuits

The sources of the following circuits are contained in the Sources section, which begins on page 675. The figure number in the box of each circuit correlates to the entry in the Sources section.

PEST REPELLER

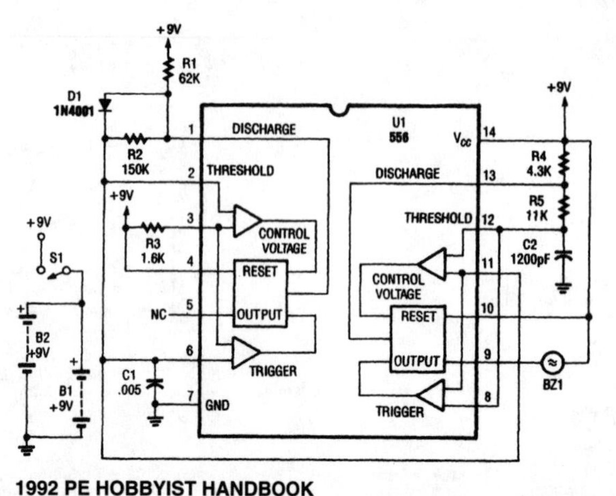

The two timers in the bug repeller have some interesting characteristics. Both of them have their thresholds externally set; the oscillator on the left has a 50% duty cycle and the oscillator on the right acts as a VCO.

1992 PE HOBBYIST HANDBOOK

FIG. 73-1

ULTRASONIC PEST REPELLER

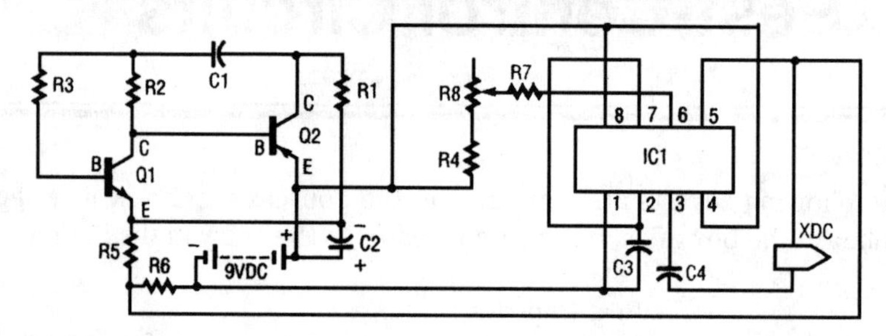

1991 PE HOBBYIST HANDBOOK

FIG. 73-2

This circuit uses two transistors and one IC (555 timer IC) to produce a pulsating ultrasonic frequency. Transistors Q1 and Q2 are connected in a direct-coupled oscillator. The frequency of that oscillator is set by capacitor C1. The oscillator output is taken from the emitter of Q2 to pin 7 of IC1. Transistor Q1 is an npn transistor, and Q2 is a pnp transistor. The signal of pin 7 on IC1 causes the output signal appearing on pin 3 to be modulated or varied by the audio frequency developed by Q1 and Q2. The IC itself is connected as a stable multivibrator with a frequency that is determined by C3. Capacitor C3 sets the basic frequency to be well above the human hearing range (ultrasonic). The combined modulated ultrasonic frequency appears on pin 3 of IC1, where it is coupled by capacitor C4 to the piezoelectric transducer.

C1, C2	0.1-μF Mylar Capacitor	R2	3.3-MΩ Resistor
C2	1-μF Electrolytic Capacitor	R3, R6	10-kΩ Resistor
C3	0.001-μF Mylar Capacitor	R4, R5	100-Ω Resistor
IC1	555 timer IC	R7	18-kΩ Resistor
Q1	2N3904 Transistor	R8	Potentiometer
Q2	2N3906 Transistor	XDC	Piezoelectric Transducer Disc
R1	4.7-kΩ Resistor	Misc	IC Socket, 9-V Snap, PC Board

74

Phase Shifter Circuits

The sources of the following circuits are contained in the Sources section, which begins on page 675. The figure number in the box of each circuit correlates to the entry in the Sources section.

LONG-TAILED PAIR PHASE-SPLITTER

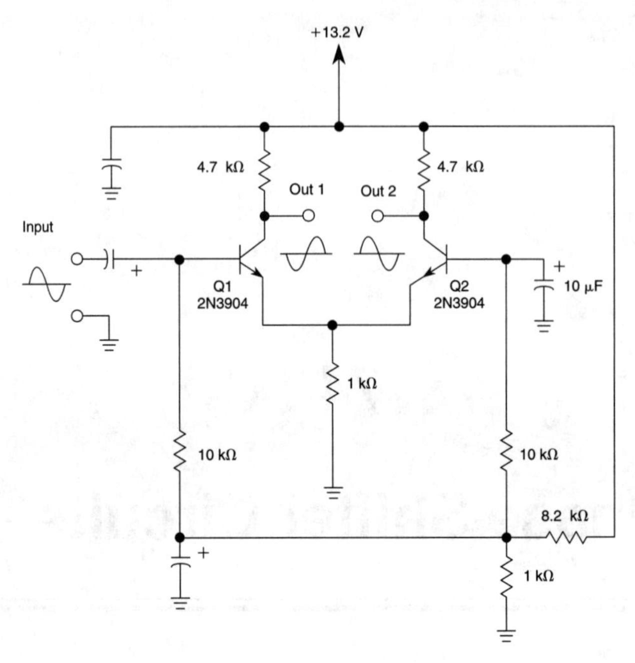

WILLIAM SHEETS ***FIG. 74-1***

The single-phase input produces out-of-phase outputs at the collectors of Q1 and Q2.

PHASE-SPLITTER CIRCUIT

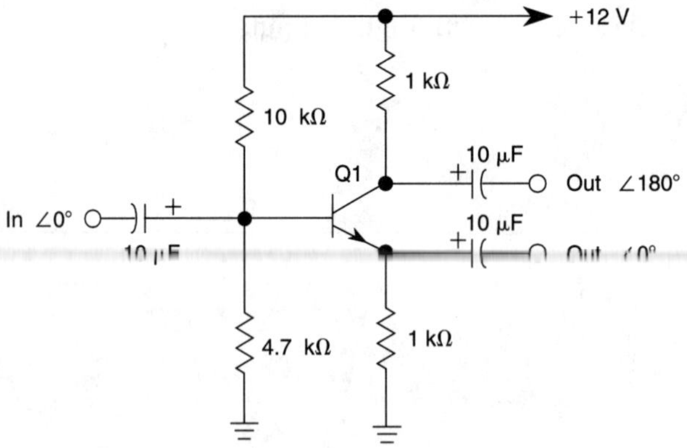

Q1: 2N2222, etc.

WILLIAM SHEETS ***FIG. 74-2***

This phase splitter uses a 2N2222 (or other general purpose npn transistor) to achieve outputs that are 180° out of phase.

PHASE SHIFTER WITH EIGHT OUTPUTS

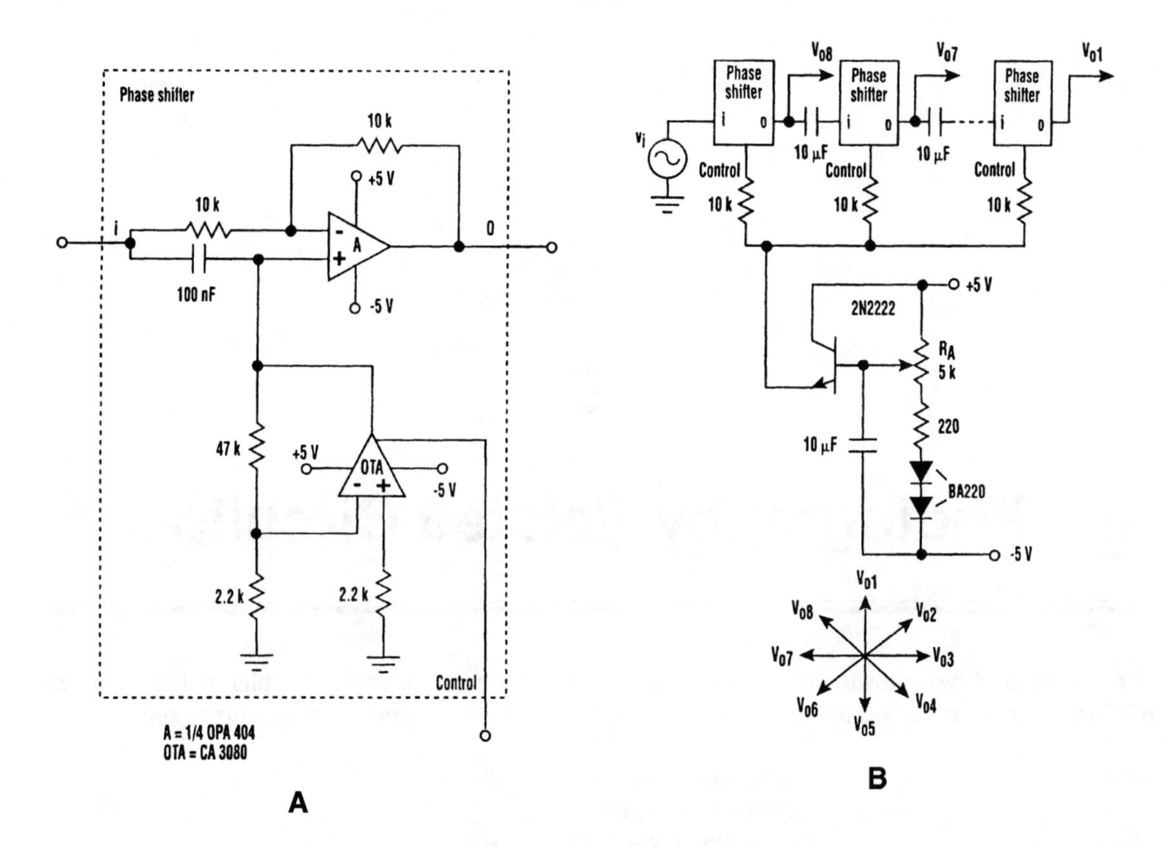

ELECTRONIC DESIGN *FIG. 74-3*

The circuit consists of eight cascaded identical cells, each cell being a dc-controlled active phase shifter. Because the dc control is common for all shifters, the circuit is adjusted by trimming R_A so that the phase difference between V_{01} and V_i is zero. As a result, each shifter will introduce a phase difference of exactly π/r. The eight signals for PSK are available at the op amps' outputs.

Phase accuracy is acceptable for 1%-tolerance resistors and 5%-tolerance 100-nF capacitors. Also, the amplitude of V_i (which is a 1700-Hz sine wave), should not exceed 1 V.

75

Photography Related Circuits

The sources of the following circuits are contained in the Sources section, which begins on page 675. The figure number in the box of each circuit correlates to the entry in the Sources section.

Time-Delay Flash-Trigger Circuit
Photo Flash Slave Unit
Enlarging Light Meter
Photo Strobe
Darkroom Timer
Photo Strobe Slave Trigger
Strobe Light
Enlarger Exposure Meter

TIME-DELAY FLASH-TRIGGER CIRCUIT

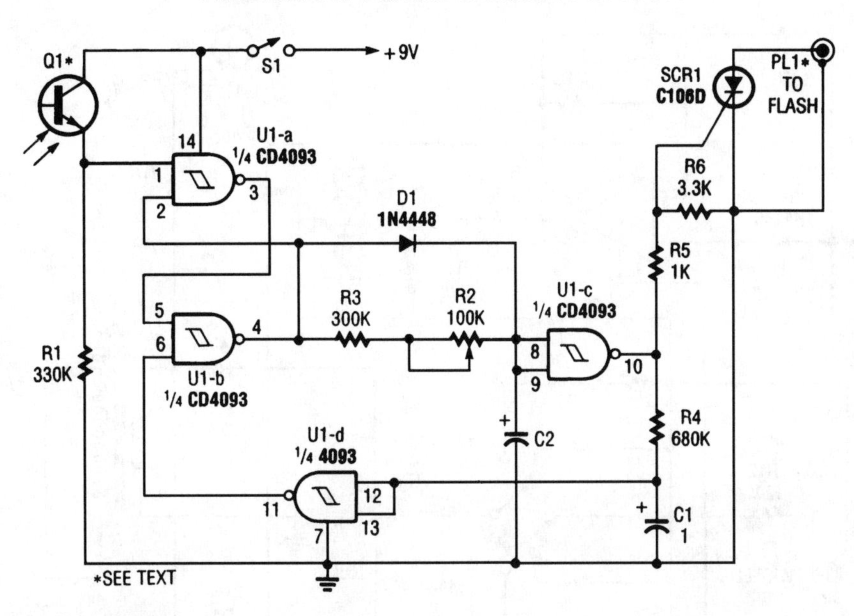

1992 PE HOBBYIST HANDBOOK ***FIG. 75-1***

The circuit is built around a single 4093 quad 2-input NAND Schmitt trigger. Two gates from that quad package (U1-a and U1-b) are configured as a set-reset flip-flop.

PHOTO FLASH SLAVE UNIT

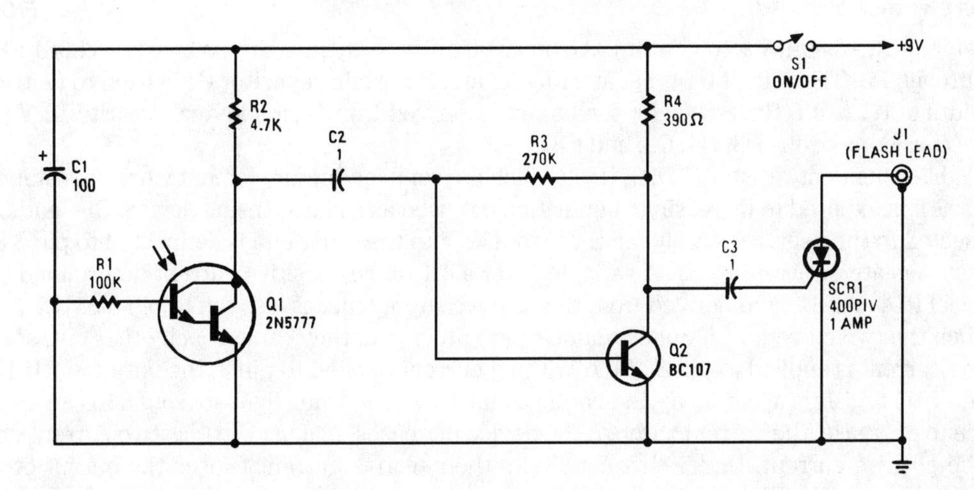

POPULAR ELECTRONICS ***FIG. 75-2***

Phototransistor Q1 receives a light pulse from a photoflash unit. The pulse is ac-coupled to amplifier Q2. It then triggers SCR1, which triggers a flash unit that is connected to J1.

ENLARGING LIGHT METER

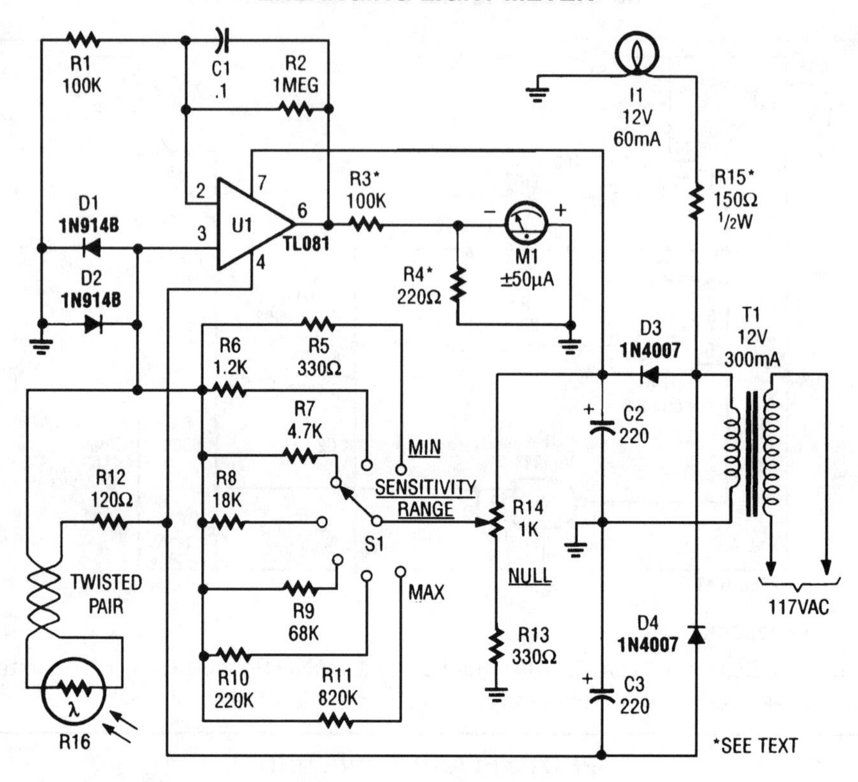

POPULAR ELECTRONICS *FIG. 75-3*

Meter M1, a +/–50-μA zero-center D'Arsonval meter movement is driven by U1, a TL081 FET op amp, through R3. The gain of U1 is set at 11 by R1 and R2, while capacitor C1 is used to restrict the bandwidth of U1 to 1.6 Hz. Power for the circuit is derived from a simple dual-polarity 12-V power supply (consisting of T1, D3, D4, C2, and C3).

A light-dependent resistor (LDR), R16 (which is a semiconductor element whose resistance decreases as it is exposed to increasing illumination), is used as a light-sensing device. One end of R16 is connected to the negative supply rail through R12, and the other end is connected to pin 3 of U1, applying a negative current to U1. A variable (over a 4:1 range) positive current determined by the settings of R14 and S1 (and derived from the positive supply rail) is also fed to pin 3 of U1.

When the two currents (of opposite polarities) are equal, they cancel each other out, so effectively no current is applied to pin 3 of U1. With no current applied to pin 3, the output of U1 is zero and meter M1 registers accordingly, indicating a null. However, when light striking R16 causes its resistance to decrease, the current through the device increases, making the negative current greater than the positive current. Under that condition, the negative current causes the output of U1 to swing negative, causing the pointer to swing in the negative direction.

That indicates that the light intensity must be reduced by using a smaller lens opening on the enlarger (smaller f/stop). The opposite occurs if the light is too dim. Lamp 11, a 12-V 60-mA "grain of wheat" unit, is used to illuminate the meter scale, and R15 is used to limit the meter's illumination to a faint glow that is just bright enough so that the face of M1 can be plainly seen in a photo darkroom.

ENLARGING LIGHT METER (*Cont.*)

Resistors R3 and R4 should be selected for the meter used. With a dual supply of +/–12 V, U1 produces an output voltage of 10 V peak-to-peak. The resistance of R3 can be found by dividing the peak voltage (i.e., 10/2) by the full-scale meter current (in amps); i.e., $R_3 = (10/2)/0.0005 = 100{,}000\ \Omega$. R4, the shunt resistor, should be selected to have a value equal to the meter's internal resistance.

PHOTO STROBE

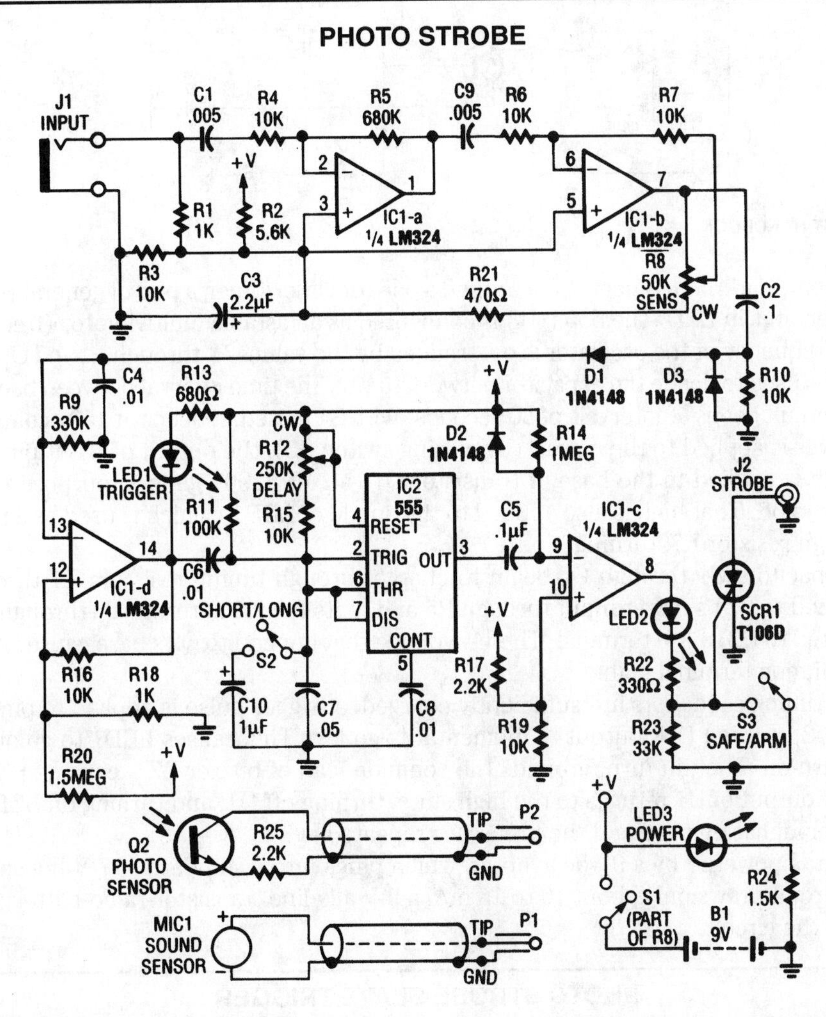

ELECTRONICS NOW

FIG. 75-4

Sound or light sensors connected to J2 produce a voltage that is amplified by IC1-a and IC1-b. A positive trigger voltage that is developed by D1 and D3 and amplified by IC1-d, drives IC2 and IC1 to trigger SCR1. SCR1 is connected to a strobe. This device is handy for photographic purposes to take pictures of events that involve sound, such as impacts, etc.

DARKROOM TIMER

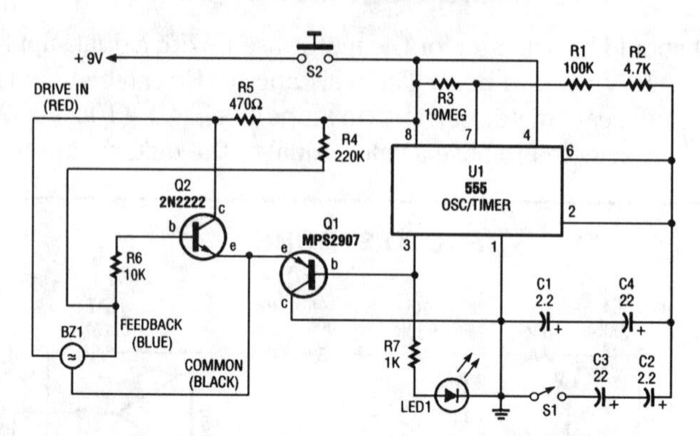

1991 PE HOBBYIST HANDBOOK ***FIG. 75-5***

The electronic darkroom timer is built around a 555 oscillator/timer, a pair of general-purpose transistors, a buzzer, and an LED. The 555 (U1) is configured as an astable multivibrator (free-running oscillator). The frequency of the oscillator is determined by the values R_1 through R_3 and C_1 through C_4.

Switch S1 is used to divide the capacitor network to vary the time interval between beeps; when S1 is closed, the circuit beeps at intervals of 30 seconds. With S1 closed, it beeps at 15-second intervals.

When power is applied to the circuit (by closing switch S2), the output of U1 at pin 3 is initially high. That high is applied to the base of transistor Q1 (an MPS2907 general-purpose pnp device), keeping it turned off. That high is also applied to the anode of LED1 (which is used as a power on indicator) through resistor R7, turning it on.

Timing capacitors C1 through C5 begin to charge through timing resistors R1 through R3. dc voltage is applied to BZ1's driver input through R5 and to its feedback terminal (through R4), which is also connected to Q2's base terminal. The $V+$ voltage that applied to Q2's base causes it to turn on, tying BZ1's common terminal high.

When the timing capacitors are sufficiently charged, a trigger pulse is applied to pin 2 (the trigger input) of U2, causing U1's output to momentarily go low. This causes LED1 to go out and transistor Q1 to turn on. That, in turn, grounds the common lead of buzzer BZ1, causing BZ1 to sound. Afterward, the output of U1 returns to the high state, turning off Q1, and turning on LED1, until another time interval has elapsed and the process is repeated.

The circuit is powered by a 9 Vac adapter, which plugs into a standard 117-V household outlet. Because the circuit draws only about 10 to 15 mA, a 9-V alkaline transistor-radio-battery can also be used to power the circuit.

PHOTO STROBE SLAVE TRIGGER

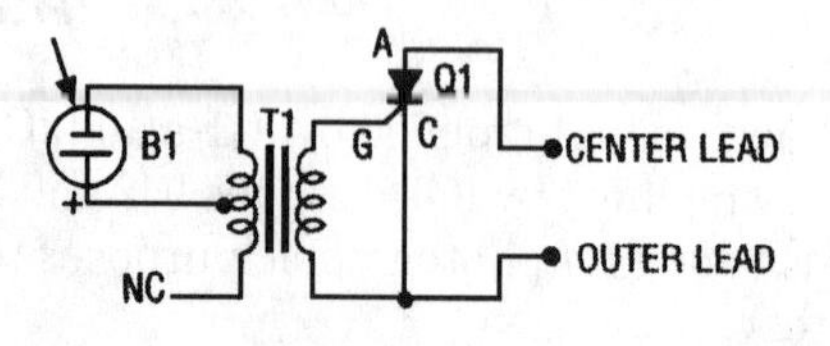

1991 PE HOBBYIST HANDBOOK ***FIG. 75-6***

The photo strobe slave trigger circuit uses a solar cell and an SCR to flash any strobe when you trigger your "master" strobe. The tiny solar cell produces a very small voltage when light falls on its surface.

STROBE LIGHT

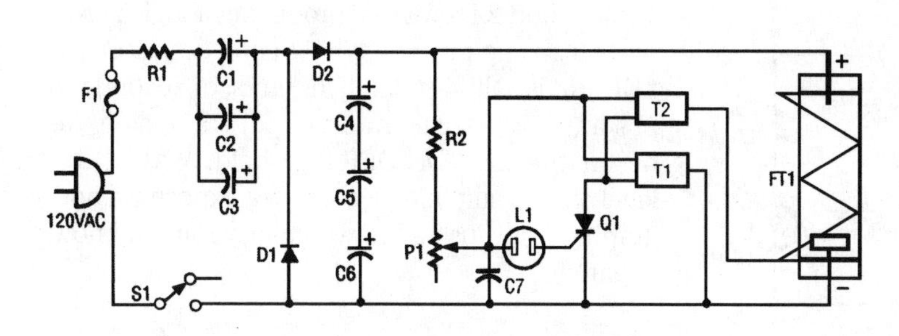

C1,C2,C3 ... 10 μF 160V Electrolytic Capacitor
C4,C5,C6 ... 160 μF 200V Electrolytic Capacitor
C7 .... 0.5 μF 250V Mylar Capacitor
D1, D2 .. 1N4004 Diodes
F1 .... 1 Amp Pigtail Fuse
FT1 .... Giant Xenon Strobe Tube
L1 .... Neon Lamp
P1 .... 10 Meg Potentiometer
Q1 .... 106D1 SCR
R1 .... 20 ohm 10 Watt Power Resistor
R2 .... 270K 1/4 Watt Resistor
S1 .... Slide Switch
T1, T2 .. Trigger Coil

1991 PE HOBBYIST HANDBOOK ***FIG. 75-7***

This strobe light operates from standard 120-Vac power. R1 limits the amount of current applied to the voltage doubler stage, which is comprised of C1, C2, C3, D1, D2, C4, C5, and C6. Capacitors C1, C2, and C3 are connected in parallel and form a capacitance of 30 μF at 160 V. Capacitors C4, C5, and C6 are connected in series and form an equivalent capacitor of about 53 μF at 480 V. Diodes D1 and D2 not only rectify the ac voltage, but also complete the voltage doubler stage, which converts the incoming 120 Vac to the appropriately 300 V that are required by the xenon strobe tube.

The next stage of the circuit is the neon relaxation oscillator and trigger stage. This stage is made up of R2, P1, C7, L1, Q1, T1, and T2. As the storage capacitor (made up of C4, C5, and C6) reaches its full-capacity charge, the voltage divider (made up of R2 and P1) applies voltage to capacitor C7. As C7 charges up, it reaches a threshold voltage level, SCR Q1. When Q1 has a positive pulse on its gate, it fires (causes a short from anode to cathode). That firing action discharges most of the energy stored in C7 into trigger transformers T1 and T2 (which have secondaries connected in series to developer 8 kV). The frequency of the 8-kV pulses is determined by the setting of P1 and the value of C_7. Because C7 is a fixed capacitor, only the setting of P1 adjusts the flash rate in this circuit.

As soon as an 8-kV pulse is applied from the secondary of T2 (trigger wire) to the trigger lead of FT1, it discharges storage capacitors C4, C5, and C6, which causes it to ionize (flash). The cycle then repeats itself until the power is removed from the circuit board by turning "off" S1 or removing the line cord.

ENLARGER EXPOSURE METER

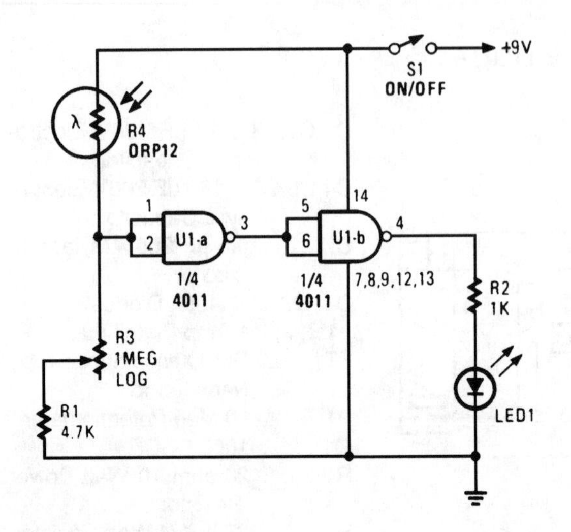

POPULAR ELECTRONICS

FIG. 75-8

Two gates of a 4011 are used as a comparator. When the resistance of R4 decreases the voltage at pin 1 and 2 increases, producing a logic zero at pin 3, causing pin 4 to go high and activating the LED. R3 is calibrated in light units, or seconds exposure time. To calibrate, set pot R3 so as to just be on the LED ON/OFF threshold. With a light level that is suitable to correctly expose a photographic print, use a known enlarger and a known negative.

76

Piezo Circuits

The sources of the following circuits are contained in the Sources section, which begins on page 675. The figure number in the box of each circuit correlates to the entry in the Sources section.

CMOS PIEZO DRIVER

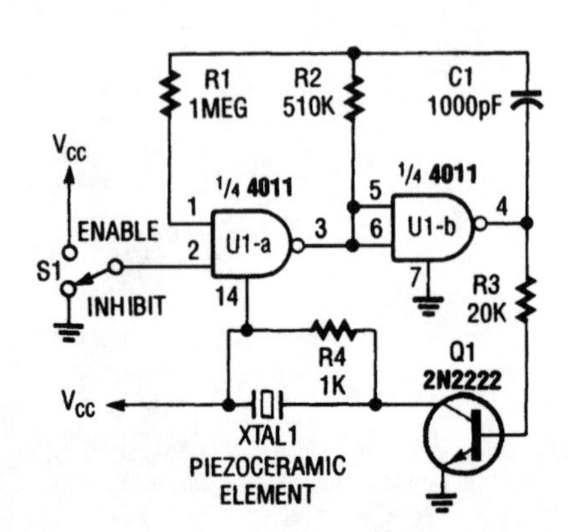

POPULAR ELECTRONICS *FIG. 76-1*

A CMOS-gate and transistor buffer can be used as an effective driver for a piezoelectric transducer.

CMOS PIEZO DRIVER USING 4049

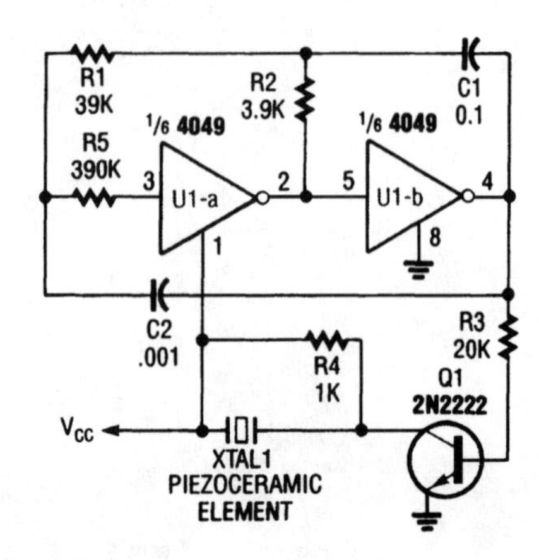

FIG 76-2

This circuit uses a 4049 IC to drive a 2N2222 switching transistor. The transistor drives crystal 1 a piezo transducer.

PIEZO DRIVER

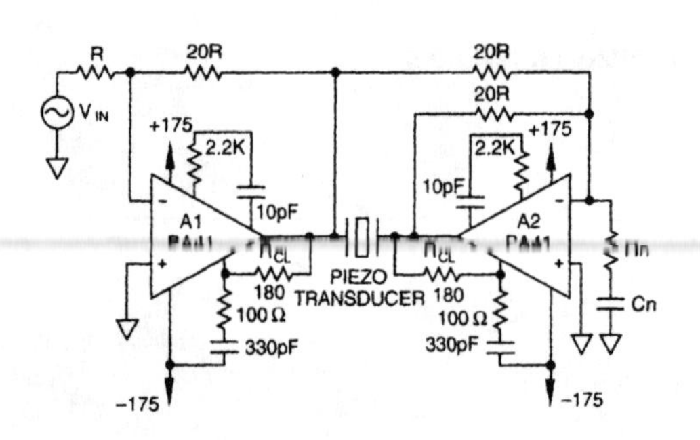

ELECTRONIC DESIGN *FIG. 76-3*

Using a PA41 from Apex Microtechnology, this monolithic amplifier is capable of 350-V operation and delivers 660 V p-p in a bridge circuit.

PIEZO MICROPOSITIONER DRIVER

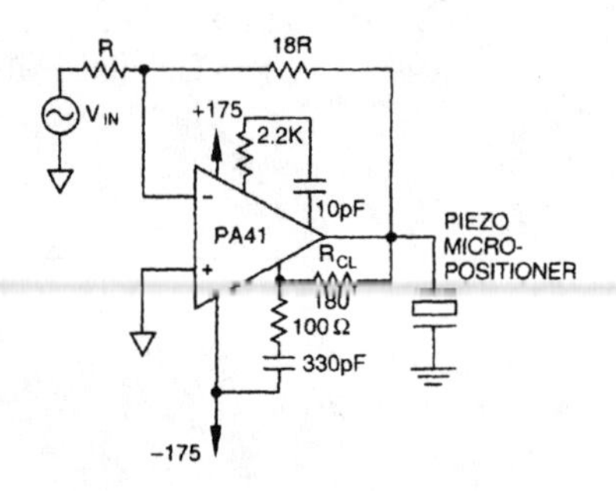

ELECTRONIC DESIGN *FIG. 76-4*

The PA41 from Apex Microtechnology is used here to drive a piezoelectric micropositioner. The drive voltage is less than 20 V p-p at input.

555 OSCILLATOR FOR DRIVING A PIEZO TRANSDUCER

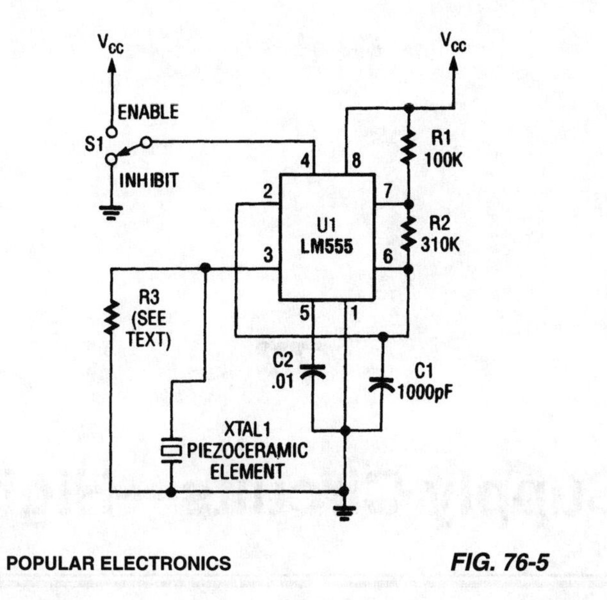

POPULAR ELECTRONICS

FIG. 76-5

A 555-timer oscillator is perhaps one of the most popular circuits for driving a piezoelectric transducer.

77

Power Supply Circuits—High Voltage

The sources of the following circuits are contained in the Sources section, which begins on page 675. The figure number in the box of each circuit correlates to the entry in the Sources section.

HIGH-VOLTAGE dc GENERATOR

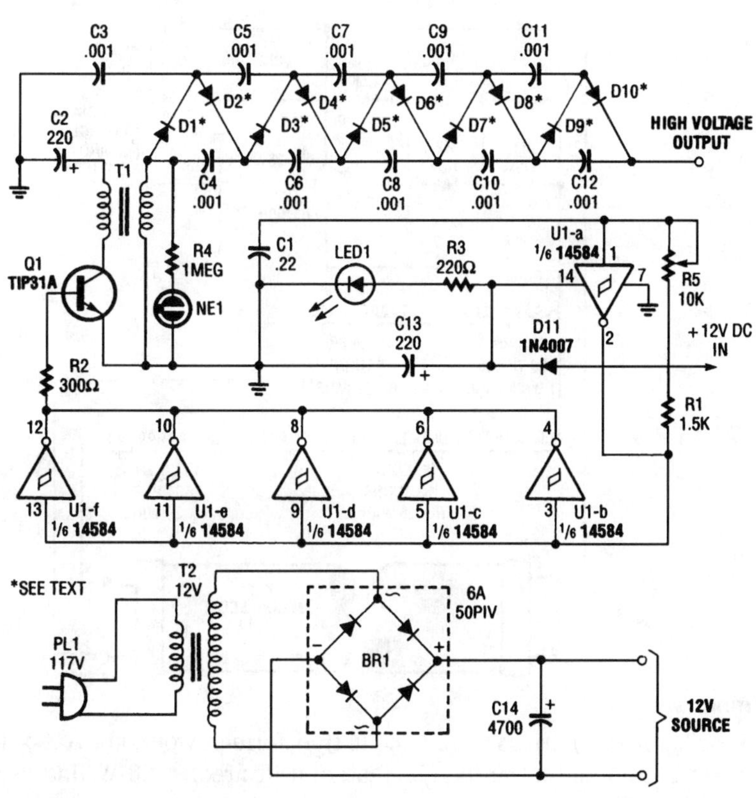

1990 PE HOBBYIST HANDBOOK ***FIG. 77-1***

In the miniature high-voltage dc generator, the input to the circuit, taken from a 12-Vdc power supply, is magnified to provide a 10,000-Vdc output causing a pulsating signal, of opposite polarity, to be induced in T1's secondary winding.

The pulsating dc output at the secondary winding of T1 (ranging from 800 to 1000 V) is applied to a 10-stage voltage-multiplier circuit, which consists of D1 through D10, and C3 through C12. The multiplier circuit increased the voltage 10 times, producing an output of up to 10,000 Vdc. The multiplier accomplishes its task by charging the capacitors (C3 through C12); the output is a series addition of the voltages on all the capacitors in the multiplier.

In order for the circuit to operate efficiently, the frequency of the square wave, and therefore the signal applied to the multiplier, must be considered. The output frequency of the oscillator (U1-a) is set by the combined values of R_1, R_5, and C_1 (which with the values specified is approximately 15 kHz). Potentiometer R5 is used to fine tune the output frequency of the oscillator. The higher the frequency of the oscillator, the lower the capacitive reactance in the multiplier.

Light-emitting diode LED1 serves as an input-power indicator, and neon lamp NE1 indicates an output at the secondary of T1. A good way to get the maximum output at the multiplier is to connect an oscilloscope to the high-voltage output of the multiplier, via a high-voltage probe, and adjust potentiometer R5 for the maximum voltage output.

FLUORESCENT TUBE POWER SUPPLY

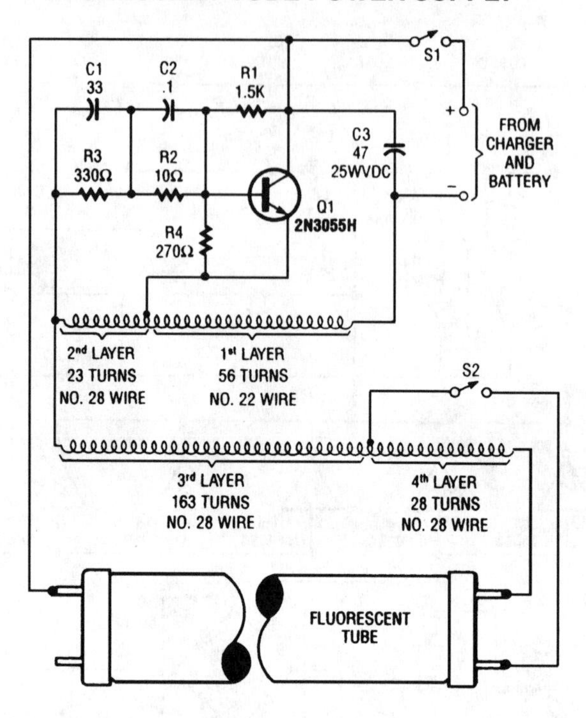

POPULAR ELECTRONICS ***FIG. 77-2***

A 2N3055 oscillator (Q1) drives a homemade transformer, wound on a 5⁄16 × 1⅞" ferrite rod. S2 is used as a filament switch and it can be eliminated, if desired. A 20-W fluorescent tube is recommended. The supply is 12 V.

PHOTOMULTIPLIER SUPPLY

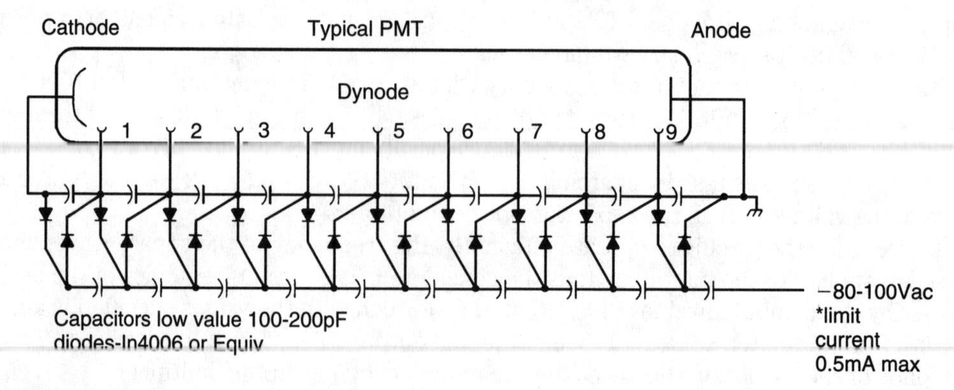

73 AMATEUR RADIO TODAY ***FIG. 77-3***

A Cockcroft-Walton voltage multiplier supplies the stepped voltage required for the dynodes of the PMT without the power-wasting voltage-divider resistor string that is traditionally used.

NEGATIVE VOLTAGE SUPPLY

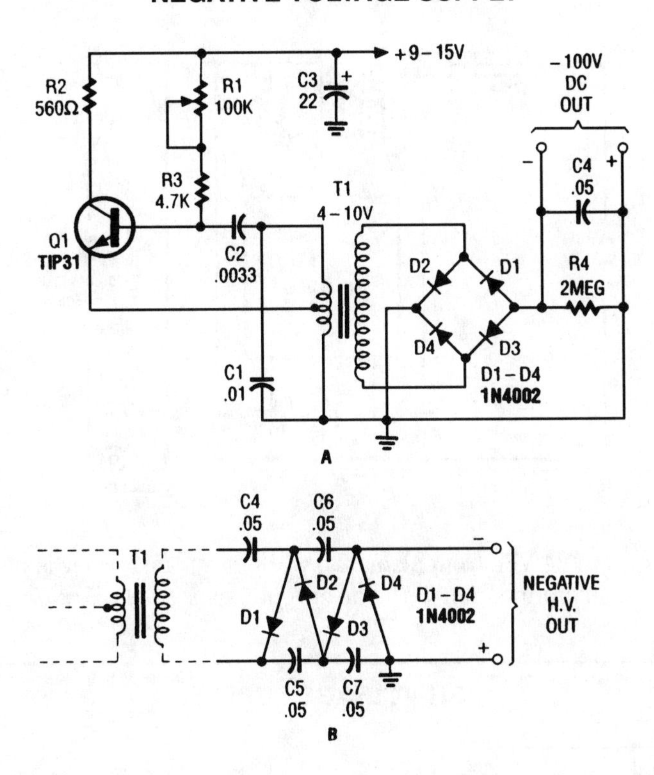

POPULAR ELECTRONICS ***FIG. 77-4***

The combination Hartley oscillator/step-up transformer shown in A can generate significant negative high voltage, especially if the voltage output of the transformer is multiplied by the circuit.

PHOTOMULTIPLIER CIRCUIT

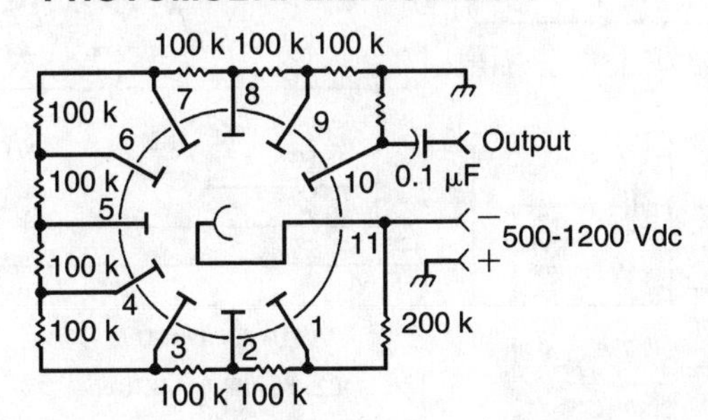

73 AMATEUR RADIO TODAY ***FIG. 77-5***

This circuit is typical of the way that a photomultiplier tube is used. The circuit shown is ac coupled, but if dc coupling is needed, the capacitor can be omitted and a suitable interfacing method used. A typical tube is the widely available 931/931A.

SINGLE-CHIP dc SUPPLY FOR 120–TO 240-Vac OPERATION

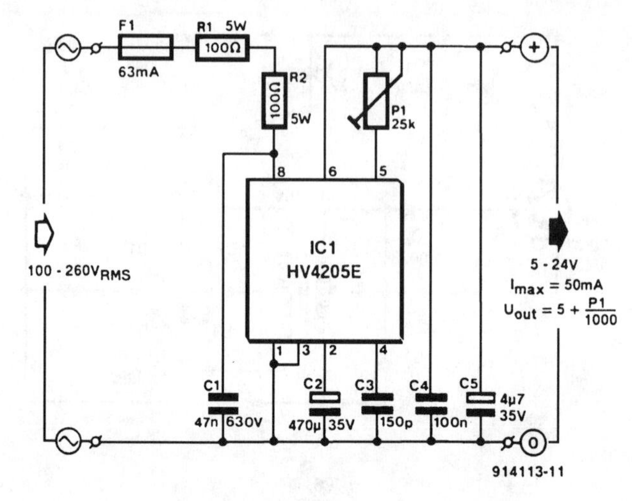

ELEKTOR ELECTRONICS *FIG. 77-6*

Direct derivation of 5 to 24 Vdc from ac mains, without a transformer is possible with this circuit. Note that a direct mains connection to the dc output exists. *Suitable safety precautions must be taken.*

HIGH-VOLTAGE SUPPLY

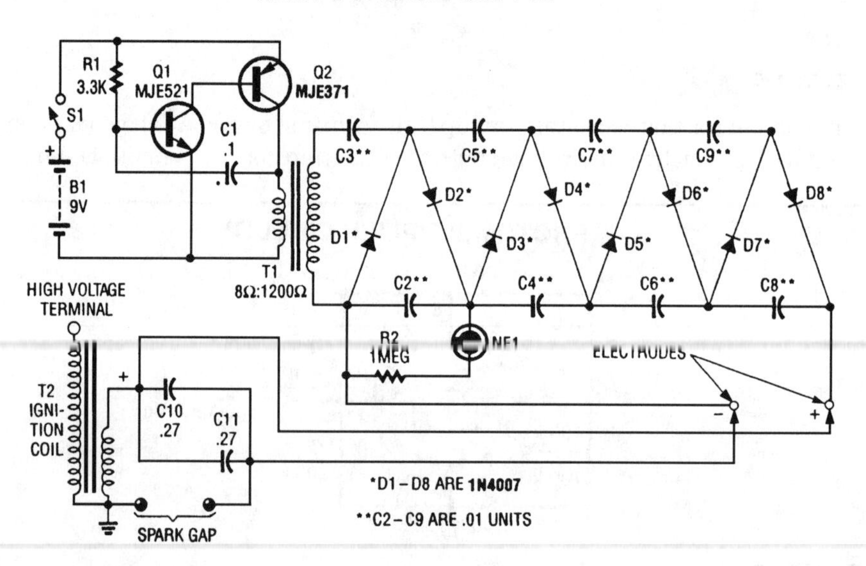

POPULAR ELECTRONICS *FIG. 77-7*

This circuit uses a transistor oscillator and a voltage multiplier to charge C10 and C11 to a high voltage. When the spark gap breaks down, T2 produces a high-voltage pulse via the capacitance discharge of C10 and C11 into its primary. T2 is an auto ignition coil.

COLD-CATHODE FLUORESCENT-LAMP POWER SUPPLY

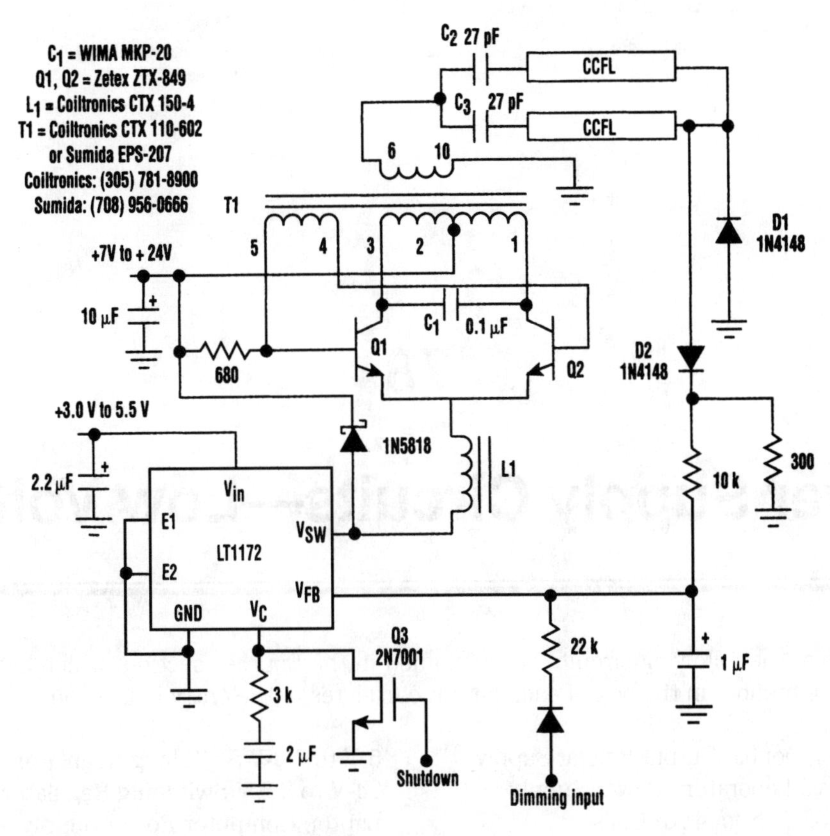

ELECTRONIC DESIGN *FIG. 77-8*

This circuit is a 92%-efficient power supply for cold-cathode fluorescent lamps (CCFLs), which are used to backlight LCD in portable equipment. The efficiency depends heavily on the component types, particularly C1, Q1, Q2, L1, and T1, whose manufacturers are noted.

78

Power Supply Circuits—Low Voltage

The sources of the following circuits are contained in the Sources section, which begins on page 675. The figure number in the box of each circuit correlates to the entry in the Sources section.

TRACKING DOUBLE-OUTPUT BIPOLAR SUPPLY

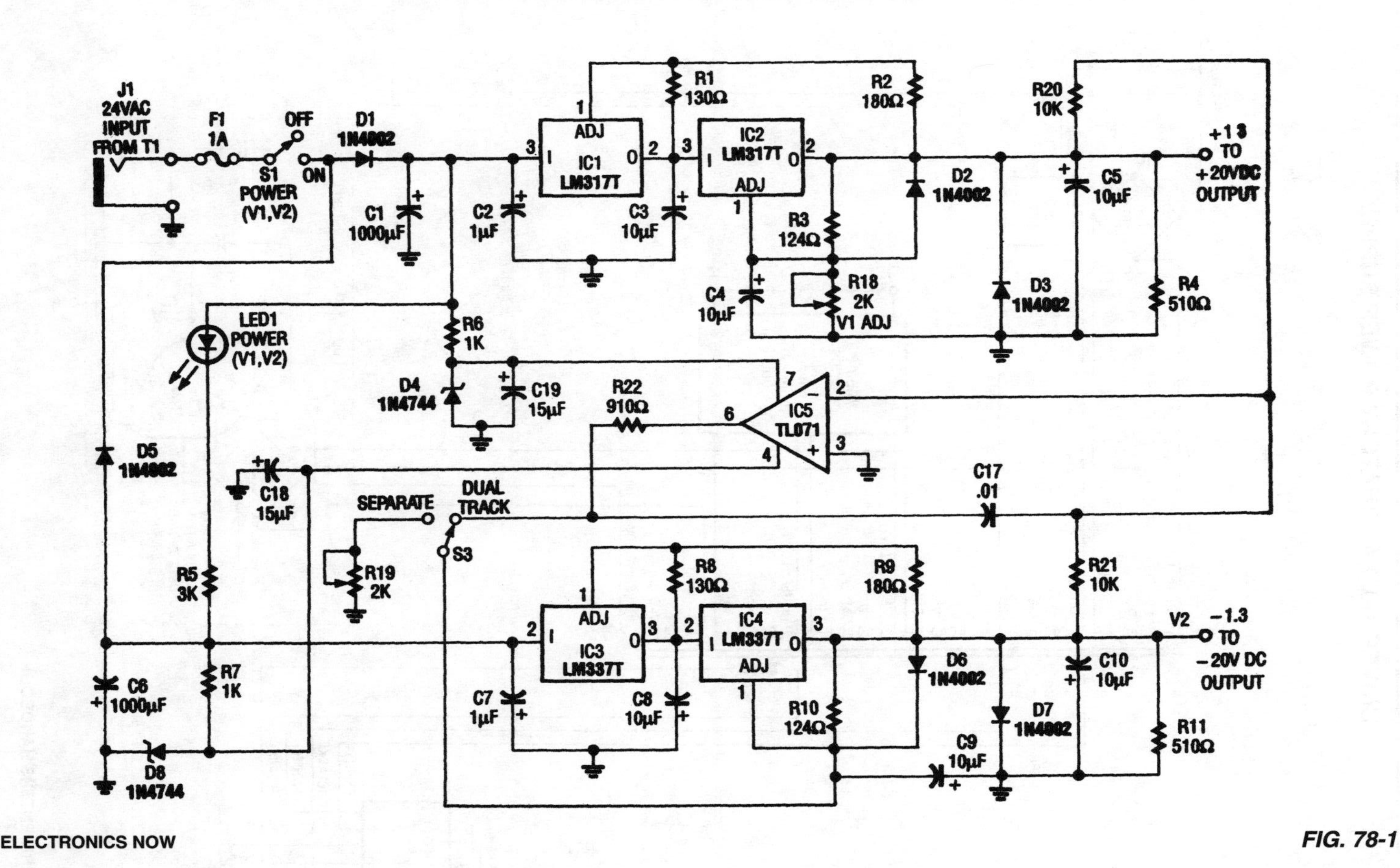

ELECTRONICS NOW

FIG. 78-1

This circuit is useful for a bench supply in the lab. Separate or tracking operation is possible. The regulators should be properly heatsinked. T1 is a 24-Vac wall transformer of suitable current capacity.

UNIVERSAL LABORATORY POWER SUPPLY

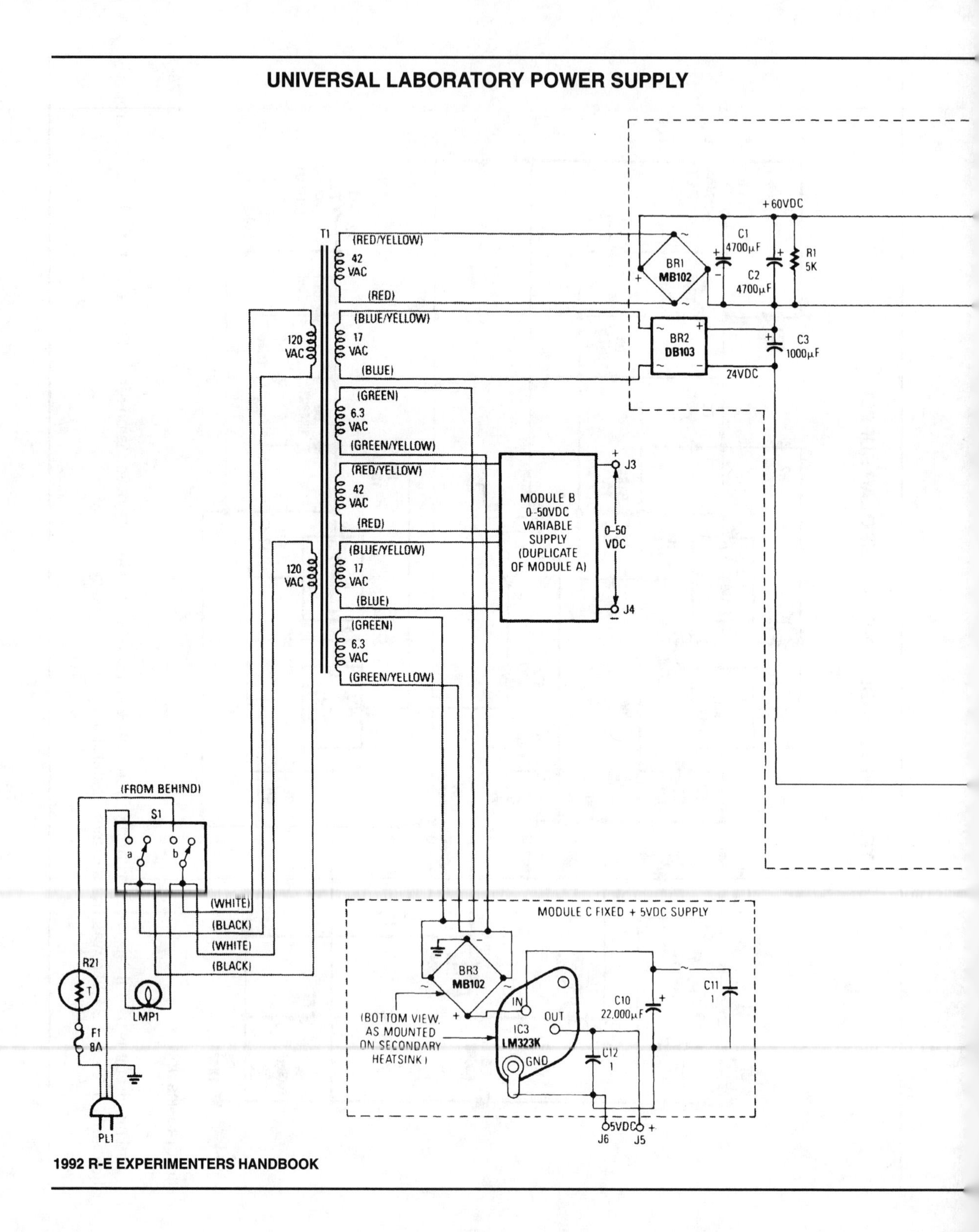

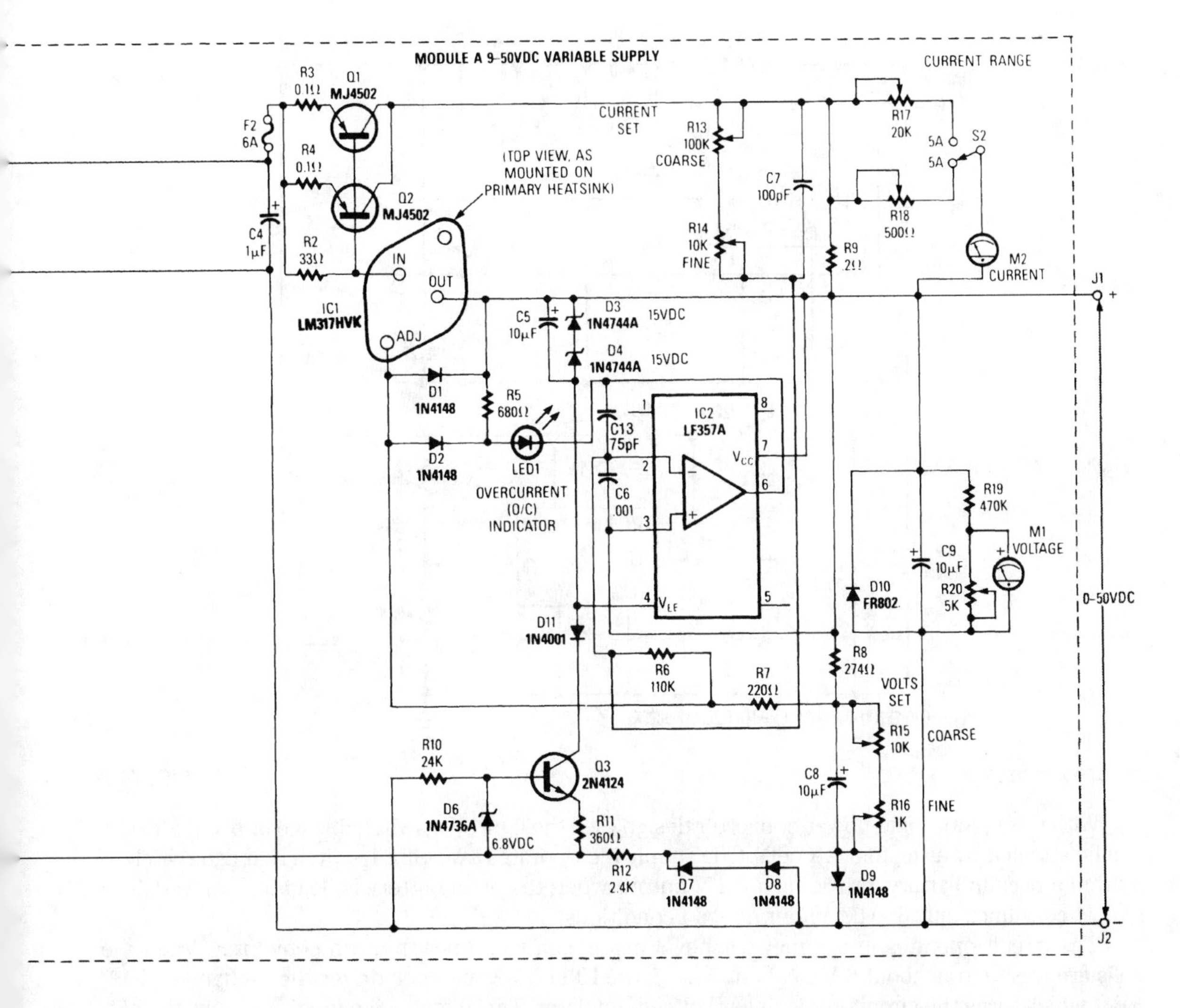

The value of the design lies in the use of IC1, an LM317HVK adjustable series-pass voltage regulator, for broad-range performance remainder supplies voltage-setting and current-limiting functions. The input to IC1 comes from the output of BR1, which is filtered by C1 and C2 to about +60 Vdc, and the input for current-sense comparator IC2 comes from BR2, which also acts as a negative bias supply for regulation down to ground. The output voltage is determined by:

$$(V_{OUT} - 1.25 + 1.3)/(R_{15} + R_{16}) = 1.25/R_8.$$

Thus, the maximum value from each variable supply board is:

$$V_{OUT} = (1.25/R_8) \times (R_{15} + R_{16}) = 50.18 \text{ Vdc}.$$

FIG. 78-2

+5 V/+3.6 V FROM 4 AA CELLS

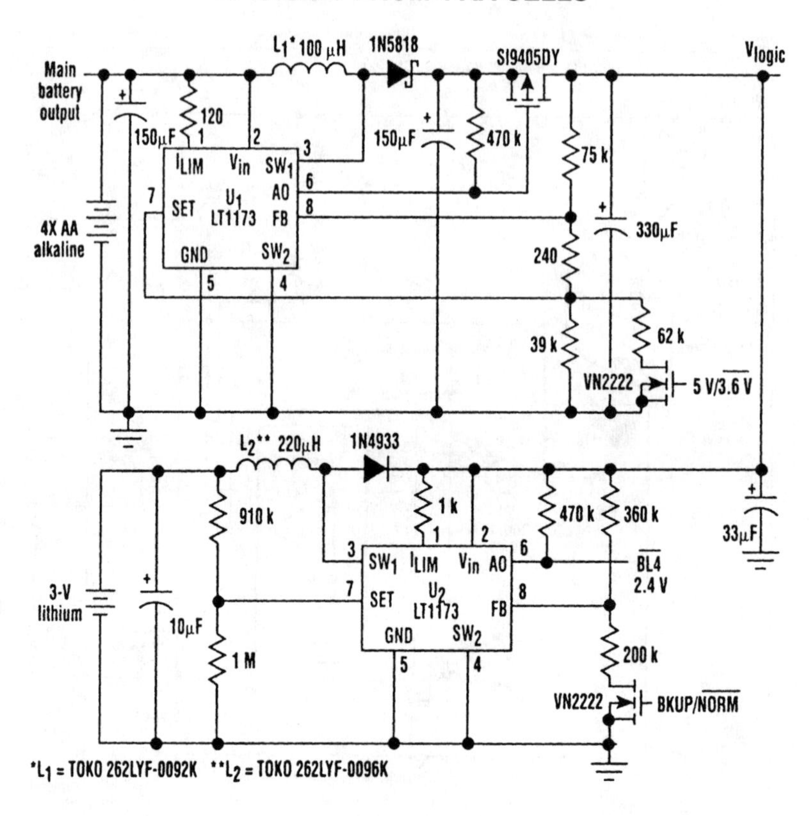

ELECTRONIC DESIGN *FIG. 78-3*

With this unique logic-power-converter design (see the figure), a switchable 3.6 or 5 V at 200 mA can be attained by using four AA cells. The supply incorporates a MOSFET switch that can switch to a lithium backup battery, providing a 3.4-V output when the main battery is dead or removed. The supply consumes only 380 μA under no-load conditions.

The circuit operates in a somewhat novel mode as a step-up/step-down converter. When the cells are fresh (from about 6 V to about 5.2 V), the LT1173's gain block drives the p-channel MOSFET, which turns the circuit into a linear voltage regulator. This might seem inefficient, but the batteries are quick to drop from 6 V to 5 V. With a 5-V input, the efficiency (for the 3.6-V output) is 3.6/5 or 72%, which is reasonable. As the battery-pack drops in voltage, efficiency increases, reaching greater than 90% with a 4.2-V input.

At a point below a 4-V input, the circuit switches to step-up mode. This mode squeezes the batteries for all of their available energy. In this case, efficiency runs between 83% at approximately a 4-V input to 73% at a 2.5-V input.

The supply can deliver 200 mA over its entire operational range. In its linear mode of operation, the supply has no current spikes that, because of the fairly high internal resistance of the alkaline cells, can reduce battery life. The topology delivers over 9.3 hours of 3.6-V 200-mA output power, compared to just 7 hours using the traditional flyback topology that is used in other designs.

INDUCTORLESS SWITCHING REGULATOR

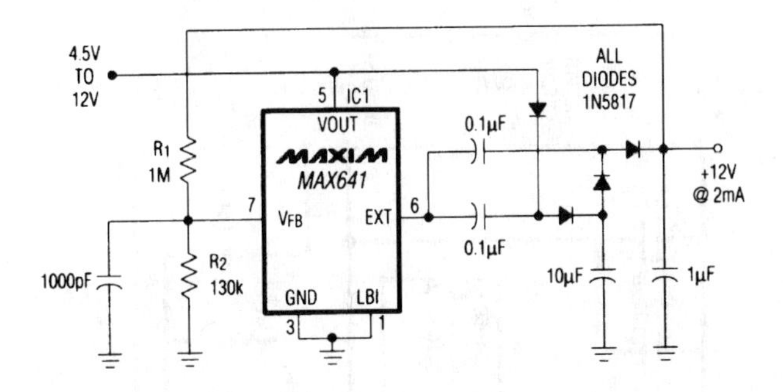

Substituting the diode-capacitor network shown for an inductor allows this switching-regulator IC to deliver 2mA at comparable line and load regulation, with somewhat reduced efficiency.

A

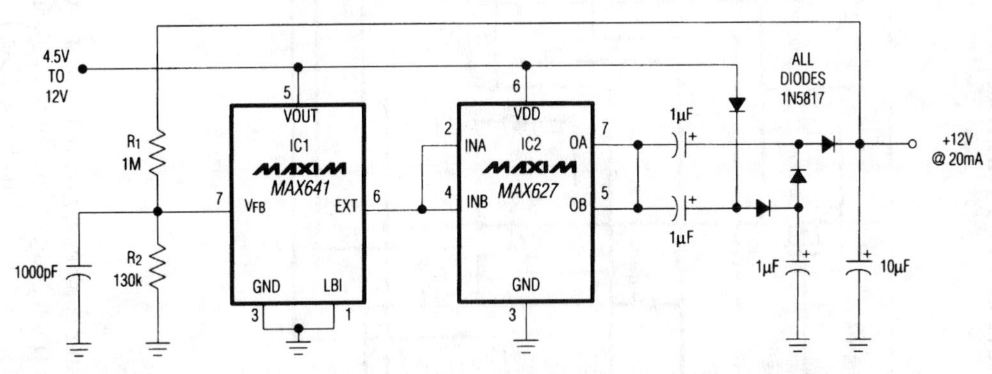

Introducing an MOS driver (IC_2) enables the Figure 1 circuit to deliver as much as 20mA.

B

MAXIM ENGINEERING JOURNAL

FIG. 78-4

In conventional applications, switching-regulator ICs regulate V_{OUT} by controlling the current through an external inductor. The IC in A, however, driving a diode-capacitor network in place of the inductor, offers comparable performance for small loads. The network can double, triple, or quadruple the input voltage.

Feedback from the R1/R2 voltage divider enables IC1 to set the regulated-output level. (As shown, the circuit derives 12 V from a 5- to 12-V input and provides as much as 2 mA of output current.) Adding a noninverting MOS driver (B) boosts the available output current to 20 mA. Substituting the diode-capacitor network shown for an inductor allows this switching-regulator IC to deliver 2 mA at comparable line and load regulation, with somewhat reduced efficiency.

SINGLE LTC POWER SUPPLY

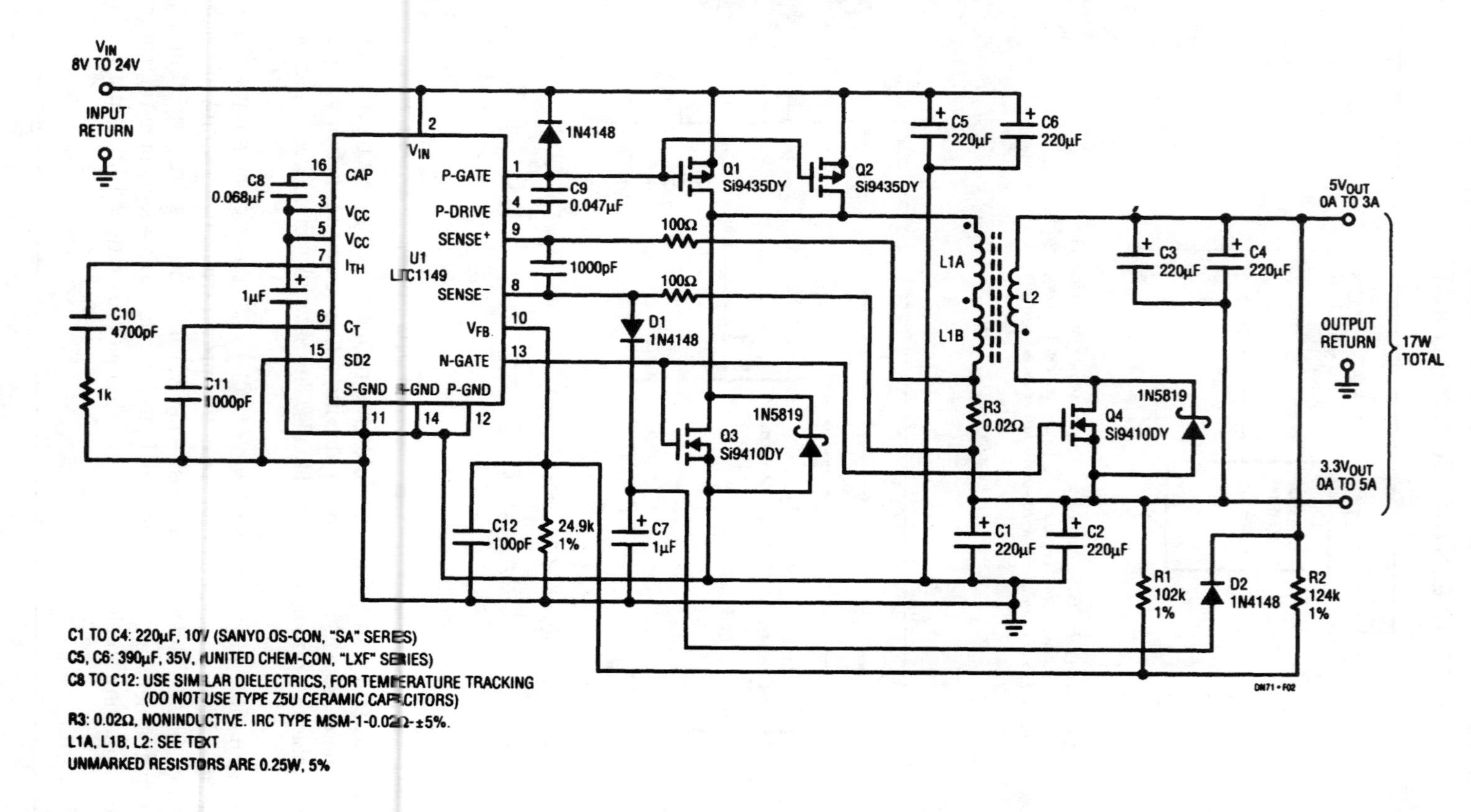

LINEAR TECHNOLOGY

FIG. 78-5

One LTC 1149 synchronous switching regulator can deliver both 3.3- and 5-V outputs. The design's simplicity, low cost, and high efficiency make it a strong contender for portable, battery-powered applications. The circuit described accepts input voltages from 8 to 24 V, to power any combination of 3.3-V and 5-V loads totalling 17 W or less. For input voltages in the 8-V to 16-V range, the LTC1148 may be used, reducing both quiescent current and cost.

CONFIGURABLE POWER SUPPLY

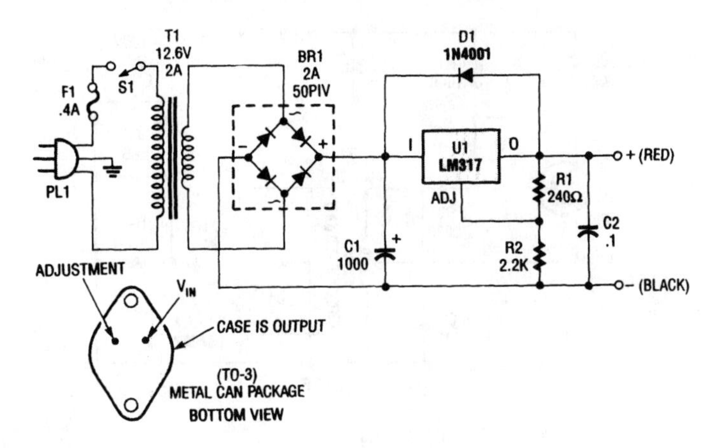

POPULAR ELECTRONICS

FIG. 78-6

The adjustable supply can easily be reconfigured by altering the value of V_2 and beefing up some other components, as is necessary.

The output voltage is given by $V_{OUT} = 1.25\ (1 + R_2/R_1)$. R_2 can be changed, as is necessary.

COMBINATION VOLTAGE AND CURRENT REGULATOR

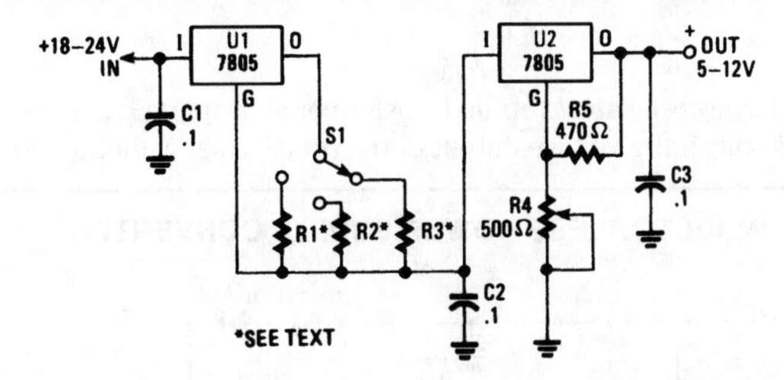

POPULAR ELECTRONICS

FIG. 78-7

This voltage-regulator/current-limiter combination can be made from two 7805 regulators as shown. R1, R2, and R3 should be selected for a 5-V drop at the maximum allowable current limit. S1 selects one of the three current values. Do not forget that U1 requires 5 mA to operate and this means that the minimum current limit setting should be 10 mA or more ($R_1 = 1.25$ kΩ). Resistor values are as follows:

$$R_x\ (\text{k}\Omega) = \frac{5 \text{ volts}}{(\textit{current limit} \text{ mA} - 5 \text{ mA})}$$

For 100 mA,

$$R_x = \frac{5}{100\text{–}5} = \frac{5}{95} \text{ k}\Omega \text{ or } 52.5\ \Omega$$

HV POWER SUPPLY WITH 9-TO 15-Vdc INPUT

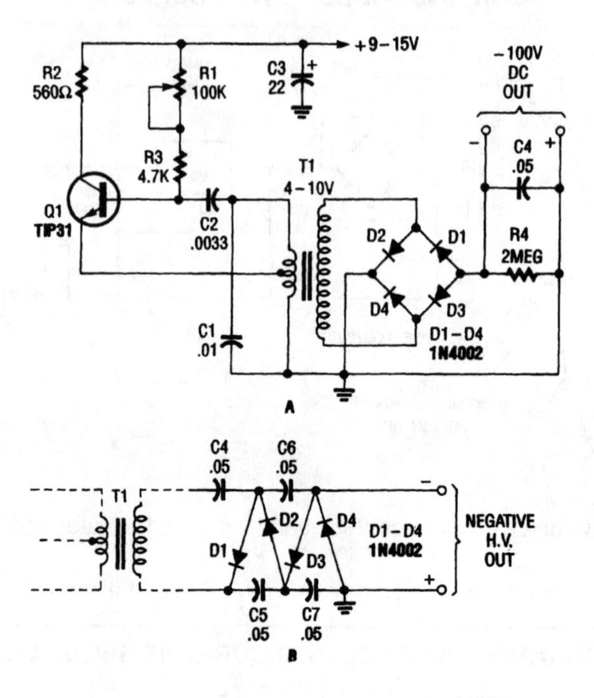

POPULAR ELECTRONICS

FIG. 78-8

The combination Hartley oscillator/step-up transformer shown in A can generate significant negative high voltage, especially if the voltage output of the transformer is multiplied by the circuit in B.

INDUCTORLESS POWER SUPPLY CONVERTER

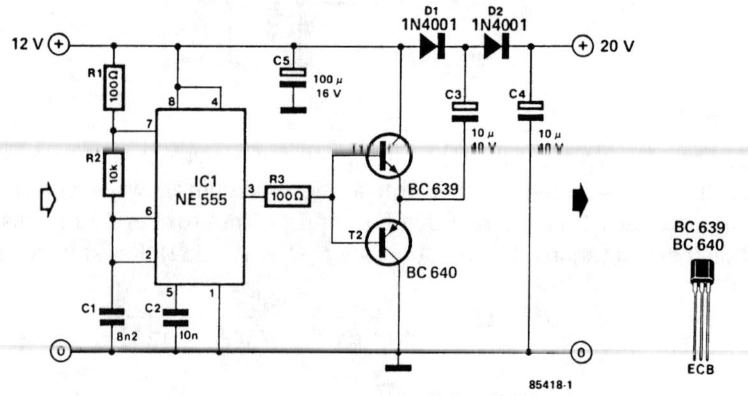

303 CIRCUITS

FIG. 78-9

Using a 555 timer and voltage doubler, this circuit will supply ≥50mA at 20 Vdc. T1 and T2 act as power amplifiers to drive the voltage doubler. Frequency of operation is approximately 8.5 kHz.

SIMPLE NEGATIVE SUPPLY FOR LOW-CURRENT APPLICATIONS

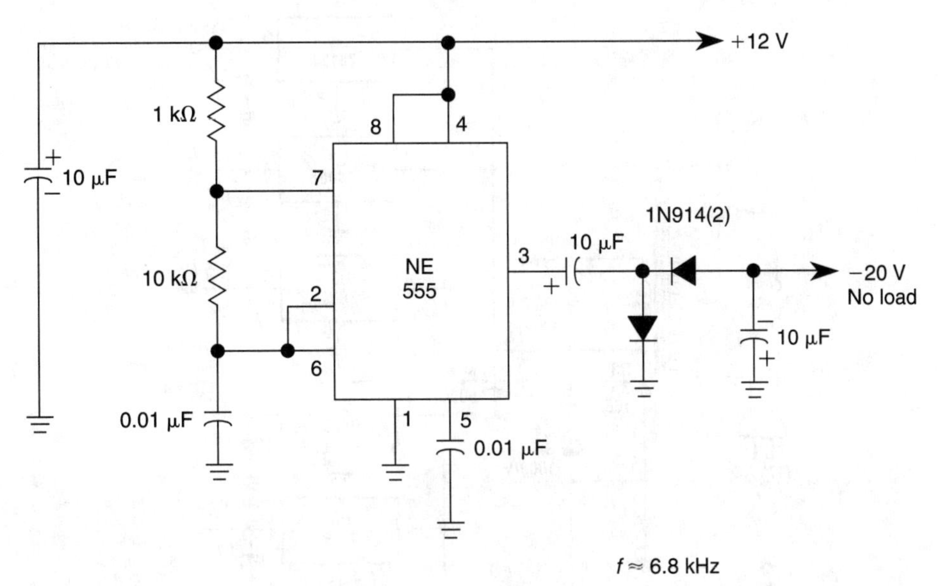

WILLIAM SHEETS

FIG. 78-10

This dc negative-voltage generator based on the 555 produces a negative output voltage equal to approximately 2x the dc supply voltage.

INVERTING POWER SUPPLY

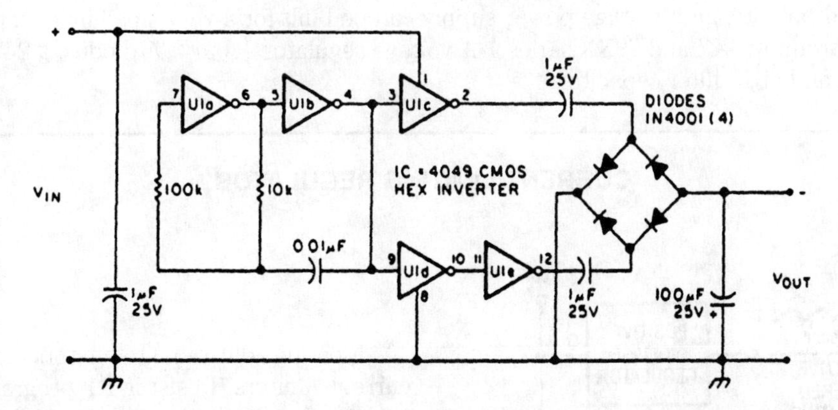

73 AMATEUR RADIO TODAY

FIG. 78-11

This circuit will provide a negative dc voltage that is approximately equal to the positive input voltage at no load and about 3 V less at 10 mA load. V_{IN} is from +5 to +15 Vdc. Do not exceed 15 V or U1 might be damaged.

MULTIVOLTAGE POWER SUPPLY

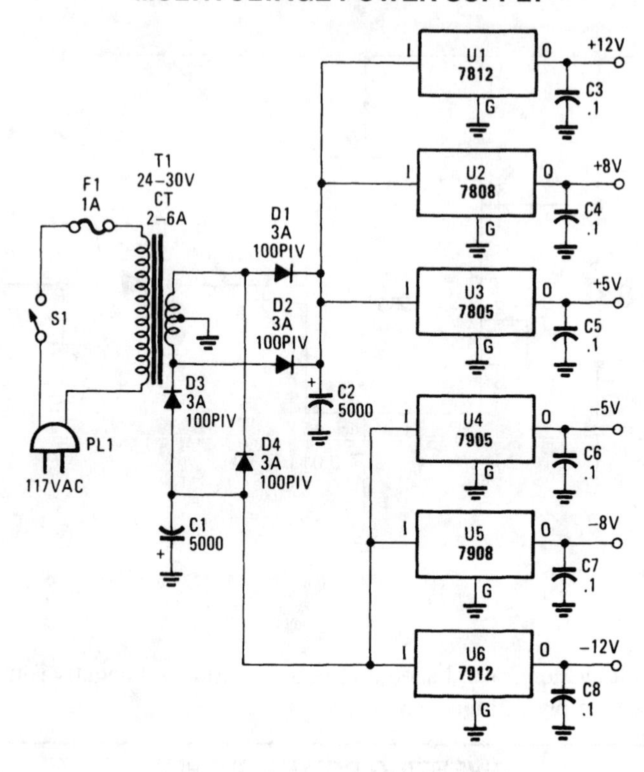

POPULAR ELECTRONICS

FIG. 78-12

This dual-polarity, multivoltage power supply can be built for a very small investment. The circuit is built around 78XX and 79XX series 1-A voltage regulators, four 3-A diodes, a 24–30-V 2–6-A transformer, and eight filter capacitors.

CURRENT-LIMITING REGULATOR

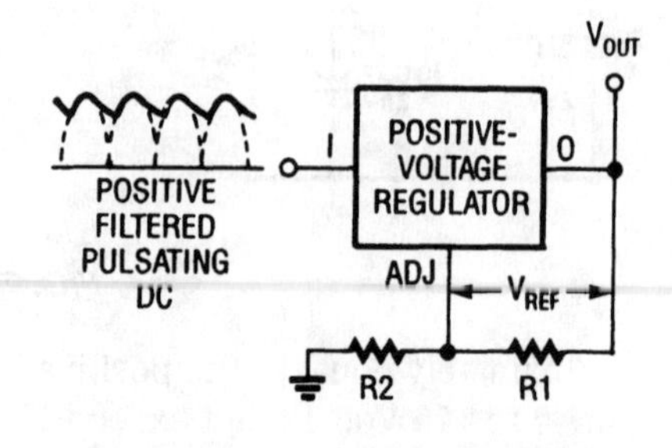

Floating adjustable regulators can be used as current limiters. Resistor R1 programs the current flowing through R2.

1993 ELECTRONICS HOBBYISTS HANDBOOK

FIG. 78-13

NEON LAMP DRIVER FOR 5- TO 15-V SUPPLIES

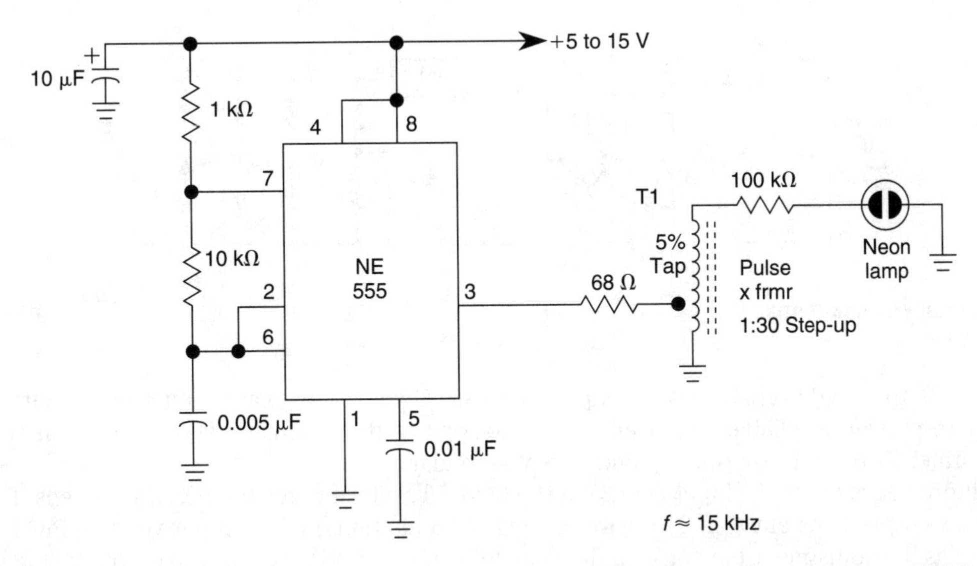

WILLIAM SHEETS

FIG. 78-14

This neon-lamp driver based on the 555 T1 can be wound on an old TV flyback transformer core.

13.8-Vdc 2-A REGULATED POWER SUPPLY

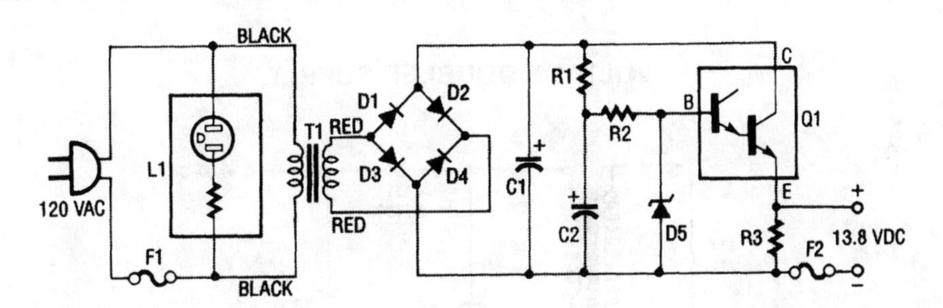

1991 PE HOBBYIST HANDBOOK

FIG. 78-15

This regulated power supply consists of step-down transformer T1, a full-wave rectifier bridge (D1 through D4), and a filtering regulator circuit made up of C1, C2, R1, R2, R3, D5, and Q1. When 120 Vac is provided, the neon-lamp assembly L1 lights up, and transformer T1 changes 120 Vac to about 28 Vac. The rectifier bridge, D1 through D4, rectifies the ac into pulsating dc, which is then filtered by C1. Capacitor C1 acts as a storage capacitor. Zener diode D5 keeps the voltage constant across the base of Darlington regulator Q1, causing constant voltage across resistor R3 and the (+) and (–) output terminals, where the load is connected. Fuse F2 is used to open ("blow"), if the current through the output terminals is too high. Make sure to take proper precautions when using projects powered by 120 Vac.

0- TO 12-V, 1-A VARIABLE POWER SUPPLY

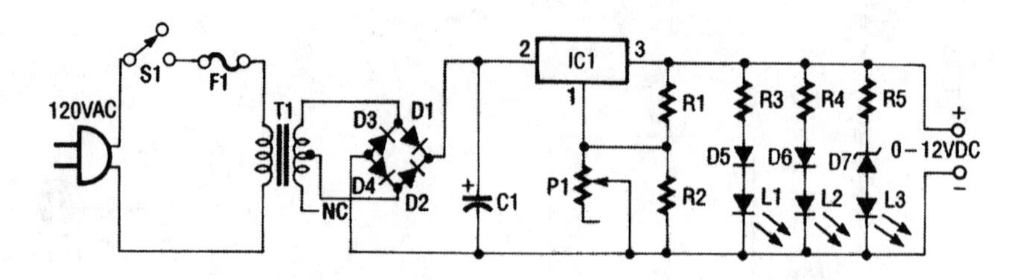

1991 PE HOBBYIST HANDBOOK

FIG. 78-16

This 0- to 12-Vdc variable power supply uses an IC voltage regulator and a heavy-duty transformer to provide a reliable dc power supply. Looking at the schematic shown, you can see that transformer T1 has a 120-V primary and a 28-V secondary.

Filtered dc is fed to the input (pin 2) of the LM317T voltage regulator, IC, which keeps the voltage at its output constant (pin 3) regardless (within limitations) of the input voltage. Pin 1 of the LM317T is the adjustment pin. Varying the voltage on pin 1 (via P1) varies the output voltage.

Diodes D5 through D7 and LEDs L1 through L3 give an approximate indication of the output voltage. Each LED/diode path has a limiting resistor to limit the current to a level that is safe for the LED.

VOLTAGE DOUBLER SUPPLY

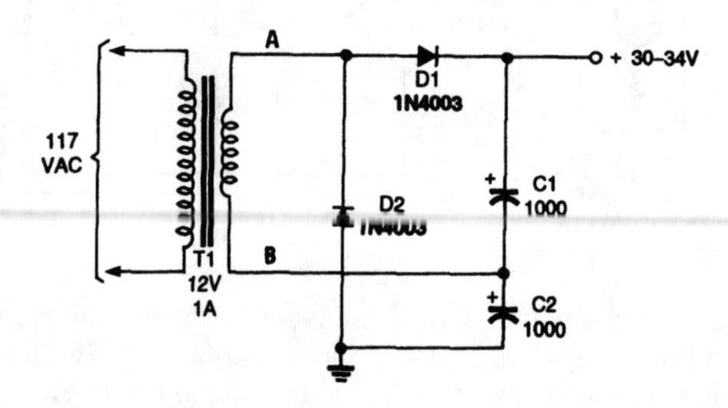

POPULAR ELECTRONICS

FIG. 78-17

The voltage doubler is built around a pair of diodes (D1 and D2) and a pair of capacitors (C1 and C2) that are fed from, in this case, a 12-V, 1-A step-down transformer (T1).

ADJUSTABLE 20-V SUPPLY

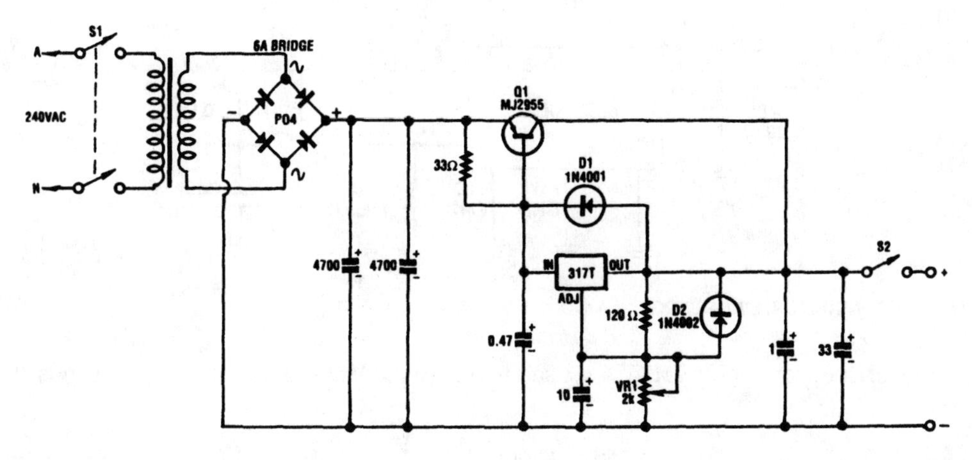

SILICON CHIP

FIG. 78-18

This circuit can deliver 3 A or more and a maximum dc voltage of a little over 20 V. It is designed around the readily available LM317T adjustable 3-terminal regulator and has a pnp power transistor to boost the current output.

The transformer has an 18-V secondary rated at 6 A; this feeds to bridge rectifier and two 4700-μF capacitors to yield around 25 Vdc. This voltage is fed to the emitter of the MJ2955 transistor and to the input of the LM317 via a 33-Ω resistor.

SWITCHING REGULATOR CONVERTER

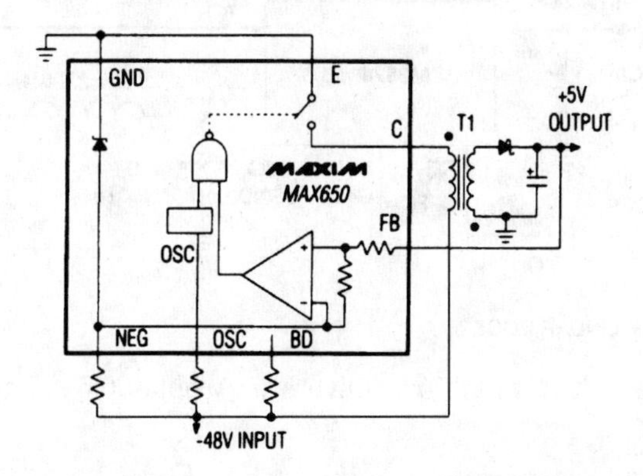

MAXIM ENGINEERING JOURNAL

FIG. 78-19

The Max650 switching regulator produces a regulated 5 V from large negative voltages, such as the –48 V found on telephone lines. The resulting power supply operates with several external components, including a transformer, and it delivers 250 mA. The device includes a 140-V 250-mA pnp transistor, short-circuit protection, and all necessary control circuitry.

5-V TO 3.3-V SWITCHING REGULATOR

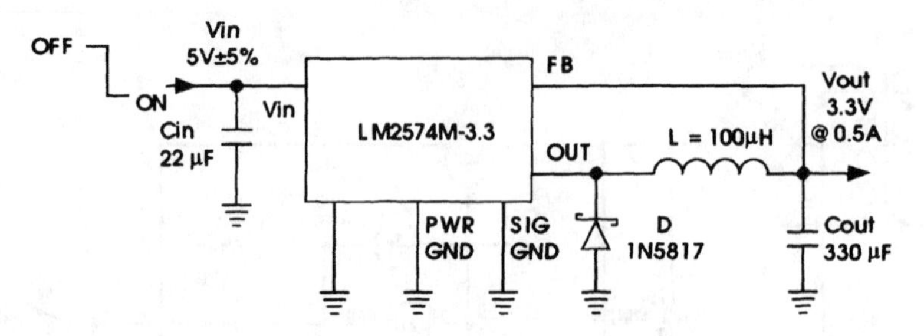

NATIONAL SEMICONDUCTOR, LINEAR EDGE

FIG. 78-20

A National Semiconductor LM2574 is used to derive 3.3 V at 0.5 A from a 5-V logic bus. The duty cycle is:

$$\frac{V_{\mathrm{OUT}} + V_D - V_{IND}}{V_{\mathrm{IN}} - V_{SAT} + V_D - 2\,V_{IND}}$$

V_D = diode drop (0.39)
V_{IND} = inductor dc drop
V_{SAT} = saturation voltage of LM2574 (0.9 V typical)

This circuit should be useful to derive 3.3 V for logic devices from existing +5-V buses.

24-V TO 3.3-V SWITCHING REGULATOR

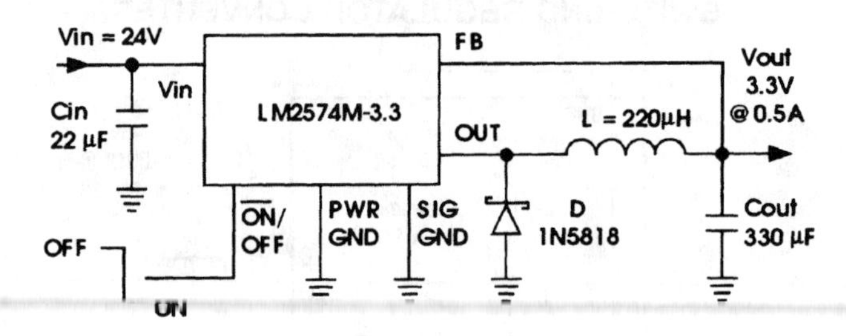

NATIONAL SEMICONDUCTOR, LINEAR EDGE

FIG. 78-21

The National Semiconductor LM2574 delivers 3.3 V out at 0.5 A from a 24-V source. The duty cycle is:

$$\frac{V_{\mathrm{OUT}} + V_D - V_{IND}}{V_{\mathrm{IN}} - V_{SAT} + V_D - 2\,V_{IND}}$$

V_D = diode drop (0.39)
V_{IND} = inductor dc drop
V_{SAT} = saturation voltage of LM2574 (0.9 V typical)

LAPTOP COMPUTER POWER SUPPLY

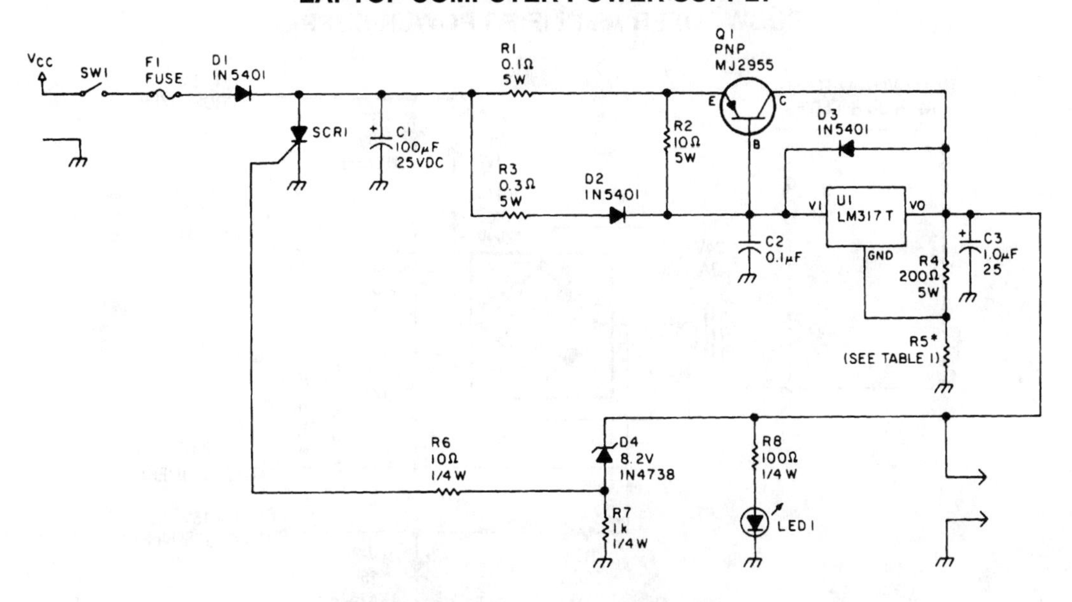

R5 Resistor Value	Voltage Out
750Ω	5V
910Ω	6V
1.2K	8V
1.5K	9V
1.8K	10V
2.0K	12V
2.7K	15V
3.3K	18V
3.6K	20V
4.3K	24V

Note: Any output voltage value greater than 10V requires a higher input voltage than 13.6V. In addition capacitor working voltage ratings will have to be increased accordingly. Allow a minimum of 2.5 times the voltage expected to appear across the capacitor as a standard for the working voltage.

Table 1. Resistor value/voltage matchup.

73 AMATEUR RADIO TODAY ***FIG. 78-22***

A laptop computer supply that has 9-V output, crowbar overvoltage protection, and operates from a 12-V supply is shown above. The supply voltage should be at least 3.6 V above the expected output voltage. Q1 should be heatsinked appropriately. R5 should have a value of 1.5 kΩ for 9-V output. Table 1 gives values for other voltages.

SUBWOOFER AMPLIFIER POWER SUPPLY

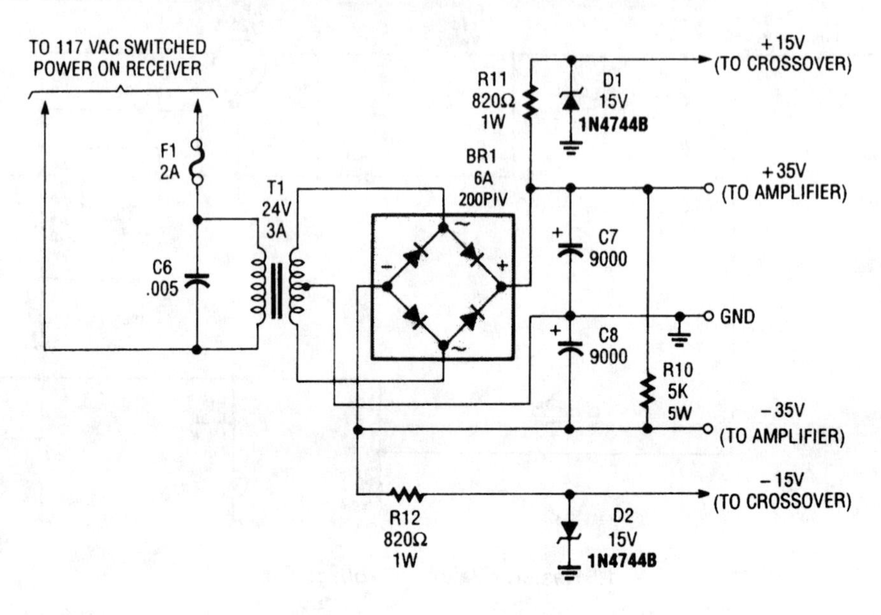

POPULAR ELECTRONICS *FIG. 78-23*

Although intended to power a 100-W low-frequency amplifier, this power supply should handle many mono or stereo amplifiers in the medium power range that require ±30 to 35 V.

DUAL VOLTAGE-RECTIFIER CIRCUIT

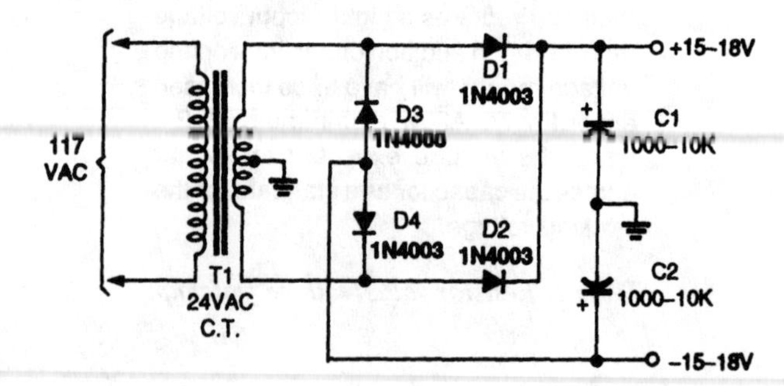

POPULAR ELECTRONICS *FIG. 78-24*

This stepped-up dual voltage supply provides ±15 to ±18 V unregulated.

DUAL AUDIO AMPLIFIER POWER SUPPLY

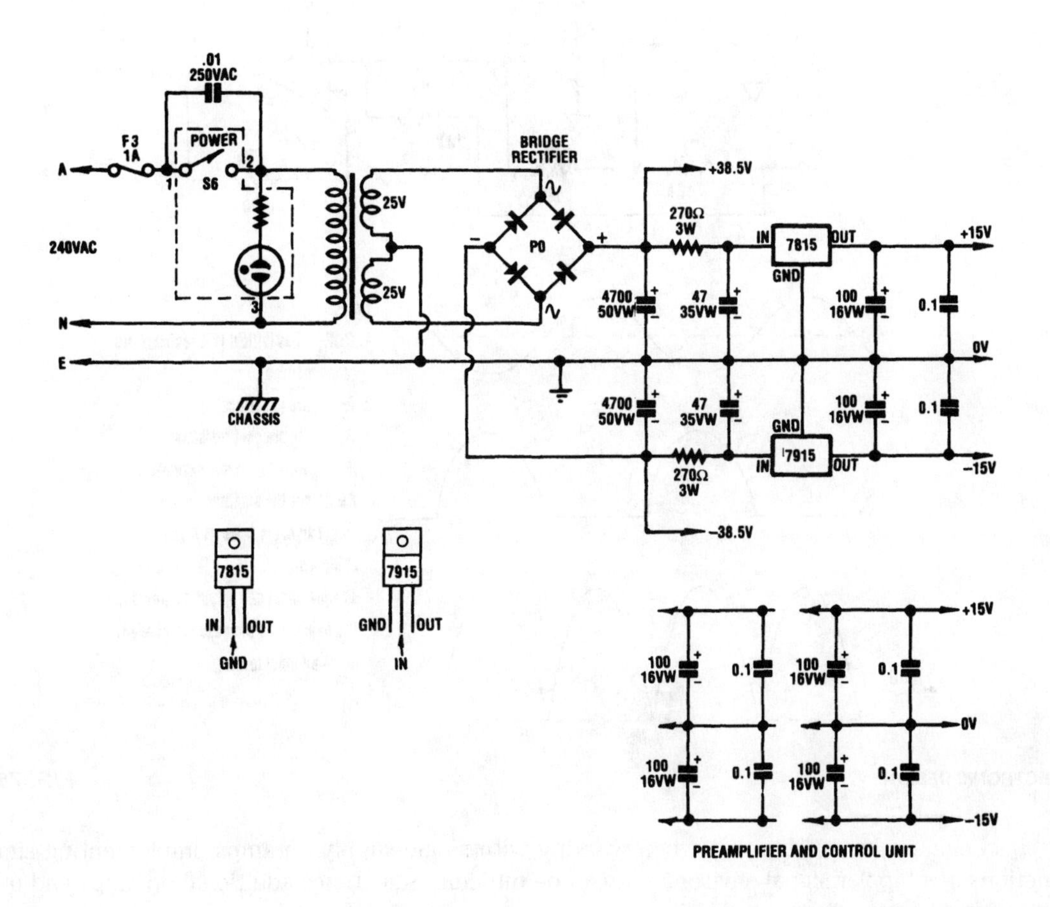

SILICON CHIP

FIG. 78-25

A dual audio amplifier that will deliver 50 W per channel is shown in the schematic. It includes preamp and tone controls, and also includes a headphone amplifier. The circuit depicts the power supply that supplies ±38.5 V and ±15 V regulated for the dual 50 watter.

DIODELESS RECTIFIER

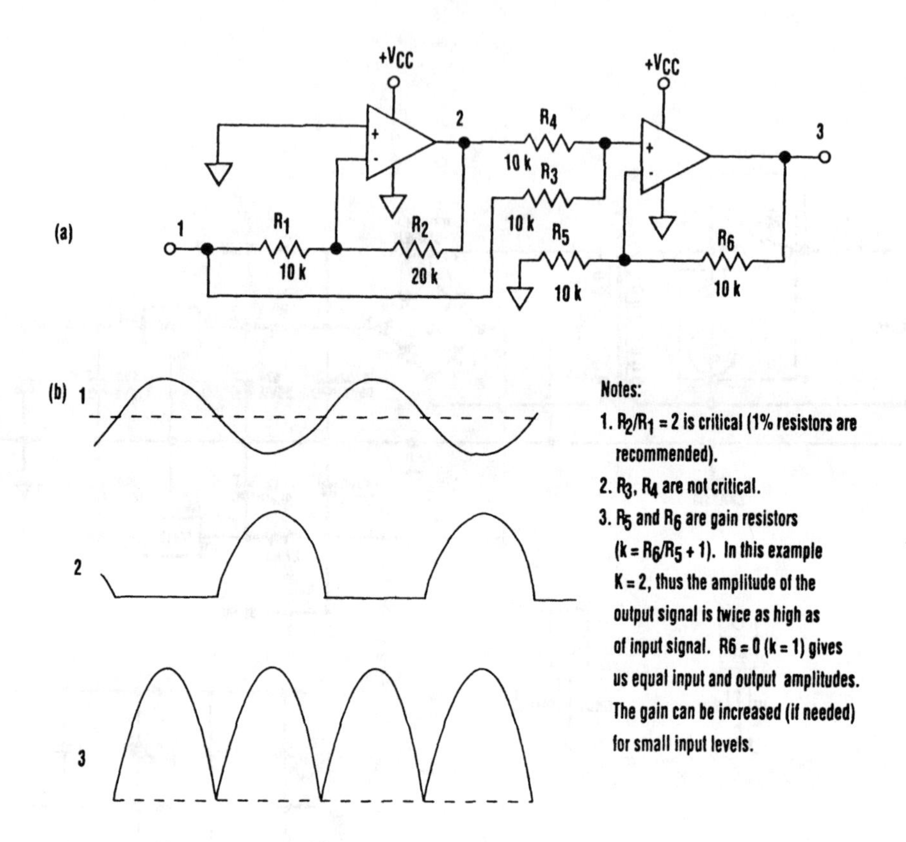

(a)
+VCC
+VCC
1
2
3
R1
10 k
R2
20 k
R4
10 k
R3
10 k
R5
10 k
R6
10 k
(b) 1
2
3
Notes:
1. R2/R1 = 2 is critical (1% resistors are recommended).
2. R3, R4 are not critical.
3. R5 and R6 are gain resistors (k = R6/R5 + 1). In this example K = 2, thus the amplitude of the output signal is twice as high as of input signal. R6 = 0 (k = 1) gives us equal input and output amplitudes. The gain can be increased (if needed) for small input levels.

ELECTRONIC DESIGN *FIG. 78-26*

It's common knowledge that when working with single-supply op amps, implementing simple functions in a bipolar signal environment can be difficult. Sometimes additional op amps and other electronic components are required.

Taking that into consideration, can any advantage be attained from this mode? The answer lies in this simple circuit (A). Requiring no diodes, the circuit is a high-precision full-wave rectifier with a high-frequency limitation equalling that of the op amps themselves. Look at the circuit's timing diagram (B) to see the principle of operation.

The first amplifier rectifies negative input levels with an inverting gain of 2 and turns positive levels to zero. The second amp, a noninverting summing amplifier, adds the inverted negative signal from the first amplifier to the original input signal. The net result is the traditional waveform produced by full-wave rectification.

In spite of the limitation on the input signal amplitude (it must be less than $V_{CC}/2$), this circuit can be useful in a variety of setups.

REGULATOR LOSS CUTTER

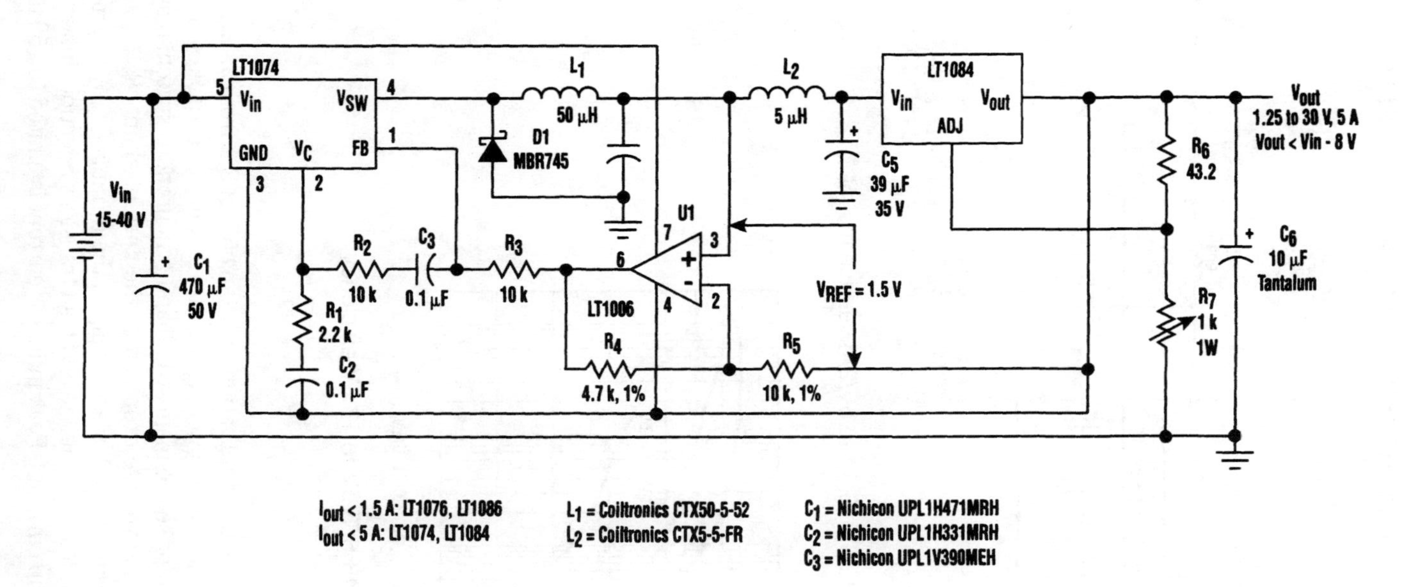

ELECTRONIC DESIGN

FIG. 78-27

Large input-to-output voltage differentials, caused by wide input voltage variations, reduce a linear regulator's efficiency and increase its power dissipation. A switching preregulator can reduce this power dissipation by minimizing the voltage drop across an adjustable linear regulator to a constant 1.5-V value.

The circuit operates the LT1084 at slightly above its dropout voltage. To minimize power dissipation, a low-dropout linear regulator was chosen. The LT1084 functions as a conventional adjustable linear regulator with an output voltage that can be varied from 1.25 to 30 V.

Without the preregulator (for a 40-V input and a 5-V output at 5 A), it would be virtually impossible to find a heatsink large enough to dissipate enough energy to keep the linear-regulator junction temperature below its maximum value. With the preregulator technique, however, the linear regulator will dissipate only 7.5 W under worst-case loading conditions for the entire input-voltage range of 15 to 40 V. Even under a short-circuit fault condition, the 1.5-V drop across the LT1084 is maintained.

SYNCHRONOUS STEPDOWN SWITCHING REGULATOR WITH 90% EFFICIENCY

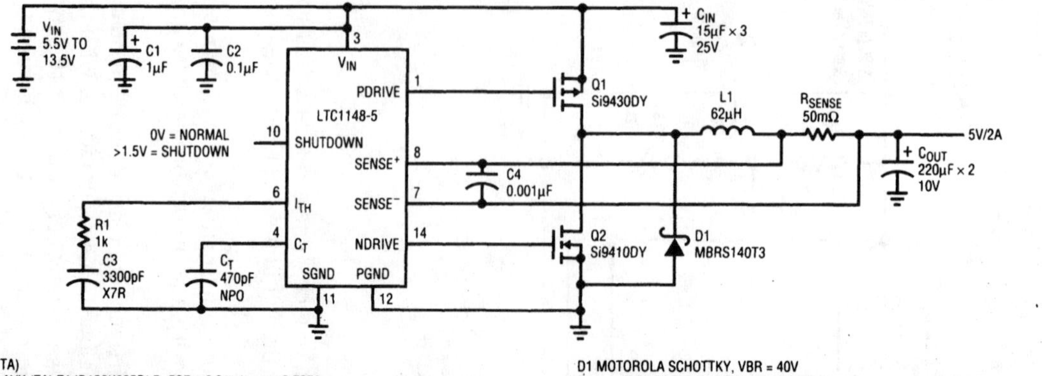

C1(TA)
C_{IN} AVX (TA) TAJD156K025RLR, ESR = 0.3Ω, I_{RMS} = 0.707A
C_{OUT} AVX (TA) TAJE227K010RLR, ESR = 0.08Ω, I_{RMS} = 1.4A
Q1 SILICONIX PMOS, BVDSS = 20V, RDS_{ON} = 0.1Ω, C_{RSS} = 400pF, Q_G = 50nC
Q2 SILICONIX NMOS, BVDSS = 30V, RDS_{ON} = 0.05Ω, C_{RSS} = 160pF, Q_G = 30nC

D1 MOTOROLA SCHOTTKY, VBR = 40V
R_{SENSE} IRC LR2512-01-R050J P_D = 1W
L1 COILTRONICS CTX62-2-MP, DCR = 0.035Ω, MPP CORE (THROUGH HOLE)
L1-1 COILTRONICS CTX02-11715-2, DCR = 0.11Ω, FERRITE CORE (SURFACE MOUNT)
ALL OTHER CAPACITORS ARE CERAMIC

A LTC1148 (5.5V-13.5V to 5V/2A) surface mount

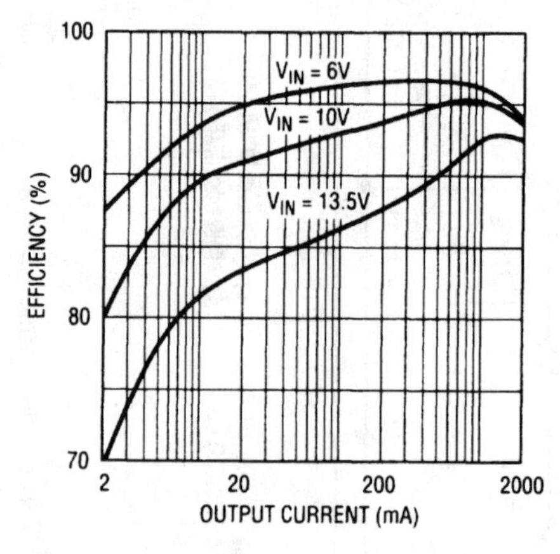

B LTC 1148-5: 5.5V to 13.5V efficiency

LINEAR TECHNOLOGY *FIG. 78-28*

A shows a typical LTC1148 surface-mount application providing 5 V at 2 A from an input voltage of 5.5 V to 13.5 V. The operating efficiency, shown in B, peaks at 97% and exceeds 90% from 10 mA to 2 A with a 10-V input. Q1 and Q2 comprise the main switch and synchronous switch, respectively, and inductor current is measured via the voltage drop across the current shunt. R_{SENSE} is the key component used to set the output current capability according to the formula I_{OUT} = 100 mV/R_{SENSE}. The advantages of current control include excellent line and load transient rejection, inherent short-circuit protection and controlled startup currents. Peak inductor current is limited to 150 mV/R_{SENSE} or 3 A for the circuit in A.

±5- TO ±35-V TRACKING POWER SUPPLY

OUTPUT VOLTAGE IS VARIABLE FROM ±5V TO ±35V.

NEGATIVE OUTPUT TRACKS POSITIVE OUTPUT TO WITHIN THE RATIO OF R6 TO R7.

POPULAR ELECTRONICS *FIG. 78-29*

This supply is designed to operate from a ±40-V nominal unregulated power source (bridge rectifier, etc.).

8-V FROM 5-V REGULATOR

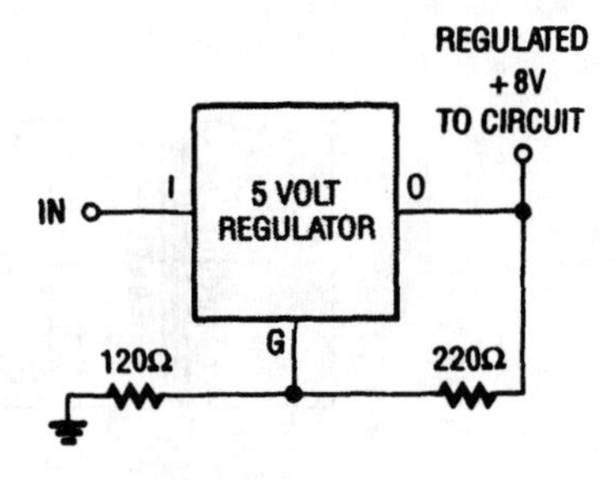

POPULAR ELECTRONICS *FIG. 78-30*

If you have trouble locating an 8-V regulator, although they are commonly available, a 5-V unit can replace it by connecting the regulator, as is shown here.

+1.5-V SUPPLY FOR ZN416E CIRCUITS

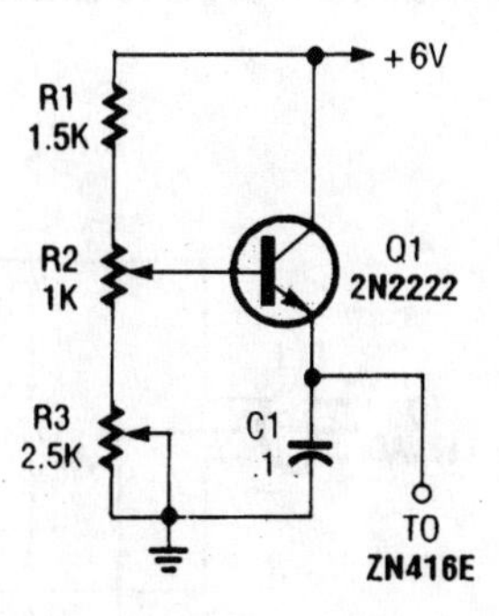

POPULAR ELECTRONICS *FIG. 78-31*

This regulator can be used with a +6-V source to supply ZN416E low-voltage TRF radio-receiver IC the necessary +1.5 V. R3 sets output voltage.

ANTIQUE RADIO dc FILAMENT SUPPLY

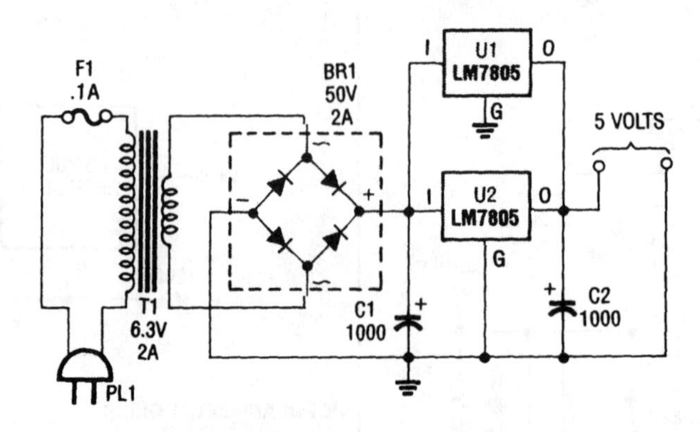

POPULAR ELECTRONICS

FIG. 78-32

This dc supply is great for operating battery-powered antique radios, because it is designed to prevent harming the tube filaments. The circuit is useful for powering filaments of 00-A, 01-A, 112A, and 71A tubes, which require 5V at 250 mA.

INEXPENSIVE ISOLATION TRANSFORMER (IMPROMPTU SETUP)

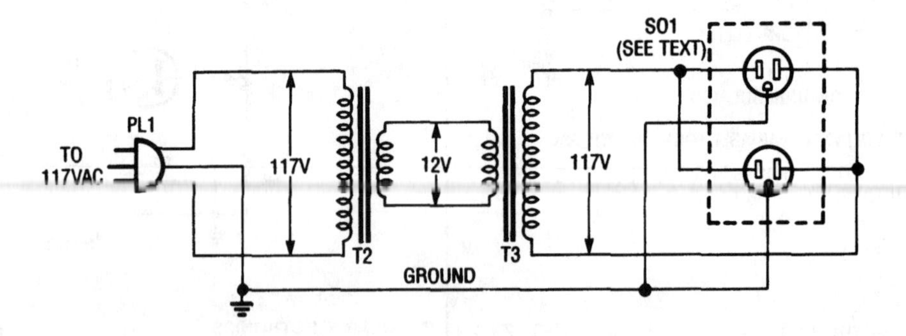

1993 ELECTRONICS HOBBYISTS HANDBOOK

FIG. 78-33

Using two 12-V filament or power transformers, an impromptu isolation transformer can be made for low-power (under 50 W) use in testing or servicing. SO1 is an ordinary, duplex ac receptable. Use heavy-wire connections between the 12-V windings because several amperes can flow.

5-V UPS

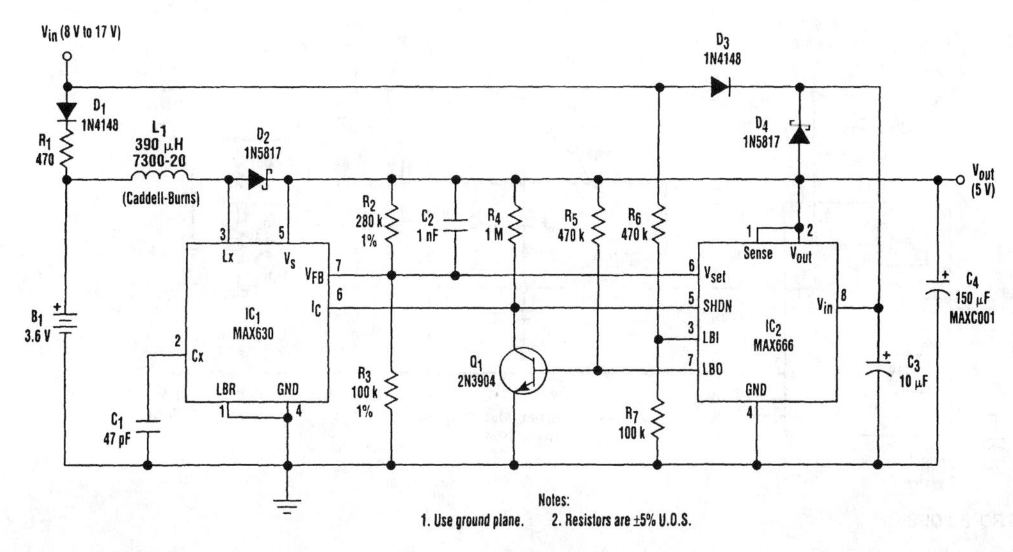

ELECTRONIC DESIGN ***FIG. 78-34***

A 9-V wall adapter supplies V_{IN}. IC2 contains a low-battery detector circuit that senses V_{IN} by means of R6 and R7. The detector output (pin 7) drives an inverter (Q1), which in turn drives the shut-down inputs I_C of IC1 and SHDN of IC2. These inputs have opposite-polarity active levels. The common feedback resistors, R2 and R3 enable both regulators to sense the output voltage, V_{OUT}.

When IC2 shuts down, its output turns off. However, when IC1 shuts down, the whole chip assumes a low-power state and draws under 1 μA. L1, D2, C1, C2, R2, and R3 are part of the 250-mW switching regulator. Diodes D3 and D4 wire-OR the power connection to IC2, and C3 improves the linear regulator's load regulation.

+5-V SUPPLY

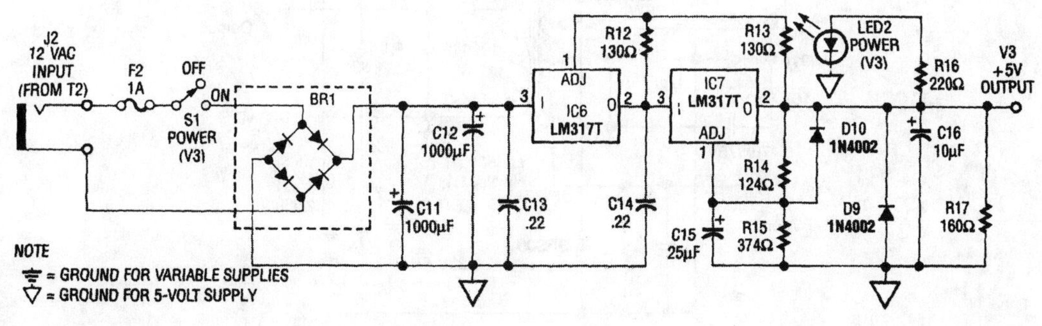

ELECTRONICS NOW ***FIG. 78-35***

The power supply shown is designed to operate from a wall transformer. This circuit can be used in conjunction with a variable supply to test circuits in the lab, etc. T2 is a 12-V wall transformer.

ADD 12-V OUTPUT TO 5-V BUCK REGULATOR

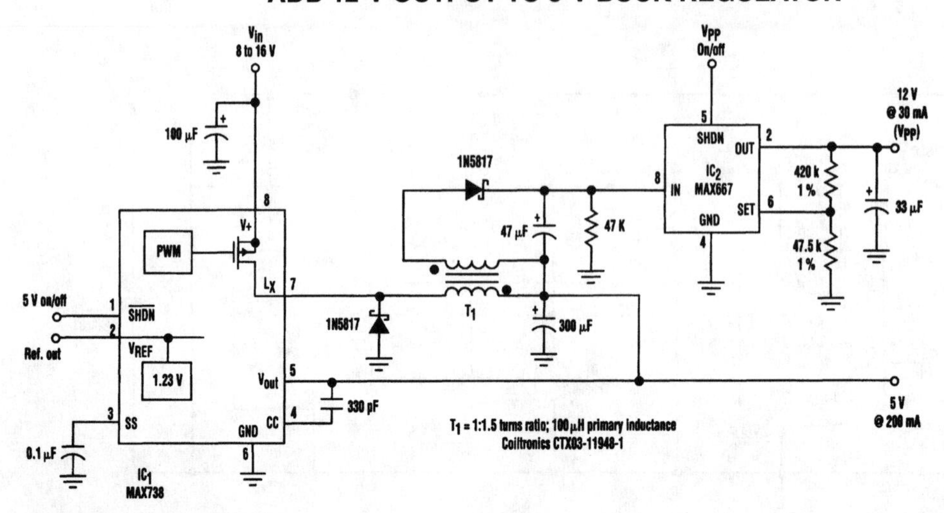

ELECTRONIC DESIGN **FIG. 78-36**

By adding a flyback winding to a buck-regulator switching converter (see the figure), which is essentially a 5-V supply with a 200-mA output capability, a 12-V output (V_{pp}) can be produced. The flyback winding on the main inductor (forming transformer T1) enables an additional low-dropout linear regulator (IC2) to create the 12-V output voltage that's needed to program EEPROMs. The required input voltage is 8 to 16 V.

TELECOM CONVERTER –48 V TO +5 V @ 1 A

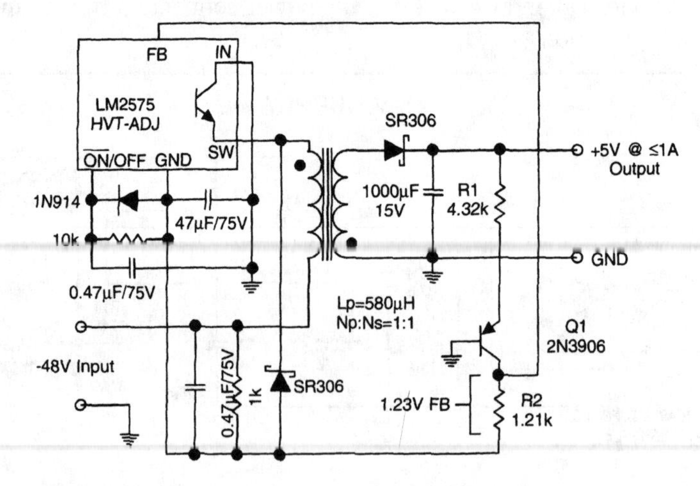

NATIONAL SEMICONDUCTOR, LINEAR EDGE **FIG. 78-37**

The circuit supplies 1 A at +5 V from the –48-V supply commonly used in telephone equipment. The National Semiconductor LM2575 is a simple switching regulator.

79

Probe Circuits

The sources of the following circuits are contained in the Sources section, which begins on page 675. The figure number in the box of each circuit correlates to the entry in the Sources section.

SIMPLE VOLTAGE PROBE

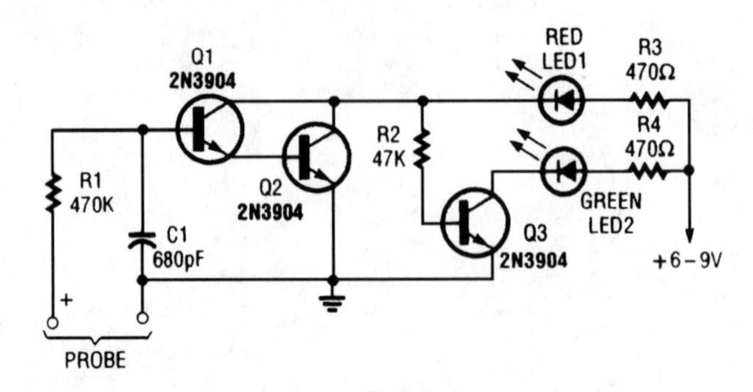

POPULAR ELECTRONICS *FIG. 79-1*

This simple voltage probe can be helpful in checking and troubleshooting solid-state circuitry.

ac VOLTAGE PROBE

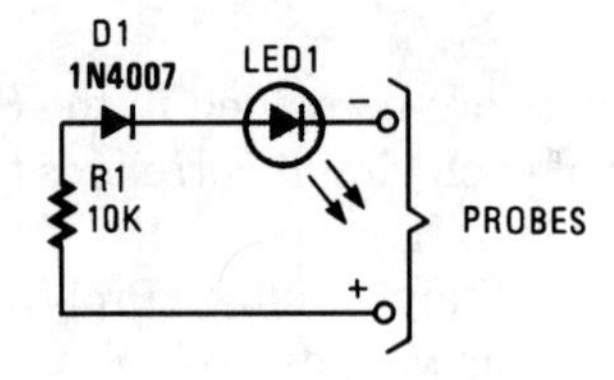

POPULAR ELECTRONICS *FIG. 79-2*

This simple probe can save your life by warning you of live circuitry. It's ideal for times when more than one person is working on a device.

80

Protection Circuits

The sources of the following circuits are contained in the Sources section, which begins on page 675. The figure number in the box of each circuit correlates to the entry in the Sources section.

SPEAKER PROTECTOR

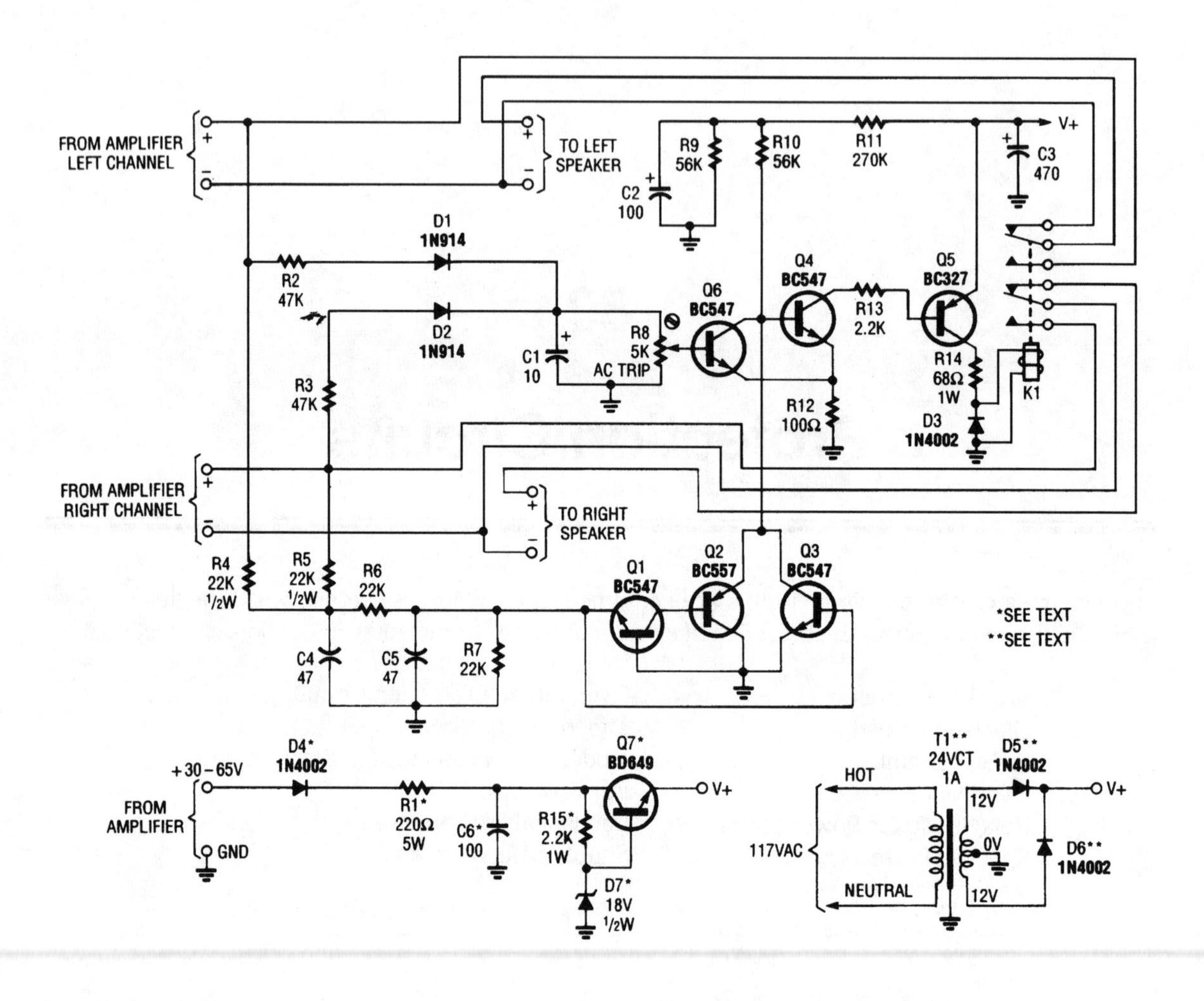

POPULAR ELECTRONICS ***FIG. 80-1***

Most of the transistors in this speaker protector function as switches. Normally, Q4, Q5, and K1 are on and the speakers are connected to the amplifier. However, if a large dc voltage appears at an amplifier output, either Q3, or Q1 and Q2 turn on, biasing Q4 off. That action turns Q5 off, de-energizes the relay, and disconnects the speakers from the amplifier. Components D1, D2, and Q6 form the overdrive-protection circuit.

ELECTRONIC FUSE

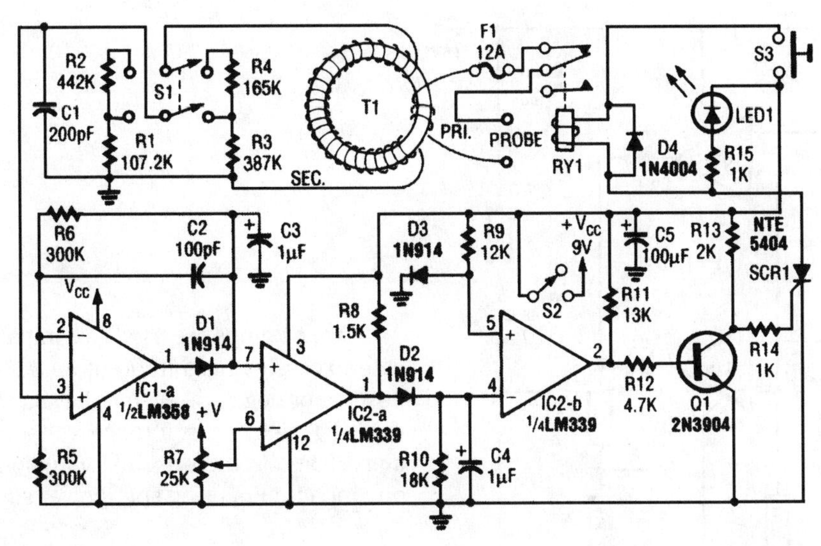

RADIO-ELECTRONICS *FIG. 80-2*

Basically, this circuit is an adjustable electronic circuit breaker, containing a toroidal transformer that senses 60-Hz load current. T1 has a two-turn winding for primary, and 100 turns of #30 gauge wire for the secondary. A high-low range switch selects 0.1 to 6 A or 1 to 12 A. The primary winding of T1 carries full load current and voltage; should be suitably insulated, as should be RY1.

SAFETY CIRCUIT

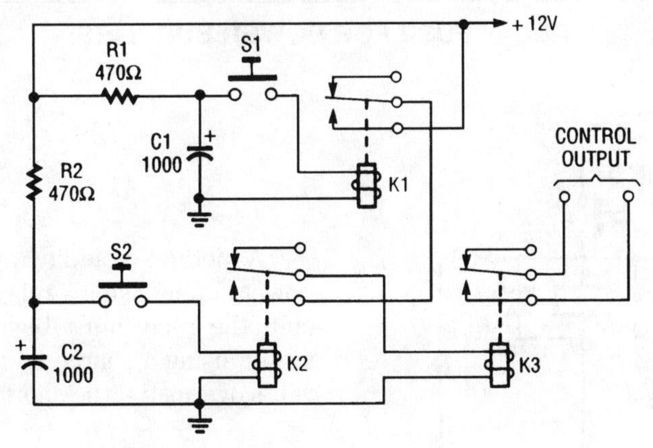

POPULAR ELECTRONICS *FIG. 80-3*

Because of the finite hold-on time of delay circuits R1/C1 and R2/C2, both S1 and S2 must be pressed at the same time to power up the load.

OVERLOAD INDICATOR

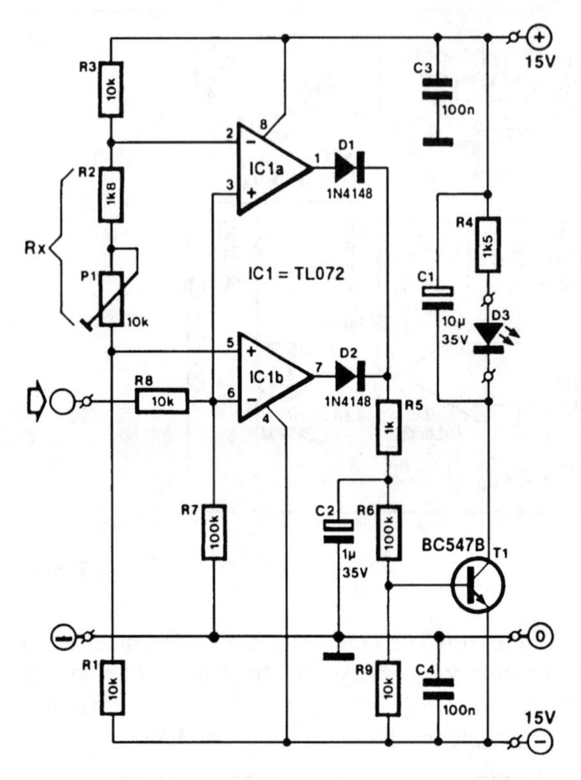

ELEKTOR ELECTRONICS

FIG. 80-4

Two op amps are used as comparators to indicate excessive magnitude of an AF signal, either positive or negative, even if the signal is asymmetrical. P1 sets the reference voltage for both op amps. This circuit is useful for audio-amplifier and op-amp circuits using split power supplies.

RELAY FUSE FOR POWER SUPPLIES

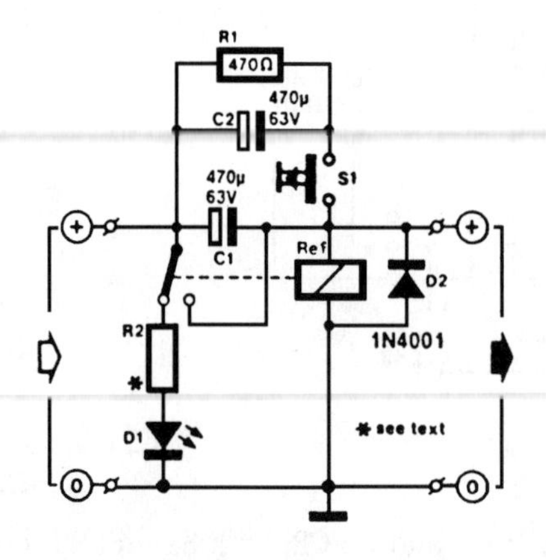

ELEKTOR ELECTRONICS

FIG. 80-5

A method of adding overload protection to a power supply using a relay is shown. In each circuit, the relay must be reset by a momentary switch using a charge on capacitor C2. This prevents overload if the short still exists.

SPEAKER PROTECTOR

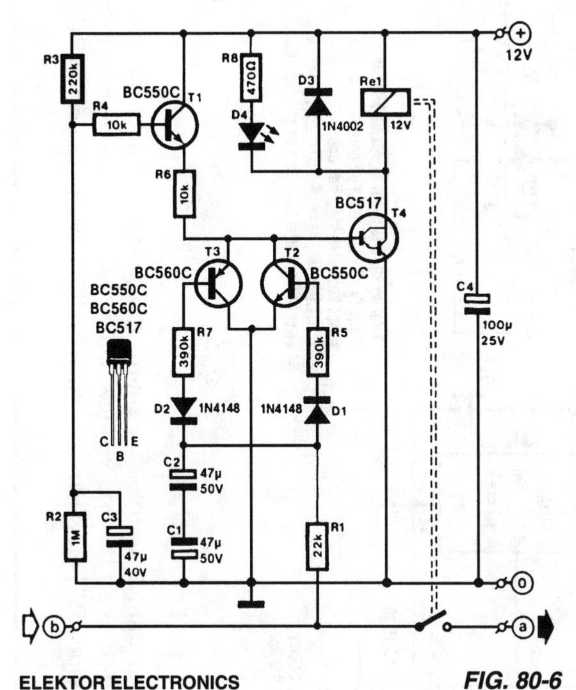

ELEKTOR ELECTRONICS

FIG. 80-6

A speaker system can be protected against amplifier failure when dc voltages (on speaker line a-b) are sensed by the circuit. Either positive or negative dc voltages are sensed. A relay opens in this case, removing the dc from the speakers. About 12 V at 50 mA is needed to power the circuit, depending on the relay.

MODEM PROTECTOR

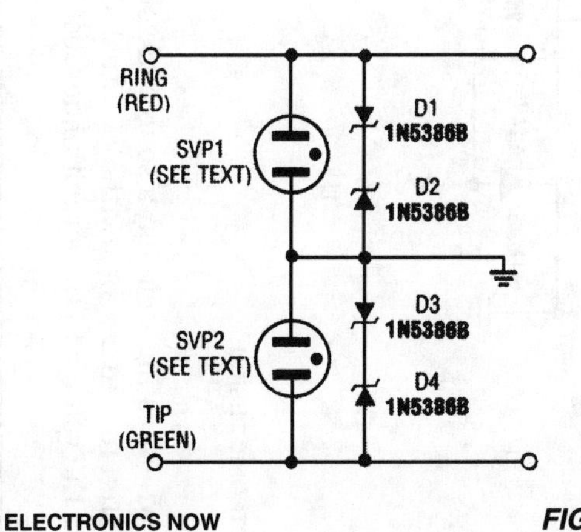

ELECTRONICS NOW

FIG. 80-7

This protector uses surge voltage protectors rated at 230-V breakdown. An effective ground should be used.

OVERVOLTAGE PROTECTION CIRCUIT

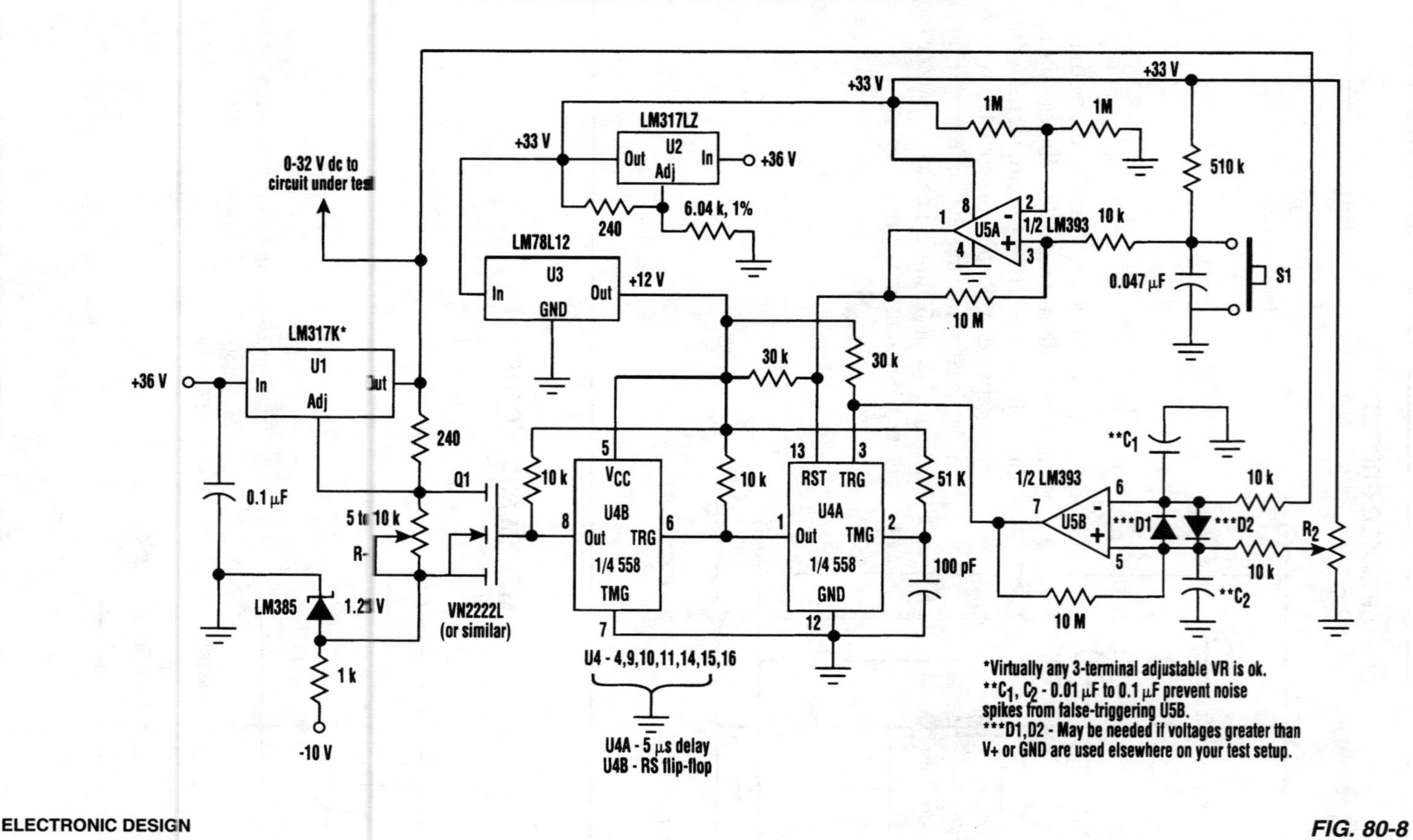

ELECTRONIC DESIGN

FIG. 80-8

When testing a circuit, a source of voltage that is variable and has overvoltage shutdown is very useful. In this circuit, R1 is adjusted to 1 to 2 V below the eventual shutdown threshold. R2 sets the trip voltage. When this voltage is reached, the circuit shuts the voltage to the circuit under test down. To reset, reduce R1 below trip threshold and depress reset switch S1.

TIMED SAFETY CIRCUIT

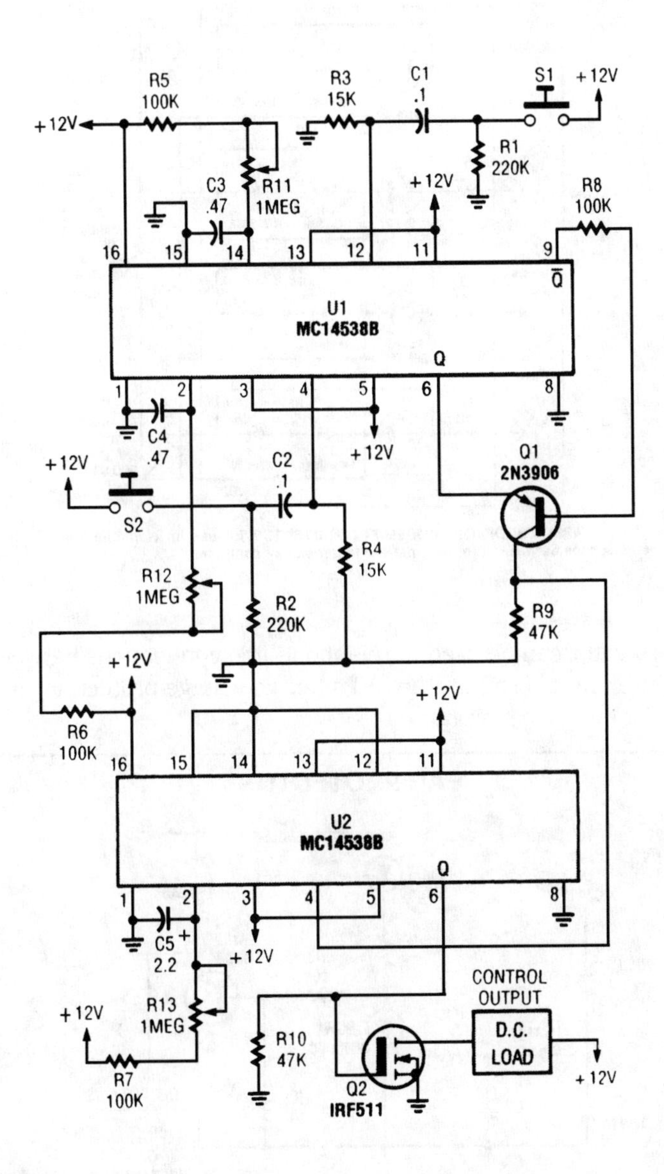

POPULAR ELECTRONICS ***FIG. 80-9***

When S1 is closed, pin 9 of U1 goes low, turning on Q1 for a preset period. If S2 is closed during this period, Q2 is turned on for a preset period. R11 and R13 set the two time periods.

MODEM/FAX PROTECTOR FOR TWO COMPUTERS

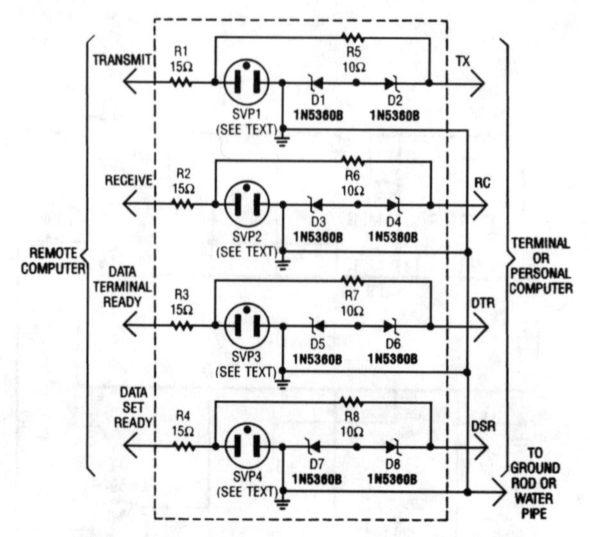

VARIATION OF THE MODEM/FAX PROTECTOR for use in telephone line connections between PC or terminal and larger distant computer.

ELECTRONICS NOW ***FIG. 80-10***

This modem/fax protector can be used in telephone-line connections between a PC or a terminal and a distant computer. In this circuit, the SVPs (surge voltage protectors) are rated at 230 V. A good ground is a must for effective operation.

EAR PROTECTOR

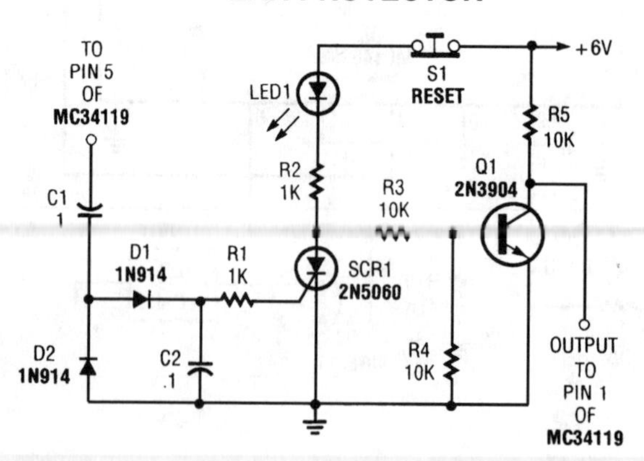

POPULAR ELECTRONICS ***FIG. 80-11***

The ear protector is actually a peak audio-detector/shutdown circuit that disables the amplifier through its chip-disable input when the output volume of an amplifier reaches the set level. The circuit, although intended for the MC34119 amplifier, should work with similar IC devices or applications.

LOUDSPEAKER PROTECTOR

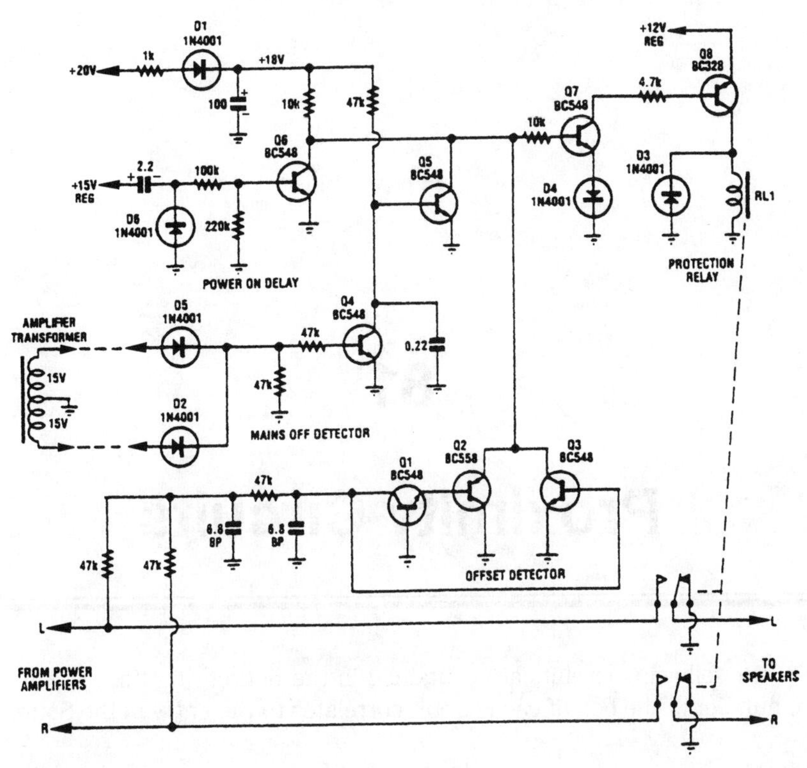

SILICON CHIP

FIG. 80-12

Transistors Q1, Q2, and Q3 monitor the two outputs of the stereo amplifier. If the offsets exceed ±2 V, Q7 is turned off, which turns off Q8 and the normally on relay. Diodes D2 and D5, together with Q4, provide a mains voltage monitor. As soon as the ac input voltage disappears, as when the amplifier is turned off, Q4 turns off and Q5 turns on. This turns off Q7, Q8, and the relay. Hence, the loudspeakers are disconnected immediately after the amplifier is turned off.

SIMPLE SAFETY CIRCUIT

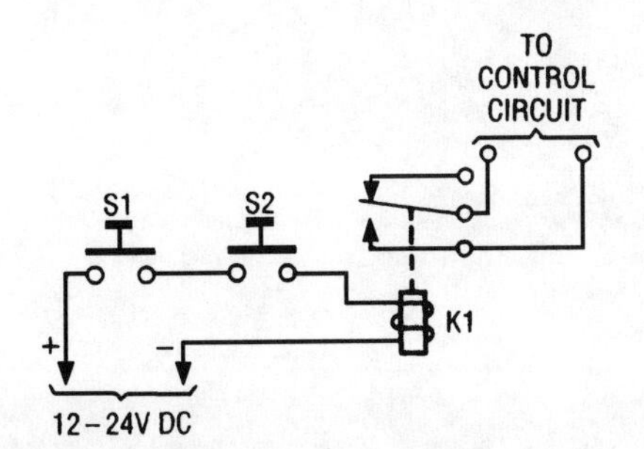

The simple two-hand safety-control switch shown here is little more than two pushbutton switches connected in series; both must be depressed in order to energize the relay.

POPULAR ELECTRONICS

FIG. 80-13

81

Proximity Circuits

The sources of the following circuits are contained in the Sources section, which begins on page 675. The figure number in the box of each circuit correlates to the entry in the Sources section.

PROXIMITY ALARM I

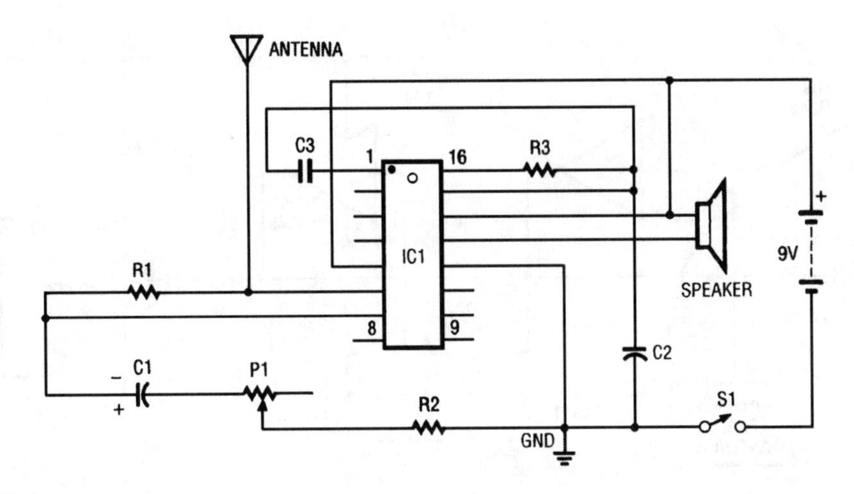

1991 PE HOBBYIST HANDBOOK

FIG. 81-1

IC1 contains several oscillators and an amplifier. The low-frequency audio-signal oscillator is used to supply an input to the amplifier. That signal is the audio tone that is amplified, then supplied to the speaker by the amplifier.

The high-frequency oscillator is purposely set to be very unstable. It is dormant or "off" until the resistor-capacitor (RC) network is changed. The resistance (*R*) in this case is made up of R2 and P1. As the resistance of P1 is decreased, the unit becomes more sensitive (more unstable), and less capacitance (*C*) is needed to cause the oscillator to oscillate.

The capacitance required is provided by C2 and by any capacitance introduced via the antenna loop. When you come near that loop, your inherent body capacitance causes the high-frequency oscillator to begin to oscillate, which then causes the low-frequency oscillator to be "switched on" internally. Once the alarm is sounding, the IC is designed so that it "latches", that is, it stays on until the power to it is switched off.

C1	1-µF Axial Capacitor
C2	27-pF Silver Mica Capacitor
C3	0.1-µF Mylar Capacitor
IC1	CM1001N IC
P1	50-kΩ Trimmer Resistor
R1	75-kΩ Resistor
R2	200-Ω Resistor
R3	100-kΩ Resistor
S1	SPDT Switch
Spk	Small Speaker
Misc	IC Socket, Battery Snap, Ground Plate, Wire, PC Board

PROXIMITY ALARM II

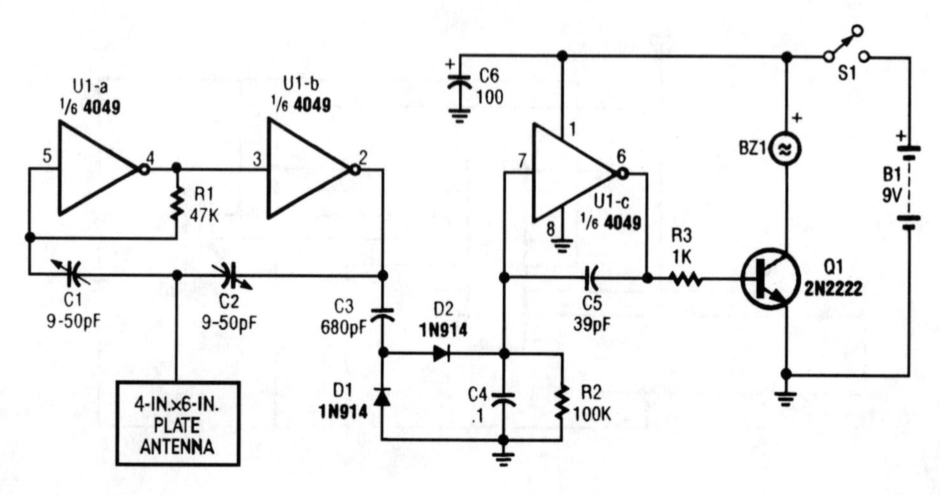

POPULAR ELECTRONICS

FIG. 81-2

A CMOS logic gate is used to make up this circuit. When an object is near the antenna, the change in oscillator output is detected by D1 and D2 and amplified by U1C, which drives Q1, sounding alarm BZ1.

82

Pulse-Generator Circuits

The sources of the following circuits are contained in the Sources section, which begins on page 675. The figure number in the box of each circuit correlates to the entry in the Sources section.

ADD-ON PULSE GENERATOR

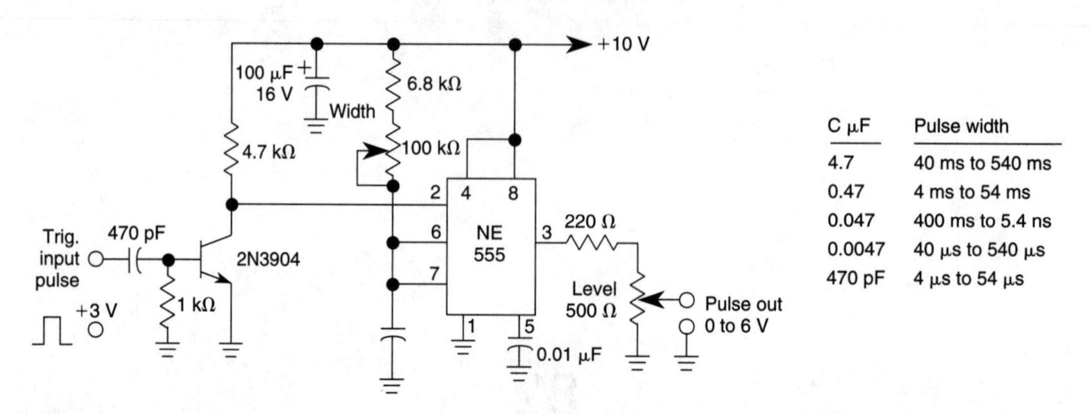

C µF	Pulse width
4.7	40 ms to 540 ms
0.47	4 ms to 54 ms
0.047	400 ms to 5.4 ns
0.0047	40 µs to 540 µs
470 pF	4 µs to 54 µs

WILLIAM SHEETS *FIG. 82-1*

This pulse generator can supplement a standalone pulse generator. Using a transistor and a 555 timer, pulse widths of <5 µs to 500 µs can be produced. The value of C_3 is approximately found from the formula:

C_3 µF ≈ $1.1 \times 10^{-5}\, T$ where T is the shortest pulse width (µs) desired in a 10:1 range (T should be greater than 5 µs)

The capacitor values and consequent pulse width range are shown.

PULSE GENERATOR

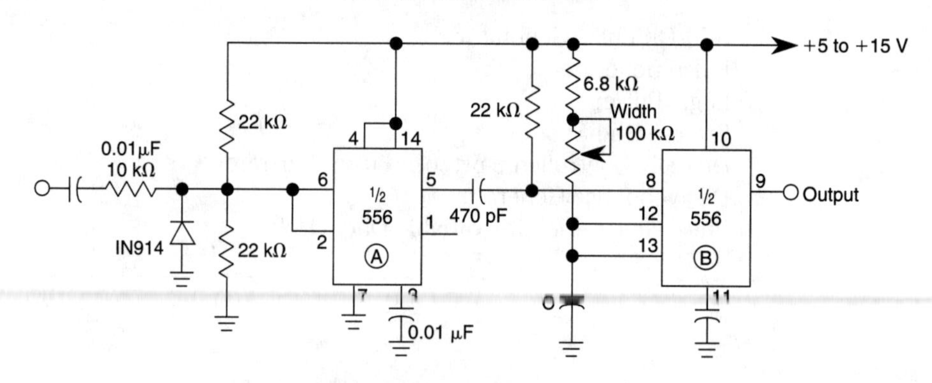

Pulsewidth $T \approx 1.1\, RC$

In this circuit $T \approx 7.4 \times 10^{-3}\, C_{\mu F}$ to $0.117 \times C_{\mu F}$ seconds with $C = 0.1\mu F$ $T = 740$ µs to 11.7 ms

WILLIAM SHEETS *FIG. 82-2*

By using a 556 dual timer with IC1A acting as a waveshaper and IC1B as a pulse generator, a 10:1 range of pulse widths can be generated.

A sine wave can be used to trigger this circuit.

LOGIC PULSER

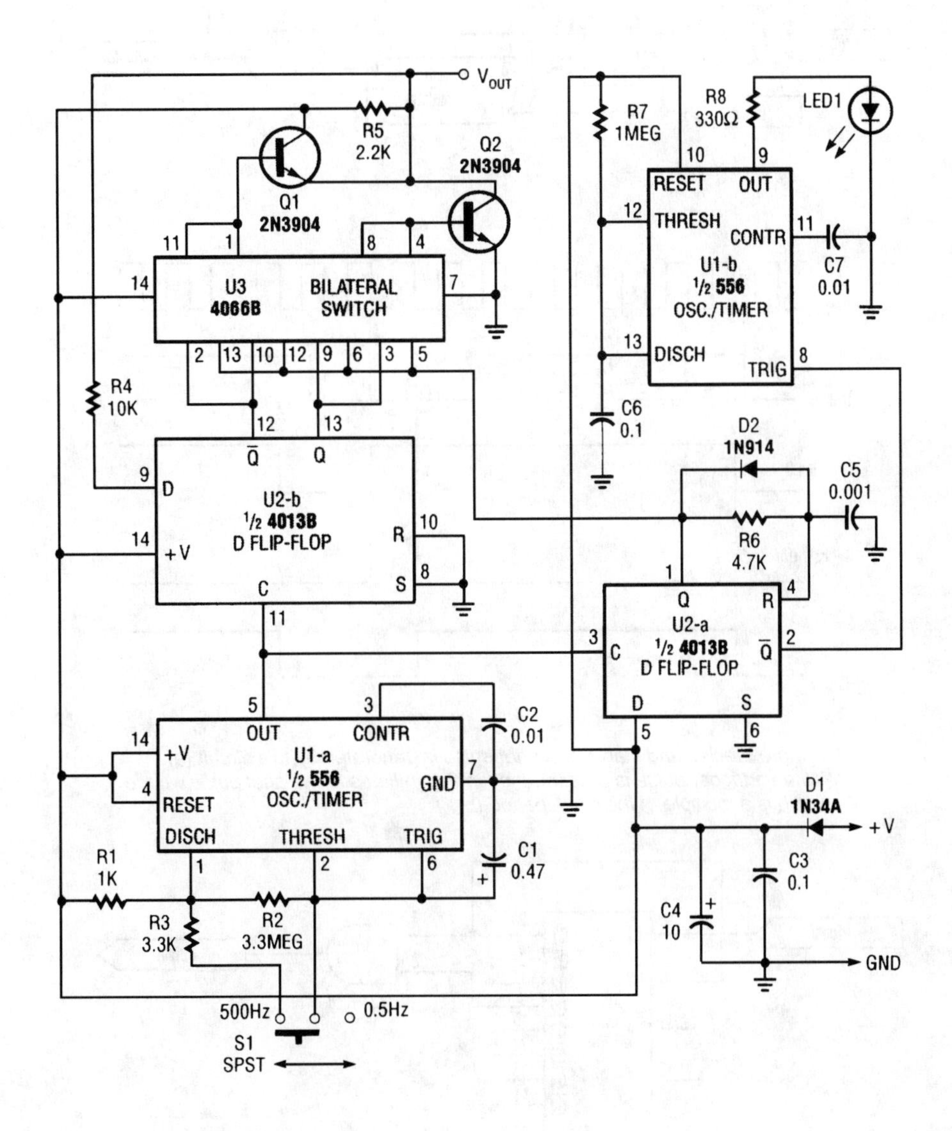

1990 PE HOBBYIST HANDBOOK ***FIG. 82-3***

The logic pulser generates pulses at 500 Hz or 0.5 Hz. When the pulser's tip connects to an input that is already being driven high or low, the pulser senses the logic state and automatically pulses the input briefly to the opposite state.

PRECISE ONE-SHOT

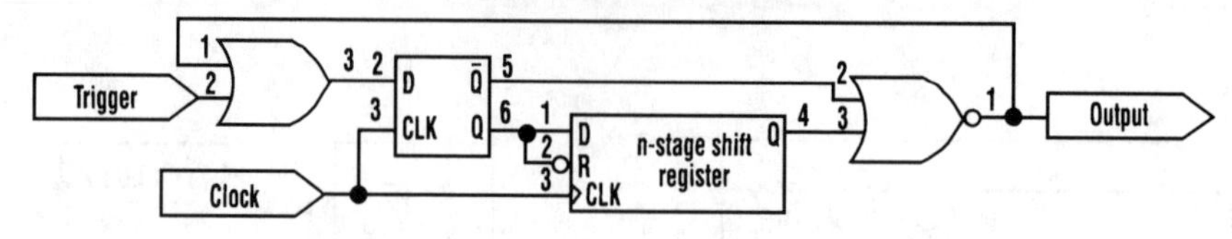

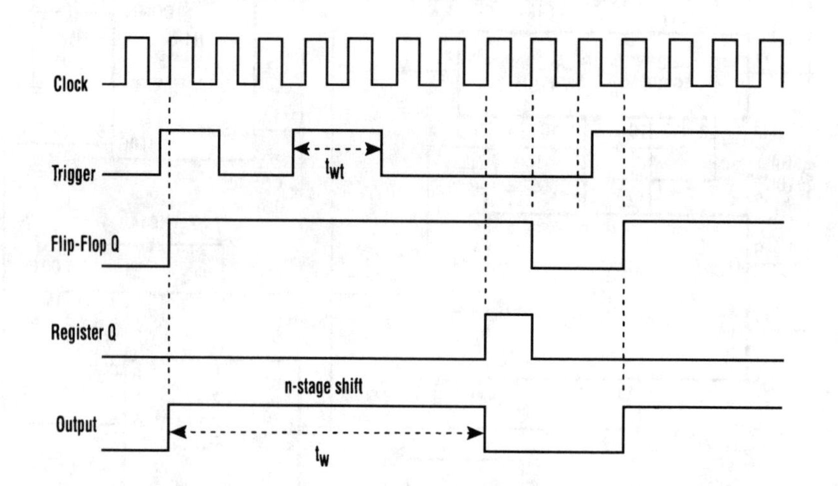

A more precise and stable one-shot pulse is generated by this circuit (a). When a trigger pulse is present, the flip-flop initiates a one-shot pulse whose width is a multiple of the clock period (b).

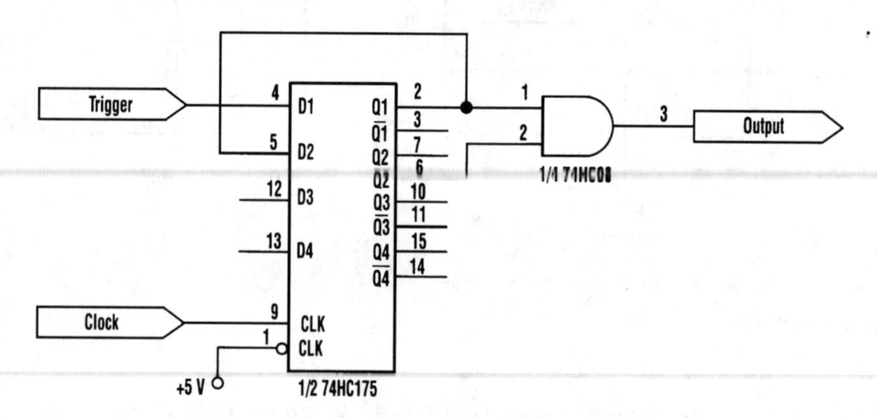

This simple one-shot circuit has a pulse width of one clock period and is more precise and stable than a multivibrator.

ELECTRONIC DESIGN **FIG. 82-4**

This approach uses a flip-flop, a shift register, and two gates (A). Before the one-shot pulse, the output of the NOR gate is 0. Consequently, the data input of the D-type flip-flop is equivalent to the trigger. When a trigger pulse is present, the flip-flop initiates the one-shot pulse, and the n-stage shift register controls the pulse width, t_w, which is a multiple of the clock's period (B).

The precision of the one-shot pulse is determined by the clock period, which is inversely proportional to its frequency. For the circuit to work properly, the width of the trigger pulse, t_{wt}, should be greater than one clock period.

The OR gate masks the trigger's effect when the circuit is generating the desired pulse. The net result is a circuit that functions as a nonretriggerable multivibrator.

When the pulse needs to be only one-clock-period wide, the circuit can be simplified. All that's required are two D-type flip-flops and an AND gate. However, despite its simplicity, this circuit generates a more stable and precise one-shot pulse than a multivibrator.

DIGITALLY CONTROLLED SAWTOOTH PULSE GENERATOR

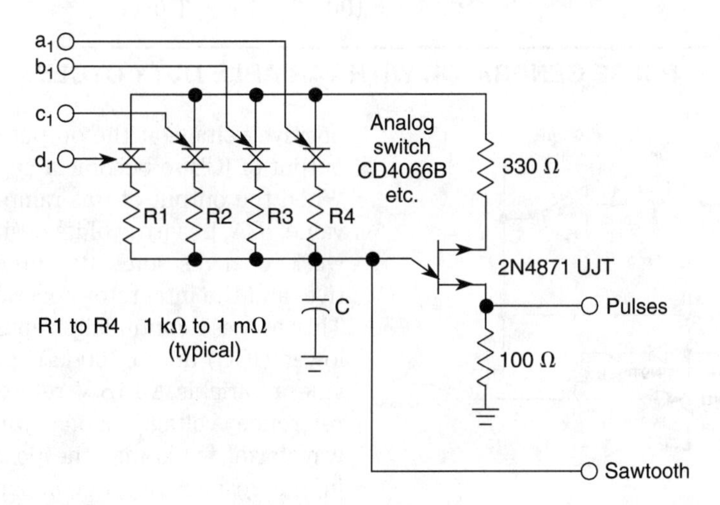

WILLIAM SHEETS

FIG. 82-5

Use of an analog switch as shown allows digital control of a UJT oscillator.

DELAYED PULSE GENERATOR

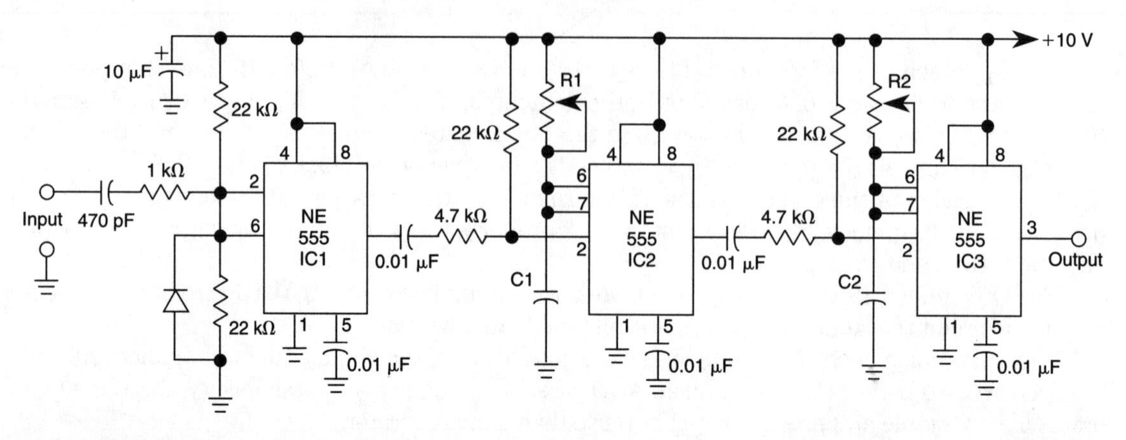

WILLIAM SHEETS *FIG. 82-6*

Three 555 IC timers are used in this circuit to construct a simple delayed-pulse generator. IC1 acts as a waveform shaper to produce a rectangular waveform. IC2 produces a delaying pulse to trigger IC3 on the trailing edge of the delaying pulse. R1 controls delay time and R2 controls pulse width. As much as a 10:1 range can be generated.

Delay: $C1 = 1.1 \times 10^{-5}$ T delay c µF
Pulse: $C2 = 1.1 \times 10^{-5}$ T pulse T µsec

PULSE GENERATOR WITH VARIABLE DUTY CYCLE

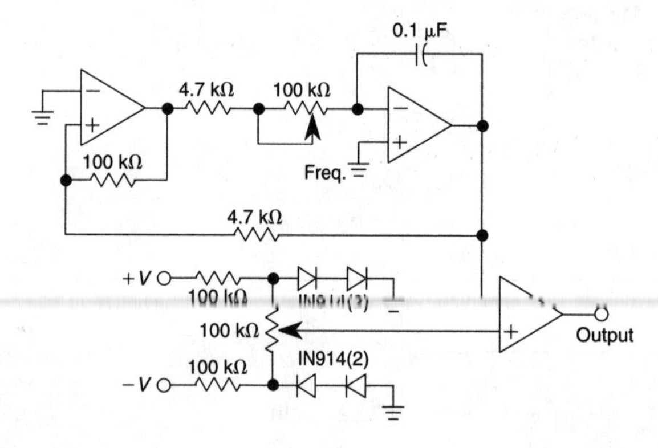

WILLIAM SHEETS *FIG. 82-7*

Using only one IC and six passive components, this pulse generator has a frequency range of 400 to 4000 Hz and an adjustable duty cycle of 1 to 99%. A threshold detector (ICA) and an integrator (ICB) generate a triangular waveform. A positive voltage at the output of ICA causes the output of ICB to become a negative-going ramp. When the output of this ramp reaches a certain value, ICA, by virtue of its positive-feedback network, changes state; its output becomes negative, and the integrator generates positive ramp. This process continually repeats. A voltage follower (ICC) and a 100-kΩ potentiometer provide a variable ±0.18-V reference voltage. This reference voltage, along with the triangular waveform, feeds into the positive and negative inputs, respectively, of comparator ICD. You can set the comparator's trip voltage at any point on the triangular waveform; ICD's output changes at that point. Varying the reference voltage alters the duty cycle of the comparator's output by adjusting the potentiometer at the negative input of the integrator, thereby varying the integration time without altering the duty cycle.

83

Receiver Circuits

The sources of the following circuits are contained in the Sources section, which begins on page 675. The figure number in the box of each circuit correlates to the entry in the Sources section.

SIMPLE DIRECT-CONVERSION RECEIVER FOR 160 TO 20 M

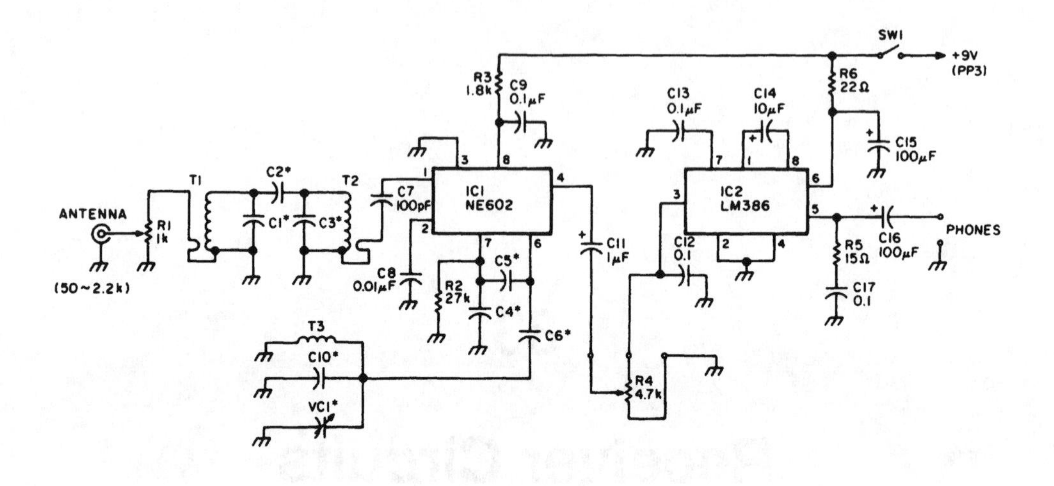

Table. Component Values for Different Bands

Band	C1	C2	C3	T1	T2
160	220 pF	10 pF	220 pF	BKXN-K3333R	BKXN-K3333R
80	47 pF	3 pF	47 pF	BKXN-K3333R	BKXN-K3333R
40	100 pF	8.2 pF	100 pF	BKXN-K3334R	BKXN-K3334R
30	47 pF	3 pF	47 pF	BKXN-K3334R	BKXN-K3334R
20	100 pF	3 pF	100 pF	BKXN-K3335R	BKXN-K3335R

VC1 + C10	C4	C5	C6	T3
All Sections + 100 pF	0.001 μF	0.001 μF	560 pF	BKXN-K3333R
All Sections + 100 pF	0.001 μF	0.001 μF	560 pF	BKXN-K3334R
1 Section + 47 pF	560 pF	560 pF	270 pF	BKXN-K4173AO
1 Section + 68 pF	680 pF	680 pF	220 pF	BKXN-K3335R
1 Section + 68 pF	220 pF	220 pF	68 pF	BKXN-K3335R

73 AMATEUR RADIO TODAY *FIG. 83-1*

Note that T1 and T2 are TOKO, including part numbers for the coils T1 and T2. The direct-conversion receiver shown uses a double-tuned input network made from readily available TOKO coils. IC1, an NE602, acts as a VFO and mixer, with the output being an IF frequency in the audio range. IC2 is an audio amplifier, R4 is a volume control.

27.145-MHz NBFM RECEIVER

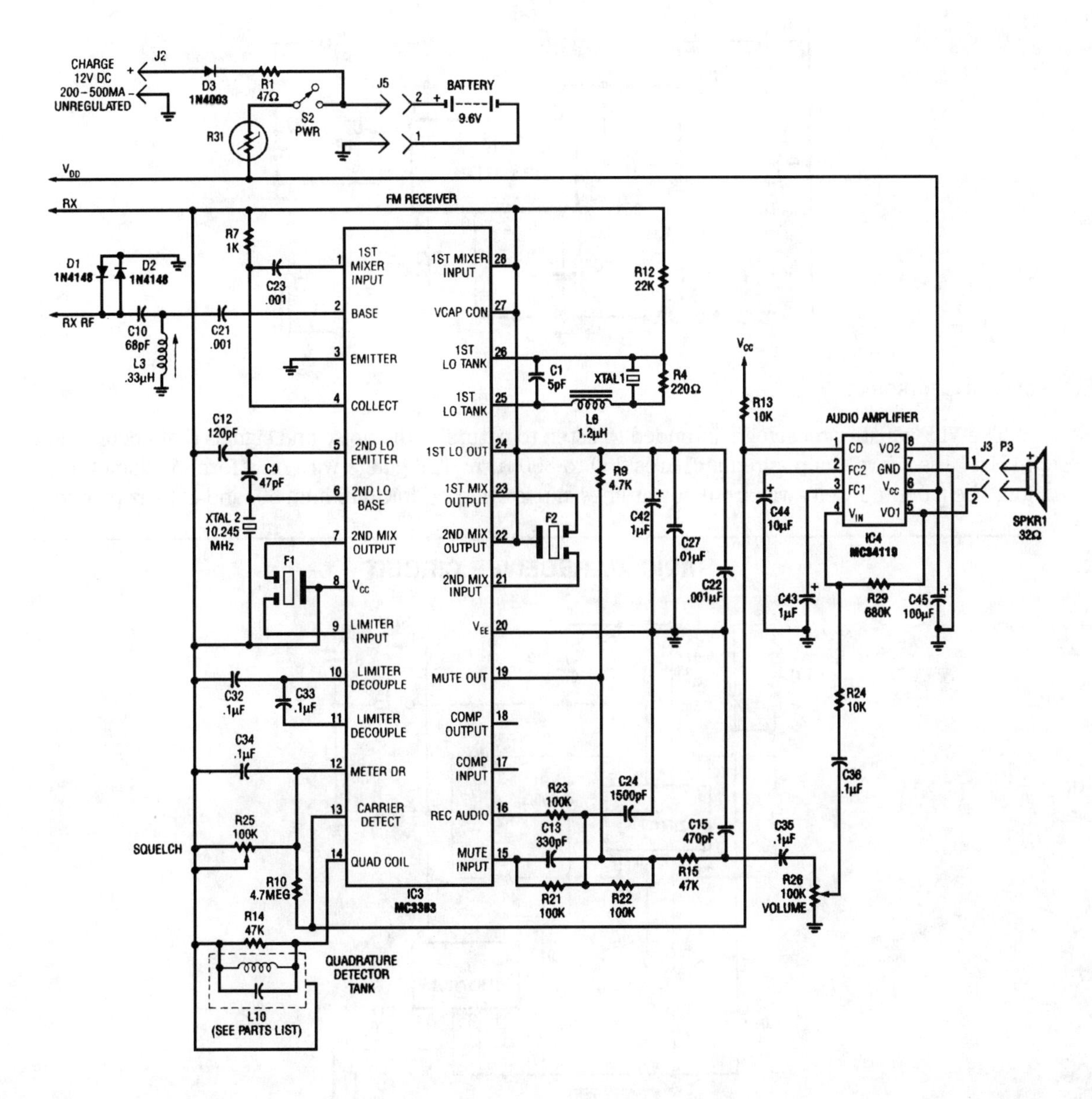

ELECTRONICS NOW *FIG. 83-2*

Using a Motorola MC3363 LSI one-chip FM receiver, the circuit is a dual-conversion FM receiver with a 10.7-MHz IF chain. IC4 provides power to drive a small speaker.

VLF WHISTLER RECEIVER

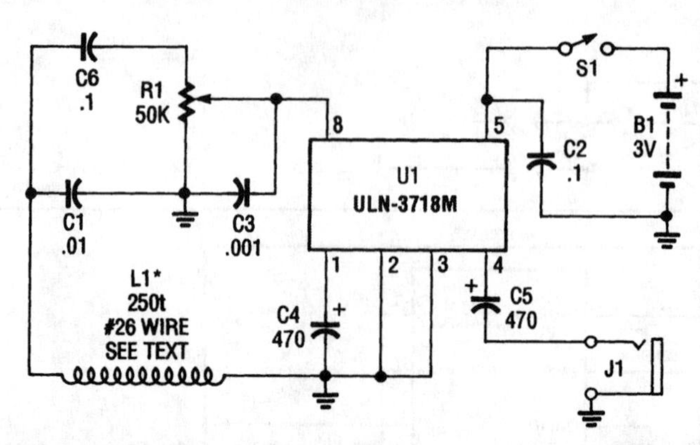

POPULAR ELECTRONICS **FIG. 83-3**

The VLF whistler receiver is intended to listen to natural radio noise and signals that occur below 20 kHz. L1 is a large loop antenna that is 250 to 300 turns #26 gauge wire on a form 3' diameter. L1 should be mounted well away from power lines and is oriented for minimum 60- and 120-Hz pickup.

BASIC AM RECEIVER CIRCUIT

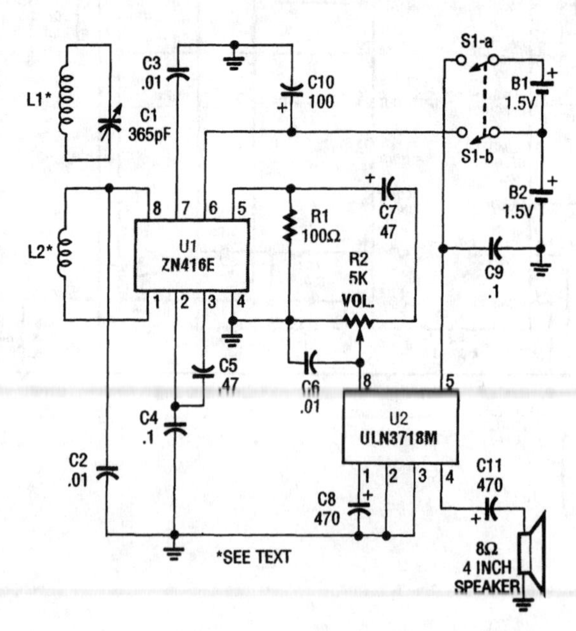

POPULAR ELECTRONICS **FIG. 83-4**

Using a single ZN416E IC and a ULN3718M, this simple TRF receiver can drive a loudspeaker. Two 1.5-V cells power the circuit.

SIMPLE 1.5-V AM BROADCAST RECEIVER

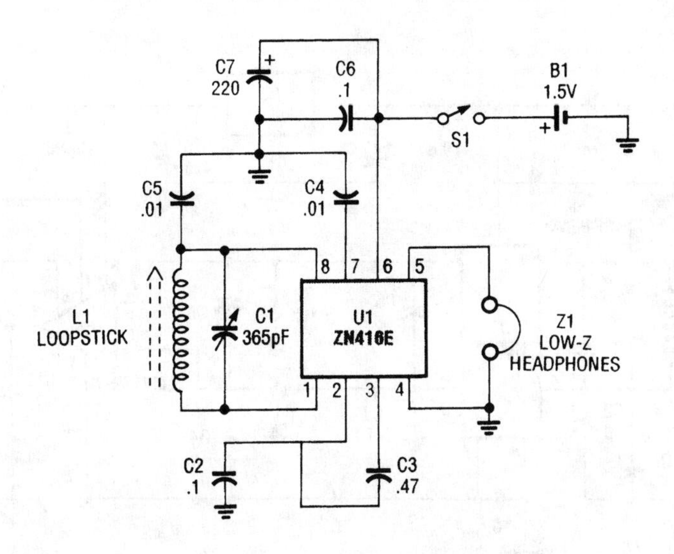

POPULAR ELECTRONICS

FIG. 83-5

This receiver uses the ZN416E made by GEC Plessey. The tuning is via C1.

CMOS LINE RECEIVER

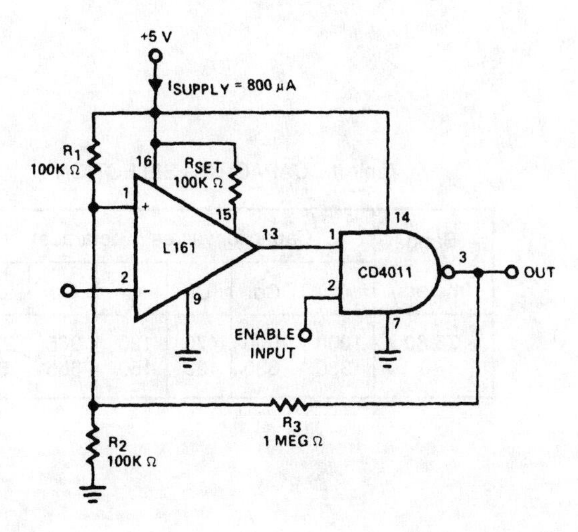

INTEGRATED CIRCUITS DATA BOOK

FIG. 83-6

This circuit will interface a line input to CMOS. The supply current is >1 mA at +5 V.

NE602 DIRECT-CONVERSION RECEIVER

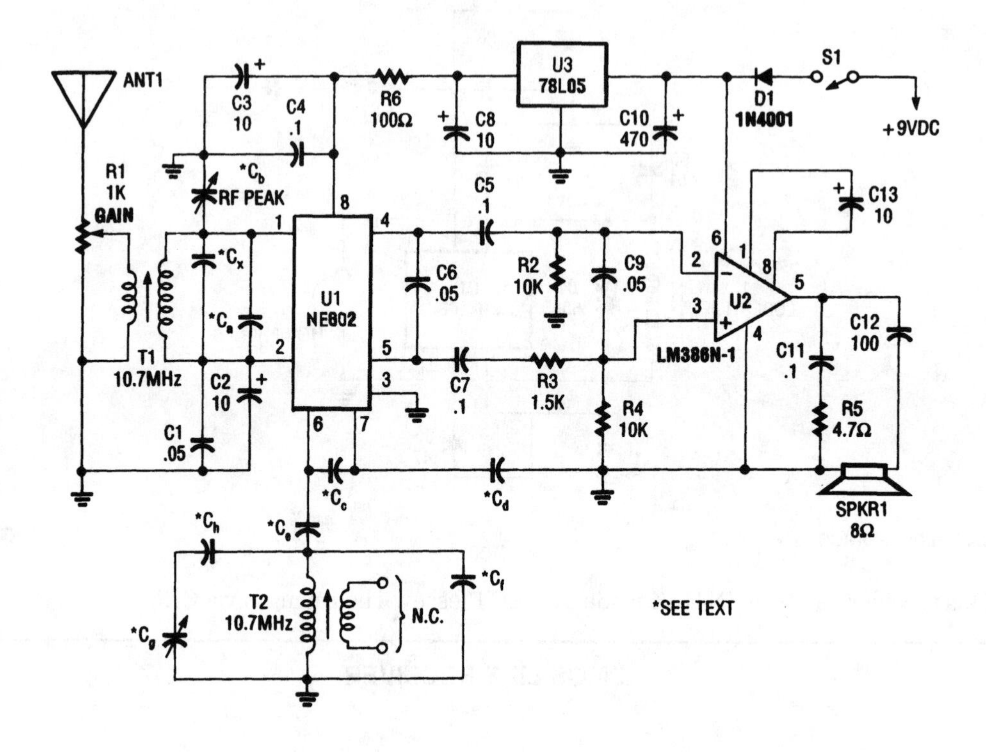

Table 1-- CAPACITOR SELECTION

Band	Capacitor values (picofarads)					
(meters)	Cc	Cd	Ce	Cf	Cg	Ch
75/80	1000	1000	470	120	365	270
40	330	330	120	150	365	68

POPULAR ELECTRONICS ***FIG. 83-7***

An NEC602 is used as a mixer with a zero IF frequency output. U2 acts as an audio amplifier. This receiver is primarily for SSB and CW signals. T1 and T2 are 10.7-MHz IF coils used in AM/FM transistorized radios, etc. or in any similar indicator.

80- AND 40-M CW/SSB RECEIVER

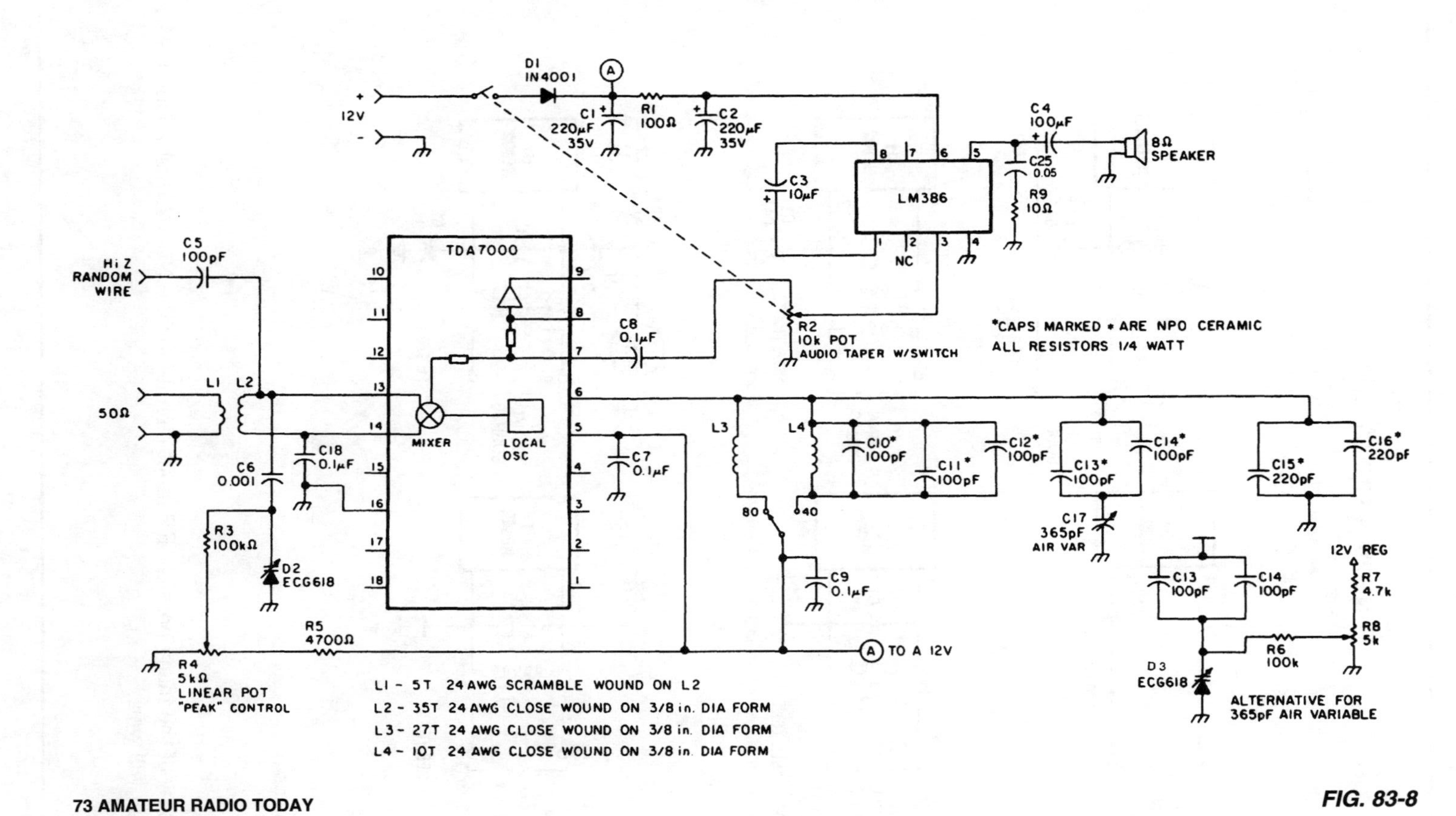

73 AMATEUR RADIO TODAY

FIG. 83-8

This direct-conversion receiver uses a TDA7000 IC and it drives an LM386 audio amplifier. The TDA7000 is used for its mixer and L.O. section. The frequency control can be either with an air variable capacitor or a varactor diode.

NE602 RF INPUT CIRCUITS

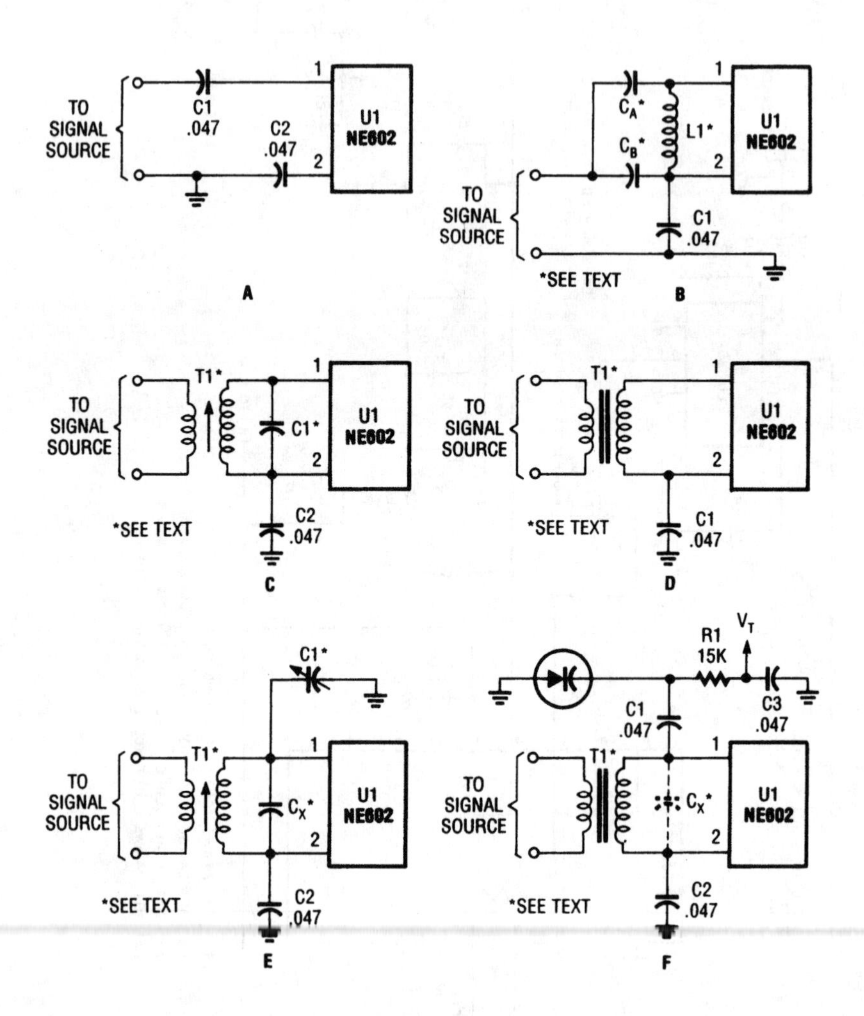

POPULAR ELECTRONICS *FIG. 83-9*

Here are a few of the many possible RF input circuits for the NE602. Just about any tuned or broadband circuit will work.

SUPER-SIMPLE SHORTWAVE RECEIVER

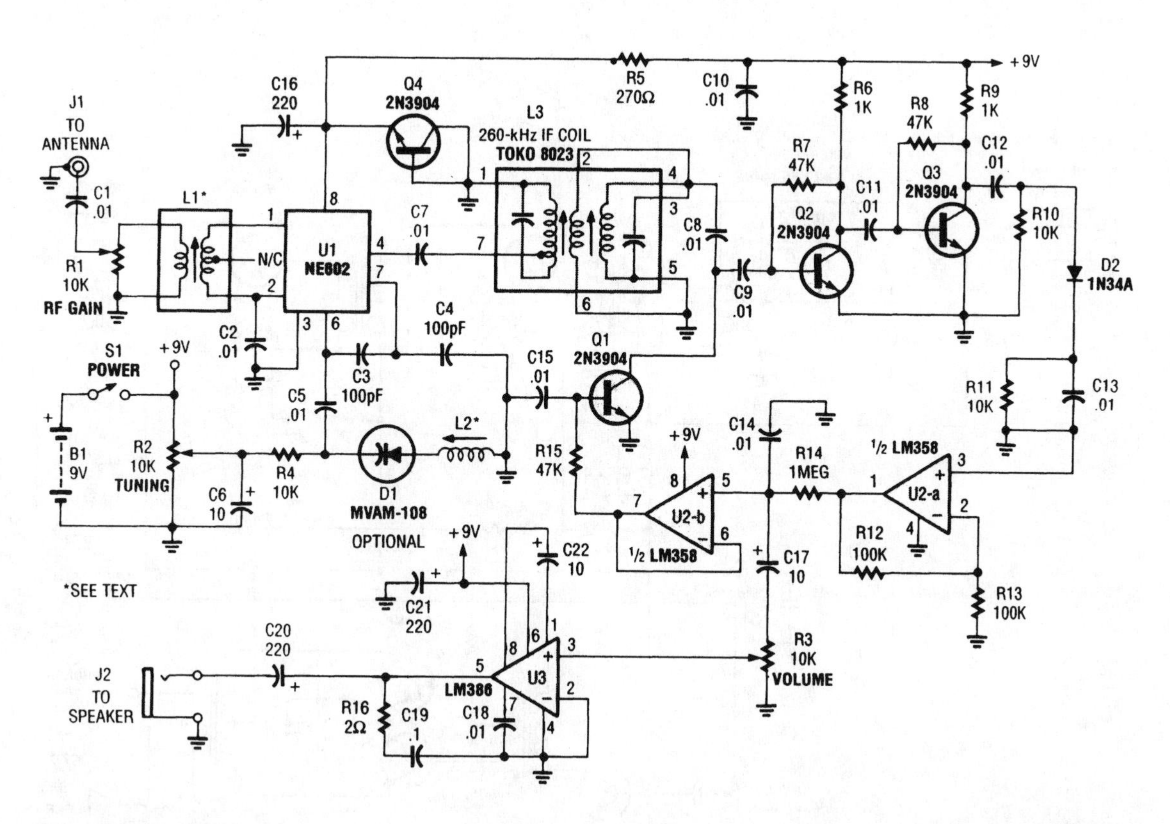

POPULAR ELECTRONICS

FIG. 83-10

Integrated circuit U1 (an NE602 double-balanced mixer) is a combination oscillator and frequency mixer. Signals from the antenna input (at J1) are fed through dc-blocking capacitor C1 to the RF-gain control, R1, and fed to the input of U1 at pins 1 and 2.

The local-oscillator frequency, which varies with the settings of R2 and L2, is mixed internally within U1, resulting in an output. The mixer output at pin 4 of U1 is applied to a tunable 260-kHz band-pass intermediate-frequency (IF) transformer, L3, through dc-blocking capacitor C7. Therefore, signals that are roughly 260 kHz above and below the local-oscillator frequency are passed while others are effectively blocked. The IF frequencies are now amplified by Q2 and Q3. The AM audio signal is detected by D2 and its associated components, which bypass the RF signals, and leave only the audio signals. The signals are preamplified by U1-a (half of an LM358 dual op amp). The audio is then boosted to speaker level by the LM386 low-voltage audio power amplifier, U3.

TRANSISTORIZED AM RADIO

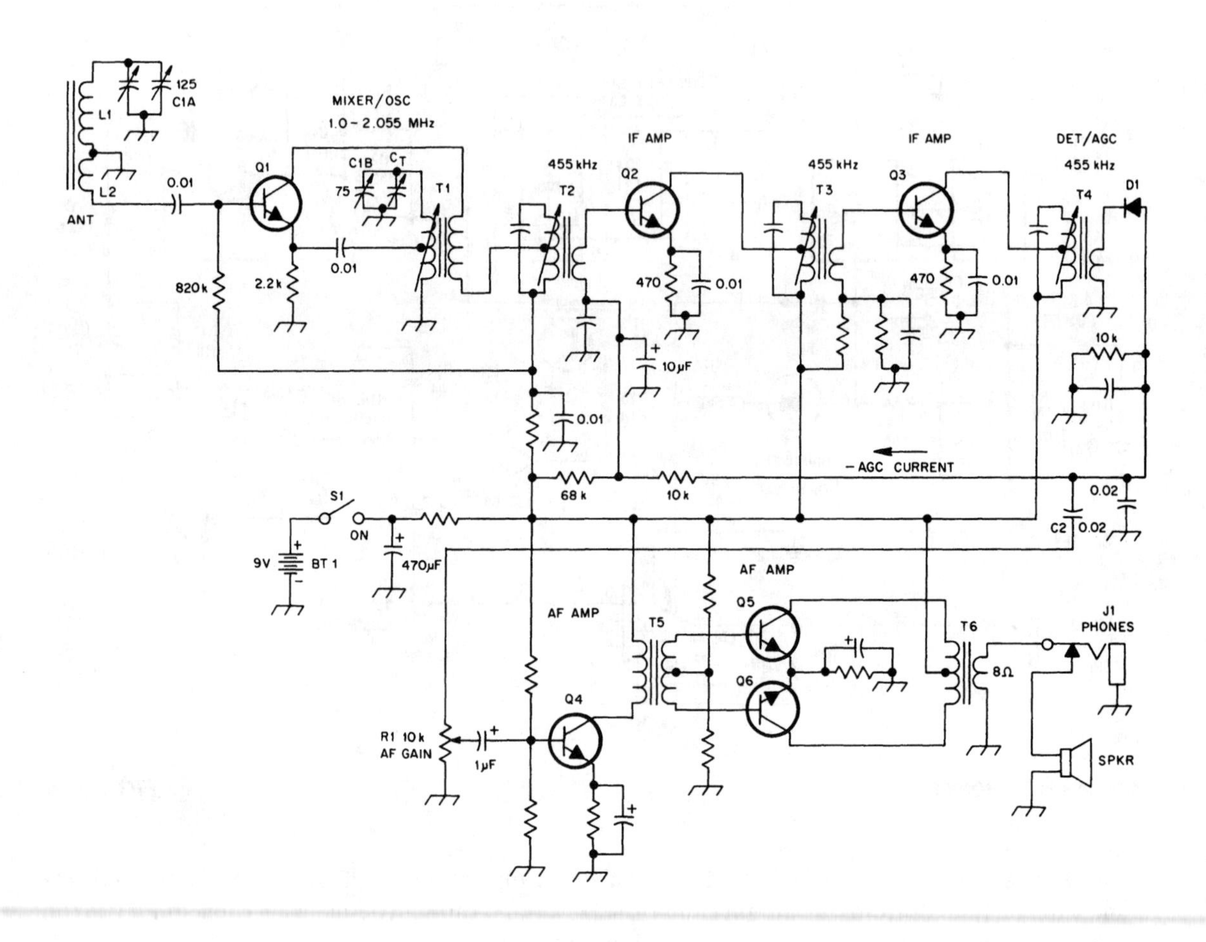

QST ***FIG. 83-11***

Shown is a schematic of a typical transistor AM radio. This circuit uses npn transistors. The circuit is "generic;" therefore, no specific values are given for some components. This circuit is for reference, to serve as a starting point for experimenters.

NE602 SUPERHET FRONT END

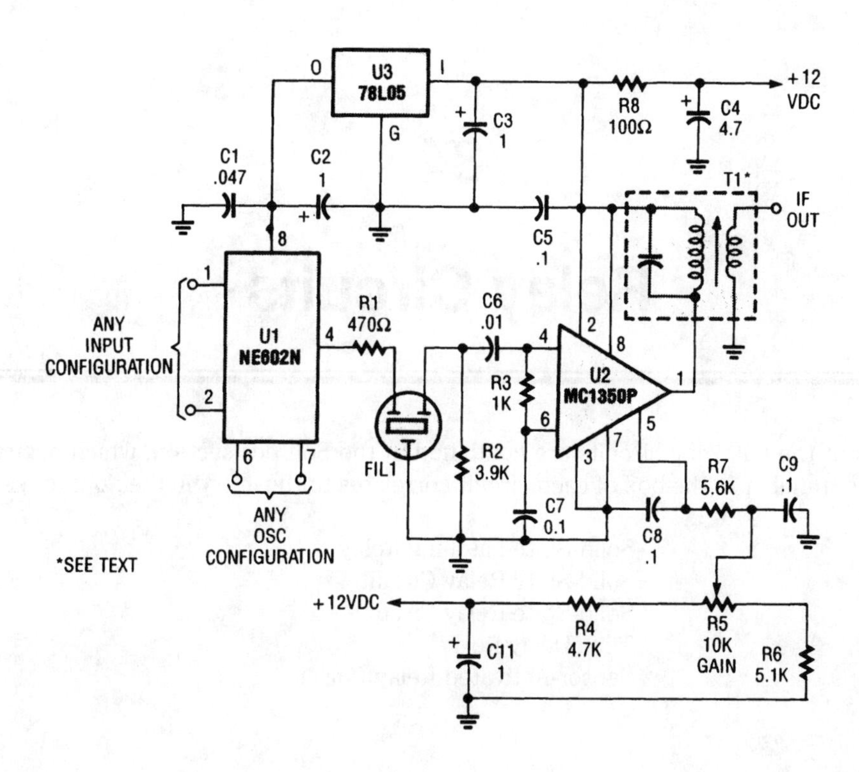

POPULAR ELECTRONICS

FIG. 83-12

By using an NE602 with a filter and an MC1350P IC, a front end and an IF system for a basic superheterodyne receiver can be built with few parts. T1 is any suitable IF transformer for 262 kHz, 455 kHz, 10.7 MHz, etc.

84

Relay Circuits

The sources of the following circuits are contained in the Sources section, which begins on page 675. The figure number in the box of each circuit correlates to the entry in the Sources section.

SOLID-STATE LATCHING RELAY

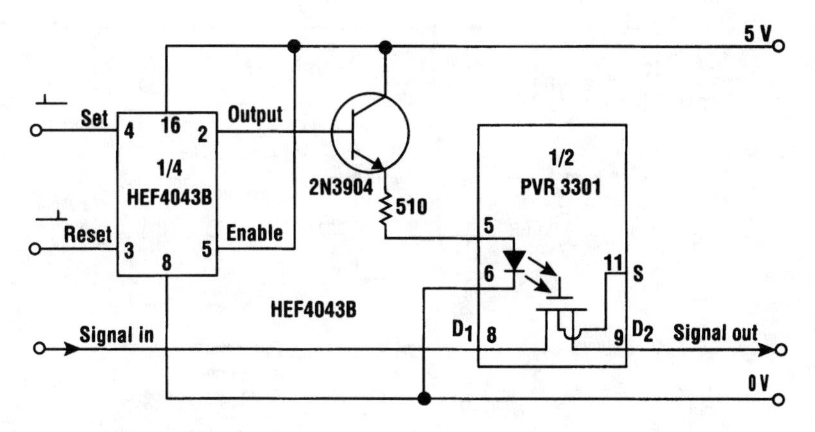

ELECTRONIC DESIGN *FIG. 84-1*

This simple circuit provides a solid-state equivalent of the electromechanical latching relay (see the figure). What's more, the switching is clean, highly resistant to vibration and shock, and isn't sensitive to magnetic fields or position.

The circuit operates as follows: a set pulse to the 4043 RS latch takes its output high and turn on the 2N3904 transistor. Current will then flow through the photovoltaic relay's LED and the resistance between D1 and D2 will fall from several gigaohms to less than 30 Ω. The PVR will remain in this state until a reset pulse is received by the 4043 RS latch.

SOLID-STATE RELAY CIRCUIT

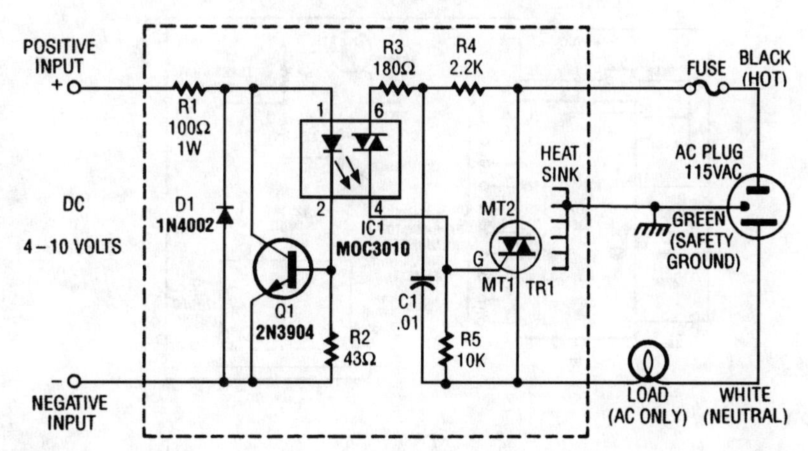

RADIO ELECTRONICS *FIG. 84-2*

R1 limits input current while Q1 acts as a current sink to protect IC1. D1 serves as a polarity protector. IC1 provides a triac output to trigger the main triac, TR1.

SOLID-STATE RELAY CIRCUITS

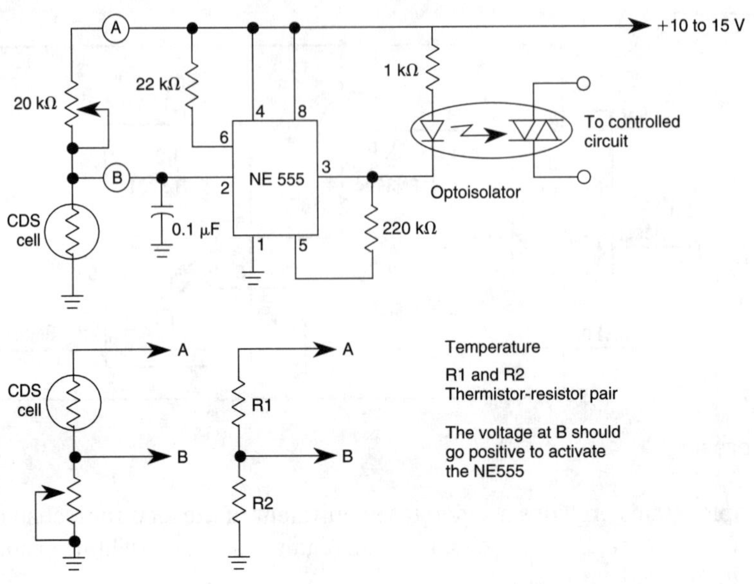

WILLIAM SHEETS

FIG. 84-3

This dark-activated relay switch can be used to turn on walkway or other outdoor lighting at dusk. By using alternate connections to A and B, increasing illumination, high and low temperatures can be sensed.

TIME DELAY RELAY

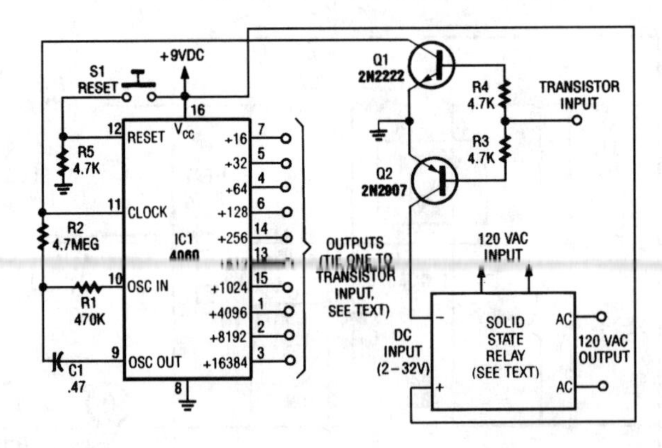

ELECTRONICS NOW

FIG. 84-4

Using a 4060 CMOS binary divider and built-in clock oscillator, a long-duration timer can be made very simply. The solid-state relay can be sized for your application, and can be replaced with a mechanical relay if a suitable power supply is available. With the components shown, a 4.5-Hz clock frequency is generated. Divided outputs are available from ÷ 4 to 16384 (about 4 hours).

SENSOR-ACTIVATED RELAY PULSER

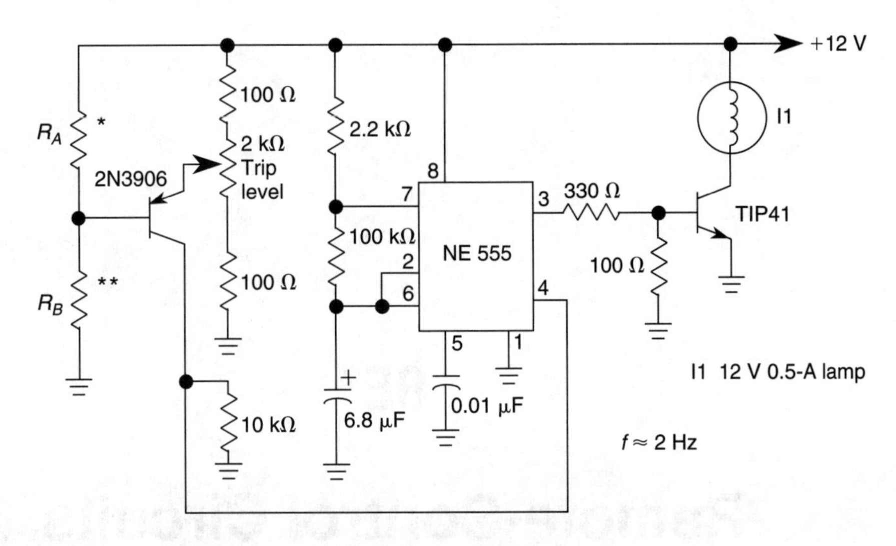

Either R_A or R_B can be sensors, as desired. A decrease in R_B or an increase in R_A will cause the NE555 to flash I1. R_A and R_B should be ≤100 kΩ max.

WILLIAM SHEETS ***FIG. 84-5***

A sensor turns on Q1 to activate the low-frequency 555 oscillator, which pulses LAMP I1. Sensor may be sensitive to changes in light or temperature.

85

Remote-Control Circuits

The sources of the following circuits are contained in the Sources section, which begins on page 675. The figure number in the box of each circuit correlates to the entry in the Sources section.

REMOTE-CONTROL TRANSMITTER

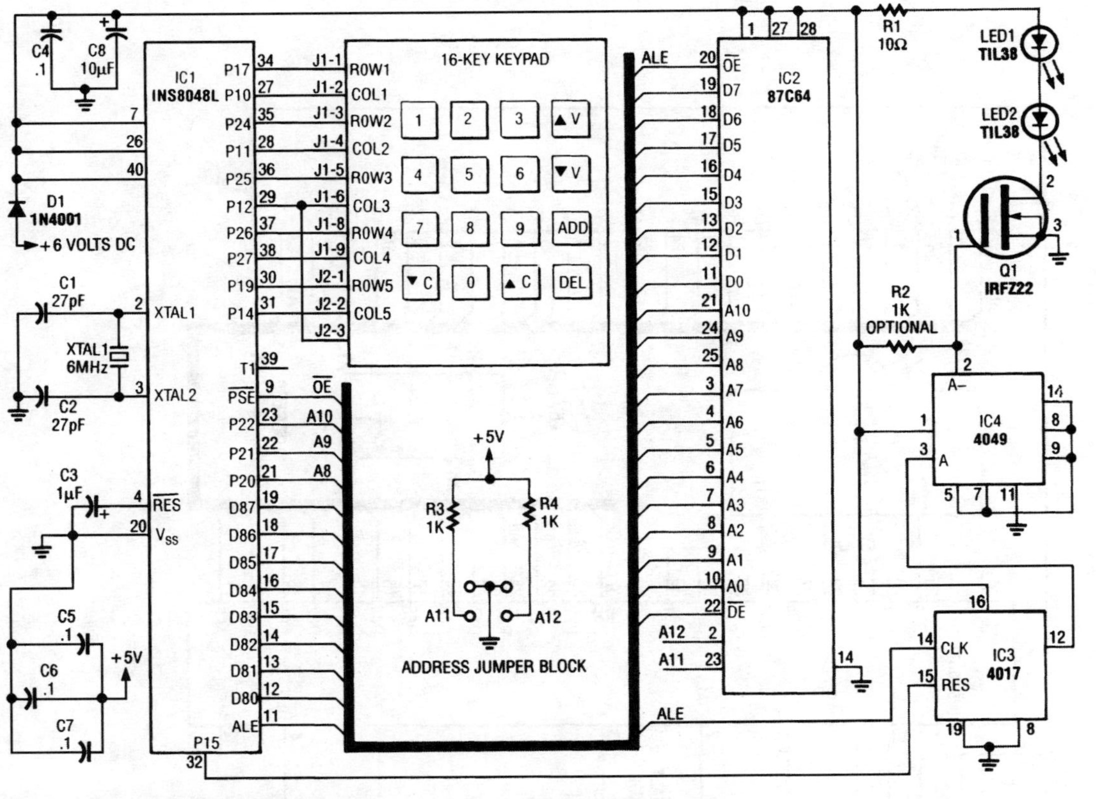

IR TRANSMITTER SCHEMATIC. The 40-kHz carrier is derived by dividing IC1's oscillator frequency (6 MHz) by 15, to get 400 kHz, which is divided by 10 by IC3.

ELECTRONICS NOW

FIG. 85-1

This transmitter sends an FM signal in the 88-to 108-MHz range, with a tone of 19 kHz. This can be used to activate the FM MPX pilot carrier indicator, which can be interfaced to external devices. L4 is for use with a 15 CM wire antenna. L1 is 9 turns of #26 enamelled wire on a ¼-W 10-kΩ resistor (carbon type), L2 is 2 turns wound over L1. L3 is 7 turns of #26 enamelled wire on a 10-kΩ ¼-W resistor.

REMOTE-CONTROL RECEIVER

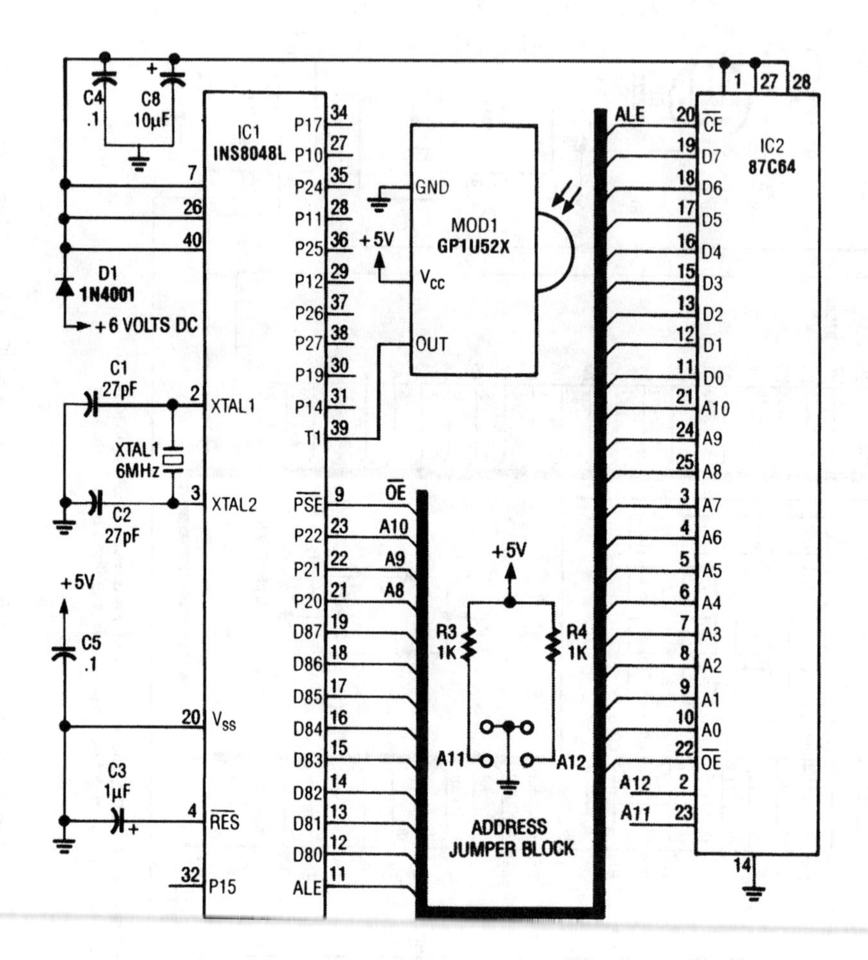

ELECTRONICS NOW *FIG. 85-2*

This circuit is based on the Sharp GP1U52X IR module and INS8048L microprocessor. The GP1U52X is a hybrid IC/infrared detector that provides a strong clean signal for later filtering and demodulation.

INTERFACE CIRCUITS FOR THE REMOTE-CONTROL TRANSMITTER

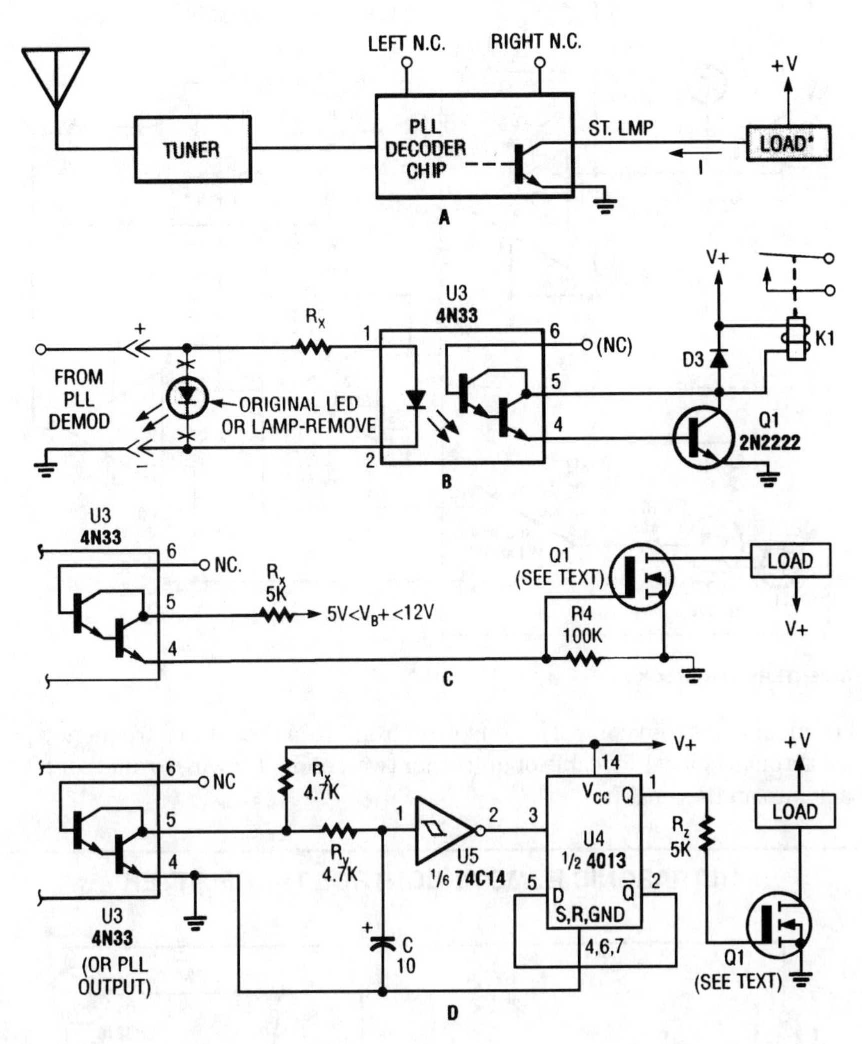

POPULAR ELECTRONICS ***FIG. 85-3***

Shown here are several possible interface circuits that can be used with the remote-control transmitter. The one in A illustrates a typical FM stereo MUX decoder with a load connected directly to the open-collector output of a TA7343 PLL. The circuit in B illustrates an optoisolator-coupler output driving a 12-V relay coil via a general-purpose transistor. C shows the gate of an N-channel power MOSFET connected to the output of a 4N33. The final circuit, D, is a toggle flip-flop that allows push-on/push-off control.

REMOTE-CONTROL EXTENDER

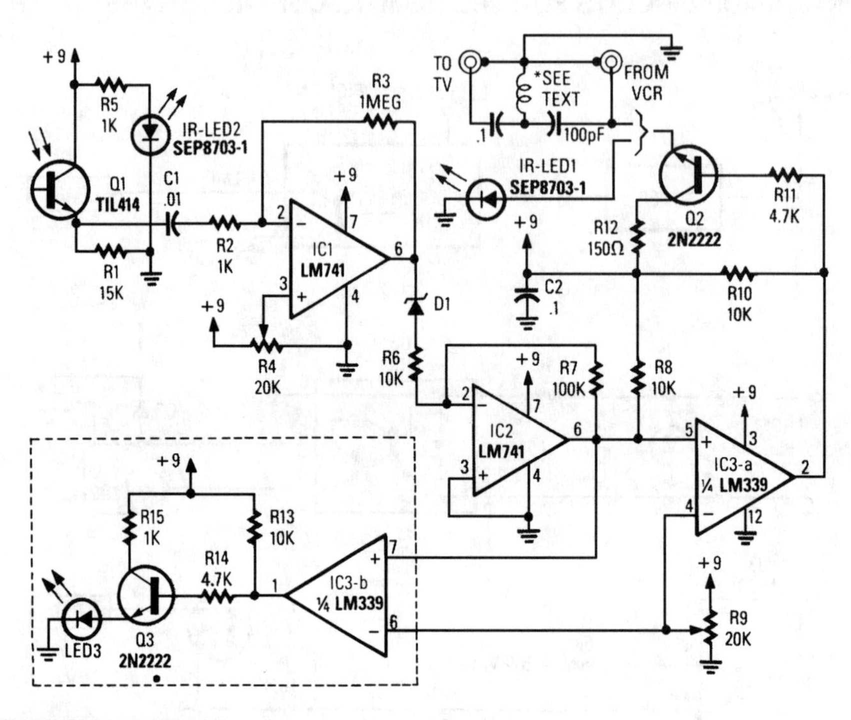

1991 R-E EXPERIMENTERS HANDBOOK *FIG. 85-4*

A signal from an IR remote control is converted from IR radiation to a frequency pulse that can be transmitted through coaxial TV cable or any other two-conductor wire to another room, where it's converted back into an IR signal.

ULTRASONIC REMOTE-CONTROL TRANSMITTER

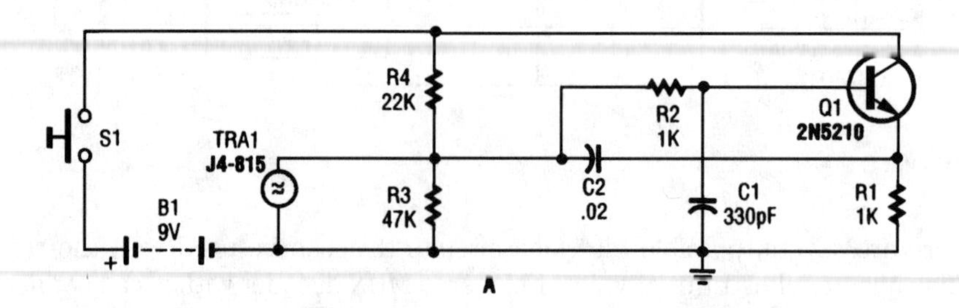

POPULAR ELECTRONICS *FIG. 85-5*

A GC Electronic P/N J4-815 ultrasonic transducer is used in this 40-kHz transmitter for remote-control application.

REMOTE-CONTROL TRANSMITTER

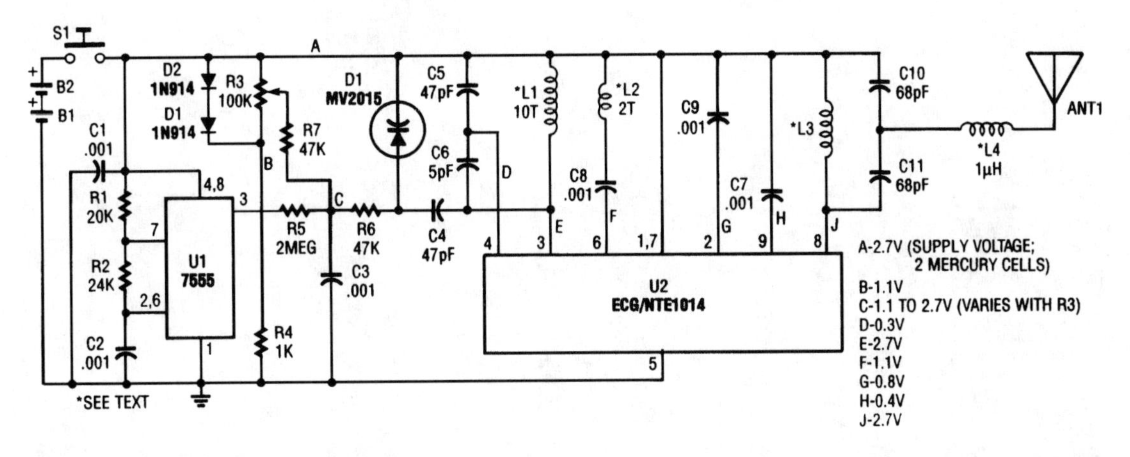

POPULAR ELECTRONICS ***FIG. 85-6***

This transmitter can be used for a variety of purposes. An INS8048L microprocessor generates various codes depending on keypad presses. The codes are modulated on a 40-kHz carrier. Q1 drives IR LEDs LED1 and LED2.

ULTRASONIC REMOTE-CONTROL RECEIVER

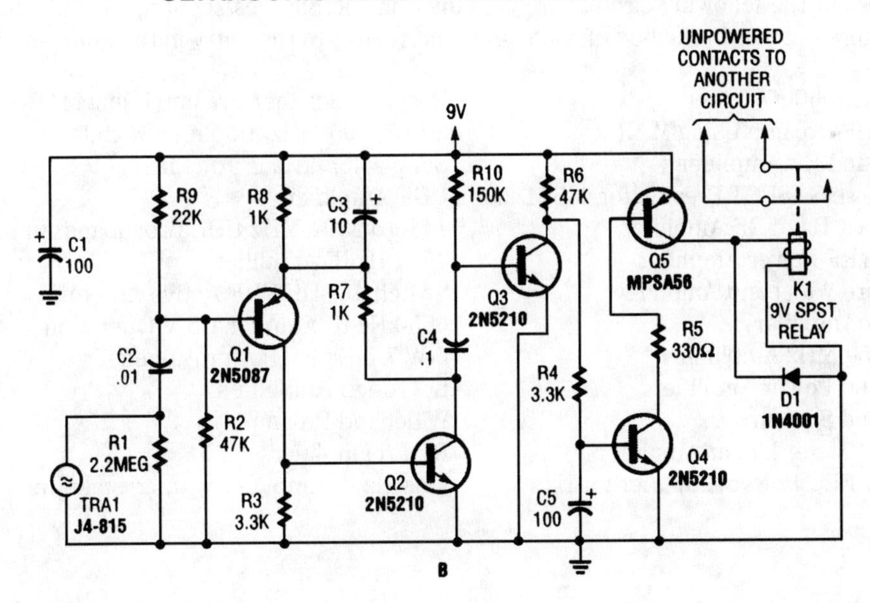

POPULAR ELECTRONICS ***FIG. 85-7***

A GC Electronics P/N J4-815 transducer is used to receive 40-kHz acoustic remote-control signals. The receiver drives a relay for control of another circuit.

86

RF Amplifier Circuits

The sources of the following circuits are contained in the Sources section, which begins on page 675. The figure number in the box of each circuit correlates to the entry in the Sources section.

HF Preamplifier
VHF/UHF Preamp Using MAR-x
Broadband RF Amplifier
Low-Noise GASFET Preamp for 435 MHz
Broadcast-Band RF Amplifier
70-MHz RF Power Amplifier
Miniature Wideband Amplifier
30-MHz Amplifier
20 W 150 MHz Amplifier
Wideband Power Amplifier
TV Sound System
10-W 10-Meter Linear Amplifier
2-Meter FET Power Amplifier for HTs
Receiver/Scanner Preamp Using MAR-1 MMIC
20-W 1296-MHz Amplifier Module
Simple 455-kHz IF Amplifier
UHF Amplifier
144- to 2304-MHz UHF Broadband Amplifier
455-kHz IF Amplifier
Switchable HF/VHF Active Antenna
455-kHz IF Amp for 1.5-V Operation
5-W 7-MHz RF Power Amplifier
LC Tuned Amplifiers
Wideband Preamp
RF Preamplifiers
45-MHz IF Amplifier with crystal filter

HF PREAMPLIFIER

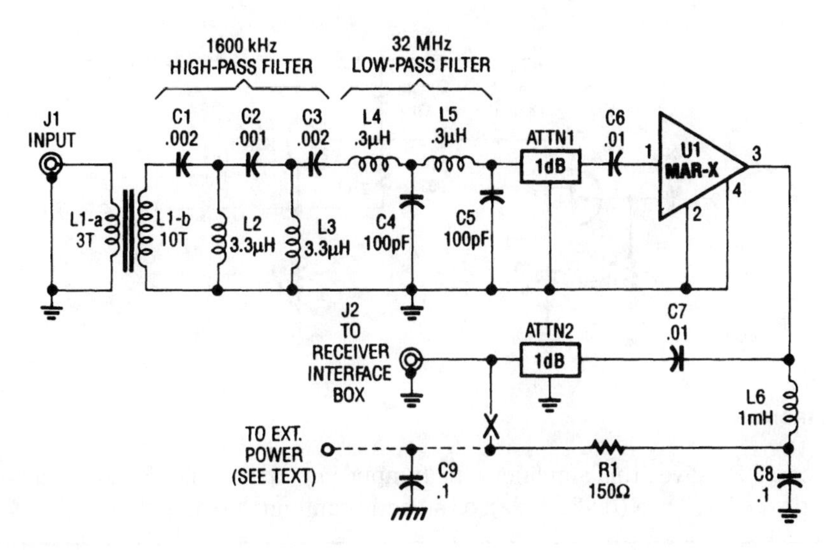

POPULAR ELECTRONICS

FIG. 86-1

This HF SW receiver preamplifier is comprised of a broadband toroidal transformer (L1-a and L1-b), a complex LC network (comprised of a 1600-kHz, high-pass filter and a 32-MHz, low-pass filter), L2 and L3 (26 turns of #26 enameled wire wound on an Amidon Associates T-50-2, red, toroidal core), a pair of resistive attenuators (ATTN1 and ATTN2), and of course, the MAR-x device. External power for the preamp can be 9 to 12 Vdc. R1 can be increased in value for higher voltages.

VHF/UHF PREAMP USING MAR-x

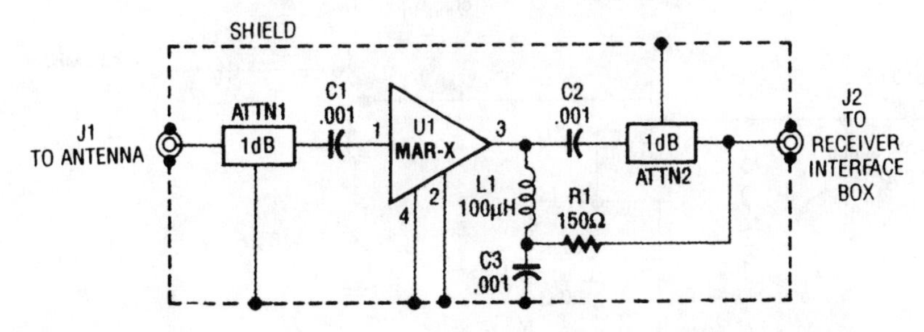

POPULAR ELECTRONICS

FIG. 86-2

The MAR-x preamp shown will cover up to 1.5 or 2 GHz with the correct MAR-x IC. ATTN1 should be omitted for low noise-figure applications. ATTN1 and ATTN2 provide a means of limiting possible termination range, for less chance of device instability.

BROADBAND RF AMPLIFIER

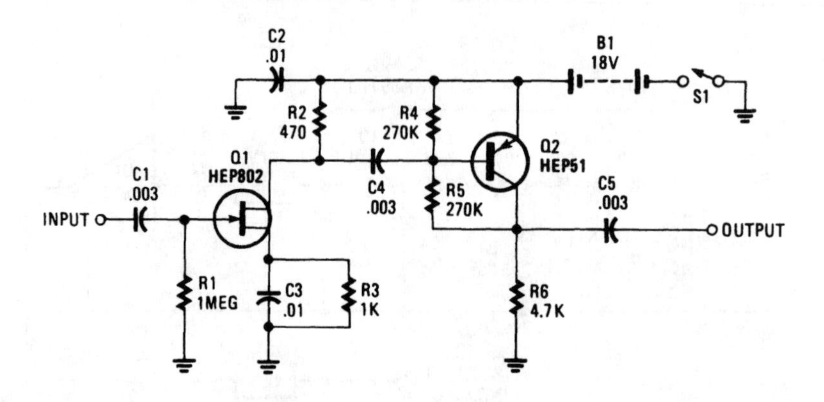

POPULAR ELECTRONICS *FIG. 86-3*

The use of a FET gives this amplifier a high input impedance. The bandwidth should be adequate for LW through HF use (dc-30 MHz), as an active antenna preamplifier.

LOW-NOISE GASFET PREAMP FOR 435 MHz

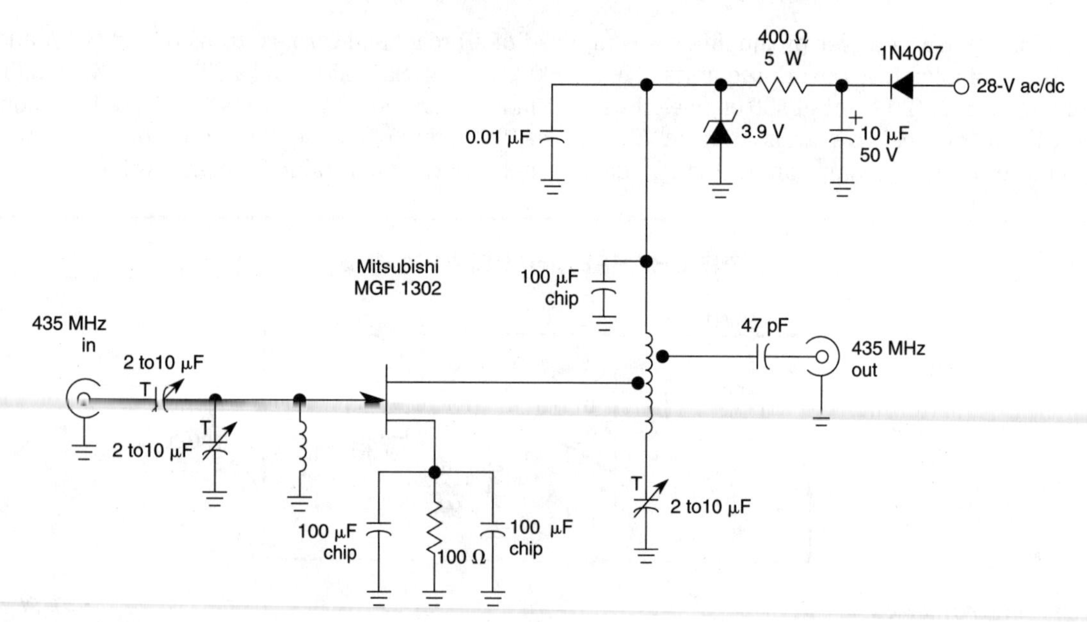

WILLIAM SHEETS *FIG. 86-4*

This circuit is a low-noise preamplifier for the 435-MHz amateur satellite frequencies. The circuit uses a Mitsubishi MGF1302. A 28-Vdc source is shown, although by changing the 400-Ω 5-W resistor lower voltages can be used.

BROADCAST-BAND RF AMPLIFIER

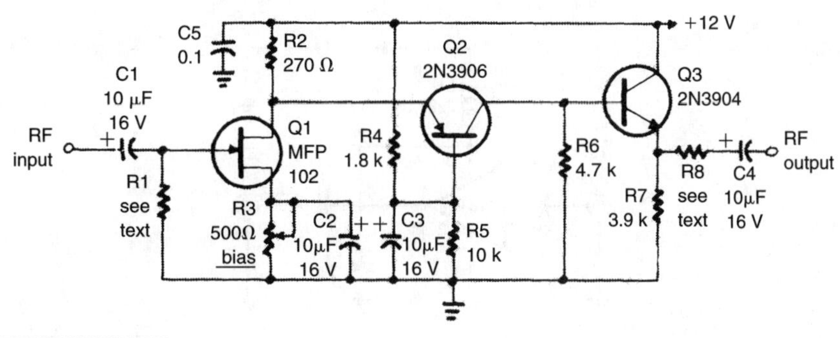

R-E EXPERIMENTERS HANDBOOK *FIG. 86-5*

The circuit has a frequency response that ranges from 100 Hz to 3 MHz; the gain is about 30 dB. Field-effect transistor Q1 is configured in the common-source self-biased mode; optional resistor R1 allows you to set the input impedance to any desired value. Commonly, it will be 50 Ω. The signal is then direct-coupled to Q2, a common-base circuit that isolates the input and output stages and provides the amplifier's exceptional stability. Last, Q3 functions as an emitter-follower, to provide low output impedance (about 50 Ω). If you need higher output impedance, include resistor R8. It will affect impedance according to this formula: $R_8 \approx R_{OUT} - 50$. Otherwise, connect output capacitor C4 directly to the emitter of Q3.

70-MHz RF POWER AMPLIFIER

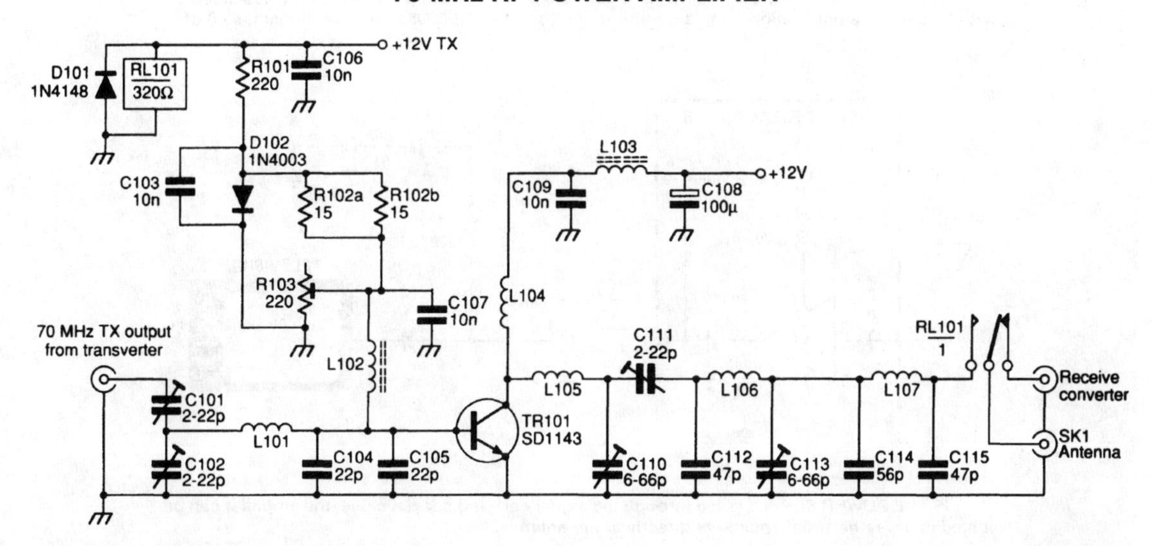

PRACTICAL WIRELESS *FIG. 86-6*

The SD1143 transistor provides a gain of about 14 dB in this circuit. It uses the fact that a 175-MHz device has a much higher gain when used at lower frequencies. The amplifier was originally designed to be used with a transverter. The output is 8 to 10 W for a 300- to 500-mW input.

MINIATURE WIDEBAND AMPLIFIER

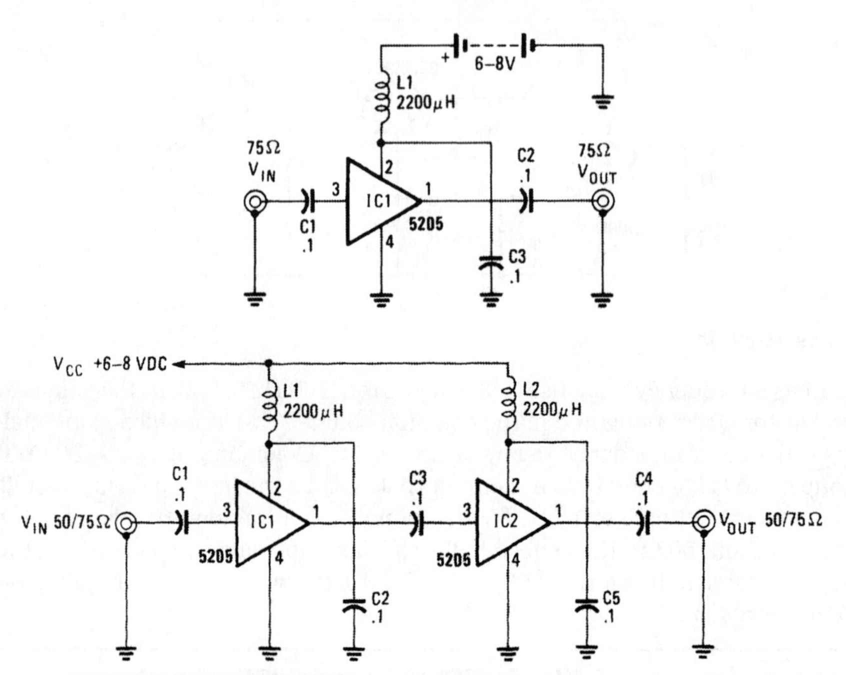

SINCE THE NE5205 FUNCTIONS as a gain block, two or more can be easily cascaded to provide additional amplification. In this circuit, which uses two NE5205s, the overall gain is 40 dB.

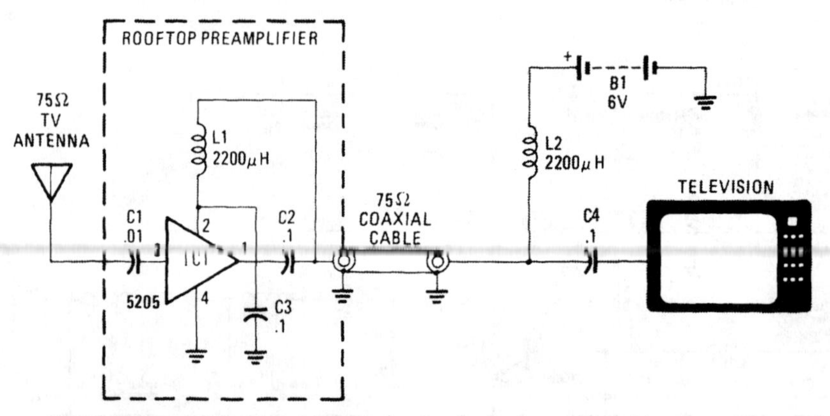

IF THE POWER SUPPLY is fed through the signal-carrying coaxial cable, the amplifier can be mounted in a weatherproof enclosure directly at the antenna.

R-E EXPERIMENTERS HANDBOOK *FIG. 86-7*

Except for the coupling and decoupling capacitors, IC1 is a complete wideband amplifier that has a fixed gain of 20 dB to 450 MHz. No external compensation is required.

30-MHz AMPLIFIER

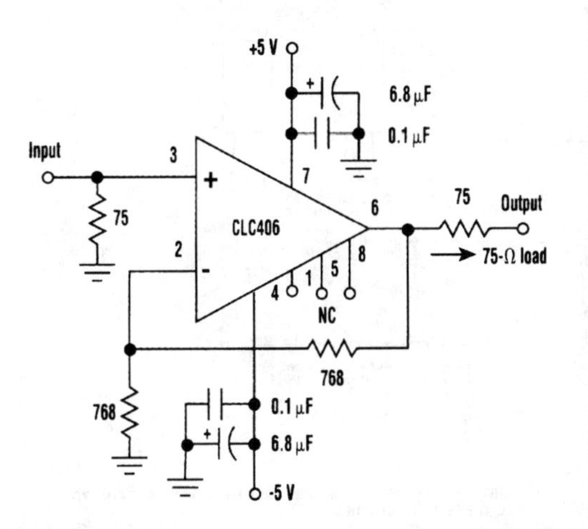

ELECTRONIC DESIGN *FIG. 86-8*

Using a CLC406 op amp, this video amplifier has a voltage gain of +2 and is flat to 30 MHz. The circuit should be useable in video switching and interfacing applications.

20-W 450-MHz AMPLIFIER

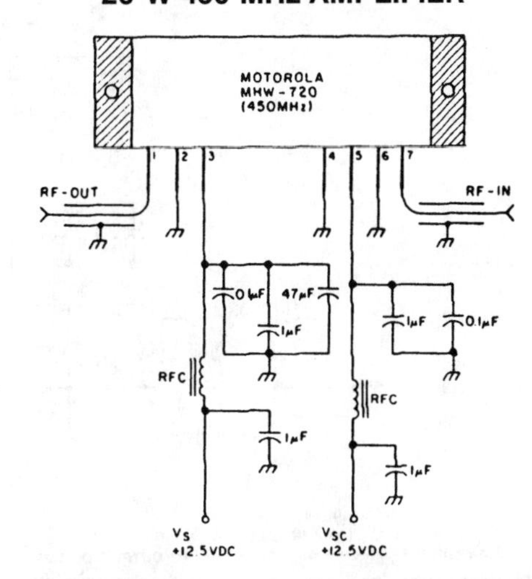

73 AMATEUR RADIO *FIG. 86-9*

Delivering 20-W output, this amplifier has a gain of 21 dB at 450 MHz. A 12-V supply powers this circuit.

WIDEBAND POWER AMPLIFIER

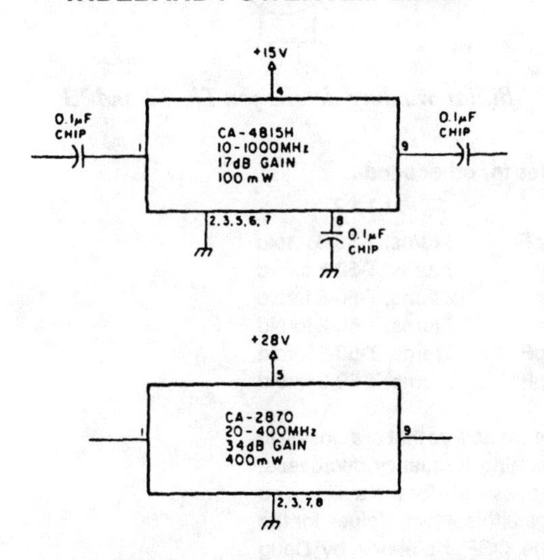

73 AMATEUR RADIO *FIG. 86-10*

Using TRW P/N CA-815H, a 17-dB gain amplifier that delivers 100 mW over 10 to 1000 MHz can be constructed. The CA-2870 will yield 0.4 W with 34-dB gain from 20 to 400 MHz.

TV SOUND SYSTEM

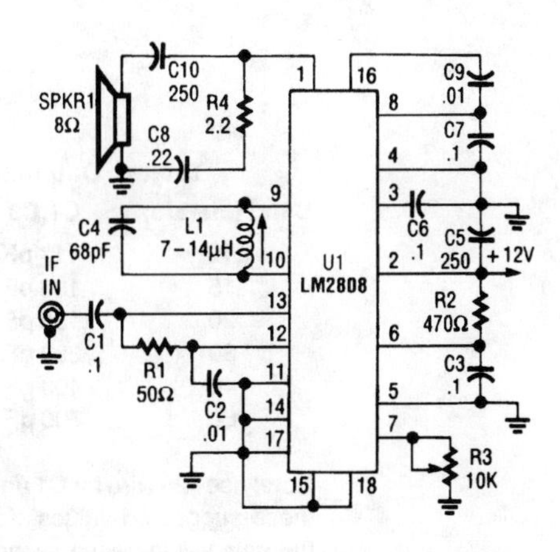

POPULAR ELECTRONICS *FIG. 86-11*

An LM2808 performs IF amplification of the 4.5-MHz sound subcarrier, limiting, detection, and audio amplification. If the center frequency must be changed, then change L1/C4. Audio output is 0.5 W. R3 is the volume control.

10-W 10-METER LINEAR AMPLIFIER

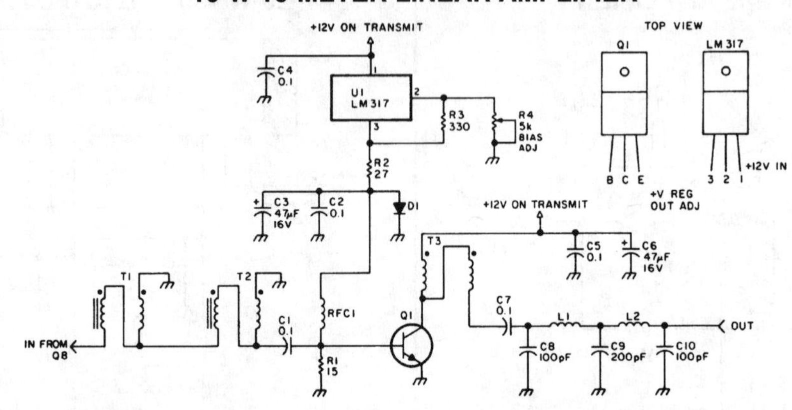

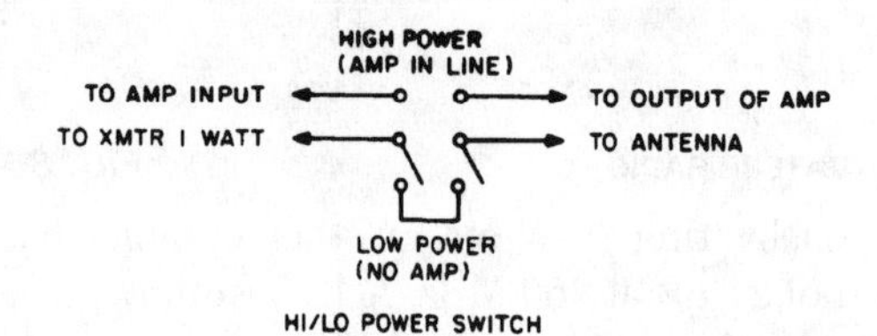

A double-pole double-throw switch can be used to switch the amplifier in and out of the circuit.

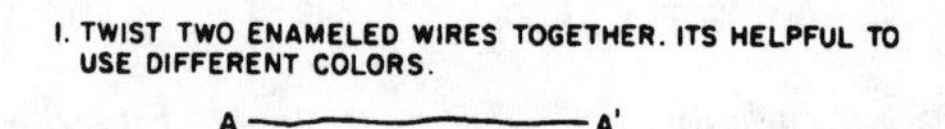

2. WIND THE A, B PAIR AROUND TOROID THE RIGHT NUMBER OF TURNS. 3. SOLDER END B TO END A'.

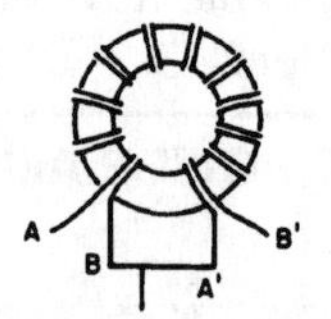

Bifilar winding details for T1, T2 and T3.

Table 1. Output filter values for other bands.

Band (meters)	C1,C3	C2	L1,L2
12	117 pF	220 pF	8 turns, T-50-6 toroid
15	138 pF	270 pF	9 turns, T-50-6 toroid
20	138 pF	420 pF	12 turns, T-50-6 toroid
30	289 pF	579 pF	12 turns, T-50-2 toroid
40	400 pF	800 pF	14 turns, T-50-2 toroid
80	700 pF	1415 pF	19 turns, T-50-2 toroid

Note: use #26 wire for C1 and C2. Use capacitors that are closest to these suggested values. As the operating frequency decreases, the gain will increase as well as the possibility for instability. You may have to use RC feedback to negate this effect. Values for the above table were obtained from the QRP Notebook by Doug DeMaw.

73 AMATEUR RADIO TODAY ***FIG. 86-12***

This linear amplifier delivers 10-W PEP output with 1.25-W drive on 10 m. T1, T2, and T3 are 10 turns of bifilar windings on an FT-50-43 toroidal core. The transformers are broadband. Filters for other bands, if desired, are shown.

2-METER FET POWER AMPLIFIER FOR HTs

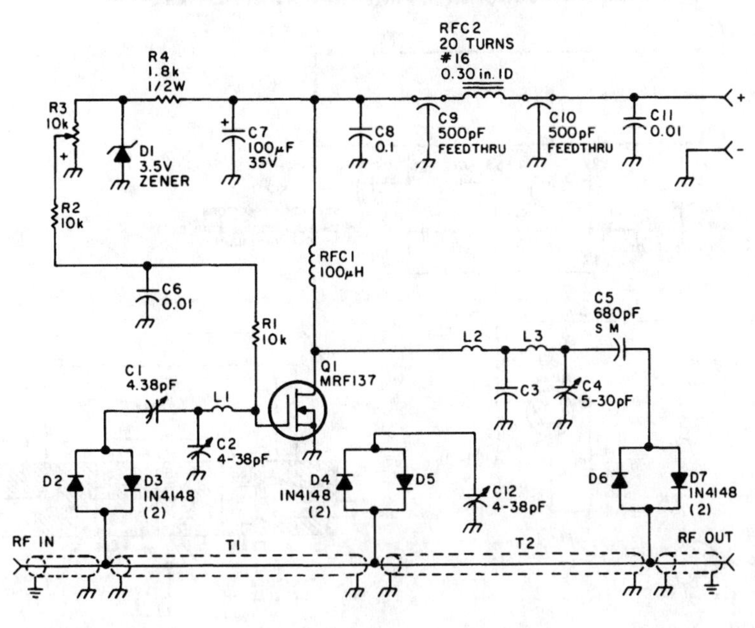

73 AMATEUR RADIO TODAY ***FIG. 86-13***

Using a power MOSFET, this amplifier can boast a 2-W handie-talkie power level to around 10 W on 2 meters. A transmission-line RF switch is used for T/R switching.

RECEIVER/SCANNER PREAMP USING MAR-1 MMIC

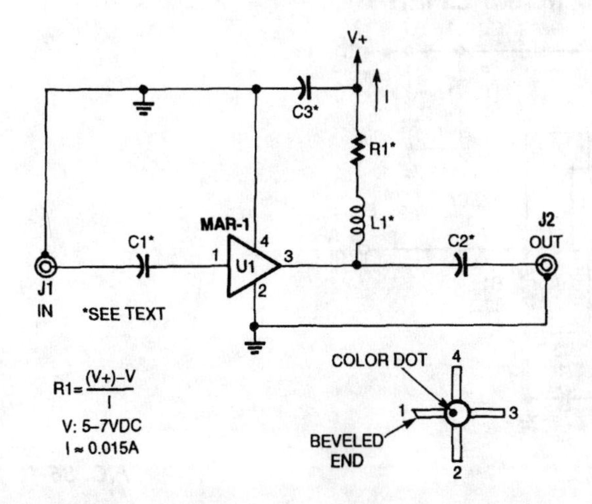

The low-cost Mini-Circuits MAR-X series of chips offer the RF builder a real advantage, with their inherent 50-Ω input and output impedances (needed for RF systems). An MAR-1-based receiver/scanner preamplifier is shown. C1 and C2 are chip capacitors. Use 0.01 μF for HF, 0.001 for VHF, and 100 pF for above 100 MHz, depending on the low-frequency limit that you desire. C3 can be a ceramic disc of 0.01 μF or 0.001 μF, depending on frequency range. L1 is an RF choke that is suitable for the frequency range that you desire (0.1 to 10 μH).

POPULAR ELECTRONICS ***FIG. 86-14***

20-W 1296-MHz AMPLIFIER MODULE

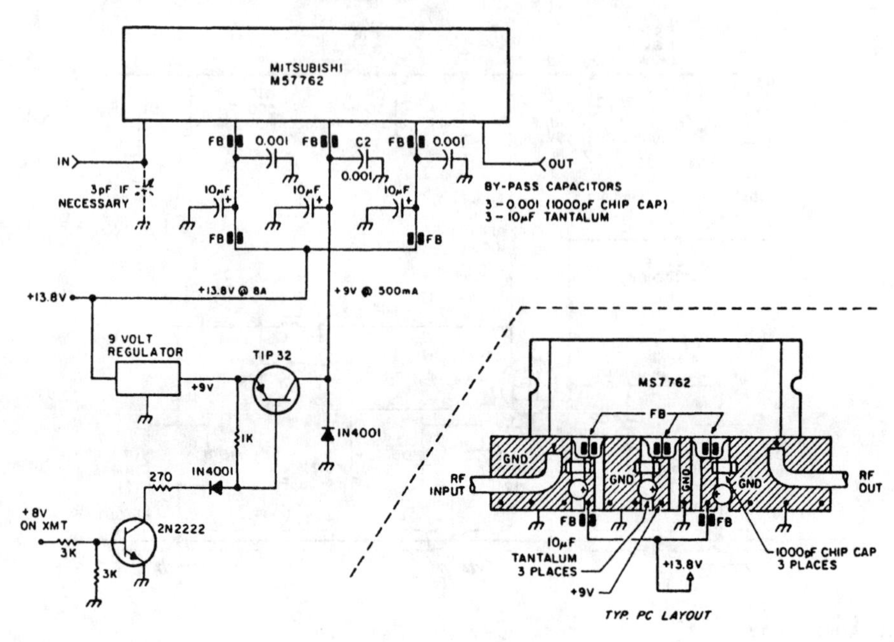

73 AMATEUR RADIO ***FIG. 86-15***

Using a Mitsubishi M57762 amplifier module, this amplifier delivers 20-W output on 1296 MHz. A single 12-V nominal power supply can be used.

SIMPLE 455-kHz IF AMPLIFIER

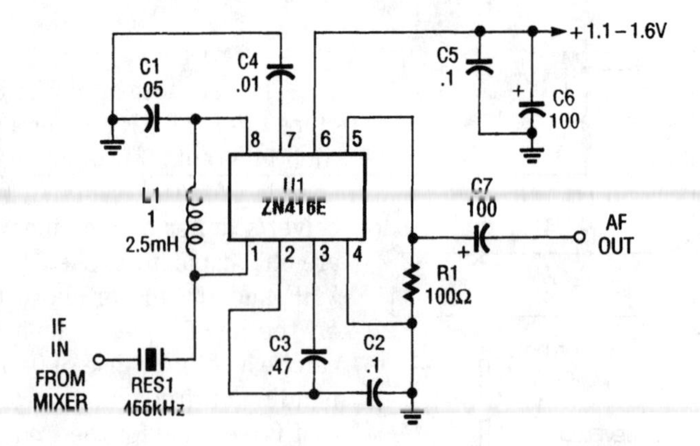

POPULAR ELECTRONICS ***FIG. 86-16***

The ZN416E can be configured as a simple 455-kHz IF amplifier. In this case, the circuit's center frequency and bandwidth are set by RES1 (a Murata CSB455E ceramic resonator).

UHF AMPLIFIER

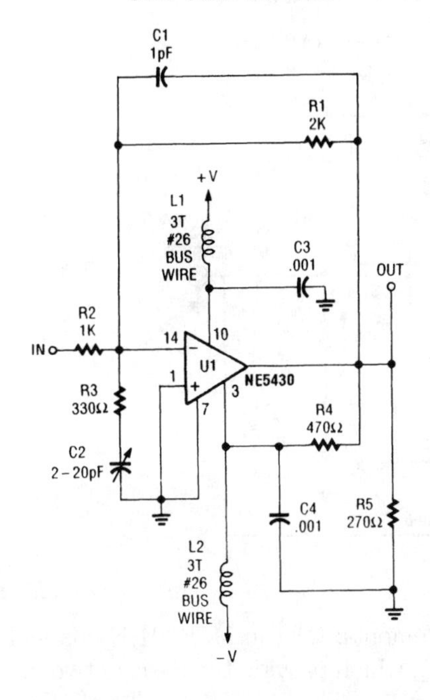

POPULAR ELECTRONICS *FIG. 86-17*

144- TO 2304-MHz UHF BROADBAND AMPLIFIER

Table 1.

Device	Max. mA	Normal Current mA.	Approx. Gain 1-GHz
MAR-1	40	20–30 mA	18 dB
MAR-2	60	30–40 mA	13 dB
MAR-3	70	30–50 mA	12 dB
MAR-4	85	50–70 mA	8 dB
MAR-6	50	15–25 mA	17 dB
MAR-7	60	25–40 mA	13 dB
MAR-8	65	30–50 mA	23 dB

Table 2.
MMIC Amplifier Performance

144 MHz	18.2 dB	2.7 dB N/F
220 MHz	18.3 dB	2.6 dB N/F
432 MHz	16.5 dB	2.8 dB N/F
902 MHz	15.0 dB	2.9 dB N/F
1296 MHz	13.0 dB	3.5 dB N/F
2304 MHz	8.8 dB	4.2 dB N/F

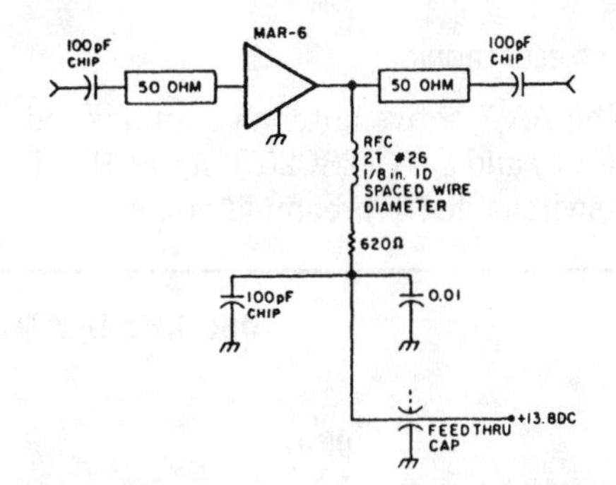

73 AMATEUR RADIO *FIG. 86-18*

Based on an MAR-6 preamp, this circuit yields low noise figures and useful gain for the 144-MHz to 2304-MHz amateur bands.

455-kHz IF AMPLIFIER

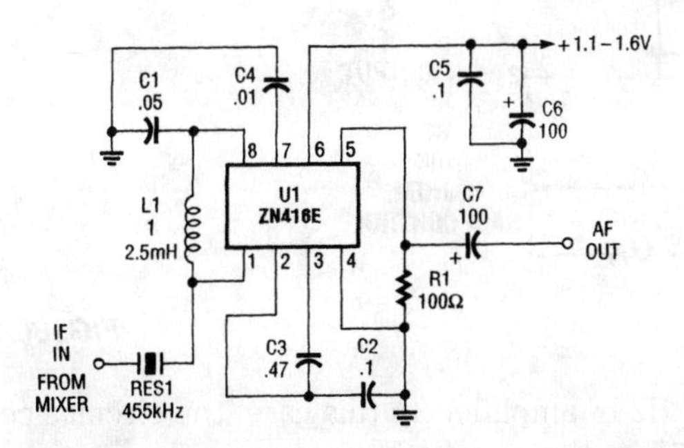

POPULAR ELECTRONICS *FIG. 86-19*

Up to 60 dB of gain at 455 kHz is available with the MC1350P. RES1 is a ceramic resonator, LC, or crystal filter. Keep the leads to pins, 1, 2, 3, and 7 short.

SWITCHABLE HF/VHF ACTIVE ANTENNA

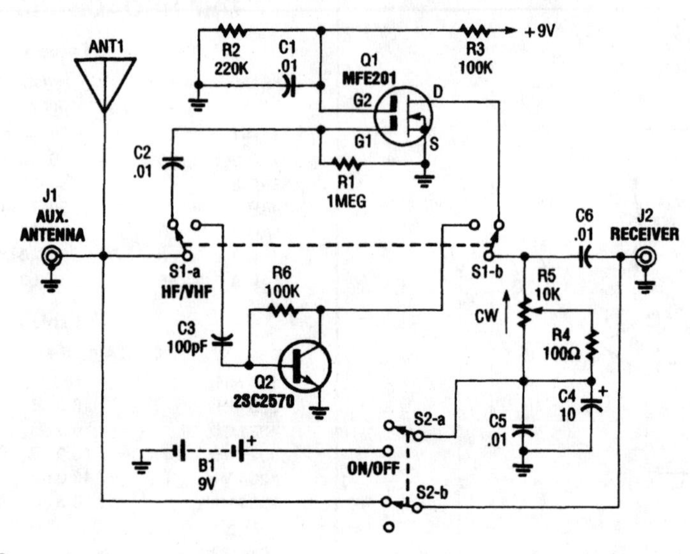

POPULAR ELECTRONICS *FIG. 86-20*

The AA-7 active antenna contains only two active elements: Q1 (an MFE201 N-channel dual-gate FET) and Q2 (a 2SC2570 npn VHF silicon transistor), which provide the basis of two independent, switchable RF preamplifiers.

455-kHz IF AMP FOR 1.5-V OPERATION

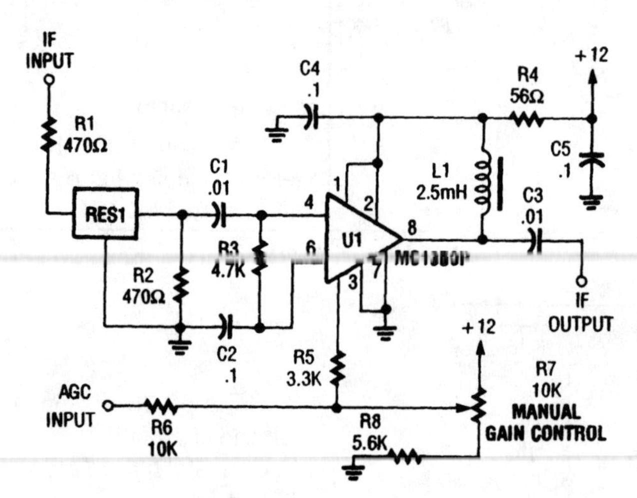

POPULAR ELECTRONICS *FIG. 86-21*

The ZN416E can be configured as a simple 455-kHz IF amplifier. In this case, the circuit's center and bandwidth are set by RES1 (a Murata CSB455E ceramic resonator).

5-W 7-MHz RF POWER AMPLIFIER

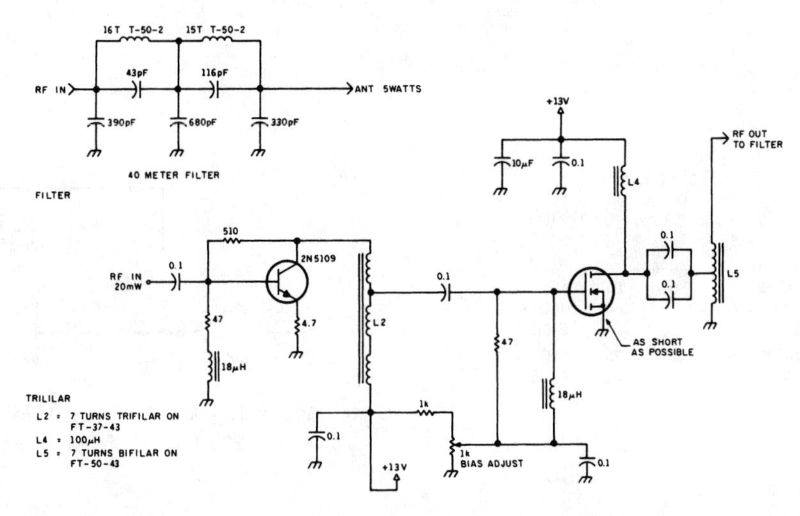

73 AMATEUR RADIO TODAY **FIG. 86-22**

The circuit shown will produce up to 5-W RF output in the 40-m (7 MHz) amateur band. The coils shown are wound on toroidal cores (Armdon Associates Inc.). The part numbers are given in the schematic. The circuit requires about 20-mW drive and a 13-V supply.

LC TUNED AMPLIFIERS

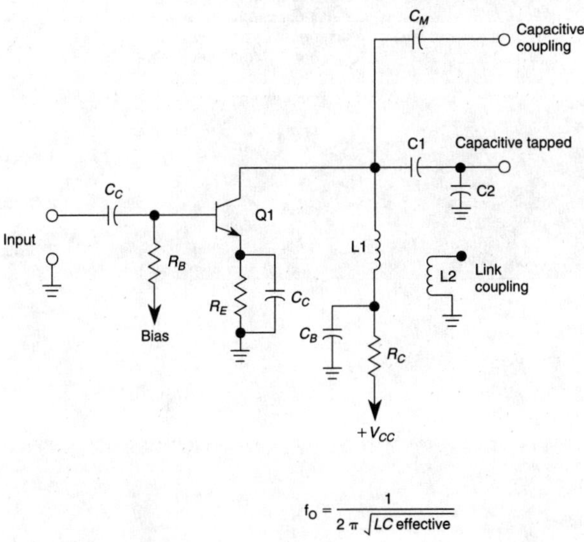

$$f_o = \frac{1}{2\pi\sqrt{LC\text{ effective}}}$$

WILLIAM SHEETS **FIG. 86-23**

This basic tuned LC amplifier can be used with three output coupling methods. They are capacitive coupling output, capacitive tapped output, or link-coupled output.

WIDEBAND PREAMP

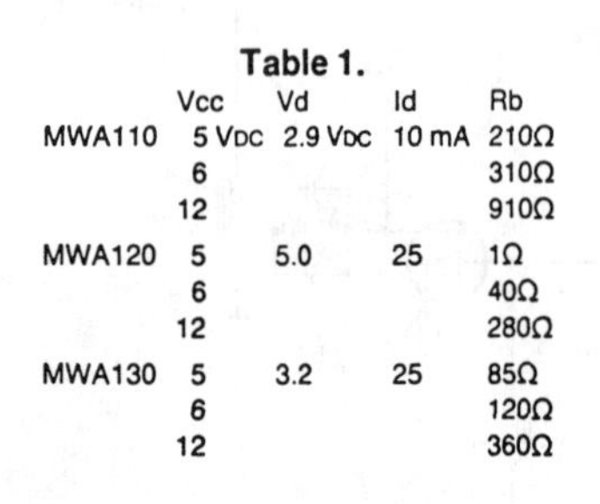

Table 1.

	Vcc	Vd	Id	Rb
MWA110	5 VDC	2.9 VDC	10 mA	210Ω
	6			310Ω
	12			910Ω
MWA120	5	5.0	25	1Ω
	6			40Ω
	12			280Ω
MWA130	5	3.2	25	85Ω
	6			120Ω
	12			360Ω

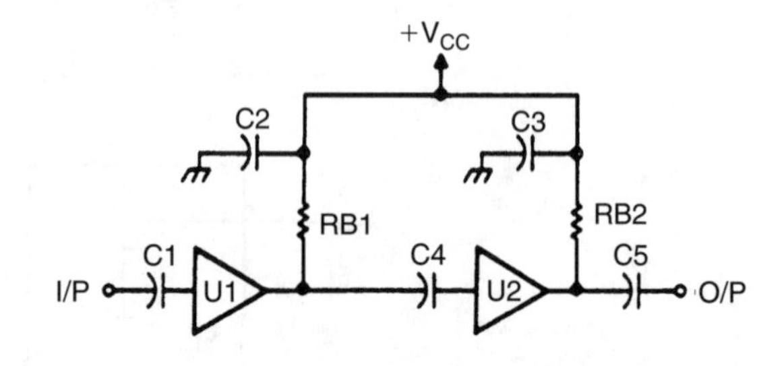

V_{CC} = 12 Vdc; C1 to C5 = 0.1µF; RB1 = 910Ω; RB2 = 280Ω; U1 = MWA110; U2 = MWA120

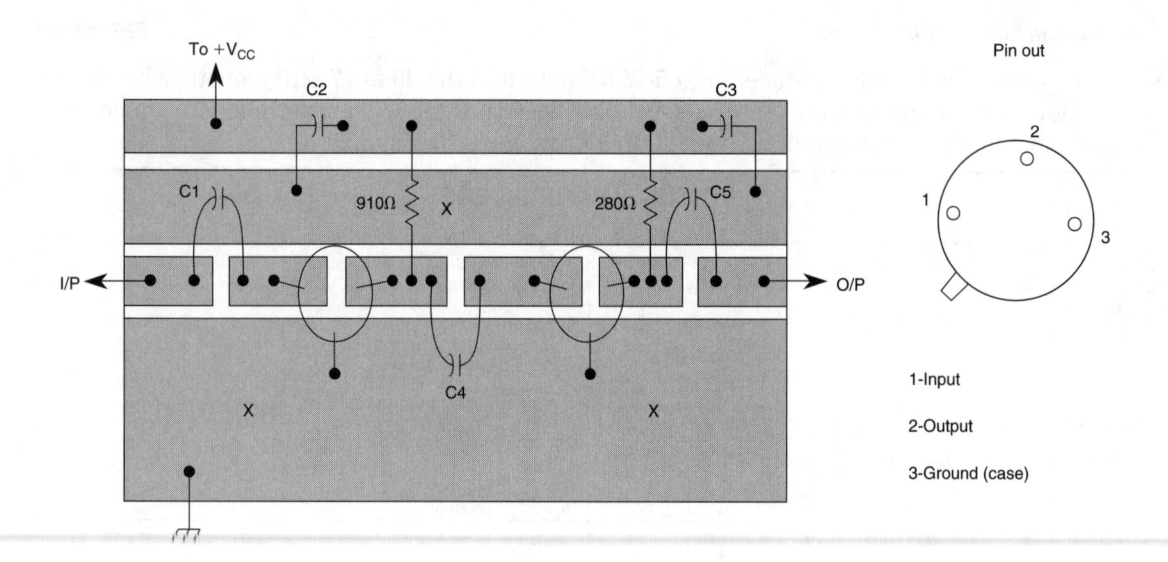

PC board layout (shading represents copper) and parts layout. "X" is the feedthrough wire to the gound plane. All capacitors are 0.1 µF. Keep all leads short.

73 AMATEUR RADIO ***FIG. 86-24***

Motorola MWA 110, 120, or 130 are wideband amplifier ICs. This wideband preamp circuit can be used in many applications. Keep the leads short when constructing the circuitry.

RF PREAMPLIFIERS

TABLE 1—MAR-X CAPABILITIES

DEVICE	MAX. FREQ. (MHz)	GAIN (100/50/1000 MHz)	N.F.	COLOR
MAR-1	1,000	18.5/17.5/15.5	5	Brown
MAR-2	2,000	13/12.8/12.5	6.5	Red
MAR-3	2,000	13/12.8/12.5	6	Orange
MAR-4	1,000	8.2/8.2/8	7	Yellow
MAR-6	2,000	20/19/16	2.8	White
MAR-7	2,000	13.5/13.1/12.5	5	Violet
MAR-8	1,000	33/28/23	3.5	Blue

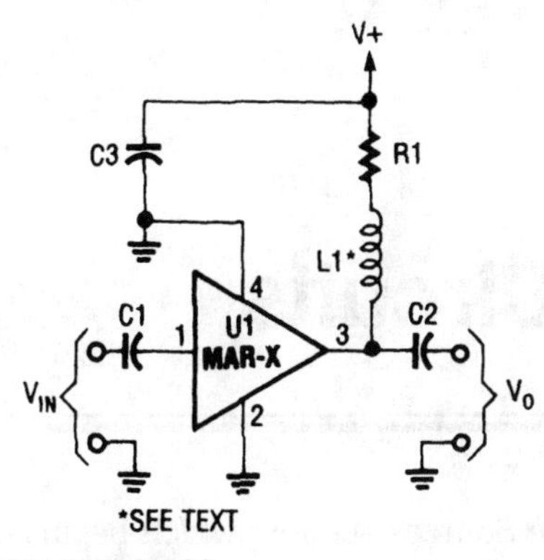

In this basic MAR-x-based circuit, both the input and output are comprised of a single dc-blocking capacitor (C1 and C2 for the input and output, respectively). The dc power-supply network (comprised of L1 and R1) is attached to the MAR-x via the RF-output terminal (lead 3).

POPULAR ELECTRONICS

FIG. 86-25

45-MHz IF AMPLIFIER WITH CRYSTAL FILTER

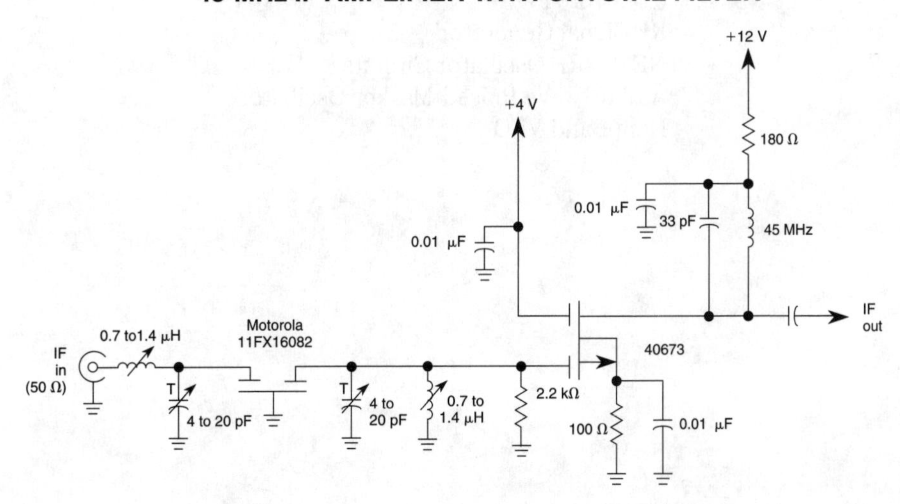

WILLIAM SHEETS

FIG. 86-26

A 40673 dual-gate MOSFET is matched to a crystal filter at 45 MHz. The filter impedance is around 2kΩ. The + 4-V source can be made variable for gain control (about +4 to −4V.)

87

RF Oscillator Circuits

The sources of the following circuits are contained in the Sources section, which begins on page 675. The figure number in the box of each circuit correlates to the entry in the Sources section.

6.5-MHz VFO

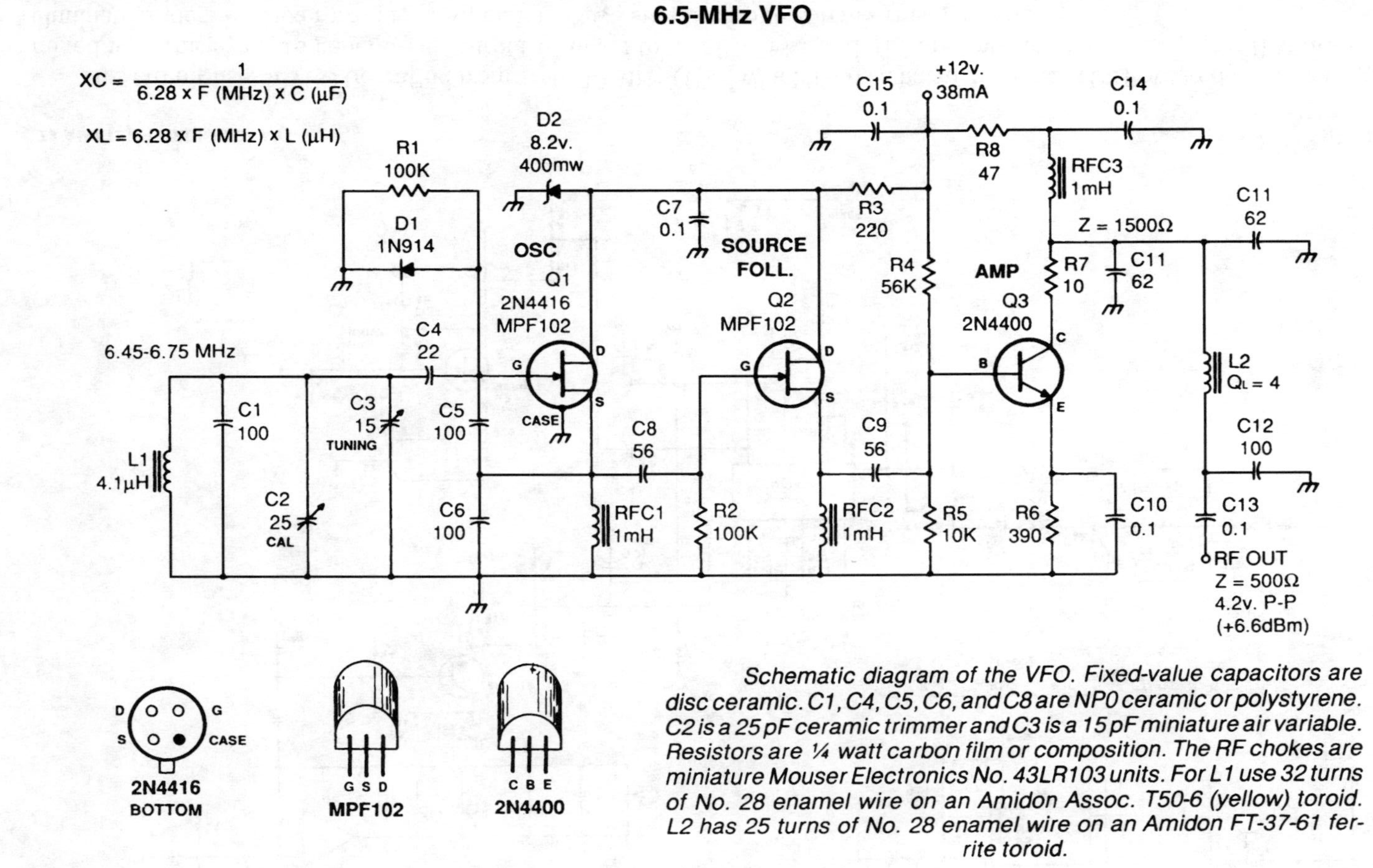

Schematic diagram of the VFO. Fixed-value capacitors are disc ceramic. C1, C4, C5, C6, and C8 are NP0 ceramic or polystyrene. C2 is a 25 pF ceramic trimmer and C3 is a 15 pF miniature air variable. Resistors are ¼ watt carbon film or composition. The RF chokes are miniature Mouser Electronics No. 43LR103 units. For L1 use 32 turns of No. 28 enamel wire on an Amidon Assoc. T50-6 (yellow) toroid. L2 has 25 turns of No. 28 enamel wire on an Amidon FT-37-61 ferrite toroid.

QST

FIG. 87-1

Fixed-value capacitors are disc ceramics. C1, C4, C5, C6, and C8 are NPO ceramic or polystyrene. C2 is a 25-pF ceramic trimmer and C3 is a 15-pF miniature air variable capacitor. The resistors are ¼-W carbon film or composition. The RF chokes are miniature Mouser Electronics No. 43LR103 units. For L1, use 32 turns of #28 enamel wire on an Amidon Assoc. T50-6 (yellow) toroid. L2 has 25 turns of #28 enamel wire on an Amidon Ft-37-61 ferrite toroid.

RF SIGNAL GENERATOR

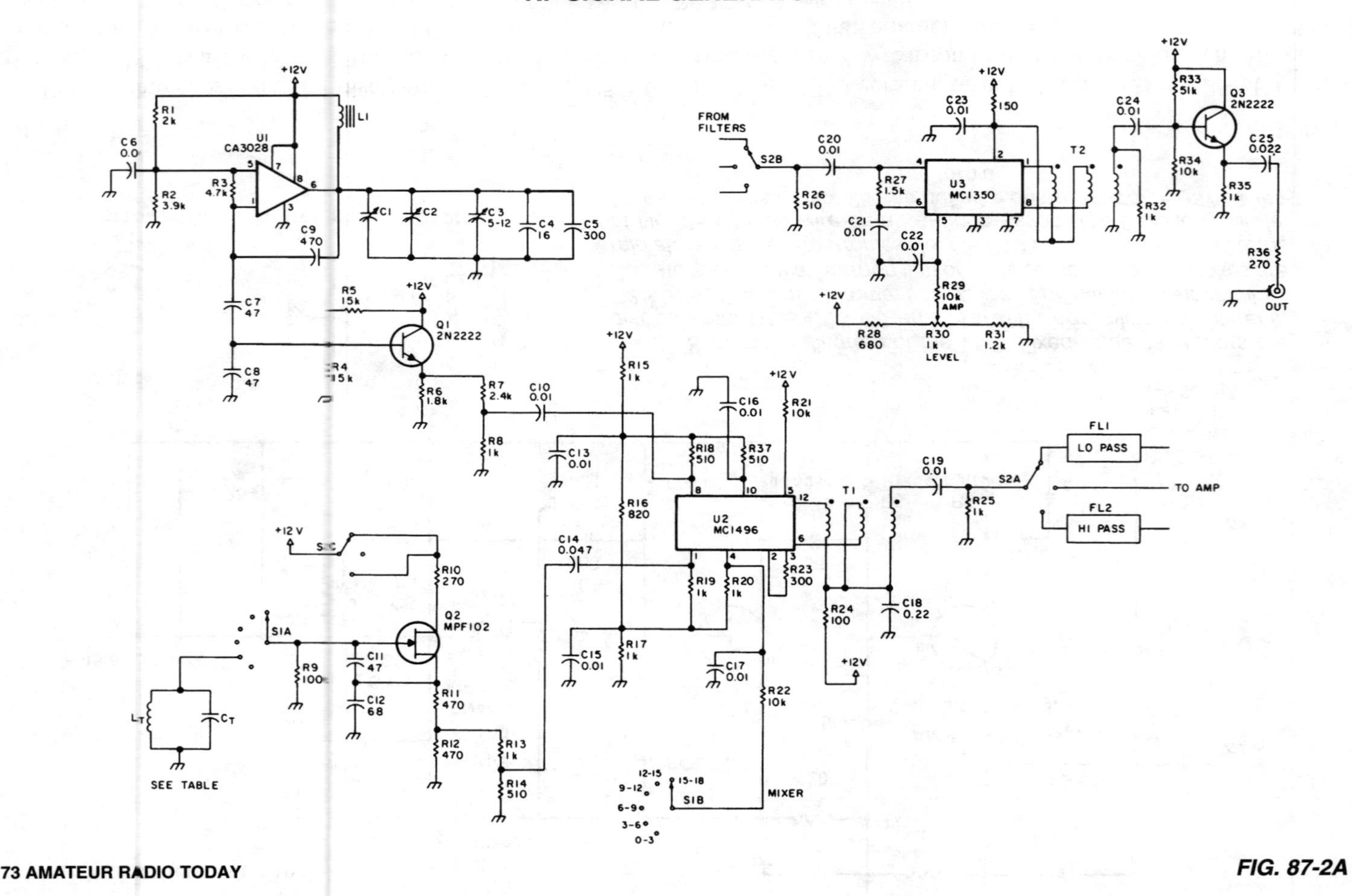

73 AMATEUR RADIO TODAY

FIG. 87-2A

This circuit uses a VFO operating from 15 to 18 MHz (U1), which feeds a balanced mixer (U2). A fixed oscillator signal is mixed with this signal to generate an output from 0.4 to 33 MHz. FL1 and FL2 are low- and high-pass filters that are used to eliminate undesired mixer products. Amplifier U3/Q3 supplies up to 200 mV rms to the output jack.

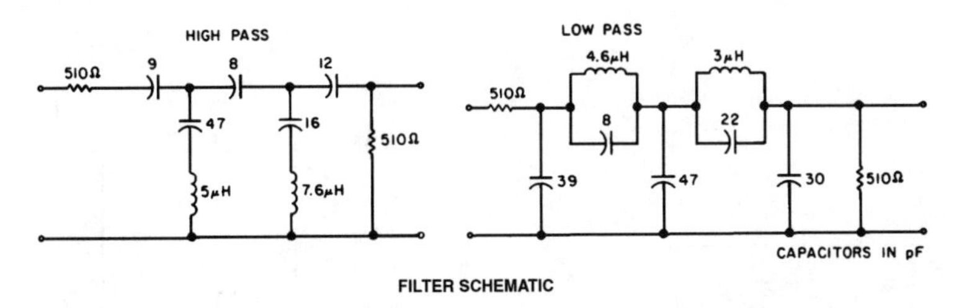

FILTER SCHEMATIC

73 AMATEUR RADIO TODAY

FIG. 87-2B

NE602 RF OSCILLATOR CIRCUITS

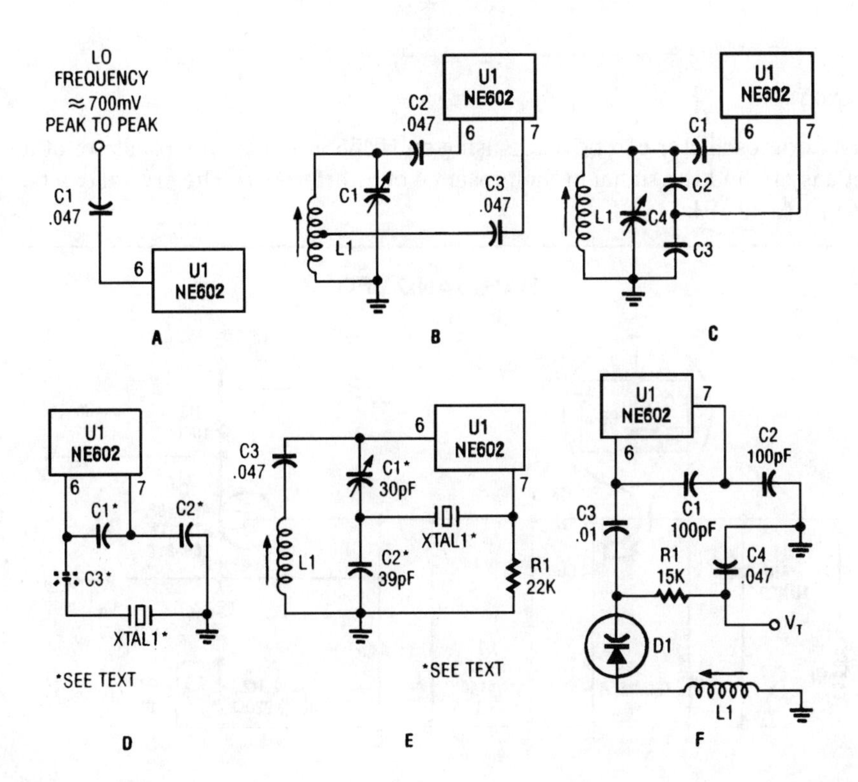

POPULAR ELECTRONICS

FIG. 87-3

Just about any standard oscillator (such as a Colpitts or Hartley configuration) can be used to generate the LO (local oscillator) frequency needed by the NE602.

A SHORTWAVE PULSED-MARKER OSCILLATOR

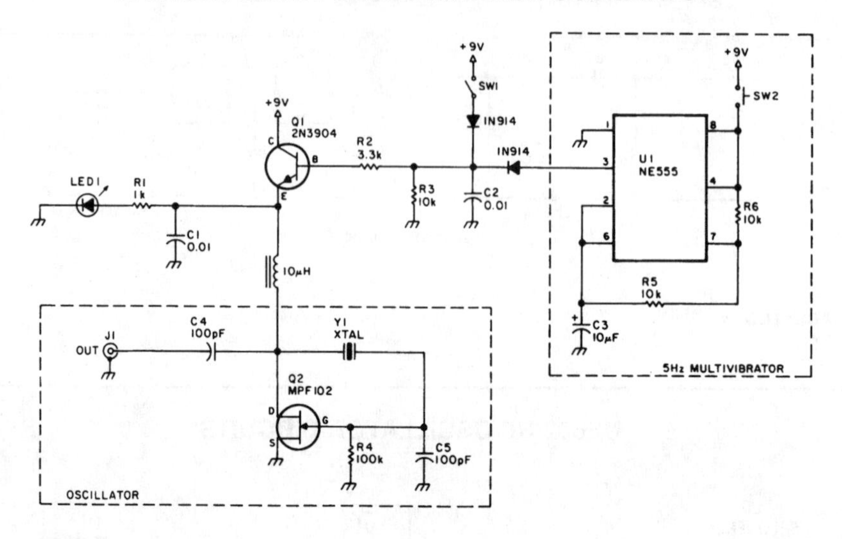

73 AMATEUR RADIO TODAY ***FIG. 87-4***

A useful marker oscillator can be made using an NE555 to pulse the oscillator at an audio rate. This makes it easy to find the signal in the presence of interference. The crystal can be any suitable frequency from 1 to 30 MHz.

HAM BAND VFO

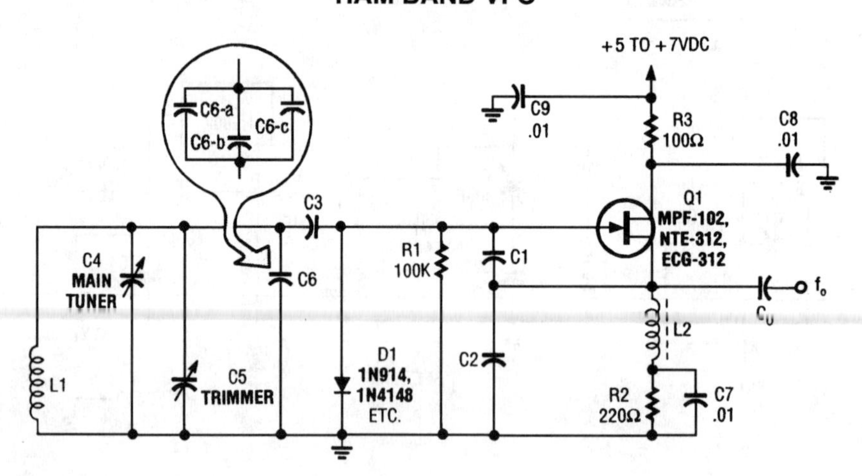

POPULAR ELECTRONICS ***FIG. 87-5***

This basic VFO for the 3- to 6-MHz range is commonly used in amateur applications, using a Colpitts circuit. For 5 to 5.5 MHz, $C_1 = C_2 = 70$ pF and for 3.5 to 4.0 MHz, use 1000 pF. C3 is typically 10 to 220 pF, depending on the frequency. C4, C5, and C6, together with C3, determine the frequency along with L1. C6 can be made up of several smaller values, paralleled to get the exact required value.

88

Sample-and-Hold Circuits

The sources of the following circuits are contained in the Sources section, which begins on page 675. The figure number in the box of each circuit correlates to the entry in the Sources section.

SAMPLE-AND-HOLD CIRCUIT I

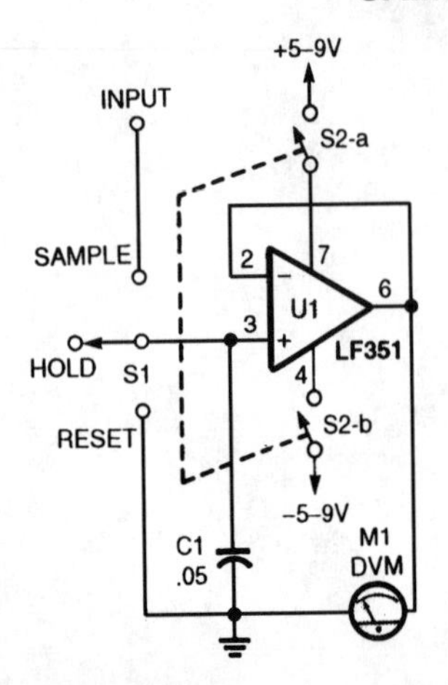

This circuit demonstrates the principle of the sample-and-hold circuit. S1 can be replaced by electronic switches (FET, etc.) in an actual application.

POPULAR ELECTRONICS

FIG. 88-1

SAMPLE-AND-HOLD CIRCUIT II

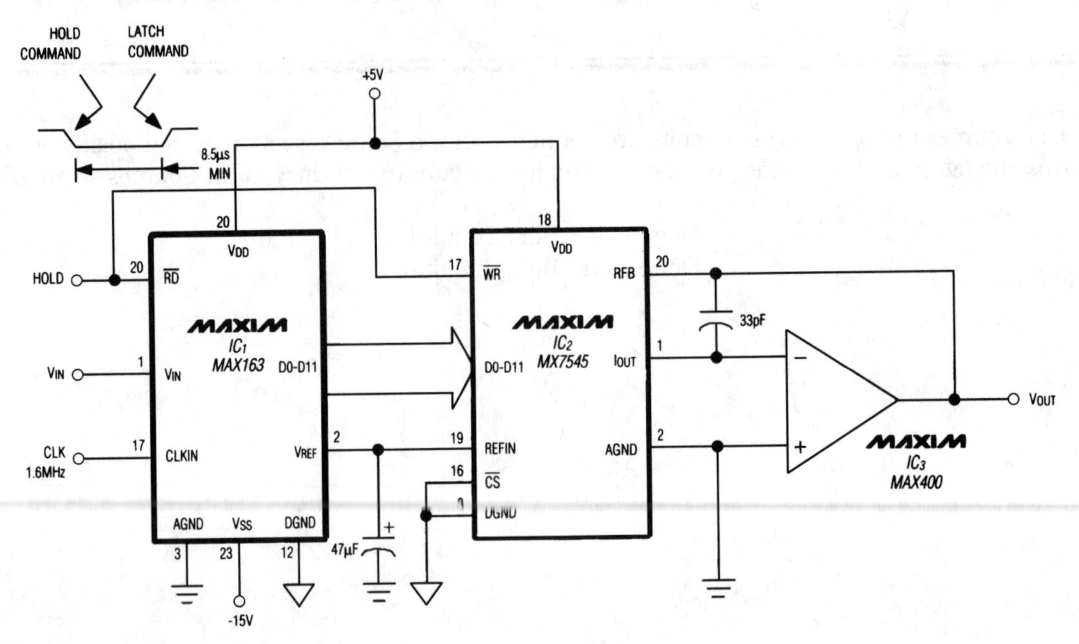

MAXIM ENGINEERING JOURNAL

FIG. 88-2

Driving a D/A converter with an A/D converter provides an overall analog-hold function, which though limited in output resolution, offers zero voltage droop and infinite hold time. The A/D converter shown (IC1) includes a 12-bit compatible track/hold at its input. The track/hold specifies a 6-MHz full-power bandwidth, a 30-ns aperture delay, and a 50-ps aperture jitter. The direct connections shown allow the D/A converter to reconstruct signal levels within the input range of 0 to 5 V.

89

SCA Circuit

The source of the following circuit is contained in the Sources section, which begins on page 675. The figure number in the box of the circuit correlates to the entry in the Sources section.

Subcarrier Adapter for FM Tuners

SUBCARRIER ADAPTER FOR FM TUNERS

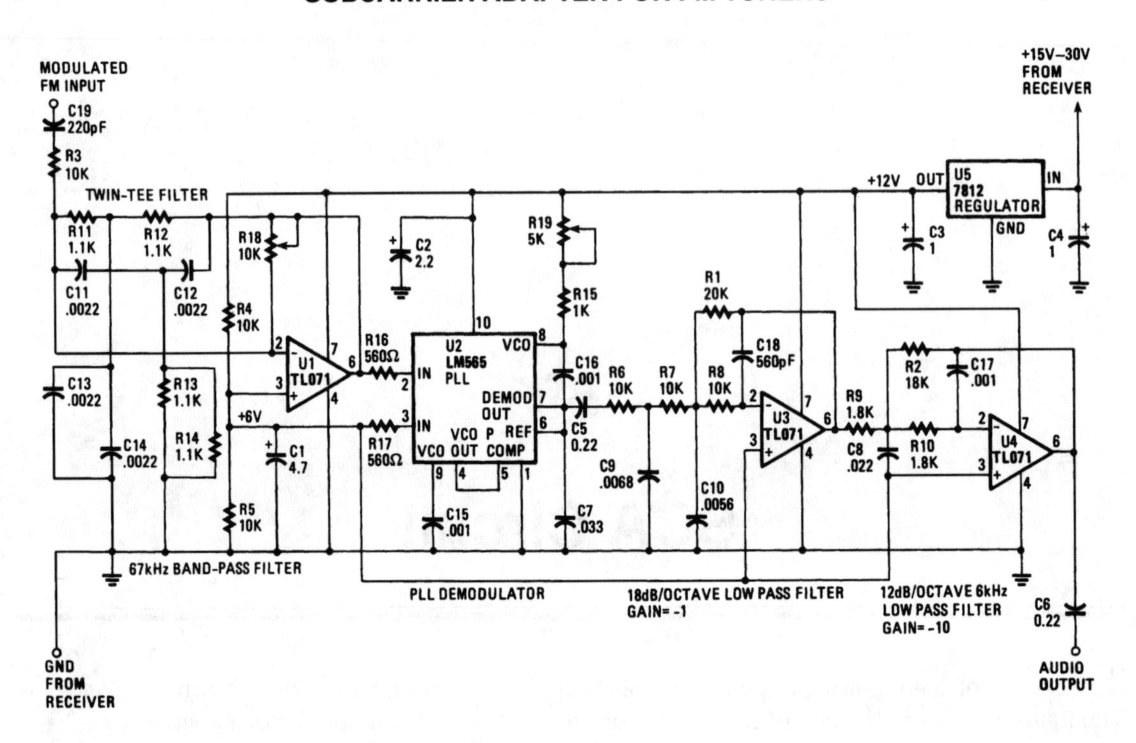

1990 PE HOBBYIST HANDBOOK ***FIG. 89-1***

Op amp U1 and its associated components comprise the 67-kHz bandpass filter. A twin-T network, comprised of four 1100-Ω resistors and four 0.0022-μF capacitors, is connected in the feedback network of the op amp. That gives some gain at 67 kHz and heavy attenuation for frequencies above and below that frequency.

An additional passive filter at the input to the twin-T network (containing a 220-pF capacitor and a 10,000-Ω resistor) provides some additional roll-off for frequencies below 67 kHz.

In practice, the bandpass-filter action covers a frequency range of about 10 kHz above and below the 67-kHz center frequency. Resistor R18 sets the gain of the bandpass-filter stage.

Integrated-circuit U2 is a National LM565 phase-locked loop that modulates the 67-kHz frequency-modulated (FM) signal from U1. The LM565 PLL consists of a voltage-controlled oscillator (VCO) set to 67 kHz, and a comparator that compares the incoming frequency-modulated 67-kHz signal at pin 2 with the VCO signal that is fed into pin 5.

The output of the comparator represents the phase difference between the incoming signal and the VCO signal. Therefore, the output is the audio modulated by the subcarrier. A treble deemphasis of 150 μs is provided by a 0.033-μF capacitor (at pin 7).

The free-running VCO frequency is determined by the 0.001-μF capacitor at pin 9 and by the resistance between the positive rail and pin 8 (100 Ω in series with R19). Variable-resistor R19 adjusts the oscillator frequency (also known as the *center frequency*) so that the incoming signal is within the lock range of the PLL.

90

Shutdown Circuits

The sources of the following circuits are contained in the Sources section, which begins on page 675. The figure number in the box of each circuit correlates to the entry in the Sources section.

RESETTABLE SHUTDOWN CIRCUITS

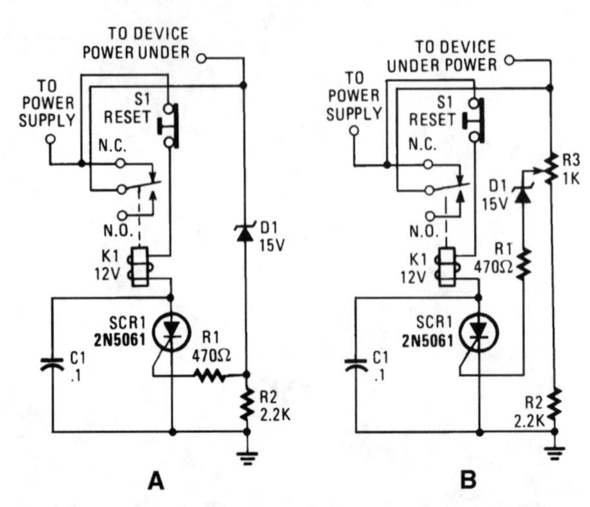

POPULAR ELECTRONICS *FIG. 90-1*

If your circuits experience frequency overvoltage conditions, continually replacing blown fuses can get pretty expensive. However, this shutdown circuit overcomes that deficiency by replacing the fuse with a relay and a low-current SCR.

When the input voltage rises above the threshold set by the Zener diode (D1), a current of sufficient magnitude is applied to the gate of SCR1, which turns it on. That draws current through the relay coil and energizes it, which swings its commutator to its normally open contact, and disrupts power to the circuit under power. Switch S1, a normally closed pushbutton switch, is used to reset the circuit; it does so by interrupting power to the relay. When S1 is pressed, the relay's wiper arm returns to the normally closed position, restoring power to the connected circuit.

If you deal with a number of circuits that have different burn-out levels, try the circuit in B. That circuit variation, a variable trip-point shutdown circuit, allows you to adjust the shutdown threshold to whatever level you desire. The circuit adjustment allows for the 30% variance in the trip point. The zener diode should be selected to have a voltage rating that is slightly lower than the minimum desired threshold voltage.

SHUTDOWN CIRCUIT

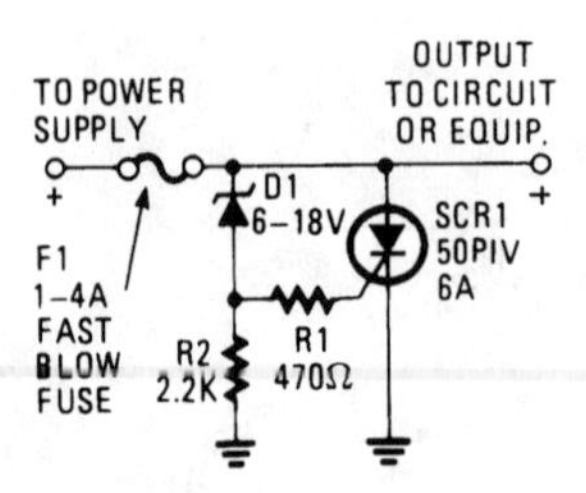

Many modern devices have shutdown circuits that are designed to remove power from the device under power when the voltage rises above a predetermined threshold. This one blows a fuse to protect the device under power.

POPULAR ELECTRONICS *FIG. 90-2*

91

Sine-Wave Oscillator Circuits

The sources of the following circuits are contained in the Sources section, which begins on page 675. The figure number in the box of each circuit correlates to the entry in the Sources section.

HIGHLY STABLE 60-Hz SINE-WAVE SOURCE

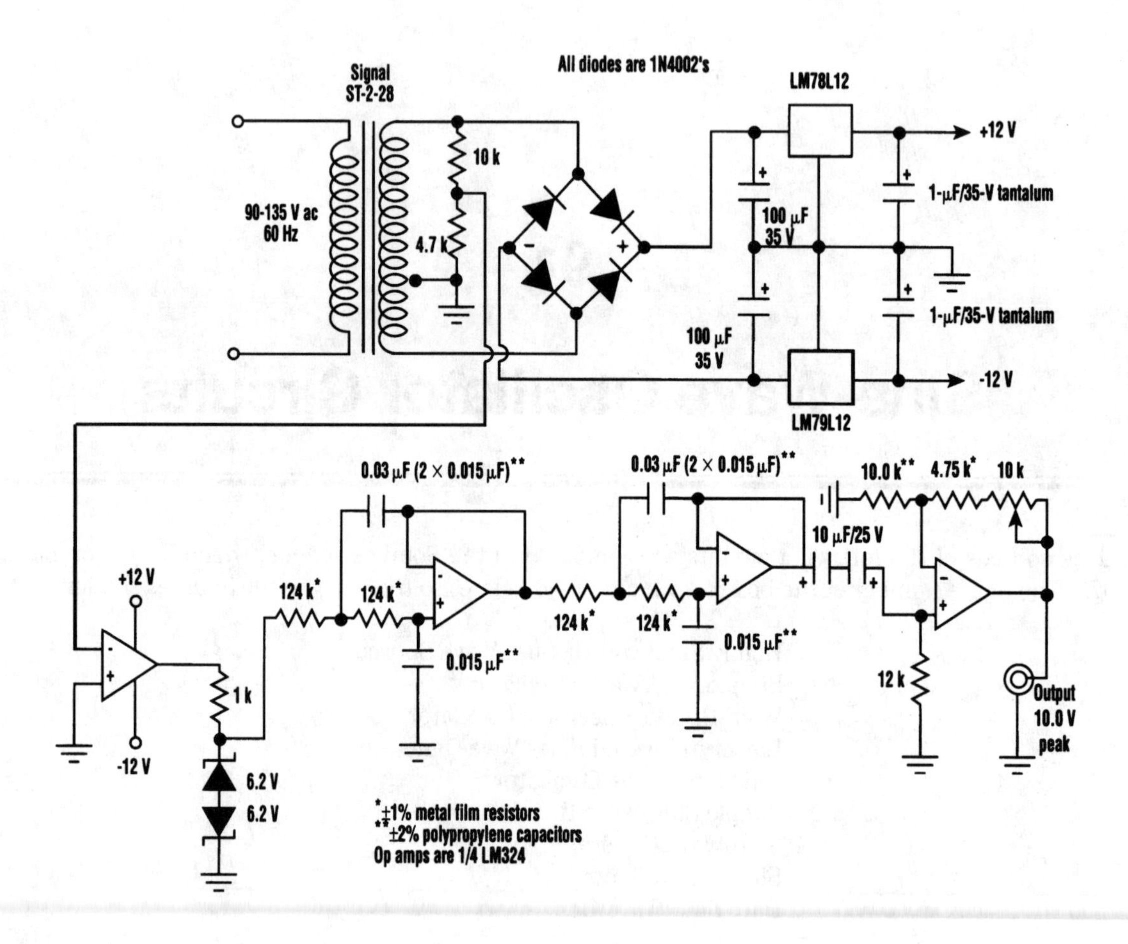

ELECTRONIC DESIGN

FIG. 91-1

A highly-stable 60-Hz sine wave can be delivered with this circuit, which offers a different and much simpler approach to gaining a stable amplitude. Capacitor coupling the last stage removes any dc component caused by unequal zener voltages in the clipping circuit that follows the comparator.

SIMPLE SINE-WAVE OSCILLATOR

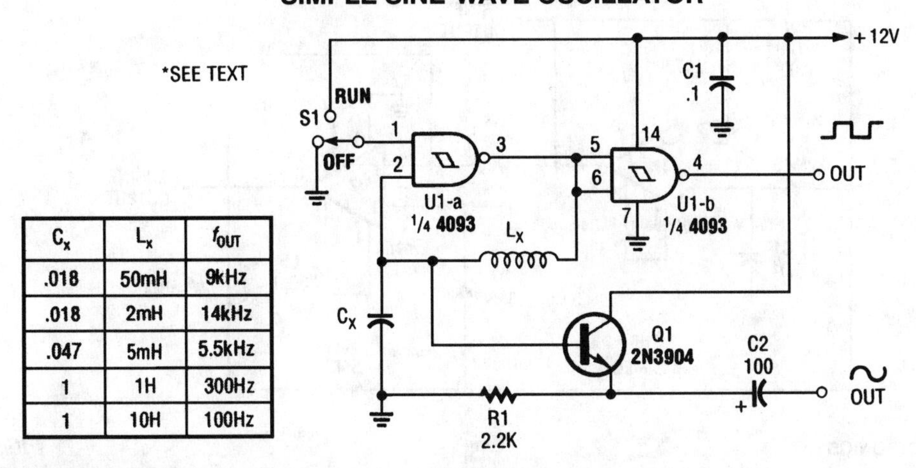

C_x	L_x	f_{OUT}
.018	50mH	9kHz
.018	2mH	14kHz
.047	5mH	5.5kHz
1	1H	300Hz
1	10H	100Hz

POPULAR ELECTRONICS *FIG. 91-2*

Using an LC circuit, this CMOS oscillator generates sine waves.

WIEN-BRIDGE SINE-WAVE OSCILLATOR

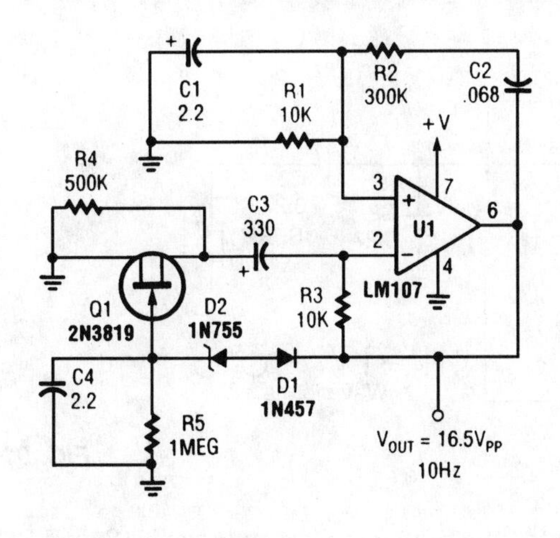

POPULAR ELECTRONICS *FIG. 91-3*

This Wien-bridge sine-wave oscillator uses a 2N3819 as an amplitude stabilizer. The 2N3819 acts as a variable-resistance element in the Wien bridge.

BATTERY-POWERED SINE-WAVE GENERATOR

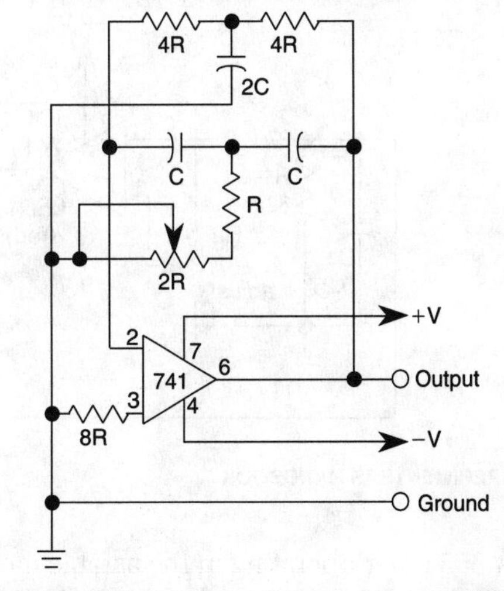

ELECTRONICS NOW *FIG. 91-4*

The quality of the sine wave depends on how closely you match the components in the twin-T network in the op amp's feedback loop.

$$f = \frac{1}{2\pi RC}$$

1-Hz SINE-WAVE OSCILLATOR

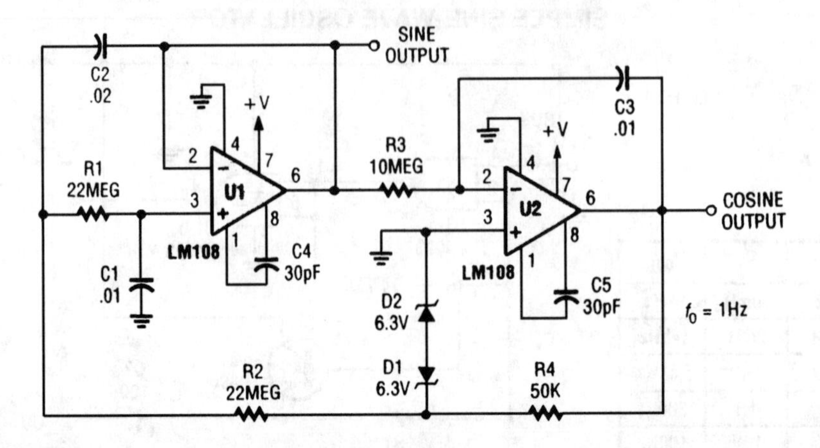

POPULAR ELECTRONICS *FIG. 91-5*

This circuit produces a 1-Hz sine wave using two op amps. A single-chip dual op amp could be used as well.

SIMPLE SINE-WAVE GENERATOR

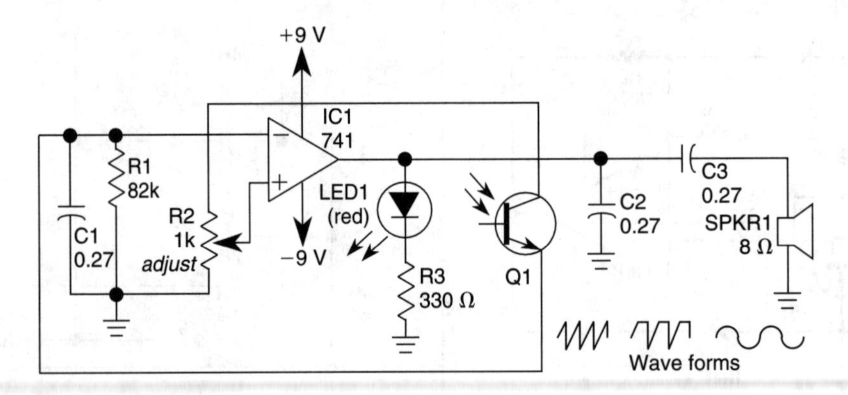

R-E EXPERIMENTERS HANDBOOK *FIG. 91-6*

A 555 timer operating in the astable mode generates the driving pulses and two 4518 dual BCD (binary coded decimal) counters provide the square waves. A TL081 op amp serves as an output buffer-amplifier, and potentiometers R1 and R2 are used in order to control the pulse's frequency and amplitude, respectively.

The output-frequency range can be varied by changing C_X. For example, a value of 0.1 µF gives a range from about 0.1 to 30 Hz, and a value of 470 pF gives a range from about 10 Hz to 1.5 kHz. The maximum output frequency is 30 kHz.

SINE-WAVE GENERATOR

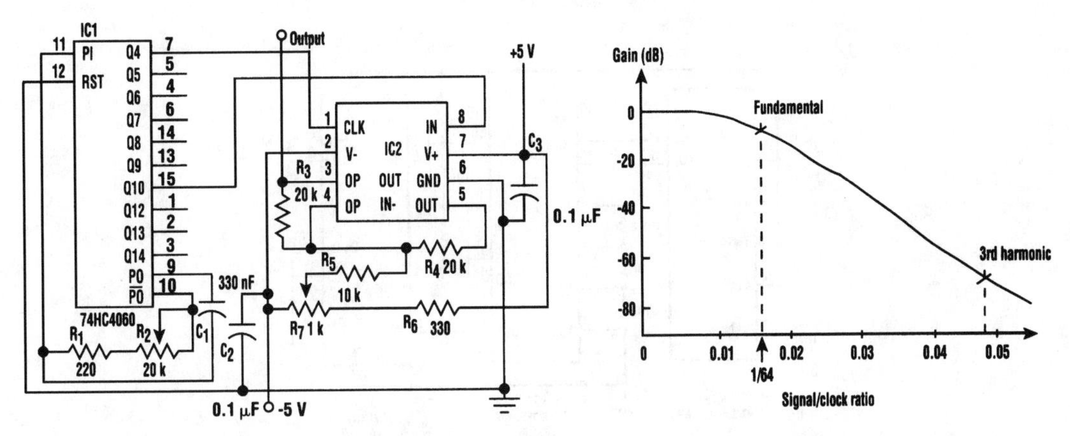

ELECTRONIC DESIGN

FIG. 91-7

In this circuit, a square wave is filtered by a high-order low-pass filter so that a –3-dB frequency will eliminate most harmonics of the waveform. As a result, the filter outputs a fundamental sine wave. This method is applied to generate a sine wave by using a switched-capacitor filter (MAX292) (see the figure). This circuit offers wide frequency range (0.1 Hz to 25 kHz), low distortion, and constant output amplitude throughout the whole frequency range.

SINE-WAVE SHAPER

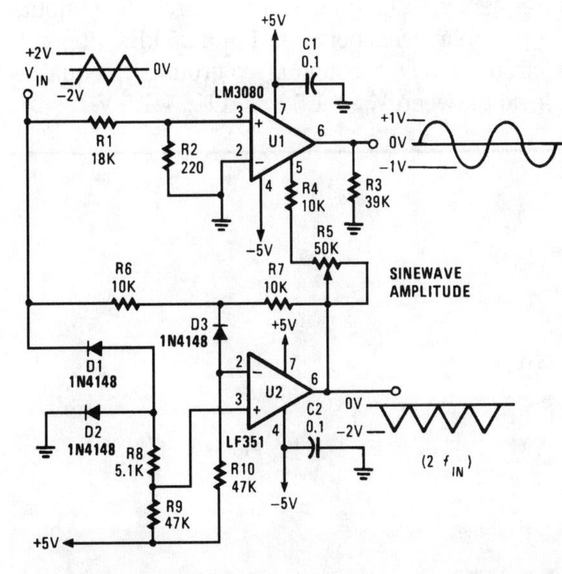

POPULAR ELECTRONICS

FIG. 91-8

Unlike most sine-wave shapers, this circuit is temperature stable. It varies the gain of a transconductance amplifier to transform an input triangle wave into a good sine-wave approximation.

PURE SINE-WAVE GENERATOR

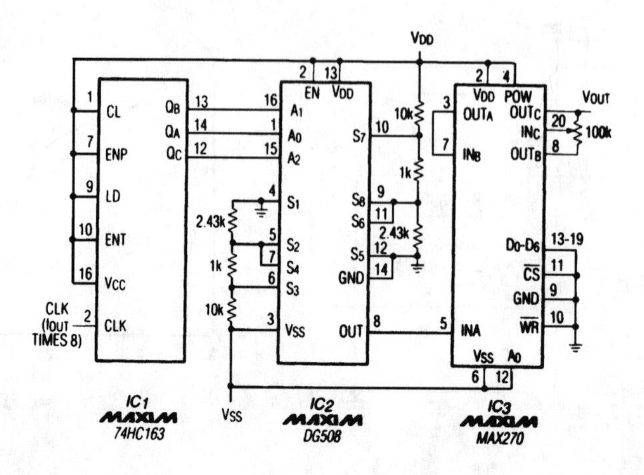

MAXIM ENGINEERING JOURNAL *FIG. 91-9*

A TTL counter, an 8-channel analog multiplexer, and a fourth-order low-pass filter can generate 10- to 25-kHz sine waves with a THD better than –80 dB. The circuit cascades the two second-order, continuous-time Sallen-Key filters within IC3 to implement the fourth-order low-pass filter.

To operate the circuit, choose the filter's cutoff frequency, f_C, by tying IC3's D_0 through D_6 inputs to 5 V or ground. The cutoff frequency can be at 128 possible levels between 1 and 25 kHz, depending on those seven digital input levels. Because the circuit ties D_0 through D_6 to ground, f_C equals 1 kHz. The 100-kΩ potentiometer adjusts the output level between V_{DD} – 1.5 V and V_{SS} + 1.5 V.

92

Sound- and Voice-Controlled Circuits

The sources of the following circuits are contained in the Sources section, which begins on page 675. The figure number in the box of each circuit correlates to the entry in the Sources section.

VOCAL STRIPPER

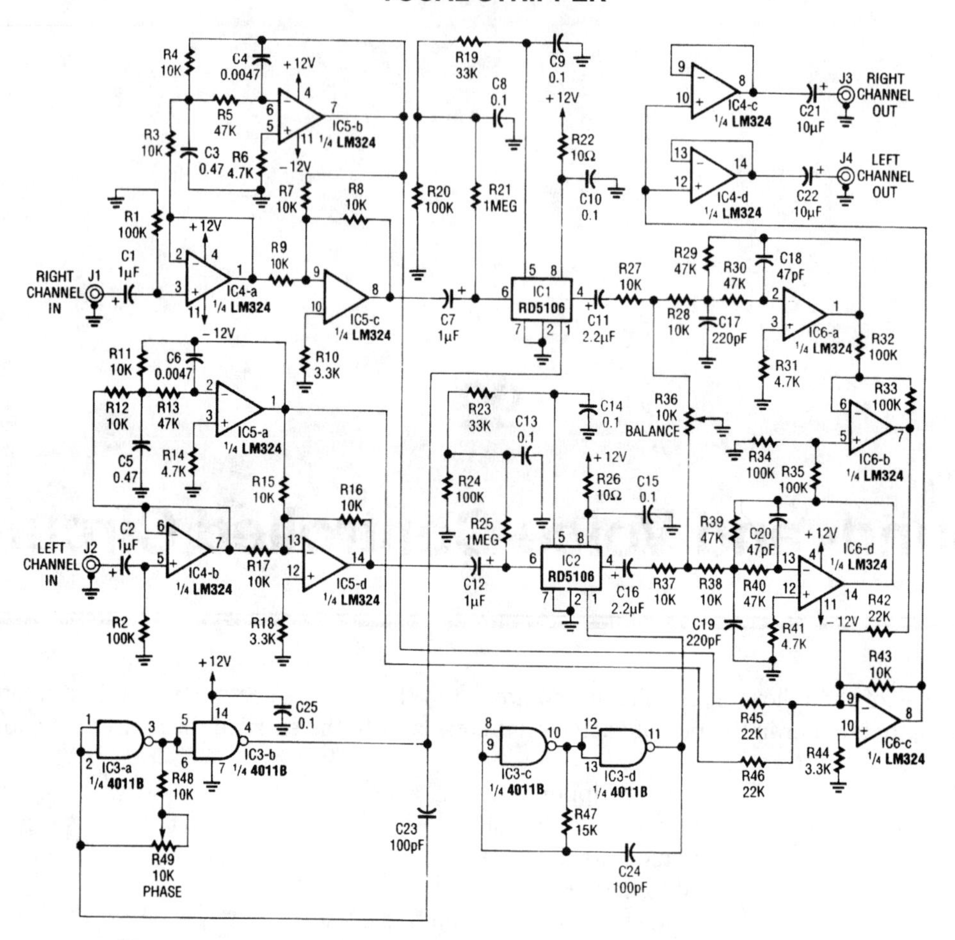

FIG. 92-1A

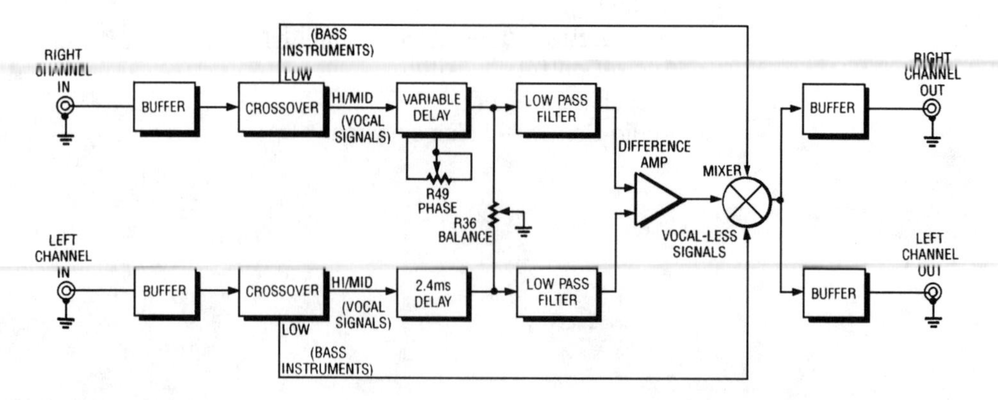

FIG. 92-1B

VOCAL STRIPPER (*Cont.*)

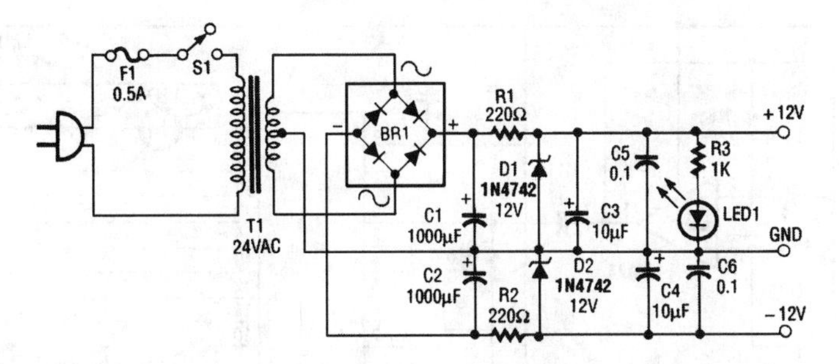

FIG. 92-1C

Right- and left-channel signals pass through 1C4-a and -b buffer amps into active crossover IC5; low frequencies are sent to the IC6-c mixer, and middle and high frequencies are sent to the analog delay lines of 1C1 and 1C2. That output passes through 1C6-a and -d to filter high-frequency sample steps. IC6-b signals are remixed with low frequencies by IC6-c and are sent to final out via IC4-c and -d buffers.

One channel (R) is a variable-delay circuit, using an analog bucket-brigade device and a variable clock frequency. This is compared in amplitude and phase to the L channel (fixed delay). The local can therefore be nulled out via R36.

SLEEP-MODE CIRCUIT

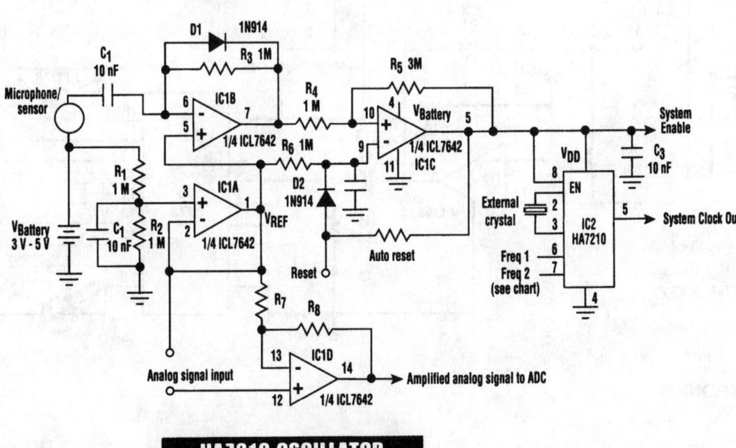

HA7210 OSCILLATOR CONTROL INPUTS

Enable	Freq 1	Freq 2	Output range
1	1	1	10 kHz-100 kHz
1	1	0	100 kHz-1 MHz
1	0	1	1 MHz-5 MHz
1	0	0	5 MHz-10 MHz+
0	X	X	High impedance

ELECTRONIC DESIGN

FIG. 92-2

The HA7210 oscillator IC combines with an ICL7642 quad CMOS op amp to produce a sleep-mode control circuit. The circuit is put into the sleep mode with a logic high applied to the Reset input or with an RC timer for automatic reset. The system is awakened by a signal from the microphone/sensor.

SONIC KALEIDOSCOPE

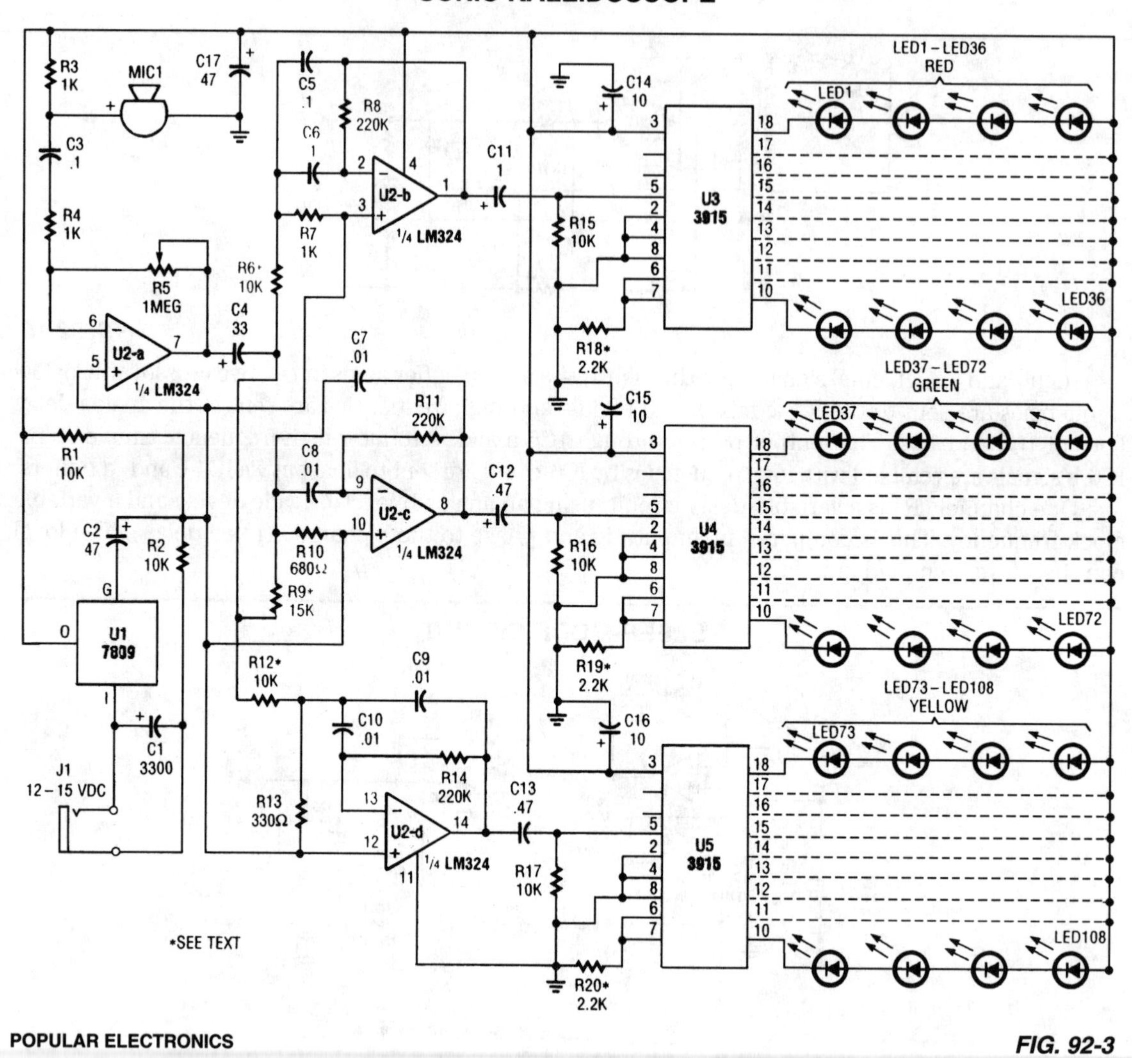

POPULAR ELECTRONICS FIG. 92-3

The microphone input, MIC1, is fed through C3 and R4 to inverting amplifier U2-a; the gain of U2-a is controlled by potentiometer R5. The output of U2-a is fed through C4 to the remaining op-amps (U2-b, U2-c, U2-d), which are all configured as band-pass filters. Each filter is tuned to pass a different range of frequencies by its resistor/capacitor combination. With the values shown, U2-b, U2-c, and U2-d have center frequencies of roughly 100, 1000 and 1500 Hz, respectively.

Resistors R6, R9, R12 control the bandwidth and gain of their respective filter circuits, and can range in value from 10 to 15 kΩ. The output of U2-b is capacitively coupled via C11 to the input of U3, with R15 serving as the load resistor for U2-b. That resistor also keeps U3's outputs from "floating" in the absence of a signal. Connected as shown, U3 uses its own internal voltage reference to make a full-scale display of 1.2 V.

SONIC KALEIDOSCOPE (*Cont.*)

Each of the nine outputs of U3 (output 1 is not used) sinks four, series-connected (red) LEDs. Op amps U2-c and U2-d are similarly connected to U4 and U5, respectively, driving green and yellow LED strings. Resistors R18, R19, and R20 control the brightness of their corresponding LED arrays, and they must be adjusted accordingly; different colors of LEDs usually vary in brightness. A lower value of resistance will make the LEDs glow brighter.

Power for the circuit is supplied by a 500 mA, 12–15-Vdc wall-pack transformer, via J1. The output of the transformer is filtered by C1 and is regulated by U1; regulation is necessary to keep power-line ripple from affecting the display. The supply pins of U2 through U5 are bypassed by capacitors C14 through C17 to further ensure stability. An on/off switch was deemed unnecessary because the power supply should be unplugged when the unit is not in use.

AUTOMATIC FADER

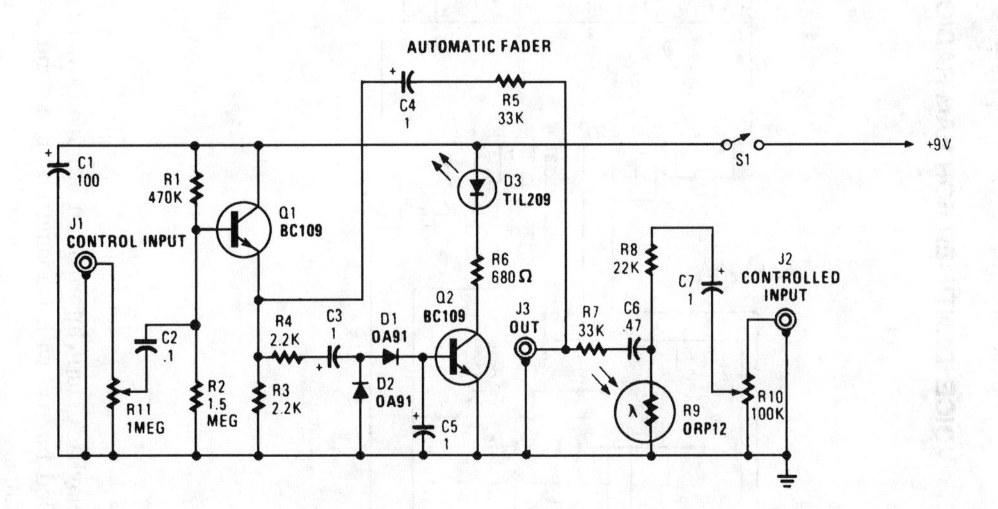

POPULAR ELECTRONICS

FIG. 92-4

In this circuit, audio fed to the control channel is amplified and rectified by D1 and D2. This dc level activates LED D3 via Q2. The light from D3 causes R9, a light-dependent resistor to decrease resistance. As R11 (audio gain) is set higher, more audio is present at the output of Q1. Audio fed into J2 is shunted to ground via R9 and less of this audio appears at J3. Therefore, audio at J1 controls the audio level fed to J3 from J2 and produces a fade effect.

VOICE IDENTIFIER FOR HAM RADIO USE

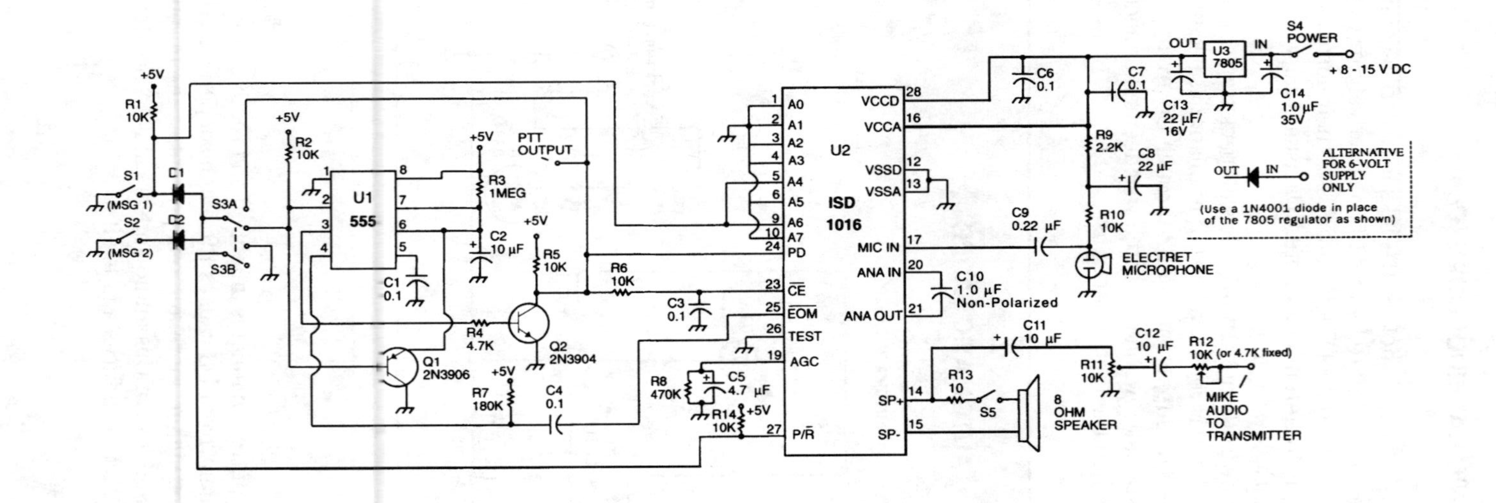

73 AMATEUR RADIO TODAY

FIG. 92-5

Using an ISD1016 audio record/playback chip (Information Storage Devices, Inc.), this circuit records and plays back messages on command. Although intended for use with transmitters, it can be used as an electronic notepad, etc. Consult the ISD1016 data sheet for other applications.

WHISTLE SWITCH

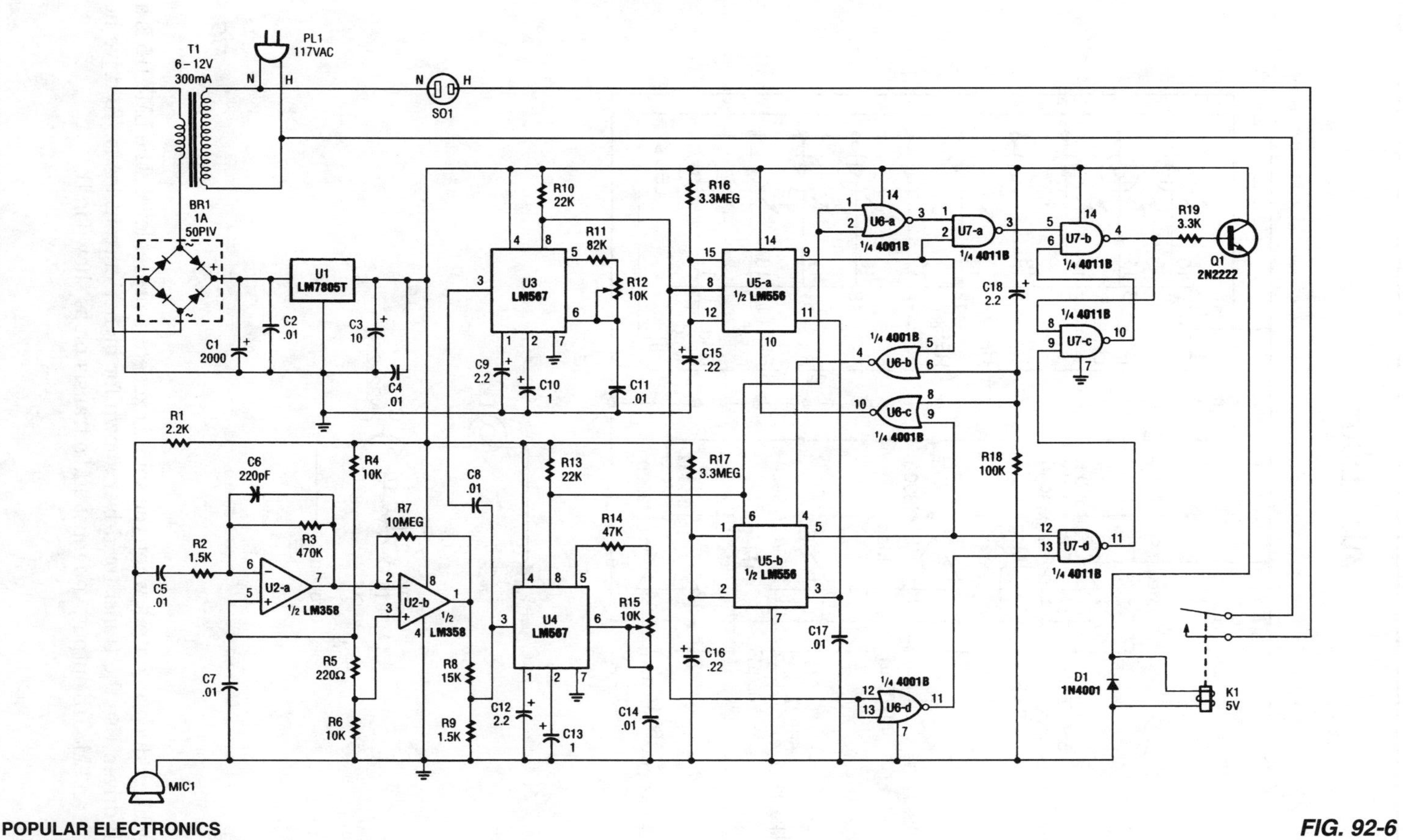

POPULAR ELECTRONICS *FIG. 92-6*

At the heart of the whistle switch are a pair of tone detectors, each of which is built around an LM567 tone decoder, which are supported by a minimum of additional components. This whistle switch is designed to respond to only two or more occurrences of a specific tone, or sequence of tones, within a specified period to prevent false triggering. Depending on the relay used, various ac loads can be controlled. Microphone MIC1 picks up the sound and U2 amplifies the signal and feeds it to tone decoders U3 and U4. These devices trigger U5-a and U5-b and the logic circuits that drive relay K1.

AUDIO LIGHT

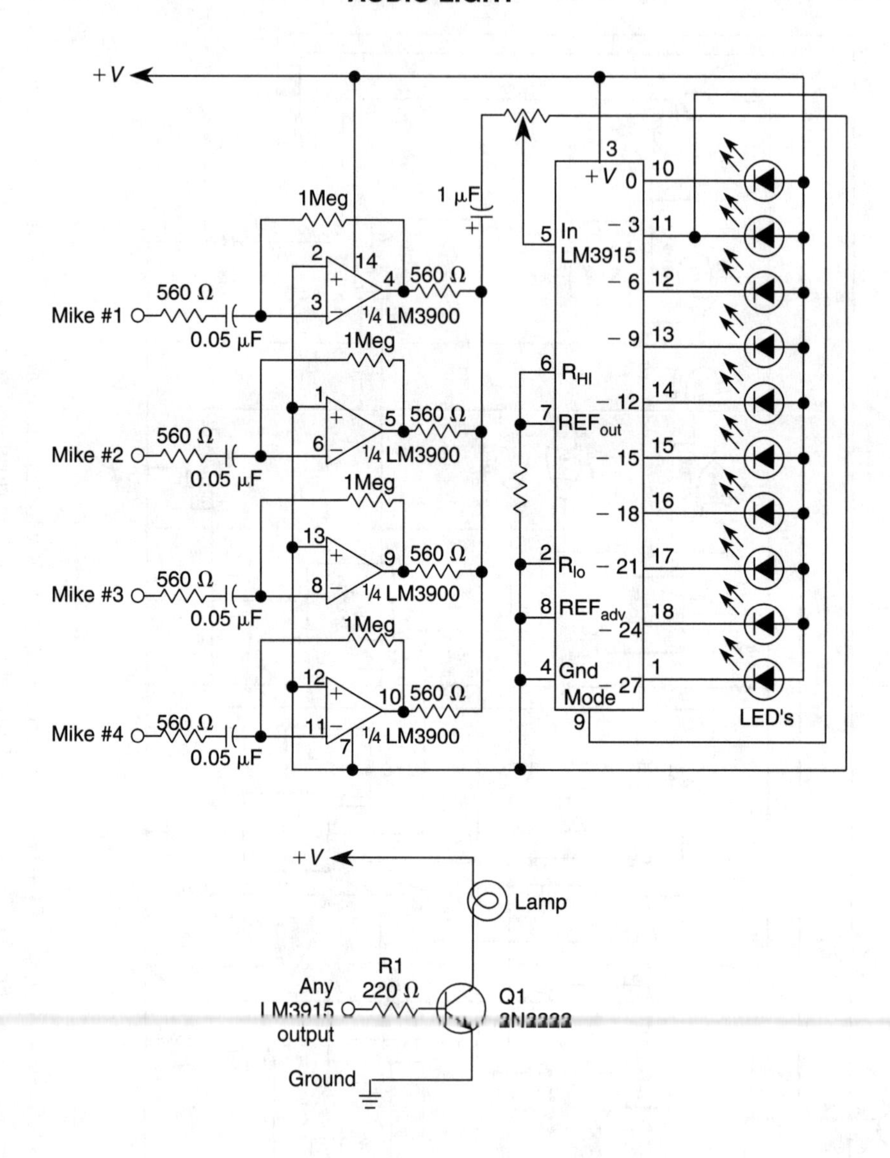

RADIO ELECTRONICS

FIG. 92-7

This circuit will produce an output when the sound exceeds a preset level. The LM3915 is a log-output bar graph driver. Use the transistor driver shown for higher current loads. To drive heavy-current loads with an LM3915 output, you must add a transistor, as shown in B.

VOICE-ACTIVATED SWITCH AND AMPLIFIER

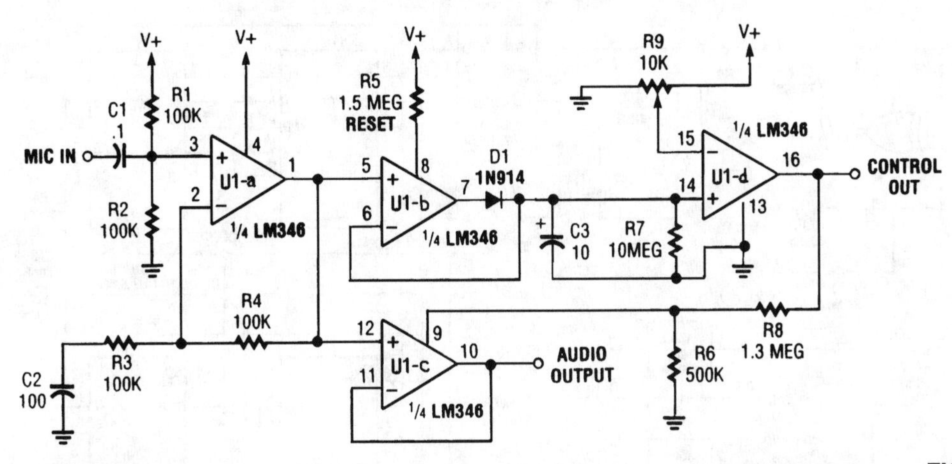

POPULAR ELECTRONICS *FIG. 92-8*

In certain applications, such as transmitter or other communications and control applications, this circuit should be useful. Both audio output and dc control outputs are provided. R9 sets the control threshold.

AUDIO-CONTROLLED SWITCH

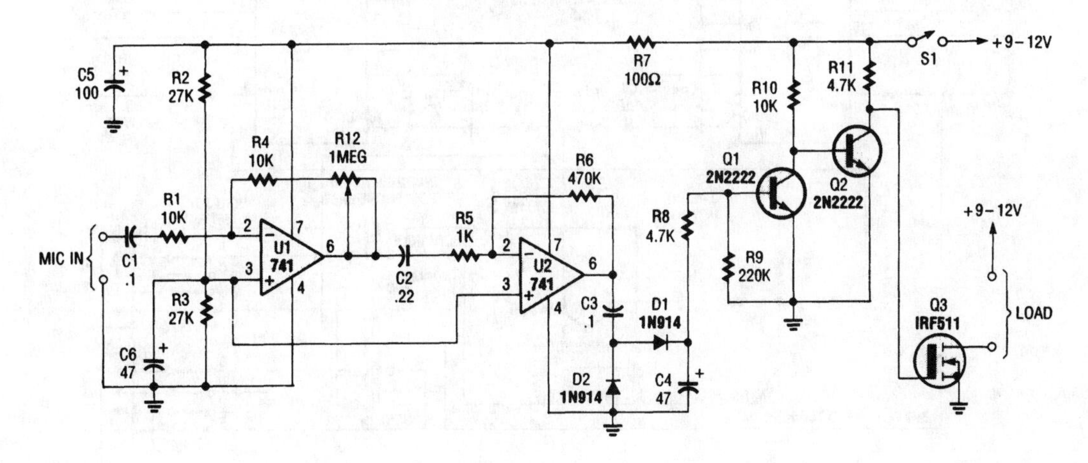

POPULAR ELECTRONICS *FIG. 92-9*

The audio-controlled switch combines a pair of 741 op amps, two 2N2222 general-purpose transistors, a hexFET, and a few support components to a circuit that can be used to turn on a tape recorder, a transmitter, or just about anything that uses sound.

SPEECH SCRAMBLER

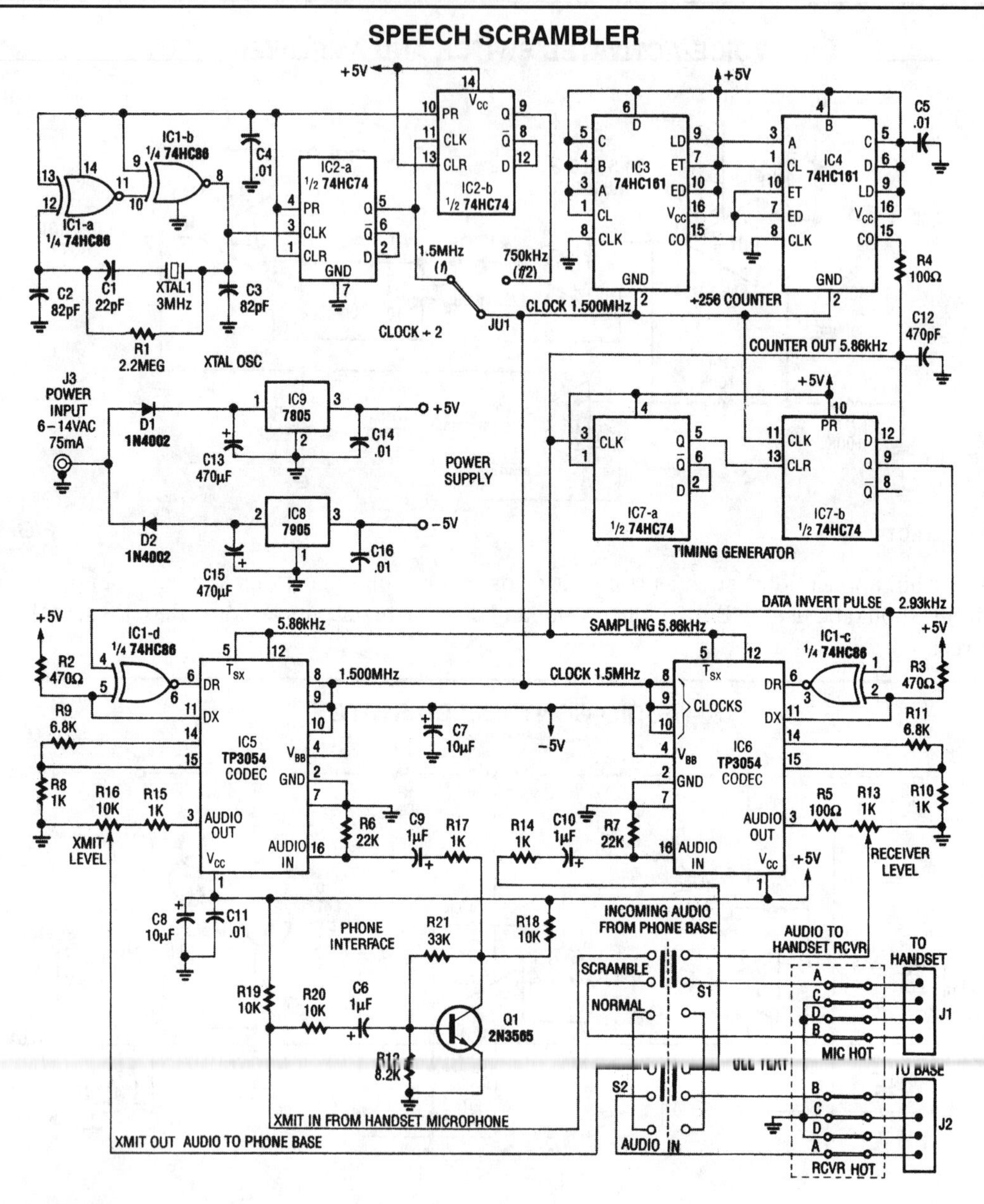

ELECTRONICS NOW *FIG. 92-10*

Using digital techniques, this circuit accomplishes the frequency-inversion algorithm via digitization of the audio, inversion of the sign of every alternate sample, and D/A conversion of the resultant data. The result is an inverted frequency spectrum. Because the circuit has two channels, this system can be used in a full duplex two-way telephone scrambler.

A complete kit of parts is available from North Country Radio, P.O. Box 53, Wykagyl Station, New Rochelle, NY 10804-0053A.

AUDIO-CONTROLLED MAINS SWITCH

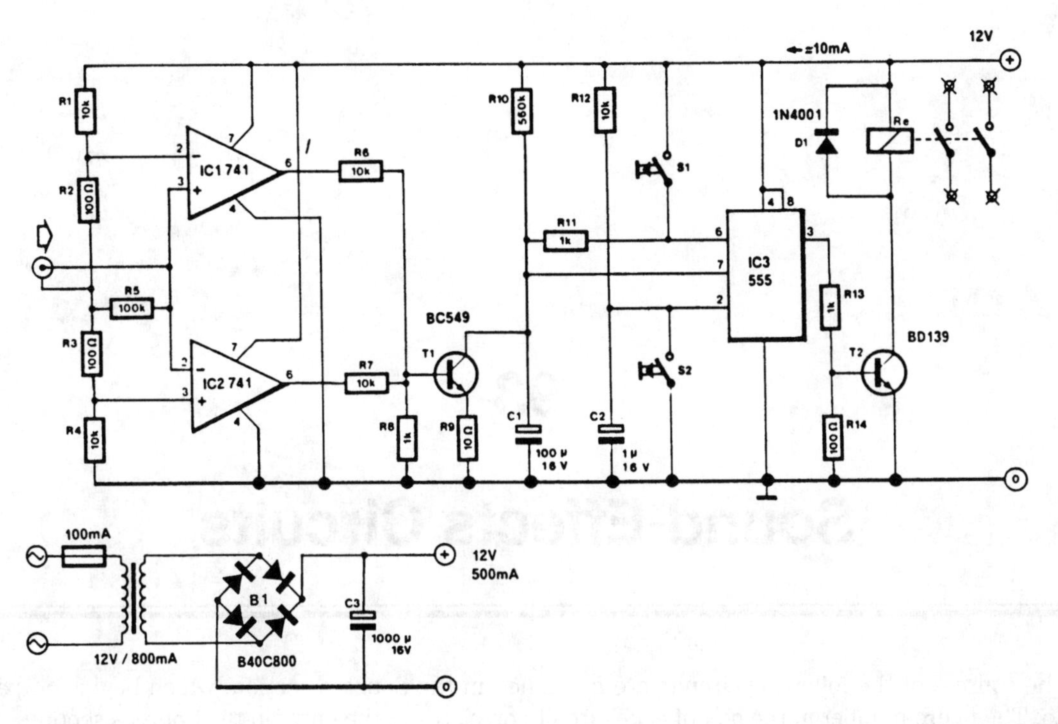

303 CIRCUITS *FIG. 92-11*

This circuit will switch off the line supply to audio or video equipment if there has been no input signal for about 2 seconds. S1 provides manual operation and S2 acts as a reset. This circuit allows for time to change a tape or compact disc. About 50 mV of audio signal is necessary.

93

Sound-Effects Circuits

The sources of the following circuits are contained in the Sources section, which begins on page 675. The figure number in the box of each circuit correlates to the entry in the Sources section.

Canary Sound Simulator
110-dB Beeper
Siren Alarm
1000-Hz Pulsed-Tone Alarm
Tone Chime
Spaceship Alarm
10-Note Sound Synthesizer
Space-Age Sound Machine
Electronic Gong
Alarm Tone Generator
Dual-Tone Sounder
Low-Level Sounder
Sound-Effects Generator
Siren
Simple Multi-Tone Generator
Siren Oscillator

CANARY SOUND SIMULATOR

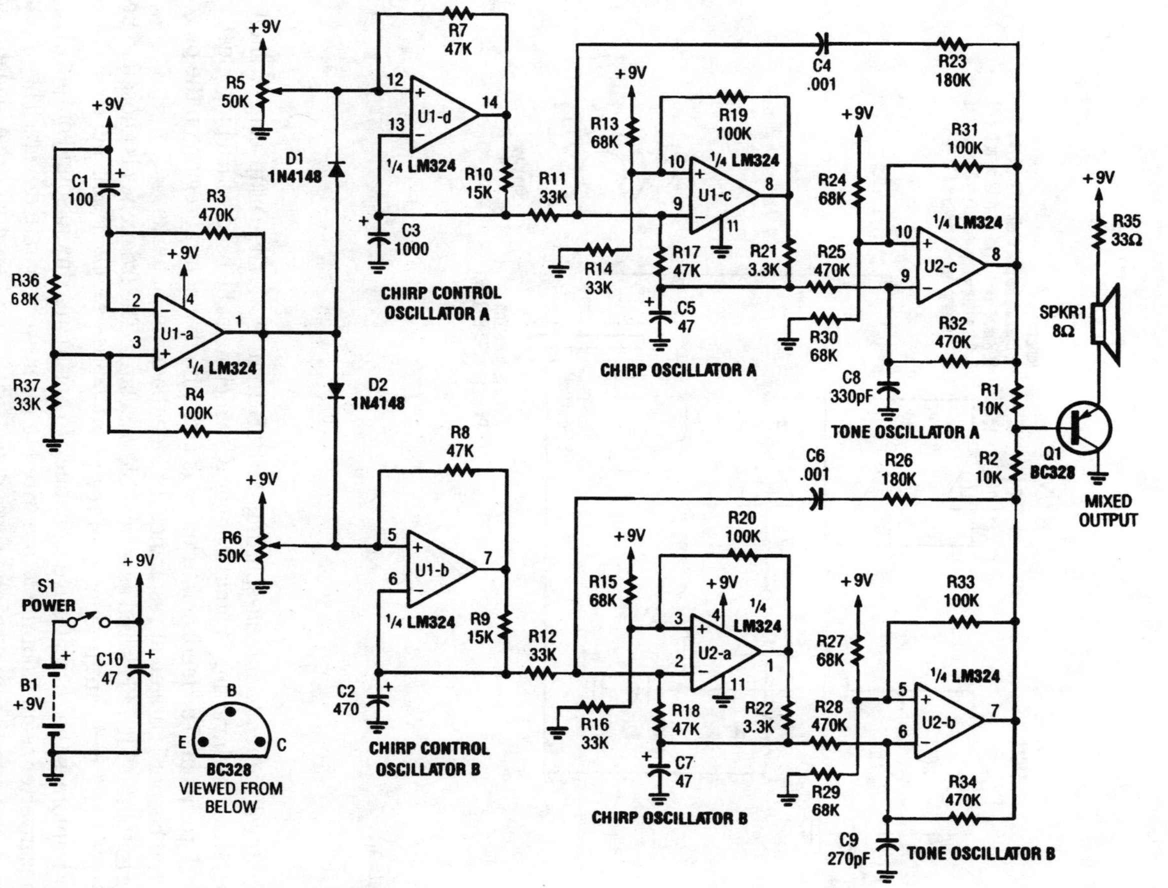

POPULAR ELECTRONICS

FIG. 93-1

This circuit generates the sound of two canaries singing in a cage. Two LM324 quad amps make up seven oscillators. One oscillator is an on/off control, the other six generate the sounds of two canaries. A 9-V supply powers the circuit.

110-dB BEEPER

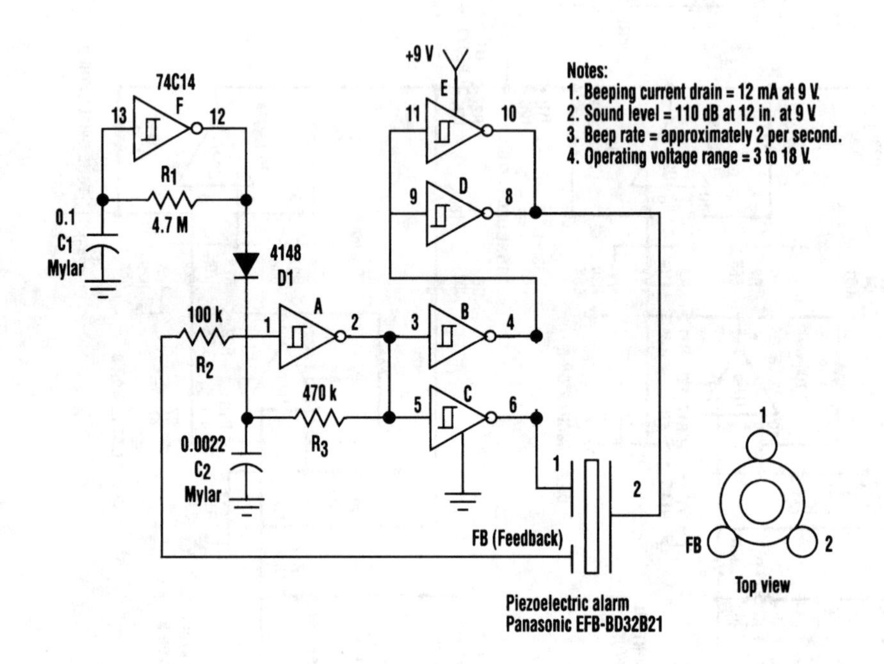

ELECTRONIC DESIGN *FIG. 93-2*

This circuit will generate an ear-splitting 110 dB from 9 V. The setup uses a single 74C14 (CD40106B) CMOS hex inverting Schmitt-trigger IC, which must be used with a piezoelectric device with a feedback terminal. The feedback terminal is attached to a central region on the piezoelectric wafer. When the beeper is driven at resonance, the feedback signal peaks.

One inverter of the 74C14 is wired as an astable oscillator. The frequency is chosen to be 5 times lower than the 3.2 kHz resonant frequency of the piezoelectric device. Feedback from the third pin of the beeper reinforces the correct drive frequency to ensure maximum sound output.

Four other inverter sections of the IC are wired to form two separate drivers. The output of one section is cross-wired to the input of the second section. The differential drive signal that results produces about 18-V p-p when measured across the beeper. The last inverter section is wired as a second astable oscillator with a frequency of about 2 Hz. It gates the main oscillator on and off through a diode. For a continuous tone, the modulation circuit can be deleted.

SIREN ALARM

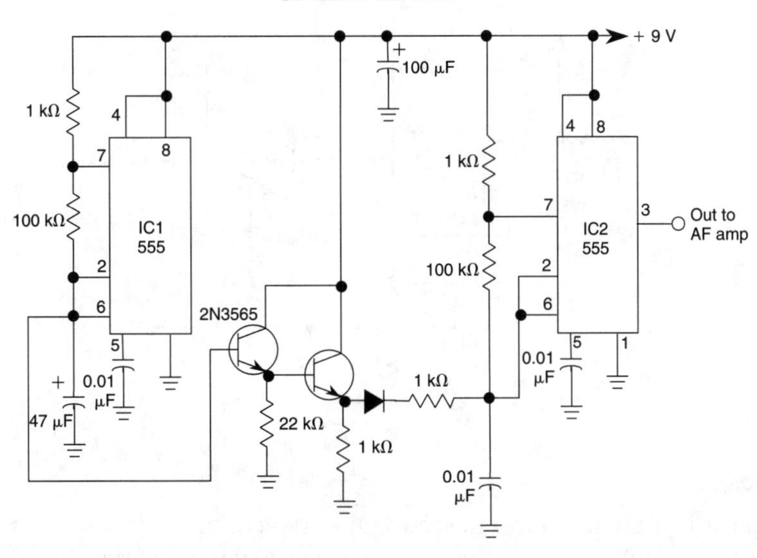

WILLIAM SHEETS *FIG. 93-3*

The ramp voltage from the low frequency oscillator IC1 modulates IC2 thereby producing a rising and falling tone like the siren wail of police cars.

1000-Hz PULSED-TONE ALARM

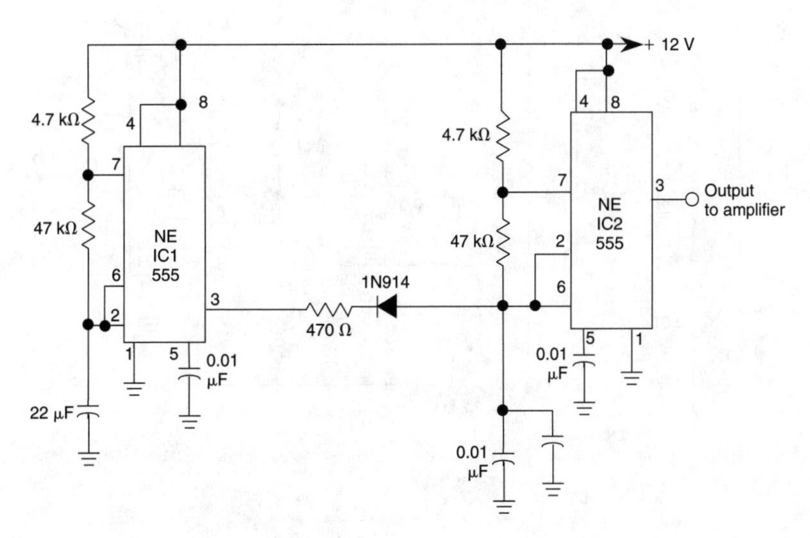

WILLIAM SHEETS *FIG. 93-4*

IC1 generates a pulse that modulates the 1000-Hz tone generated by IC2. This circuit can be used to generate warning or alert signals.

TONE CHIME

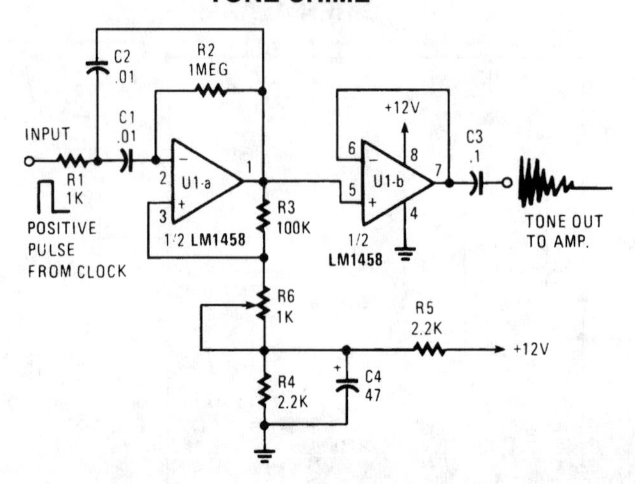

POPULAR ELECTRONICS **FIG. 93-5**

A positive pulse input to R1 causes the active filter U1-a to "ring." If the gain is set too high (R6), the circuit will oscillate. R6 controls the positive feedback and the *Q* of the circuit. C1 and C2 can be changed to adjust the tone frequency.

SPACESHIP ALARM

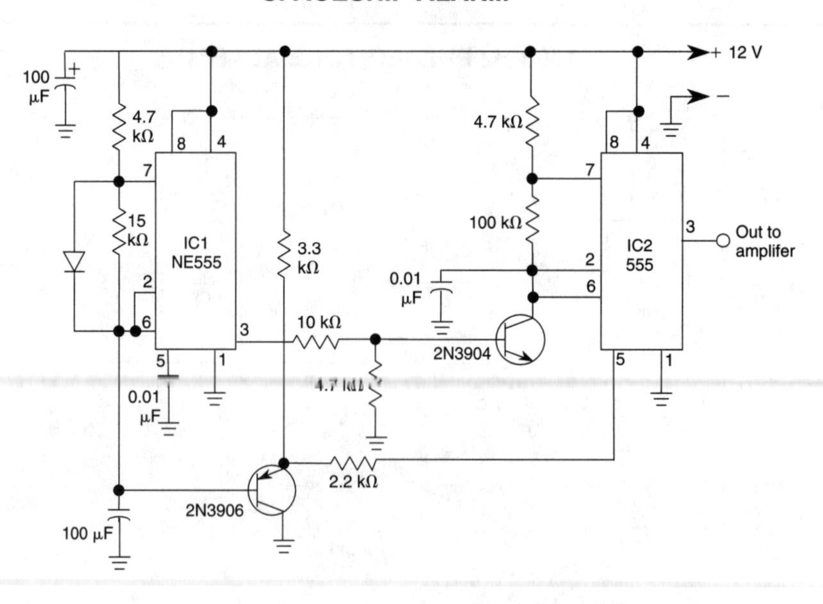

WILLIAM SHEETS **FIG. 93-6**

By using two 555 timers this circuit produces a low frequency tone that rises to a high frequency tone in a little over 1 second. Then the sound stops for about 0.3 seconds, thereafter the cycle repeats. To produce the alarm sound of the Star Trek spaceship.

10-NOTE SOUND SYNTHESIZER

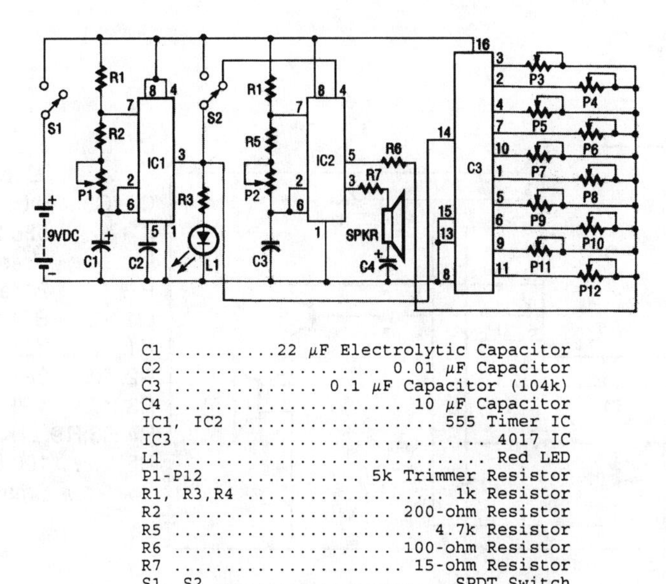

C1	22 μF Electrolytic Capacitor
C2	0.01 μF Capacitor
C3	0.1 μF Capacitor (104k)
C4	10 μF Capacitor
IC1, IC2	555 Timer IC
IC3	4017 IC
L1	Red LED
P1-P12	5k Trimmer Resistor
R1, R3,R4	1k Resistor
R2	200-ohm Resistor
R5	4.7k Resistor
R6	100-ohm Resistor
R7	15-ohm Resistor
S1, S2	SPDT Switch

1991 PE HOBBYIST HANDBOOK ***FIG. 93-7***

As shown, three ICs are used to produce the sounds. IC1 is a 555 timer that generates clock pulses. It is configured as an astable multivibrator. The frequency of the clock pulses is set by trimmer potentiometer P1. These clock pulses are coupled to the input of IC3 (a 4017 CMOS Johnson counter) on its clock input pin 14. Each clock pulse causes IC3 to shift a "high" to each of its output pins in sequence. A trimmer resistor, which can be adjusted to set a different frequency for each note, is connected to each of IC3's output pins. One side of each of the trimmers is connected to pin 5 (the control voltage pin) of IC2.

IC2, another 555 timer IC, creates the tone; the overall pitch of the tone can be varied by P2. As the output sequences from the 4017, that tone, which is changed in frequency by each output shift is applied to a small speaker from pin 3 of IC2. An LED, which flashes with each clock pulse, is connected to pin 3 of IC1. Switch S2 is used to vary the sound between "flowing" and distinct notes.

SPACE-AGE SOUND MACHINE

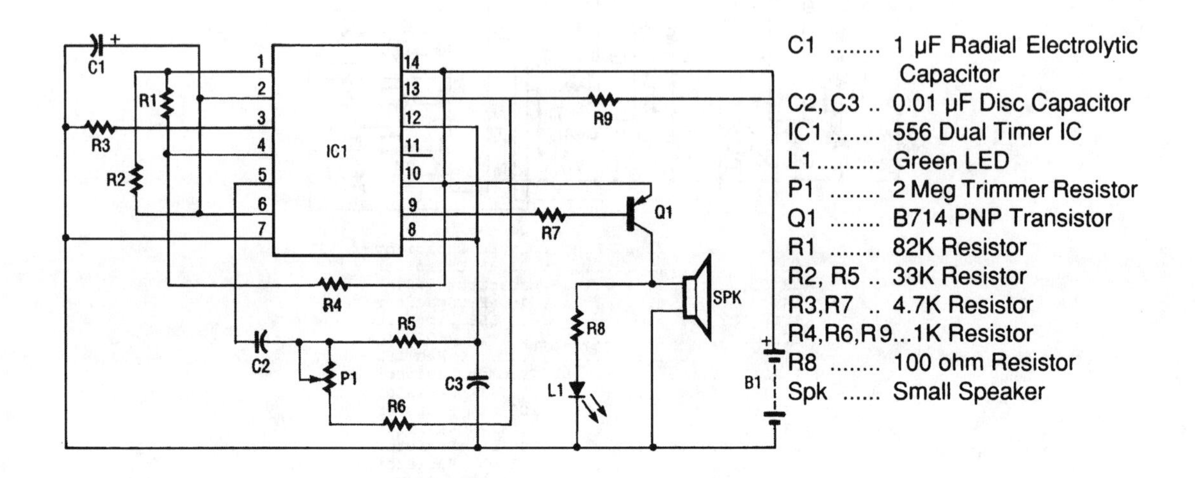

PE HOBBYIST HANDBOOK *FIG. 93-8*

The space-age sound device uses a 556 dual-times IC to produce a phasor sound. That IC is actually two 555 timer ICs in one 14-pin package, as shown in the schematic. Each timer inside the 556 is connected in an astable multivibrator mode.

The first timer has its frequency set by R1, R2, and C1. Its output appears on pin 5 and it is coupled through C2 and R5 into the trigger input of the second timer. The second timer has an adjustable frequency that is controlled by P1, R6, and C3.

In the second timer, the first frequency mixes with the second frequency and produces the phasor-like sounds. The output of the second timer, which has the two signals mixed together, is brought from pin 9 through limiting resistor R7 to the input of Q1. The function of pnp germanium power transistor Q1 is to amplify the signal to the level that is needed to drive the speaker. The green LED, L1, converts electrons directly into visible photons (light) in time with the pulses from the speaker. The purpose of resistor R8 is to limit the current through the LED to a safe level.

ELECTRONIC GONG

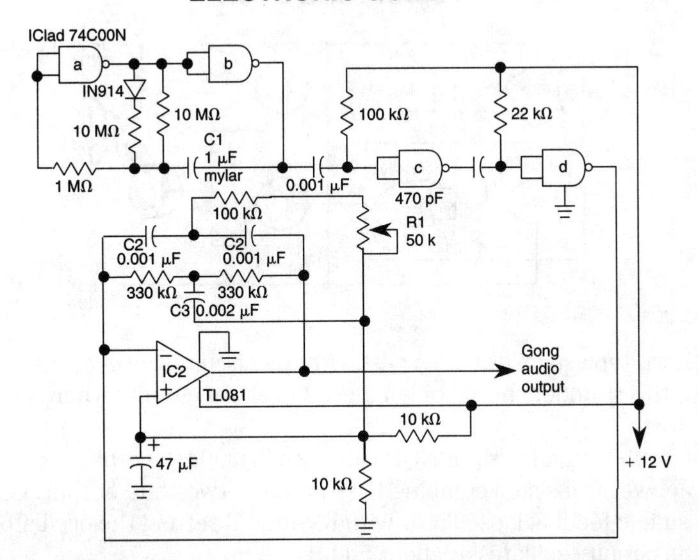

WILLIAM SHEETS *FIG. 93-9*

The electronic gong is comprised of an oscillator (built around half of a 74COON quad 2-input NAND gate), an active twin-T filter (built around a TLO81), and will drive an audio amplifier IC such as an LM386N. Pulses from astable multivibrator IC1 cause the twin-tee active filter U2 to ring, producing a damped sinusoidal output. C1 varies rate and C2-C3 vary gong frequency. Adjust R1 for best "tone" sound.

ALARM TONE GENERATOR

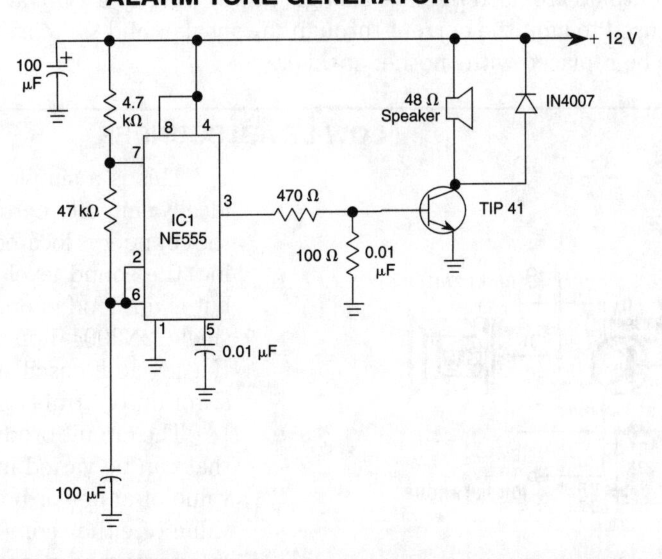

WILLIAM SHEETS *FIG. 93-10*

In this alarm tone generator, a TIP41 transistor is used as a speaker driver. R1, R2, and C1 determines the frequency which is 1400 Hz with the values shown.

DUAL-TONE SOUNDER

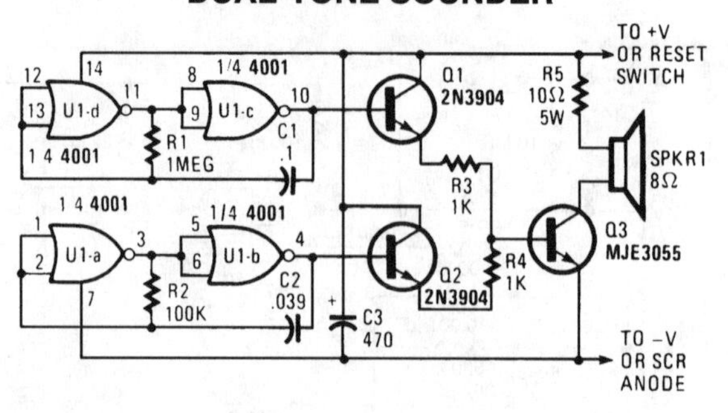

POPULAR ELECTRONICS *FIG. 93-11*

An outside horn-type speaker works best with the circuit. However, such devices require a great deal of power, so this sounder should only be used in alarm circuits where at least a 6-A SCR is used as the sounder driver.

A single CMOS 4001 quad 2-input NOR gate, two 2N3904 general-purpose npn transistors, and a single MJE3055 power transistor combine to generate a two-tone output. Gates U1-a and U1-b are configured as a simple feedback oscillator with R2 and C2 setting the oscillator's frequency. With the values shown, the circuit oscillates at about 500 Hz.

Gates U1-c and U1-d are connected in a similar oscillator circuit, but they operate at a much lower frequency. The oscillator frequencies (and thus the tones that they produce) can be altered by increasing or decreasing the values of R_1 and C_1 for the low-frequency oscillator and R_2 and C_2 for the high-frequency oscillator. Decreasing the values of those components will increase the frequency; increasing their values will decrease the frequency.

The two oscillator outputs are connected to separate amplifiers (configured as emitter followers), whose outputs are used to drive a single power transistor (Q3, an MJE3055). A 10-Ω, 5-W resistor, R5, is used to limit the current through the speaker and Q3 to a safe level. To boost the sound level, R5 can be replaced with another speaker.

LOW-LEVEL SOUNDER

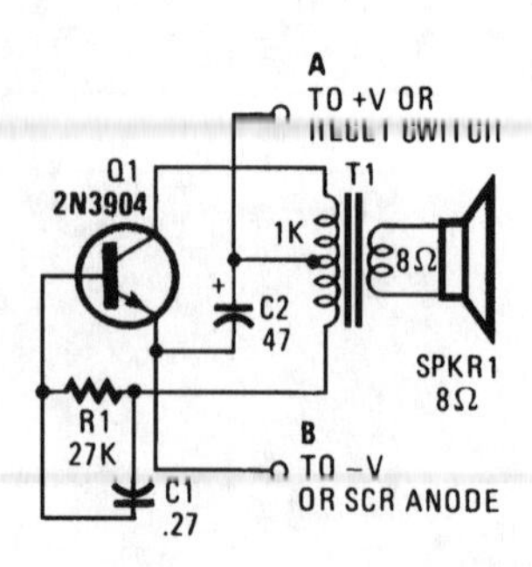

POPULAR ELECTRONICS *FIG. 93-12*

This is a simple low-level noise maker that's ideally suited to certain alarm applications. When the sounder is located in another part of the building, the sound level is loud enough to be heard, but is not loud enough to warn off an intruder. A single 2N3904 npn transistor is connected in a Hartley audio oscillator, with a 1 kΩ to 8-Ω transistor-output transformer doing double duty.

The circuit produces a single-frequency tone that can be varied in frequency by changing the value of either or both R_1 and C_1. Increasing the value of either component will lower the output frequency and decreasing their values will raise the frequency. Don't go below 4.7 kΩ for R1 because you could easily destroy Q1.

SOUND-EFFECTS GENERATOR

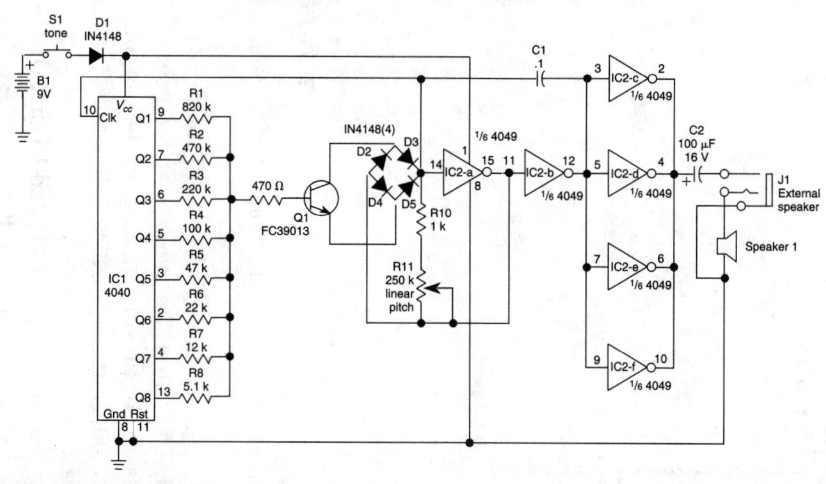

1989 R-E EXPERIMENTERS HANDBOOK ***FIG. 93-13***

The circuit consists of four parts: a binary counter, a D/A converter, a VCO, and an audio output amplifier. The speed at which the counter counts depends on the frequency of the output of the VCO, which in turn is determined by the output of the counter. That feedback loop gives this circuit its characteristic output.

The initial frequency of oscillation is determined by potentiometer R11. The VCO first oscillates at a relatively low frequency, and it gradually picks up speed as the control voltage supplied by the D/A converter increases.

The D/A converter is simply the group of resistors R1 through R8. When none of IC1's outputs is active, little current will flow into the base of Q1, so the VCO's control voltage will be low. As more and more counter outputs become active, base current increases, and so does the VCO's frequency of oscillation.

The VCO itself is composed of IC2-a, IC2-b, and Q1; the timing network is D1 through D4, C1, R10, and R11. The diode bridge functions basically as a voltage-controlled resistor. The buffer amplifier is made up of the four remaining gates from IC2, all wired in parallel. The volume is sufficient for experimental purposes, but you might want to add an amplifier, speaker, or both.

SIREN

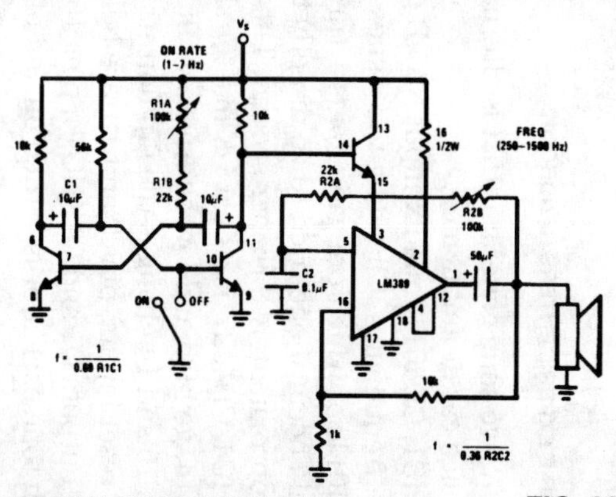

An LM380 audio IC is configured as a feedback audio oscillator. A transistor astable modulates this oscillator at a low frequency, which produces a siren tone.

NATIONAL SEMICONDUCTOR ***FIG. 93-14***

ALTERNATE TONE ALARM

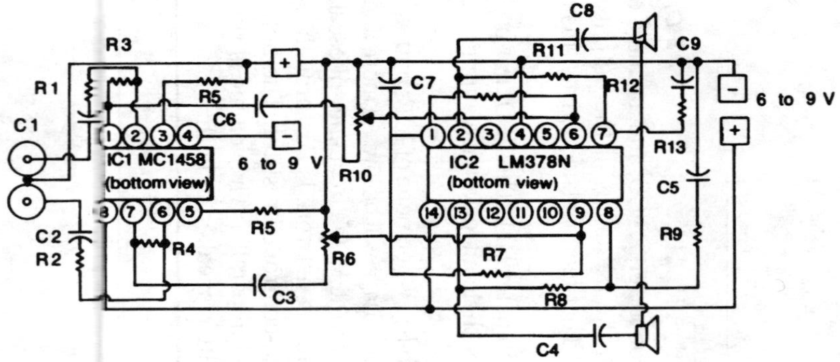

IC1	MC1458 dual op amp
IC2	dual 4-watt amplifier
R1,R2	10 kΩ
R3,R4	1 MΩ
R5, R14	2.2 kΩ
R10,R6	500 kΩ potentiometers
R11,R7	100 kΩ
R8, R12	
R13,R9	4.7 kΩ

C1,C2	.1 μF
C3,C6	.1 μF
C4,C8	200 μF, 15 V
C7	60 μF, 10 V
C9,C5	6 μF
Spks	8 μF, 12-inch

Batteries (2 required or bipolar supply)

1989 R-E EXPERIMENTERS HANDBOOK

FIG. 93-15

A two-tone generator that is alternately switched ON provides a high/low output as might be heard from a traffic vehicle like a police car or ambulance.

IC1, CD4011, quad 2-input NAND gate is a two-tone oscillator in which each side, pins 1 through 7 and 8 through 13 set the tone frequencies. Changing the values of C_2 and C_1 determines the high/low tones. The outout frequencies are coupled to IC2, CD4011, of which one side (pins 1 through 6) acts as a buffer. The buffer is necessary to prevent loading on the outputs that would occur if one tried to go directly to the LM386 amplifier. The other side of IC2, pins 8 through 13, is a slow pulse oscillator of approximately 8 Hz per second. The output at pin 10 is connected to IC4 as a clock.

IC4, CD4027, is a dual J-K master-slave flip-flop that is wired to perform as a toggle switch in which Q1 and 15, and Q1 (NOT) pin 14, go high and low alternately (flip-flop). The clock input from IC2 pin 10 is connected to pin 13 of IC4, and the outputs at pins 15 and 14 changes the flip/flop state with each positive pulse transition. The CD4027 functions in toggle mode when the set and reset inputs, pins 9 and 12, are held low or grounded. Also, J-K inputs, pins 10 and 11, must be held high or to the positive. The outputs Q1 and Q1 (NOT), pins 15 and 14 are connected to pins 13 and 1 respectively of IC1 that enables or disables. Thus, each tone oscillator is turned on and off alternately. IC3 is a straightforward low-voltage audio amplifier.

SIREN OSCILLATOR

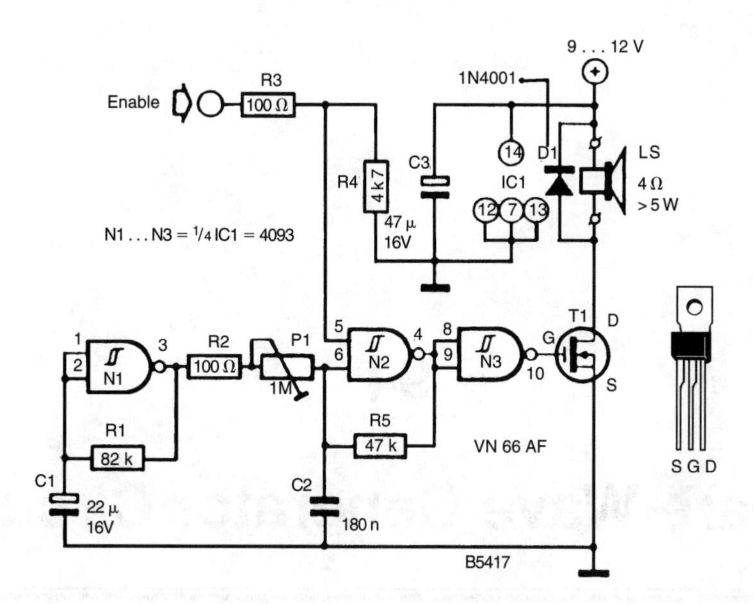

303 CIRCUITS ***FIG. 93-16***

A CD4093 chip and a few components make up a siren oscillator, which drives power MOSFET T1. A 4-Ω speaker is driven directly from this device. The siren is enabled by a logic high applied to the ENABLE input.

94

Square-Wave Generator Circuits

The sources of the following circuits are contained in the Sources section, which begins on page 675. The figure number in the box of each circuit correlates to the entry in the Sources section.

SQUARE-WAVE OSCILLATOR

SQUAREWAVE OSCILLATOR

POPULAR ELECTRONICS *FIG. 94-1*

An op amp with positive feedback generates a square wave. The period of the oscillator is determined by R3 and C1.

$$T = T_1 + T_2 \approx 0.69 \times 2\ (R_3C_1) \qquad T_1 = T_2$$

SCHMITT TRIGGER OR SINE-TO-SQUARE-WAVE CONVERTER

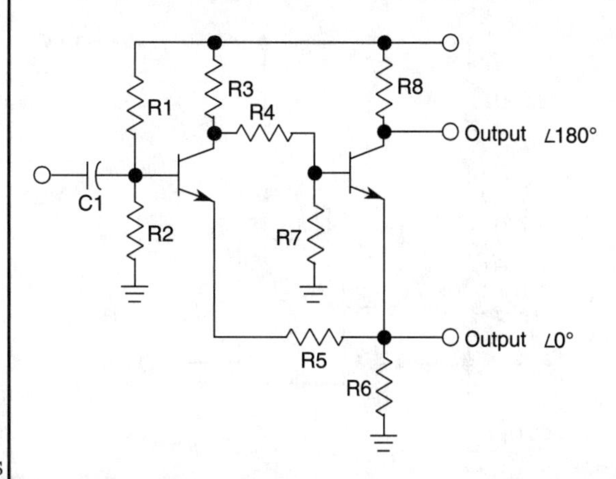

WILLIAM SHEETS *FIG. 94-2*

This sine-wave triggered circuit produces two square-wave outputs that are 180° out of phase.

60-Hz SQUARE-WAVE GENERATOR

SQUARE-WAVE GENERATOR

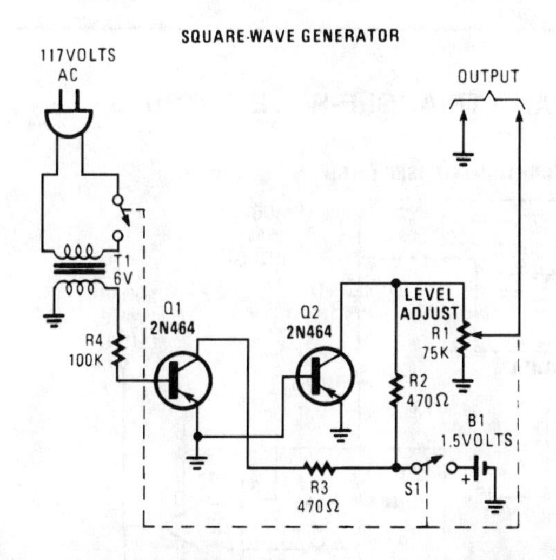

POPULAR ELECTRONICS *FIG. 94-3*

This generator circuit uses an overdriven amplifier to produce a 60-Hz square wave from the 60-Hz ac line. The circuit can be used in line-operated applications as a clock source.

SQUARE-WAVE OSCILLATOR

SQUAREWAVE OSCILLATOR

POPULAR ELECTRONICS *FIG. 94-4*

Positive feedback is via R3 and R4 and R1 and C1 determine period.

VARIABLE-FREQUENCY SQUARE-WAVE GENERATOR

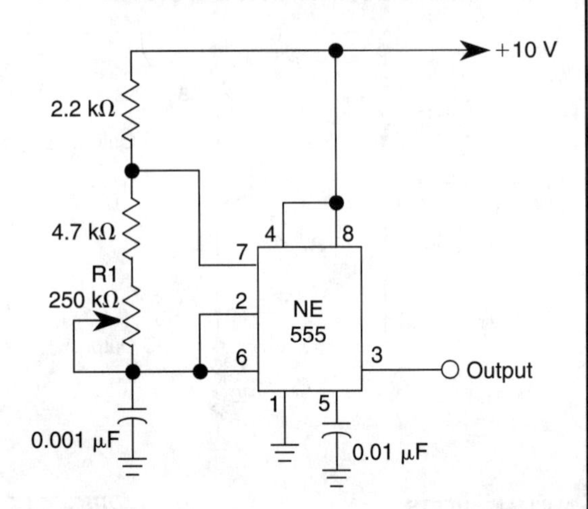

WILLIAM SHEETS

FIG. 94-5

This simple square-wave generator produces a variable frequency output of 2800 Hz to 80 kHz with the values shown. Frequency is adjusted with potentiometer R1.

SCHMITT TRIGGER SINE-/SQUARE-WAVE GENERATOR

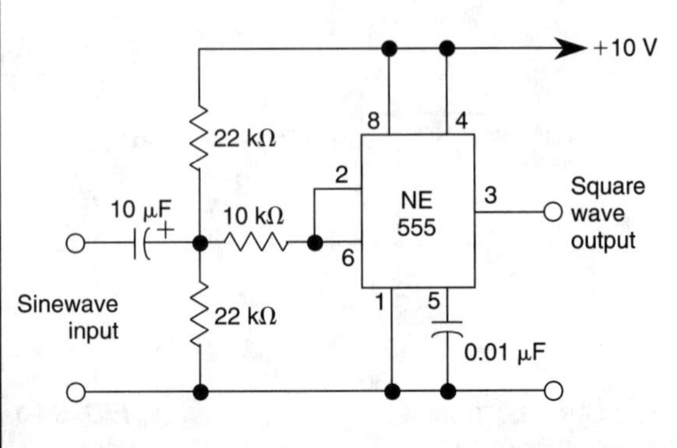

WILLIAM SHEETS

FIG. 94-6

A sine wave input can produce a square wave output by this Schmitt trigger circuit based on a 555 IC.

10-Hz TO 10-kHz VCO WITH SQUARE- AND TRIANGLE-WAVE OUTPUTS

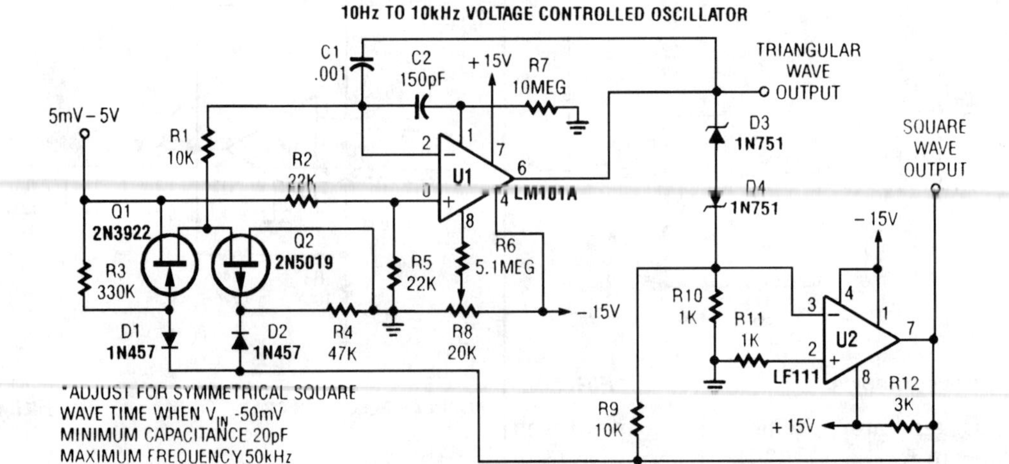

POPULAR ELECTRONICS

FIG. 94-7

95

Stepper Motor Circuits

The sources of the following circuits are contained in the Sources section, which begins on page 675. The figure number in the box of each circuit correlates to the entry in the Sources section.

BIPOLAR STEPPER MOTOR DRIVE CIRCUIT

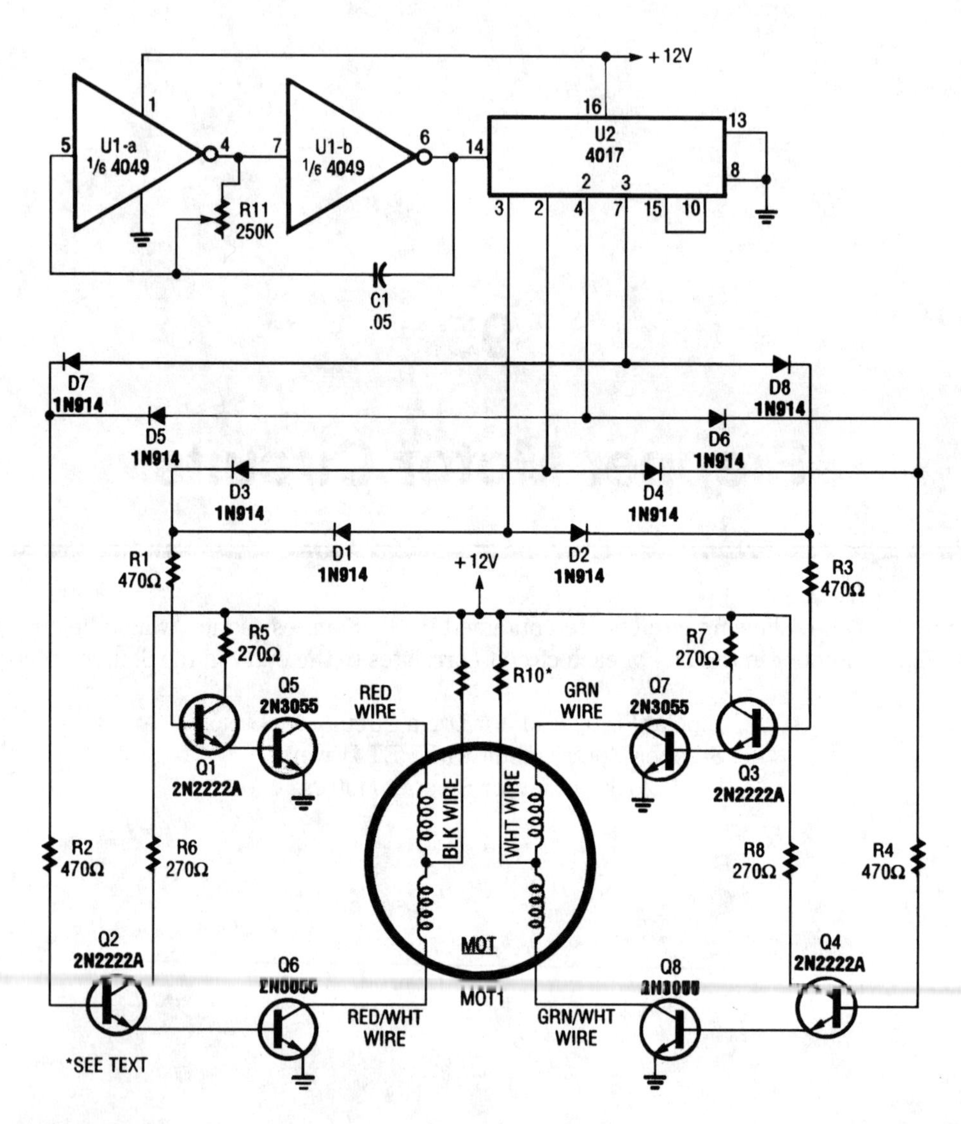

POPULAR ELECTRONICS ***FIG. 95-1***

A 4017 decade counter/divider driven from a low-frequency oscillator (U1-a and U1-b) is used to drive transistor switches to sequence the windings, as is needed. MOT1 is a 12-V stepper motor. R9 and R10 are selected for the motor's current rating. A 3.3-Hz signal from U1 will cause the motor to run at 1 rpm, a 33-Hz signal will result in 10 rpm, etc.

STEPPER MOTOR CIRCUIT WITH FET DRIVERS

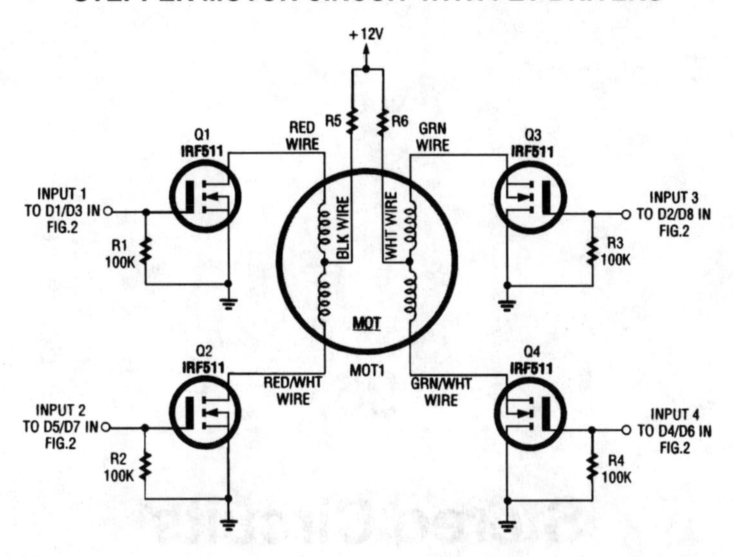

POPULAR ELECTRONICS *FIG. 95-2*

This motor-driver circuit replaces the eight bipolar transistors of the previous circuit with four IFR511 power hexFET's (Q1 through Q4).

DUAL CLOCK CIRCUIT FOR STEPPER MOTORS

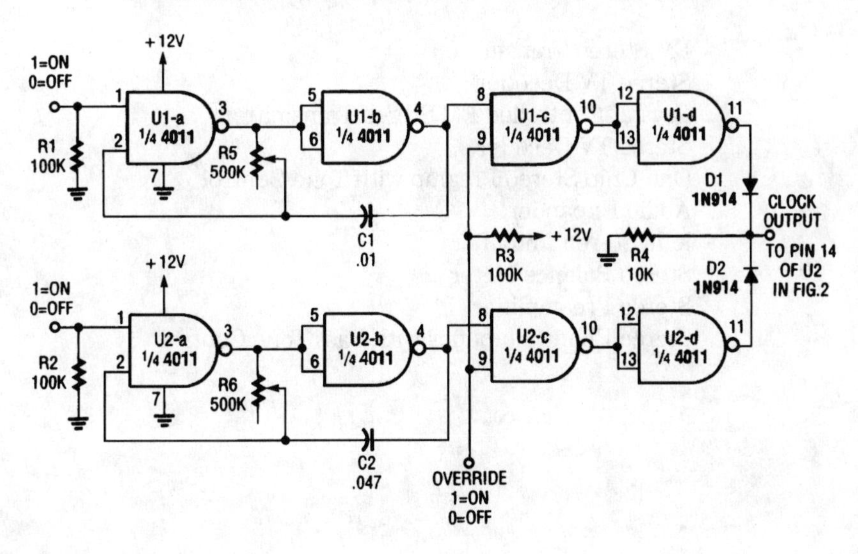

POPULAR ELECTRONICS *FIG. 95-3*

This oscillator can be used to drive a stepper motor circuit at two preset speeds with override to shut the motors off.

96

Stereo Circuits

The sources of the following circuits are contained in the Sources section, which begins on page 675. The figure number in the box of each circuit correlates to the entry in the Sources section.

FM STEREO TRANSMITTER

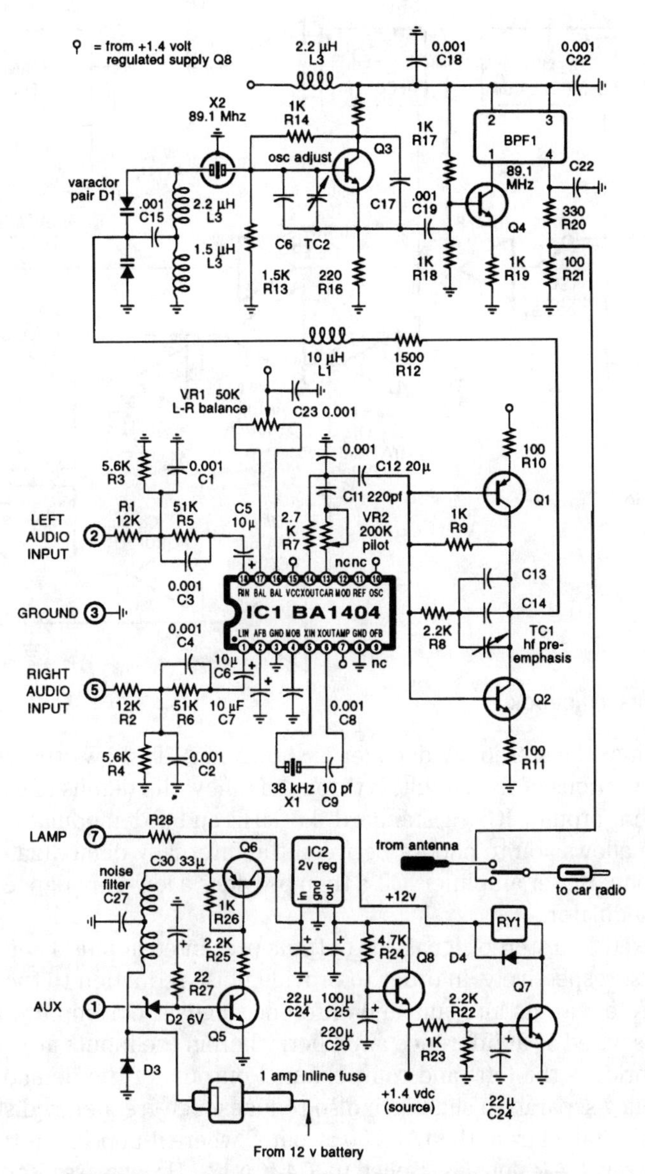

RADIO ELECTRONICS ***FIG. 96-1***

A BA1404 IC is used to generate a complete FM MPX signal. The chip contains all of the necessary circuitry. C1 and R3, and R4 and C4 provide pre-emphasis. The transmitter runs on a single AA cell. L3 is 3 turns of #20 wire on a 3/16" drill (for a form). L3 is 1/4" long. L4 is 4 turns #20 wire on 3/16" drill bit, spaced to 3/8". If monophonic operation is wanted, omit C5 and the 38-kHz oscillator components.

STEREO TV DECODER

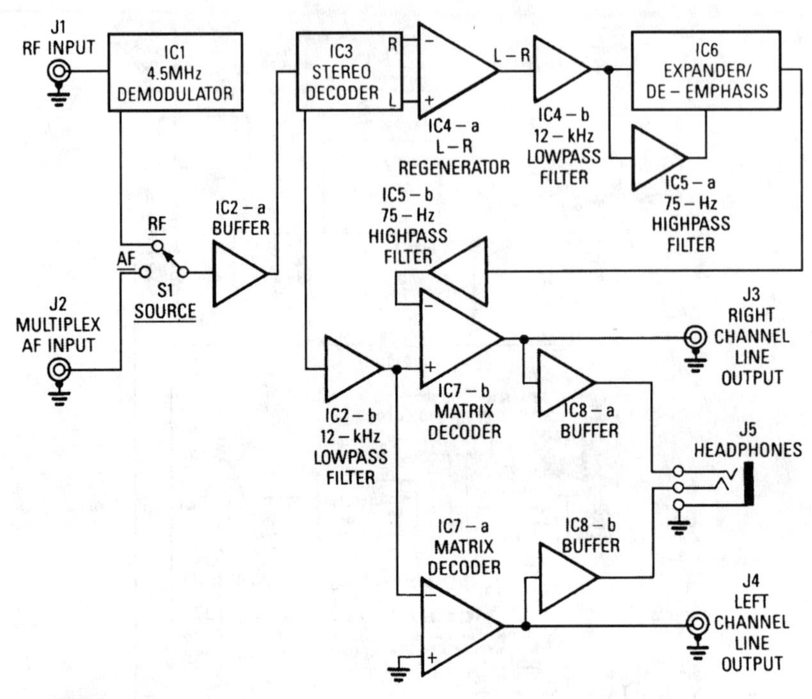

1989 R-E EXPERIMENTERS HANDBOOK ***FIG. 96-2***

A block diagram of the stereo-TV decoder is shown in A. It shows the overall relationships between the separate sections of the circuit; B through E show the details of each subsection. The decoder section centers around IC1, a standard 4.5-MHz audio demodulator. The output of IC1 is routed to S1, which allows you to choose between the internally demodulated signal and an externally demodulated one. Buffer amplifier IC2-a then provides a low-impedance source to drive IC3, an LM1800 stereo demodulator.

When IC3 is locked on a stereo signal, the outputs presented at pins 4 and 5 are discrete left- and right-channel signals, respectively. In order to provide noise reduction to the $L - R$ signal, you must recombine the discrete outputs into sum and difference signals. Op amp IC4-a is used to regenerate the $L - R$ signal. It is wired as a difference amplifier, wherein the inputs are summed together ($+L - R$). Capacitor C18 bridges the left- and right-channel outputs of the demodulator. Although it decreases high-frequency separation slightly, it also reduces high-frequency distortion.

The $L + R$ signal is taken from the LM1800 at pin 2, where it appears at the output of an internal buffer amplifier. The raw $L - R$ signal is applied to IC4-b, a 12-kHz lowpass filter. The $L + R$ signal is also fed through a 12-kHz low pass filter in order to keep the phase shift undergone by both signals equal.

Next, the $L - R$ signal is fed to Q2. It allows you to add a level control to the $L - R$ signal path; it provides a low source impedance for driving the following circuits, and it inverts the signal 180°. Inversion is necessary to compensate for the 180° inversion in the compander.

Next comes the expander stage. At the collector of Q2 is a 75-µs de-emphasis network (R27 and C29) that functions just like the network that is associated with Q1. Note that Q2 feeds both Q3 and

STEREO TV DECODER (*Cont.*)

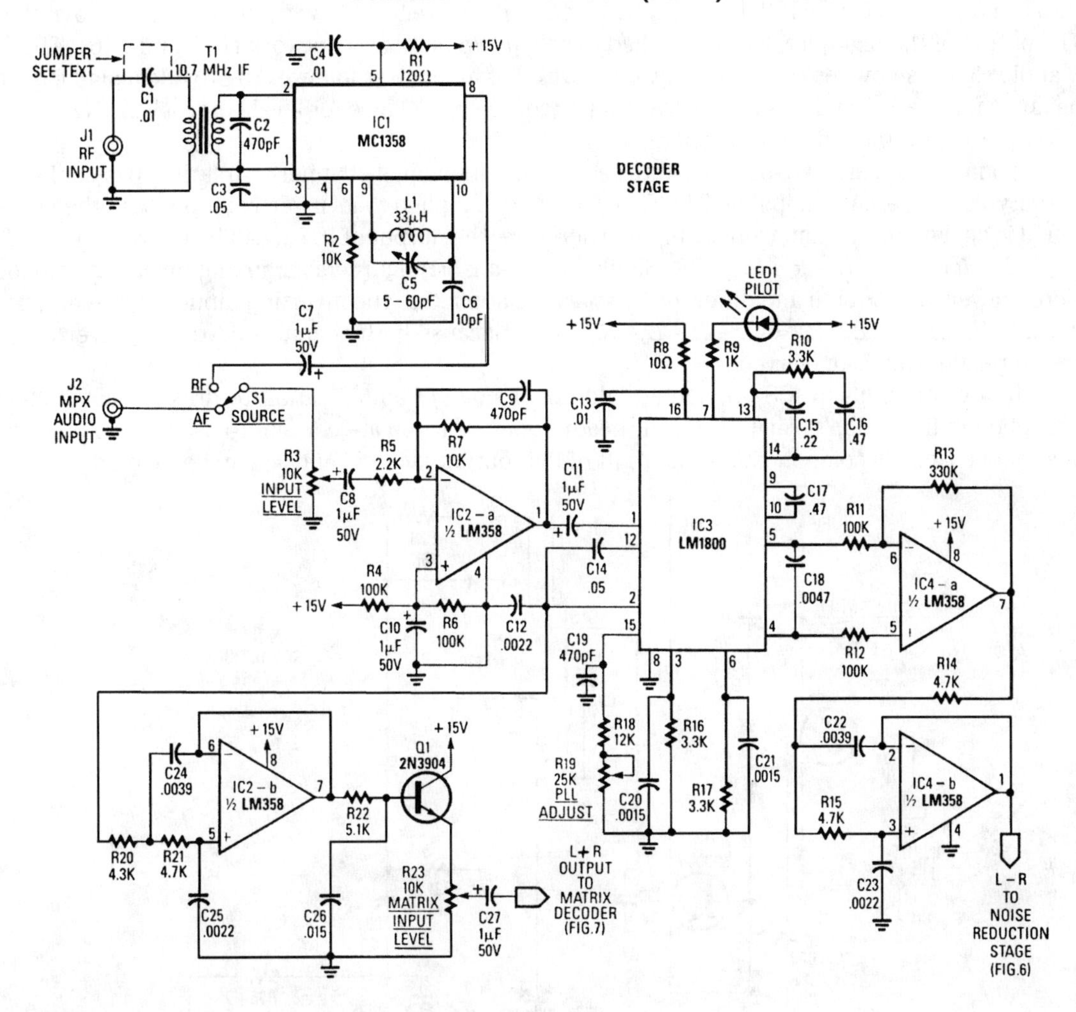

THE DECODER STAGE converts the multiplexed audio signal into L + R and L – R signals.

IC5-a, a –12-dB per octave high-pass filter. The output of that filter drives the rectifier input of IC6, an NE570. The 75-Hz high-pass filter at the rectifier input helps to prevent hum, 60-Hz sych buzz, and other low-frequency noise in the $L - R$ signal from causing pumping or breathing.

The NE570 contains an on-board op amp; its inverting input is available directly at pin 5 and via a 20-kΩ series resistor at pin 6. The 18-kΩ resistor (R30) combines with the internal resistor and C32 (0.01 μF) to form a first-order filter with a 390-μs time constant. Because the internal op amp operates in the inverting mode, the $-(L - R)$ signal is restored to the proper $(L - R)$ form.

The output of the expander drives another 75-Hz high-pass filter, but this one is a third-order type that provides –18 dB per octave rolloff. It is used to keep low-frequency noise from showing up at the output of the decoder. At this point, the $(L - R)$ signal has been restored, more or less, to the condition it was in before it was dBx companded at the transmitter.

STEREO TV DECODER (*Cont.*)

The $L + R$ signal from IC3 is fed to a 12-kHz low-pass filter, IC2-b, with a –12 dB per octave slope. The output of the high-pass filter is applied to a 75 μs de-emphasis network (R22 and C26). The $L + R$ audio signal is now restored properly. Q1 is wired as an emitter follower to provide a high load impedance for the de-emphasis network and a low source impedance for level control R23. Next, the $L + R$ signal is fed to the matrix decoder.

Op amps IC7-a and IC7-b are used to recover the individual channels. First, IC7-b is configured as unity-gain difference amplifier. The $(L + R)$ signal is applied to its inverting input, and the $(L - R)$ signal is applied to the noninverting input. Therefore, the output of IC7-b can be expressed as $-(L + R) + (L - R) = -L + L - R - R = -2R$. Similarly, IC7-a is configured as a mixing inverting amplifier. Here, however, both sum and difference signals are applied to the inverting input. So, the output of IC7-a is $(L + R) - (L - R) = -L - R - L + R = -2L$. Because both channels have been inverted, the stereo relationship is preserved.

The two op amps in IC8 provide an additional stage of amplification to drive a pair of stereo headphones. If you don't plan to use your headphones, or if you are content to use only your stereo's headphone jack, all components to the right of line-output jacks J3 and J4 can be deleted.

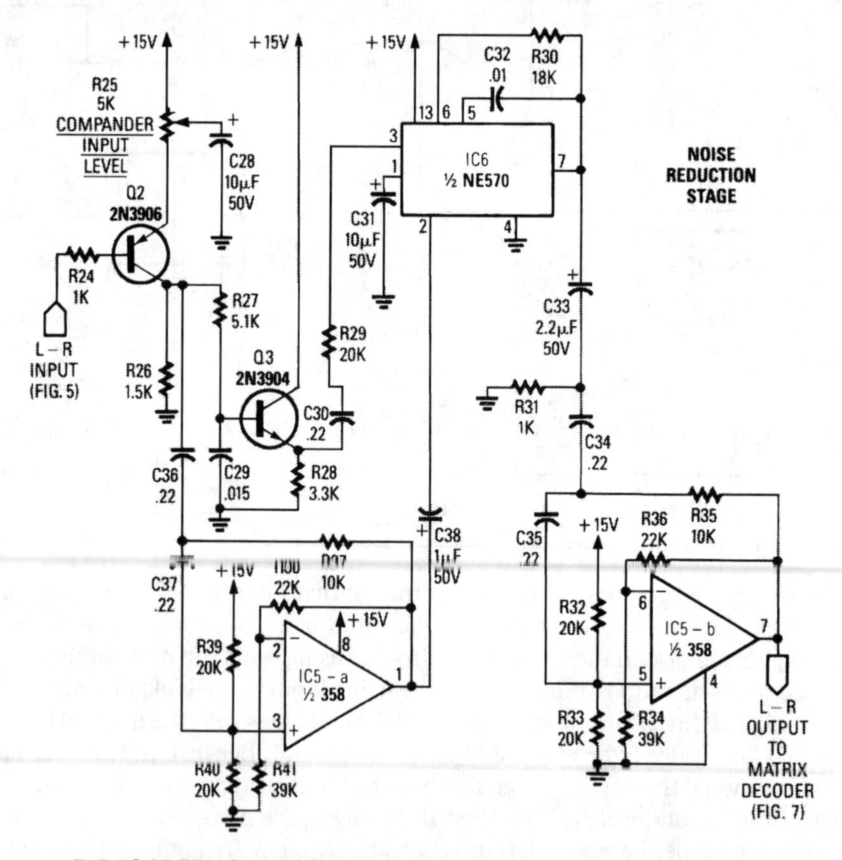

THE NOISE-REDUCTION STAGE de-compands the L−R signal, and emulates dbx-style processing. As described elsewhere in this article (see box), true dbx processing is not currently possible in a home-built circuit due to the inavailability of the dbx IC's.

STEREO TV DECODER (*Cont.*)

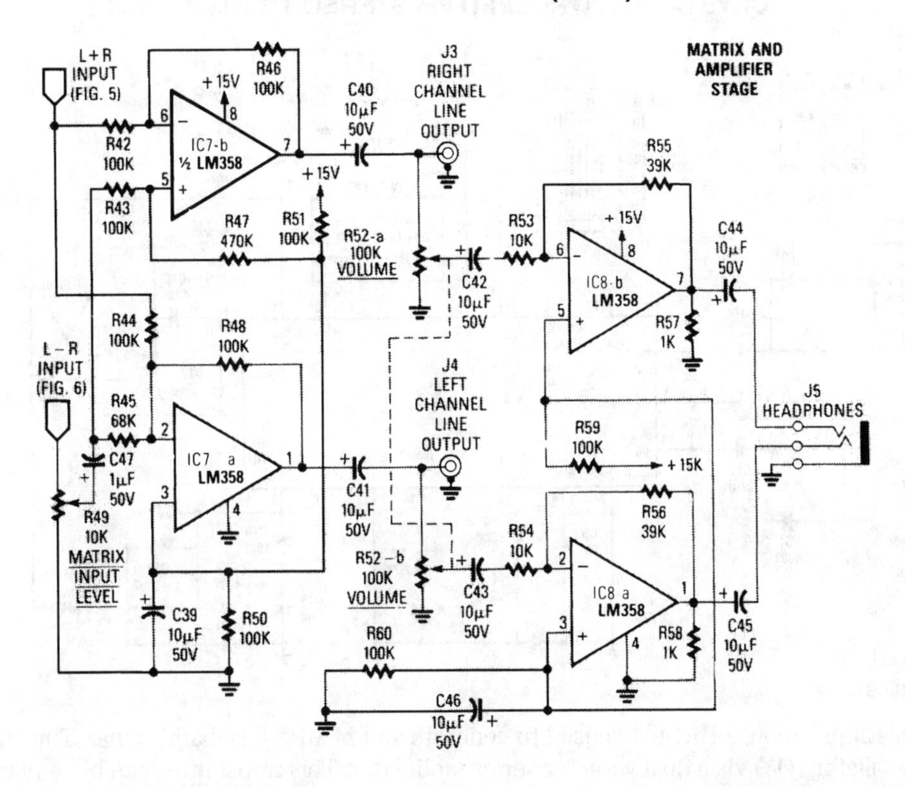

THE MATRIX STAGE separates the L + R and L − R signals into the left- and right-channel components. Op-amp IC8 and associated components provide an optional headphone output. If you do not wish to drive a pair of headphones, or plan to use your amplifier's headphone jack for that purpose, all components to the right of jacks J3 and J4 can be deleted.

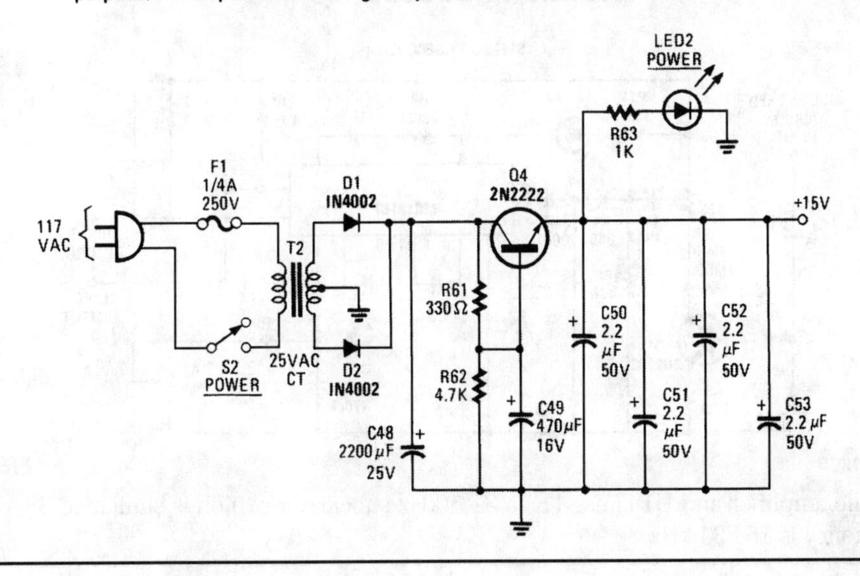

CRYSTAL-CONTROLLED FM STEREO TRANSMITTER

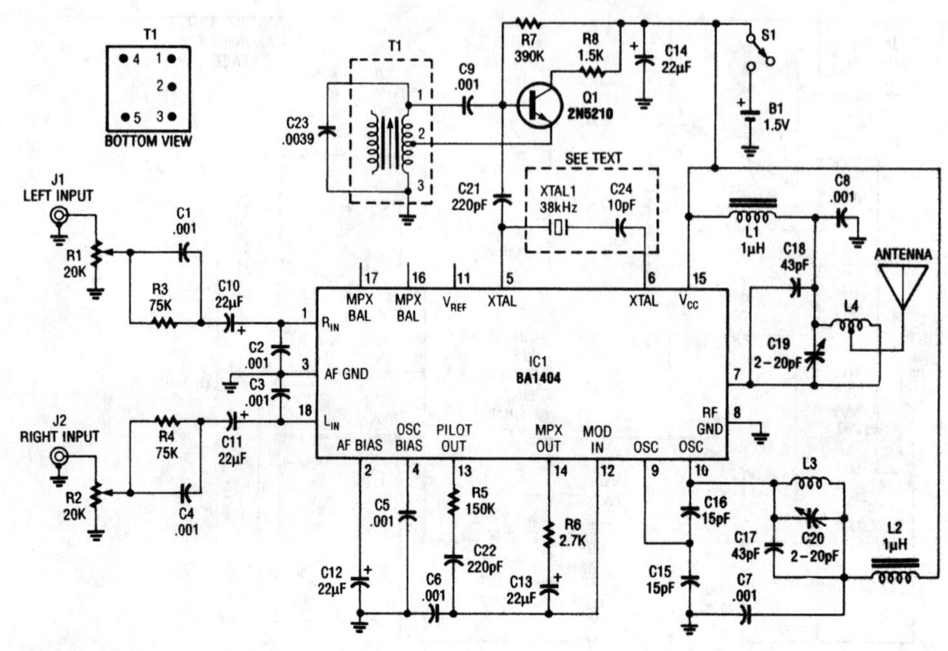

ELECTRONICS NOW *FIG. 96-3*

In this application, a BA1404 is used to generate an FM MPX baseband signal. This modulates a crystal oscillator (Q3) via a dual varactor series modulator. This transmitter can be to play CD audio on an existing FM auto radio.

STEREO TV DECODER

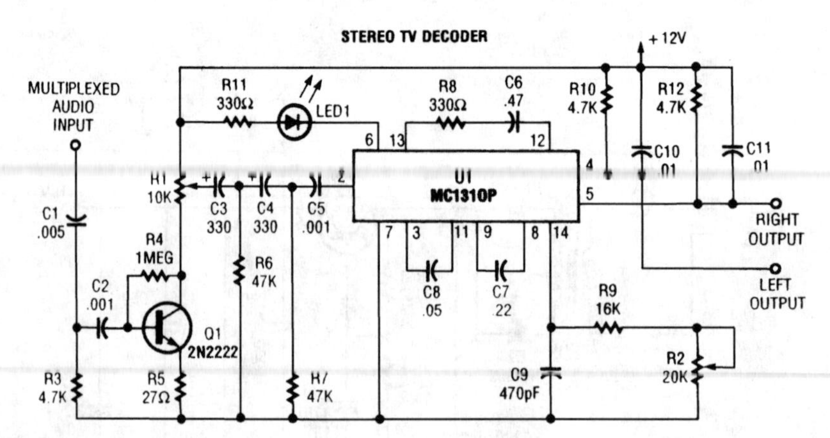

POPULAR ELECTRONICS *FIG. 96-4*

Q1 is an audio amplifier and U1 is used as a 31.5-kHz subcarrier, which is similar to 38-kHz FM MPX. Pilot frequency is 15.734 kHz.

ONE CHIP STEREO PREAMP WITH TONE CONTROL

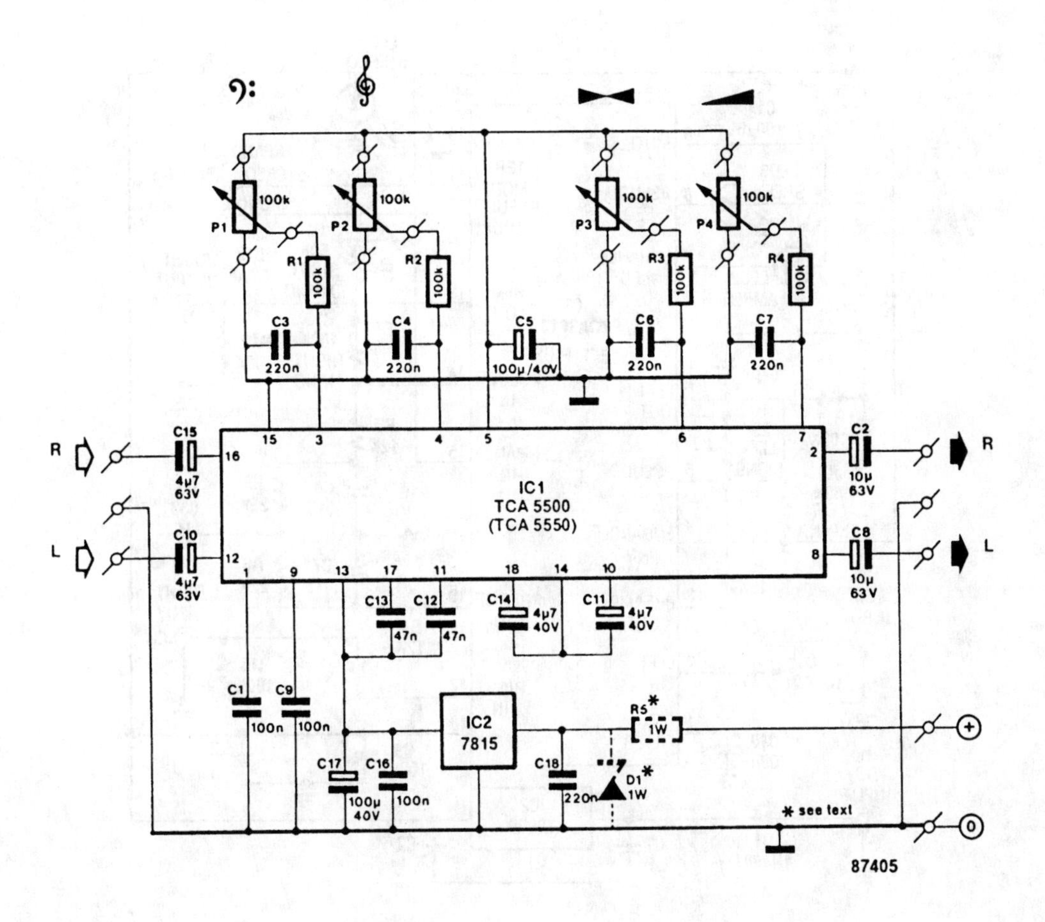

303 CIRCUITS

FIG. 96-5

A Motorola TCA5500 or TCA5550 can provide a stereo preamplifier system with tone controls. This circuit provides a gain of about 10X, a 14-dB tone-control range, a 75-dB volume control range, and it can operate from 8 to 18 Vdc. IC2 provides 15 V for IC1, and the input of IC2 can be supplied from the power amplifier's power supply (+) rail. D1 and R5 should be used if over 30 V input will be used.

AUDIO EXPANDER

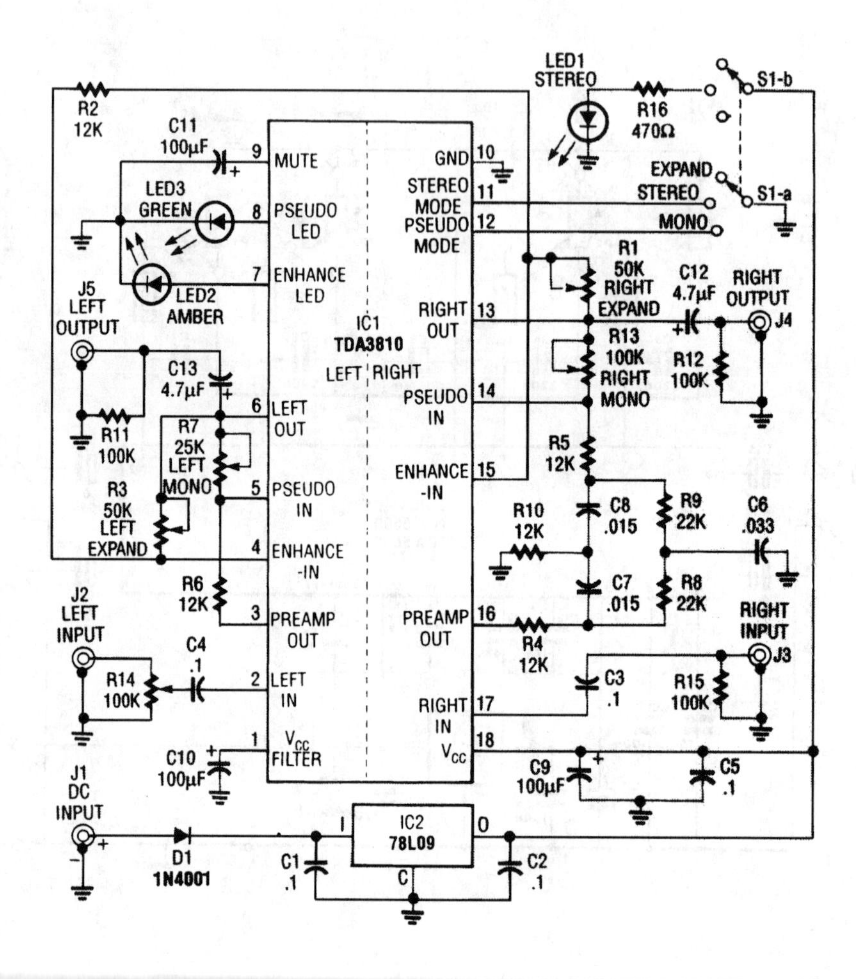

ELECTRONICS NOW

FIG. 96-6

This audio processor is based on the Signetics/Philips TDA3810N stereo, spatial, pseudo-stereo processor, IC. This processor uses a Philips TDA3810IC device, and it functions as an expander, pseudo stereo processor, and audio enhancer. Pseudo stereo is obtained by routing various frequencies to each channel via active filters.

MINI STEREO AMPLIFIER

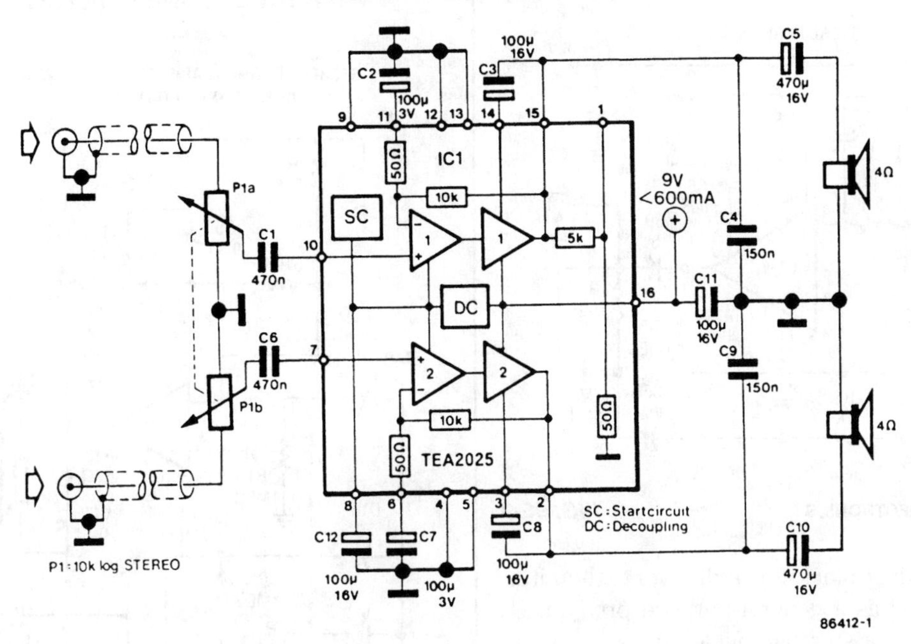

303 CIRCUITS *FIG. 96-7*

Using a Thomson TEA2025, this stereo amplifier provides 1 W per channel into 4 Ω with a 9-V supply. Input sensitivity is 25 mV p-p for full output. Note that pins 4, 5, 12, and 13 of IC1 should be effectively grounded to a ground plane and heatsinked.

STEREO BALANCE METER

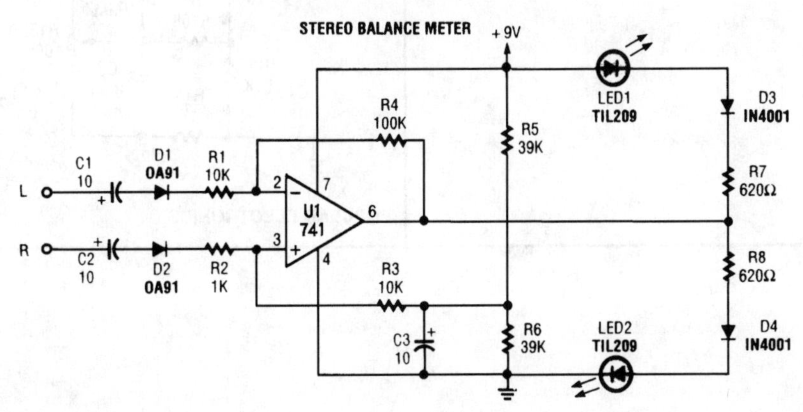

POPULAR ELECTRONICS *FIG. 96-8*

When L & R signals are equal, no output is present from U1, and pin 6 is at a steady 4.5 V. Unbalanced audio causes the LEDs to vary in brightness, which causes a difference that corresponds to unbalance between channels.

STEREO PREAMPLIFIER

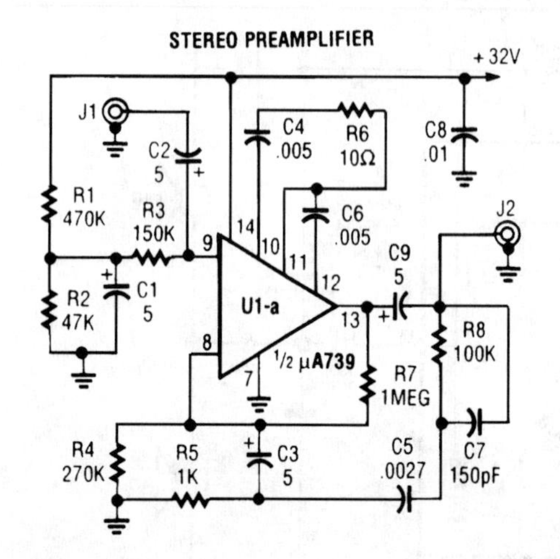

POPULAR ELECTRONICS *FIG. 96-9*

A building block for audio work, the circuit can be used as a general-purpose preamp. Use two circuits for stereo applications.

STEREO PHONO AMPLIFIER WITH BASS TONE CONTROL

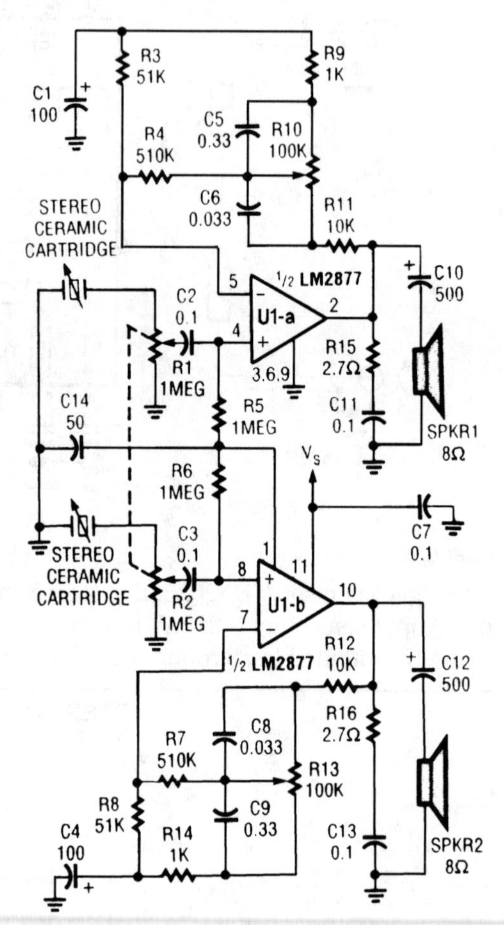

POPULAR ELECTRONICS *FIG. 96-10*

97

Switching Circuits

The sources of the following circuits are contained in the Sources section, which begins on page 675. The figure number in the box of each circuit correlates to the entry in the Sources section.

SIMPLE VIDEO/AUDIO SWITCHER

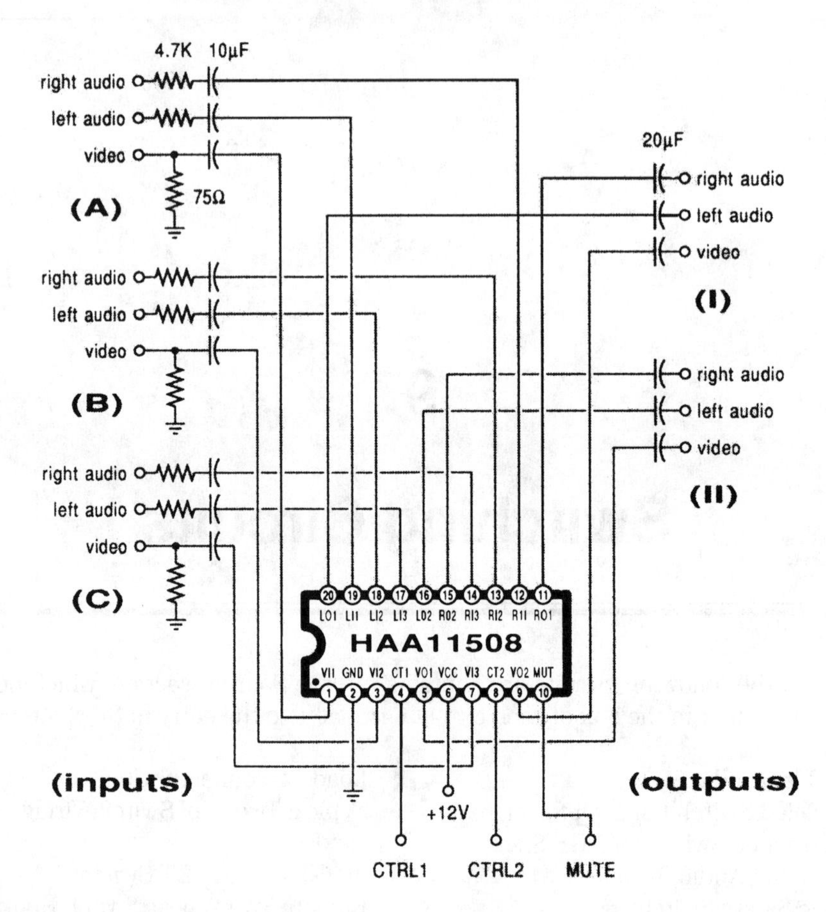

RADIO ELECTRONICS *FIG. 97-1*

This channel selector selects video and stereo audio from any one of three different sources. The circuit should be constructed on a PC board with plenty of ground plane to minimize noise.

dc-CONTROLLED SWITCH USING OPTOISOLATOR

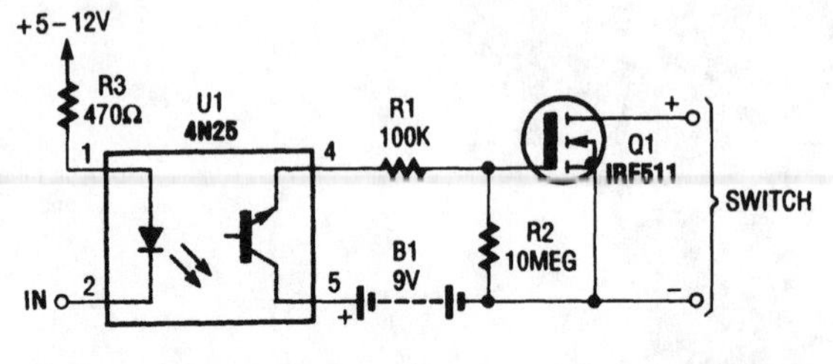

POPULAR ELECTRONICS *FIG. 97-2*

This dc-controlled switch uses an optoisolator/coupler, U1, to electrically isolate the input signal from the output-control device.

WIDEBAND VIDEO SWITCH FOR RGB SIGNALS

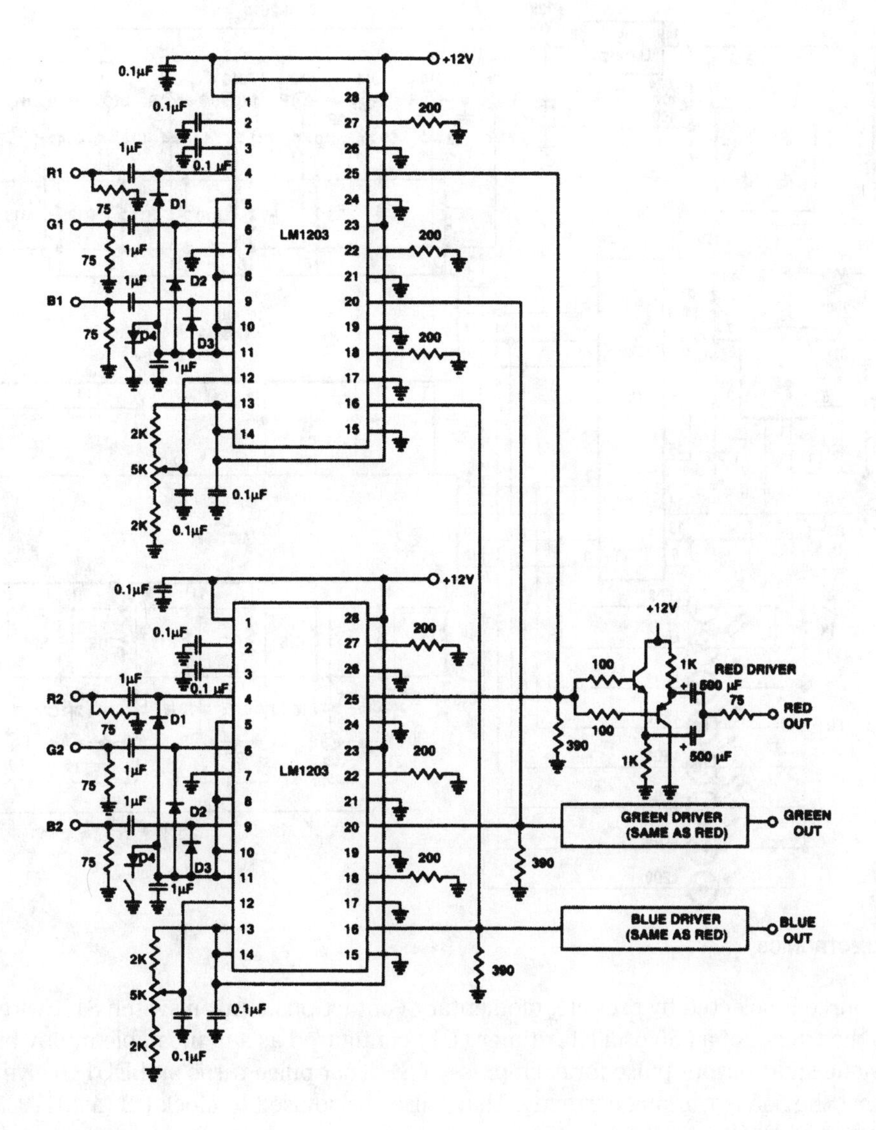

NATIONAL SEMICONDUCTOR

FIG. 97-3

The switch shown selects 1 to 2 inputs and uses a National LM1203. The slew rate is 4-V p-p into 390 Ω in 5 to 7 ns.

EIGHT-CHANNEL AUDIO SWITCHER

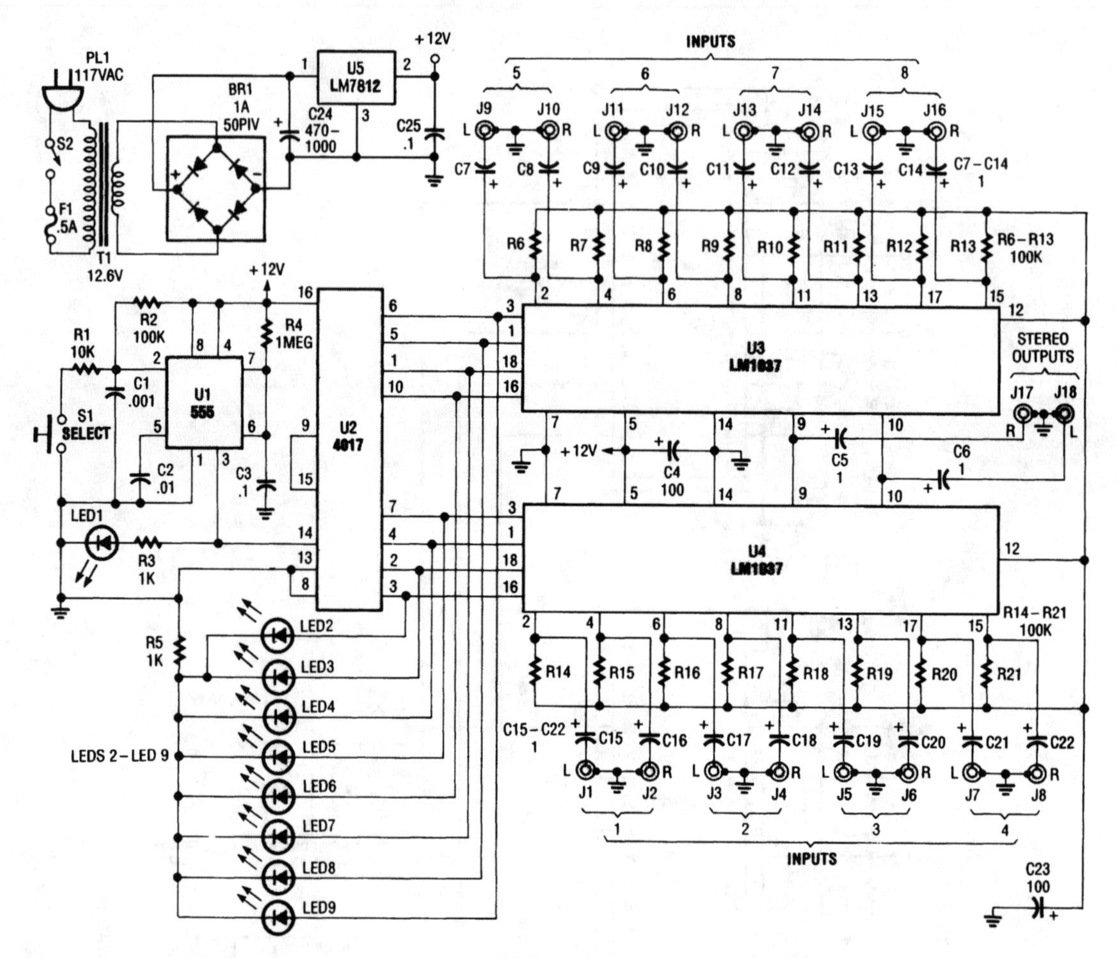

POPULAR ELECTRONICS

FIG. 97-4

This source is selected by pressing momentary-contact pushbutton switch S1. Switch S1 is connected to the trigger of a 555 oscillator/timer (U1) configured as a monostable multivibrator, which generates one short output pulse for each press of S1. That pulse turns on LED1 to give a visible indication that the 555 is working correctly. That pulse is also used to clock U2 (a 4017 CMOS divide-by-1-counter/divider).

Both LED1 and its associated current-limiting resistor R3 are optional and can be left out of the finished project without any affect on circuit operation. The 4017 advances by one clock pulse each time S1 is pressed, turning on its corresponding output. Pin 9 (corresponding to output 8) of U2 is directly connected to its own reset terminal at pin 15. This allows the counter to count from zero to seven, and then reset to zero on the eighth count.

EIGHT-CHANNEL AUDIO SWITCHER (*Cont.*)

Pin 13, the enable input of U2, is tied to ground to allow the counter to operate. Outputs zero through seven are connected to eight indicator LEDs and the control pins of the two LM1037s (U3 and U4). When an output is selected, its LED lights and the corresponding control input on the LM1037 is brought high.

The LM1037 has extremely high-impedance inputs and low-impedance outputs, so interconnection between various types and brands of equipment should not be a problem. That, together with a wide-frequency response and low distortion, makes it ideal for use with good-quality, home-entertainment systems. The prototype of the audio switcher has a usable frequency response of from just a few hertz to over 100 kHz.

Power for the switcher is provided by a rather simple circuit. Because the switcher only draws between 20 and 30 mA, a simple circuit using the popular 7812 or 78L12 (a low-power version) voltage regulator works quite well.

ELECTRONIC SAFETY SWITCH

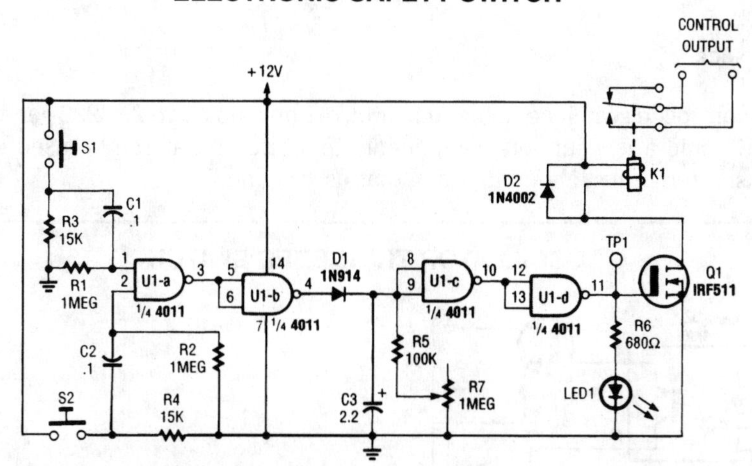

The electronic safety-control is built around a 4011 quad two-input NAND gate and an IRF511 hexFET.

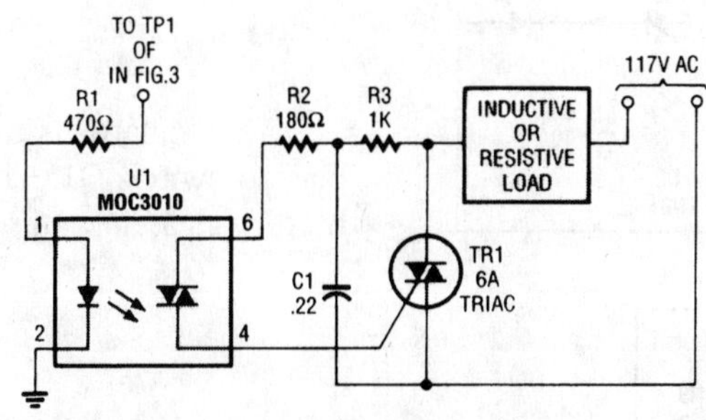

The relay-replacement circuit (shown here) can be used to operate inductive or resistive loads.

POPULAR ELECTRONICS **FIG. 97-5**

S1 and S2 must be depressed within 200 ms of each other to activate K1. The hold time is adjustable via R7. S1 and S2 overlap time can be changed by changing C1 and C2 or R1 and R2.

AUDIO-CONTROLLED SWITCH

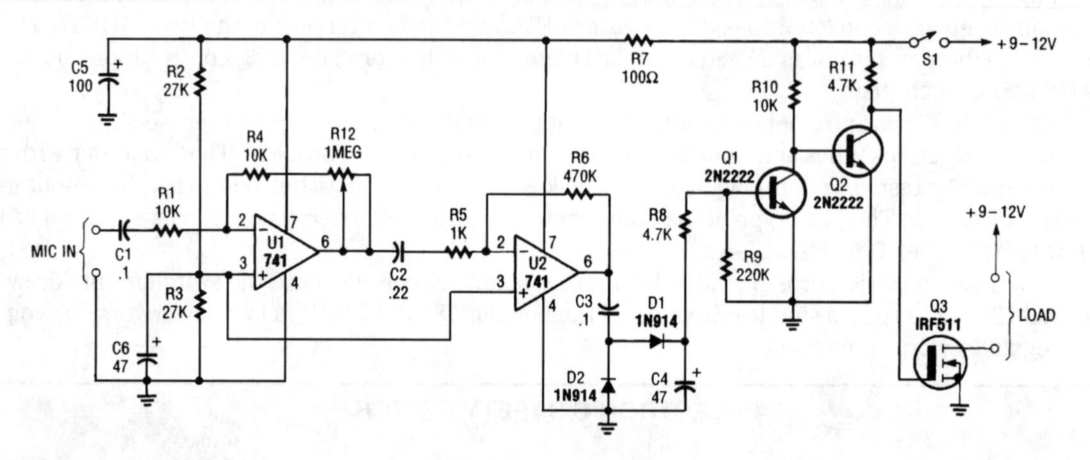

POPULAR ELECTRONICS *FIG. 97-6*

This audio-controlled switch combines a pair of 741 op amps, two 2N2222 general-purpose transistors, a hexFET, and a few support components to a circuit that can be used to turn on a tape recorder, a transmitter, or just about anything that uses sound.

OSCILLATOR TRIGGERED SWITCH

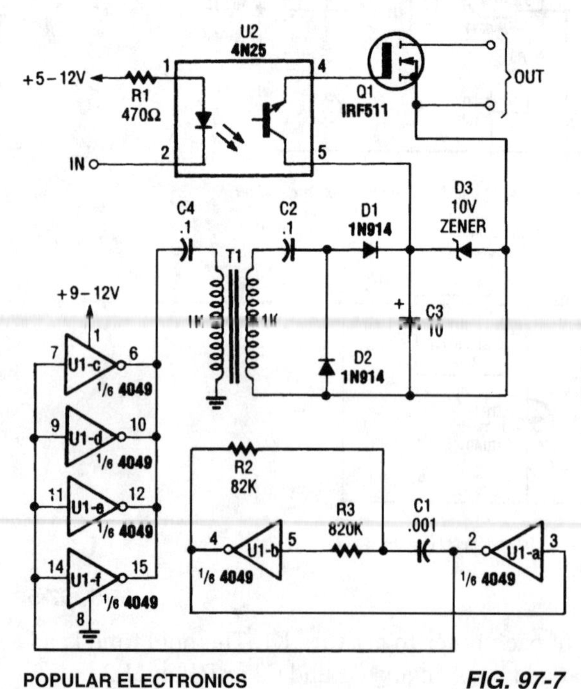

POPULAR ELECTRONICS *FIG. 97-7*

An oscillator is used here to generate a 0 V bias to switch Q1. This removes the need for a battery as a bias source.

LOAD-DISCONNECT SWITCH

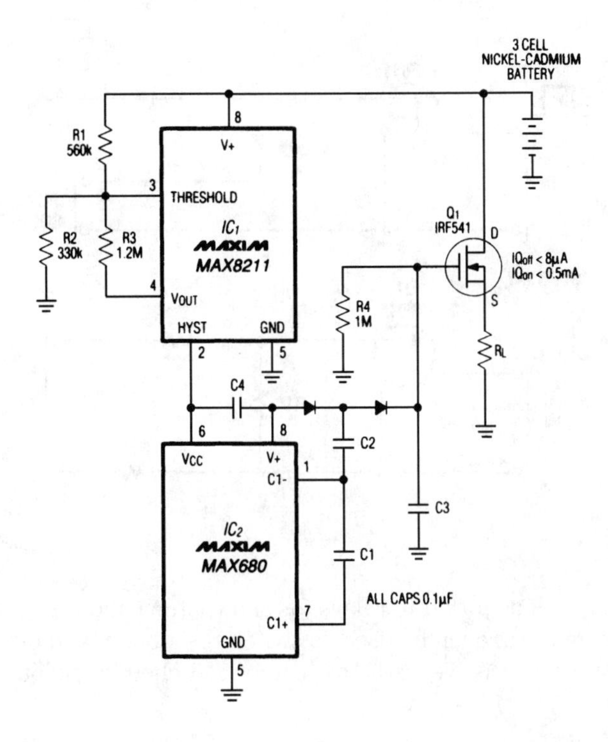

MAXIM ENGINEERING JOURNAL

FIG. 97-8

Deep discharge can damage a rechargeable battery. By disconnecting the battery from its load, this circuit halts battery discharge at a predetermined level of declining terminal voltage. Transistor Q1 acts as the switch. The overall circuit draws about 500 μA when the switch is closed and about 8 μA when the switch is open.

TYPICAL TWO-WAY SWITCH WIRING

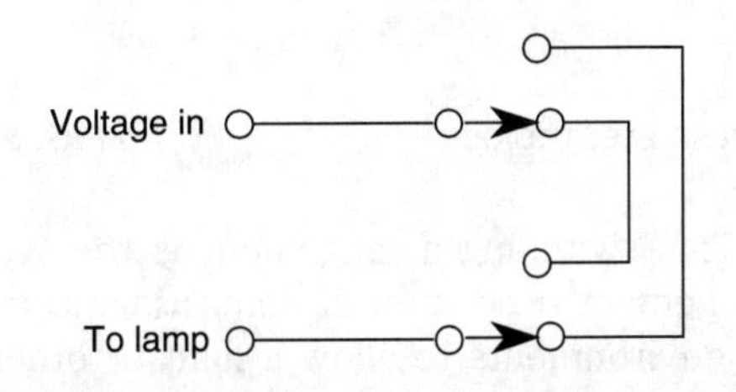

When the light is off, it can be turned on with either switch. When it's on, it can be turned off with either switch.

ELECTRONICS NOW

FIG. 97-9

HEXFET SWITCH

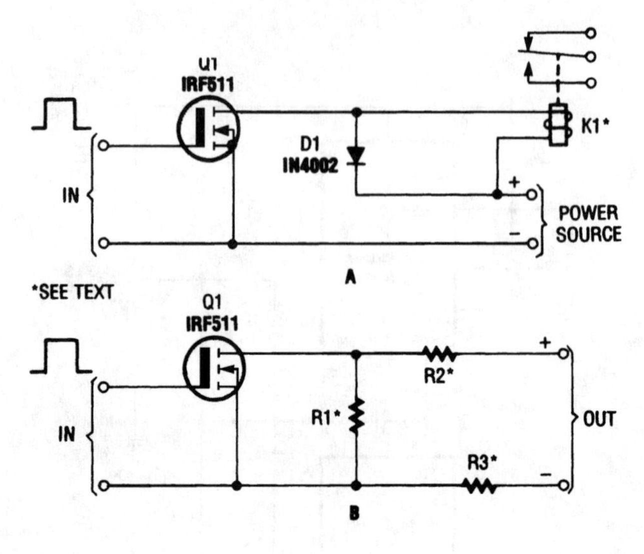

POPULAR ELECTRONICS *FIG. 97-10*

The hexFET can switch dc power to relays (as shown in A), motors, lamps, and numerous other devices. That arrangement can even be used to switch resistors in and out of a circuit, as shown in B. R1, R2, and R3 represent resistive loads that can be switched in and out of the circuit.

dc-CONTROLLED FET SWITCH

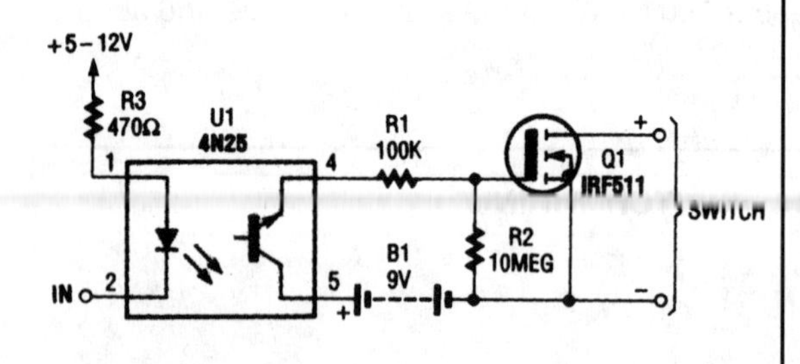

POPULAR ELECTRONICS *FIG. 97-11*

This dc-controlled switch uses an optoisolator/coupler, U1, to electrically isolate the input signal from the output-control device.

REMOTE TWO WAY ac SWITCH HOOKUP

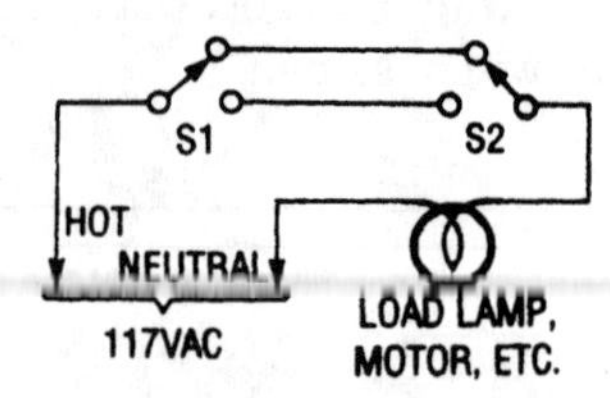

POPULAR ELECTRONICS *FIG. 97-12*

This switching arrangement is the type of arrangement used in both domestic and industrial environments to allow a light or other ac-operated device to be controlled from more than one location.

DUAL-CONTROL HEXFET SWITCH

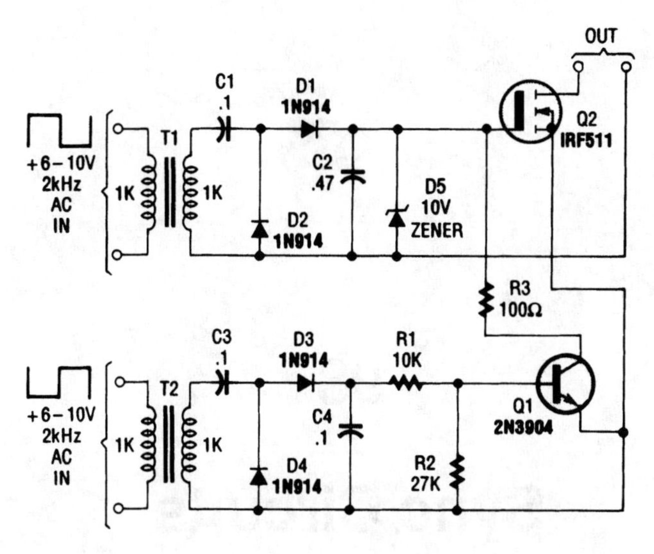

POPULAR ELECTRONICS

FIG. 97-13

This dual-control switch uses two 6 to 10-Vac sources to trigger the circuit on and off; one source for each function.

98

Sync Circuits

The sources of the following circuits are contained in the Sources section, which begins on page 675. The figure number in the box of each circuit correlates to the entry in the Sources section.

SYNC GATING CIRCUIT

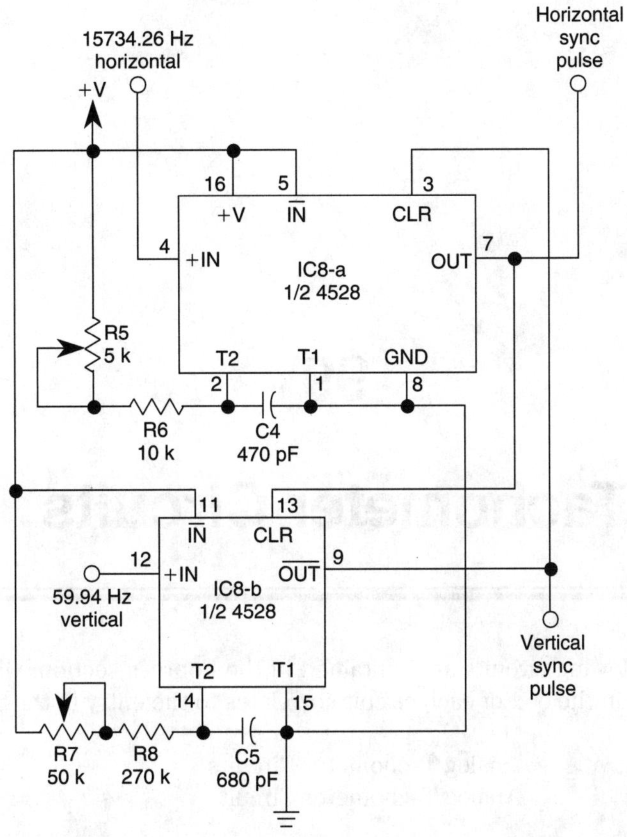

RADIO ELECTRONICS *FIG. 98-1*

This circuit guarantees that only one type of sync pulse is generated at a time. During vertical sync periods, horizontal sync is disabled.

SYNC COMBINER

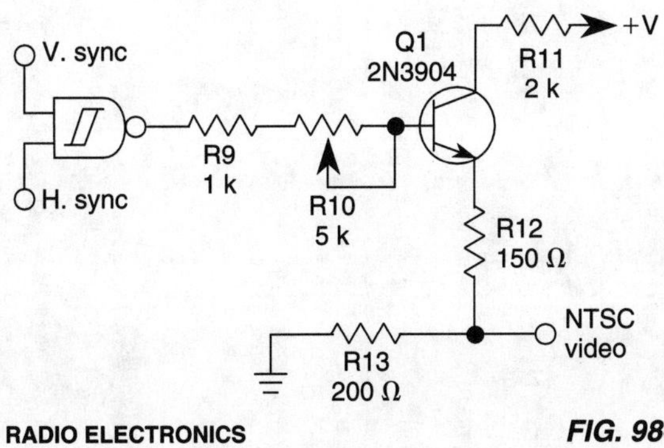

This circuit combines H and V sync signals at TTL or CMOS levels and produces an NTSC video sync output.

RADIO ELECTRONICS *FIG. 98-2*

99

Tachometer Circuits

The sources of the following circuits are contained in the Sources section, which begins on page 675. The figure number in the box of each circuit correlates to the entry in the Sources section.

ANALOG TACHOMETER CIRCUITS

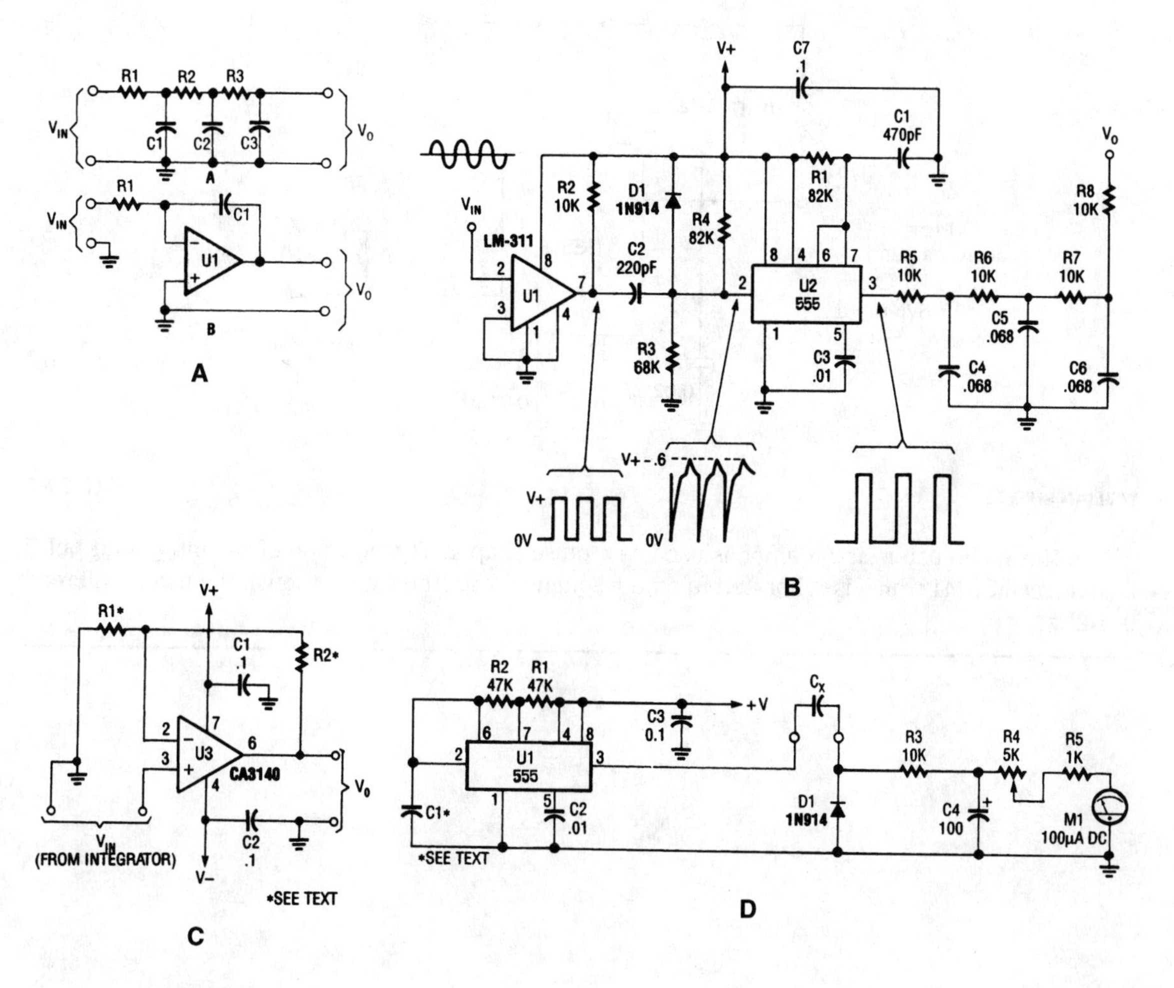

POPULAR ELECTRONICS

FIG. 99-1

The four circuits shown are: a passive and active integrator, an analog tachometer, a scaling amplifier, and a capacitance meter.

In B, $T = 1.1\, R_1\, C_1$ (output pulse duration)

In C, $V_o = V_{\text{in}} \left(1 + \dfrac{R_2}{R_1}\right)$

ANALOG TACHOMETER CIRCUIT

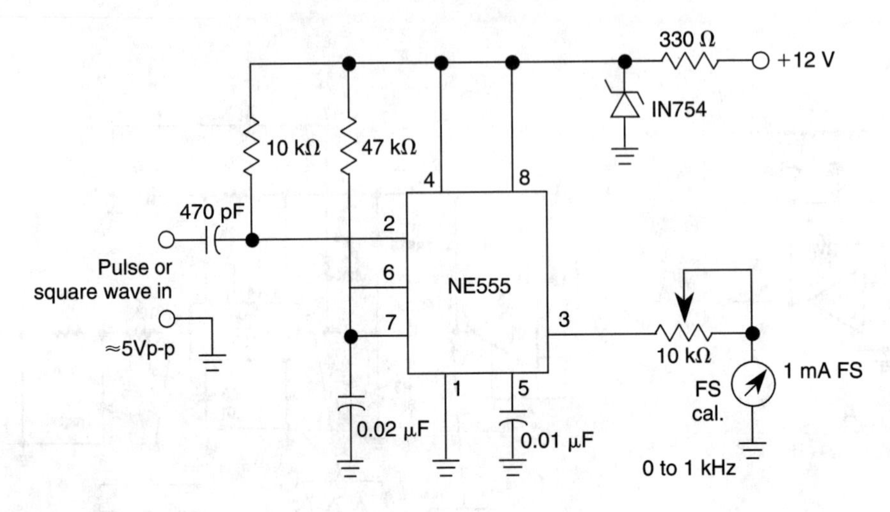

WILLIAM SHEETS

FIG. 99-2

In this tachometer circuit a 555 is used as a pulse shaper. The dc value of the integrated pulse train is read by M1 which is calibrated to read frequency. With the values shown, the meter will read 0–1 kHz.

100

Telephone-Related Circuits

The sources of the following circuits are contained in the Sources section, which begins on page 675. The figure number in the box of each circuit correlates to the entry in the Sources section.

Telephone Ringer
Automatic Telephone-Call Recording Circuit
Music on Hold
Telephone Ring Converter
Phone-In-Use Indicator
Emergency Telephone Dialer
Telephone Bell Simulator
Simple Telephone Ring Indicator
Phone-Line Interface
Music-On-Hold Box
Speakerphone Adapter
Telephone Voice-Mail Alert
Telephone Scrambler
Phone Pager
5-V Dial-Tone Circuit
Phone Pager
Alarm Dialer
Telephone Audio Interface
Caller ID Circuit
FCC Part 68 Phone Interface
Telephone Amplifier
Telephone Hold Circuit
Telephone Circuit
Telephone-Line Tester

TELEPHONE RINGER

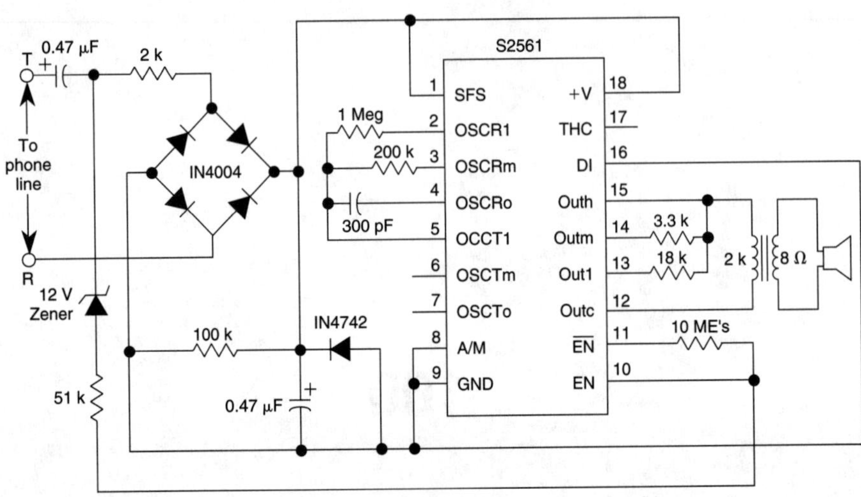

RADIO-ELECTRONICS

FIG. 100-1

Using an AMI chip P/N S2561, this telephone ringer can be powered directly off the telephone line. Audio output is about 50 mW when powered from a 10-V source.

AUTOMATIC TELEPHONE-CALL RECORDING CIRCUIT

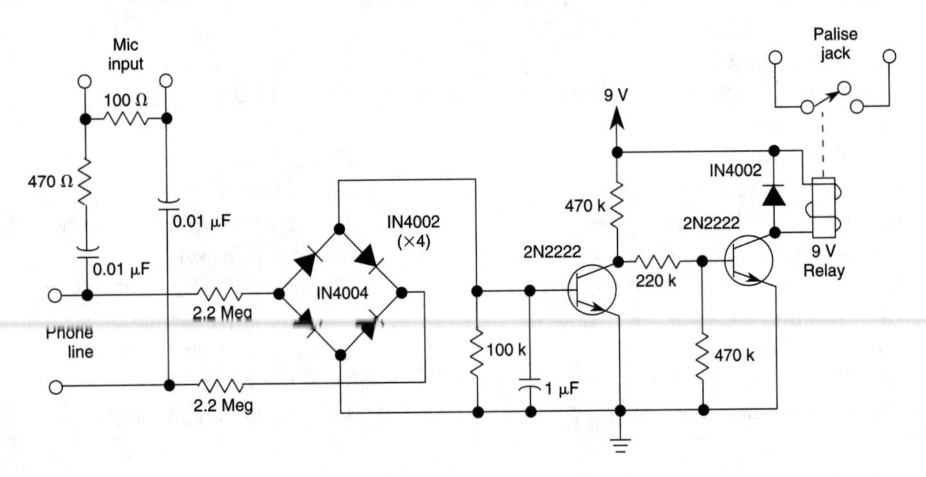

RADIO-ELECTRONICS

FIG. 100-2

The dc voltage present on a telephone line is usually around 45 to 50 V on-hook and 6 V off-hook. This circuit uses this drop in voltage to activate a relay. The relay controls a cassette tape recorder. Audio is taken off through a network to the microphone input of the cassette.

MUSIC ON HOLD

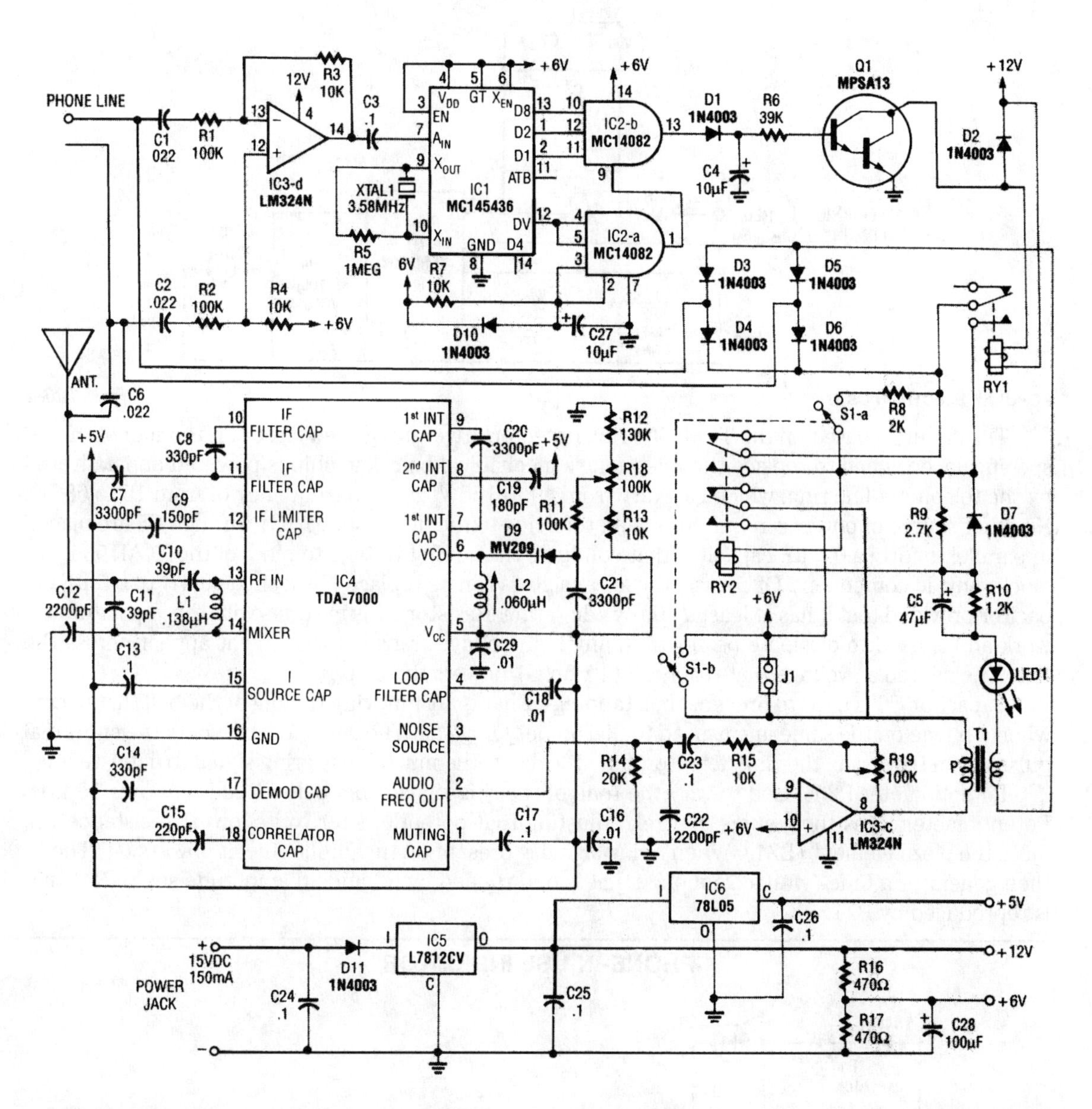

RADIO-ELECTRONICS *FIG. 100-3*

When an asterisk * is pressed on the touch-tone phone, IC1 a DTMF decoder, controls on-hold logic. Audio from the FM receiver IC4 is placed on the telephone line when a hold condition is present. RY2 is a DPDT 12-V relay. To place a caller on hold, press the asterisk button on the touch-tone phone and hang up the handset.

TELEPHONE RING CONVERTER

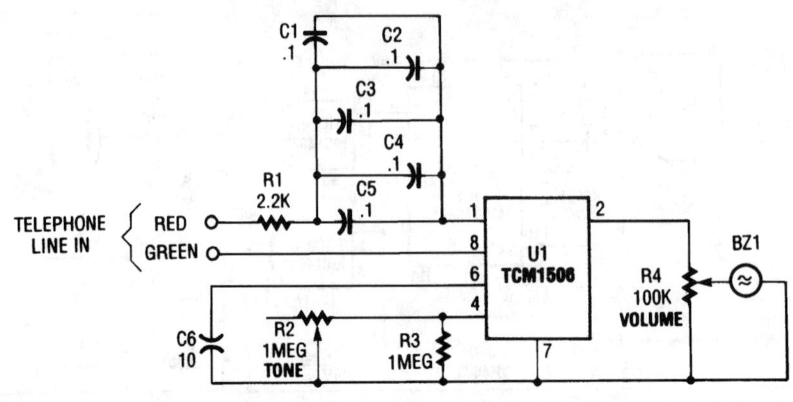

POPULAR ELECTRONICS *FIG. 100-4*

The circuit is based on the TCM1506 ring detector/driver integrated circuit. It is a monolithic IC specifically designed to replace the telephone's mechanical bell. The chip is powered and activated by the telephone-line ring, which can vary from 40 to 150 V rms at a frequency of from 15 to 68 Hz. No other source of power is required. Again, referring to the figure shown, C1 through C5 are placed in parallel to form a 0.5-μF capacitor that conducts the ac ring voltage to pin 1 of the TCM1506, but blocks any dc component. Of course, those capacitors can be replaced by a single 0.47- to 0.5-μF capacitor provided that it has at least a 400-WVdc rating. Resistor R1 is in series with the capacitor network and is used to dissipate power from any high-voltage transient that might appear across the line. The diluted ac voltage that reaches pin 1 on U1 powers the chip.

Capacitor C6 is used to prevent "bell tapping." That is an annoying ringing of the bell that occurs when a phone on the same line is used to dial an outgoing call. The capacitor prevents the short dial pulses from triggering the ring detector, but still allows the much longer ring signal to activate it.

Potentiometer R2 is used to vary the tone of the ring signal from below 100 Hz to over 15 kHz. Potentiometer R4 is the volume control; adjusting that potentiometer to its lowest resistance will mute the piezo element (BZ1). When a ring signal is present on the phone-line, it powers U1. The IC then generates a tone (with a frequency that is determined by R2 and an amplitude set by R4) that is reproduced by BZ1.

PHONE-IN-USE INDICATOR

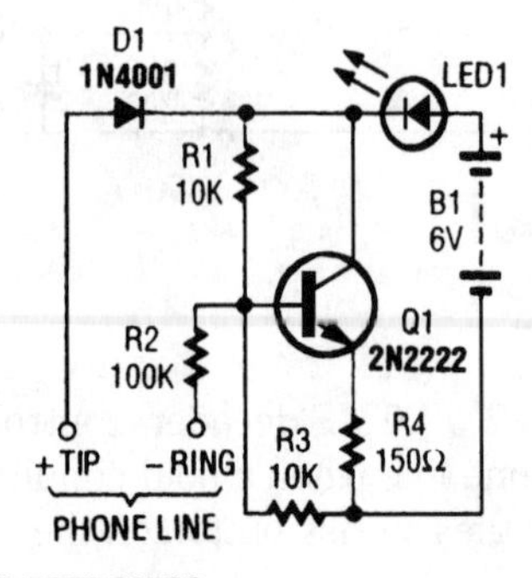

This phone-in-use indicator also indicates the presence of a ring signal. Just the thing for the hearing impaired.

POPULAR ELECTRONICS *FIG. 100-5*

EMERGENCY TELEPHONE DIALER

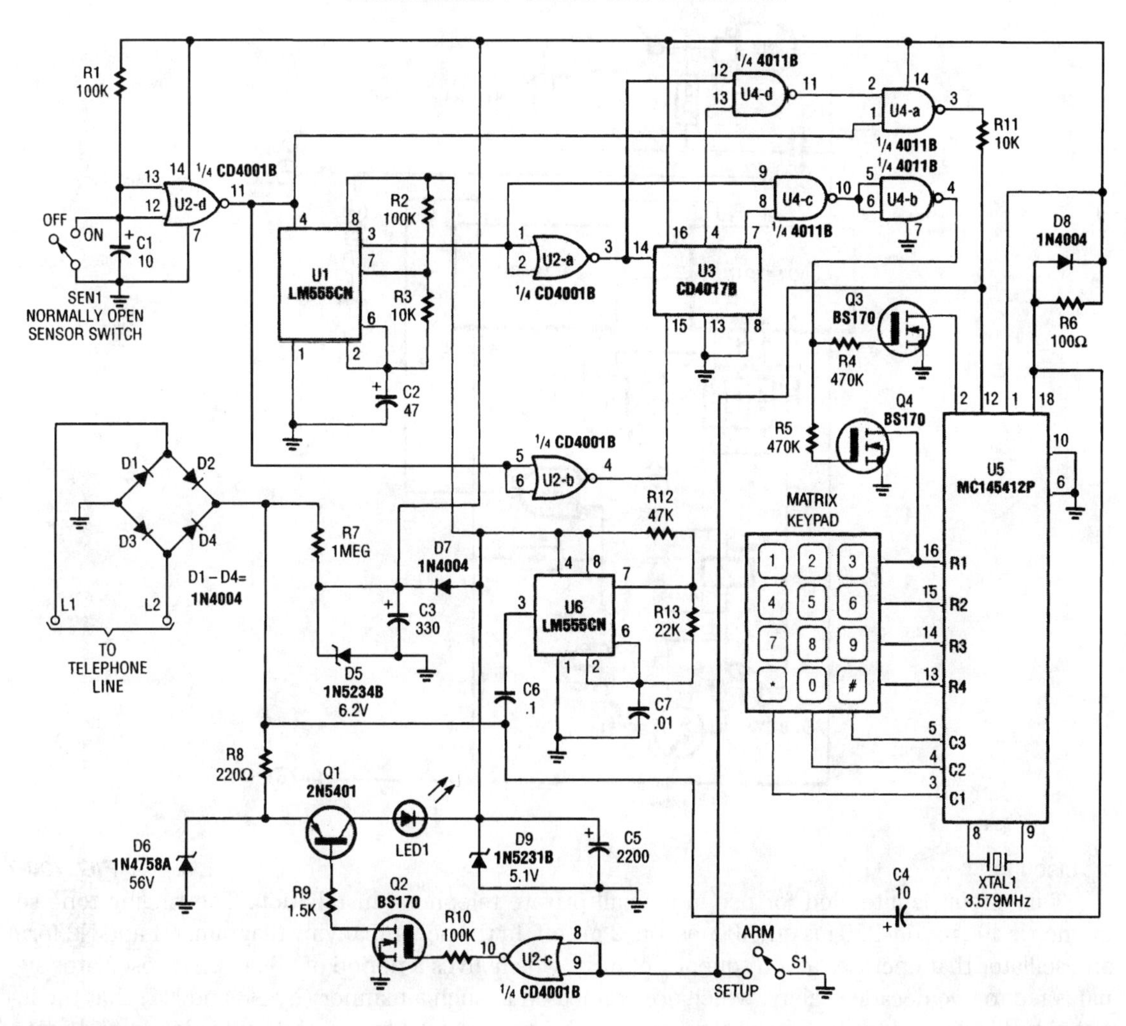

POPULAR ELECTRONICS ***FIG. 100-6***

This system will alert you or anyone chosen by automatically dialing a programmed phone number. This is accomplished by monitoring an open-loop or closed-loop sensor switch located in the protected area. When the sensor detects a problem (such as a break-in, fire, heating system failure, flood, etc.), Teleguard dials whatever telephone number has been programmed into its memory. When the phone is taken off the hook, Teleguard emits an unusual tone to alert the party on the receiving end that something is amiss.

The circuit is not hampered by busy signals when a call is placed; it automatically redials the number again and again (about once a minute) until it gets through. In addition, Teleguard can also automatically dial a number in the event of a medical emergency; for instance, where a mobility-impaired person is unable to dial the telephone. That can be accomplished by adding a "panic" switch to the circuit.

TELEPHONE BELL SIMULATOR

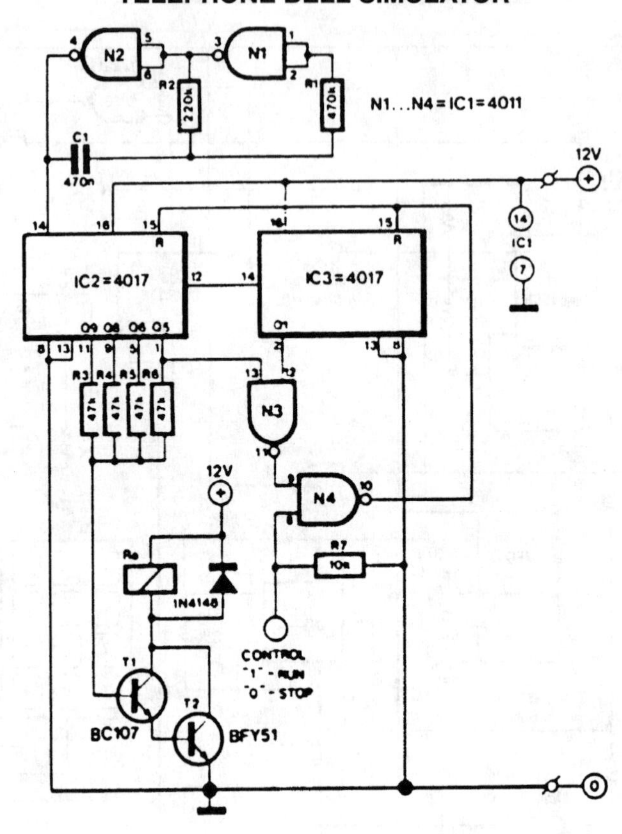

303 CIRCUITS *FIG. 100-7*

This circuit is intended for use in a small private telephone installation. The ringing tone sequence is 400 ms on, 200 ms off, 400 ms on, 2 ms off. In the accompanying diagram, N1 and N2 form an oscillator that operates at a frequency of 5 Hz, which gives a period of 200 ms. The oscillator signal is fed to two decade scalers, which are connected in such a manner (by N3 and N4) that the input signal is divided by 15. The second input of N4 can be used to switch the divider on and off by logic levels. If this facility is not used, the two inputs of N4 should be interconnected.

SIMPLE TELEPHONE RING INDICATOR

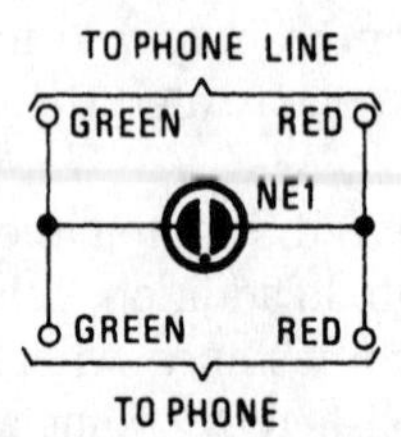

A neon lamp can easily be added to the phone line to act as a ring indicator. It's perfect for times when you can't hear the phone.

POPULAR ELECTRONICS *FIG. 100-8*

PHONE-LINE INTERFACE

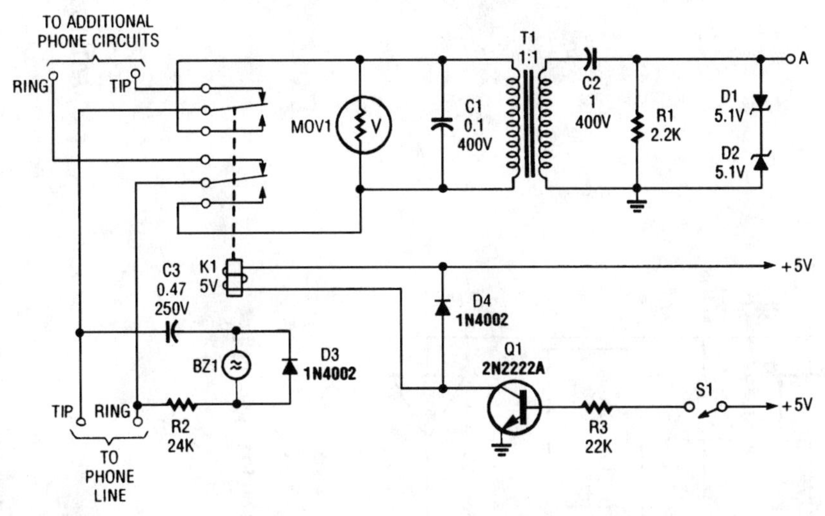

POPULAR ELECTRONICS

FIG. 100-9

This circuit should be useful for interfacing phone projects to the telephone line. It has a ringer, can interrupt the wiring, and isolates project from the phone line.

MUSIC-ON-HOLD BOX

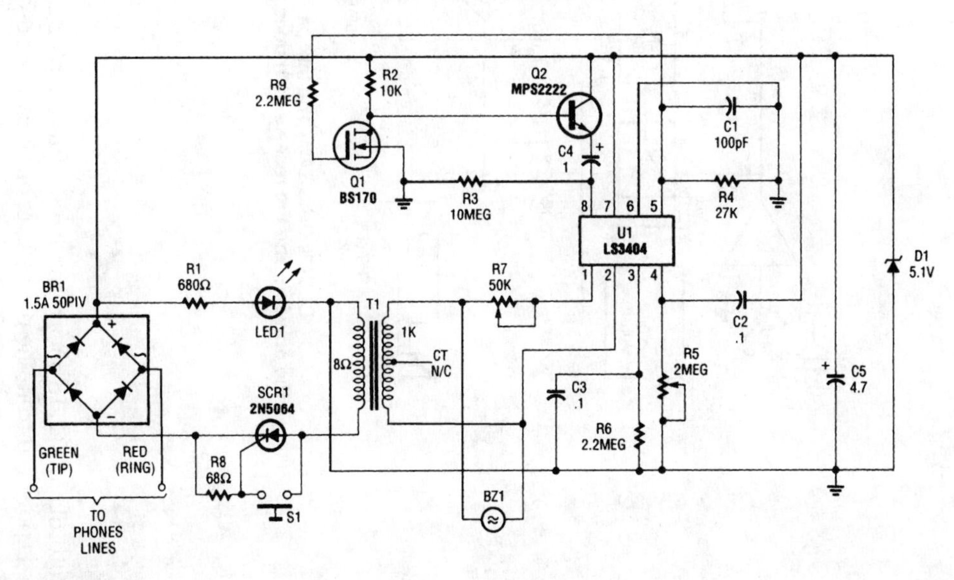

POPULAR ELECTRONICS

FIG. 100-10

U1, an LS3404 melody chip is activated when "hold" S1 is pressed, which causes SCR1 to conduct and hold the telephone line via T1, R1, and LED1. The voltage across R1 and LED1 is used to activate the melody chip. Q1 and Q2 form a restart circuit to keep the melody chip going during hold.

SPEAKERPHONE ADAPTER

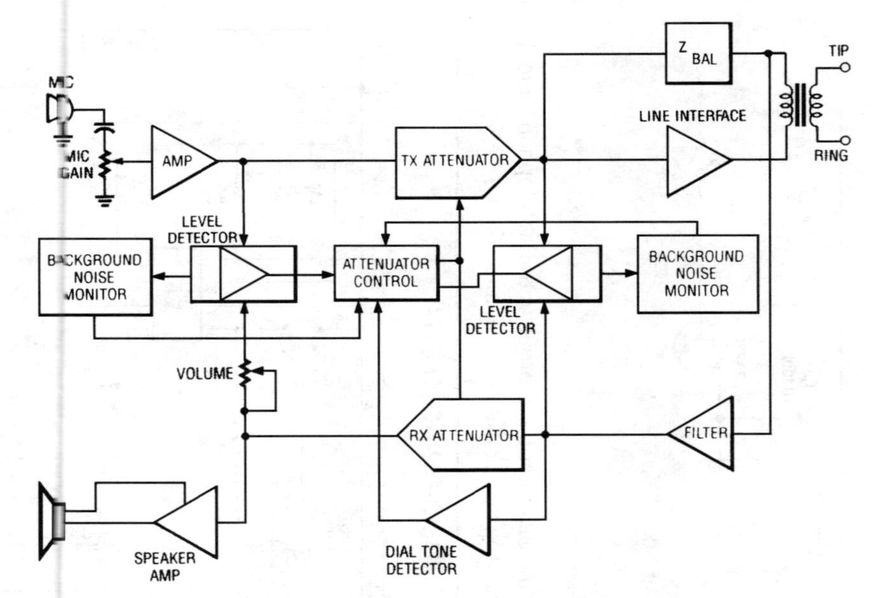

BLOCK DIAGRAM. The talk path goes left to right on the upper half of the drawing, and the receive path goes from right to left.

RADIO-ELECTRONICS

FIG. 100-11

Using a Motorola MC34118 speakerphone IC, this adapter can be used with a regular telephone to provide speaker capability. This device is powered from the phone line, but it can be powered via an external power supply if the line loop current is marginally low. An external phone is needed for ringing and dialing functions.

SPEAKERPHONE ADAPTER (*Cont.*)

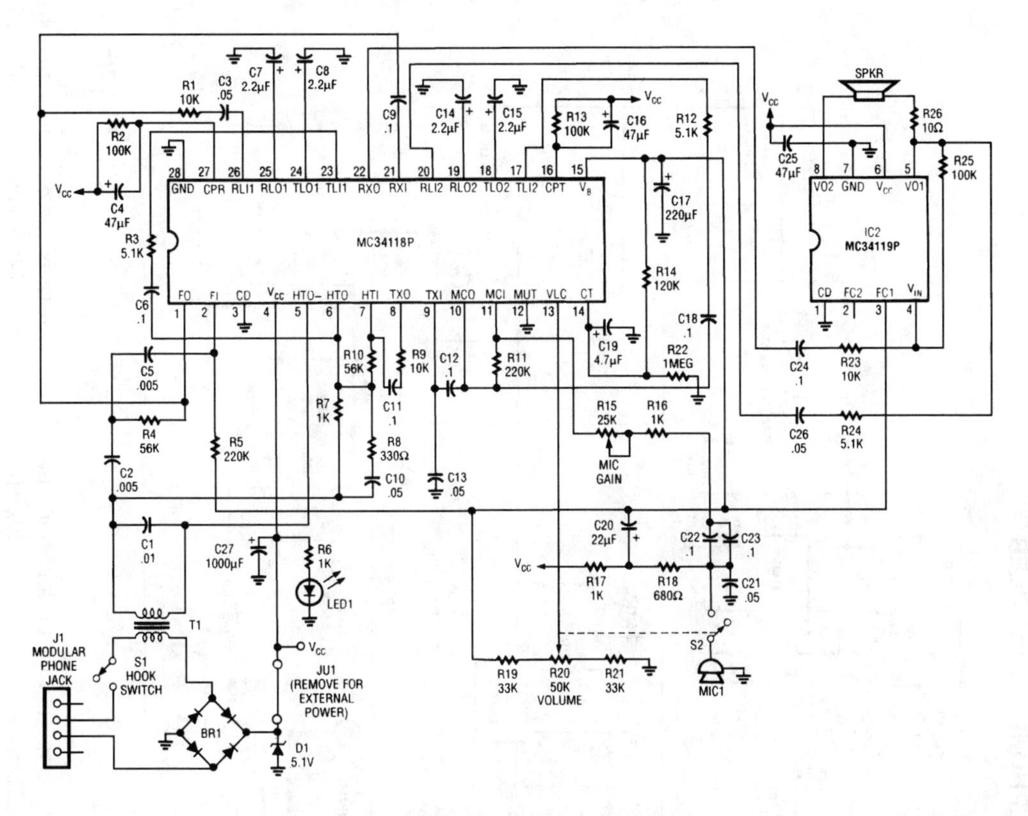

TELEPHONE VOICE-MAIL ALERT

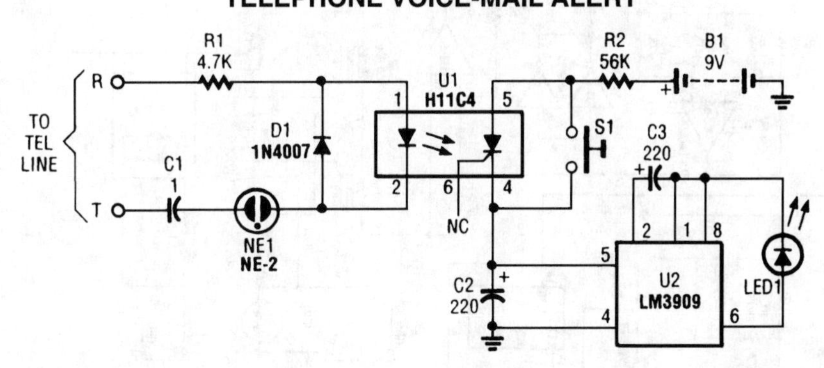

1993 ELECTRONICS HOBBYIST HANDBOOK ***FIG. 100-12***

The circuit is built around a couple of low-cost ICs: an H11C4 optoisolator/coupler with an SCR output (U1) and an LM3909 LED flasher (U2). It is connected to the phone line in the same manner as any extension phone. A ring signal on the telephone activates the optoisolator/SCR, and causes U2 to flash LED1. This flash signifies that a ring signal has been received.

TELEPHONE SCRAMBLER

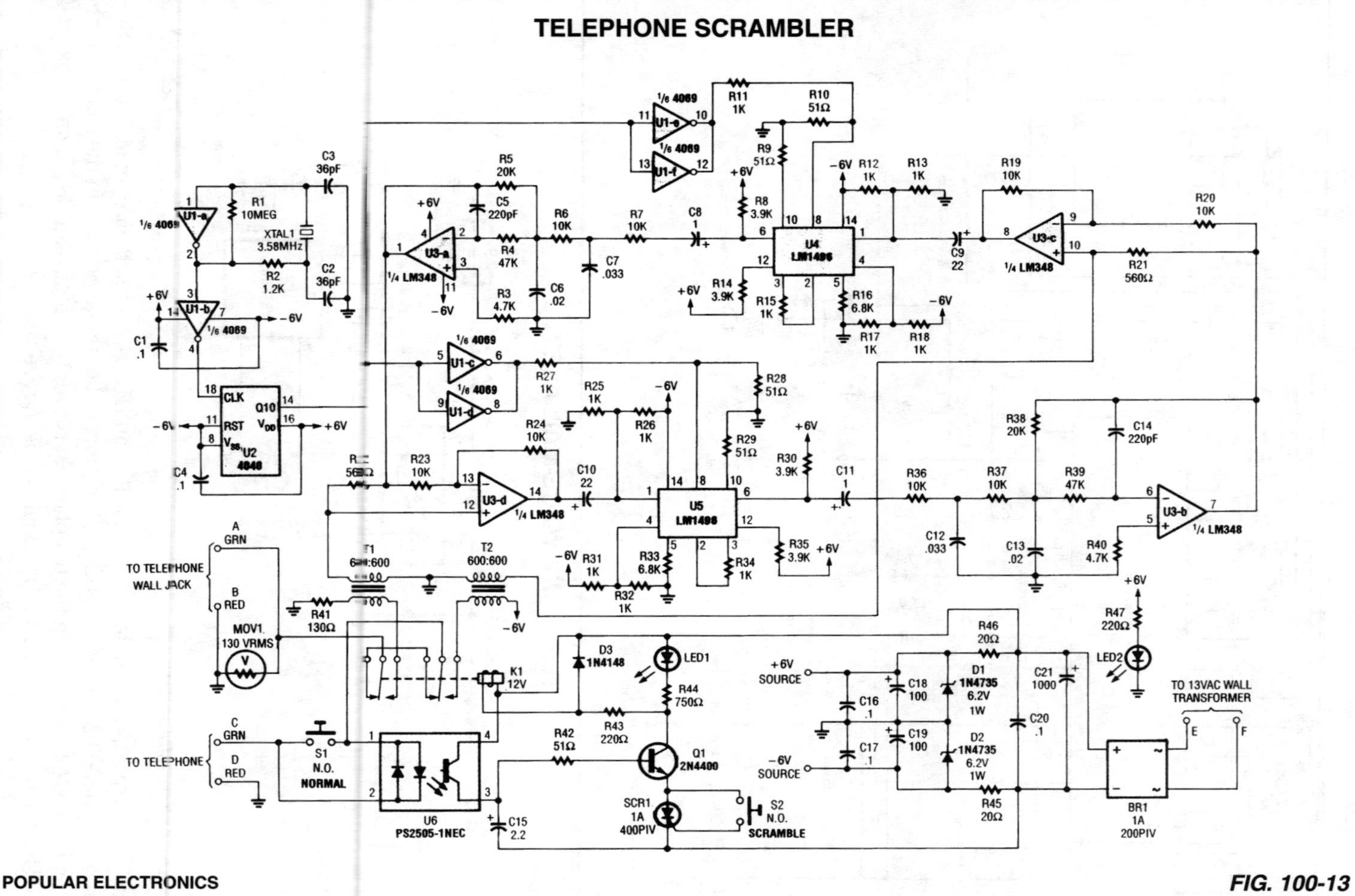

POPULAR ELECTRONICS

FIG. 100-13

Two hybrids (T1 and T2) are used to allow direct connection to a telephone line. This circuit uses the common speech-inversion algorithm where the frequency of an audio signal is inverted about a center frequency. An LM1496 balanced modulator is used to heterodyne the speech range against a 3.58-kHz signal.

PHONE PAGER

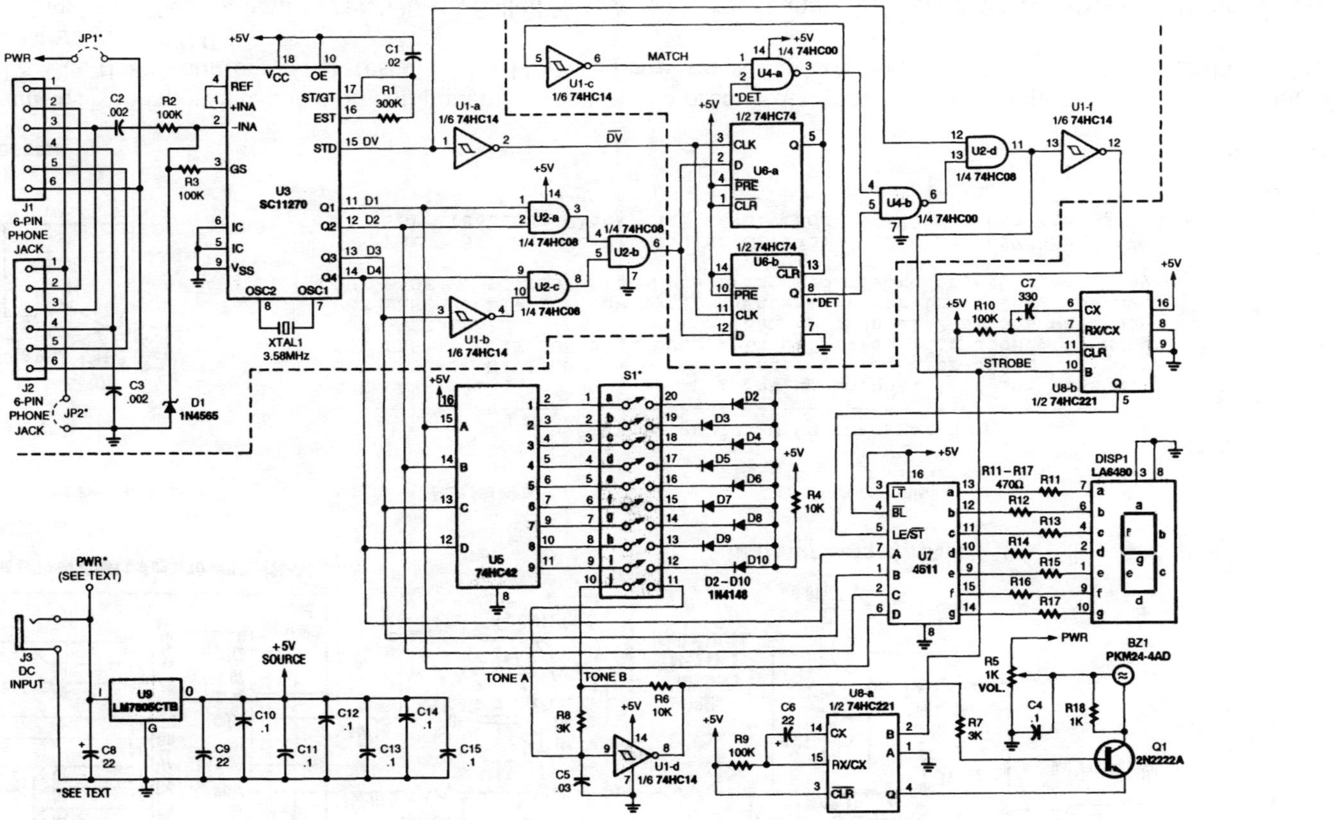

POPULAR ELECTRONICS

FIG. 100-14

This pager allows you to use your in-house phone wiring as a PA system. It uses two tone decoders to detect a particular touch-tone key. This key enables an audio amplifier.

5-V DIAL-TONE CIRCUIT

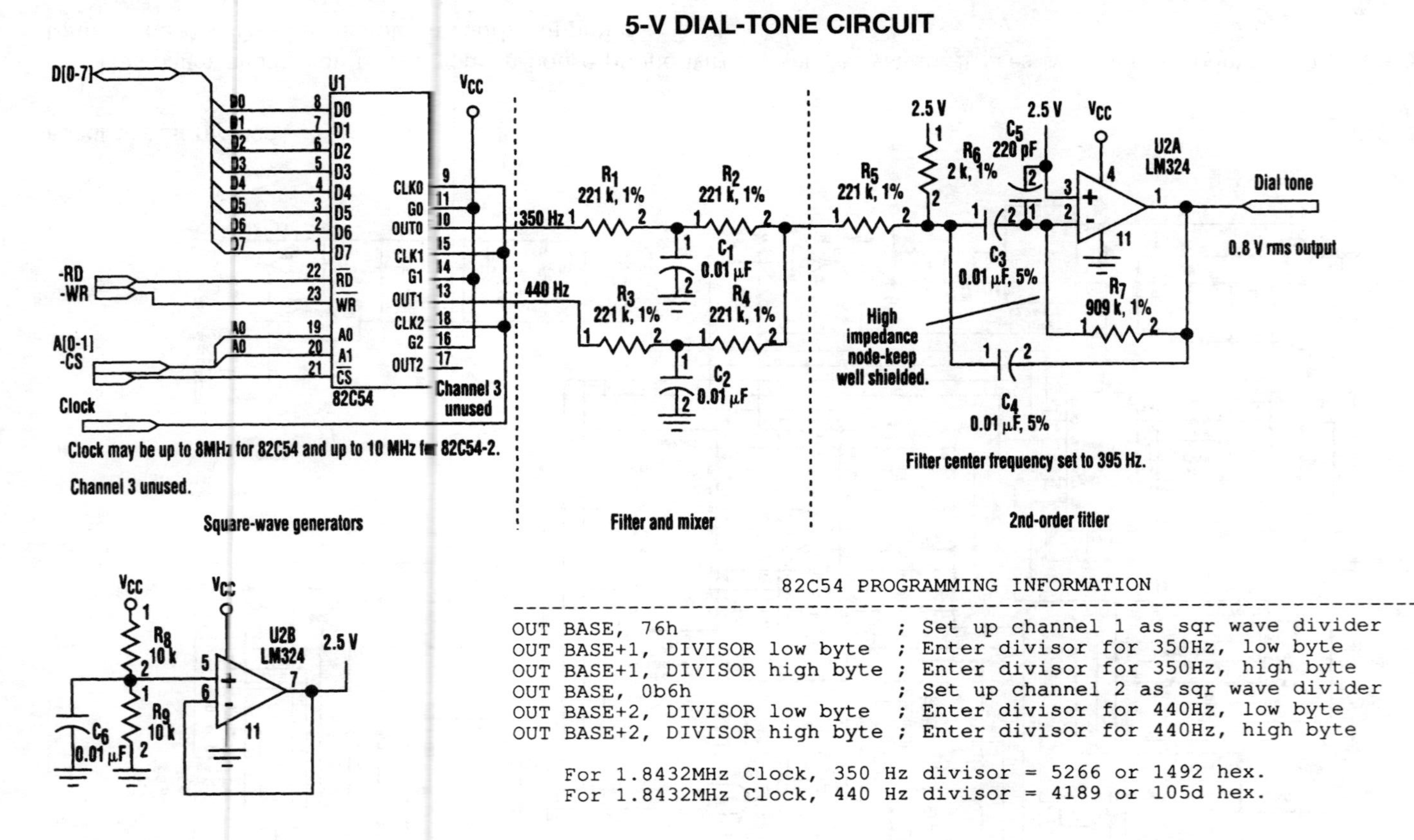

ELECTRONIC DESIGN

FIG. 100-15

This circuit uses inexpensive, common components to generate a precise dial tone for phone applications (see the figure). U1 (an Intel 82C54 timer-counter) generates 350- and 440-Hz square waves that are filtered by R_1/C_1 and R_3/C_2, and mixed together by resistors R2 and R4.

An operational amplifier configured as a 395-Hz, Sallen-Key, second-order bandpass filter (halfway between 350 and 440 Hz) removes unwanted signal harmonics. Almost any timer-counter can be used as the signal source, so long as it produces roughly square-wave outputs.

PHONE PAGER

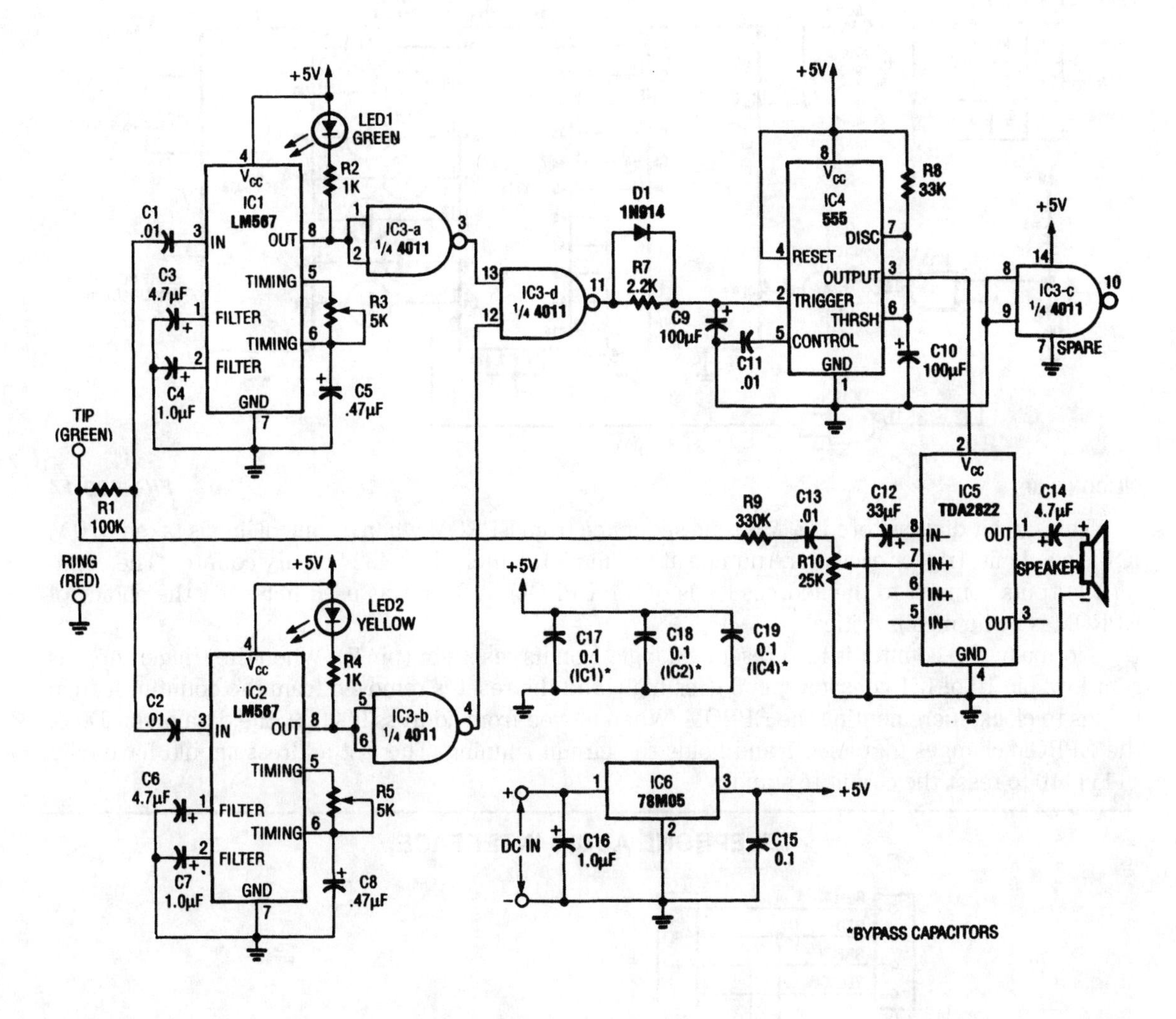

ELECTRONICS NOW *FIG. 100-16*

This pager works with DTMF phones. It displays a number and sounds an alert as the number on the display corresponds to a specific message.

ALARM DIALER

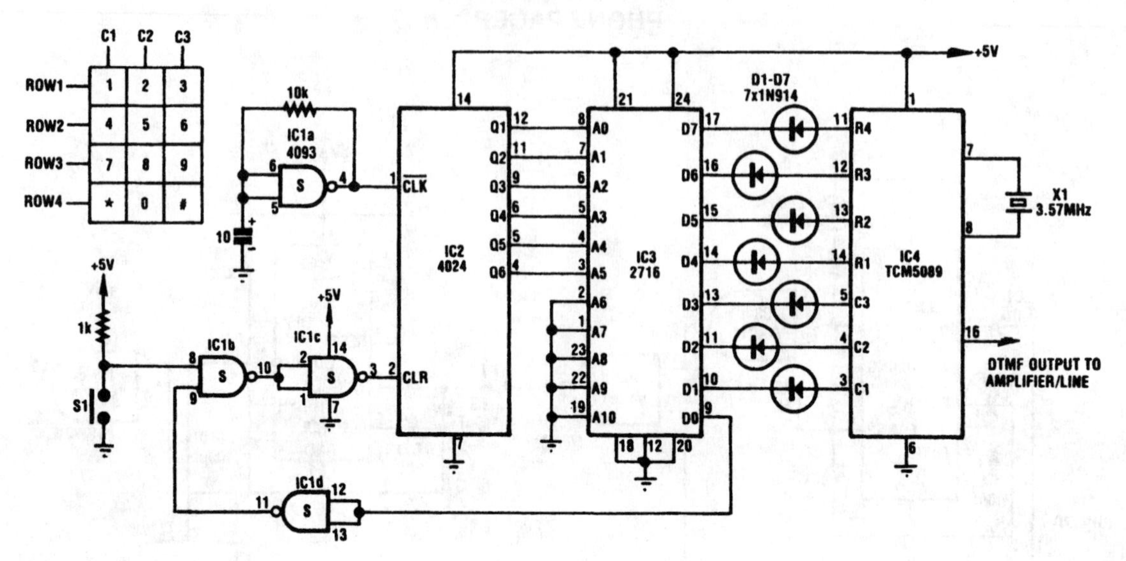

SILICON CHIP *FIG. 100-17*

This circuit dials a stored DTMF tone sequence from EPROM when a control line is taken to 0 V. IC1 is a Schmitt trigger oscillator, running at around 2 Hz. It clocks a 4024 binary counter. The counter's outputs connect to the address leads of the EPROM. A 2716 was used here, but the choice of EPROM is by no means critical.

Normally, the counter is held reset by a logic 1 on its reset pin (pin 2). When the trigger input is sent low, pin 10 of IC1 goes low, pin 3 goes high, and the reset is removed from the counter. It then begins to clock, incrementing the EPROM. When moved from address 000000, the data on bit D0 of the EPROM changes to a logic 1 and holds the circuit running. The last address should have data 11111110 to reset the circuit to standby.

TELEPHONE AUDIO INTERFACE

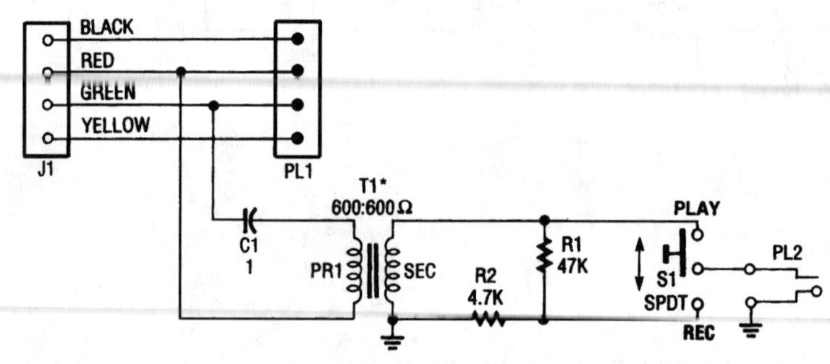

POPULAR ELECTRONICS *FIG. 100-18*

Used to record and play back tapes via the phone lines, this simple circuit has an audio level switch (S1).

CALLER ID CIRCUIT

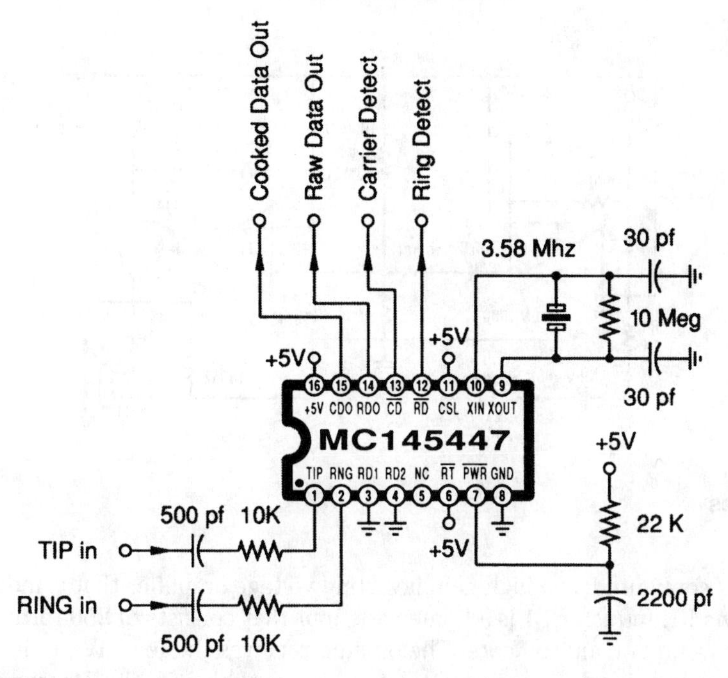

RADIO-ELECTRONICS *FIG. 100-19*

This caller ID circuit uses the Motorola MC145447 IC chip. This service must be available from your local phone company in order for this circuit to be used.

FCC PART 68 PHONE INTERFACE

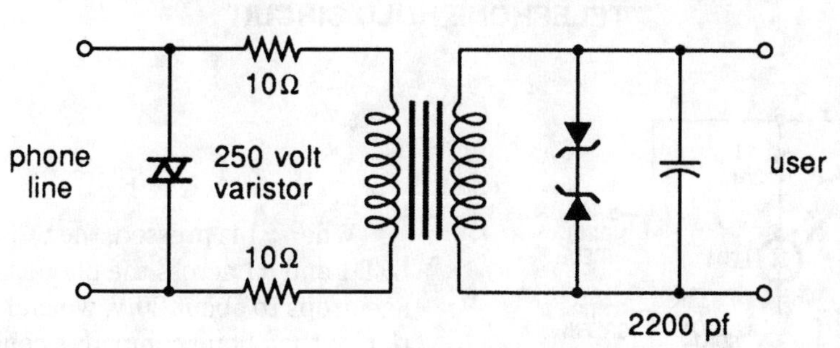

RADIO-ELECTRONICS *FIG. 100-20*

An FCC Part 68 interface is required any time you connect any circut of your own to the phone line.

TELEPHONE AMPLIFIER

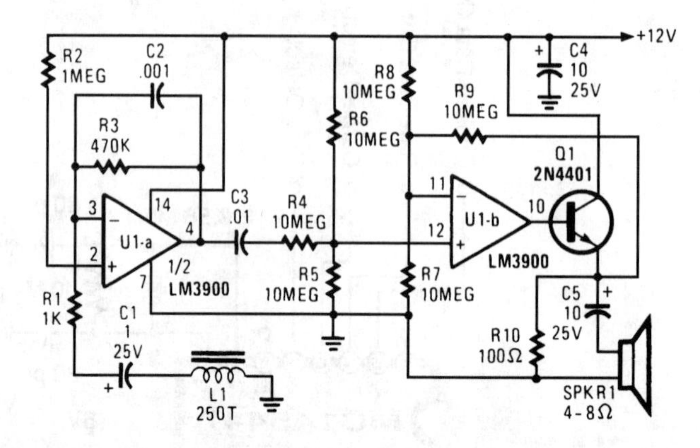

POPULAR ELECTRONICS

FIG. 100-21

Section U1-a is configured as a high-gain inverting voltage amplifier that is inductively coupled to the phone line via L1. Inductor L1 is a homemade unit that consists of 250 turns of fine, enamel-coated wire that is wound on an iron core. The op amp receives the few mV produced by L1 via C1 and R1 and amplifies the signal. Capacitor C1 acts as the negative-feedback component that limits the circuit's high-frequency gain, while R3 limits the low-frequency gain. Resistor R3 is particularly important because without it, the amplifier would saturate.

Op amp U1-b is configured as a difference amplifier. It receives a signal from U1-a via C3 and R4 and amplifies the difference between it and half of the supply voltage. Transistor Q1 is configured as a common-collector amplifier ensuring sufficient signal to drive the speaker. Capacitor C5 is used to remove any dc component provided by transistor Q1.

TELEPHONE HOLD CIRCUIT

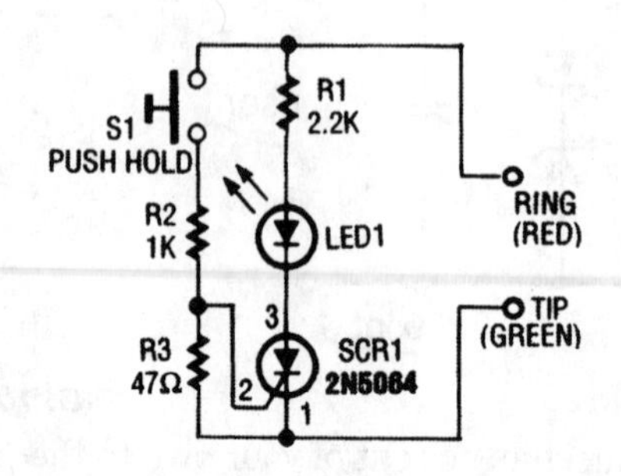

When S1 is pressed, the SCR fires, and places LED1 and R1 across the phone line. The line voltage drops to about 20 V, which holds the connection to the phone company's central office.

ELECTRONICS NOW

FIG. 100-22

TELEPHONE CIRCUIT

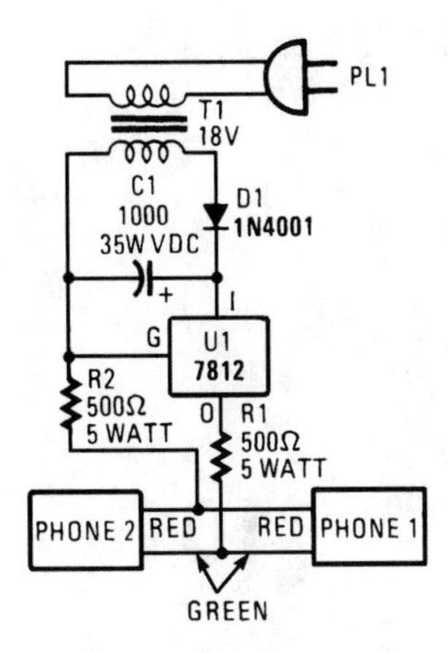

POPULAR ELECTRONICS

FIG. 100-23

This circuit is useful for checking out old telephones by providing them with the dc voltage that they require for operation.

TELEPHONE-LINE TESTER

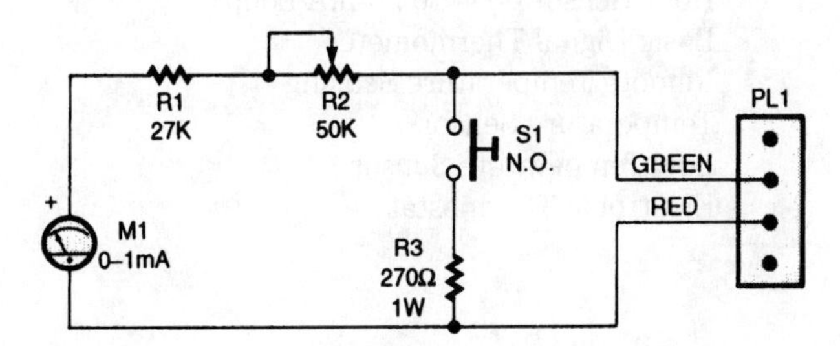

POPULAR ELECTRONICS

FIG. 100-24

The telephone-line tester consists of nothing more than a meter (that's used to measure line voltage in the on- and off-hook state), three resistors (one of which is variable), a pushbutton switch, and a modular telephone connector. When the circuit is connected to the telephone line, a meter reading of 5 to 10 V (when S1 is pressed) indicates that the line is okay.

101

Temperature-Related Circuits

The sources of the following circuits are contained in the Sources section, which begins on page 675. The figure number in the box of each circuit correlates to the entry in the Sources section.

TEMPERATURE COMPENSATION ADJUSTER

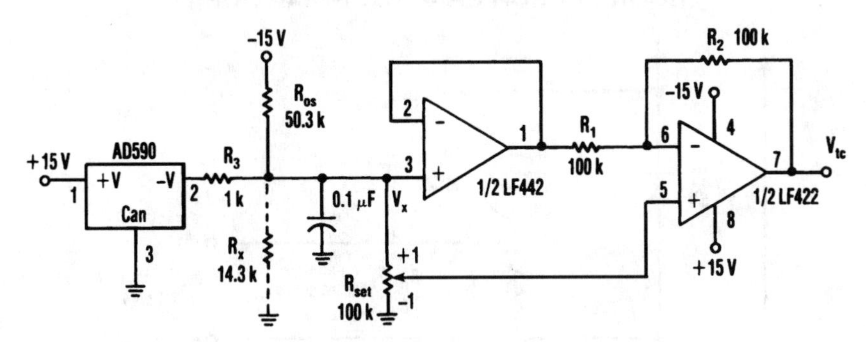

ELECTRONIC DESIGN **FIG. 101-1**

The circuit shown delivers +10 to –10 mV°/C output using an Analog Devices' AD590 temperature transducer. R_x is a scaling resistor.

THERMOMETER FOR 5-V OPERATION

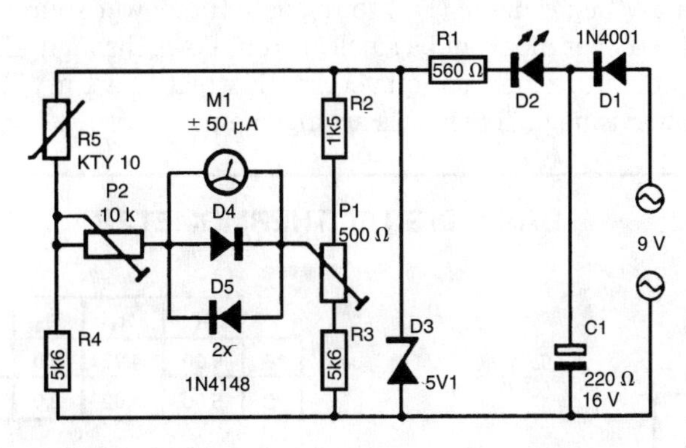

303 CIRCUITS **FIG. 101-2**

At the heart of this simple circuit is the well-known type KTY10 temperature sensor from Siemens. This silicon sensor is essentially a temperature-dependent resistor that is connected as one arm in a bridge circuit here. Preset P1 functions to balance the bridge at 0°C. At that temperature, moving coil meter M1 should not deflect, i.e., the needle is in the center position. Temperature variations cause the bridge to be unbalanced, and hence produce a proportional indication on the meter. Calibration at, say, 20°C is carried out with the aid of P2.

The bridge is fed from a stabilized 5.1-V supply, based on a temperature-compensated zener-diode. It is also possible to feed the thermometer from a 9-V battery, provided D1–D3, R1 and C1 are replaced with a Type 78L05 voltage regulator, because this is more economic as regards to current consumption.

HOOK SENSOR ON 4- TO 20-mA LOOP

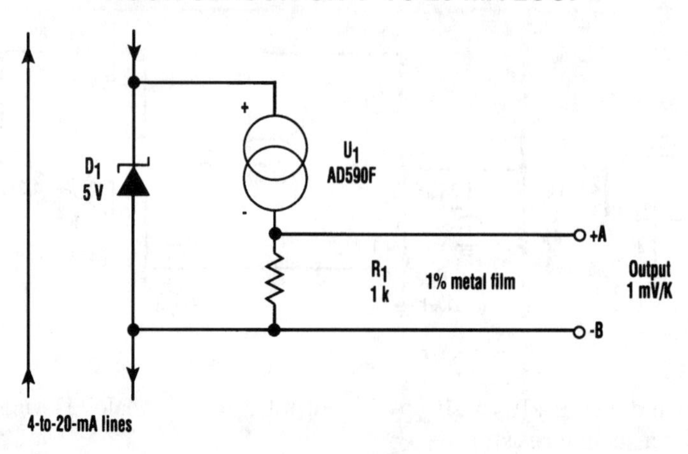

ELECTRONIC DESIGN *FIG. 101-3*

Here's an effective for a temperature sensor to receive power from a 4-to-20 mA loop without actually affecting the loop current (see the figure). This particular temperature sensor IC (AD590F) conducts 1 μA/K when powered by a supply in the range of 4 V to 40 Vdc.

The scheme uses a 5-V Zener diode (D1) to regulate the power source for AD590F. Most of the current flows through the Zener diode and a small current flows through AD590F. A high-impedance device can read the temperature information across R1, which is a 1 mV/K in the range of –55°C to 150°C. The waste of power is negligible in this arrangement.

BASIC DIGITAL THERMOMETER

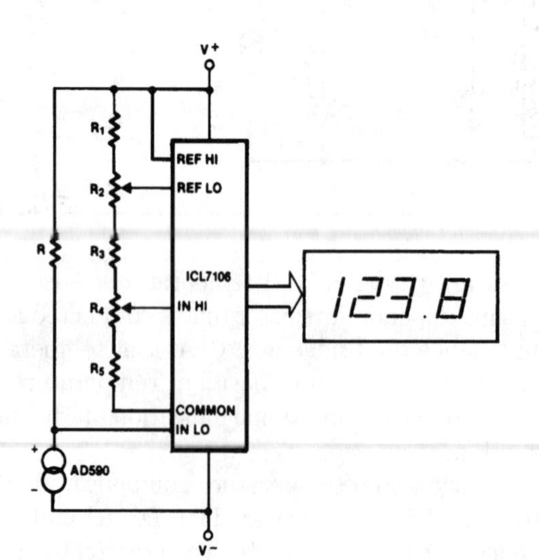

	R	R_1	R_2	R_3	R_4	R_5
°F	9.00	4.02	2.0	12.4	10.0	0
°C	5.00	4.02	2.0	5.11	5.0	11.8

$\sum_{n=1}^{5} R_n = 28k\Omega$ nominal

All values in kΩ

The ICL7106 has a V_{IN} span of ±2.0V, and a V_{CM} range of (V^+ −0.5) Volts to (V^- +1) Volts; R is scaled to bring each range within V_{CM} while not exceeding V_{IN}. V_{REF} for both scales is 500mV. Maximum reading on the Celsius range is 199.9°C, limited by the (short-term) maximum allowable sensor temperature. Maximum reading on the Fahrenheit range is 199.9°F (93.3°C), limited by the number of display digits. See note next page.

INTERSIL *FIG. 101-4*

REMOTE TEMPERATURE SENSING

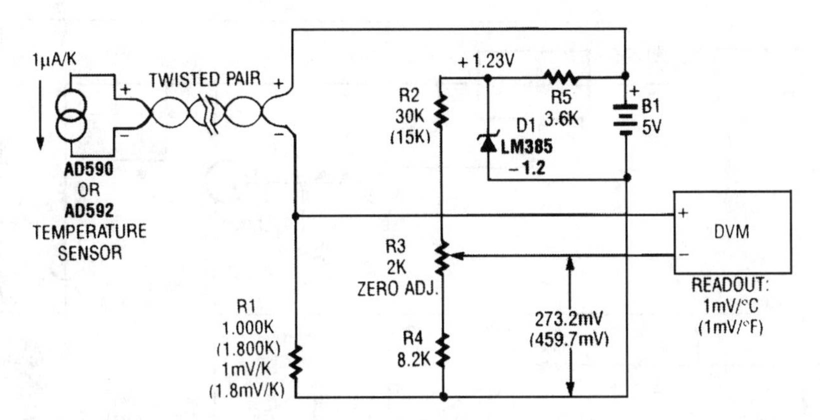

RADIO-ELECTRONICS

FIG. 101-5

An AD590 or AD592 makes it easy to transmit temperature data over a pair of wires. The circuit produces 1mV/°C (or 1mV/°F using the values in parentheses).

TEMPERATURE SENSOR

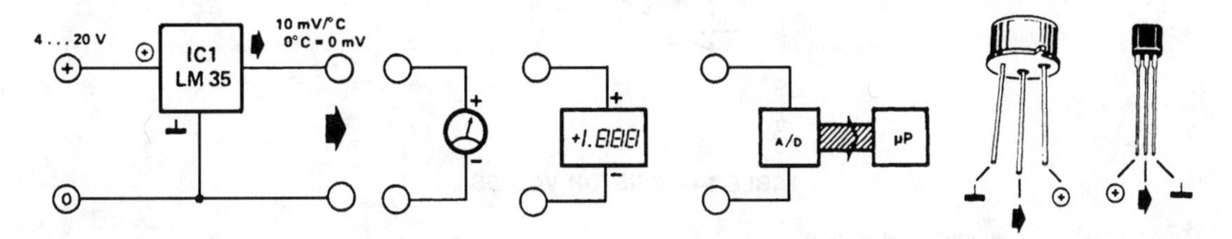

303 CIRCUITS

FIG. 101-6

The LM35 temperature sensor provides an output of 10 mV/°C for every degree Celsius over 0°C. At 20°C the output voltage is 20 × 10 = 200 mV. The circuit consumes 60 μA. The load resistance should not be less than 5 kΩ. A 4- to 20-V supply can be used.

LOW TEMPERATURE SENSOR

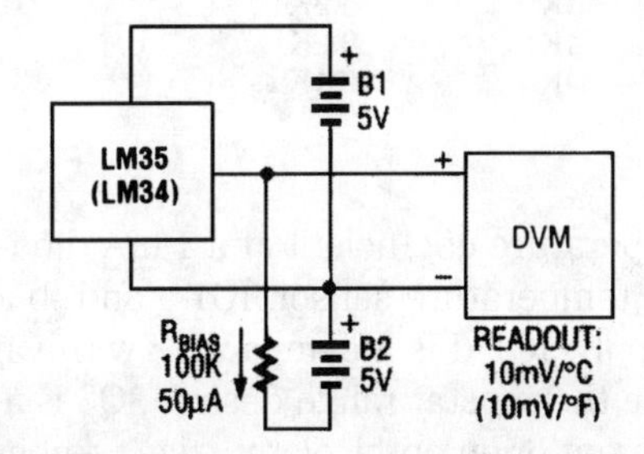

A negative bias current can produce the offset needed for below-zero readings using the LM34 or LM35 temperature sensor.

RADIO-ELECTRONICS

FIG. 101-7

ELECTRONIC THERMOSTAT

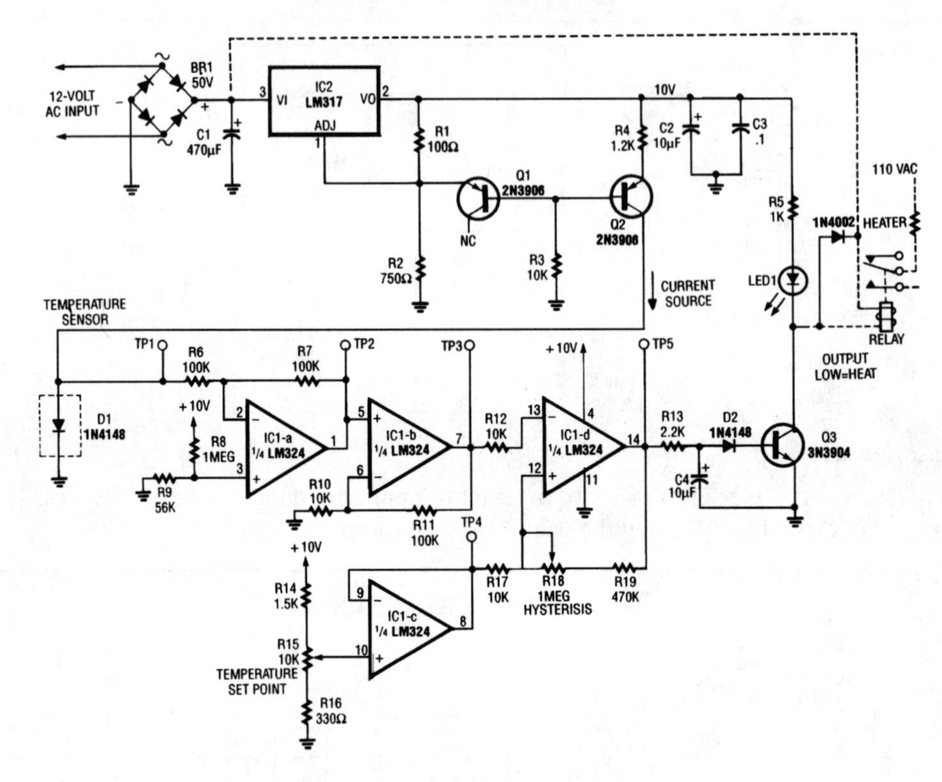

TABLE 1—RESISTOR VALUES

Temperature Range (Degrees C)	R14	R15	R16
−50 to −30	10K	1K	330Ω
−30 to −10	9.1K	1K	1.2K
−10 to 15	8.2K	1K	2.2K
15 to 35	7.5K	1K	3.3K
35 to 55	6.2K	1K	4.3K
55 to 75	5.1K	1K	5.1K
75 to 95	4.3K	1K	6.2K
95 to 115	3.3K	1K	6.8K
115 to 135	2.2K	1K	8.2K
135 to 155	1.2K	1K	9.1K

RADIO-ELECTRONICS ***FIG. 101-8***

A diode, such as a IN4148, has a typical –2m V/°C temperature coefficient at a 1 mA diode current. Q1 and Q2 form a constant current source. D1 is the temperature sensor. IC1-a and -b are dc amplifiers, with IC1-c a temperature reference voltage supply. IC1-d is a comparator with variable hysteresis. R14, R15, and R16 are chosen depending on the thermostat range desired. Q3 is a relay driver (2N3904). The relay used should handle the load current or an optoisolator triac combination can be used.

102

Timer Circuits

The sources of the following circuits are contained in the Sources section, which begins on page 675. The figure number in the box of each circuit correlates to the entry in the Sources section.

REFLEX TIMER

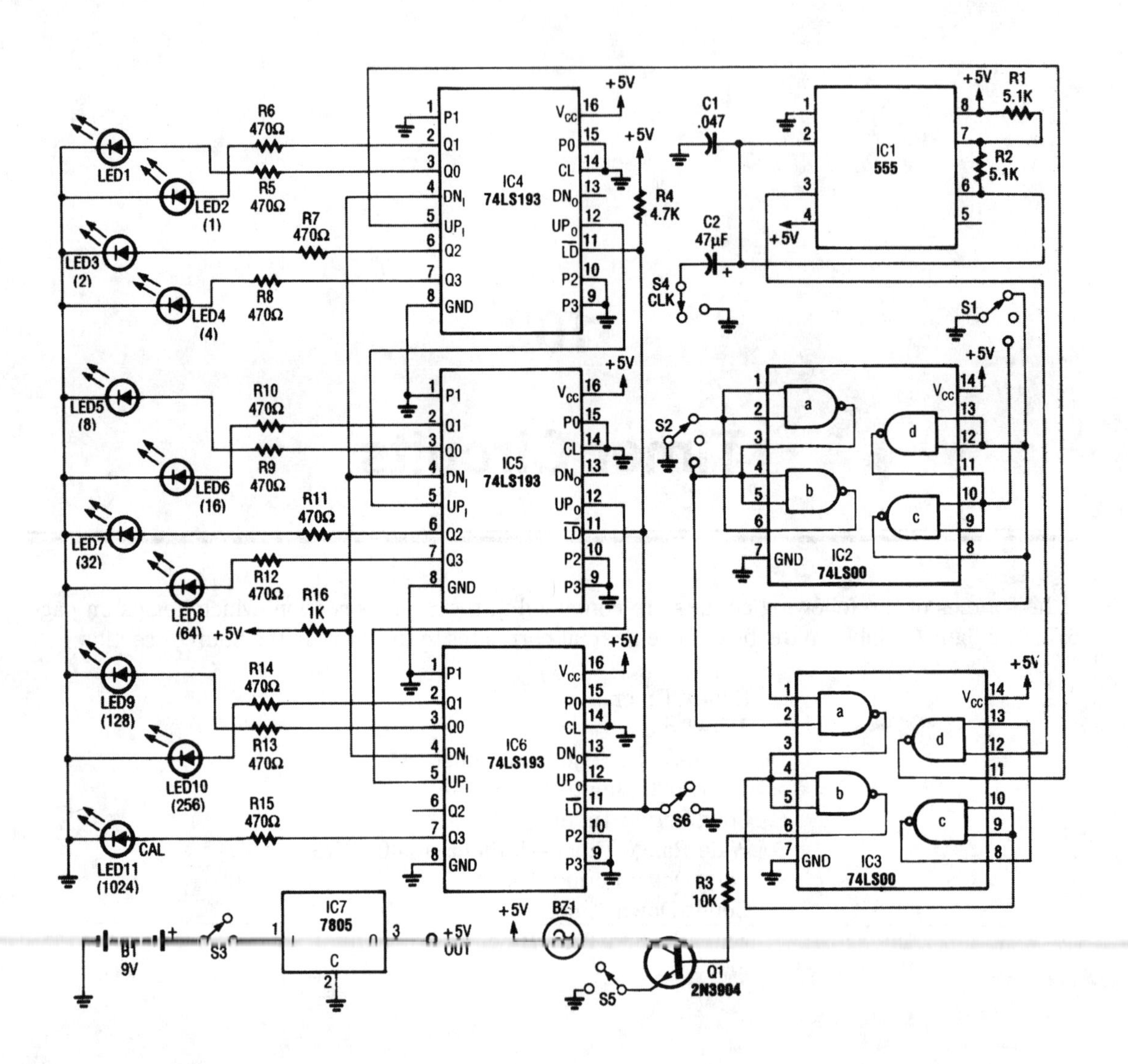

ELECTRONICS NOW

FIG. 102-1

This timer circuit uses a 555 IC timer and three 74LS193 counters to drive an LED display. S1 is activated by one person, who turns on piezo buzzer BZ1 via Q1 and also starts the clock; S1 is activated by the other person being timed. This shuts off the timer, and the number of LEDs lit indicate, in binary form, the elapsed time.

TELE-TIMER

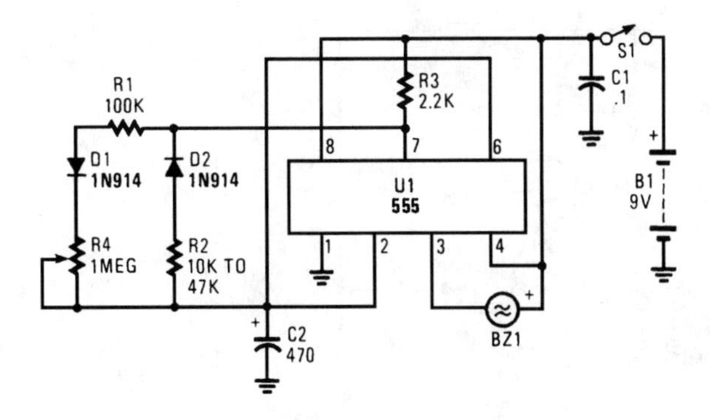

POPULAR ELECTRONICS **FIG. 102-2**

The circuit is built around a 555 oscillator/timer. The circuit provides two time periods. The long-running time period is adjustable from about 1 to 10 minutes, and the short time period is preset to about three seconds.

Here's how the dual timer operates. When the power is switched on, C2 begins to charge through R3, R1, D1, and R4 to start the long-term timer period. When the voltage across C2 reaches the 555's internal switching point, the long-term timer times out, discharging C2 through R2, D2, and pin 7 of the 555. During that time, pin 3 of the 555 is pulled to ground, activating the piezo sounder.

To set the short time period to about four seconds, use a 10 k resistor for R2, and for about twenty seconds use a 47 k resistor. The timing capacitor, C2, should be a good-quality, low-leakage unit.

THREE-STAGE SEQUENTIAL TIMER

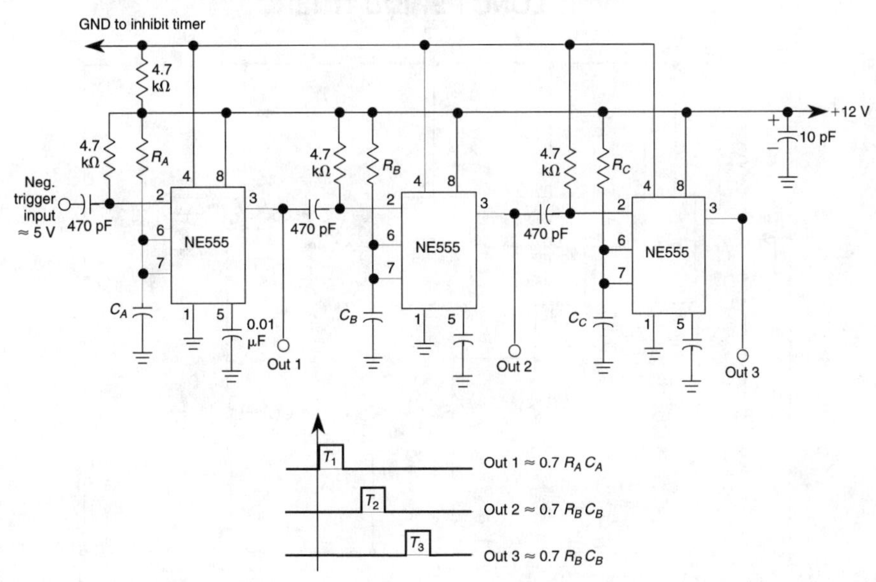

WILLIAM SHEETS **FIG. 102-3**

By using three 555 ICs, three sequential pulses can be generated. Output 3 can be connected back to trigger input to achieve astable operation.

2- TO 2000-MINUTE TIMER

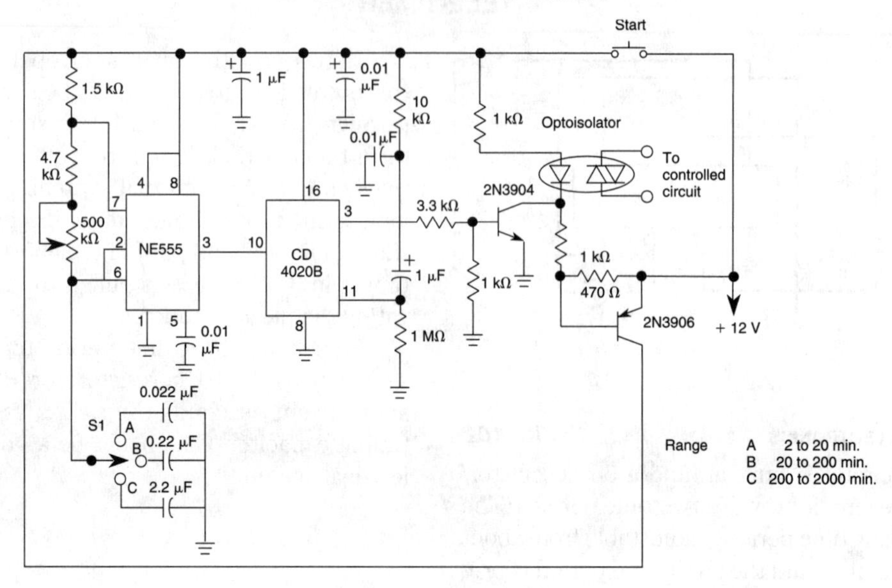

WILLIAM SHEETS ***FIG. 102-4***

A CD4020B divider is used to divide the frequency of a 555 timer. S1 is a range selector switch that provides timing intervals of 2–20, 20–200 and 200–2,000 minutes.

LONG PERIOD TIMER

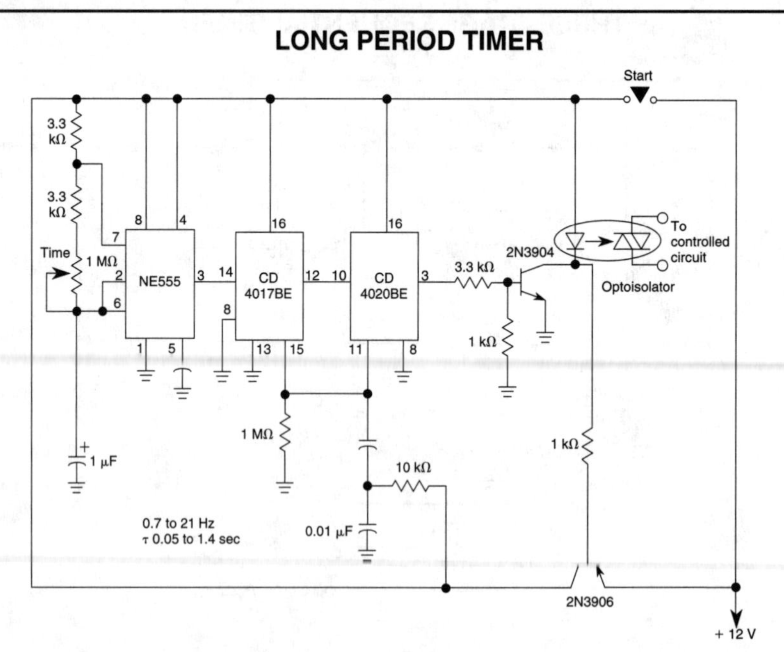

WILLIAM SHEETS ***FIG. 102-5***

By using a 555 timer, time intervals of 150 minutes 60 hours can be generated with this circuit.

WIDE-RANGE TIMER–1MINUTE TO 400 HRS

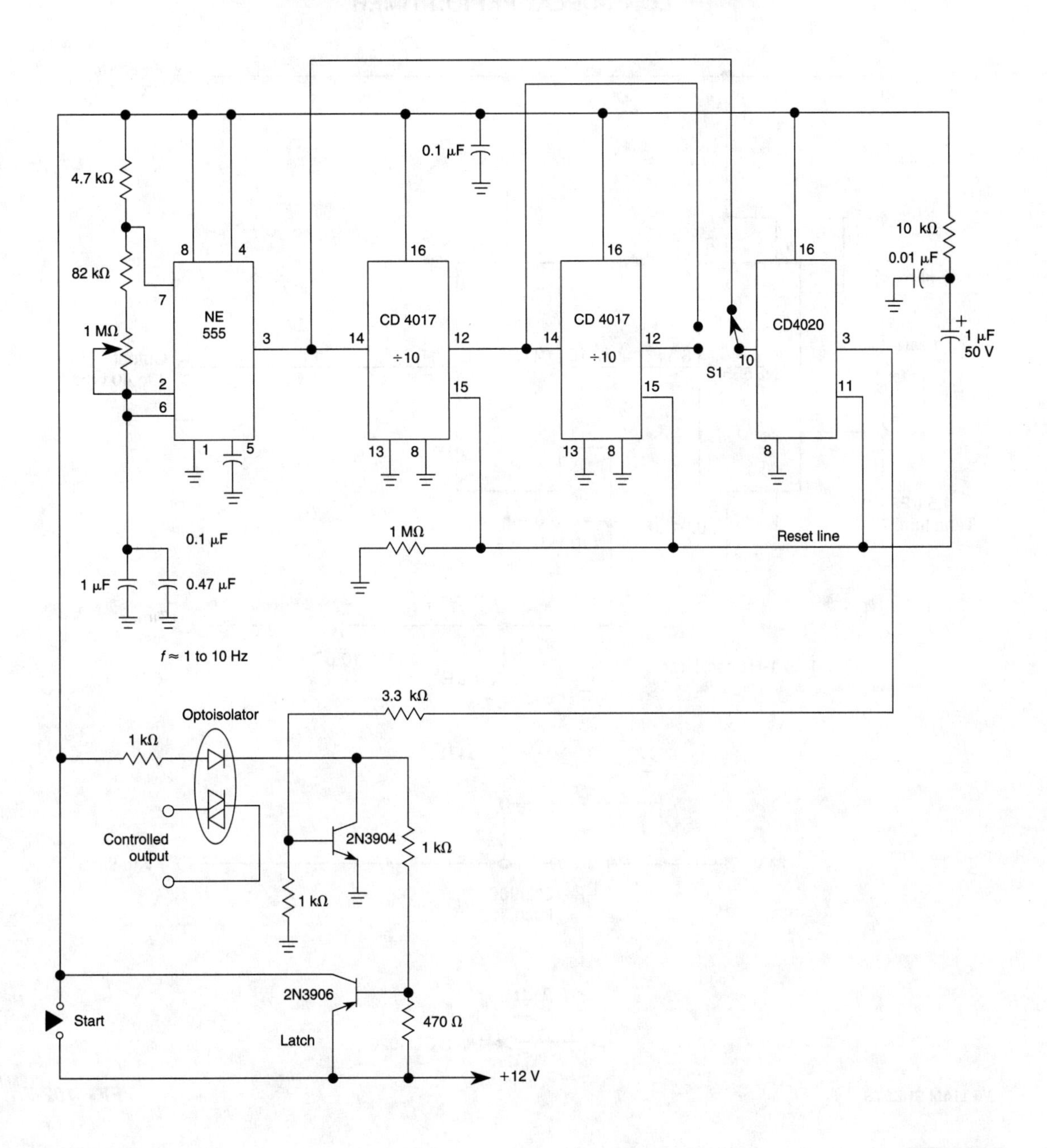

WILLIAM SHEETS

FIG. 102-6

This ultra wide range timer uses a 555 timer base, two 4017Bs and a 4020B that act as frequency dividers that can be switched in and out. S1 is a SP3T range switch.

LONG-DELAY-PERIOD TIMER

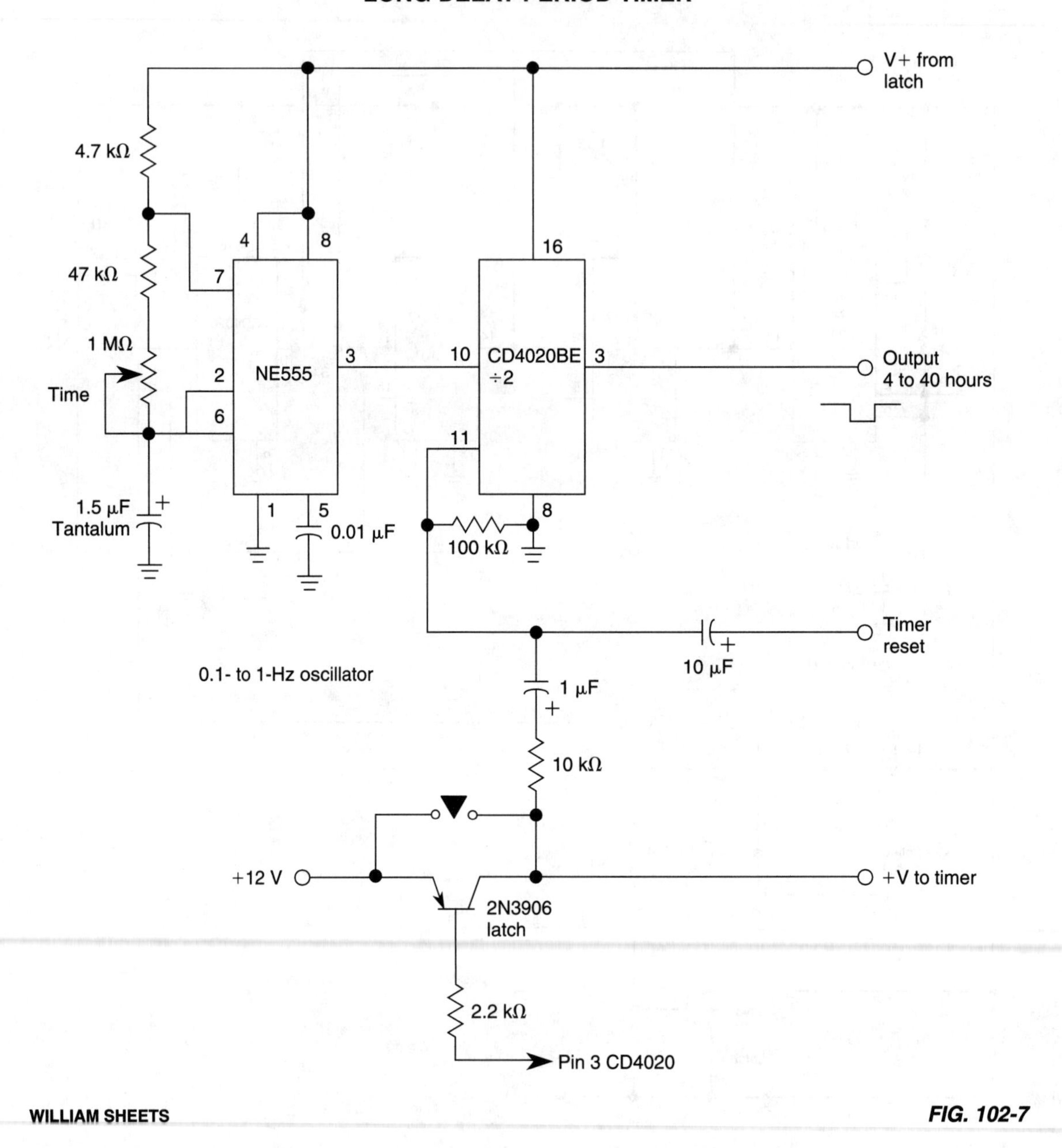

WILLIAM SHEETS

FIG. 102-7

This method of obtaining a 4 to 40 hour timing period from a 555 IC can be further expanded to produce even longer delays with equal accuracy.

COUNT-DOWN TIMER

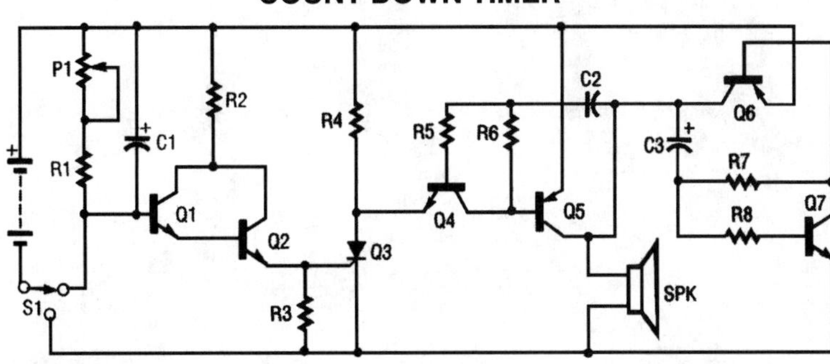

1991 PE HOBBYIST HANDBOOK *FIG. 102-8*

C1	100-µF Electrolytic Capacitor	R3	33-kΩ Resistor
C2	0.0047-µF Mylar Capacitor	R4	200 Ω Resistor
C3	1-µF Electrolytic Capacitor	R5	2.2-kΩ Resistor
P1	2-MΩ Trimmer Resistor	R6	220-kΩ Resistor
Q1, Q2, Q4, Q7	2N3904 Transistor	R7	2.2-MΩ Resistor
Q3	106 SCR	R8	7.5-kΩ Resistor
Q5, Q6	2N3906 Transistor	S1	SPDT Slide Switch
R1	1-MΩ Resistor	SPK	Small Speaker
R2	10-kΩ Resistor	Misc	PC Board, 9-V Snap Wire

With switch S1 in the off position, as shown, battery voltage is applied across timing-capacitor C1, which stays charged while the rest of the circuitry has no power supplied to it. Transistor Q1, and thus transistors Q2 through Q4, are kept in an off condition as long as C1 has a sufficient charge.

EXTENDED ON-TIME TIMER

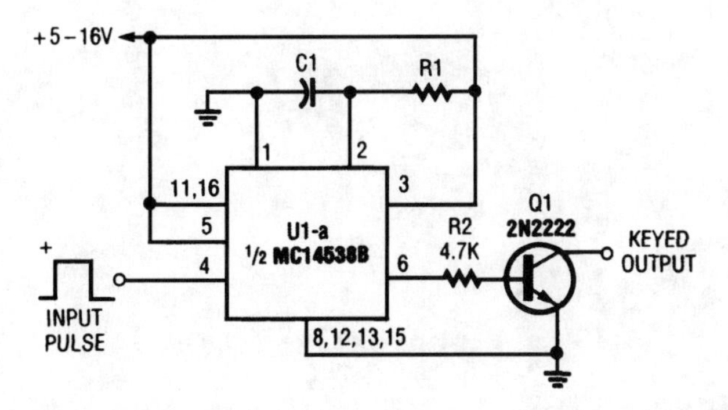

POPULAR ELECTRONICS *FIG. 102-9*

Half of a Motorola MC14538B dual, precision, retriggerable monostable multivibrator is used to form an extended on-time timer circuit. That type of circuit can be used as a switch debouncer. Such circuits are often used in digital circuitry, where each and every bounce of a switch contact is seen as a separate digital input.

The delay on time (established by C1 and R1) is easily set using the formula, $C_1 \times R_1 = T$, where C_1 is in microfarads, R_1 is in megohms, and T is in seconds.

103

Tone Circuits

The sources of the following circuits are contained in the Sources section, which begins on page 675. The figure number in the box of each circuit correlates to the entry in the Sources section.

REPEATER-TONE BURST GENERATOR

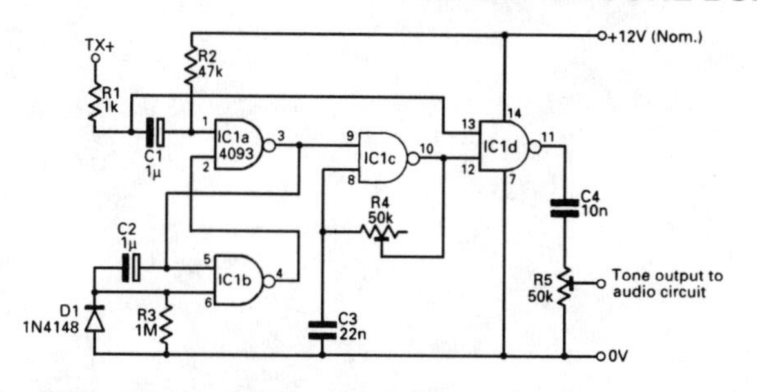

Fig. 1: The circuit, based on a single c.m.o.s chip and a few other components.

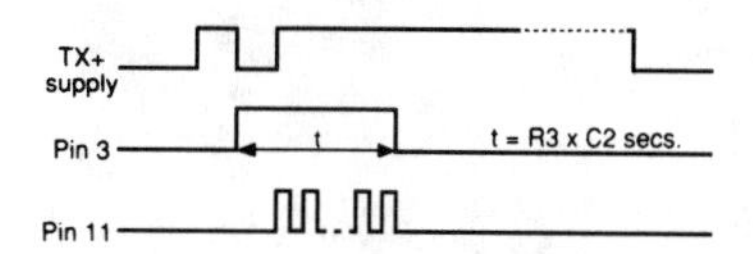

Fig. 2: Pulse and timing diagram, see the text for more details.

PRACTICAL WIRELESS

FIG. 103-1

Integrated circuit gates IC1-a and IC1-b form a monostable, whose time constant is determined by C2 and R3. When the transmitter is dekeyed (and then almost immediately rekeyed) point TX+ goes low and takes pin 1 low for a short time. This triggers the start of the timing period controlled by C_2/R_3. The capacitor C2, charges via R3 until the trigger point of gate IC1-b is reached. At this point, the monostable changes state and pin 3 goes low again. On the prototype, this time was about 700 ms. The pulse occurs each time after dekeying and it is normally inaudible. If, however, point TX+ goes high again (as in immediate rekeying) the monostable is still in the enabled state and the oscillations of IC1-c are present in the transmission. During this time period, the buffer gate, IC1-d, is enabled and the tone is therefore passed to the output.

TWO-TONE ENCODER

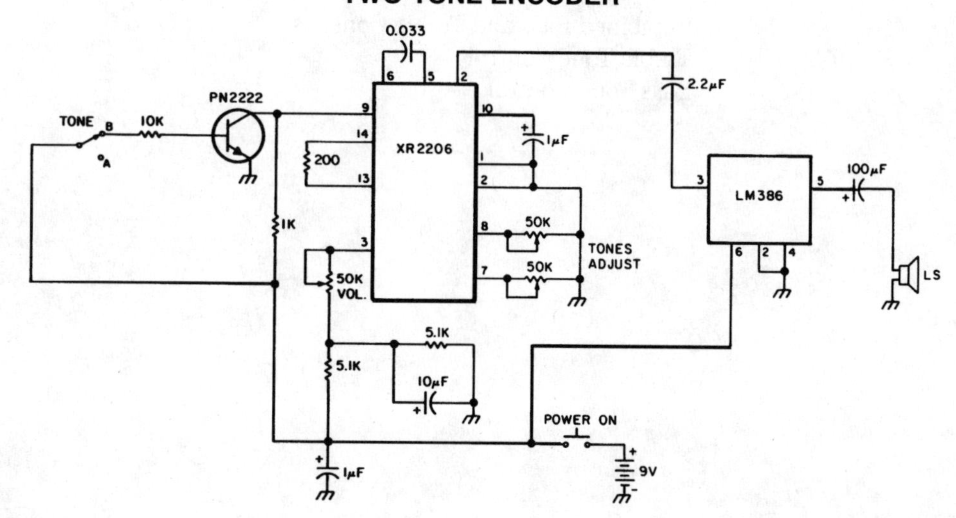

73 AMATEUR RADIO

FIG. 103-2

Using an XR2206 oscillator, this circuit can generate two audio tones. Switching between tones can be done with a logic level to either the base of the PN2222 or pin 9 of the XR2206.

104

Tone-Control Circuits

The sources of the following circuits are contained in the Sources section, which begins on page 675. The figure number in the box of each circuit correlates to the entry in the Sources section.

COMBINED BASS AND TREBLE CONTROLS

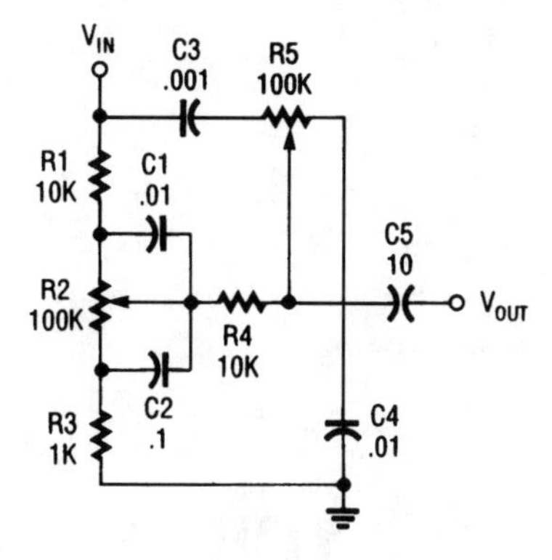

1993 ELECTRONICS HOBBYIST HANDBOOK ***FIG. 104-1***

Bass and treble circuits can be combined to form a two-control tone-adjust circuit, as shown here.

TREBLE TONE CONTROL

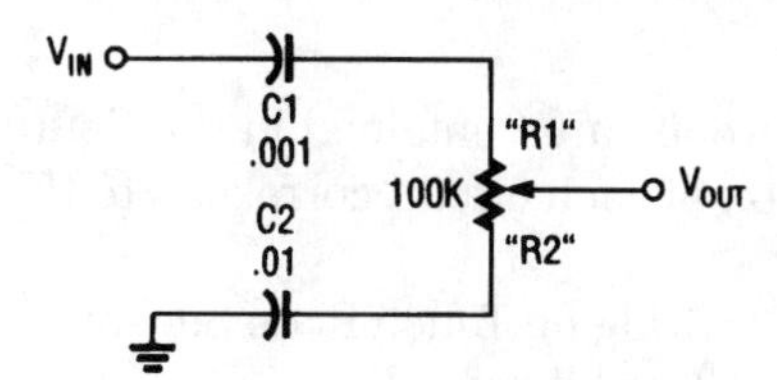

1993 ELECTRONICS HOBBYIST HANDBOOK ***FIG. 104-2***

The treble control has capacitors placed in series with the potentiometer.

BASS TONE CONTROL

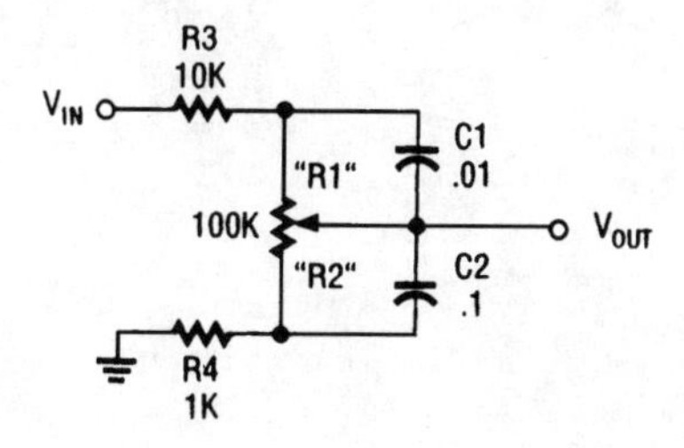

1993 ELECTRONICS HOBBYIST HANDBOOK ***FIG. 104-3***

The frequency dependence of the capacitor's impedance permits this circuit to boost the bass frequencies.

105

Touch-Control Circuits

The sources of the following circuits are contained in the Sources section, which begins on page 675. The figure number in the box of each circuit correlates to the entry in the Sources section.

Bridging Touch Plate Sensor
Touch Switch I
Touch Switch II
Touch On-Only Switch
Latching Touch Switch
Single Plate Touch Sensor

BRIDGING TOUCH PLATE SENSOR

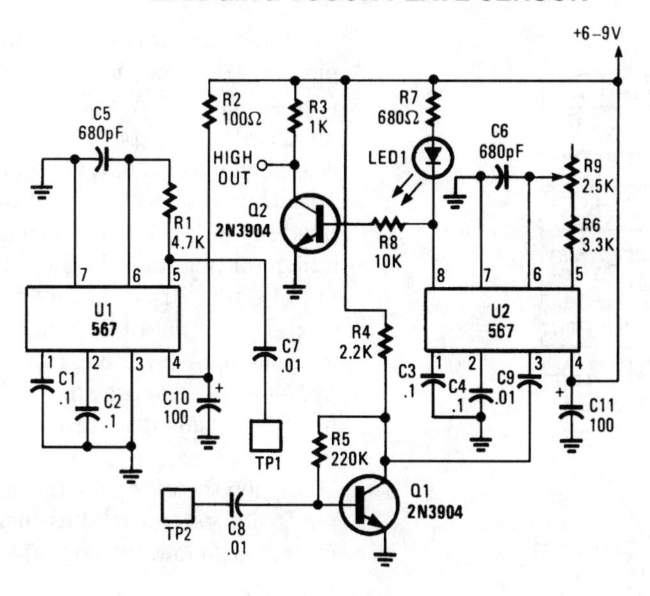

POPULAR ELECTRONICS *FIG. 105-1*

In this circuit, two 567 tone decoders are used. One is an oscillator, the other is a detector. Bridging TP1 and TP2 causes U2 to receive U1's signal, which causes pin 8 of U2 to go low. This action lights LED1 and drives the output of Q2 high.

TOUCH SWITCH I

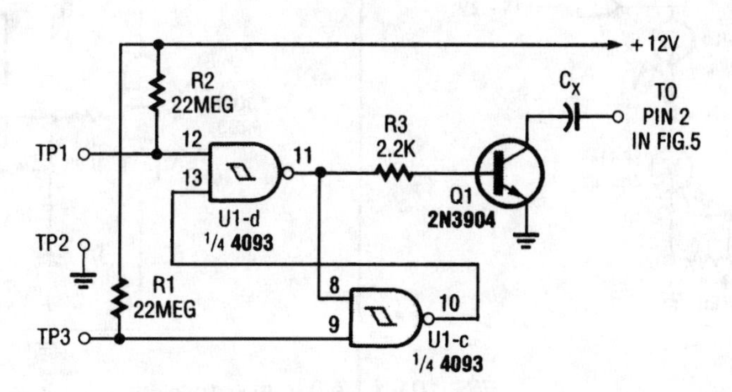

POPULAR ELECTRONICS *FIG. 105-2*

Two NAND Schmitt triggers are connected in a flip-flop configuration to produce a bridged touch-activated switch.

TOUCH SWITCH II

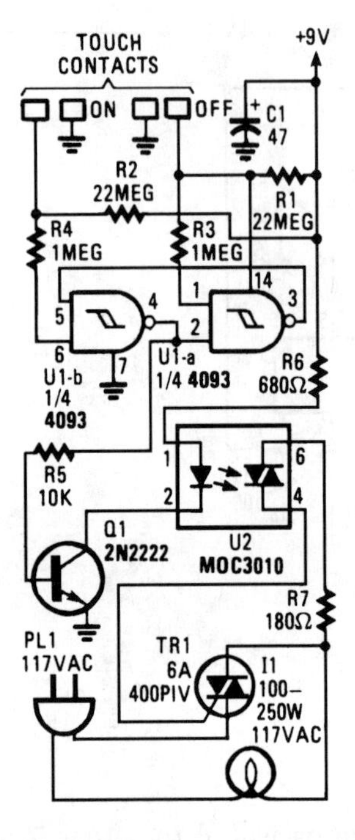

POPULAR ELECTRONICS FIG. 105-3

When the touch-on contacts are bridged, pin 6 of U1-b goes low, which forces its output (the set output) at pin 4 to go high. That high divides along two paths: in one path, the output is applied to pin 2 of U1-a, which causes its output at pin 3 to go low. That low is, in turn, applied to pin 5 of U1-b, which latches the gate in a high output state. In the other path, the output of U1-b is used to drive Q1. When Q1 turns on, U2's internal LED lights, which turns on its internal, light-sensitive, triac-driver (diac) output element. The triac driver feeds gate current to TR1, causing it to turn on, and light the lamp (I1).

When the off contact is bridged, U1-a's output switches and latches high, causing U1-b's output to go low, turning off the lamp.

TOUCH ON-ONLY SWITCH

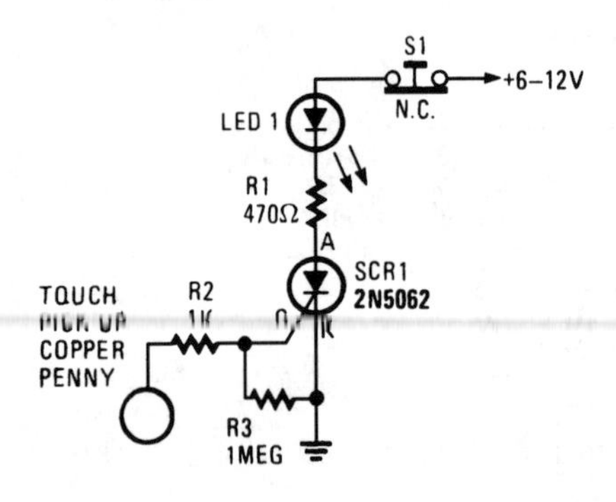

HANDS-ON ELECTRONICS FIG. 105-4

This touch on-only switch can be triggered into conduction by electrical means, and can only be reset by way of a mechanical switch. When the touch terminal is contacted by a finger, the SCR turns on and illuminates LED1.

LATCHING TOUCH SWITCH USING CD4066B

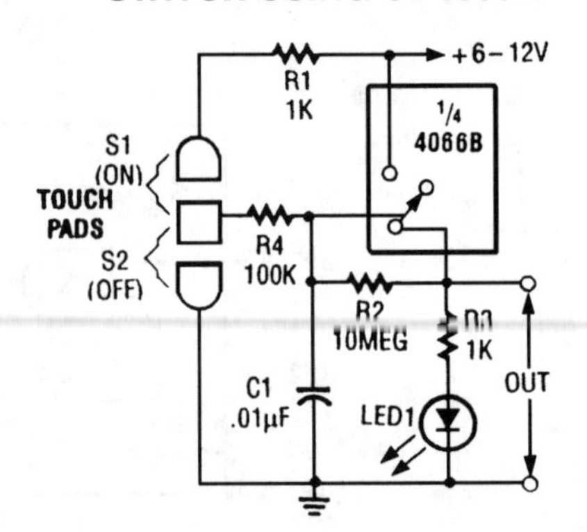

RADIO-ELECTRONICS FIG. 105-5

When touch switch S1 is activated, R4 is driven high, and the control voltage goes high, which latches the switch. When S2 is activated, R4 goes low and the control voltage goes low, which deactivates the switch.

SINGLE-PLATE TOUCH SENSOR

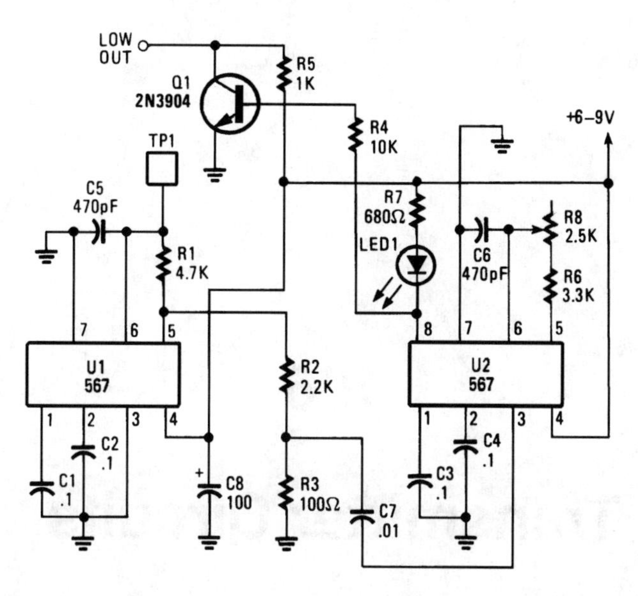

POPULAR ELECTRONICS *FIG. 105-6*

This system operates on the principle that capacitance loading of an oscillator will lower its frequency. When a foreign body comes into contact with touch plate, the frequency of U1 is lowered. This removes the oscillator signal from U1 from U2's passband, which causes U2 to lose lock, turns off the LED, and causes the collector of Q1 to go low.

106

Transmitter Circuits

The sources of the following circuits are contained in the Sources section, which begins on page 675. The figure number in the box of each circuit correlates to the entry in the Sources section.

27.125-MHz NBFM TRANSMITTER

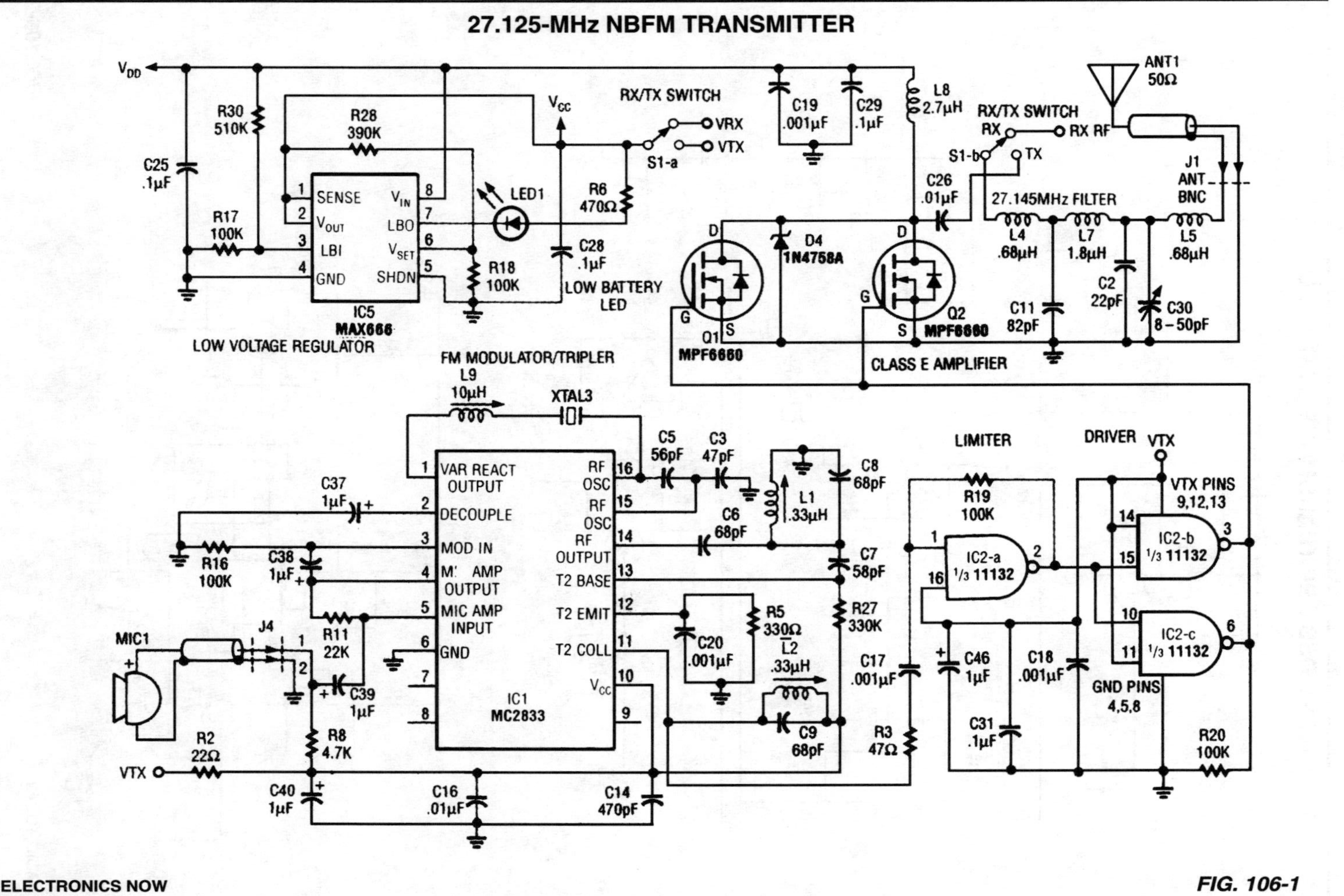

ELECTRONICS NOW

FIG. 106-1

Using a Motorola MC2833 one-chip FM transmitter, a few support components, and an MPF6660 FET RF amp, this transmitter delivers about 3 W into a 50-Ω load. It is capable of operation over about 29 to 32 MHz with the components shown.

10-M DSB QRP TRANSMITTER WITH VFO

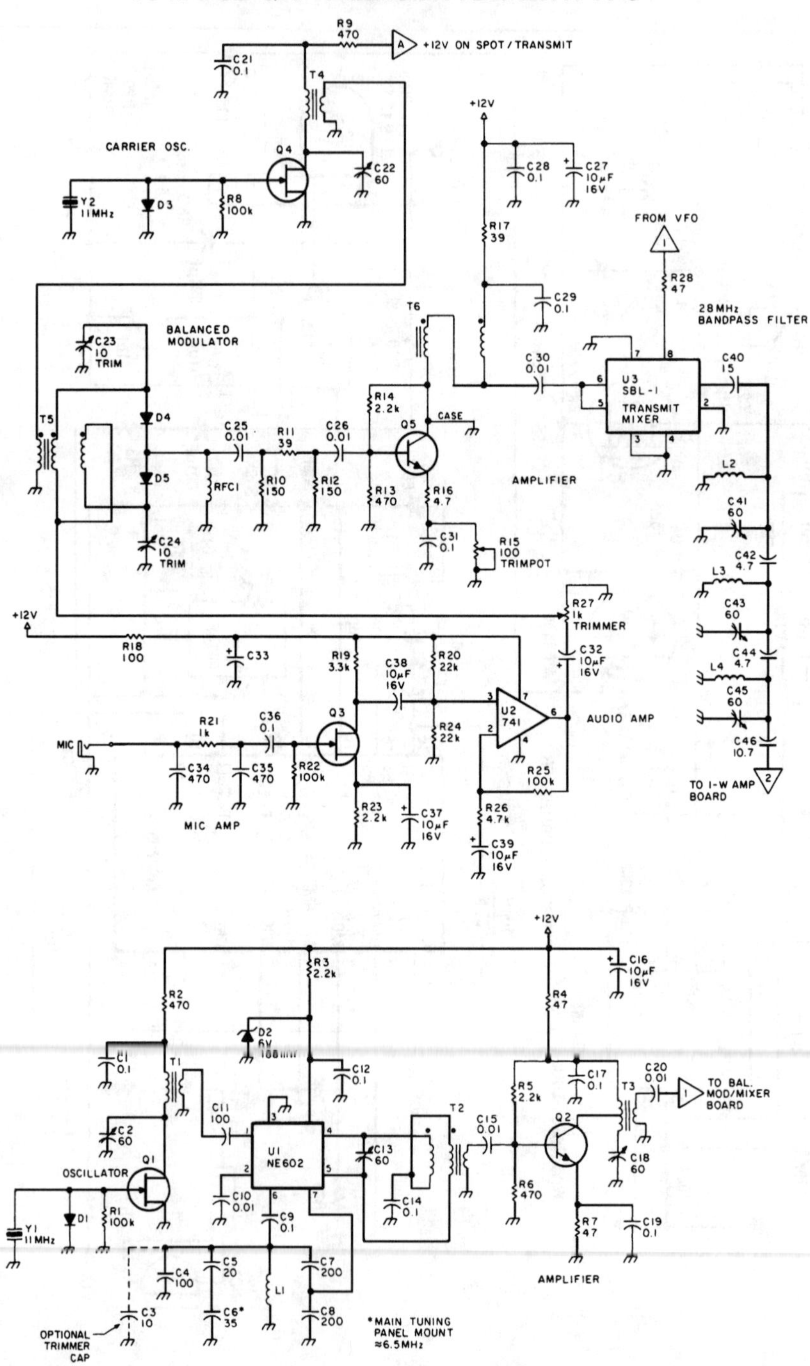

73 AMATEUR RADIO TODAY

FIG. 106-2

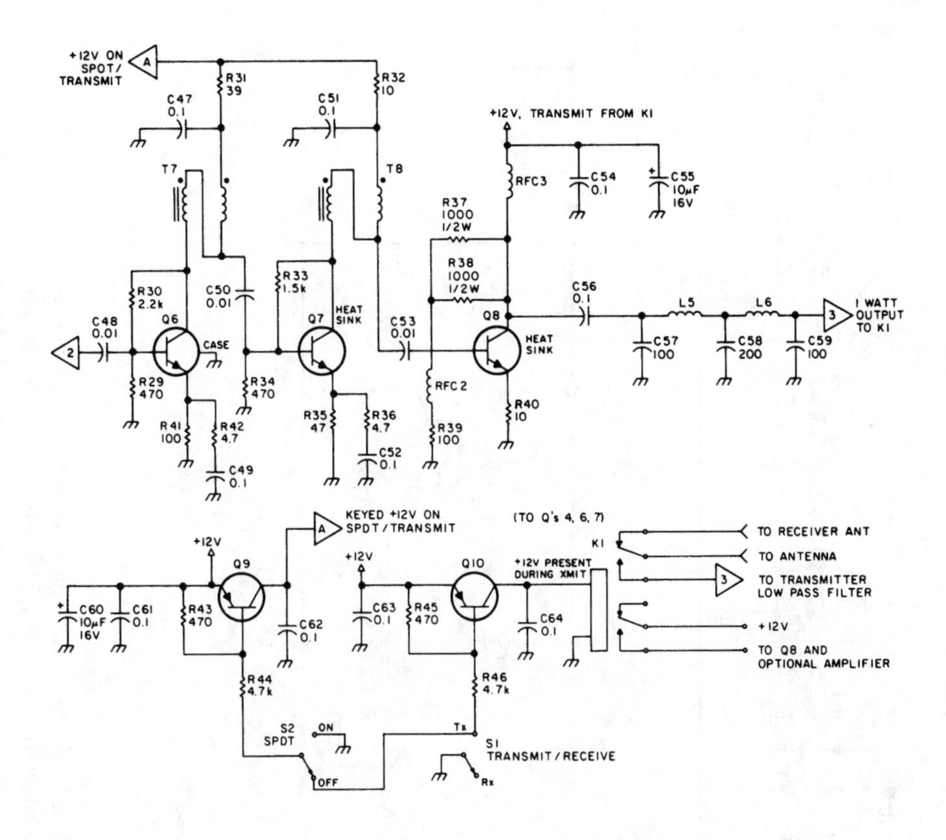

The three schematics represent three building blocks for a 10-meter SSB transmitter. Or these blocks can be used separately as circuit modules for other transmitters. The VFO board uses an FET transmittal oscillator, the VFO signal is mixed in an NE602 mixer and is amplified by Q2 to a level sufficient to drive an SBL-1 mixer in the transmit mixer stage (+7 to +10 dBm). In the balance mixer/modulator board, an 11-MHz crystal oscillator drives a diode balanced mixer. Audio for modulation purposes is also fed to this mixer. The DSB signal feeds a 28-MHz BPF. The 1-W amplifier board consists of a 3-stage amplifier and transmit/receive switching circuitry.

ATV JR TRANSMITTER 440 MHz

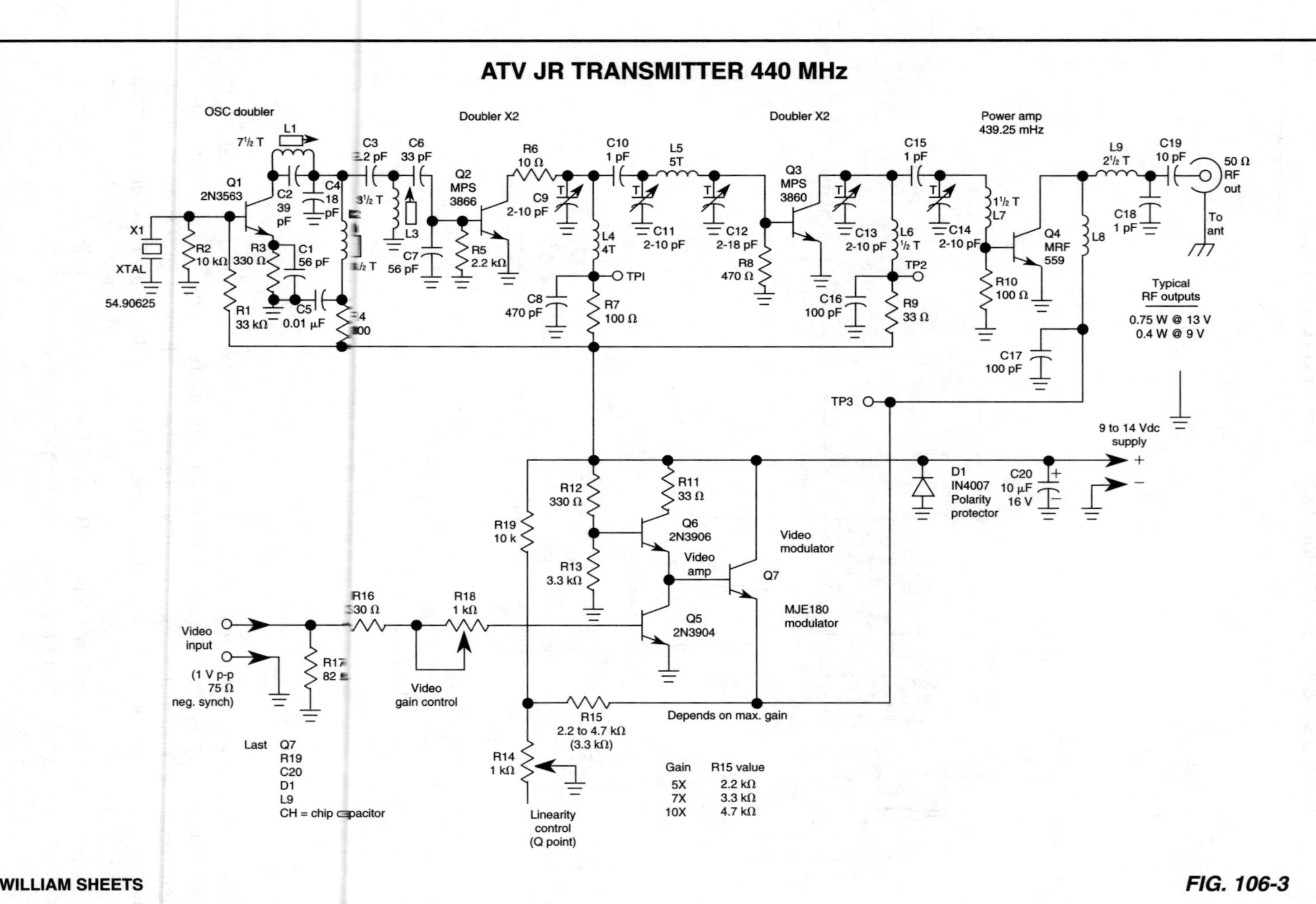

WILLIAM SHEETS *FIG. 106-3*

This low-power video transmitter is useful for R/C applications, surveillance, or amateur radio applications. Seven transistors are used in a crystal oscillator-multiplier RF power amplifier chain, and a high-level video modulator. A 9- to 14-Vdc supply is required. Output is 0.4 to 1.2 W, depending on supply voltage. A complete kit of parts is available from North Country Radio, P.O. Box 53, Wykagyl Station, New Rochelle, NY 10804-00530

6-W ECONOMY MORSE-CODE TRANSMITTER FOR 7 MHz

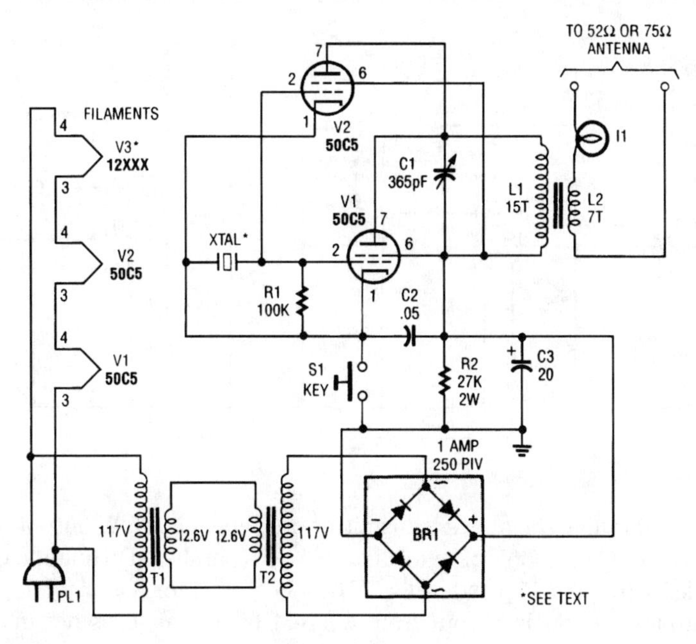

POPULAR ELECTRONICS ***FIG. 106-4***

The vacuum tube is still alive and useful in some applications, as in this CW transmitter. The circuit was built in old-fashioned breadboard style on a wooden base. Old table radios are a good source of parts for this circuit. V3 is used as a ballast resistor—a 75-Ω or 100-Ω 5-W resistor could be substituted. L1 is 15 turns of hookup wire on a ⅞" form 2" long. L2 is 7 turns of the same wire. L2 is wound over L1. Be careful as up to 160 V is present on V1 and V2.

SIMPLE FM TRANSMITTER

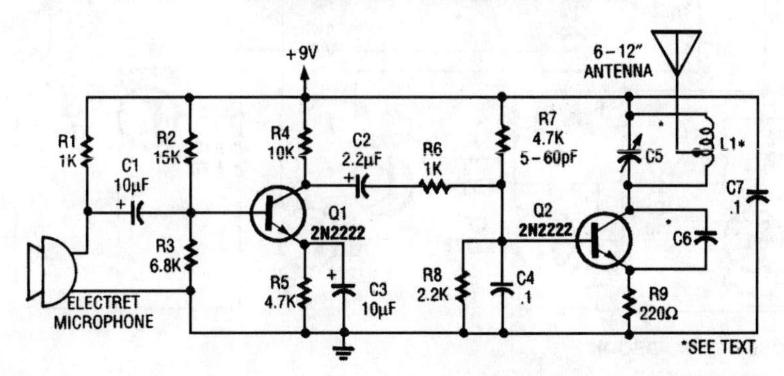

RADIO-ELECTRONICS ***FIG. 106-5***

Running from a 9-V battery, this transmitter can be used as a wireless microphone with an ordinary 88- to 108-MHz FM broadcast receiver. Keep the antenna length under 12 inches to comply with FCC limits. L1 is 6 turns of #24 wire wound around a pencil or a ¼" form, with turns spaced 1 wire diameter. C6 is a gimmick capacitor of about 1 pF.

VACUUM-TUBE LOW-POWER 80/40-METER TRANSMITTER

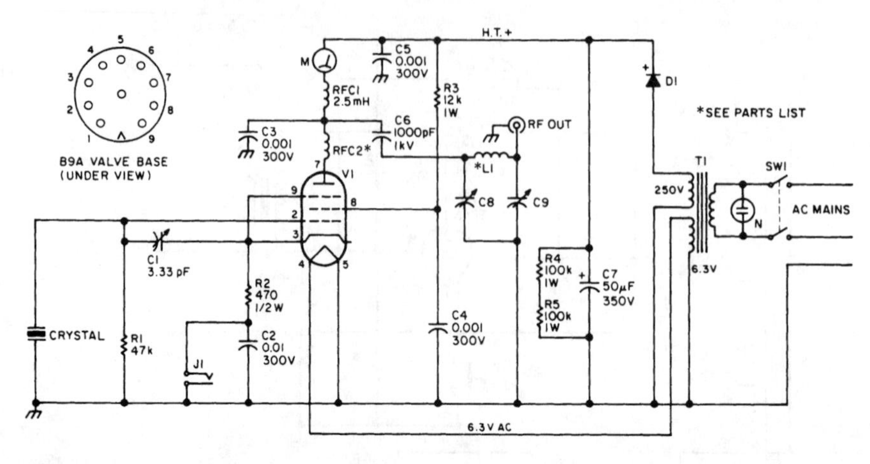

73 AMATEUR RADIO TODAY ***FIG. 106-6***

Using a 6BW6 vacuum tube, the above transmitter delivers about 5 W output. C1 is adjusted for cleanest CW note. C8 and C9 are 365 pF and dual-365 pF (paralleled) tuning capacitors. L1 is 35 turns of #24 enamelled wire on a 1" plastic tube. FT-243 crystals for 3.5 or 7 MHz are used. Do not use this circuit to produce a 7-MHz output from a 3.5-MHz crystal—it is not intended to "double over" crystal frequencies.

TRACKING TRANSMITTER

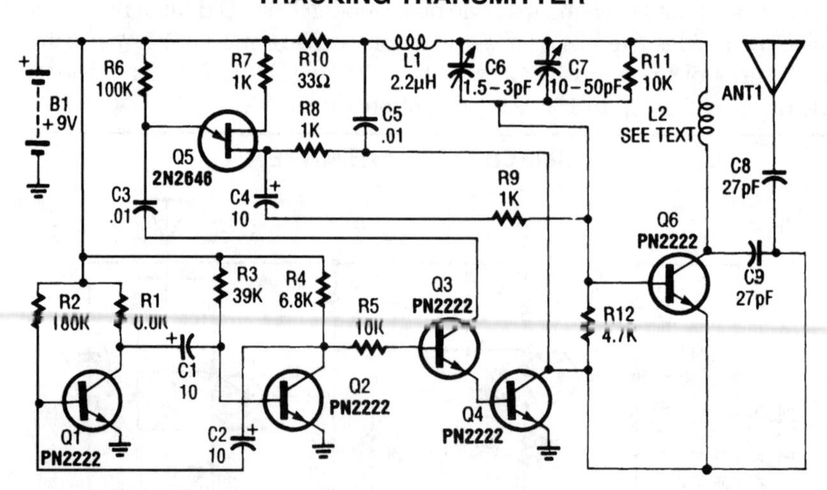

1993 ELECTRONICS HOBBYIST HANDBOOK ***FIG. 106-7***

This tracking transmitter consists of four distinct subassemblies; a free-running multivibrator, a transmit switch, an audio-tone generator, and an FM transmitter. The multivibrator (which produces a pulse width with a pulse separation of 1500 ms) is built around Q1 and Q2. The multivibrator output is coupled through R5 to the base of Q3, whose emitter feeds Q4, which controls the circuit's transmitter section.

49-MHz FM TRANSMITTER

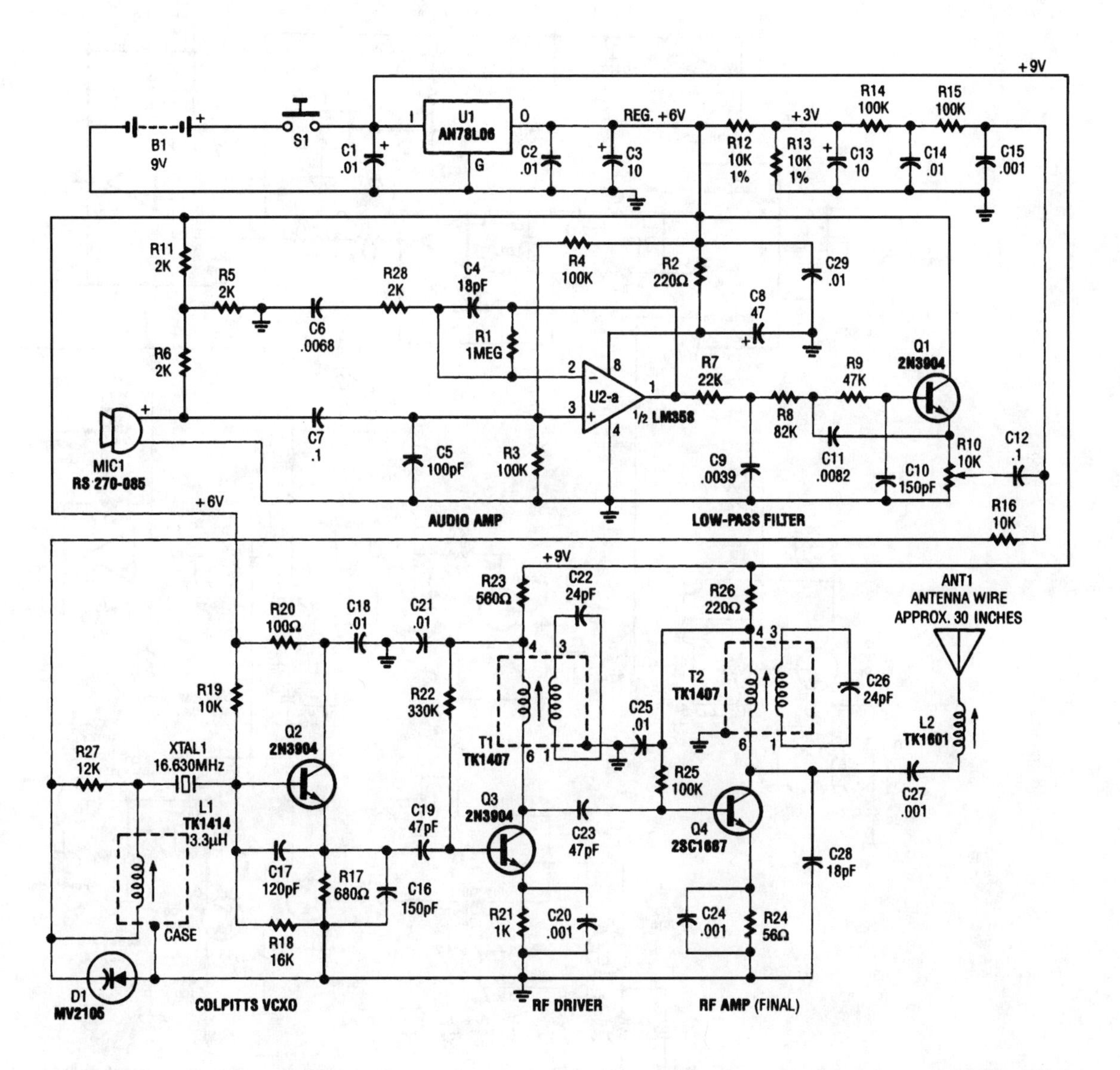

1993 ELECTRONICS HOBBYIST HANDBOOK ***FIG. 106-8***

This 49-MHz FM transmitter consists of an audio amplifier, a low-pass filter, three RF stages, and a regulated-dc power supply. The output is about 16 mW into a 50-Ω load. This transmitter can be used in many 49-MHz applications, such as in a baby monitor, cordless telephone, or in conjunction with a scanner as a one-way voice link.

QRP TRANSCEIVER FOR 18, 21, AND 24 MHz

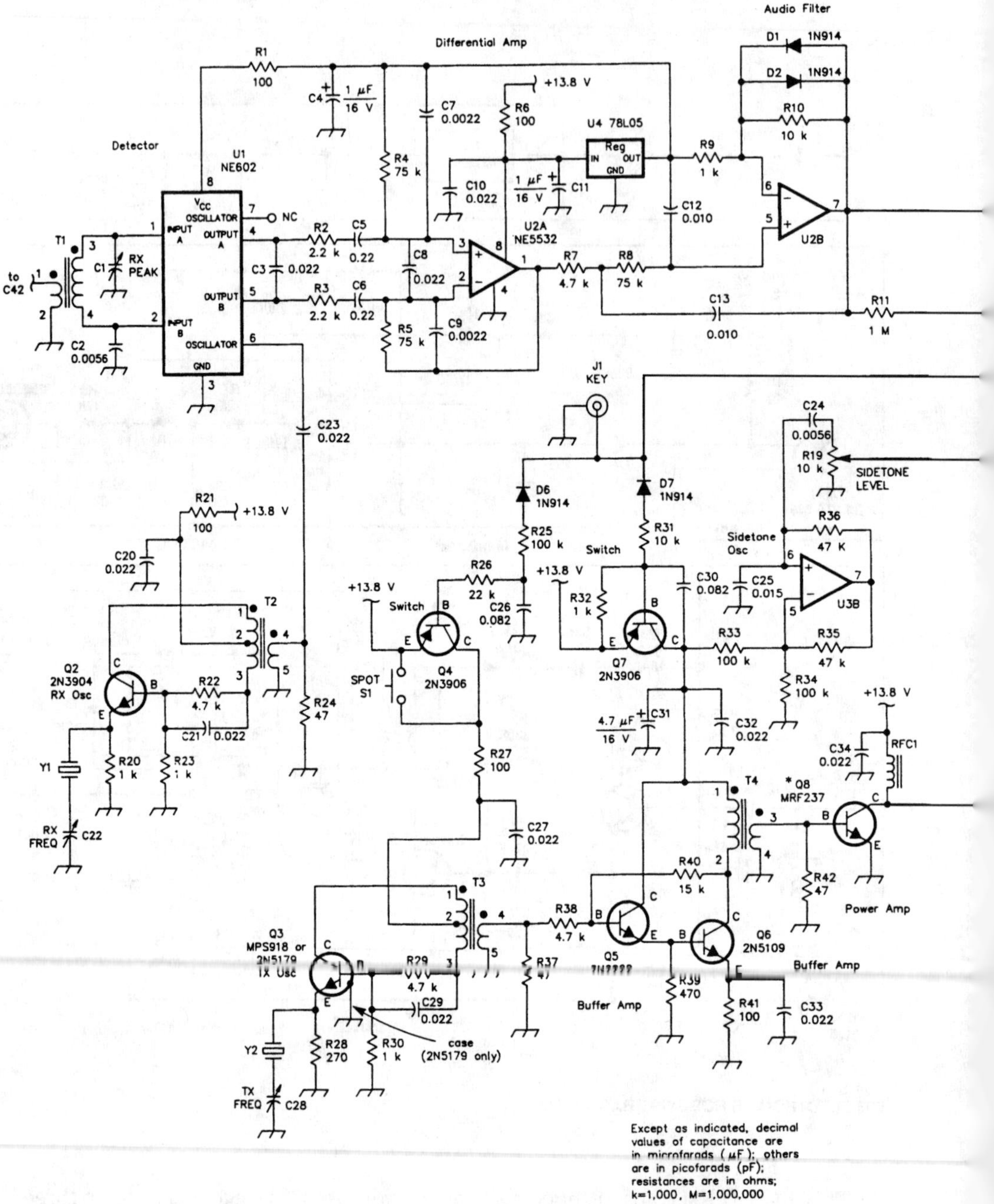

Except as indicated, decimal values of capacitance are in microfarads (μF); others are in picofarads (pF); resistances are in ohms; k=1,000, M=1,000,000

*Heat sink required; see text
● = phasing
SM = silver mica

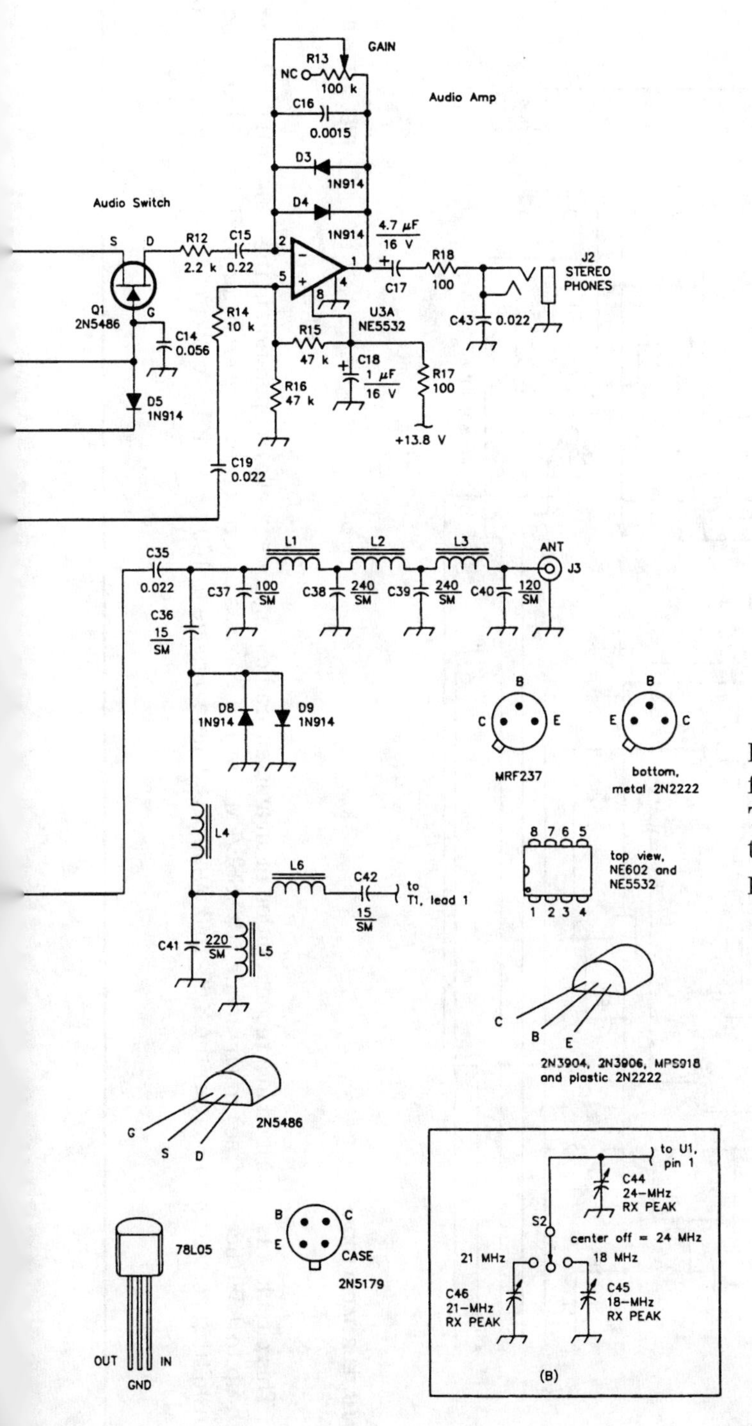

This CW transceiver has 1.25 to 4 W RF output, a direct-conversion receiver, full break-in, and SW sidetone generation. The power supply is 13.8 V, which makes this transceiver suitable for mobile or portable operation.

FIG. 106-9

1750-METER TRANSVERTER

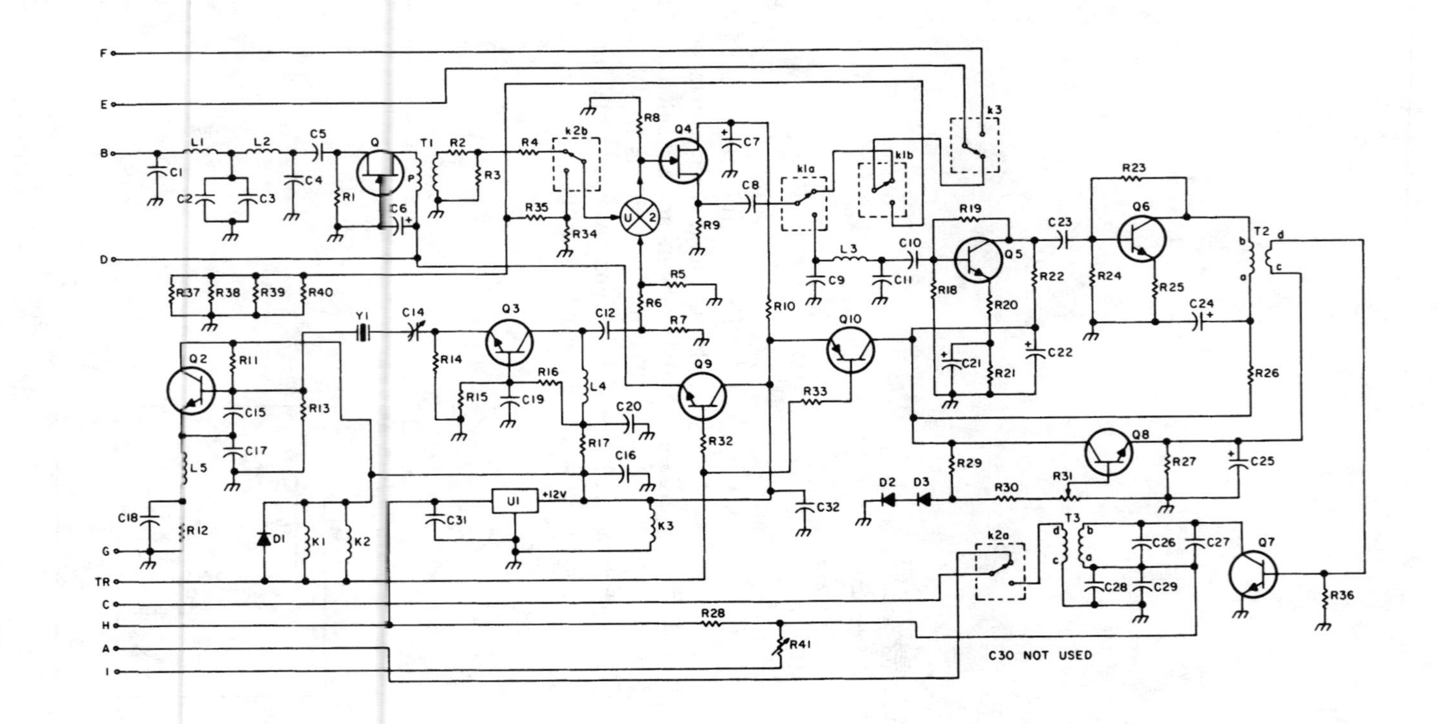

73 AMATEUR RADIO TODAY

FIG. 106-10

This circuit was described in a recent edition of an amateur radio magazine. It allows operation in the 160- to 190-kHz band with up to 1 W (license free) in any mode (CW/SSB/FM, etc.). It consists of a receiving converter for 5 kHz to 450 kHz and a transmitting converter to convert the 3.66- to 3.69-MHz (80 meter) range to 160 to 190 kHz. A 12- to 24-V power supply can be used.

10-METER DSB TRANSMITTER

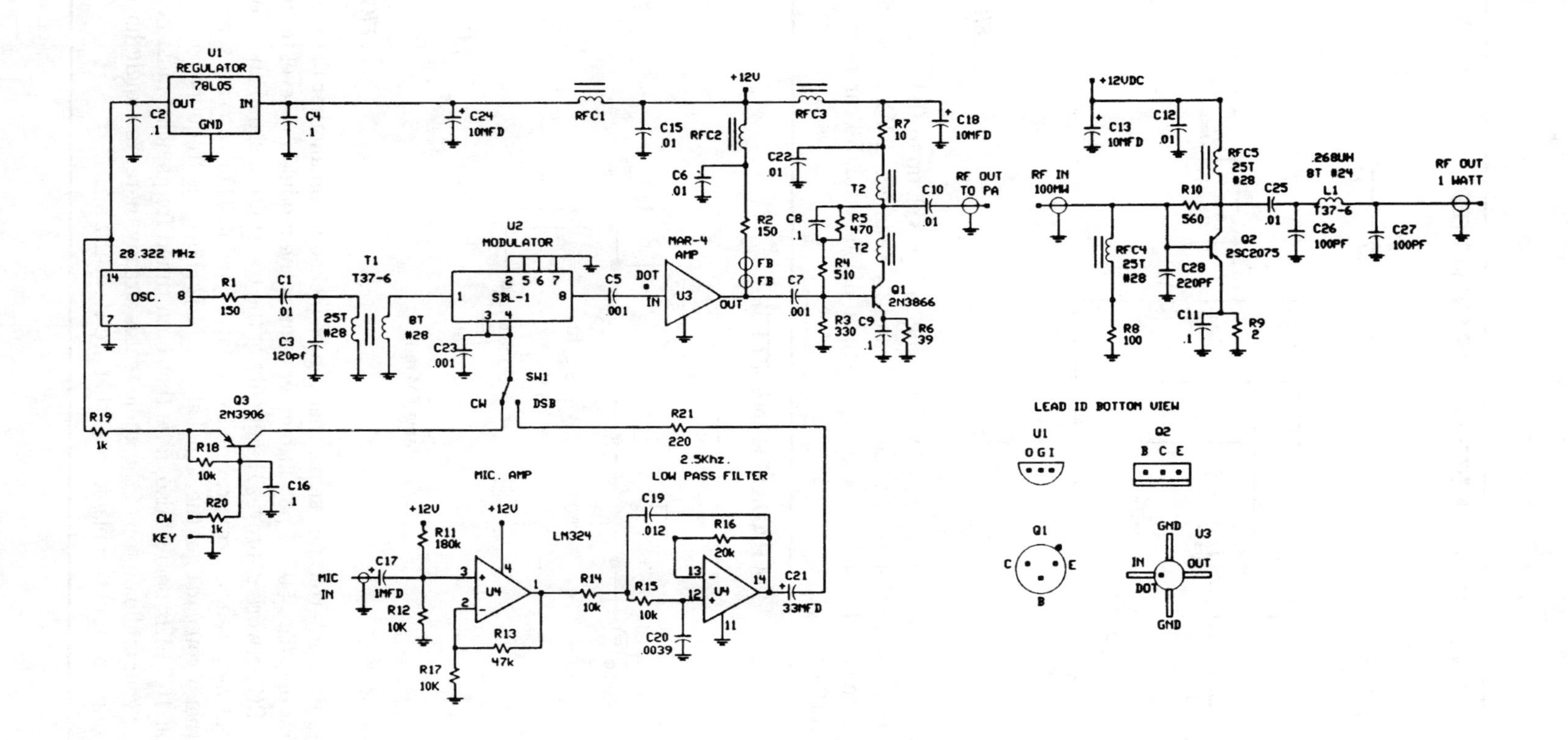

73 AMATEUR RADIO TODAY ***FIG. 106-11***

A DSB transmitter is much cheaper to build than an SSB transmitter because no filter or phasing networks are required. This circuit produces up to 1-W output on the 10-meter band. The frequency 28.322 MHz is used, which is a commonly available clock frequency crystal. CW operation is also provided. A doubly balanced mixer assembly is used as a modulator and CW keyer.

LOW-POWER 40-METER CW TRANSMITTER

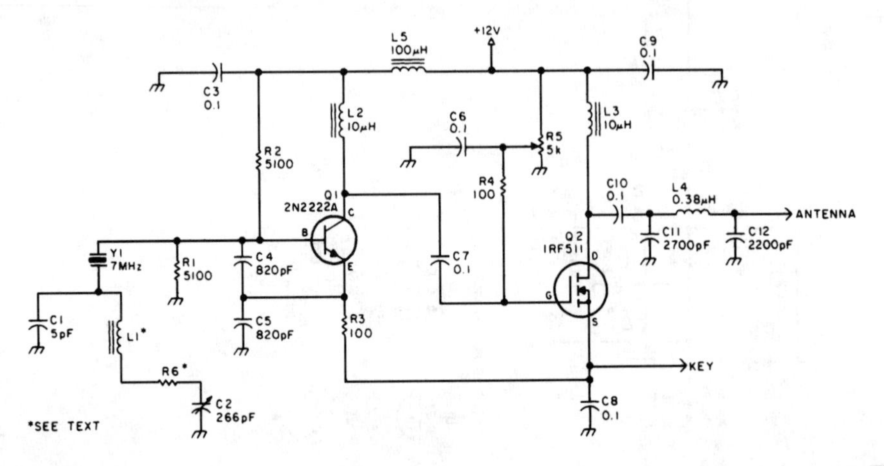

73 AMATEUR RADIO TODAY ***FIG. 106-12***

This CW transmitter has an output of up to 3 W. By using 24 V on Q2, up to 10 W output can be obtained. If a 24-V supply is used, Q1 must not see more than 12 V. Connect 12 V between junctions C3, R2 and L2, and remove L5. L1 should be a low-*Q* 18- to 20-μH inductor. R6 can be used (up to 47 Ω) to reduce the *Q* further.

FM RADIO TRANSMITTER

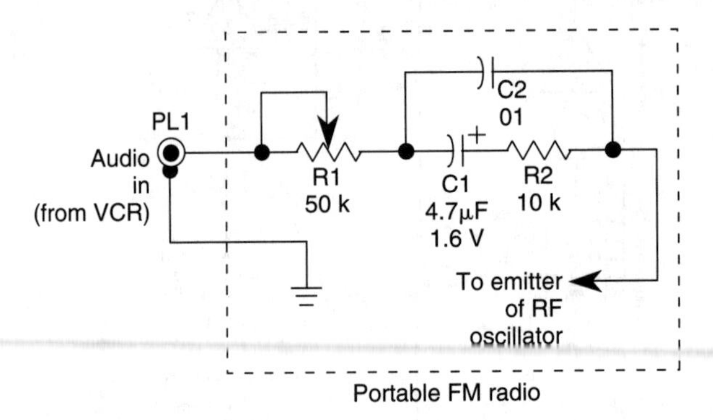

R-E EXPERIMENTERS HANDBOOK ***FIG. 106-13***

An FM radio generates an interference signal that can be picked up on another FM radio tuned 10.7 MHz above the first one. The 50-kΩ potentiometer adjusts the modulation level to maximum without distortion. The RC network improves the fidelity of the transmitted signal and provides dc isolation. The component values shown are provided as a starting point. They can vary somewhat for different radios. Note that if you can't get the signal at 10.7 MHz above the frequency setting of the first radio, try tuning at 10.7 MHz below. Also, note that both tuned frequencies must be unused. Otherwise, you will hear your audio on top of the audio that is already there. You might have to play with both frequencies until you find two blank spots that are 10.7 MHz apart.

LOW-POWER 20-METER CW TRANSMITTER

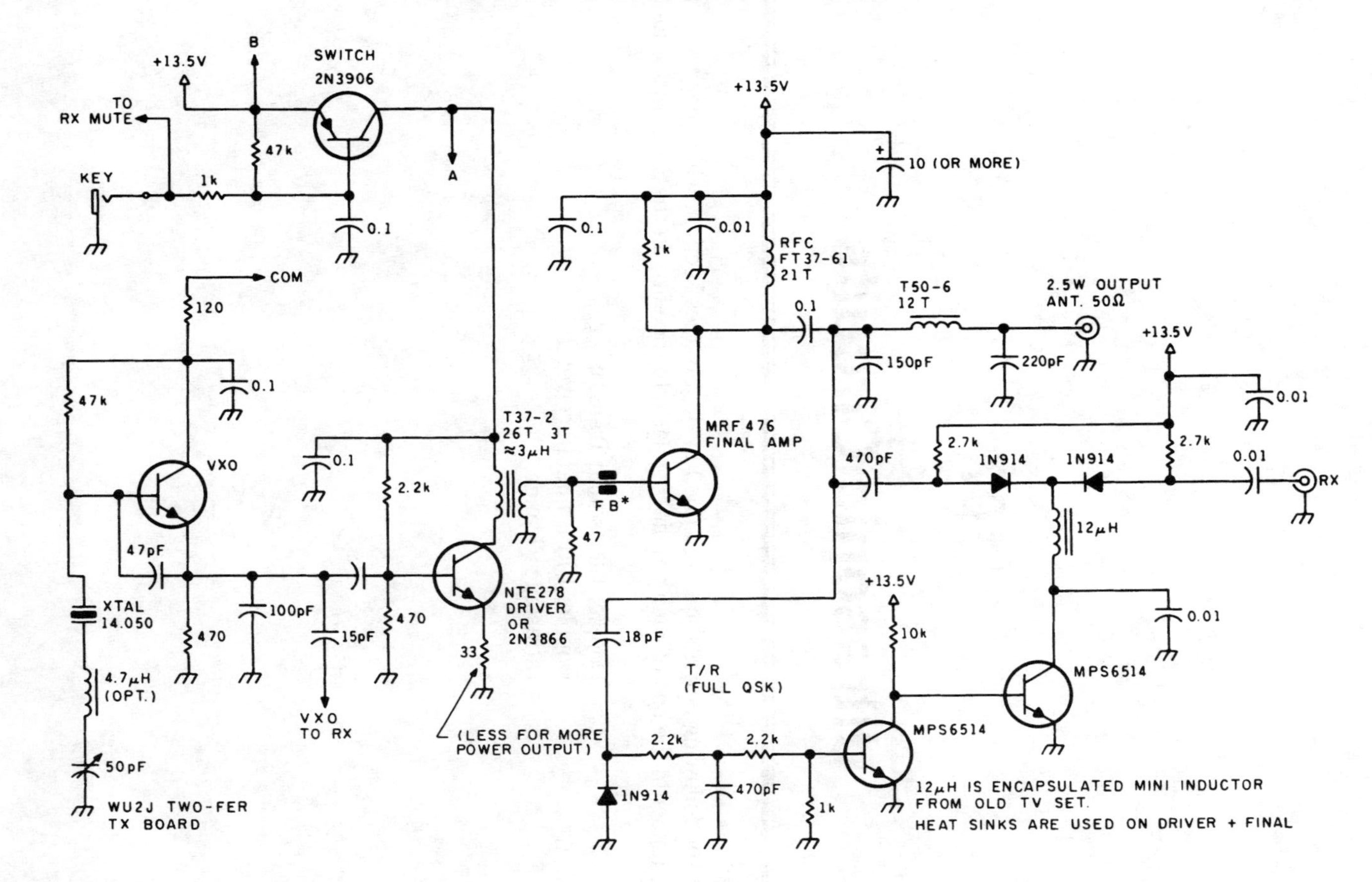

73 AMATEUR RADIO TODAY

FIG. 106-14

The transmitter has a VXO circuit to drive an amplifier that is keyed. The keyed amplifier drives an MRF 476 final amplifier, which delivers about 2-W output. A solid-state T-R switch is included for the receiver. The parts values shown are for the 20-meter band.

107

Ultrasonic Circuits

The sources of the following circuits are contained in the Sources section, which begins on page 675. The figure number in the box of each circuit correlates to the entry in the Sources section.

Doppler Ultrasound Transmitter
Doppler Ultrasound Receiver
Ultrasonic Cleaner

DOPPLER ULTRASOUND TRANSMITTER

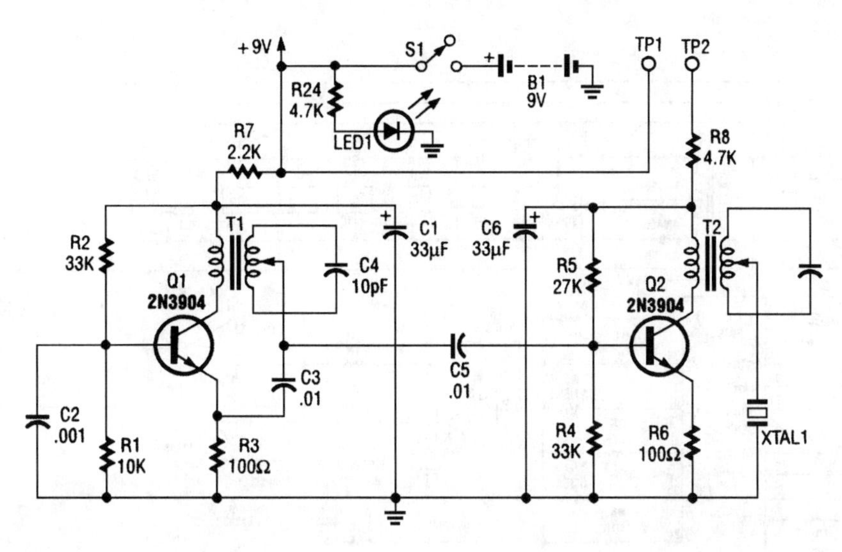

RADIO-ELECTRONICS *FIG. 107-1*

The 2.25-MHz oscillator Q1 drives amplifier Q2 and XTAL1, an ultrasonic transducer. The transducer is a lead zirconate-titanate type. Taps on T1 and T2 provide low-impedance drive points.

DOPPLER ULTRASOUND RECEIVER

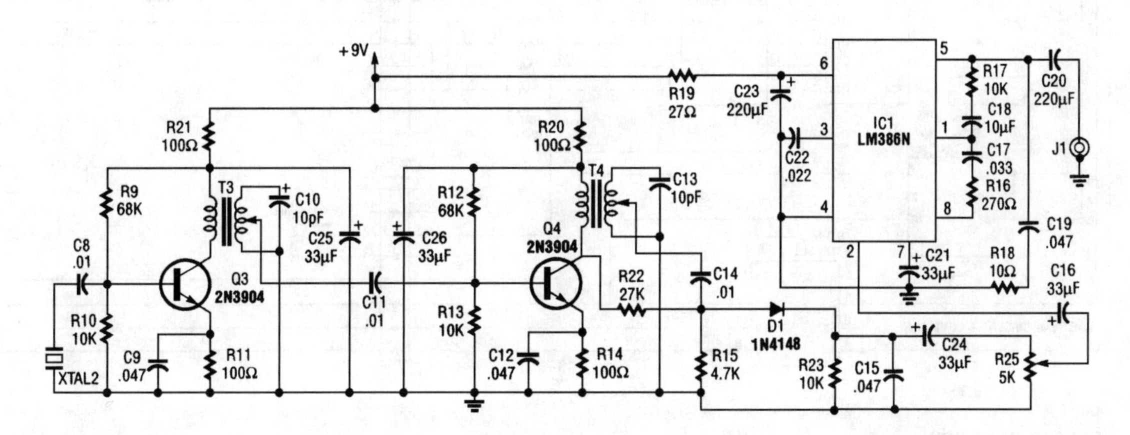

RADIO-ELECTRONICS *FIG. 107-2*

XTAL1 drives amplifier Q3/Q4, which is tuned to 2.25 MHz. The detected signal is fed to audio amplifier IC1. A 9-V supply is used. The circuit operates at 2.25 MHz and is designed to be used with an ultrasonic sound transmitter at this frequency.

ULTRASONIC CLEANER

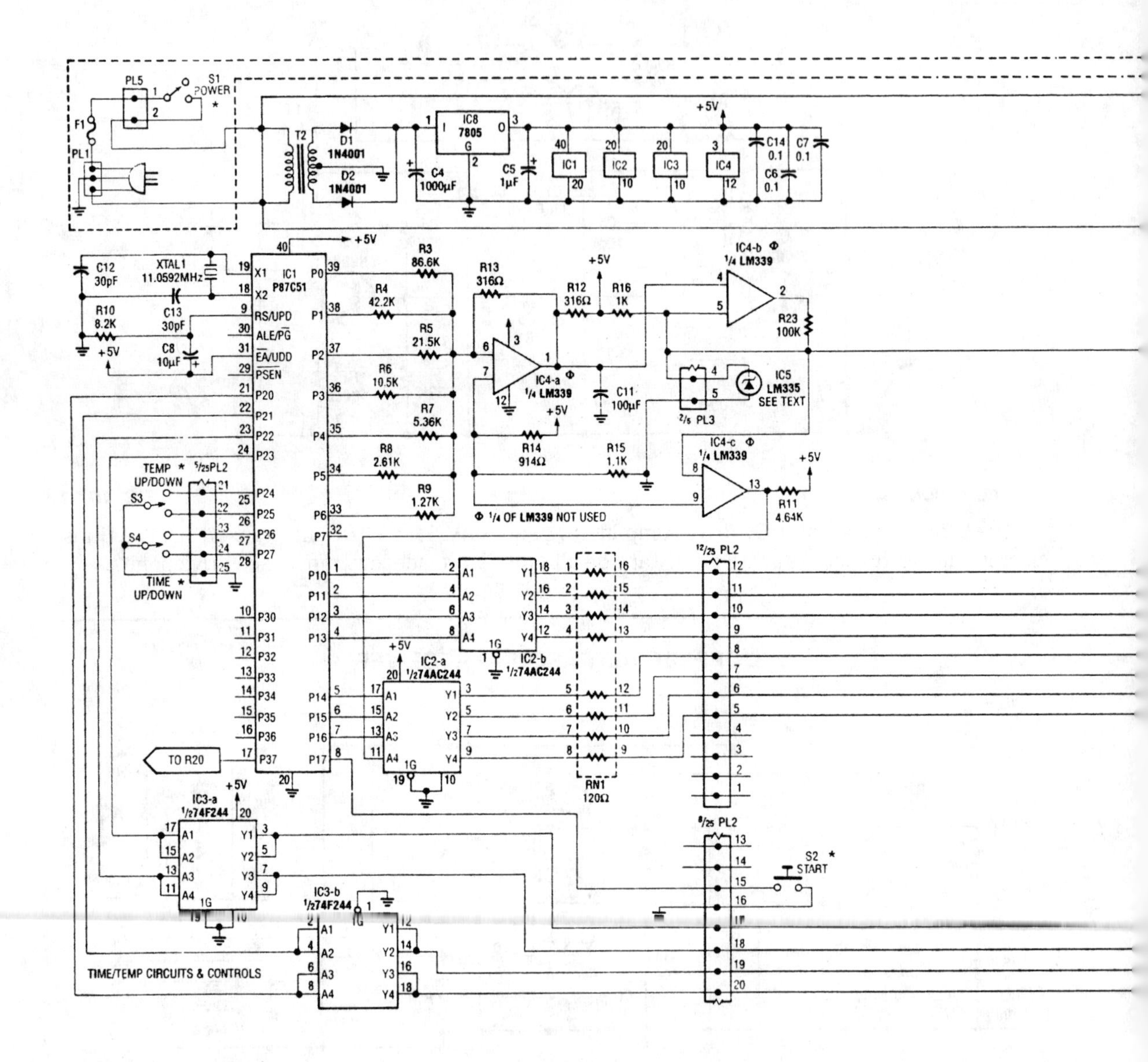

ELECTRONICS NOW

An ultrasonic cleaner is useful to clean certain items. This circuit uses a microcontroller to control timing and give a digital readout, but only the basic oscillator can be used, if desired. RES1, RES2 are piezoelectric transducers driven by power oscillator Q1. Q1 is powered by a bridge rectifier-capacitor input filter that operates directly off the ac line. The frequency is 40 to 60 kHz.

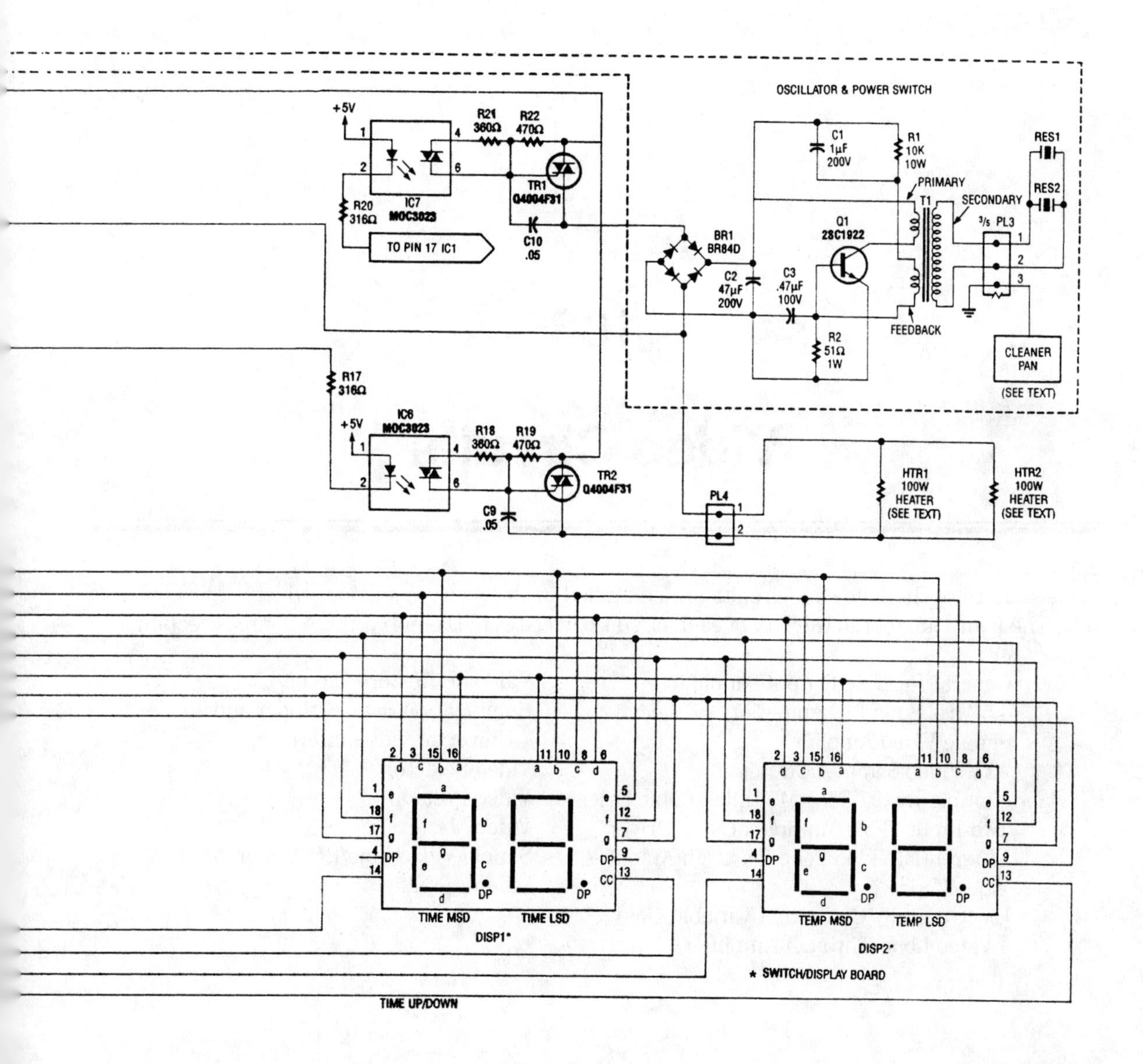

OSCILLATOR & POWER SWITCH
+5V
R21 360Ω
R22 470Ω
IC7 MOC3023
R20 316Ω
TR1 Q4004F31
TO PIN 17 IC1
C10 .05
C1 1μF 200V
R1 10K 10W
PRIMARY
SECONDARY
RES1
RES2
T1
Q1 2SC1922
BR1 BR84D
3/s PL3
C2 47μF 200V
C3 .47μF 100V
FEEDBACK
R2 51Ω 1W
CLEANER PAN
(SEE TEXT)
R17 316Ω
IC6 MOC3023
R18 360Ω
R19 470Ω
TR2 Q4004F31
C9 .05
PL4
HTR1 100W HEATER (SEE TEXT)
HTR2 100W HEATER (SEE TEXT)
TIME MSD
TIME LSD
DISP1*
TEMP MSD
TEMP LSD
DISP2*
* SWITCH/DISPLAY BOARD
TIME UP/DOWN

FIG. 107-3

108

Video Circuits

The sources of the following circuits are contained in the Sources section, which begins on page 675. The figure number in the box of each circuit correlates to the entry in the Sources section.

- General-Purpose Output Amplifier
- 4.5-MHz Sound IF Amplifier
- Simple Video Amplifier
- ATV Video Sampler Circuit
- Multiple-Input Video Multiplex Cable Driver
- Two-Input Video Multiplex Cable Driver
- Differential Video Loop-Through Amplifier
- Video Fader
- Electronically Controlled Variable Gain Video Loop-Through Amplifier
- Video dc-Restore Circuit
- Combination Sync Stripper and Universal Video Interface
- Video Selector
- Video Preamp
- Video Master
- Simple Video Line/Bar Generator
- Video Amplifier

GENERAL-PURPOSE OUTPUT AMPLIFIER

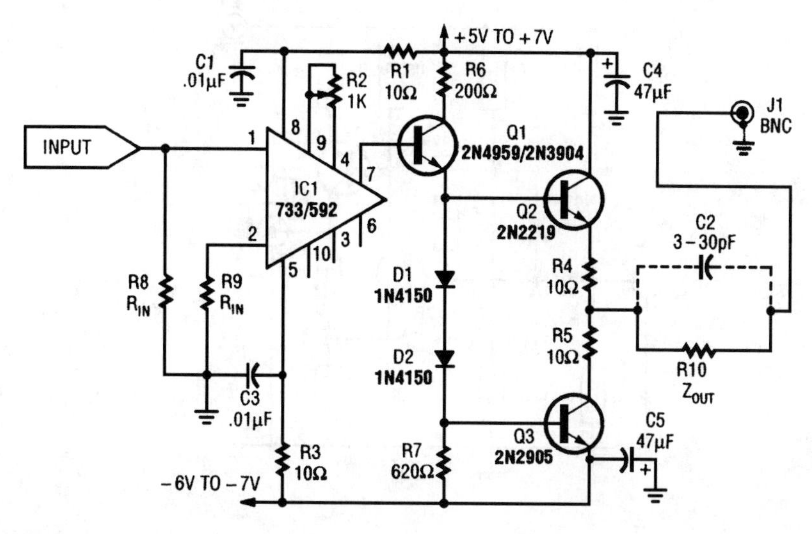

RADIO-ELECTRONICS *FIG. 108-1*

This general-purpose amplifier has a bandwidth of approximately 20 MHz and it uses an LM733/NE592 video amp IC. This circuit can be used as a line driver or as a LAN line driver.

4.5-MHz SOUND IF AMPLIFIER

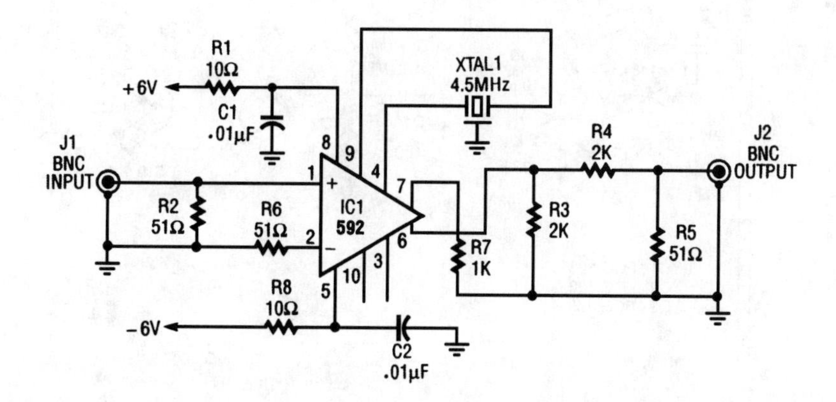

RADIO-ELECTRONICS *FIG. 108-2*

An NE592 is used as a 4.5-MHz amplifier sound subcarrier in video applications. XTAL1 is a 4.5-MHz crystal or ceramic resonator.

SIMPLE VIDEO AMPLIFIER

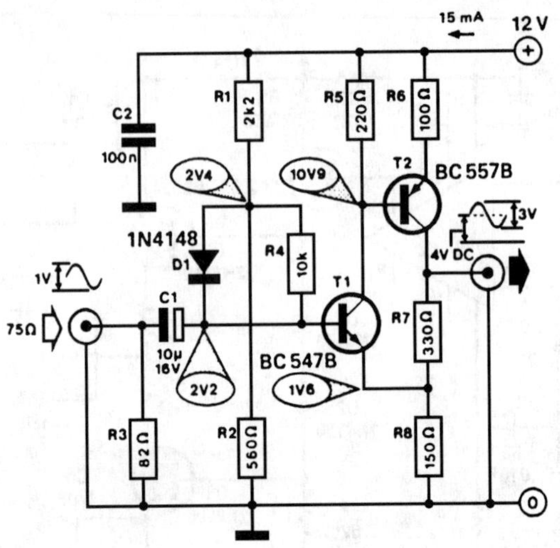

303 CIRCUITS

FIG. 108-3

Useful for interfacing B/W TV sets with a camera or computer, this amplifier has a bandwidth of ≥10 MHz and a gain of 3X.

ATV VIDEO SAMPLER CIRCUIT

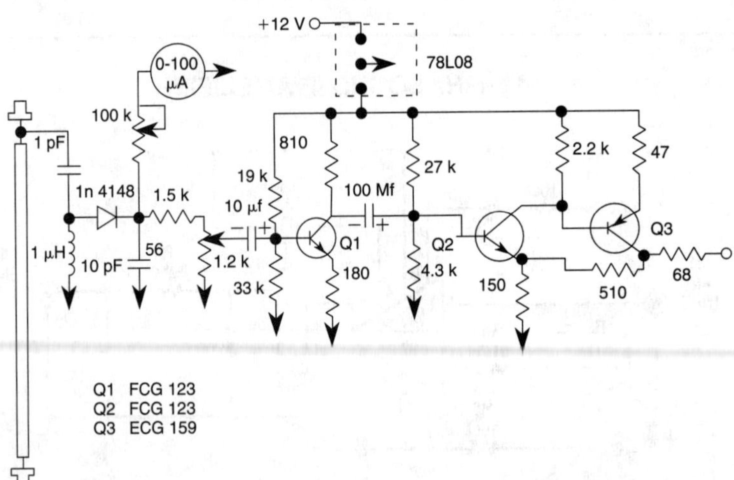

SPEC-COM

FIG. 108-4

This unit picks up your ATV signal by sampling the transmission line with negligible insertion loss. It uses 2 "N" connectors for input and output connections. A BNC connector is used on the video output. The detected output is connected to your monitor and scope so that you can accurately adjust your transmitter for proper video and synch levels. Two different models are provided. Both have relative power output meters, but one has greater accuracy. There are two PC controls, one for video level and the other for power output.

MULTIPLE-INPUT VIDEO MULTIPLEX CABLE DRIVER

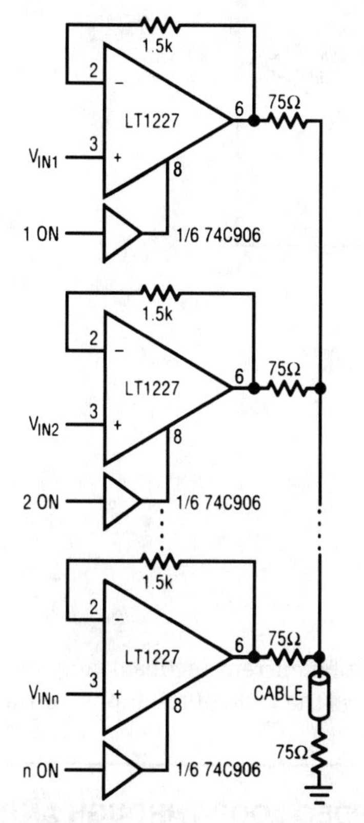

LINEAR TECHNOLOGY *FIG. 108-5*

Using a Linear Technology LT1227, the multiplex video amp uses logic levels to turn on and off selected inputs.

TWO-INPUT VIDEO MULTIPLEX CABLE DRIVER

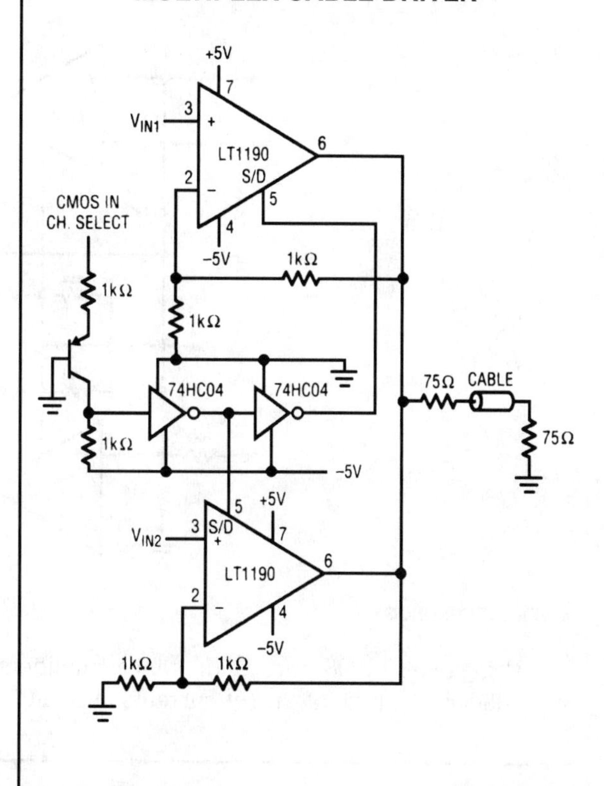

LINEAR TECHNOLOGY *FIG. 108-6*

CMOS logic levels select one of two video inputs with this circuit. The op amps are Linear Technology LT1190s.

DIFFERENTIAL VIDEO LOOP-THROUGH AMPLIFIER

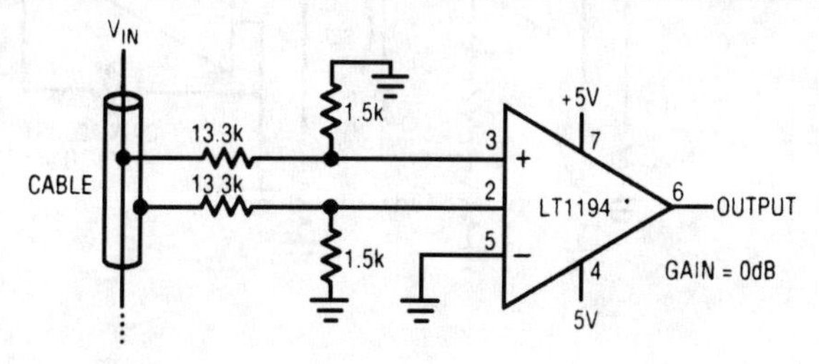

LINEAR TECHNOLOGY *FIG. 108-7*

An LT1194 is used as a differential amplifier for video applications, where low cable loading is needed.

VIDEO FADER

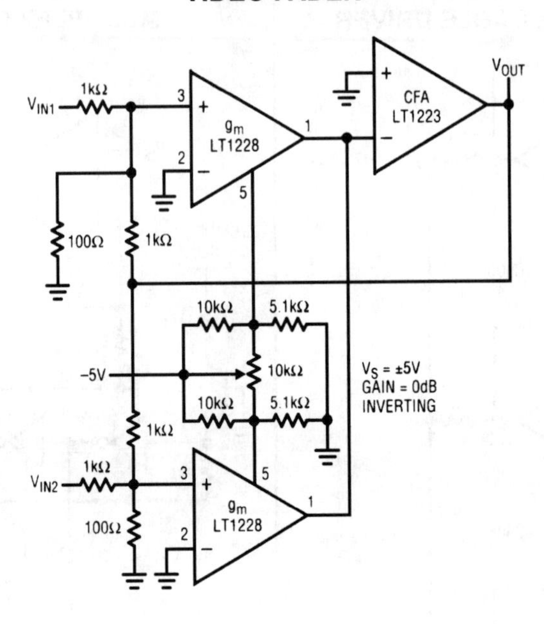

LINEAR TECHNOLOGY *FIG. 108-8*

Using two LT1228 transconductance amplifiers in front of a current feedback amplifier forms a video fader. The ratio of the set currents into pin 5 determines the ratio of the inputs at the output.

ELECTRONICALLY CONTROLLED VARIABLE-GAIN VIDEO LOOP-THROUGH AMPLIFIER

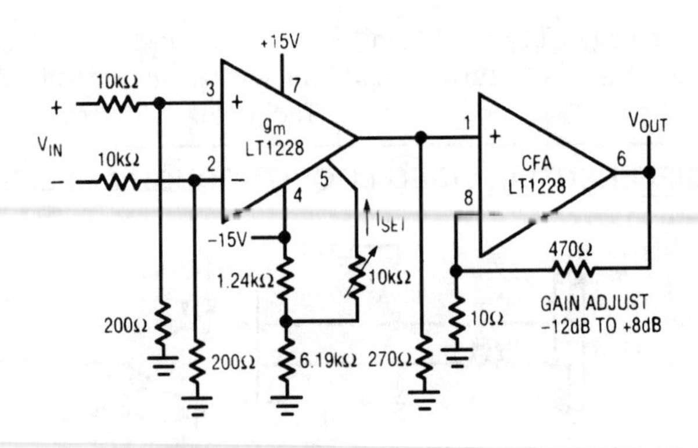

LINEAR TECHNOLOGY *FIG. 108-9*

An LT1228 transconductance amplifier is used in this application. The gain is adjustable from −12 to +8 dB.

VIDEO dc RESTORE CIRCUIT

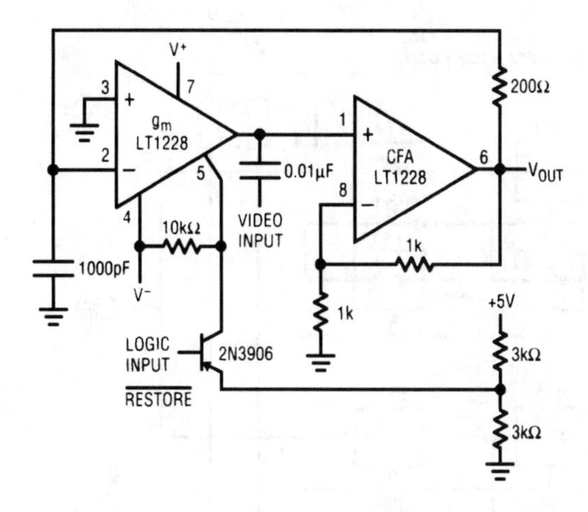

LINEAR TECHNOLOGY **_FIG. 108-10_**

This circuit restores the black level of a monochrome composite video signal to 0 V at the beginning of every horizontal line. This circuit is also useful with CCD scanners to set the black level.

COMBINATION SYNC STRIPPER AND UNIVERSAL VIDEO INTERFACE

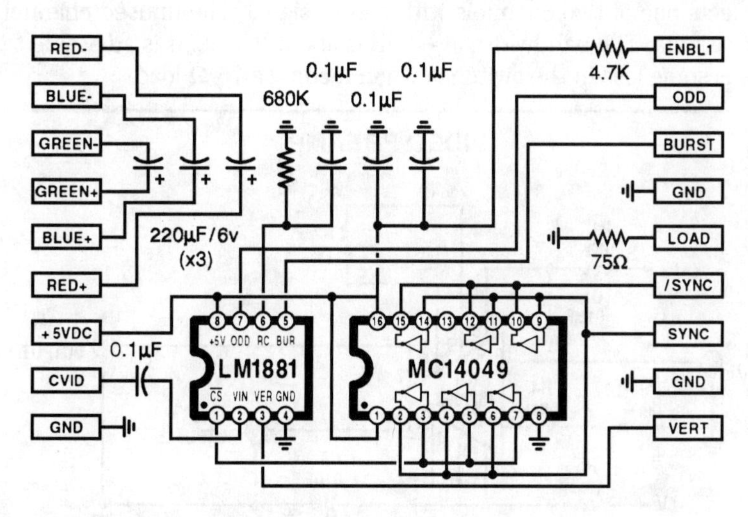

RADIO-ELECTRONICS **_FIG. 108-11_**

This combination sync stripper and universal video interface can solve a lot of problems for you, including Super-Nintendo-to-anything interfacing, video overlay and scope TV frame locking. Kits, fully tested units, and custom cable assemblies are available through Redmond Cable. This unit uses an LM1881 (NS) synch separator IC.

VIDEO SELECTOR

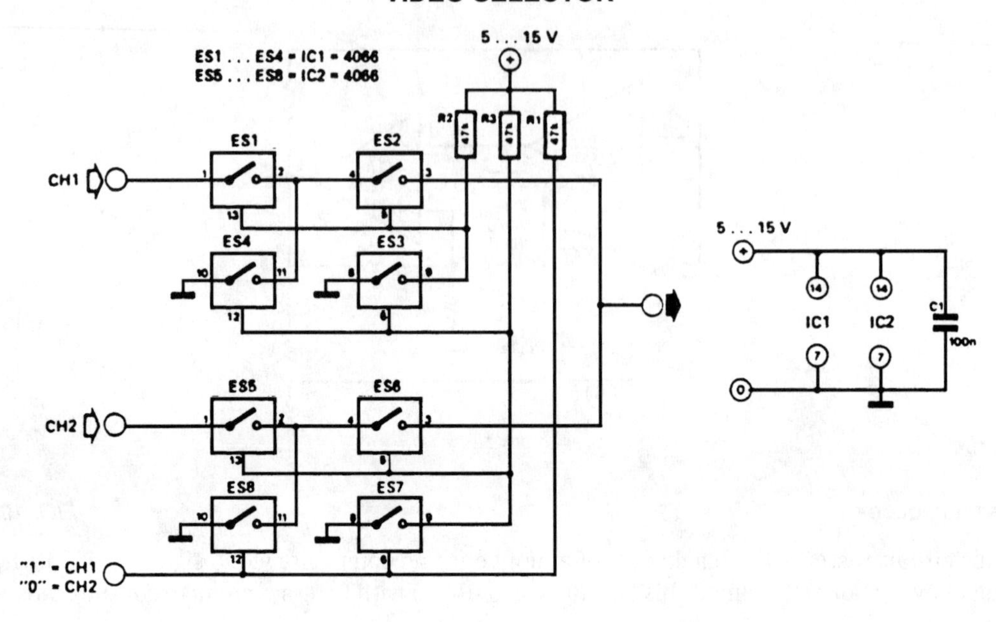

303 CIRCUITS

FIG. 108-12

This circuit selects one of two channels with a logic signal. The unused channel is shorted out, which minimizes crosstalk. The bandwidth at –3 dB is about 8 MHz. It is advisable to buffer this circuit because there is some loss in the switches when feeding a 75-Ω load.

VIDEO PREAMP

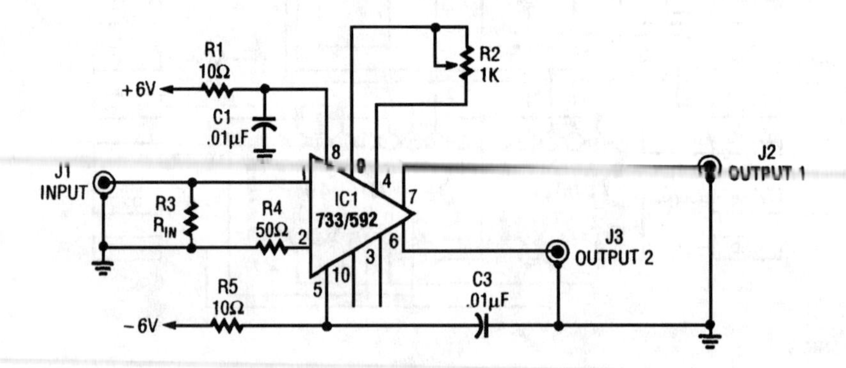

RADIO-ELECTRONICS

FIG. 108-13

An NE592 or LM733 is used as a general-purpose video amplifier in this schematic. J2 and J3 provide two anti-phase outputs. R2 is a gain control. The bandwidth is about 100 MHz.

VIDEO MASTER

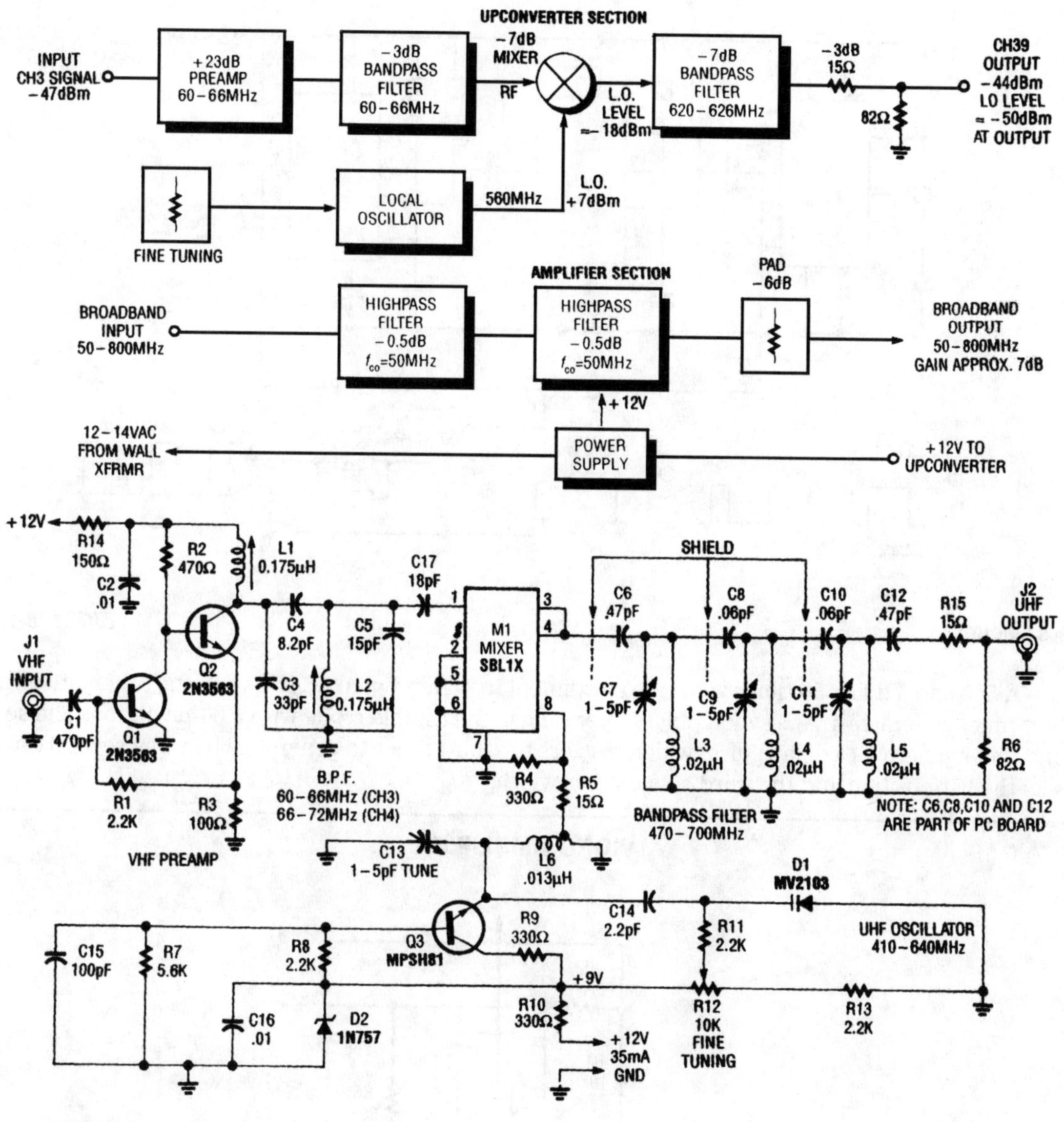

ELECTRONICS NOW

FIG. 108-14

The video master consists of a series of converters that place all your video sources on unused UHF channels, which then combines them with normal TV channels (terrestrial or cable into one cable). That one cable can then feed several TV sets for whole-house coverage. The desired video source is selected with the TV set's tuner. All of the TV's remote-control features are retained.

A complete kit of parts is available from North Country Radio, P.O. Box 53, Wykagyl Station, New Rochelle, NY 10804-0053A.

SIMPLE VIDEO LINE/BAR GENERATOR

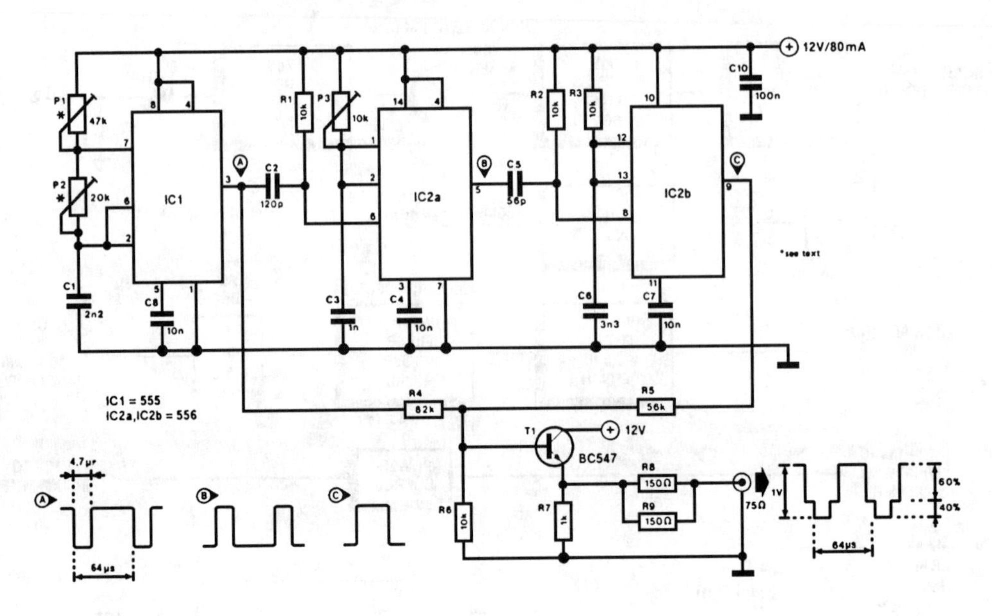

303 CIRCUITS

FIG. 108-15

A 555 and a dual 556 timer generate a rudimentary video signal, as shown in the schematic. The first timer generates 4.7-µs synch pulses operating in the astable mode with a 64-µs period. The second timer generates a delay pulse, which triggers the third timer to generate a bar. The second timer sets the bar position and the third sets the bar width.

VIDEO AMPLIFIER

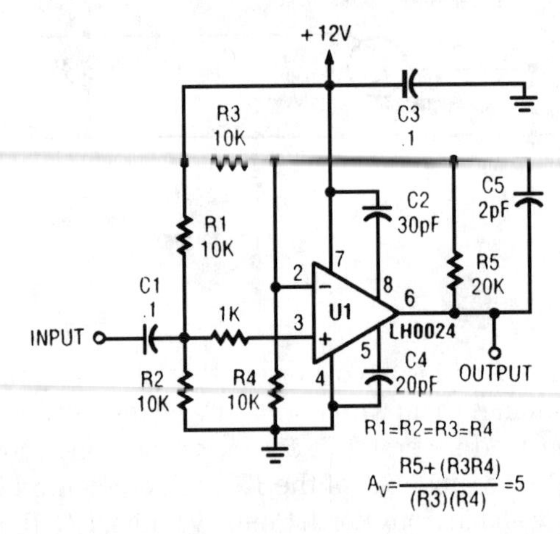

$$A_V = \frac{R5 + (R3R4)}{(R3)(R4)} = 5$$

POPULAR ELECTRONICS

FIG. 108-16

109

Voltage-Controlled Oscillator Circuits

The sources of the following circuits are contained in the Sources section, which begins on page 675. The figure number in the box of each circuit correlates to the entry in the Sources section.

SINUSOIDAL 3-Hz TO 300-kHz VCO

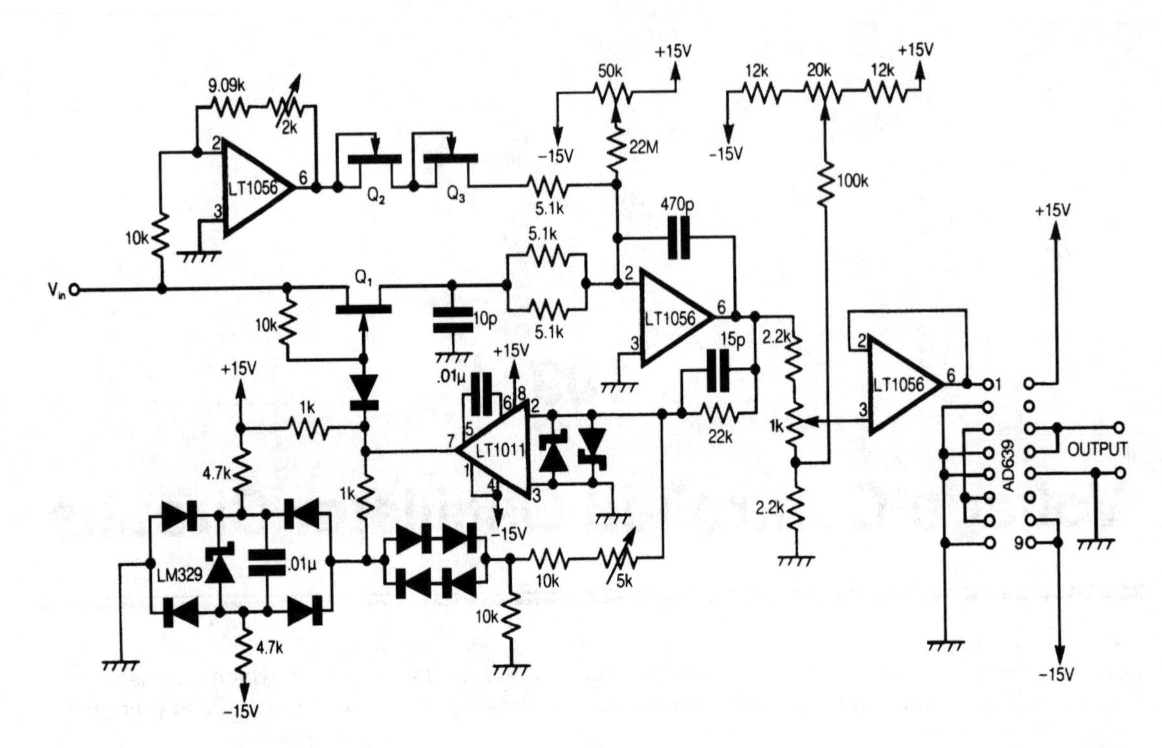

ELECTRONIC ENGINEERING

FIG. 109-1

This circuit uses Analog Devices' AD639 universal trigonometric function generator to convert a triangle waveform, the basic waveform of the VCO itself, into a very low-distortion sine wave.

By using the AD639 in its frequency tripler mode [2], the frequency range 3 Hz to 300 kHz is now covered. The circuit has been drawn here so that the oscillator loop, consisting of Q1, the integrator and the LT1011 comparator, is clearly shown.

When Q1 is off, the input amplifier, which is adjusted to have a gain of exactly –1, pulls a current V_{IN}/R, where R is 5.1 kΩ in series with two JFETs, and Q2 and Q3, out of the virtual earth of the integrator. The output of the integrator thus rises at a rate of V_{IN}/CR, where C = 470 pF. At a level that can be adjusted by the 5-kΩ potentiometer, the comparator flips and turns on Q1.

A current of exactly $2V_{IN}/R$, is now supplied to the virtual earth of the integrator because there are now two 5.1-kΩ resistors in parallel and only a single JFET in between the virtual earth and V_{in}. The integrator output now falls at a rate of V_{IN}/CR and the cycle repeats. Any offset in the current to the virtual earth of the integrator, due to circuit board leakage, etc., can be corrected by adjusting the 50-kΩ potentiometer. It follows that the symmetry of the triangle wave at the integrator output can be corrected by adjusting the 2-kΩ potentiometer, and the 50-kΩ potentiometer at VLF, and the frequency can be trimmed with the 5-kΩ potentiometer.

SINUSOIDAL 3-Hz TO 300-kHz VCO (*Cont.*)

The 1-kΩ potentiometer variable is adjusted to give the input level to the AD639 needed to drive it over ±270° and so produce a sinusoidal output at three times the frequency of the triangle-wave input. Offset correction for the AD639 is made at the input to the voltage follower by means of the 20-kΩ potentiometer.

Once a symmetric triangle wave has been obtained by adjusting the 2-kΩ and 50-kΩ potentiometers, and the correct frequency of 100 kHz has been set for V_{IN} = 10 V, by adjusting the 5-kΩ potentiometer, the triple-frequency sine-wave output can be set up by adjustment of the 1-kΩ and 20-kΩ potentiometers.

This is best done by triggering the CRO from the triangle wave, and then viewing at least three complete cycles of output. Having adjusted for a clean-looking sine wave, the final adjustment of the 1-kΩ and 20-kΩ potentiometers should be made on a single sinusoidal cycle display, using internal trigger so that the three slightly different parts of the output cycle lie one upon the other and can be made to merge. Q1, Q2, and Q3 are 2N4391s, the two Schottky diodes are 5082–2810, and the other nine diodes are 1N914.

All device power supply pins should be decoupled with 0.33 μF. Resistors associated with the inputs of the devices should be 1% high-stability parts.

SIMPLE TL082 VCO

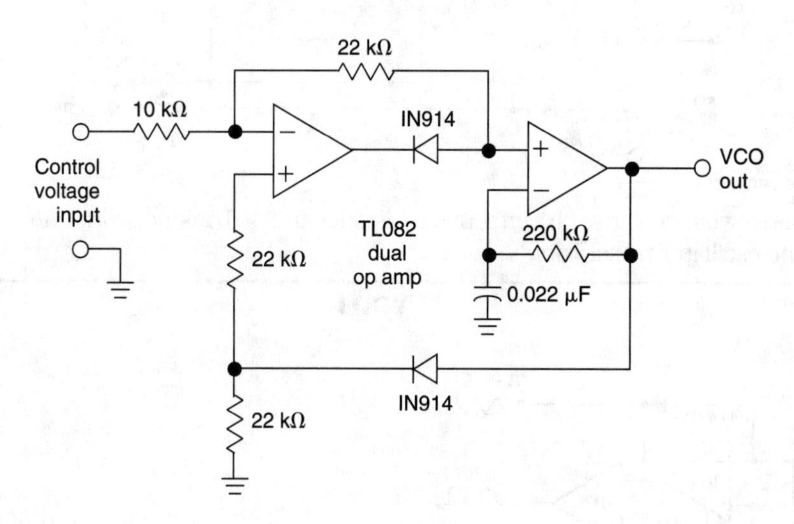

WILLIAM SHEETS

FIG. 109-2

This circuit uses a dual operational amplifier (TL082) to form a voltage-controlled oscillator (VCO). With the component values shown, the output-frequency range is 100 Hz to 10 kHz when the input control voltage is between 0.05 and 10 V.

10-Hz TO 10-kHz 3-DECADE VCO

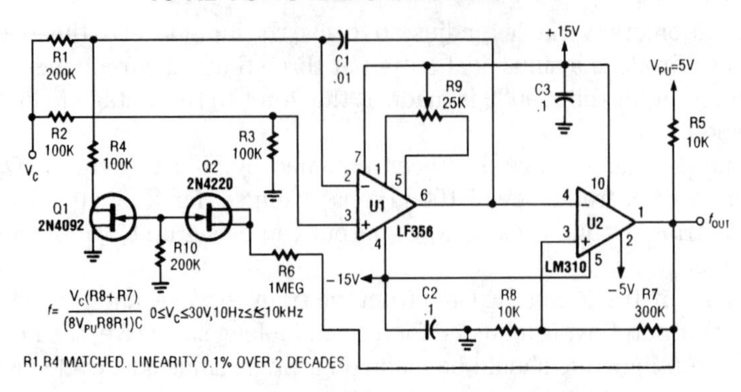

POPULAR ELECTRONICS **FIG. 109-3**

SINE-WAVE VCO

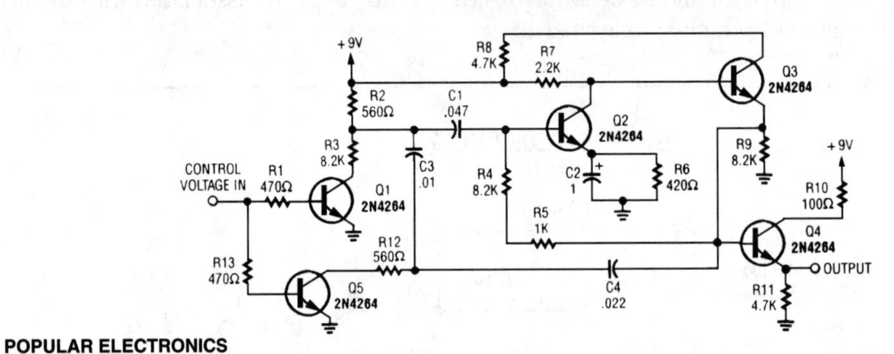

POPULAR ELECTRONICS **FIG. 109-4**

A dc control voltage varies the effective resistance in feedback network C4/C3/C1 and R12/R3. Q2/Q3 are the oscillator transistors.

VCO I

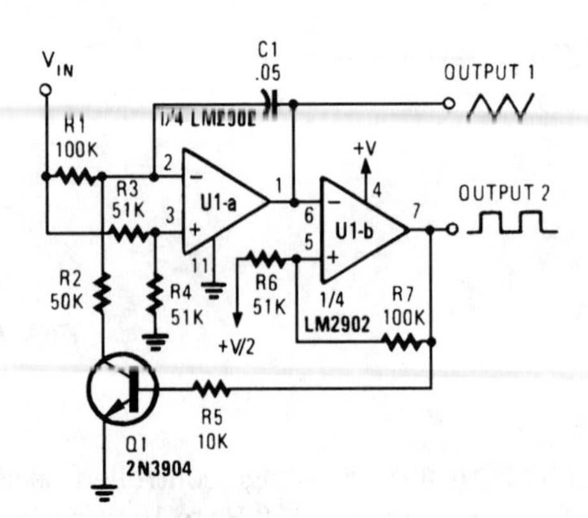

This circuit gives both triangle- and square-wave outputs. The frequency range is determined by C1.

POPULAR ELECTRONICS **FIG. 109-5**

VCO II

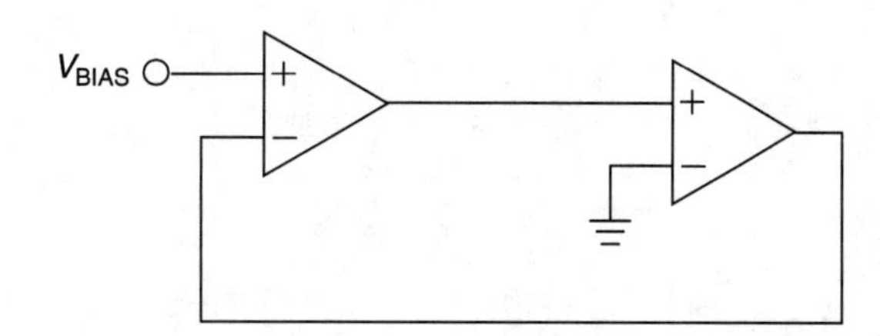

WILLIAM SHEETS

FIG. 109-6

The output frequency of this simple low-cost active voltage-controlled oscillator circuit is based upon the inherent frequency dependent characteristics of our operational amplifier.

The oscillator circuit shown uses a TL082 op amp. When power is applied, the circuit generates a sinusoidal wave. The frequency of oscillation can be changed by varying the bias supply.

110

Voltage Converter/Inverter Circuits

The sources of the following circuits are contained in the Sources section, which begins on page 675. The figure number in the box of each circuit correlates to the entry in the Sources section.

dc/dc CONVERTER

303 CIRCUITS *FIG. 110-1*

This low-power converter will supply about 100 mW of dc to a load and it is useful to isolate or derive dc voltages. It operates at around 200 kHz. L1 is wound on a 22-mm diameter × 13-mm high pot core with #32 magnet wire. The primary is 80 turns and the secondary is 80 turns (for 12-V nominal output). The two windings should be insulated for the expected voltage difference between input and output in insulation applications.

SIMPLE dc/ac INVERTER

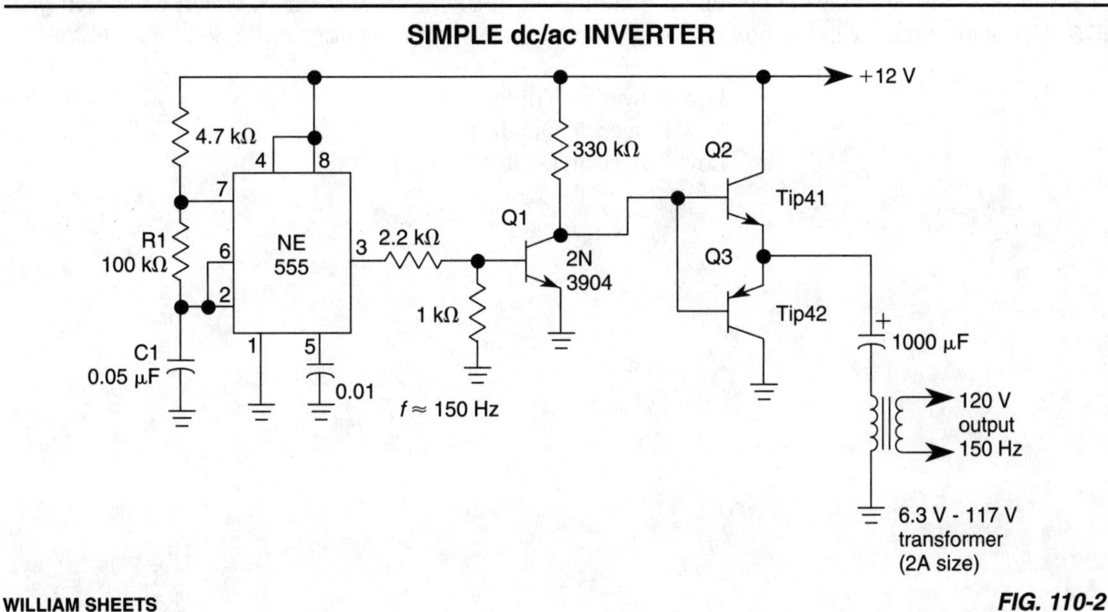

WILLIAM SHEETS *FIG. 110-2*

This dc-to-ac inverter is based on the popular 555. A 555 oscillator circuit drives a buffer amplifier consisting of Q1, Q2, and Q3. The circuit operates at 150 to 160 Hz. T1 can be a 6.3-V or 12.6-V filament transformer as applicable. The frequency can be changed by changing the values of R1 and/or C1.

111

Voltage Multiplier Circuits

The sources of the following circuits are contained in the Sources section, which begins on page 675. The figure number in the box of each circuit correlates to the entry in the Sources section.

LOW-POWER dc TRIPLER

+12 V
10 μF
1 kΩ
4
8
7
NE
555
22 kΩ
22 μF
3
2
6
22 μF
22 μF
0.01 μF
1
5
0.01 μF
22 μF
≈ + 34 V
out

WILLIAM SHEETS ***FIG. 111-1***

This dc voltage-tripler circuit based on the 555 can produce a dc output voltage equal to approximately 3× the dc supply voltage.

LOW-POWER dc QUADRUPLER

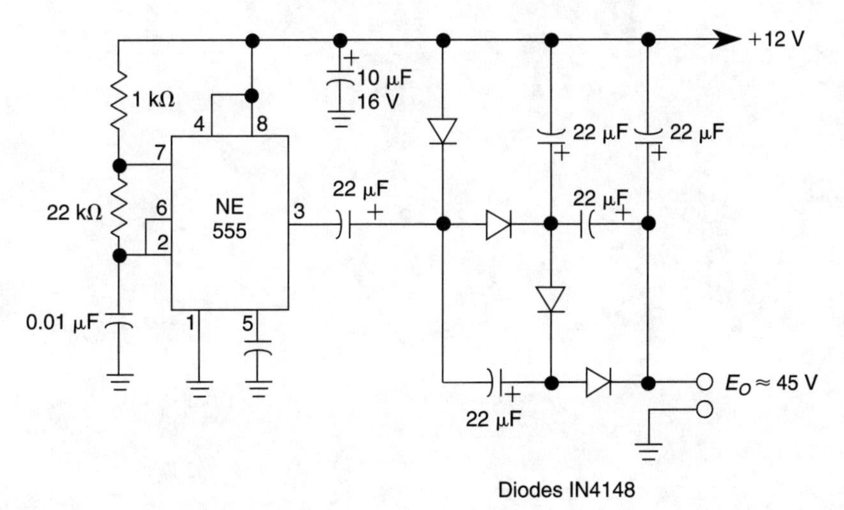

WILLIAM SHEETS ***FIG. 111-2***

This dc voltage-quadrupler circuit based on the 555 can produce a dc output voltage equal to approximately 4× the dc supply voltage.

LOW-POWER dc DOUBLER

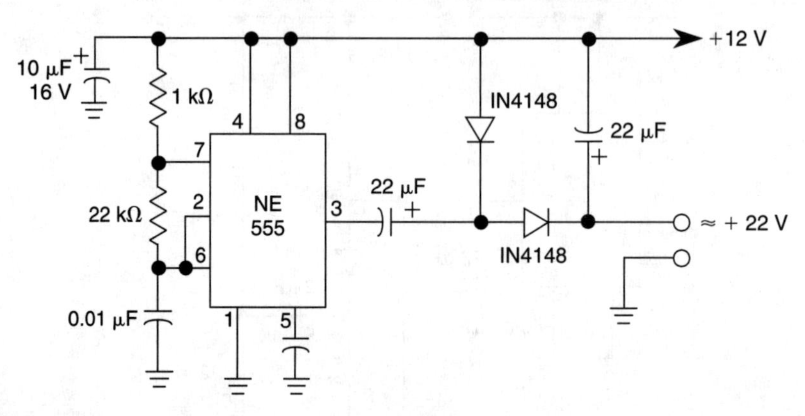

WILLIAM SHEETS

FIG. 111-3

This dc voltage-doubler circuit based on the 555 can produce a dc output voltage equal to approximately 2× the dc supply voltage.

112

Window Comparator and Discriminator Circuits

The sources of the following circuits are contained in the Sources section, which begins on page 675. The figure number in the box of each circuit correlates to the entry in the Sources section.

WINDOW COMPARATOR

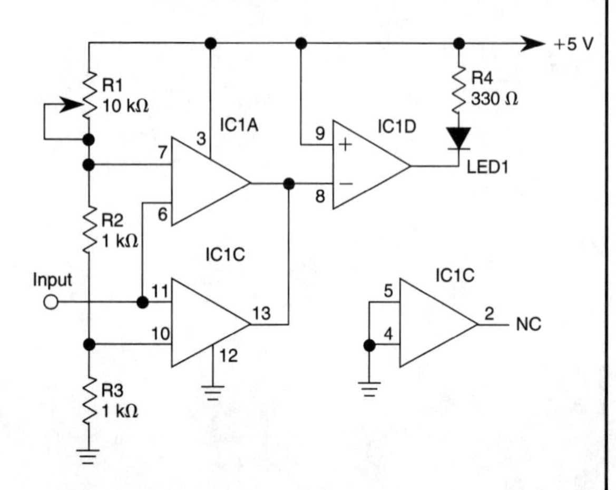

WILLIAM SHEETS *FIG. 112-1*

IC1-c functions as a noninverting comparator, and IC1-a operates as an inverting comparator. Potentiometer R1 and fixed resistors R2 and R3 form a divider chain that delivers slightly different voltages to the two comparators. These voltages define the upper and lower limits of the circuit's switching "window," which can be changed easily by varying R2 and R3. The LED glows only when the input voltage falls within the window region.

MULTIPLE-APERTURE WINDOW DISCRIMINATOR

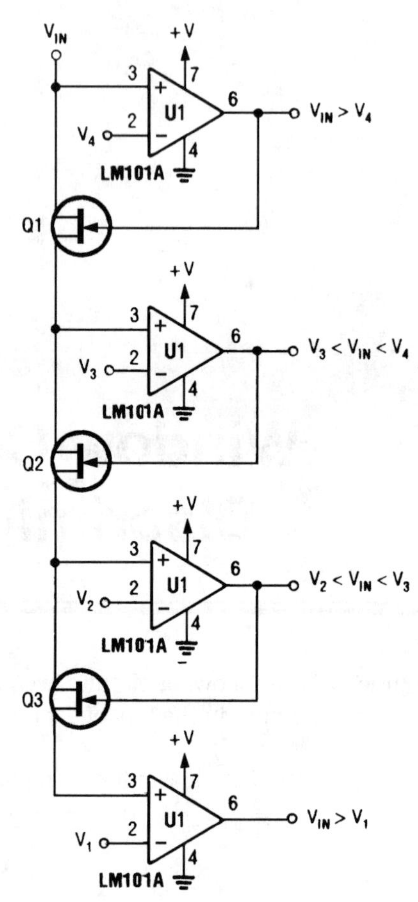

POPULAR ELECTRONICS *FIG. 112-2*

V1 through V4 are reference voltages that are derived from separate sources or from a common voltage divider.

Sources

Chapter 1

Fig. 1-1. Reprinted with permission from Popular Electronics, 1/92, p. 80. (c) Copyright Gernsback Publications, Inc., 1992.

Fig. 1-2. Reprinted with permission from Popular Electronics, 1/92, p. 80. (c) Copyright Gernsback Publications, Inc., 1992.

Fig. 1-3. Reprinted with permission from Popular Electronics, 1/92, p. 79. (c) Copyright Gernsback Publications, Inc., 1992.

Fig. 1-4. Reprinted with permission from Popular Electronics, 1/92, p. 79. (c) Copyright Gernsback Publications, Inc., 1992.

Fig. 1-5. Reprinted with permission from Popular Electronics, 1/92, p. 79. (c) Copyright Gernsback Publications, Inc., 1992.

Fig. 1-6. Reprinted with permission from Popular Electronics, 2/92, pp. 65-66. (c) Copyright Gernsback Publications, Inc., 1992.

Fig. 1-7. Reprinted with permission from Popular Electronics, 11/93, p. 53. (c) Copyright Gernsback Publications, Inc., 1993.

Fig. 1-8. Reprinted with permission from PE Hobbyist Handbook, 1992, pp. 93-94. (c) Copyright Gernsback Publications, Inc., 1992.

Fig. 1-9. William Sheets.

Fig. 1-10. Reprinted with permission from Popular Electronics, 2/92, pp. 70-71. (c) Copyright Gernsback Publications, Inc., 1992.

Fig. 1-11. Reprinted with permission from Popular Electronics, 12/92, p. 68. (c) Copyright Gernsback Publications, Inc., 1992.

Fig. 1-12. William Sheets.

Fig. 1-13. Reprinted with permission from PE Hobbyist Handbook, 1991, pp. 31-32. (c) Copyright Gernsback Publications, Inc., 1991.

Fig. 1-14. Reprinted with permission from Popular Electronics, 8/92, p. 71. (c) Copyright Gernsback Publications, Inc., 1992.

Fig. 1-15. Reprinted with permission from Popular Electronics, 2/92, p. 66. (c) Copyright Gernsback Publications, Inc., 1992.

Fig. 1-16. William Sheets.

Fig. 1-17. Reprinted with permission from Electronics Now, 7/92, p. 66. (c) Copyright Gernsback Publications, Inc., 1992.

Fig. 1-18. William Sheets.

Fig. 1-19. William Sheets.

Fig. 1-20. William Sheets.

Fig. 1-21. William Sheets.

Fig. 1-22. William Sheets.

Fig. 1-23. William Sheets.

Fig. 1-24. Reprinted with permission from Popular Electronics, 3/93, p. 42. (c) Copyright Gernsback Publications, Inc., 1993.

Fig. 1-25. Reprinted with permission from PE Hobbyist Handbook, 1991, pp. 19-20. (c) Copyright Gernsback Publications, Inc., 1991.

Fig. 1-26. Reprinted with permission from Popular Electronics, 1/92, p. 78. (c) Copyright Gernsback Publications, Inc., 1992.

Chapter 2

Fig. 2-1. Reprinted with permission from Popular Electronics, Fact Card 255, (c) Copyright Gernsback Publications, Inc.

Fig. 2-2. Reprinted with permission from Popular Electronics, Fact Card 254, (c) Copyright Gernsback Publications, Inc.

Fig. 2-3. Reprinted with permission from Popular Electronics, Fact Card 254, (c) Copyright Gernsback Publications, Inc.

Fig. 2-4. Reprinted with permission from Popular Electronics, Fact Card 253, (c) Copyright Gernsback Publications, Inc.

Fig. 2-5. Reprinted with permission from Popular Electronics, Fact Card 254, (c) Copyright Gernsback Publications, Inc.

Fig. 2-6. Reprinted with permission from Popular Electronics, Fact Card 254, (c) Copyright Gernsback Publications, Inc.

Fig. 2-7. Reprinted with permission from Popular Electronics, Fact Card 253, (c) Copyright Gernsback Publications, Inc.

Fig. 2-8. William Sheets.

Fig. 2-9. Reprinted with permission from Popular Electronics, Fact Card 253, (c) Copyright Gernsback Publications, Inc.

Fig. 2-10. William Sheets.

Fig. 2-11. William Sheets.

Fig. 2-12. William Sheets.

Fig. 2-13. William Sheets.

Fig. 2-14. William Sheets.

Fig. 2-15. Reprinted with permission from Electronics Now, 7/92, p. 36. (c) Copyright Gernsback Publications, Inc., 1992.

Fig. 2-16. Reprinted with permission from Popular Electronics, Fact Card 255, (c) Copyright Gernsback Publications, Inc.

Fig. 2-17. William Sheets.

Fig. 2-18. William Sheets.

Fig. 2-19. Reprinted with permission from Popular Electronics, Fact Card 223, (c) Copyright Gernsback Publications, Inc.

Fig. 2-20. William Sheets.

Fig. 2-21. William Sheets.

Fig. 2-22. Reprinted with permission from Popular Electronics, Fact Card 264, (c) Copyright Gernsback Publications, Inc.

Fig. 2-23. Reprinted with permission from Radio-Electronics, 6/92, p. 59. (c) Copyright Gernsback Publications, Inc., 1992.

Fig. 2-24. Reprinted with permission from 73 Amateur Radio Today, 4/92, p. 71.

Fig. 2-25. William Sheets.

Fig. 2-26. Reprinted with permission from National Semiconductor, Linear Edge, Spring 1992.

Fig. 2-27. Reprinted with permission from Popular Electronics, Fact Card 206, (c) Copyright Gernsback Publications, Inc.

Fig. 2-28. Reprinted with permission from Popular Electronics, 9/93, p. 47. (c) Copyright Gernsback Publications, Inc., 1993.

Chapter 3

Fig. 3-1. Reprinted with permission from Electronic Design, 3/93, p. 67.

Fig. 3-2. Reprinted with permission from 73 Amateur Radio Today, 7/92, p. 42.

Chapter 4

Fig. 4-1. Reprinted with permission from Popular Electronics, 12/91, p. 63. (c) Copyright Gernsback Publications, Inc., 1991.

Fig. 4-2. Reprinted with permission from Elector Electronics USA, 10/92, p. 14.

Fig. 4-3. Reprinted with permission from 73 Amateur Radio Today, 5/90, p. 47.

Fig. 4-4. Reprinted with permission from Practical Wireless, 6/91, p. 36.

Fig. 4-5. Reprinted with permission from Elektor Electronics, 12/91, pp. 88-89.

Fig. 4-6. Reprinted with permission from Popular Electronics, 6/93, p. 55. (c) Copyright Gernsback Publications, Inc., 1993.

Fig. 4-7. Reprinted with permission from 73 Amateur Radio Today, 7/92, p. 34.

Fig. 4-8. Reprinted with permission from Electronics Hobbyists Handbook, 1993, p. 89.

Fig. 4-9. Reprinted with permission from 73 Amateur Radio Today, 10/92, p. 28.

Chapter 5

Fig. 5-1. Reprinted with permission from PE Hobbyist Handbook, 1991, pp. 65-66. (c) Copyright Gernsback Publications, Inc., 1991.

Fig. 5-2. Reprinted with permission from Electronics Now, 11/92, p. 42. (c) Copyright Gernsback Publications, Inc., 1992.

Fig. 5-3. Reprinted with permission from Electronics Now, 12/92, p. 14. (c) Copyright Gernsback Publications, Inc., 1992.
Fig. 5-4. Reprinted with permission from Silicon Chip.
Fig. 5-5. Reprinted with permission from Electronics Now, 11/92, p. 43. (c) Copyright Gernsback Publications, Inc., 1992.
Fig. 5-6. Reprinted with permission from Electronics Now, 11/92, p. 41. (c) Copyright Gernsback Publications, Inc., 1992.
Fig. 5-7. Reprinted with permission from 303 Circuits, p. 42.
Fig. 5-8. Reprinted with permission from Electronics Now, 11/92, p. 39. (c) Copyright Gernsback Publications, Inc., 1992.
Fig. 5-9. Reprinted with permission from Electronics Now, 11/92, p. 39. (c) Copyright Gernsback Publications, Inc., 1992.
Fig. 5-10. Reprinted with permission from R-E Experimenters Handbook, 1987, p. 74.
Fig. 5-11. Reprinted with permission from Popular Electronics, 1/92, p. 36. (c) Copyright Gernsback Publications, Inc., 1992.
Fig. 5-12. Reprinted with permission from Popular Electronics, Fact Card 243, (c) Copyright Gernsback Publications, Inc.
Fig. 5-13. Reprinted with permission from Radio-Electronics, 1/92, p. 35. (c) Copyright Gernsback Publications, Inc., 1992.
Fig. 5-14. Reprinted with permission from Popular Electronics, 4/92, p. 69. (c) Copyright Gernsback Publications, Inc., 1992.
Fig. 5-15. Reprinted with permission from Electronic Design, 3/89, p. 100.

Chapter 6

Fig. 6-1. Reprinted with permission from 303 Circuits, p. 22.
Fig. 6-2. Reprinted with permission from 303 Circuits, p. 10.
Fig. 6-3. William Sheets.
Fig. 6-4. Reprinted with permission from 303 Circuits, p. 40.
Fig. 6-5. Reprinted with permission from Popular Electronics, 6/92, p. 68. (c) Copyright Gernsback Publications, Inc., 1992.
Fig. 6-6. Reprinted with permission from 303 Circuits, pp. 12-13.
Fig. 6-7. William Sheets.
Fig. 6-8. Reprinted with permission from R-E Experimenters Handbook, 1987, p. 74.
Fig. 6-9. Reprinted with permission from Popular Electronics, Fact Card 267, (c) Copyright Gernsback Publications, Inc.
Fig. 6-10. Reprinted with permission from Popular Electronics, Fact Card 267, (c) Copyright Gernsback Publications, Inc.
Fig. 6-11. Reprinted with permission from R-E Experimenters Handbook, 1992, p. 37.
Fig. 6-12. Reprinted with permission from Popular Electronics, Fact Card 263, (c) Copyright Gernsback Publications, Inc.

Chapter 7

Fig. 7-1. Reprinted with permission from 73 Amateur Radio Today, 8/92, p. 36.
Fig. 7-2. William Sheets.
Fig. 7-3. Reprinted with permission from RF Communications Handbook, 1989, pp. 2-14.

Chapter 8

Fig. 8-1. Reprinted with permission from Popular Electronics, 9/92, p. 33. (c) Copyright Gernsback Publications, Inc., 1992.
Fig. 8-2. Reprinted with permission from Popular Electronics, 3/92, p. 75. (c) Copyright Gernsback Publications, Inc., 1992.
Fig. 8-3. William Sheets.
Fig. 8-4. Reprinted with permission from Electronics Now, 9/93, p. 63. (c) Copyright Gernsback Publications, Inc., 1993.
Fig. 8-5. William Sheets.
Fig. 8-6. Reprinted with permission from Silicon Chip, p. 27.
Fig. 8-7. Reprinted with permission from PE Hobbyist Handbook, 1990, pp. 86-87. (c) Copyright Gernsback Publications, Inc., 1990.
Fig. 8-8. Reprinted with permission from Electronics Now, 11/92, p. 59. (c) Copyright Gernsback Publications, Inc., 1992.
Fig. 8-9. Reprinted with permission from Electronics Hobbyist Handbook, 1993, p. 22.
Fig. 8-10. Reprinted with permission from Popular Electronics, 10/93, p. 64. (c) Copyright Gernsback Publications, Inc., 1993.
Fig. 8-11. Reprinted with permission from PE Hobbyist Handbook, p. 73. (c) Copyright Gernsback Publications, Inc.
Fig. 8-12. Reprinted with permission from Popular Electronics, 3/93, p. 62. (c) Copyright Gernsback Publications, Inc., 1993.

Fig. 8-13. Reprinted with permission from Popular Electronics, 11/93, p. 72. (c) Copyright Gernsback Publications, Inc., 1993.
Fig. 8-14. Reprinted with permission from Popular Electronics, 3/92, p. 73. (c) Copyright Gernsback Publications, Inc., 1992.
Fig. 8-15. Reprinted with permission from Popular Electronics, 3/92, p. 75. (c) Copyright Gernsback Publications, Inc., 1992.
Fig. 8-16. Reprinted with permission from Popular Electronics, 3/92, p. 75. (c) Copyright Gernsback Publications, Inc., 1992.
Fig. 8-17. Reprinted with permission from Popular Electronics, 3/92, p. 75. (c) Copyright Gernsback Publications, Inc., 1992.
Fig. 8-18. Reprinted with permission from Popular Electronics, 12/92, p. 71. (c) Copyright Gernsback Publications, Inc., 1992.
Fig. 8-19. William Sheets.
Fig. 8-20. Reprinted with permission from Radio-Electronics, 12/91, p. 75. (c) Copyright Gernsback Publications, Inc., 1991.
Fig. 8-21. Reprinted with permission from Radio-Electronics, 5/92, p. 82. (c) Copyright Gernsback Publications, Inc., 1992.
Fig. 8-22. Reprinted with permission from Popular Electronics, 4/92, p. 71. (c) Copyright Gernsback Publications, Inc., 1992.
Fig. 8-23. William Sheets.
Fig. 8-24. Reprinted with permission from R-E Experimenters Handbook, 1989, p. 158.

Chapter 9

Fig. 9-1. Reprinted with permission from Popular Electronics, 6/93, p. 76. (c) Copyright Gernsback Publications, Inc., 1993.
Fig. 9-2. Reprinted with permission from R-E Experimenters Handbook, 1992, p. 122.
Fig. 9-3. Reprinted with permission from Popular Electronics, 4/92, p. 71 & 88. (c) Copyright Gernsback Publications, Inc., 1992.
Fig. 9-4. Reprinted with permission from National Semiconductor, Linear Edge, Summer 1992.
Fig. 9-5. Reprinted with permission from Radio-Electronics, 5/92, p. 12. (c) Copyright Gernsback Publications, Inc., 1992.

Chapter 10

Fig. 10-1. Reprinted with permission from PE Hobbyist Handbook, 1991, pp. 44-45. (c) Copyright Gernsback Publications, Inc., 1991.
Fig. 10-2. Reprinted with permission from Electronics Now, 7/92, pp. 57-62. (c) Copyright Gernsback Publications, Inc., 1992.
Fig. 10-3. Reprinted with permission from Popular Electronics, Fact Card 198. (c) Copyright Gernsback Publications, Inc.
Fig. 10-4. Reprinted with permission from 73 Amateur Radio Today, 5/92, p. 26.
Fig. 10-5. Reprinted with permission from Electronic Design, 7/93, p. 78.
Fig. 10-6. Reprinted with permission from Popular Electronics, Fact Card 198. (c) Copyright Gernsback Publications, Inc.
Fig. 10-7. Reprinted with permission from Popular Electronics, 11/91, p. 20. (c) Copyright Gernsback Publications, Inc., 1991.
Fig. 10-8. Reprinted with permission from Linear Technology, Design Note 60.
Fig. 10-9. Reprinted with permission from PE Hobbyist Handbook, 1991, pp. 63-64. (c) Copyright Gernsback Publications, Inc., 1991.
Fig. 10-10. Reprinted with permission from Elektor Electronics, 12/91, p. 72.
Fig. 10-11. Reprinted with permission from Elektor Electronics USA, 12/91, p. 36.

Chapter 11

Fig. 11-1. Reprinted with permission from Popular Electronics, 8/93, p. 79. (c) Copyright Gernsback Publications, Inc., 1993.
Fig. 11-2. Reprinted with permission from Popular Electronics, 6/93, p. 70. (c) Copyright Gernsback Publications, Inc., 1993.
Fig. 11-3. Reprinted with permission from Popular Electronics, 8/93, p. 79. (c) Copyright Gernsback Publications, Inc., 1993.
Fig. 11-4. Reprinted with permission from Popular Electronics, 8/93, p. 79. (c) Copyright Gernsback Publications, Inc., 1993.
Fig. 11-5. Reprinted with permission from National Semiconductor, Linear Edge, Issue #5.
Fig. 11-6. Reprinted with permission from Electronics Now, 9/92, p. 96. (c) Copyright Gernsback Publications, Inc., 1992.
Fig. 11-7. William Sheets.

Chapter 12

Fig. 12-1. Reprinted with permission from Electronics Hobbyist Handbook, 1993, p. 58.
Fig. 12-2. Reprinted with permission from Electronics Hobbyist Handbook, 1993, p. 59.

Chapter 13

Fig. 13-1. Reprinted with permission from PE Hobbyist Handbook, 1992, p. 49. (c) Copyright Gernsback Publications, Inc.

Chapter 14

Fig. 14-1. Reprinted with permission from Popular Electronics, 3/93, p. 44. (c) Copyright Gernsback Publications, Inc., 1993.

Fig. 14-2. Reprinted with permission from 73 Amateur Radio Today, 5/92, p. 20.

Fig. 14-3. Reprinted with permission from 73 Amateur Radio Today, 5/92, p. 18.

Fig. 14-4. Reprinted with permission from Popular Electronics, Fact Card 206, (c) Copyright Gernsback Publications, Inc.

Fig. 14-5. William Sheets.

Fig. 14-6. Reprinted with permission from Popular Electronics, 10/93, p. 73. (c) Copyright Gernsback Publications, Inc., 1993.

Fig. 14-7. Reprinted with permission from Popular Electronics, Fact Card 206, (c) Copyright Gernsback Publications, Inc.

Chapter 15

Fig. 15-1. Reprinted with permission from PE Hobbyist Handbook, 1991, pp. 24-26. (c) Copyright Gernsback Publications, Inc., 1991.

Chapter 16

Fig. 16-1. Reprinted with permission from Popular Electronics, 3/92, p. 60. (c) Copyright Gernsback Publications, Inc., 1992.

Fig. 16-2. Reprinted with permission from Electronics Now, 7/92, p. 51. (c) Copyright Gernsback Publications, Inc., 1992.

Fig. 16-3. Reprinted with permission from Electronic Design, 3/93, pp. 67-68.

Chapter 17

Fig. 17-1. Reprinted with permission from Elektor Electronics, 12/91, pp. 78-79.

Fig. 17-2. Reprinted with permission from Electronics Hobbyist Handbook, 1993, p. 84.

Fig. 17-3. Reprinted with permission from Electronic Design.

Fig. 17-4. Reprinted with permission from Popular Electronics, 11/91, p. 20. (c) Copyright Gernsback Publications, Inc., 1991.

Fig. 17-5. Reprinted with permission from Popular Electronics, 6/93, p. 72. (c) Copyright Gernsback Publications, Inc., 1993.

Chapter 18

Fig. 18-1. Reprinted with permission from Popular Electronics, 12/91, p. 58. (c) Copyright Gernsback Publications, Inc., 1991.

Fig. 18-2. Reprinted with permission from PE Hobbyist Handbook, 1990, pp. 34-36. (c) Copyright Gernsback Publications, Inc., 1990.

Fig. 18-3. Reprinted with permission from Electronic Design, 5/92, p. 91.

Fig. 18-4. Reprinted with permission from Popular Electronics, 6/92, p. 57. (c) Copyright Gernsback Publications, Inc., 1992.

Fig. 18-5. Reprinted with permission from R-E Experimenters Handbook, 1992, p. 92.

Fig. 18-6. Reprinted with permission from Radio-Electronics, 3/92, p. 50. (c) Copyright Gernsback Publications, Inc., 1992.

Fig. 18-7. Reprinted with permission from 303 Circuits, p. 197.

Fig. 18-8. Reprinted with permission from Radio-Electronics, 2/92, p. 42. (c) Copyright Gernsback Publications, Inc., 1992.

Fig. 18-9. Reprinted with permission from Popular Electronics, 12/91, p. 58. (c) Copyright Gernsback Publications, Inc., 1991.

Fig. 18-10. Reprinted with permission from National Semiconductor, Linear Edge, Issue #4, Summer 1992.

Fig. 18-11. Reprinted with permission from 73 Amateur Radio Today, 2/93, p. 28.

Fig. 18-12. William Sheets.

Fig. 18-13. Reprinted with permission from 73 Amateur Radio Today, 3/92, p. 24.

Fig. 18-14. Reprinted with permission from Popular Electronics, 7/92, p. 70. (c) Copyright Gernsback Publications, Inc., 1992.

Fig. 18-15. Reprinted with permission from Electronics Now, 9/92, p. 79. (c) Copyright Gernsback Publications, Inc., 1992.

Fig. 18-16. Reprinted with permission from National Semiconductor, Linear Edge, Spring 1992.

Fig. 18-17. Reprinted with permission from Popular Electronics, Fact Card 259, (c) Copyright Gernsback Publications, Inc.

Fig. 18-18. Reprinted with permission from Popular Electronics, Fact Card 257, (c) Copyright Gernsback Publications, Inc.

Fig. 18-19. Reprinted with permission from Linear Technology Corporation, 1993, Design Note 69.

Chapter 19

Fig. 19-1. Reprinted with permission from Electronics Hobbyist Handbook, 1993, p. 47.

Fig. 19-2. Reprinted with permission from PE Hobbyist Handbook, 1990, p. 101. (c) Copyright Gernsback Publications, Inc., 1990.

Chapter 20

Fig. 20-1. Reprinted with permission from RF Design, 5/92, p. 80.

Fig. 20-2. William Sheets.

Fig. 20-3. Reprinted with permission from Electronic Design, 11/92, p. 61.

Fig. 20-4. Reprinted with permission from Electronics Now, 12/92, p. 12. (c) Copyright Gernsback Publications, Inc., 1992.

Fig. 20-5. Reprinted with permission from Radio-Electronics, 2/92, p. 89. (c) Copyright Gernsback Publications, Inc., 1992.

Fig. 20-6. Reprinted with permission from 73 Amateur Radio Today, 5/92, p. 64.

Fig. 20-7. Reprinted with permission from 73 Amateur Radio Today, 8/92, p. 48.

Fig. 20-8. Reprinted with permission from 73 Amateur Radio Today, 1/92, p. 22.

Fig. 20-9. Reprinted with permission from 73 Amateur Radio Today, 7/92, p. 60.

Fig. 20-10. William Sheets.

Fig. 20-11. Reprinted with permission from Popular Electronics, Fact Card 229, (c) Copyright Gernsback Publications, Inc.

Chapter 21

Fig. 21-1. Reprinted with permission from Electronics Hobbyist Handbook, 1993, p. 78.

Fig. 21-2. Reprinted with permission from National Semiconductor, Linear Edge, Issue #5.

Fig. 21-3. Reprinted with permission from Popular Electronics, Fact Card 257, (c) Copyright Gernsback Publications, Inc.

Fig. 21-4. Reprinted with permission from National Semiconductor, Linear Edge, Issue #5.

Chapter 22

Fig. 22-1. Reprinted with permission from Silicon Chip.

Fig. 22-2. Reprinted with permission from Electronics Now, 12/92, p. 49. (c) Copyright Gernsback Publications, Inc., 1992.

Chapter 23

Fig. 23-1. Reprinted with permission from Elektor Electronics, 12/91, p. 81.

Chapter 24

Fig. 24-1. Reprinted with permission from PE Hobbyist Handbook, 1992, p. 70. (c) Copyright Gernsback Publications, Inc., 1992.

Fig. 24-2. Reprinted with permission from Electronics Now, 12/92, p. 61. (c) Copyright Gernsback Publications, Inc., 1992.

Fig. 24-3. Reprinted with permission from Electronics Now, 12/92, p. 65. (c) Copyright Gernsback Publications, Inc., 1992.

Fig. 24-4. Reprinted with permission from Electronic Design, 5/92, p. 95.

Fig. 24-5. Reprinted with permission from Linear Technology, Design Note 61.

Fig. 24-6. Reprinted with permission from Linear Technology, Design Note 61.

Fig. 24-7. Reprinted with permission from Linear Technology, Design Note 61.

Fig. 24-8. Reprinted with permission from Popular Electronics, Fact Card, 270, (c) Copyright Gernsback Publications, Inc.

Fig. 24-9. Reprinted with permission from Popular Electronics, Fact Card, 269, (c) Copyright Gernsback Publications, Inc.

Fig. 24-10. Reprinted with permission from Popular Electronics, Fact Card, 270, (c) Copyright Gernsback Publications, Inc.

Fig. 24-11. Reprinted with permission from 73 Amateur Radio Today, 8/89, p. 48.

Chapter 25

Fig. 25-1. Reprinted with permission from PE Hobbyist Handbook, 1992, pp. 63-64. (c) Copyright Gernsback Publications, Inc., 1992.

Fig. 25-2. Reprinted with permission from 303 Circuits, p. 14.

Fig. 25-3. Reprinted with permission from R-E Experimenters Handbook, 1989, p. 58.

Fig. 25-4. Reprinted with permission from R-E Experimenters Handbook, 1989, p. 57.

Fig. 25-5. Reprinted with permission from R-E Experimenters Handbook, 1989, p. 57.

Fig. 25-6. Reprinted with permission from Elector Electronics USA, 12/91, p. 36.

Chapter 26

Fig. 26-1. Reprinted with permission from Radio-Electronics, 5/92, p. 72. (c) Copyright Gernsback Publications, Inc., 1992.

Fig. 26-2. Reprinted with permission from Radio-Electronics, 5/92, p. 71. (c) Copyright Gernsback Publications, Inc., 1992.

Fig. 26-3. Reprinted with permission from Electronic Design, 8/92, p. 70.

Fig. 26-4. Reprinted with permission from Radio-Electronics, 5/92, p. 69. (c) Copyright Gernsback Publications, Inc., 1992.

Fig. 26-5. Reprinted with permission from Radio-Electronics, 5/92, p. 69. (c) Copyright Gernsback Publications, Inc., 1992.

Fig. 26-6. Reprinted with permission from Radio-Electronics, 5/92, p. 70. (c) Copyright Gernsback Publications, Inc., 1992.

Fig. 26-7. Reprinted with permission from Radio-Electronics, 5/92, p. 69. (c) Copyright Gernsback Publications, Inc., 1992.

Fig. 26-8. Reprinted with permission from Radio-Electronics, 5/92, p. 70. (c) Copyright Gernsback Publications, Inc., 1992.

Fig. 26-9. Reprinted with permission from Radio-Electronics, 5/92, p. 71. (c) Copyright Gernsback Publications, Inc., 1992.

Fig. 26-10. Reprinted with permission from Radio-Electronics, 5/92, p. 70. (c) Copyright Gernsback Publications, Inc., 1992.

Fig. 26-11. Reprinted with permission from Radio-Electronics, 5/92, p. 71. (c) Copyright Gernsback Publications, Inc., 1992.

Fig. 26-12. Reprinted with permission from Radio-Electronics, 5/92, p. 70. (c) Copyright Gernsback Publications, Inc., 1992.

Chapter 27

Fig. 27-1. Reprinted with permission from Popular Electronics, 9/93, p. 42. (c) Copyright Gernsback Publications, Inc., 1993.

Fig. 27-2. Reprinted with permission from 303 Circuits, p. 266.

Fig. 27-3. Reprinted with permission from Popular Electronics, 6/93, p. 63. (c) Copyright Gernsback Publications, Inc., 1993.

Chapter 28

Fig. 28-1. Reprinted with permission from R-E Experimenters Handbook, 1992, p. 65.

Fig. 28-2. Reprinted with permission from R-E Experimenters Handbook, 1992, p. 65.

Chapter 29

Fig. 29-1. Reprinted with permission from 73 Amateur Radio Today, 3/92, p. 44.

Fig. 29-2. Reprinted with permission from 73 Amateur Radio Today, 1/92

Fig. 29-3. Reprinted with permission from 73 Amateur Radio Today, 5/90, p. 80.

Fig. 29-4. Reprinted with permission from 73 Amateur Radio Today, 9/90, p. 9.

Fig. 29-5. Reprinted with permission from Popular Electronics, 11/93, p. 73. (c) Copyright Gernsback Publications, Inc., 1993.

Chapter 30

Fig. 30-1. Reprinted with permission from Popular Electronics, Fact Card, 226, (c) Copyright Gernsback Publications, Inc.

Fig. 30-2. Reprinted with permission from Popular Electronics, Fact Card, 226, (c) Copyright Gernsback Publications, Inc.

Fig. 30-3. Reprinted with permission from Popular Electronics, Fact Card, 226, (c) Copyright Gernsback Publications, Inc.

Fig. 30-4. Reprinted with permission from Popular Electronics, Fact Card, 227. (c) Copyright Gernsback Publications, Inc.

Fig. 30-5. Reprinted with permission from Popular Electronics, Fact Card, 227. (c) Copyright Gernsback Publications, Inc.

Fig. 30-6. Reprinted with permission from Popular Electronics, Fact Card, 227. (c) Copyright Gernsback Publications, Inc.

Fig. 30-7. Reprinted with permission from Popular Electronics, Fact Card, 227. (c) Copyright Gernsback Publications, Inc.

Fig. 30-8. Reprinted with permission from Popular Electronics, Fact Card, 228. (c) Copyright Gernsback Publications, Inc.

Fig. 30-9. Reprinted with permission from Popular Electronics, Fact Card, 228. (c) Copyright Gernsback Publications, Inc.

Fig. 30-10. Reprinted with permission from Popular Electronics, Fact Card, 228. (c) Copyright Gernsback Publications, Inc.

Fig. 30-11. Reprinted with permission from Popular Electronics, Fact Card, 228. (c) Copyright Gernsback Publications, Inc.

Fig. 30-12. Reprinted with permission from Popular Electronics, Fact Card, 225. (c) Copyright Gernsback Publications, Inc.
Fig. 30-13. Reprinted with permission from Popular Electronics, Fact Card, 224. (c) Copyright Gernsback Publications, Inc.
Fig. 30-14. Reprinted with permission from Popular Electronics, Fact Card, 223. (c) Copyright Gernsback Publications, Inc.
Fig. 30-15. Reprinted with permission from Popular Electronics, Fact Card, 231. (c) Copyright Gernsback Publications, Inc.
Fig. 30-16. Reprinted with permission from Popular Electronics, Fact Card, 231. (c) Copyright Gernsback Publications, Inc.
Fig. 30-17. Reprinted with permission from Electronic Design, 2/93, p. 75.
Fig. 30-18. Reprinted with permission from Electronics Now, 8/93, p. 73. (c) Copyright Gernsback Publications, Inc., 1993.
Fig. 30-19. Reprinted with permission from Electronics Now, 8/93, p. 72. (c) Copyright Gernsback Publications, Inc., 1993.
Fig. 30-20. Reprinted with permission from Electronics Now, 4/93, p. 72. (c) Copyright Gernsback Publications, Inc., 1993.
Fig. 30-21. Reprinted with permission from Electronics Now, 8/93, p. 70. (c) Copyright Gernsback Publications, Inc., 1993.
Fig. 30-22. Reprinted with permission from Popular Electronics, 6/92, p. 68. (c) Copyright Gernsback Publications, Inc., 1992.
Fig. 30-23. Reprinted with permission from Electronics Now, 8/93, p. 71. (c) Copyright Gernsback Publications, Inc., 1993.
Fig. 30-24. Reprinted with permission from Popular Electronics, 6/92, p. 67. (c) Copyright Gernsback Publications, Inc., 1992.
Fig. 30-25. Reprinted with permission from Electronics Now, 8/93, p. 71. (c) Copyright Gernsback Publications, Inc., 1993.
Fig. 30-26. Reprinted with permission from Electronics Now, 8/93, p. 71. (c) Copyright Gernsback Publications, Inc., 1993.
Fig. 30-27. Reprinted with permission from Linear Technology Corporation, 1993, Advertisement, Circle No. 51.
Fig. 30-28. Reprinted with permission from Electronics Now, 8/93, p. 70. (c) Copyright Gernsback Publications, Inc., 1993.
Fig. 30-29. William Sheets.
Fig. 30-30. William Sheets.
Fig. 30-31. William Sheets.
Fig. 30-32. William Sheets.
Fig. 30-33. William Sheets.
Fig. 30-34. William Sheets.
Fig. 30-35. Reprinted with permission from Electronic Design, 7/92, p. 62.
Fig. 30-36. Reprinted with permission from 303 Circuits, p. 185.
Fig. 30-37. Reprinted with permission from National Semiconductor, Linear Edge, Summer 1992.
Fig. 30-38. Reprinted with permission from Popular Electronics, Fact Card 224. (c) Copyright Gernsback Publications, Inc.
Fig. 30-39. Reprinted with permission from Popular Electronics, Fact Card 231. (c) Copyright Gernsback Publications, Inc.
Fig. 30-40. Reprinted with permission from Popular Electronics, Fact Card 242. (c) Copyright Gernsback Publications, Inc.
Fig. 30-41. William Sheets.

Chapter 31

Fig. 31-1. Reprinted with permission from R-E Experimenters Handbook, 1989, p. 159.
Fig. 31-2. Reprinted with permission from Popular Electronics, 12/91, p. 80. (c) Copyright Gernsback Publications, Inc., 1991.
Fig. 31-3. William Sheets.
Fig. 31-4. William Sheets.
Fig. 31-5. Reprinted with permission from PE Hobbyist Handbook, 1991, p. 10. (c) Copyright Gernsback Publications, Inc., 1991.
Fig. 31-6. William Sheets.
Fig. 31-7. Reprinted with permission from Radio-Electronics, 11/89, p. 12. (c) Copyright Gernsback Publications, Inc., 1989.
Fig. 31-8. Reprinted with permission from R-E Experimenters Handbook, p. 28.

Chapter 32

Fig. 32-1. Reprinted with permission from Electronic Engineering, 9/89, p. 30.

Chapter 33

Fig. 33-1. Reprinted with permission from Electronic Design, 7/93, p. 76.
Fig. 33-2. Reprinted with permission from Popular Electronics, Fact Card 254. (c) Copyright Gernsback Publications, Inc.

Fig. 33-3. Reprinted with permission from 73 Amateur Radio Today, 1/92, p. 28.

Fig. 33-4. Reprinted with permission from Popular Electronics, Fact Card 253. (c) Copyright Gernsback Publications, Inc.

Fig. 33-5. Reprinted with permission from Popular Electronics, 11/91, p. 22. (c) Copyright Gernsback Publications, Inc., 1991.

Fig. 33-6. Reprinted with permission from Elektor Electronics, 3/92, p. 58.

Fig. 33-7. William Sheets.

Fig. 33-8. William Sheets.

Fig. 33-9. Reprinted with permission from Popular Electronics, Fact Card 230. (c) Copyright Gernsback Publications, Inc.

Fig. 33-10. William Sheets.

Fig. 33-11. Reprinted with permission from Popular Electronics, Fact Card 257. (c) Copyright Gernsback Publications, Inc.

Fig. 33-12. Reprinted with permission from Popular Electronics, Fact Card 243. (c) Copyright Gernsback Publications, Inc.

Fig. 33-13. Reprinted with permission from Electronic Design, 7/93, p. 76.

Fig. 33-14. William Sheets.

Fig. 33-15. Reprinted with permission from Popular Electronics, Fact Card 258. (c) Copyright Gernsback Publications, Inc.

Chapter 34

Fig. 34-1. Reprinted with permission from Electronics Hobbyist Handbook, 1993, p. 33.

Fig. 34-2. Reprinted with permission from PE Hobbyist Handbook, 1991, p. 47. (c) Copyright Gernsback Publications, Inc., 1991.

Fig. 34-3. Reprinted with permission from PE Hobbyist Handbook, 1991, pp. 36-37. (c) Copyright Gernsback Publications, Inc., 1991.

Chapter 35

Fig. 35-1. Reprinted with permission from Popular Electronics, 11/93, p. 33. (c) Copyright Gernsback Publications, Inc., 1993.

Fig. 35-2. Reprinted with permission from Electronics Now, 7/93, p. 40. (c) Copyright Gernsback Publications, Inc., 1993.

Chapter 36

Fig. 36-1. Reprinted with permission from Popular Electronics, Fact Card, 268. (c) Copyright Gernsback Publications, Inc.

Chapter 37

Fig. 37-1. Reprinted with permission from Popular Electronics, 7/92, pp. 42-43. (c) Copyright Gernsback Publications, Inc., 1992.

Fig. 37-2. Reprinted with permission from PE Hobbyist Handbook, 1991, p. 59. (c) Copyright Gernsback Publications, Inc., 1991.

Chapter 38

Fig. 38-1. Reprinted with permission from Electronics Hobbyist Handbook, 1993, p. 81.

Fig. 38-2. Reprinted with permission from Popular Electronics, 7/93, p. 75. (c) Copyright Gernsback Publications, Inc., 1993.

Chapter 39

Fig. 39-1. Reprinted with permission from Popular Electronics, 6/92, p. 39. (c) Copyright Gernsback Publications, Inc., 1992.

Fig. 39-2. Reprinted with permission from Popular Electronics, 1/92, p. 24. (c) Copyright Gernsback Publications, Inc., 1992.

Fig. 39-3. Reprinted with permission from Radio-Electronics, 10/89, p. 43. (c) Copyright Gernsback Publications, Inc., 1989.

Fig. 39-4. Reprinted with permission from Popular Electronics, 1/92, p. 24. (c) Copyright Gernsback Publications, Inc., 1992.

Fig. 39-5. William Sheets.

Fig. 39-6. William Sheets.

Fig. 39-7. Reprinted with permission from Popular Electronics, 12/93, p. 32. (c) Copyright Gernsback Publications, Inc., 1993.

Fig. 39-8. Reprinted with permission from Popular Electronics, 12/93, p. 32. (c) Copyright Gernsback Publications, Inc., 1993.

Fig. 39-9. Reprinted with permission from PE Hobbyist Handbook, 1991, pp. 75-77. (c) Copyright Gernsback Publications, Inc., 1991.

Fig. 39-10. Reprinted with permission from Radio-Electronics, 10/89, p. 43. (c) Copyright Gernsback Publications, Inc., 1989.

Fig. 39-11. Reprinted with permission from Popular Electronics, 3/93, p. 43. (c) Copyright Gernsback Publications, Inc., 1993.

Fig. 39-12. Reprinted with permission from Electronics Now, 5/93, p. 12. (c) Copyright Gernsback Publications, Inc., 1993.

Fig. 39-13. Reprinted with permission from Electronics Now, 3/93, p. 83. (c) Copyright Gernsback Publications, Inc., 1993.

Chapter 40

Fig. 40-1. Reprinted with permission from Popular Electronics, 4/92, p. 88. (c) Copyright Gernsback Publications, Inc., 1992.

Fig. 40-2. Reprinted with permission from Popular Electronics, 4/92, pp. 70-71. (c) Copyright Gernsback Publications, Inc., 1992.

Chapter 41

Fig. 41-1. Reprinted with permission from National Semiconductor, Linear Edge, Spring 1992.

Fig. 41-2. Reprinted with permission from National Semiconductor, Linear Edge, Summer 1992.

Chapter 42

Fig. 42-1. Reprinted with permission from Popular Electronics, Fact Card 255. (c) Copyright Gernsback Publications, Inc.

Chapter 43

Fig. 43-1. Reprinted with permission from Popular Electronics, 4/92, p. 67. (c) Copyright Gernsback Publications, Inc., 1992.

Fig. 43-2. Reprinted with permission from Popular Electronics, 8/91, p. 75. (c) Copyright Gernsback Publications, Inc., 1991.

Fig. 43-3. Reprinted with permission from Popular Electronics, 8/92, p. 76. (c) Copyright Gernsback Publications, Inc., 1992.

Chapter 44

Fig. 44-1. Reprinted with permission from Radio-Electronics, 2/92, p. 66. (c) Copyright Gernsback Publications, Inc., 1992.

Fig. 44-2. Reprinted with permission from Popular Electronics, Fact Card 265. (c) Copyright Gernsback Publications, Inc.

Fig. 44-3. Reprinted with permission from 73 Amateur Radio Today, 4/89. p. 87. (c) Copyright Gernsback Publications, Inc., 1989.

Fig. 44-4. Reprinted with permission from Popular Electronics, 6/93, p. 55. (c) Copyright Gernsback Publications, Inc., 1993.

Fig. 44-5. Reprinted with permission from Radio-Electronics, 7/90, p. 64. (c) Copyright Gernsback Publications, Inc., 1990.

Chapter 45

Fig. 45-1. Reprinted with permission from Electronics Now, 10/92, p. 76. (c) Copyright Gernsback Publications, Inc., 1992.

Fig. 45-2. William Sheets.

Fig. 45-3. Reprinted with permission from Popular Electronics, Fact Card 267. (c) Copyright Gernsback Publications, Inc.

Fig. 45-4. Reprinted with permission from Popular Electronics, Fact Card 266. (c) Copyright Gernsback Publications, Inc.

Chapter 46

Fig. 46-1. Reprinted with permission from Electronics Hobbyist Handbook, 1993, p. 14.

Chapter 47

Fig. 47-1. Reprinted with permission from Electronic Design, 10/93, p. 73.

Fig. 47-2. Reprinted with permission from 73 Amateur Radio Today, 7/92, p. 62.

Fig. 47-3. Reprinted with permission from R-E Experimenters Handbook, 1992, p. 122.

Fig. 47-4. William Sheets.

Fig. 47-5. Reprinted with permission from PE Hobbyist handbook, 1991, pp. 42-43. (c) Copyright Gernsback Publications, Inc., 1991.

Fig. 47-6. Reprinted with permission from Popular Electronics, 6/93, p. 78. (c) Copyright Gernsback Publications, Inc., 1993.

Chapter 48

Fig. 48-1. Reprinted with permission from PE Hobbyist Handbook, 1991, pp. 89-90. (c) Copyright Gernsback Publications, Inc., 1991.

Chapter 49

Fig. 49-1. Reprinted with permission from Popular Electronics, Application Circuit 215. (c) Copyright Gernsback Publications, Inc.

Fig. 49-2. Reprinted with permission from Popular Electronics, Application Circuit 215. (c) Copyright Gernsback Publications, Inc.

Fig. 49-3. Reprinted with permission from Electronics Now, 12/92, p. 60. (c) Copyright Gernsback Publications, Inc., 1992.

Fig. 49-4. Reprinted with permission from Electronics Now, 12/92, p. 59. (c) Copyright Gernsback Publications, Inc., 1992.

Fig. 49-5. William Sheets.

Fig. 49-6. William Sheets.

Fig. 49-7. Reprinted with permission from Popular Electronics, 3/92, p. 71. (c) Copyright Gernsback Publications, Inc., 1992.

Fig. 49-8. Reprinted with permission from Popular

Electronics, 3/92, p. 71. (c) Copyright Gernsback Publications, Inc., 1992.

Chapter 50

Fig. 50-1. Reprinted with permission from PE Hobbyist Handbook, 1991, pp. 79-80. (c) Copyright Gernsback Publications, Inc., 1991.

Fig. 50-2. Reprinted with permission from PE Hobbyist Handbook, 1990, pp. 45-47. (c) Copyright Gernsback Publications, Inc., 1990.

Fig. 50-3. Reprinted with permission from Electronics Hobbyist Handbook, 1993, p. 64.

Fig. 50-4. Reprinted with permission from 303 Circuits, p. 238.

Fig. 50-5. Reprinted with permission from Electronics Design, 5/92, p. 93.

Fig. 50-6. Reprinted with permission from Radio-Electronics, 7/90, p. 65. (c) Copyright Gernsback Publications, Inc., 1990.

Fig. 50-7. Reprinted with permission from PE Hobbyist Handbook, 1991, p. 82. (c) Copyright Gernsback Publications, Inc., 1991.

Fig. 50-8. Reprinted with permission from Radio-Electronics, 7/90, p. 64. (c) Copyright Gernsback Publications, Inc., 1990.

Fig. 50-9. Reprinted with permission from 303 Circuits, p. 251.

Fig. 50-10. Reprinted with permission from Popular Electronics, 11/93, p. 71. (c) Copyright Gernsback Publications, Inc., 1993.

Fig. 50-11. Reprinted with permission from Elektor Electronics, 12/91, p. 87.

Fig. 50-12. Reprinted with permission from Radio-Electronics, 10/89, p. 12. (c) Copyright Gernsback Publications, Inc., 1989.

Fig. 50-13. Reprinted with permission from Electronic Design, 5/92, p. 93.

Fig. 50-14. Reprinted with permission from 303 Circuits, p. 63.

Chapter 51

Fig. 51-1. Reprinted with permission from Popular Electronics, 8/92, p. 73. (c) Copyright Gernsback Publications, Inc., 1992.

Fig. 51-2. William Sheets.

Fig. 51-3. William Sheets.

Fig. 51-4. William Sheets.

Fig. 51-5. William Sheets.

Fig. 51-6. William Sheets.

Fig. 51-7. William Sheets.

Fig. 51-8. William Sheets.

Fig. 51-9. Reprinted with permission from Popular Electronics, 2/92, p. 90. (c) Copyright Gernsback Publications, Inc., 1992.

Fig. 51-10. William Sheets.

Fig. 51-11. William Sheets.

Fig. 51-12. William Sheets.

Fig. 51-13. Reprinted with permission from Popular Electronics, 3/93, p. 43. (c) Copyright Gernsback Publications, Inc., 1993.

Fig. 51-14. William Sheets.

Fig. 51-15. William Sheets.

Chapter 52

Fig. 52-1. Reprinted with permission from PE Hobbyist Handbook, pp. 93-94.

Fig. 52-2. Reprinted with permission from Popular Electronics, 6/92, p. 33. (c) Copyright Gernsback Publications, Inc., 1992.

Chapter 53

Fig. 53-1. Reprinted with permission from Popular Electronics, 12/91, p. 22. (c) Copyright Gernsback Publications, Inc., 1991.

Fig. 53-2. Reprinted with permission from Popular Electronics, 12/91, p. 18. (c) Copyright Gernsback Publications, Inc., 1991.

Chapter 54

Fig. 54-1. Reprinted with permission from Electronic Design, 2/93, p. 83.

Fig. 54-2. Reprinted with permission from 73 Amateur Radio Today, 11/92, p. 34.

Fig. 54-3. Reprinted with permission from Popular Electronics, Fact Card 258. (c) Copyright Gernsback Publications, Inc.

Chapter 55

Fig. 55-1. Reprinted with permission from Radio-Electronics, 12/91, pp. 31-36. (c) Copyright Gernsback Publications, Inc., 1991.

Fig. 55-2. Reprinted with permission from Radio-Electronics, 12/91, p. 48. (c) Copyright Gernsback Publications, Inc., 1991.

Fig. 55-3. William Sheets.

Fig. 55-4. Reprinted with permission from Popular Electronics, 2/92, pp. 53-54. (c) Copyright Gernsback Publications, Inc., 1992.

Fig. 55-5. Reprinted with permission from R-E Experimenters Handbook, 1989, p. 39.

Fig. 55-6. Reprinted with permission from Radio-Electronics, 5/92, p. 52. (c) Copyright Gernsback Publications, Inc., 1992.

Fig. 55-7. Reprinted with permission from Electronics Now, 7/93, p. 45. (c) Copyright Gernsback Publications, Inc., 1993.

Fig. 55-8. Reprinted with permission from Radio-Electronics, 8/86, p. 42. (c) Copyright Gernsback Publications, Inc., 1986.

Fig. 55-9. Reprinted with permission from Popular Electronics, 4/92, p. 53. (c) Copyright Gernsback Publications, Inc., 1992.

Fig. 55-10. Reprinted with permission from Popular Electronics, 12/91, p. 26. (c) Copyright Gernsback Publications, Inc., 1991.

Fig. 55-11. Reprinted with permission from Popular Electronics, 9/92, p. 72. (c) Copyright Gernsback Publications, Inc., 1992.

Fig. 55-12. William Sheets.

Fig. 55-13. Reprinted with permission from R-E Experimenters Handbook, 1989, p. 101.

Fig. 55-14. Reprinted with permission from Popular Electronics, 3/93, p. 42. (c) Copyright Gernsback Publications, Inc., 1993.

Fig. 55-15. William Sheets.

Fig. 55-16. Reprinted with permission from Electronics Hobbyist Handbook, 1993, p. 90.

Fig. 55-17. Reprinted with permission from Electronic Design, 4/93, p. 94.

Fig. 55-18. Reprinted with permission from Radio-Electronics, 12/91, p. 51. (c) Copyright Gernsback Publications, Inc., 1991.

Fig. 55-19. Reprinted with permission from Popular Electronics, 5/92, p. 75. (c) Copyright Gernsback Publications, Inc., 1992.

Fig. 55-20. Reprinted with permission from 303 Circuits, p. 308.

Fig. 55-21. Reprinted with permission from Electronics Now, 12/92, p. 64. (c) Copyright Gernsback Publications, Inc., 1992.

Fig. 55-22. William Sheets.

Fig. 55-23. Reprinted with permission from Popular Electronics, 3/93, p. 42. (c) Copyright Gernsback Publications, Inc., 1993.

Fig. 55-24. Reprinted with permission from Popular Electronics, Fact Card 110. (c) Copyright Gernsback Publications, Inc.

Fig. 55-25. Reprinted with permission from Electronic Design, 4/93, p. 56.

Fig. 55-26. Reprinted with permission from Popular Electronics, Fact Card 221. (c) Copyright Gernsback Publications, Inc.

Fig. 55-27. Reprinted with permission from Popular Electronics, Fact Card 221. (c) Copyright Gernsback Publications, Inc.

Fig. 55-28. Reprinted with permission from Popular Electronics, 11/93, p. 42. (c) Copyright Gernsback Publications, Inc., 1993.

Fig. 55-29. Reprinted with permission from Popular Electronics, 9/93, p. 45. (c) Copyright Gernsback Publications, Inc., 1993.

Fig. 55-30. Reprinted with permission from 303 Circuits, p. 187.

Fig. 55-31. Reprinted with permission from 73 Amateur Radio Today, 5/92, p. 62.

Fig. 55-32. Reprinted with permission from Electronics Now, 8/93, p. 73. (c) Copyright Gernsback Publications, Inc., 1993.

Fig. 55-33. Reprinted with permission from Popular Electronics, 11/91, p. 18. (c) Copyright Gernsback Publications, Inc., 1991.

Fig. 55-34. Reprinted with permission from R-E Experimenters Handbook, 1989, pp. 156-157.

Fig. 55-35. Reprinted with permission from Popular Electronics, 3/93, p. 75. (c) Copyright Gernsback Publications, Inc., 1993.

Fig. 55-36. Reprinted with permission from R-E Experimenters Handbook, 1992, p. 31.

Fig. 55-37. Reprinted with permission from Electronic Design, 5/92, p. 92.

Fig. 55-38. Reprinted with permission from Popular Electronics, 9/92, p. 72. (c) Copyright Gernsback Publications, Inc., 1992.

Fig. 55-39. Reprinted with permission from 73 Amateur Radio Today, 1/92, p. 38.

Fig. 55-40. Reprinted with permission from Popular Electronics, 3/93, p. 42. (c) Copyright Gernsback Publications, Inc., 1993.

Fig. 55-41. Reprinted with permission from 303 Circuits, p. 218.

Fig. 55-42. William Sheets.

Fig. 55-43. Reprinted with permission from PE Hobbyist Handbook, 1991, p. 14. (c) Copyright Gernsback Publications, Inc., 1991.

Fig. 55-44. William Sheets.

Fig. 55-45. Reprinted with permission from Electronic Design, 3/93.

Fig. 55-46. Reprinted with permission from Popular Electronics, 9/93, p. 46. (c) Copyright Gernsback Publications, Inc., 1993.

Fig. 55-47. Reprinted with permission from Popular Electronics, Fact Card, 198. (c) Copyright Gernsback Publications, Inc.

Fig. 55-48. Reprinted with permission from Popular Electronics, Fact Card, 221. (c) Copyright Gernsback Publications, Inc.

Fig. 55-49. Reprinted with permission from Popular Electronics, 3/93, p. 73. (c) Copyright Gernsback Publications, Inc., 1993.

Chapter 56

Fig. 56-1. Reprinted with permission from 303 Circuits, pp. 249-250.

Fig. 56-2. Reprinted with permission from Popular Electronics, 3/89, p. 69. (c) Copyright Gernsback Publications, Inc., 1989.

Fig. 56-3. Reprinted with permission from PE Hobbyist Handbook, 1991, pp. 71-72. (c) Copyright Gernsback Publications, Inc., 1991.

Chapter 57

Fig. 57-1. Reprinted with permission from Popular Electronics, 12/92, pp. 53-54. (c) Copyright Gernsback Publications, Inc., 1992.

Fig. 57-2. Reprinted with permission from Electronics Hobbyist Handbook, 1992, p. 93.

Fig. 57-3. Reprinted with permission from Popular Electronics, 3/93, p. 36. (c) Copyright Gernsback Publications, Inc., 1993.

Fig. 57-4. Reprinted with permission from Popular Electronics, 10/92, pp. 39-40. (c) Copyright Gernsback Publications, Inc., 1992.

Fig. 57-5. William Sheets.

Fig. 57-6. Reprinted with permission from 303 Circuits, p 265.

Fig. 57-7. Reprinted with permission from Popular Electronics, 11/93, p. 55. (c) Copyright Gernsback Publications, Inc., 1993.

Fig. 57-8. Reprinted with permission from R-E Experimenters Handbook, 1989, pp. 38-39.

Fig. 57-9. Reprinted with permission from Radio-Electronics, 1/92, p. 82. (c) Copyright Gernsback Publications, Inc., 1992.

Fig. 57-10. Reprinted with permission from Electronic Engineering, 11/89, pp. 21-22.

Fig. 57-11. Reprinted with permission from R-E Experimenters Handbook, pp. 118-120.

Fig. 57-12. Reprinted with permission from Electronic Design, 8/92, p. 70.

Fig. 57-13. Reprinted with permission from Electronics Now, 7/92, p. 10. (c) Copyright Gernsback Publications, Inc., 1992.

Fig. 57-14. Reprinted with permission from Popular Electronics, 12/92, p. 70. (c) Copyright Gernsback Publications, Inc., 1992.

Fig. 57-15. Reprinted with permission from 73 Amateur Radio Today, 8/92, p. 48.

Fig. 57-16. Reprinted with permission from Maxim Engineering Journal, Volume 4, pp. 11-12.

Fig. 57-17. Reprinted with permission from Radio-Electronics, 10/89, p. 13. (c) Copyright Gernsback Publications, Inc., 1989.

Fig. 57-18. Reprinted with permission from Elektor Electronics, 3/92, p.20.

Fig. 57-19. Reprinted with permission from Popular Electronics, Fact Card 223. (c) Copyright Gernsback Publications, Inc.

Fig. 57-20. Reprinted with permission from Popular Electronics, Fact Card 257. (c) Copyright Gernsback Publications, Inc.

Fig. 57-21. Reprinted with permission from Popular Electronics, Fact Card 259. (c) Copyright Gernsback Publications, Inc.

Fig. 57-22. Reprinted with permission from Popular Electronics, 11/93, p. 80. (c) Copyright Gernsback Publications, Inc., 1993.

Fig. 57-23. Reprinted with permission from 73 Amateur Radio Today, 3/92, p. 8.

Fig. 57-24. Reprinted with permission from Electronic Design, 4/93, p. 93.

Fig. 57-25. Reprinted with permission from Popular Electronics, 11/91, p. 18. (c) Copyright Gernsback Publications, Inc., 1991.

Fig. 57-26. Reprinted with permission from Popular Electronics, 11/91, p. 18. (c) Copyright Gernsback Publications, Inc., 1991.

Fig. 57-27. Reprinted with permission from Electronic Design, 7/92, p. 59.

Fig. 57-28. Reprinted with permission from Popular Electronics, 6/93, p. 55. (c) Copyright Gernsback Publications, Inc., 1993.

Fig. 57-29. Reprinted with permission from Electronic Design, 10/93, p. 74.

Fig. 57-30. Reprinted with permission from 73 Amateur Radio Today, 4/89, p. 87.

Fig. 57-31. William Sheets.

Fig. 57-32. Reprinted with permission from Popular Electronics, 9/92, p. 72. (c) Copyright Gernsback Publications, Inc., 1992.

Fig. 57-33. Reprinted with permission from Precision Monolithics Inc., 1981. Full Line Catalog, pp. 6-59.

Fig. 57-34. William Sheets.

Fig. 57-35. Reprinted with permission from Linear Databook, 1986, pp. 8-12.

Fig. 57-36. Reprinted with permission from QST, 3/89, p. 36.

Fig. 57-37. Reprinted with permission from 73 Amateur Radio Today, 2/93, p. 46. (c) Copyright Gernsback Publications, Inc., 1993.

Fig. 57-38. Reprinted with permission from 73 Amateur Radio Today, 2/93, p. 48. (c) Copyright Gernsback Publications, Inc., 1993.

Fig. 57-39. Reprinted with permission from Electronics Now, 10/92, p. 80. (c) Copyright Gernsback Publications, Inc., 1992.

Fig. 57-40. Reprinted with permission from Electronic Design, 5/92, p. 94.

Fig. 57-41. William Sheets.

Fig. 57-42. Reprinted with permission from QST, 3/89, p.35.

Fig. 57-43. Reprinted with permission from 73 Amateur Radio Today, 2/93, p. 46.

Fig. 57-44. Reprinted with permission from Popular Electronics, 3/92, p. 70. (c) Copyright Gernsback Publications, Inc., 1992.

Chapter 58

Fig. 58-1. Reprinted with permission from Silicon Chip, p. 56.

Fig. 58-2. Reprinted with permission from Electronics Now, 10/93, p. 12. (c) Copyright Gernsback Publications, Inc., 1993.

Fig. 58-3. William Sheets.

Fig. 58-4. William Sheets.

Fig. 58-5. Reprinted with permission from Popular Electronics, Fact Card 264. (c) Copyright Gernsback Publications, Inc.

Chapter 59

Fig. 59-1. Reprinted with permission from RF Design, 3/93, pp. 92-93.

Fig. 59-2. William Sheets.

Fig. 59-3. Reprinted with permission from RF Design, 3/93, p. 92.

Chapter 60

Fig. 60-1. Reprinted with permission from Silicon Chip, p. 46.

Fig. 60-2. Reprinted with permission from Popular Electronics, 10/92, pp. 31-32. (c) Copyright Gernsback Publications, Inc., 1992.

Fig. 60-3. Reprinted with permission from Popular Electronics, 10/93, p. 72. (c) Copyright Gernsback Publications, Inc., 1993.

Fig. 60-4. Reprinted with permission from R-E Experimenters Handbook, p.41.

Chapter 61

Fig. 61-1. Reprinted with permission from Popular Electronics, Fact Card 198. (c) Copyright Gernsback Publications, Inc.

Fig. 61-2. Reprinted with permission from Popular Electronics, 9/92, p. 75. (c) Copyright Gernsback Publications, Inc., 1992.

Fig. 61-3. Reprinted with permission from PE Hobbyist Handbook, p. 12. (c) Copyright Gernsback Publications, Inc.

Chapter 62

Fig. 62-1. Reprinted with permission from Popular Electronics, 10/93, p. 31. (c) Copyright Gernsback Publications, Inc., 1993.

Chapter 63

Fig. 63-1. Reprinted with permission from Radio-Electronics, 7/90, p. 66. (c) Copyright Gernsback Publications, Inc., 1990.

Fig. 63-2. Reprinted with permission from Radio-Electronics, 2/92, p. 12. (c) Copyright Gernsback Publications, Inc., 1992.

Fig. 63-3. Reprinted with permission from Popular Electronics, 6/93, p. 73.

Fig. 63-4. Reprinted with permission from Apex Microtechnology Corporation.

Fig. 63-5. Reprinted with permission from Popular Electronics, 3/92, p. 72. (c) Copyright Gernsback Publications, Inc., 1992.

Fig. 63-6. Reprinted with permission from Popular Electronics, 6/93, p. 73. (c) Copyright Gernsback Publications, Inc., 1993.

Chapter 64

Fig. 64-1. William Sheets.

Chapter 65

Fig. 65-1. William Sheets.

Fig. 65-2. William Sheets.

Fig. 65-3. William Sheets.

Fig. 65-4. William Sheets.

Fig. 65-5. William Sheets.

Fig. 65-6. William Sheets.

Fig. 65-7. William Sheets.

Fig. 65-8. Reprinted with permission from Popular Electronics, Fact Card 259. (c) Copyright Gernsback Publications, Inc.

Fig. 65-9. Reprinted with permission from Electronics Now, 10/92, p. 69. (c) Copyright Gernsback Publications, Inc., 1992.

Fig. 65-10. Reprinted with permission from Popular Electronics, Fact Card 268. (c) Copyright Gernsback Publications, Inc.

Chapter 66

Fig. 66-1. Reprinted with permission from Electronics Hobbyist Handbook, 1993, p. 17.

Fig. 66-2. Reprinted with permission from Electronics Now, 6/93, p. 47. (c) Copyright Gernsback Publications, Inc., 1993.

Fig. 66-3. Reprinted with permission from Elektor Electronics, 3/92, p. 15.

Fig. 66-4. Reprinted with permission from Electronics Now, 11/92, p. 63. (c) Copyright Gernsback Publications, Inc., 1992.

Fig. 66-5. Reprinted with permission from Elektor Electronics, 3/92, p. 14.

Fig. 66-6. Reprinted with permission from Radio-Electronics, 7/90, p. 66. (c) Copyright Gernsback Publications, Inc., 1990.

Fig. 66-7. Reprinted with permission from Radio-Electronics, 3/92, p. 77. (c) Copyright Gernsback Publications, Inc., 1992.

Chapter 67

Fig. 67-1. Reprinted with permission from 303 Circuits, p. 312.

Chapter 68

Fig. 68-1. Reprinted with permission from Popular Electronics, Fact Card 242. (c) Copyright Gernsback Publications, Inc.

Fig. 68-2. Reprinted with permission from 73 Amateur Radio Today, 11/92, p. 12.

Fig. 68-3. Reprinted with permission from 73 Amateur Radio Today, 11/92, p. 12.

Fig. 68-4. Reprinted with permission from 73 Amateur Radio Today, 11/92, p. 12.

Chapter 69

Fig. 69-1. Reprinted with permission from Electronic Design, 1/93, p. 116.

Fig. 69-2. Reprinted with permission from Electronic Design, 4/93, p. 93.

Fig. 69-3. Reprinted with permission from Popular Electronics, Fact Card 253. (c) Copyright Gernsback Publications, Inc.

Fig. 69-4. Reprinted with permission from Electronic Design, 3/89, p. 100.

Fig. 69-5. William Sheets.

Fig. 69-6. Reprinted with permission from Electronic Design, 1/93, p. 63.

Fig. 69-7. Reprinted with permission from Maxim Engineering Journal, Volume 3, p. 17.

Fig. 69-8. Reprinted with permission from Maxim Engineering Journal, Volume 3, p. 28.

Chapter 70

Fig. 70-1. Reprinted with permission from Popular Electronics, 3/93, p. 45. (c) Copyright Gernsback Publications, Inc., 1993.

Fig. 70-2. Reprinted with permission from Electronic Engineering, 8/93, p. 18.

Fig. 70-3. Reprinted with permission from Popular Electronics, 10/92, pp. 55-56. (c) Copyright Gernsback Publications, Inc., 1992.

Fig. 70-4. Reprinted with permission from Popular Electronics, 10/92, p. 56. (c) Copyright Gernsback Publications, Inc., 1992.

Fig. 70-5. William Sheets.

Fig. 70-6. Reprinted with permission from Popular Electronics, 9/93, p. 76. (c) Copyright Gernsback Publications, Inc., 1993.

Fig. 70-7. Reprinted with permission from Electronics Now, 11/92, p. 14. (c) Copyright Gernsback Publications, Inc., 1992.

Fig. 70-8. William Sheets.

Chapter 71

Fig. 71-1. Reprinted with permission from 73 Amateur Radio Today, 2/93, p. 48.

Fig. 71-2. Reprinted with permission from Popular Electronics, Fact Card 229. (c) Copyright Gernsback Publications, Inc.

Fig. 71-3. Reprinted with permission from Popular Electronics, Fact Card 230. (c) Copyright Gernsback Publications, Inc.

Fig. 71-4. Reprinted with permission from QST, 2/89, pp. 33-35.

Fig. 71-5. Reprinted with permission from Popular Electronics, 11/91, p. 21. (c) Copyright Gernsback Publications, Inc., 1991.

Fig. 71-6. Reprinted with permission from Radio-Electronics, 12/91, p. 12. (c) Copyright Gernsback Publications, Inc., 1991.

Fig. 71-7. William Sheets.

Fig. 71-8. William Sheets.

Fig. 71-9. William Sheets.

Fig. 71-10. William Sheets.

Fig. 71-11. Reprinted with permission from 303 Circuits, p. 323.

Fig. 71-12. Reprinted with permission from 73 Amateur Radio Today, 7/92, p. 59.

Fig. 71-13. Reprinted with permission from Popular Electronics, 2/92, pp. 29-31. (c) Copyright Gernsback Publications, Inc., 1992.

Fig. 71-14. Reprinted with permission from Popular Electronics, 12/93, p. 70. (c) Copyright Gernsback Publications, Inc., 1993.

Fig. 71-15. Reprinted with permission from Popular Electronics, 12/93, p. 71. (c) Copyright Gernsback Publications, Inc., 1993.

Fig. 71-16. William Sheets.

Fig. 71-17. Reprinted with permission from Popular Electronics, Fact Card 260. (c) Copyright Gernsback Publications, Inc.

Fig. 71-18. William Sheets.

Fig. 71-19. Reprinted with permission from Popular Electronics, 8/93, p. 79. (c) Copyright Gernsback Publications, Inc., 1993.

Fig. 71-20. Reprinted with permission from Popular Electronics, 12/93, p. 68. (c) Copyright Gernsback Publications, Inc., 1993.

Fig. 71-21. Reprinted with permission from Popular Electronics, 11/91, p. 21. (c) Copyright Gernsback Publications, Inc., 1991.

Fig. 71-22. Reprinted with permission from Popular Electronics, 12/93, p. 70. (c) Copyright Gernsback Publications, Inc., 1993.

Chapter 72

Fig. 72-1. Reprinted with permission from Radio-Electronics, 6/92, p. 60. (c) Copyright Gernsback Publications, Inc., 1992.

Fig. 72-2. Reprinted with permission from Popular Electronics, 12/91, p. 77. (c) Copyright Gernsback Publications, Inc., 1991.

Fig. 72-3. William Sheets.

Fig. 72-4. William Sheets.

Fig. 72-5. Reprinted with permission from Electronics Now, 7/92, p. 88. (c) Copyright Gernsback Publications, Inc., 1992.

Chapter 73

Fig. 73-1. Reprinted with permission from PE Hobbyist Handbook, 1992, p. 61. (c) Copyright Gernsback Publications, Inc., 1992.

Fig. 73-2. Reprinted with permission from PE Hobbyist Handbook, 1991, pp. 69-70. (c) Copyright Gernsback Publications, Inc., 1991.

Chapter 74

Fig. 74-1. William Sheets.

Fig. 74-2. William Sheets.

Fig. 74-3. Reprinted with permission from Electronic Design, 1/93, p. 62. (c) Copyright Gernsback Publications, Inc., 1993.

Chapter 75

Fig. 75-1. Reprinted with permission from PE Hobbyist Handbook, 1992, p. 41. (c) Copyright Gernsback Publications, Inc., 1992.

Fig. 75-2. Reprinted with permission from Popular Electronics, Fact Card 198. (c) Copyright Gernsback Publications, Inc.

Fig. 75-3. Reprinted with permission from Popular Electronics, 4/92, p. 31. (c) Copyright Gernsback Publications, Inc., 1992.

Fig. 75-4. Reprinted with permission from Electronics Now, 11/92, p. 32. (c) Copyright Gernsback Publications, Inc., 1992.

Fig. 75-5. Reprinted with permission from PE Hobbyist Handbook, 1991, p. 38. (c) Copyright Gernsback Publications, Inc., 1991.

Fig. 75-6. Reprinted with permission from PE Hobbyist Handbook, 1991, p. 73. (c) Copyright Gernsback Publications, Inc., 1991.

Fig. 75-7. Reprinted with permission from PE Hobbyist Handbook, 1991, p. 54. (c) Copyright Gernsback Publications, Inc., 1991.

Fig. 75-8. Reprinted with permission from Popular Electronics, Fact Card 198. (c) Copyright Gernsback Publications, Inc.

Chapter 76

Fig. 76-1. Reprinted with permission from Popular Electronics, 5/92, p. 60. (c) Copyright Gernsback Publications, Inc., 1992.

Fig. 76-2. Reprinted with permission from Popular Electronics, 5/92, p. 60. (c) Copyright Gernsback Publications, Inc., 1992.

Fig. 76-3. Reprinted with permission from Apex Microtechnology Corporation.

Fig. 76-4. Reprinted with permission from Apex Microtechnology Corporation.

Fig. 76-5. Reprinted with permission from Popular Electronics, 5/92, p. 60. (c) Copyright Gernsback Publications, Inc., 1992.

Chapter 77

Fig. 77-1. Reprinted with permission from PE Hobbyist Handbook, 1990, p. 92. (c) Copyright Gernsback Publications, Inc., 1990.

Fig. 77-2. Reprinted with permission from Popular Electronics, 6/93, p. 77. (c) Copyright Gernsback Publications, Inc., 1993.

Fig. 77-3. Reprinted with permission from 73 Amateur Radio Today, 7/92, p. 62.

Fig. 77-4. Reprinted with permission from Popular Electronics, 6/93, p. 77. (c) Copyright Gernsback Publications, Inc., 1993.

Fig. 77-5. Reprinted with permission from 73 Amateur Radio Today, 7/92, p. 62.

Fig. 77-6. Reprinted with permission from Elektor Electronics, 12/91, p. 94.

Fig. 77-7. Reprinted with permission from Popular Electronics, 7/93, p. 76. (c) Copyright Gernsback Publications, Inc., 1993.

Fig. 77-8. Reprinted with permission from Electronic Design, 6/93, p. 76.

Chapter 78

Fig. 78-1. Reprinted with permission from Electronics Now, 10/93, p. 53. (c) Copyright Gernsback Publications, Inc., 1993.

Fig. 78-2. Reprinted with permission from R-E Experimenters Handbook, p. 60.

Fig. 78-3. Reprinted with permission from Electronic Design, 2/93, p. 71.

Fig. 78-4. Reprinted with permission from Maxim Engineering Journal, Volume 3, p. 16.

Fig. 78-5. Reprinted with permission from Linear Technology, Design Note 72.

Fig. 78-6. Reprinted with permission from Popular Electronics, 6/93, p. 48. (c) Copyright Gernsback Publications, Inc., 1993.

Fig. 78-7. Reprinted with permission from Popular Electronics, 3/92, p. 72. (c) Copyright Gernsback Publications, Inc., 1992.

Fig. 78-8. Reprinted with permission from Popular Electronics, 6/93, p. 77. (c) Copyright Gernsback Publications, Inc., 1993.

Fig. 78-9. Reprinted with permission from 303 Circuits, p. 283.

Fig. 78-10. Reprinted with permission from Electronics Now, 12/92, p. 66. (c) Copyright Gernsback Publications, Inc., 1992.

Fig. 78-11. Reprinted with permission from 73 Amateur Radio Today, 7/92, p. 60.

Fig. 78-12. Reprinted with permission from Popular Electronics, 5/92, p. 73. (c) Copyright Gernsback Publications, Inc., 1992.

Fig. 78-13. Reprinted with permission from PE Hobbyist Handbook, 1993, p. 93. (c) Copyright Gernsback Publications, Inc., 1993.

Fig. 78-14. William Sheets.

Fig. 78-15. Reprinted with permission from PE Hobbyist Handbook, 1991. (c) Copyright Gernsback Publications, Inc., 1991.

Fig. 78-16. Reprinted with permission from PE Hobbyist Handbook, 1991, pp. 28-29. (c) Copyright Gernsback Publications, Inc., 1991.

Fig. 78-17. Reprinted with permission from Popular Electronics, 11/93, p. 71. (c) Copyright Gernsback Publications, Inc., 1993.

Fig. 78-18. Reprinted with permission from Silicon Chip, pp. 63-64.

Fig. 78-19. Reprinted with permission from Maxim Engineering Journal, Volume 4, p. 19.

Fig. 78-20. Reprinted with permission from National Semiconductor, Linear Edge, Issue #5.

Fig. 78-21. Reprinted with permission from National Semiconductor, Linear Edge, Issue #5.

Fig. 78-22. Reprinted with permission from 73 Amateur Radio Today, 3/92, p. 54.

Fig. 78-23. Reprinted with permission from Popular Electronics, 1/92, p. 37. (c) Copyright Gernsback Publications, Inc., 1992.

Fig. 78-24. Reprinted with permission from Popular Electronics, 11/93, p. 71. (c) Copyright Gernsback Publications, Inc., 1993.

Fig. 78-25. Reprinted with permission from Silicon Chip, p. 10.

Fig. 78-26. Reprinted with permission from Electronic Design, 2/93, p. 72.

Fig. 78-27. Reprinted with permission from Electronic Design, 8/93, p. 84.

Fig. 78-28. Reprinted with permission from Linear Technology, Design Note 68.

Fig. 78-29. Reprinted with permission from Popular Electronics, Fact Card 260. (c) Copyright Gernsback Publications, Inc.

Fig. 78-30. Reprinted with permission from Popular Electronics, 11/93, p. 54.

Fig. 78-31. Reprinted with permission from Popular Electronics, 8/93, p. 88. (c) Copyright Gernsback Publications, Inc., 1993.

Fig. 78-32. Reprinted with permission from Popular Electronics, 3/93, p. 74. (c) Copyright Gernsback Publications, Inc., 1993.

Fig. 78-33. Reprinted with permission from PE Hobbyist Handbook, 1993, p. 61. (c) Copyright Gernsback Publications, Inc., 1993.

Fig. 78-34. Reprinted with permission from Electronic Design, 2/93, pp. 75-76.

Fig. 78-35. Reprinted with permission from Electronics Now, 10/93, p. 54. (c) Copyright Gernsback Publications, Inc., 1993.

Fig. 78-36. Reprinted with permission from Electronic Design, 4/93, p. 54.

Fig. 78-37. Reprinted with permission from National Semiconductor, Linear Edge, Spring 1992.

Chapter 79

Fig. 79-1. Reprinted with permission from Popular Electronics, 9/92, p. 71. (c) Copyright Gernsback Publications, Inc., 1992.

Fig. 79-2. Reprinted with permission from Popular Electronics, 11/91, p. 22. (c) Copyright Gernsback Publications, Inc., 1991.

Chapter 80

Fig. 80-1. Reprinted with permission from Popular Electronics, 3/92, p. 42. (c) Copyright Gernsback Publications, Inc., 1992.

Fig. 80-2. Reprinted with permission from Radio-Electronics, 12/91, p. 63. (c) Copyright Gernsback Publications, Inc., 1991.

Fig. 80-3. Reprinted with permission from Popular Electronics, 9/93, p. 69. (c) Copyright Gernsback Publications, Inc., 1993.

Fig. 80-4. Reprinted with permission from Elektor Electronics, 12/91, p. 73.

Fig. 80-5. Reprinted with permission from Elektor Electronics, 12/91, p. 72.

Fig. 80-6. Reprinted with permission from Elektor Electronics, 12/91, p. 85.

Fig. 80-7. Reprinted with permission from Electronics Now, 12/92, p. 45. (c) Copyright Gernsback Publications, Inc., 1992.

Fig. 80-8. Reprinted with permission from Electronic Design, 7/93, p. 87.

Fig. 80-9. Reprinted with permission from Popular Electronics, 9/93, p. 71. (c) Copyright Gernsback Publications, Inc., 1993.

Fig. 80-10. Reprinted with permission from Electronics Now, 12/92, p. 46. (c) Copyright Gernsback Publications, Inc., 1992.

Fig. 80-11. Reprinted with permission from Popular Electronics, 10/93, p. 73. (c) Copyright Gernsback Publications, Inc., 1993.

Fig. 80-12. Reprinted with permission from Silicon Chip, p. 64.

Fig. 80-13. Reprinted with permission from Popular Electronics, 9/93, p. 69. (c) Copyright Gernsback Publications, Inc., 1993.

Chapter 81

Fig. 81-1. Reprinted with permission from PE Hobbyist Handbook, 1991, pp. 85-86. (c) Copyright Gernsback Publications, Inc., 1991.

Fig. 81-2. Reprinted with permission from Popular Electronics, Fact Card 270. (c) Copyright Gernsback Publications, Inc.

Chapter 82

Fig. 82-1. William Sheets.

Fig. 82-2. William Sheets.

Fig. 82-3. Reprinted with permission from PE Hobbyists Handbook, 1990, p. 120. (c) Copyright Gernsback Publications, Inc., 1990.

Fig. 82-4. Reprinted with permission from Electronic Design 1/93, p. 61.

Fig. 82-5. William Sheets.

Fig. 82-6. William Sheets.

Fig. 82-7. William Sheets.

Chapter 83

Fig. 83-1. Reprinted with permission from 73 Amateur Radio Today, 10/91, p. 8.

Fig. 83-2. Reprinted with permission from Electronics Now, 10/92, p. 37. (c) Copyright Gernsback Publications, Inc., 1992.

Fig. 83-3. Reprinted with permission from Popular Electronics, 10/93, p. 74. (c) Copyright Gernsback Publications, Inc., 1993.

Fig. 83-4. Reprinted with permission from Popular Electronics, 8/93, p. 71. (c) Copyright Gernsback Publications, Inc., 1993.

Fig. 83-5. Reprinted with permission from Popular Electronics, 8/93, p. 70. (c) Copyright Gernsback Publications, Inc., 1993.

Fig. 83-6. Reprinted with permission from Integrated Circuits Data Book, 3/85, pp. 5-16.

Fig. 83-7. Reprinted with permission from Popular Electronics, 3/93, p. 79. (c) Copyright Gernsback Publications, Inc., 1993.

Fig. 83-8. Reprinted with permission from 73 Amateur Radio Today, 8/93, p. 32.

Fig. 83-9. Reprinted with permission from Popular

Electronics, 6/92, p. 55. (c) Copyright Gernsback Publications, Inc., 1992.

Fig. 83-10. Reprinted with permission from Popular Electronics, 8/93, p. 32. (c) Copyright Gernsback Publications, Inc., 1993.

Fig. 83-11. Reprinted with permission from QST, 2/89, p. 34.

Fig. 83-12. Reprinted with permission from Popular Electronics, 6/92, p. 57. (c) Copyright Gernsback Publications, Inc., 1992.

Chapter 84

Fig. 84-1. Reprinted with permission from Electronic Design, 8/92, p. 69.

Fig. 84-2. Reprinted with permission from Radio-Electronics, 5/92, p. 47. (c) Copyright Gernsback Publications, Inc., 1992.

Fig. 84-3. William Sheets.

Fig. 84-4. Reprinted with permission from Electronics Now, 3/93, p. 69. (c) Copyright Gernsback Publications, Inc., 1993.

Fig. 84-5. William Sheets.

Chapter 85

Fig. 85-1. Reprinted with permission from Electronics Now, 11/92, p. 53. (c) Copyright Gernsback Publications, Inc., 1992.

Fig. 85-2. Reprinted with permission from Electronics Now, 11/92, p. 54. (c) Copyright Gernsback Publications, Inc., 1992.

Fig. 85-3. Reprinted with permission from Popular Electronics, 8/93, p. 56. (c) Copyright Gernsback Publications, Inc., 1993.

Fig. 85-4. Reprinted with permission from R-E Experimenters Handbook, 1991, p. 30.

Fig. 85-5. Reprinted with permission from Popular Electronics, 3/93, p. 45. (c) Copyright Gernsback Publications, Inc., 1993.

Fig. 85-6. Reprinted with permission from Popular Electronics, 8/93, p. 53. (c) Copyright Gernsback Publications, Inc., 1993.

Fig. 85-7. Reprinted with permission from Popular Electronics, 3/93, p. 45. (c) Copyright Gernsback Publications, Inc., 1993.

Chapter 86

Fig. 86-1. Reprinted with permission from Popular Electronics, 6/93, p. 55. (c) Copyright Gernsback Publications, Inc., 1993.

Fig. 86-2. Reprinted with permission from Popular Electronics, 6/93, p. 56. (c) Copyright Gernsback Publications, Inc., 1993.

Fig. 86-3. Reprinted with permission from Popular Electronics, Fact Card 225. (c) Copyright Gernsback Publications, Inc.

Fig. 86-4. William Sheets.

Fig. 86-5. Reprinted with permission from R-E Experimenters Handbook, 1989, p. 156.

Fig. 86-6. Reprinted with permission from Practical Wireless, 6/91, p. 34.

Fig. 86-7. Reprinted with permission from R-E Experimenters Handbook, p. 33.

Fig. 86-8. Reprinted with permission from Electronic Design, 6/93, p. 83.

Fig. 86-9. Reprinted with permission from 73 Amateur Radio Today, 5/90, p. 78.

Fig. 86-10. Reprinted with permission from 73 Amateur Radio Today, 5/90, p. 78.

Fig. 86-11. Reprinted with permission from Popular Electronics, Fact Card 241. (c) Copyright Gernsback Publications, Inc.

Fig. 86-12. Reprinted with permission from 73 Amateur Radio Today, 11/91, pp. 52-56. (c) Copyright Gernsback Publications, Inc., 1991.

Fig. 86-13. Reprinted with permission from 73 Amateur Radio Today, 10/92, p. 20.

Fig. 86-14. Reprinted with permission from Popular Electronics, 11/93, p. 81. (c) Copyright Gernsback Publications, Inc., 1993.

Fig. 86-15. Reprinted with permission from 73 Amateur Radio, 5/90, p. 78.

Fig. 86-16. Reprinted with permission from Popular Electronics, 8/93, p. 72. (c) Copyright Gernsback Publications, Inc., 1993.

Fig. 86-17. Reprinted with permission from Popular Electronics, Fact Card 262. (c) Copyright Gernsback Publications, Inc.

Fig. 86-18. Reprinted with permission from 73 Amateur Radio, 5/90, p. 77.

Fig. 86-19. Reprinted with permission from Popular Electronics, 8/93, p. 72. (c) Copyright Gernsback Publications, Inc., 1993.

Fig. 86-20. Reprinted with permission from Popular Electronics, 3/93, p. 47. (c) Copyright Gernsback Publications, Inc., 1993.

Fig. 86-21. Reprinted with permission from Popular Electronics, 9/93, p. 83. (c) Copyright Gernsback Publications, Inc., 1993.

Fig. 86-22. Reprinted with permission from 73 Amateur Radio Today, 2/93, p. 60.

Fig. 86-23. William Sheets.

Fig. 86-24. Reprinted with permission from 73 Amateur Radio Today, 5/90, p. 31.

Fig. 86-25. Reprinted with permission from Popular Electronics, 6/93, p. 54. (c) Copyright Gernsback Publications, Inc., 1993.

Fig. 86-26. William Sheets.

Chapter 87

Fig. 87-1. Reprinted with permission from QST, 6/91, p. 18.

Fig. 87-2. Reprinted with permission from 73 Amateur Radio Today, 7/92, p. 30.

Fig. 87-3. Reprinted with permission from Popular Electronics, 6/92, p. 56. (c) Copyright Gernsback Publications, Inc., 1992.

Fig. 87-4. Reprinted with permission from 73 Amateur Radio Today, 3/92, p. 16.

Fig. 87-5. Reprinted with permission from Popular Electronics, 7/93, p. 80. (c) Copyright Gernsback Publications, Inc., 1993.

Chapter 88

Fig. 88-1. Reprinted with permission from Popular Electronics, 11/93, p. 73. (c) Copyright Gernsback Publications, Inc., 1993.

Fig. 88-2. Reprinted with permission from Maxim Journal, Vol. 3., p. 22.

Chapter 89

Fig. 89-1. Reprinted with permission from PE Hobbyist Handbook, 1990, p. 21. (c) Copyright Gernsback Publications, Inc., 1990.

Chapter 90

Fig. 90-1. Reprinted with permission from Popular Electronics, 5/92, p. 71. (c) Copyright Gernsback Publications, Inc., 1992.

Fig. 90-2. Reprinted with permission from Popular Electronics, 5/92, p. 74. (c) Copyright Gernsback Publications, Inc., 1992.

Chapter 91

Fig. 91-1. Reprinted with permission from Electronic Design, 11/92, p. 62.

Fig. 91-2. Reprinted with permission from Popular Electronics, 12/93, p. 71. (c) Copyright Gernsback Publications, Inc., 1993.

Fig. 91-3. Reprinted with permission from Popular Electronics, Fact Card 256. (c) Copyright Gernsback Publications, Inc.

Fig. 91-4. Reprinted with permission from Electronics Now, 6/93, p. 14. (c) Copyright Gernsback Publications, Inc., 1993.

Fig. 91-5. Reprinted with permission from Popular Electronics, Fact Card 256. (c) Copyright Gernsback Publications, Inc.

Fig. 91-6. Reprinted with permission from R-E Experimenters Handbook, 1989, p. 160.

Fig. 91-7. Reprinted with permission from Electronic Design, 10/93, p. 74.

Fig. 91-8. Reprinted with permission from Popular Electronics, 11/91, p. 22. (c) Copyright Gernsback Publications, Inc., 1991.

Fig. 91-9. Reprinted with permission from Maxim Engineering Journal, Vol. 4, p. 15.

Chapter 92

Fig. 92-1. Reprinted with permission from R-E Experimenters Handbook, 1992, p. 98.

Fig. 92-2. Reprinted with permission from Electronic Design, 6/93, p. 82.

Fig. 92-3. Reprinted with permission from Popular Electronics, 10/92, p. 58. (c) Copyright Gernsback Publications, Inc., 1992.

Fig. 92-4. Reprinted with permission from Popular Electronics, Fact Card 201. (c) Copyright Gernsback Publications, Inc.

Fig. 92-5. Reprinted with permission from 73 Amateur Radio Today, 11/91, p. 11.

Fig. 92-6. Reprinted with permission from Popular Electronics, 6/93, p. 59. (c) Copyright Gernsback Publications, Inc., 1993.

Fig. 92-7. Reprinted with permission from Radio-Electronics, 1/92, p. 12.

Fig. 92-8. Reprinted with permission from Popular Electronics, Fact Card 255. (c) Copyright Gernsback Publications, Inc.

Fig. 92-9. Reprinted with permission from Popular Electronics, 6/93, p. 72. (c) Copyright Gernsback Publications, Inc., 1993.

Fig. 92-10. Reprinted with permission from Electronics Now, 12/93, p. 39. (c) Copyright Gernsback Publications, Inc., 1993.

Fig. 92-11. Reprinted with permission from 303 Circuits.

Chapter 93

Fig. 93-1. Reprinted with permission from Popular Electronics, 1/92, p. 43. (c) Copyright Gernsback

Publications, Inc., 1992.

Fig. 93-2. Reprinted with permission from Electronic Design, 8/93, p. 81.

Fig. 93-3. William Sheets.

Fig. 93-4. William Sheets.

Fig. 93-5. Reprinted with permission from Popular Electronics, 12/91, p. 81. (c) Copyright Gernsback Publications, Inc., 1991.

Fig. 93-6. William Sheets.

Fig. 93-7. William Sheets.

Fig. 93-8. Reprinted with permission from PE Hobbyist, 1991, p. 77. (c) Copyright Gernsback Publications, Inc., 1991.

Fig. 93-9. William Sheets.

Fig. 93-10. William Sheets.

Fig. 93-11. Reprinted with permission from Popular Electronics, 2/92, p. 67. (c) Copyright Gernsback Publications, Inc., 1992.

Fig. 93-12. Reprinted with permission from Popular Electronics, 2/92, p. 66. (c) Copyright Gernsback Publications, Inc., 1992.

Fig. 93-13. Reprinted with permission from R-E Experimenters Handbook, 1989, p. 155.

Fig. 93-14. Reprinted with permission from National Semiconductor, Linear Applications Handbook.

Fig. 93-15. Reprinted with permission from R-E Experimenters Handbook, 1989, p. 161.

Fig. 93-16. Reprinted with permission from 303 Circuits, p. 257 (#221).

Chapter 94

Fig. 94-1. Reprinted with permission from Popular Electronics, Fact Card 223. (c) Copyright Gernsback Publications, Inc.

Fig. 94-2. William Sheets.

Fig. 94-3. Reprinted with permission from Popular Electronics, Fact Card 221. (c) Copyright Gernsback Publications, Inc.

Fig. 94-4. Reprinted with permission from Popular Electronics, Fact Card 243. (c) Copyright Gernsback Publications, Inc.

Fig. 94-5. William Sheets.

Fig. 94-6. William Sheets.

Fig. 94-7. Reprinted with permission from Popular Electronics, Fact Card 263. (c) Copyright Gernsback Publications, Inc.

Chapter 95

Fig. 95-1. Reprinted with permission from Popular Electronics, 3/93, p. 71. (c) Copyright Gernsback Publications, Inc., 1993.

Fig. 95-2. Reprinted with permission from Popular Electronics, 3/93, p. 72. (c) Copyright Gernsback Publications, Inc., 1993.

Fig. 95-3. Reprinted with permission from Popular Electronics, 3/93, p. 72. (c) Copyright Gernsback Publications, Inc., 1993.

Chapter 96

Fig. 96-1. Reprinted with permission from Radio-Electronics, 6/92, p. 71. (c) Copyright Gernsback Publications, Inc., 1992.

Fig. 96-2. Reprinted with permission from R-E Experimenters Handbook, 1989, pp. 12-15.

Fig. 96-3. Reprinted with permission from Electronics Now, 7/92, p. 33. (c) Copyright Gernsback Publications, Inc., 1992.

Fig. 96-4. Reprinted with permission from Popular Electronics, Fact Card 261. (c) Copyright Gernsback Publications, Inc.

Fig. 96-5. Reprinted with permission from 303 Circuits, p. 49.

Fig. 96-6. Reprinted with permission from Electronics Now, 3/93, p. 71. (c) Copyright Gernsback Publications, Inc., 1993.

Fig. 96-7. Reprinted with permission from 303 Circuits, p. 41.

Fig. 96-8. Reprinted with permission from Popular Electronics, Fact Card 261. (c) Copyright Gernsback Publications, Inc.

Fig. 96-9. Reprinted with permission from Popular Electronics, Fact Card 241. (c) Copyright Gernsback Publications, Inc.

Fig. 96-10. Reprinted with permission from Popular Electronics, Fact Card 262. (c) Copyright Gernsback Publications, Inc.

Chapter 97

Fig. 97-1. Reprinted with permission from Radio-Electronics, 12/90, pp. 72-73.

Fig. 97-2. Reprinted with permission from Popular Electronics, 6/93, p. 71. (c) Copyright Gernsback Publications, Inc., 1993.

Fig. 97-3. Reprinted with permission from National Semiconductor, Linear Edge, Summer 1992.

Fig. 97-4. Reprinted with permission from Popular Electronics, 12/92, p. 32. (c) Copyright Gernsback Publications, Inc., 1992.

Fig. 97-5. Reprinted with permission from Popular Electronics, 9/93, p. 70. (c) Copyright Gernsback

Publications, Inc., 1993.

Fig. 97-6. Reprinted with permission from Popular Electronics, 6/93, p. 72. (c) Copyright Gernsback Publications, Inc., 1993.

Fig. 97-7. Reprinted with permission from Popular Electronics, 6/93, p. 71. (c) Copyright Gernsback Publications, Inc., 1993.

Fig. 97-8. Reprinted with permission from Maxim Engineering Journal, Vol. 4. p. 10.

Fig. 97-9. Reprinted with permission from Electronics Now, 6/93, p. 14. (c) Copyright Gernsback Publications, Inc.

Fig. 97-10. Reprinted with permission from Popular Electronics, 6/91, p. 71. (c) Copyright Gernsback Publications, Inc., 1991.

Fig. 97-11. Reprinted with permission from Popular Electronics, 6/93, p. 71. (c) Copyright Gernsback Publications, Inc., 1993.

Fig. 97-12. Reprinted with permission from Popular Electronics, 9/92, p. 70. (c) Copyright Gernsback Publications, Inc., 1992.

Fig. 97-13. Reprinted with permission from Popular Electronics, 6/93, p. 72. (c) Copyright Gernsback Publications, Inc., 1993.

Chapter 98

Fig. 98-1. Reprinted with permission from Radio-Electronics, 6/90, p. 71.

Fig. 98-2. Reprinted with permission from Radio-Electronics, 6/90, p. 71.

Chapter 99

Fig. 99-1. Reprinted with permission from Popular Electronics, 7/92, pp. 60-61. (c) Copyright Gernsback Publications, Inc., 1992.

Fig. 99-2. William Sheets.

Chapter 100

Fig. 100-1. Reprinted with permission from Radio-Electronics, 7/90, p. 8.

Fig. 100-2. Reprinted with permission from Radio-Electronics, 10/89, p. 8.

Fig. 100-3. Reprinted with permission from Radio-Electronics, 11/91, p. 59.

Fig. 100-4. Reprinted with permission from Popular Electronics, 4/92, p. 38. (c) Copyright Gernsback Publications, Inc., 1992.

Fig. 100-5. Reprinted with permission from Popular Electronics, p. 75. (c) Copyright Gernsback Publications, Inc.

Fig. 100-6. Reprinted with permission from Popular Electronics, 9/92, pp. 38-40. (c) Copyright Gernsback Publications, Inc., 1992.

Fig. 100-7. Reprinted with permission from 303 Circuits, 226, p. 263.

Fig. 100-8. Reprinted with permission from Popular Electronics, 8/92, p. 76. (c) Copyright Gernsback Publications, Inc., 1992.

Fig. 100-9. Reprinted with permission from Popular Electronics, 9/92, p. 74. (c) Copyright Gernsback Publications, Inc., 1992.

Fig. 100-10. Reprinted with permission from Popular Electronics, 12/91, p. 53. (c) Copyright Gernsback Publications, Inc., 1991.

Fig. 100-11. Reprinted with permission from Radio-Electronics, 1/93, p. 43. (c) Copyright Gernsback Publications, Inc., 1993.

Fig. 100-12. Reprinted with permission from Electronics Hobbyist Handbook, 1993, p. 26.

Fig. 100-13. Reprinted with permission from Popular Electronics, 9/93, p. 33. (c) Copyright Gernsback Publications, Inc., 1993.

Fig. 100-14. Reprinted with permission from Popular Electronics, 11/93, p. 38. (c) Copyright Gernsback Publications, Inc., 1993.

Fig. 100-15. Reprinted with permission from Electronic Design, 8/93, p. 86.

Fig. 100-16. Reprinted with permission from Electronics Now, 5/93, p. 47. (c) Copyright Gernsback Publications, Inc., 1993.

Fig. 100-17. Reprinted with permission from Silicon Chip, p. 62.

Fig. 100-18. Reprinted with permission from Popular Electronics, 12/93, p. 62. (c) Copyright Gernsback Publications, Inc., 1993.

Fig. 100-19. Reprinted with permission from Radio-Electronics, 3/92, p. 74. (c) Copyright Gernsback Publications, Inc., 1992.

Fig. 100-20. Reprinted with permission from Radio-Electronics, 2/92, p. 81. (c) Copyright Gernsback Publications, Inc., 1992.

Fig. 100-21. Reprinted with permission from Popular Electronics, 2/92, p. 70. (c) Copyright Gernsback Publications, Inc., 1992.

Fig. 100-22. Reprinted with permission from Electronics Now, 11/92, p. 45. (c) Copyright Gernsback Publications, Inc., 1992.

Fig. 100-23. Reprinted with permission from Popular Electronics, 7/92, p. 74. (c) Copyright Gernsback Publications, Inc., 1992.

Fig. 100-24. Reprinted with permission from Popular Electronics, 11/92, p. 72. (c) Copyright Gernsback Publications, Inc., 1992.

Chapter 101

Fig. 101-1. Reprinted with permission from Electronic Design, 5/92, p. 94.

Fig. 101-2. Reprinted with permission from 303 Circuits, #229, pp. 264-265.

Fig. 101-3. Reprinted with permission from Electronic Design, 11/92, p. 62. (c) Copyright Gernsback Publications, Inc., 1992.

Fig. 101-4. Reprinted with permission from Intersil, Component Data Catalog, 1987, pp. 6-10.

Fig. 101-5. Reprinted with permission from Radio-Electronics, 3/90, p. 50. (c) Copyright Gernsback Publications, Inc., 1990.

Fig. 101-6. Reprinted with permission from 303 Circuits, #228, p. 364.

Fig. 101-6. Reprinted with permission from Radio-Electronics, 3/92, p. 50. (c) Copyright Gernsback Publications, Inc., 1992.

Fig. 101-7. Reprinted with permission from Radio-Electronics, 6/92, p. 54. (c) Copyright Gernsback Publications, Inc., 1992.

Fig. 101-8. Reprinted with permission from Radio-Electronics.

Chapter 102

Fig. 102-1. Reprinted with permission from Electronics Now, 10/92, p. 43. (c) Copyright Gernsback Publications, Inc., 1992.

Fig. 102-2. Reprinted with permission from Popular Electronics, 5/92, p. 75. (c) Copyright Gernsback Publications, Inc., 1992.

Fig. 102-3. William Sheets.

Fig. 102-4. William Sheets.

Fig. 102-5. William Sheets.

Fig. 102-6. William Sheets.

Fig. 102-7. William Sheets.

Fig. 102-8. Reprinted with permission from PE Hobbyist Handbook, 1991, p. 57. (c) Copyright Gernsback Publications, Inc., 1991.

Fig. 102-9. Reprinted with permission from Popular Electronics, 12/92, p. 68. (c) Copyright Gernsback Publications, Inc., 1992.

Chapter 103

Fig. 103-1. Reprinted with permission from Practical Wireless, 2/91, p. 49.

Fig. 103-2. Reprinted with permission from 73 Amateur Radio, 7/88, p. 14.

Chapter 104

Fig. 104-1. Reprinted with permission from Electronics Hobbyist Handbook, 1993, p. 101.

Fig. 104-2. Reprinted with permission from Electronics Hobbyist Handbook, 1993, p. 101.

Fig. 104-3. Reprinted with permission from Electronics Hobbyist Handbook, 1993, p. 100.

Chapter 105

Fig. 105-1. Reprinted with permission from Popular Electronics, 8/92, p. 74. (c) Copyright Gernsback Publications, Inc., 1992.

Fig. 105-2. Reprinted with permission from Popular Electronics, Fact Card 266. (c) Copyright Gernsback Publications, Inc.

Fig. 105-3. Reprinted with permission from Popular Electronics, 8/92, p. 74. (c) Copyright Gernsback Publications, Inc., 1992.

Fig. 105-4. Reprinted with permission from Popular Electronics, 12/93, p. 71. (c) Copyright Gernsback Publications, Inc., 1993.

Fig. 105-5. Reprinted with permission from Popular Electronics, 7/92, p. 70. (c) Copyright Gernsback Publications, Inc., 1992.

Fig. 105-6. Reprinted with permission from Hands-On Electronics, 9/87, p. 88.

Chapter 106

Fig. 106-1. Reprinted with permission from Electronics Now, 10/92, p. 36. (c) Copyright Gernsback Publications, Inc., 1992.

Fig. 106-2. Reprinted with permission from 73 Amateur Radio Today, 10/91, pp. 14-22.

Fig. 106-3. William Sheets.

Fig. 106-4. Reprinted with permission from Popular Electronics, 8/92, p. 46. (c) Copyright Gernsback Publications, Inc., 1992.

Fig. 106-5. Reprinted with permission from Radio-Electronics, 11/91, p. 85. (c) Copyright Gernsback Publications, Inc., 1991.

Fig. 106-6. Reprinted with permission from 73 Amateur Radio Today, 11/92, p. 8.

Fig. 106-7. Reprinted with permission from Electronics Hobbyist Handbook, 1993, p. 24.

Fig. 106-8. Reprinted with permission from Electronics Hobbyist Handbook, 1993, p. 52.

Fig. 106-9. Reprinted with permission from QST, 10/89, p. 25.
Fig. 106-10. Reprinted with permission from 73 Amateur Radio Today, 4/92, p. 36.
Fig. 106-11. Reprinted with permission from 73 Amateur Radio Today, 7/92, p. 20.
Fig. 106-12. Reprinted with permission from 73 Amateur Radio Today, 4/92, p. 25.
Fig. 106-13. Reprinted with permission from R-E Experimenters Handbook, 1989, p. 158.
Fig. 106-14. Reprinted with permission from 73 Amateur Radio Today, 4/93, p. 53.

Chapter 107

Fig. 107-1. Reprinted with permission from Radio-Electronics, 11/91, pp. 49-57. (c) Copyright Gernsback Publications, Inc., 1991.
Fig. 107-2. Reprinted with permission from Radio-Electronics, 11/91, p. 49. (c) Copyright Gernsback Publications, Inc., 1991.
Fig. 107-3. Reprinted with permission from Electronics Now, 3/93, p. 33. (c) Copyright Gernsback Publications, Inc., 1993.

Chapter 108

Fig. 108-1. Reprinted with permission from Radio-Electronics, 6/92, p. 61. (c) Copyright Gernsback Publications, Inc., 1992.
Fig. 108-2. Reprinted with permission from Radio-Electronics, 6/92, p. 59. (c) Copyright Gernsback Publications, Inc., 1992.
Fig. 108-3. Reprinted with permission from 303 Circuits, p. 307.
Fig. 108-4. Reprinted with permission from SpecCom, 5/91, p. 15.
Fig. 108-5. Reprinted with permission from Linear Technology Design Note #57.
Fig. 108-6. Reprinted with permission from Linear Technology Design Note #57.
Fig. 108-7. Reprinted with permission from Linear Technology Design Note #57.
Fig. 108-8. Reprinted with permission from Linear Technology Design Note #57.
Fig. 108-9. Reprinted with permission from Linear Technology Application Note #57.
Fig. 108-10. Reprinted with permission from Linear Technology Application Note #57.
Fig. 108-11. Reprinted with permission from Radio-Electronics, 4/92, p. 64. (c) Copyright Gernsback Publications, Inc., 1992.
Fig. 108-12. Reprinted with permission from 303 Circuits, #303, p. 332.
Fig. 108-13. Reprinted with permission from Radio-Electronics, 6/92, p. 59. (c) Copyright Gernsback Publications, Inc., 1992.
Fig. 108-14. Reprinted with permission from Electronics Now, 8/93, p. 39. (c) Copyright Gernsback Publications, Inc., 1993.
Fig. 108-15. Reprinted with permission from 303 Circuits, #300, p. 331.
Fig. 108-16. Reprinted with permission from Popular Electronics, Fact Card 268. (c) Copyright Gernsback Publications, Inc.

Chapter 109

Fig. 109-1. Reprinted with permission from Electronic Engineering, 9/89, p. 28.
Fig. 109-2. William Sheets.
Fig. 109-3. Reprinted with permission from Popular Electronics, Fact Card 269. (c) Copyright Gernsback Publications, Inc.
Fig. 109-4. Reprinted with permission from Popular Electronics, Fact Card #241. (c) Copyright Gernsback Publications, Inc.
Fig. 109-5. Reprinted with permission from Popular Electronics, Fact Card 224. (c) Copyright Gernsback Publications, Inc.
Fig. 109-6. William Sheets.

Chapter 110

Fig. 110-1. Reprinted with permission from 303 Circuits, p. 280.
Fig. 110-2. William Sheets.

Chapter 111

Fig. 111-1. William Sheets.
Fig. 111-2. William Sheets.
Fig. 111-3. William Sheets.

Chapter 112

Fig. 112-1. William Sheets.
Fig. 112-2. Reprinted with permission from Popular Electronics, Fact Card 259. (c) Copyright Gernsback Publications, Inc.

Index

Numbers preceded by a "I,", "II," "III,", "IV," or "V" are from *Encyclopedia of Electronic Circuits* Vol. I, II, III, IV, or V respectively.

B

D

E

F

G

H

I

J

K

L

N

O

S

T

U

V

X

Y

Z